	IB	IIB	IIIA	IVA	VA	VIA	VIIA	0
								2 He 4.00260
			5 ✓ B 10.81	6 C 12.011	7 N 14.0067	8 O 15.9994	9 F 18.9984	10 Ne 20.179
			13 Al 26.9815	14 Si 28.086	15 P 30.9738	16 S 32.06	17 Cl 35.453	18 Ar 39.948
28 Ni 58.71	29 Cu 63.546	30 Zn 65.37	31 Ga 69.72	32 Ge 72.59	33 As 74.9216	34 Se 78.96	35 Br 79.904	36 Kr 83.80
46 Pd 106.4	47 Ag 107.868	48 Cd 112.40	49 In 114.82	50 Sn 118.69	51 Sb 121.75	52 Te 127.60	53 I 126.9045	54 Xe 131.30
78 Pt 195.09	79 Au 196.9665	80 Hg 200.59	81 Tl 204.37	82 Pb 207.2	83 Bi 208.9806	84 Po (210)[b]	85 At (210)[b]	86 Rn (222)[b]

64 Gd 157.25	65 Tb 158.9254	66 Dy 162.50	67 Ho 164.9303	68 Er 167.26	69 Tm 168.9342	70 Yb 173.04	71 Lu 174.97
96 Cm (247)[b]	97 Bk (247)[b]	98 Cf (251)[b]	99 Es (254)[b]	100 Fm (253)[b]	101 Md (256)[b]	102 No (254)[b]	103 Lr (257)[b]

Atomic symbols shown in black represent metals; colored symbols represent non-metals; shaded boxes are metalloids.

GENERAL CHEMISTRY

GENERAL CHEMISTRY

RALPH S. BECKER
WAYNE E. WENTWORTH

University Of Houston

HOUGHTON MIFFLIN COMPANY •BOSTON
Atlanta •Dallas •Geneva, Illinois
Hopewell, New Jersey •Palo Alto

This book is dedicated to the memory of Mark Korry Becker
September 1, 1953–May 13, 1972

Printed in the United States of America

Library of Congress Card Number 72-5642

ISBN: 0-395-16002-2

CONTENTS

*An asterisk indicates an optional chapter or section.

PREFACE

GENERAL CHEMISTRY is written for an introductory full year course, with students of chemistry, science and engineering in mind. The book is rigorous in the sense that it is thorough. At the same time it is written at a level comprehensible to beginning college students. No knowledge of calculus and no previous exposure to chemistry are necessary for the material to be understood.

The basic ideas of chemistry are integrated throughout the text, and are related where appropriate to other disciplines. One of our objectives in writing the book was to make the content useful to all students taking a general chemistry course. To aid in accomplishing this goal, we have discussed with instructors of other sciences and engineering what topics in chemistry are important to their disciplines. We have attempted to be sensitive to their suggestions both in the text and in our selection of problems and exercises.

The content of the book is focused around the chemical background which we feel a science or engineering student needs in solving present day problems. This does not mean that every aspect of chemistry is touched on in the text. In order to keep the book reasonable in length, we have omitted certain topics which we feel are less important to beginning students. For example, there is no discussion of unit cell crystal arrangements, nor are there diagrams of complex equipment. We believe these serve no real need and often cannot be adequately understood at this stage of study. Our aim is to give the student a firm scientific basis in chemistry which will prepare him to pursue his own special interests in advanced courses.

The text presents material in as unified a manner as possible. Essential aspects of a given topic are largely covered in a single chapter. For example, all of the essential ionic and molecular interactions are presented together in Chapter Eight rather than being scattered over several chapters. By not fragmenting the initial presentation of new topics, the text helps students focus on the existence and basis of fundamental principles.

Once a new concept is introduced, it is used repeatedly throughout the book. Each chapter increases the student's repertoire, and succeeding chapters use information from earlier chapters. Students should realize that the body of knowledge is not isolated chapter by chapter, or even within the confines of the book itself. All knowledge is enriched by overlapping and cross-referencing of ideas. Present academic work, research, and industrial processes require application and integration of a variety of information.

The text sets the stage for meaningful laboratory experiments to be started early in the academic year. Chapters Two and Three begin immediately with much fundamental information which provides a basis for laboratory work. Some topics are discussed at a qualitative level in these early chapters and are treated quantitatively at appropriate places in later chapters.

We have tried to make this text truly represent *general* chemistry in that both inorganic and organic compounds are considered throughout the text. To accomplish this, the structure and classification of organic compounds have been included in Chapter Seven, which considers molecular structure and bonding in polyatomic molecules. By introducing organic compounds at this point in the text, the subsequent topics of solutions, chemical equilibrium and kinetics can be applied to organic as well as inorganic compounds — truly a *general* chemistry approach.

Coverage of some aspects of chemistry are condensed without sacrificing clarity. For example, Chapter Nine discusses all the essentials of inorganic chemistry in a general consideration of the periodic table and group properties. Transition elements, which present special additional considerations, are discussed in a separate optional chapter. Other optional chapters (and sections within chapters) have been designated with a marginal asterisk to provide flexibility of presentation. Omitting these optional sections in no way impairs the student's understanding of succeeding chapters, except in Chapter Nineteen where occasional reference is made to Chapter Eighteen.

Chapters Eighteen and Nineteen provide a genuine challenge. They present as much up-to-date thinking as possible on the methods of approach to and the status of current research. In addition, they show the importance of integrating the various aspects of chemistry with such fields as physics and biology.

Problems and exercises are located at two different places in the text and serve two different purposes. The problems are located within each chapter, where they serve to stimulate thinking and to illustrate the applications of concepts just discussed in the previous section. All problems should be worked by the student immediately after the section has been discussed in class. The exercises are located at the end of each chapter and involve or interrelate concepts from this chapter as well as from earlier chapters. Many exercises illustrate applications of chemistry to other areas of science or to practical situations. It is not necessary that all exercises be worked. Answers to *all* problems and exercises are located at the end of the text.

In summary, our principal goal has been to provide material actually needed by all students of general chemistry, to present it at a level comprehensible to the student, and to stress the concept that all knowledge relies on the integration of ideas within and between disciplines. We believe this text will be one the student will want to use throughout his college career as a reference source.

We especially want to thank the following people for their careful reading of the manuscript and for useful scientific and pedagogic suggestions: Professor Gordon Atkinson, Professor Herbert H. Richtol, Professor Rodney E. Black, and Professor John A. Ricketts. In addition, we would like to thank Nancy L. Marcus for her help in editing our preliminary manuscript, and Rose Ann Bourgeois, Cherry Day, and Julie Norris for typing the manuscript. Also we wish to express our appreciation to the following students for checking manuscript and problems: Dorothy Gray, Betty Bartschmid, Kenneth Lepow, Shen Nan Lin, Shabbir Zubari and Wu Ting-Long.

RALPH BECKER
WAYNE WENTWORTH

CHAPTER ONE

INTRODUCTION

Before we can begin any useful discussion of chemistry, we must first attempt to define what chemistry is and how it relates to other disciplines. The most general definition of chemistry is: the study of the properties of substances, the relationship of these properties to composition, and the transformation or combination of substances (chemical reaction) to form new substances. This very loose definition generalizes the primary objective of most chemical investigations.

Chemistry overlaps in many areas with other sciences; some areas in chemistry are inorganic chemistry, organic chemistry, physical chemistry, biochemistry, photochemistry, analytical chemistry, theoretical chemistry, agricultural chemistry, and petroleum chemistry. In each of these areas the particular focus of study may be certain classes of compounds, particular techniques of study, or kinds of chemical reactions. For example, inorganic chemistry is concerned with the transformations or chemical reactions of noncarbon-containing substances, organic chemistry with chemical reactions of carbon-containing compounds, and biochemistry with any chemical reactions and compounds in living systems. On the other hand, analytical chemistry is not directly concerned with the transformation of substances, but rather with the identification of a substance (qualitative analysis) and determination of the amount (quantitative analysis). This information is essential, of course, to studying chemical reactions, whether the reactions are inorganic, organic, or biochemical. We must not only have all substances properly identified but a method must also be available for the quantitative analysis of the

TABLE 1–1 Chemistry—the subareas

SUBAREA	PRIMARY FOCUS FOR INVESTIGATION
inorganic chemistry	structure, properties, and reactions of noncarbon-containing compounds
organic chemistry	structure, properties, and reactions of carbon-containing compounds
biochemistry or biological chemistry	structure, properties, and reactions of compounds in plant and animal systems
photochemistry	structure, properties, and reactivity of compounds in excited states caused by electromagnetic radiation
analytical chemistry	development of techniques and procedures for the qualitative and quantitative determination of compounds, elements, ions, or radicals
physical chemistry	study of the nature of physical and chemical properties, techniques for investigating these properties, and their relationship to geometrical and electronic structure of molecules
polymer chemistry	structure and properties of polymers and reactions to produce them
geochemistry	study of the chemical constitution of the earth's crust

substances before we can fully understand chemical reactions. In some areas emphasis is put on the investigation of molecular structure. Although this is not a direct study of chemical reactions, it ultimately leads to a better understanding of the nature of the substance and of its ability (or inability) to enter into a reaction.

Table 1–1 lists some areas of chemistry and their primary focus of investigation. Note the overlap of the areas; this list is not complete; in fact, no one area is so clearly defined that it is independent of the others. The table illustrates how broad and also fragmented the subareas of chemistry are.

1–A. WHY IS CHEMISTRY IMPORTANT?

Having an idea of what chemistry is, we now might ask, "Why is it important for anyone other than a chemist to know some chemistry?" The answer to this question becomes obvious if we consider the applications of chemistry in the other science and engineering disciplines and in everyday life. It is advantageous in today's world to have some knowledge of chemistry. All of the various bodily functions involve series of chemical reactions, although some of these are not yet known in detail. The entire digestive process is a complex series of chemical reactions. The cycle of birth, life, and death involve a myriad of unconsciously controlled chemical reactions.

Most of the food we eat has been affected by chemicals in one way or another. Chemicals are frequently used as food additives for various reasons, and the effect of these additives on health is being questioned. Insecticides have played an important role in increasing food production over the past twenty years. However, some insecticides, such as DDT, have also had adverse effects upon wildlife, and particularly upon certain birds. Experiences such as these have taught us to be more cautious in the use of insecticides. Other effective insecticides have subsequently been developed which do not appear to have these adverse effects, but continual monitoring by investigators is needed if we are to interact effectively with nature.

There has been an enormous advance in the field of pharmaceuticals, which has produced drugs and medicines that have saved the lives of millions of people. Many of these materials are purified natural products; however, many are unique compounds made through a series of synthetic reactions. As with insecticides and fertilizers, adverse side effects can occur in drug usage. A knowledge of chemistry is needed if we are to exercise caution. Tests must be developed to determine whether adverse effects can be expected. One of the controversial developments of the past few years has been oral contraceptives. Many people recognize that overpopulation is a serious threat to human life and welfare on this planet, and certainly advances in birth control measures can help in controlling this problem. Again, adverse effects have been found associated with the use of oral contraceptives. Scientists who have produced medicines must continually investigate their subsequent side effects.

The moral of this discussion is, of course, that chemistry and the use of chemicals can be either friend or foe. Survival of human life as we know it has been and may always be a struggle with nature against insects, rodents, predatory fish species, etc. Chemicals can play a most important role in this struggle; yet they should not be allowed to grossly upset the balance of nature. Upsetting this balance has, in some cases, led to the extermination of a species. Man must learn to live with nature if he is to survive, and with the assistance of chemicals he can make nature more fruitful.

Many areas of chemistry have contributed significantly to the present high standard of living of many people. For example, synthetic fabrics, plastics, and construction materials are direct results of the application of chemistry to the needs of the community. In many cases these developments have resulted from the combination of chemistry with other sciences such as biology, geology, and physics, and the various fields of engineering. It is this interaction that makes it essential for all scientists and engineers to have a basic understanding of the principles of chemistry. No one field of science or engineering can remain isolated. All fields are highly interrelated and highly dependent on one another.

1–B. CHEMISTRY AND SOCIETY

If we were to compare life as it is today with that of twenty to thirty years previously, we would quickly realize the great influence which science and engineering have had on our society. This influence has had both good and bad effects. Television, nuclear bombs and nuclear power, computers, and travel in outer space have made a tremendous impact on our thinking and on our lives today. What the future holds is anyone's guess.

The society is responsible for the demands it makes on science and technology. It is critical, therefore, that members of society understand enough science to make intelligent decisions. Chemistry and related fields have been responsible for the significant advances in the areas of plastics, petroleum chemistry, synthetic fibers, pharmaceuticals, fuels, fertilizers, insecticides, and detergents. The continuing advances are largely dependent on the needs of the community. If society demands continued research in any area, and if additional funds are made available, scientific research will proceed.

In recent years science, as well as industry, has been given a "bad name" as a result of problems which appear to be consequences of our highly technological society. Air and water pollution, noise level, waste disposal, and detrimental effects from insecticides and drugs have caused hardship to many people. These are real problems which obviously need real and immediate solutions. Some of these problems are discussed in appropriate chapters later in the text. We will describe the chemistry of many of these problems as they are understood today. With an awareness of these problems, the scientists and engineers of the future must keep these less desirable aspects of industry in mind and must solve those problems while developing new products.

The general purpose of this text is to give the science student a modern presentation of general chemistry which is not only scientifically sound but also applicable in the modern technological world. The book is intended for chemistry majors as well as other science majors; however, it is not written with only the chemists' interests in mind. Wherever possible, applications and interest in other scientific and engineering disciplines have been incorporated in this text without sacrificing an emphasis on chemical principles.

1–C. THE SCIENTIFIC METHOD

The so-called scientific method is somewhat of an idealized method or model for scientific investigation. The method consists of the following steps:

1. collection of data
2. evaluation of the data—analysis of the interdependency of the data
3. development of a *law* which describes the variation in the data or expresses it mathematically

4. development of a *theory* which is used to derive or which may explain the law.*

The theory should be based upon a fundamental model involving some basic assumptions. If the theory agrees (or disagrees) with the law, this proves (or disproves) the basic assumptions of the theoretical model. The value of a satisfactory theory is that it gives a deeper and more fundamental insight than does a law into *how* and *why* the phenomenon occurs. The theory can then be used to suggest further investigations that will more critically examine the theory and its limitations.

The development of the equation of state for an ideal gas can be used to illustrate the scientific method. The development followed the stages

1. Collection of data: Measurements of pressure P, volume V, temperature T, and moles n are made on various gases.

2. Analysis of the data: Upon analysis of the data it is seen that V decreases with increasing P but increases with increasing T and n.

3. Development of laws: Laws are written in the form of mathematical statements which describe quantitatively how the volume V varies with the variable pressure P, temperature T and number of moles n. The names of the laws acknowledge early investigators who helped formulate the laws. The symbol α means "is proportional to."

Boyle's law	$V \propto \dfrac{1}{P}$	$(n, T = \text{constants})$
Avogadro's law	$V \propto n$	$(P, T = \text{constants})$
Charles' law	$V \propto T$	$(n, P = \text{constants})$

These laws can be combined into one equation known as the *ideal gas law* or *equation of state*

$$PV = nRT$$

where $R = \text{constant}$.

4. Development of a theory: The *kinetic molecular theory* has been developed to describe the behavior of gases in terms of a fundamental model of molecules and atoms and their motion in the gas phase. Using this kinetic molecular theory, the ideal gas equation can be derived. The pertinent assumptions in the model are

   a. the gas molecules have negligible volume compared to the distance between the molecules,

   b. there are no forces of attraction between the molecules.

*There is not consistency in all texts between steps 3 and 4 in the "scientific method." Many texts outline the sequence of steps as 3. *hypothesis* is advanced to explain the experimental data 4. further testing of the hypothesis (providing it is capable of correlating the data) elevates the hypothesis to a theory 5. finally the theory is tested and found to be accepted as "truth," at which time it is called a *law*. In chemistry this sequence is not as appropriate as the one used here.

In this example, if there is agreement between the theoretical derivation (step 4) and the empirical development (step 3) of the ideal gas law, the two given assumptions are valid for any gas which obeys the ideal gas equation of state. On the contrary, disagreement of the *P, V, T, n* data with the ideal gas law would suggest that the theoretical model with its basic assumptions is not valid for the gas in question. That is, the gas does not obey the ideal gas law, because the size of the molecules and forces of attraction between the molecules cannot be neglected.

Earlier we stated that the scientific method is an *idealized* method of investigation. For anyone doing research, the scientific method can seldom be rigorously put into practice. The principal step which is missing in the sequence of four steps is the motivation for collecting the experimental data in the first place. Seldom does a scientist go into the laboratory to obtain data for the sake of simply collecting it. More likely his purpose is to prove or disprove a certain theory or aspect of a theory. Possibly he is going to investigate some compounds on the basis of an educated "hunch." In any event there has been some serious thought given to the experiment and this is one of the most important aspects of the scientific approach. The subsequent steps are sometimes followed and more likely the entire process will be repeated many times. On the basis of the first theory formed, the researcher frequently goes back to look at other data that will test the theory more critically. The theory or model may then be refined, and the process repeated again and again. It is rather naive to profess that the scientific method as outlined above is carried out in all research. In some cases, the theoretical work was in fact developed first, and the experimental work simply confirmed the theoretical prediction. Such was the case with the development of the LASER, which we discuss briefly in Chapter 19.

In this text we have not made any special effort to present chemical principles in the order of the scientific method (that is, facts, generalizations, laws, and/or theories). Rather we have structured the book so that it will be a logical and interesting approach to modern, general chemistry.

Scientific thinking is not a stereotyped procedure. In fact, science is an exciting and dynamic field. The thrill of seeing an experimental result confirm a major breakthrough in understanding is difficult to express in words. It may be impossible for you to appreciate just how exciting science can be until you have had your first experimental success. We can only encourage you to find out for yourself.

CHAPTER TWO

THE STRUCTURE AND STATES OF MATTER

What are the properties of substances in our immediate environment? What do we know about our physical world? In this chapter we will begin by discussing the microworld of small indivisible particles which we will build into atoms, then molecules and compounds. We will discuss chemical reactions in Chapter 3. The discussion of the theory of atomic and molecular structure in Chapters 4 through 7 will help you understand why compounds form and why chemical reactions occur. We will use this background of basic facts in presenting theories to explain easily visible phenomena as well as current research later in the book.

2–A. GENERAL STRUCTURE OF ATOMS—ELEMENTS

All matter is composed of discrete units called atoms. All atoms consist of two general regions: *nuclear* and *extranuclear*. The nuclear region is a generally spherical region that contains protons and neutrons (see Chapter 17 for more detail on nuclear structure). Protons are positively charged, p^+, and neutrons are neutral, n. The sum of the mass of these particles is very nearly equal to the total mass of the atom; very little mass is in the extranuclear portion. The diameter of the nucleus is approximately 10^{-12} centimeters. Nearly all of the mass of the atom is concentrated in the small nuclear volume; the nucleus is extremely dense. This density is approximately 10^8 tons per cubic centimeter (cm^3) or 1000 million tons per cubic inch. This is many orders of magnitude more dense than any known star. The density of some stars is of the order of 1 ton per cubic inch. In fact, the great density of these stars is considered to be due to the fact that they are composed of atoms from which a part of the extranuclear region has been stripped.

The extranuclear region is also spherical, and contains only one kind of particle, the electron, e^-. The mass of the electron is small, approximately 1/1840 of that of the proton or neutron. The electron has a negative charge equal in magnitude to the positive charge of the proton. The number of protons in the nucleus is equal to the number of electrons in the extranuclear region, and thus, the atom as a whole has a neutral charge. The diameter of the extranuclear sphere, approximately 10^{-8} cm, is approximately 10,000 times larger than that of the nuclear sphere. Thus the density of the extranuclear region is very small compared to that of the nuclear region. The *size* of the atom is almost entirely dependent on the volume occupied by the extranuclear region.

The *atomic number Z* is defined as the number of protons in the nucleus and because we know that the charge on an atom is neutral, we know that the atomic number must also be equal to the number of electrons in the atom.

$$Z = \text{number of protons} = \text{number of electrons}$$

The atomic number characterizes an element; all atoms of a particular element have the same atomic number. This atomic number, then, is a unique or identifying characteristic enabling us to classify atoms into groups called *elements*. The *mass number A* is the total number of protons and neutrons. It is always greater in value than the atomic number (except for hydrogen).

$$A = Z + \text{number of neutrons}$$

The atoms of any one element can have different mass numbers and these atoms are called *isotopes* of that element.

The simplest of all atoms is the hydrogen atom. A convenient way of depicting an atom with its particulate structure is shown in Figure 2–1. This picture, though

FIGURE 2–1 A. hydrogen with $A = 1$ B. hydrogen with $A = 2$ C. carbon with $A = 12$ D. carbon with $A = 14$. (See Chapter 4 for the details of how the location of electrons is represented.)

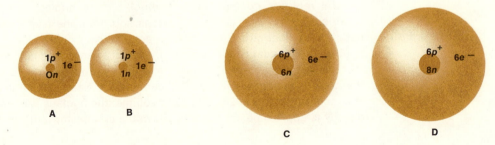

based on accepted evidence, is in no way a model. The electron cannot be positioned, and the nucleus is not a sphere with a sharply defined boundary. It is important to use pictures and models with great care. Hydrogen (Figure 2–1A) has a single proton in the nucleus and one electron. There is also an isotope of hydrogen having a mass number of 2, which is called deuterium and is shown in Figure 2–1B. The carbon atom has an isotope of $A = 12$. Since the atomic number of carbon is 6, the nucleus of this isotope contains 6 protons and 6 neutrons as shown in Figure 2–1C. Another isotope of carbon has $A = 14$. The composition of an atom of this isotope is shown in Figure 2–1D. A more detailed discussion of isotopes and their use will be given in Chapter 17.

SYMBOLS

For convenience we have already started using a kind of shorthand, for example, $Z = $ atomic number. There is also a shorthand for the names of all of the elements. In order that we can conveniently refer to the elements, we assign to each element a certain *symbol* consisting of one or two letters extracted from the name of the element. A list of all known elements, along with their symbols and atomic numbers is given in the table inside the front cover of this book. A convention has been adopted to designate the different isotopes for an element. The mass number is shown as a presuperscript and the atomic number as a presubscript. For example the isotopes of nydrogen are

$$\,^1_1H \qquad \,^2_1H \qquad \,^3_1H$$

The isotopes of carbon are

$$\,^{12}_6C \qquad \,^{13}_6C \qquad \,^{14}_6C$$

Most elements have more than one isotope. A certain fraction of each isotope exists in nature. For example the fractions of ^1_1H, ^2_1H, and ^3_1H are 0.99985, 0.00015, and 10^{-11}, respectively.

PROBLEM 2–1 Diagram the number of electrons, protons, and neutrons for the atoms $^{18}_8O$, $^{19}_9F$, ^7_3Li using Figure 2–1 as a model.

ATOMIC WEIGHT SCALE

The mass m of the proton, neutron, and electron are

$$m_{p+} = 0.16724 \quad \times 10^{-23}\ g$$
$$m_n = 0.16747 \quad \times 10^{-23}\ g$$
$$m_{e-} = 0.0000910 \times 10^{-23}\ g$$

As you can see, all of these quantities are extremely small. The mass of the atom is *approximately* the sum of the masses of these *elementary particles* in the atom. However, the mass of the actual atom is slightly less than the sum of the masses of the particles, because of the large nuclear binding forces between the protons and neutrons. This discrepancy is discussed in Chapter 17. The mass of the atoms range anywhere from 0.167×10^{-23} g for hydrogen to 39.52×10^{-23} g for uranium, U. These numbers are small, and are awkward to use in comparing relative masses of atoms. For this reason a *relative* mass scale has been devised, called the *atomic weight scale*, which arbitrarily assigns the $^{12}_{6}C$ isotope a value of 12.00000. The mass of other isotopes on this atomic weight scale is given relative to the mass of $^{12}_{6}C$. The units on this scale are amu (atomic mass units). The masses of the proton, neutron, and electron on the atomic weight scale are

$$m_{p+} = 1.00728 \text{ amu}$$

$$m_n = 1.00867 \text{ amu}$$

$$m_{e-} = 0.000549 \text{ amu}$$

Note that this atomic weight scale has been devised so that the masses of the proton and neutron are conveniently close to one. It would be more appropriate to call this an atomic mass scale since it is a relative scale of the atomic masses rather than atomic weight. Weight depends upon the gravitational attraction whereas mass is independent of any forces and is a true measure of the quantity of matter.* This scale has been conventionally called the atomic weight scale and we will retain this name despite the fact that it is a measure of relative masses. The term *atomic mass* will be used for the mass of an individual atom.

PROBLEM 2–2   The mass of $^{12}_{6}C$ is 1.99×10^{-23} g and the mass of $^{19}_{9}F$ is 3.15×10^{-23} g. What is the mass of $^{19}_{9}F$ on the atomic weight scale? [Hint: The mass of an isotope is *not* equal to the sum of the masses of the p^+, n, and e^-.]

ATOMIC WEIGHTS OF ELEMENTS

$^{1}_{1}H$ is by far the predominant isotope of hydrogen. In other cases the isotopes are more evenly distributed. For example, $^{35}_{17}Cl$, $^{37}_{17}Cl$, have the fractions 0.758, 0.242, respectively. The average mass of all the isotopes for an element on the atomic weight scale is called the *atomic weight* of that element. This is *not* a simple average like the average between two numbers. This is a weighted average, where the relative fraction of each isotope is taken into account. The atomic weight of an element is found by considering a natural mixture of isotopes. Let $f_1, f_2, \ldots$ be the fractional abundance of the different isotopes and $M_1, M_2, \ldots$ be the mass of

*Since the weight of an object depends upon the gravitational attraction, the weight of a given object is different when weighed at different altitudes on the earth's surface or on planets with different masses. The mass of an object is measured by comparing the weight of the object relative to some weight standard (a one kilogram bar of platinum at the National Bureau of Standards). Thus, the mass is independent of the force of gravity and is a true measure of the quantity of material.

these isotopes on the atomic weight scale. The weighted average of these, which is the atomic weight (M) of the element, is given by

$$M = f_1M_1 + f_2M_2 + \ldots = \Sigma f_iM_i$$

EXAMPLE 2-1

The masses of the isotopes for hydrogen $_1^1H$, $_1^2H$, and $_1^3H$ are 1.007825, 2.0140, and 3.01605 amu, respectively. The fractions f_1, f_2, f_3 are 0.99985, 0.00015, 10^{-11}. The atomic weight of hydrogen is

$$M = f_1M_1 + f_2M_2 + f_3M_3$$

$$M = (0.99985)(1.0078) + (0.0015)(2.0140) + 10^{-11}(3.0161)$$

$$M = 1.00797$$

PROBLEM 2-3

The masses of $_{17}^{35}Cl$ and $_{17}^{37}Cl$ on the atomic weight scale are 34.97 amu and 36.97 amu, respectively. The fractions of these isotopes as they occur in nature are 0.758 and 0.242. Calculate the atomic weight for chlorine.

From this point on we will not refer to the mass number of an element since the different isotopes have different numbers of neutrons. Only when we are referring to a specific isotope will the mass number be designated.

MONATOMIC IONS

An *ion* is an atom or group of atoms with positive or negative charge. Monatomic ions can be formed from single atoms by addition or removal of electrons. Take the lithium atom, Li, for example. Lithium has 3 protons in the nucleus and 3 electrons surrounding the nucleus. If it were to lose an electron, the overall charge balance would be upset and it would acquire a net positive charge.

Similarly, fluorine can accept an electron to form the negative ion F⁻.

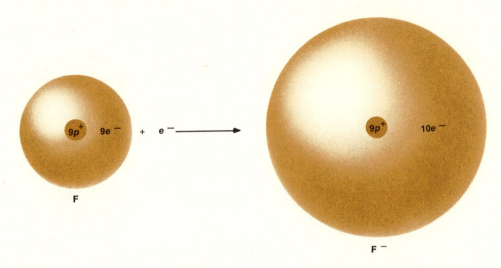

Obviously in a monatomic ion of an element the atomic number, Z, still equals the number of protons in the nucleus, but Z *does not equal* the number of electrons.

PROBLEM 2–4 How many protons and electrons are there in Fe^{3+}?

PROBLEM 2–5 What single feature characterizes an element regardless of whether it exists as an atom or an ion?

Positive ions are referred to as *cations* and negative ions as *anions*. These names arise from the direction in which these ions move in an electric field (Chapter 14). The cations migrate toward the negative electrode (cathode) whereas the anions migrate toward the positive electrode (anode). A list of some common monatomic ions and their names are given in Table 2–1.

The name of a cation (positive ion) is just the name of the element itself. If there is more than one ion possible for an element, the charge of a given ion is designated by a Roman numeral in parenthesis. For example, Cu^+ is called copper(I) ion, and Cu^{2+} is called copper(II) ion. Previously, an older nomenclature was used to distinguish between two possible (+) charges on an ion: an *-ous* ending for the lower charged ion and an *-ic* ending for the higher charged ion. For example, Cu^+ was called cupr*ous* and Cu^{2+} was called cupr*ic*. The disadvantage in using this nomenclature is that one must keep in mind the two possible (+) charges of both ions in order to name the ions. For example, chromium has two positively charged ions +2, +3: Cr^{2+} is called chrom*ous* and Cr^{3+} is called chrom*ic*. Because of the inconvenience and possible confusion in the *-ous*, *-ic* system of nomenclature, the *use of Roman numerals is preferred*. Using Roman numerals designates explicitly the charge on the ion and there is no chance for misinterpretation.

TABLE 2–1 Nomenclature of common monatomic ions

+1 CHARGE	+2 CHARGE	+3 CHARGE
H^+ hydrogen ion	Be^{2+} beryllium ion	Al^{3+} aluminum ion
Li^+ lithium ion	Fe^{2+} iron(II) ion (ferrous)	Fe^{3+} iron(III) ion (ferric)
Na^+ sodium ion	Cr^{2+} chromium(II) ion	Cr^{3+} chromium(III) ion (chromic)
K^+ potassium ion	(chromous)	
Rb^+ rubidium ion	Mg^{2+} magnesium ion	
Cu^+ Copper (I) ion	Ca^{2+} calcium ion	
(cuprous)	Sr^{2+} strontium ion	
	Ba^{2+} barium ion	
	Cu^{2+} copper(II) ion (cupric)	

−4 CHARGE	−3 CHARGE	−2 CHARGE	−1 CHARGE
C^{4-} carbide	N^{3-} nitride	O^{2-} oxide	F^- fluoride
	P^{3-} phosphide	S^{2-} sulfide	Cl^- chloride
		Se^{2-} selenide	Br^- bromide
		Te^{2-} telluride	I^- iodide
			H^- hydride

Later we will see that the Roman numeral is also used to designate the oxidation number or oxidation state of an element. At this time, it is sufficient to say that the charge on a monatomic ion is also the oxidation number of the ion. However, the oxidation number is a more general term. We cannot simply say that the oxidation number of an atom refers to the charge on an ion of that element. The determination of oxidation numbers and their application will be described in Chapter 3.

The monatomic anions have a characteristic *-ide* ending which frequently replaces the last three letters or so of the name of the element. A few examples will make this clear.

F fluorine, F^- fluor*ide*

Cl chlorine, Cl^- chlor*ide*

O oxygen, O^{2-} ox*ide*

S sulfur, S^{2-} sulf*ide*

After repeated use of these ions in writing formulas and names of chemical compounds, you will begin to learn what atoms form what ions. After a detailed discussion of atomic structure and the periodic table in Chapter 5, you will understand why an ion with a certain charge exists for a given element. Furthermore, association of an element with its position in the periodic table will facilitate your remembering the charge on the ion.

2–B. GENERAL STRUCTURE OF MOLECULES—COMPOUNDS

MOLECULAR FORMULAS

A *compound* is a substance which is composed of certain elements in definite proportions or amounts. A compound has a unique set of properties which is characteristic of that compound. The atom is the fundamental unit which characterizes an element. The fundamental unit of a compound is the *molecule*. There is one exception: ionic compounds. This class of compounds will be discussed later in the chapter.

A molecule is simply a specific group of atoms which act as a single unit because they are held tightly together by forces called chemical bonds. The molecules of any one compound are all identical; they all have the same number of atoms of a given element and they all have the same structure and size.

The number of atoms of each element in the molecule is designated by a subscript in the chemical formula for the molecule. For example, the formula for water is H_2O. The subscript 2 designates that there are two hydrogen atoms. The oxygen in H_2O has no subscript; the absence of a subscript indicates that there is only one oxygen atom in the molecule.

Another more complicated molecule is acetic acid whose chemical formula is $C_2H_4O_2$. Reading the subscripts we see that there are 2 carbon atoms, 4 hydrogen atoms, and 2 oxygen atoms in one molecule of acetic acid. All molecules of acetic acid will have exactly this same combination of atoms.

The formula for acetic acid may also be written as CH_3COOH. The formula written in this manner gives information about the geometrical structure of acetic acid molecules. An even more accurate way of writing the formula is

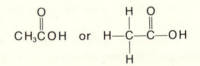

All acetic acid molecules have the same molecular structure, in addition to having the same formula. The lines in the molecular diagram represent chemical bonds. One carbon atom is bonded to three hydrogen atoms and is written as CH_3. The other carbon is bonded with *two bonds or a double bond* to one oxygen atom and with a single bond to another oxygen which is in turn bonded to a hydrogen atom. This arrangement is frequently shortened to $COOH$. The two carbon atoms are also joined by a chemical bond.

These formulas suggest the geometry of the molecule. However, they are not at all accurate because they are two-dimensional. All molecules exist and react in three dimensions. When using this shorthand notation, you must remember its limitations.

PROBLEM 2-6   Glucose is an important compound which acts as a source of energy for chemical reactions in the body. Write the chemical formula for glucose whose molecules can be represented by the diagram

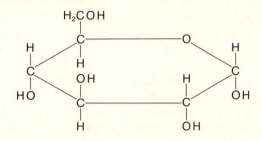

Write the number of atoms of each element in the order C, H, O.

Frequently, it is convenient to consider a certain group of atoms within the molecule as a unit (for example, CH_3 in acetic acid). Other common groups are CH_2, H_2O, NH_2, and OH. If there are more than one of these units in a molecule, parentheses are put around the unit in the formula and the number of that unit in the molecule is indicated by a subscript after the parentheses. As an example, consider the molecular formula for ethylene glycol

$$C_2H_6O_2 \quad \text{or} \quad HO(CH_2)_2OH \quad \text{or} \quad C_2H_4(OH)_2$$

There are two units of CH_2; this information is indicated by $(CH_2)_2$. The reason for using a molecular formula which sets off the CH_2 units is obvious from the geometrical structure of the molecule.

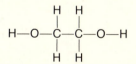

Notice that the CH_2 groups are bonded together and that each is bonded at the ends to an OH group. The molecular formula $HO(CH_2)_2OH$ gives us much more information about structure than does the formula $C_2H_6O_2$, which tells us only numbers of atoms.

PROBLEM 2-7   The geometrical structure of 1,1,1-trifluoro-4-chlorobutane is

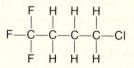

How many CH_2 and F groups are there? Write the molecular formula designating the number of CH_2 and F groups.

POLYATOMIC IONS

In the same way that atoms can be bonded together to form molecules, atoms can combine to form ions. Since these ions contain more than one atom they are called *polyatomic ions*. In general, the atoms in a polyatomic ion are held together firmly by strong bonds, and the polyatomic ion acts as a single unit analogous to a molecule. The negative charge on a polyatomic anion results from the addition of one electron or more to the bonded atoms. The hydroxide ion, OH^-, has a total of 10 electrons: 8 electrons from oxygen, one from hydrogen, and one more. There are also polyatomic cations but they are much less common than polyatomic anions. A polyatomic cation has fewer electrons than the bonded atoms. The ammonium ion, NH_4^+, has a total of 10 electrons: 7 electrons from nitrogen, one electron from each of the 4 hydrogen, and one electron removed to give a positive charge.

There are many other polyatomic ions. A few common ions are SO_4^{2-} (sulfate), NO_3^- (nitrate), PO_4^{3-} (phosphate), ClO_3^- (chlorate), H_3O^+ (hydronium), NH_4^+ (ammonium). The charge on these ions cannot be identified with any particular atom but rather with the entire polyatomic ion. A list of some common polyatomic ions is given in Table 2–2.

The nomenclature of polyatomic anions of the form XO_n is based on the number of oxygen atoms in the ion. One of the polyatomic ions which each element forms with oxygen is given a name ending in —*ate*. For example,

PO_4^{3-} phosph*ate*
AsO_4^{3-} arsen*ate*

SO_4^{2-} sulf*ate*
CrO_4^{2-} chrom*ate*
NO_3^- nitr*ate*
ClO_3^- chlor*ate*
BrO_3^- brom*ate*
IO_3^- iod*ate*

It is necessary to learn the formulas and charges of these ions since the ion to which the -*ate* ending is assigned is not defined by the number of oxygen atoms or by the charge. The systematic nomenclature of a XO_n ion is based on the number of oxygen atoms in that ion relative to the number of oxygen atoms in the ion to which the -*ate* ending has been arbitrarily assigned. When the ion has one oxygen atom fewer than the -*ate* ion, the ending is changed to -*ite*. When the ion has two oxygen atoms fewer, the prefix *hypo-* is attached to the name in addition to the -*ite* ending. When the ion contains one oxygen atom more than the -*ate* ion, the prefix *per-* is attached.

		ClO^- *hypochlorite*
PO_3^{3-} phosph*ite*	SO_3^{2-} sulf*ite*	ClO_2^- chlor*ite*
PO_4^{3-} phosph*ate*	SO_4^{2-} sulf*ate*	ClO_3^- chlor*ate*
		ClO_4^- *perchlorate*

TABLE 2–2 Nomenclature of common polyatomic ions

+1 CHARGE	−3 CHARGE	−2 CHARGE	−1 CHARGE
NH_4^+ ammonium	PO_4^{3-} phosphate	SO_3^{2-} sulfite	NO_2^- nitrite
H_3O^+ hydronium	AsO_3^{3-} arsenite	SO_4^{2-} sulfate	NO_3^- nitrate
	AsO_4^{3-} arsenate	$S_2O_3^{2-}$ thiosulfate	ClO^- hypochlorite
		CrO_4^{2-} chromate	ClO_2^- chlorite
		$Cr_2O_7^{2-}$ dichromate	ClO_3^- chlorate
		CO_3^{2-} carbonate	ClO_4^- perchlorate
		SiO_3^{2-} silicate	BrO_3^- bromate
		HPO_4^{2-} biphosphate	IO_3^- iodate
		or hydrogen	MnO_4^- permangate
		phosphate	HSO_4^- bisulfate or hydrogen sulfate
		$C_2O_4^{2-}$ oxalate	$H_2PO_4^-$ dihydrogen phosphate
		O_2^{2-} peroxide	$HCOO^-$ formate
			CH_3COO^- acetate
			OH^- hydroxide
			CN^- cyanide

Eight of the negative ions in Table 2–2 are not of the form we have just described. These are CH_3COO^-, OH^-, CN^-, $HCOO^-$, $C_2O_4^{2-}$, O_2^{2-}, $S_2O_3^{2-}$, and $Cr_2O_7^{2-}$. Refer to Table 2–2 when you need to know the names, until you can remember them.

PROBLEM 2–8   Name the following ions: IO_3^-, IO_2^-, BrO^-.

COVALENT VERSUS IONIC BONDING

We have described the structure of compounds in which the fundamental unit is the molecule. The bonding in these molecules is what we refer to as *covalent* bonding. A covalent bond involves a *sharing* of one or more pairs of electrons. The pair of electrons cannot be assigned to either atom but is associated with both atoms simultaneously. In the molecular structure diagrams we have already drawn, each line joining two atoms represents a pair of electrons making up the covalent bond.

A very different type of bond called an *ionic bond* can be formed between two ions. In this case there are no electrons shared between the atoms. To illustrate, we will consider the bond between lithium, Li, and fluorine, F. We know from the previous section that Li can form a positive ion Li^+ by removal of an electron and that F can form a negative ion F^- by addition of an electron. The two oppositely charged ions can form a bond in which the bonding results primarily from electrostatic attraction. Despite the fact that lithium and fluorine usually exist as separate

TABLE 2-3 Formulas of some ionic compounds

CHEMICAL FORMULA	IONS
$NaCl$	Na^+, Cl^-
$MgCl_2$	Mg^{2+}, Cl^-
Na_2SO_4	Na^+, SO_4^{2-}
$MgSO_4$	Mg^{2+}, SO_4^{2-}
$Mg(OH)_2$	Mg^{2+}, OH^-
$Ca_3(PO_4)_2$	Ca^{2+}, PO_4^{3-}

ions, for convenience we write the molecular formula for lithium fluoride as LiF, rather than the more descriptive Li^+F^-. Compounds in which the bonding is electrostatic are called ionic compounds.

There are numerous examples of ionic compounds. In Table 2-3 we have given the chemical formulas for a few ionic compounds and for the ions involved. Note that the number of cations and anions in an ionic compound is such as to make the compound neutral. For example, there must be $2Cl^-$ for each Mg^{2+} in $MgCl_2$ since Cl^- has only a -1 charge and Mg^{2+} has a $+2$ charge.

Ionic compounds can usually be vaporized to a gas but only at exceedingly high temperatures (on the order of 600–1500 °C). In the gas phase, the compound exists in the form of molecules, such as LiF or $NaCl$ molecules.* However, in the solid state the compound forms a crystal made up of ions. One cannot identify molecules in the solid ionic compound. In $NaCl$, the Na^+ and Cl^- exist in an ordered three-dimensional arrangement, each Na^+ surrounded by 6 Cl^- ions and each Cl^- ion surrounded by 6 Na^+ ions. Figure 2-2 illustrates the crystal network of $NaCl$. It is impossible to assign a given Na^+ ion to a single Cl^- ion. Consequently, it is inaccurate to refer to molecules when talking about the solid, crystal form of an ionic compound. For an ionic compound in the solid state, the chemical formula represents only the proportions of the ions. The chemical formula $NaCl$ designates that the Na^+ and Cl^- are in a 1:1 proportion.

The reasons for the formation of covalent or ionic bonds will be discussed in Chapter 6. An adequate explanation requires a more detailed understanding of atomic structure and properties of atoms.

PROBLEM 2-9   Write the chemical formula for the ionic compounds formed from the following ions:

a. Al^{3+} and Cl^-
b. ammonium ion and the sulfate ion
c. SiO_3^{2-} and Na^+

*Since these molecules consist essentially of ions, some people prefer to call gaseous species such as LiF *ion pairs*. In any event, the LiF and $NaCl$ in the gas phase remain as distinct diatomic units.

FIGURE 2–2 Crystal structure of NaCl. A. location of Na$^+$ and Cl$^-$ at corners of a cube B. spheres representing relative sizes of Na$^+$ and Cl$^-$.

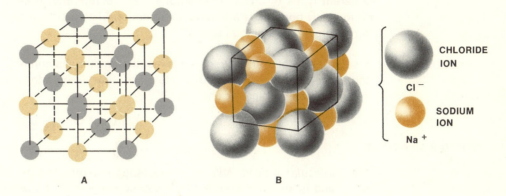

CHLORIDE
ION

Cl$^-$

SODIUM
ION

Na$^+$

A

B

COMPOUNDS VERSUS MIXTURES

It was emphasized earlier that the molecule (and thus the compound itself) is a unique species with its own properties. This set of properties is characteristic of the compound and is independent of the way in which the compound was prepared. This compound can be contrasted to a mixture, which does not contain substances in any definite or distinct proportion. The properties of a mixture will vary as the composition of the mixture varies.

To illustrate the difference between mixtures and compounds we can consider a simple laboratory experiment. We mix clean iron, Fe, dust with powdered sulfur, S. Different mixtures are formed as we vary the relative amounts of Fe and S. Properties such as color and density of the mixture vary as the composition is varied. Furthermore, the iron can be readily separated from the sulfur by physical means: since the iron is magnetic, it can be extracted with a magnet. The sulfur can be removed by combining the mixture with liquid carbon disulfide, since only sulfur will dissolve. Both the combination and the separation are physical processes and do not involve any chemical change in iron and sulfur. Iron and sulfur have not been combined chemically, in this example, to form compounds.

However, if one were to place any of the mixtures of iron and sulfur in a covered crucible and heat it to 1000 °C, the iron and sulfur would react to form a compound, FeS. In addition to the FeS, there might be excess iron or sulfur remaining since only a unique proportion of Fe and S will combine to form FeS. If the original mixture did not have this unique proportion, the excess elements will remain unchanged. The compound FeS is not magnetic and any attempt at separating Fe and S in FeS by this means will fail. Sulfur in FeS will not dissolve in carbon disulfide. Furthermore, any other *physical method* fails to separate iron from sulfur in FeS. Only a chemical reaction or a series of chemical reactions can be used to separate FeS into its components.

Heating a mixture to combine two elements directly to give a compound is not, of course, the only way to form compounds. Compounds can be formed by many different types of chemical reactions. Each compound, whether it is covalently or ionically bonded, is a unique combination of elements in definite proportions. A mixture is *not* a unique species.

PROBLEM 2–10 If sugar and water are mixed, is the resulting material a compound? Explain. Can the sugar be separated from the water. If so, how? Would the changes involved in separation be physical or chemical?

MOLECULAR WEIGHT

The *molecular weight*, *MW*, of a substance is expressed on the atomic weight scale and is simply the sum of the atomic weights of the atoms making up the compound. For ionic compounds, which do not consist of molecules, some people prefer to use the term *formula weight, FW*. In this text we will use the term molecular weight for both covalent and ionic compounds. An example or two will make the definition clear. A list of the atomic weights of the elements is given inside the front cover of the book and in the periodic table inside the back cover of the book.

EXAMPLE 2–2 *Calculate the molecular weight of water (H_2O).*

$$wt.\ of\ H = 2 \times at\ wt\ H = 2 \times 1.008 = 2.016$$
$$wt.\ of\ O = 1 \times at\ wt\ O = \underline{1 \times 16.00 = 16.00}$$

$$MW\ of\ H_2O = 18.02$$

Since there were 2 hydrogen atoms in H_2O, the weight of hydrogen in the compound was found by multiplying the atomic weight of hydrogen by 2. Similarly, there is only one oxygen atom so the atomic weight of oxygen was multiplied by one. The molecular weight is the sum of these two products.

EXAMPLE 2–3 *Calculate the molecular weight of sucrose, or common sugar, whose formula is $C_{12}H_{22}O_{11}$.*

$$wt.\ of\ C = 12 \times at\ wt\ of\ C = 12 \times 12.01 = 144.12$$
$$wt.\ of\ H = 22 \times at\ wt\ of\ H = 22 \times 1.008 = 22.18$$
$$wt.\ of\ O = 11 \times at\ wt\ of\ O = \underline{11 \times 16.00 = 176.00}$$

$$MW\ of\ C_{12}H_{22}O_{11} = 342.30$$

EXAMPLE 2–4 *Calculate the molecular weight of the ionic compound $Mg_3(PO_4)_2$.*

$$wt.\ of\ Mg = 3 \times at\ wt\ of\ Mg = 3 \times 24.31 = 72.93$$
$$wt.\ of\ P\ \ = 2 \times at\ wt\ of\ P\ \ \ = 2 \times 30.97 = 61.94$$
$$wt.\ of\ O\ \ = 8 \times at\ wt\ of\ O\ \ \ = \underline{8 \times 16.00 = 128.00}$$

$$MW\ of\ Mg_3(PO_4)_2 = 262.87$$

PROBLEM 2–11   Calculate the molecular weight of the covalent compound fructose, $C_6H_{12}O_6$ (fruit sugar).

PROBLEM 2–12   Calcium chloride readily absorbs water and is used as a drying agent. Calculate the molecular weight of the ionic compound $CaCl_2$.

PROBLEM 2–13   What are the proportions (by mass) of iron and sulfur that must be used to make FeS without having an excess of Fe or S?

PROBLEM 2–14   What proportions (by mass) of hydrogen and oxygen must be used in order to make H_2O without any excess of H or O?

PERCENTAGE COMPOSITION

Frequently we need to know how much of a specific element is contained in a compound. This information is sometimes expressed in terms of the percentage by mass of that element. In order to calculate the *percentage composition* by mass of a compound, one simply divides the total mass of each element by the molecular weight and multiplies each quotient by 100. This procedure gives the percentage composition by mass for each element.

$$\% \text{ element} = \frac{\text{no. atoms of element in formula} \times \text{at. wt.} \times 100}{\text{molecular weight}}$$

$$= \frac{\text{wt. of element} \times 100}{\text{molecular weight}}$$

An example will readily illustrate the concept.

EXAMPLE 2–5   *Calculate the percentage composition of H_2O. (You calculated the molecular weight of H_2O in Example 2–2.)*

$$\% \ H = \frac{2 \times (1.008)}{MW} \times 100 = \frac{2.016}{18.02} \times 100 = 11.2 \ \%$$

$$\% \ O = \frac{1 \times (16.00)}{MW} \times 100 = \frac{16.00}{18.02} \times 100 = 88.8 \ \%$$

PROBLEM 2–15   Calculate the percentage composition of $Mg_3(PO_4)_2$ and $C_6H_{12}O_6$.

DETERMINATION OF SIMPLEST FORMULA

Previously, we determined the percentage composition using the chemical formula for the compound. The percentage composition gives the proportions by mass of the elements in the compound and these proportional masses can be used to determine the proportions of atoms of each element in the compound.

The formula representing the proportion of atoms in the compound is called the *simplest formula*. If we represent a general compound composed of elements A, B, C, the simplest formula is

$$A_x \, B_y \, C_z$$

where x, y, z are the smallest whole numbers that designate the proportions of A, B, and C. If the compound is ionic, the simplest formula is also the chemical formula. However, if the compound is covalently bonded and exists as molecules, the simplest formula will *not* necessarily represent the molecular formula for the compound. A molecule of the compound may contain a multiple of x, y, and z atoms of A, B, and C, and the molecular weight must be known in order to determine this multiple.

The atomic weight of an element represents the average weight of an atom of that element relative to other elements. Therefore, the mass of an element in a compound divided by the atomic weight of the element is proportional to the number of atoms of that element in that compound

$$\text{number of atoms} \; \alpha \; \frac{\text{mass}}{\text{mass (relative)/atom}} \; \alpha \; \frac{\text{mass}}{\text{atomic weight}}$$

For our general compound $A_x B_y C_z$, the x, y, and z can then be expressed as

$$x \; \alpha \; \frac{m_A}{\text{at. wt. A}}$$

$$y \; \alpha \; \frac{m_B}{\text{at. wt. B}}$$

$$z \; \alpha \; \frac{m_C}{\text{at. wt. C}}$$

where m_A, m_B, and m_C are the masses of A, B, and C in some sample of the compound. If the percentage composition is known, then the values for m_A, m_B, and m_C can be taken as the masses for a 100 g sample of the compound. The values for x, y, and z are then reduced to the smallest possible whole numbers by dividing x, y, and z by the smallest number of the set. If the resulting values for x, y, or z are not whole numbers, it may be necessary to multiply them by some appropriate factor to make them whole numbers. An example showing the mathematical steps will clarify the procedure.

EXAMPLE 2-6 *Calculate the simplest formula for an iron oxide which contains 69.94 % Fe and 30.06 % O.*

$$Fe_x \, O_y$$

$$x \; \alpha \; \frac{69.94}{55.85} = 1.252$$

$$y \; \alpha \; \frac{30.06}{16} = 1.879$$

Dividing the values by the smallest number 1.252 in the set,

$$x \; \alpha \; 1.000$$

$$y \; \alpha \; 1.501$$

If we multiply both values by 2, we reduce both to whole numbers

$$x = 2$$

$$y = 3$$

The simplest formula is thus

$$Fe_2O_3$$

PROBLEM 2–16 Using the results for $C_6H_{12}O_6$ from Problem 2–15, calculate the simplest formula. Is the simplest formula equal to the molecular formula? What additional information is required to determine the molecular formula from the simplest formula?

2–C. NOMENCLATURE OF INORGANIC COMPOUNDS

In this section a systematic procedure is given for the nomenclature of all inorganic compounds except for complex ions and acids. The nomenclature of complex ions will be discussed in Chapter 16, where we also talk about transition metals and lanthanides. Acid-base reactions are discussed in Section 3–E, and the systematic nomenclature of acids will be presented then. The nomenclature in organic chemistry is discussed in Chapter 7.

COMPOUNDS NAMED ON THE BASIS OF IONIC CONSTITUENTS

We name many compounds as if their constituent atoms or groups of atoms could exist as ions. This does *not* necessarily mean that the compounds themselves are ionic. The compounds may or may not be ionic in nature, but the atoms or groups of atoms which they contain could, upon separation, exist as ions. As a simple example let us take hydrogen chloride, HCl. This is a covalent compound, but its constituents can exist as the ions H^+ and Cl^-. The name of the compound is based on the names of its constituent ions.

The names of common monatomic and polyatomic cations and anions were given in Tables 2–2 and 2–3. You should be familiar with most of these ions, their charges, and their names. Many inorganic compounds consist of only one type of cation and one type of anion. The name of the compound simply gives the name of the cation first and the anion second. The examples in Table 2–4 show the molecular formula, ionic constituents, and the name of the compound.

Since any ion has a specific charge, two ions can only combine in a unique proportion in order for the compound to be neutral. Therefore, only the names of the ions need be specified in order to unambiguously designate the molecular formula. For example, calcium phosphate consists of Ca^{2+} ions and PO_4^{3-} ions. These can only combine in the ratio 3:2 for the compound to be neutral. Hence, we know that calcium phosphate must have the formula $Ca_3(PO_4)_2$.

It should also be noted in the examples we have just given that either alternate name of a cation could be used in naming the compound. For example, $FeCl_3$ could be called iron (III) chloride or ferric chloride.

EXAMPLE 2–7   *Write chemical formulas for the following compounds: a. chromium (III) chloride and b. chromous arsenite.*

   a. Chromium (III) ion is Cr^{3+} and chloride ion is Cl^- (see Table 2–1). If the compound chromium (III) chloride is to be neutral, we must have 3 Cl^- for each Cr^{3+}. The formula is thus

$$CrCl_3$$

TABLE 2–4   Nomenclature of inorganic compounds based on ionic constituents

MOLECULAR FORMULA	CATION	ANION	NAME OF COMPOUND
HCl	H^+, hydrogen ion	Cl^-, chloride	hydrogen chloride
NaCl	Na^+, sodium ion	Cl^-, chloride	sodium chloride
$MgCl_2$	Mg^{2+}, magnesium ion	Cl^-, chloride	magnesium chloride
$Al(OH)_3$	Al^{3+}, aluminum ion	OH^-, hydroxide	aluminum hydroxide
$MgSO_4$	Mg^{2+}, magnesium ion	SO_4^{2-}, sulfate	magnesium sulfate
$Mg_3(PO_4)_2$	Mg^{2+}, magnesium ion	PO_4^{3-}, phosphate	magnesium phosphate
$FeCl_2$	Fe^{2+}, iron(II) ion (ferrous ion)	Cl^-, chloride	iron(II) chloride (ferrous chloride)
$FeCl_3$	Fe^{3+}, iron(III) ion (ferric ion)	Cl^-, chloride	iron(III) chloride (ferric chloride)
$CrAsO_4$	Cr^{3+}, chromium(III) ion (chromic ion)	AsO_4^{3-}, arsenate	chromium(III) arsenate (chromic arsenate)
$Cr(ClO)_2$	Cr^{2+}, chromium(II) ion (chromous ion)	ClO^-, hypochlorite ion	chromium(II) hypochlorite (chromous hypochlorite)
$CaHPO_4$	Ca^{2+}, calcium ion	HPO_4^{2-}, biphosphate ion (hydrogen phosphate ion)	calcium biphosphate (calcium hydrogen phosphate)

b. *Chromous ion is Cr^{2+} (chromium (II)) and arsenite ion is AsO_3^{3-} (see Tables 2–1 and 2–2). To have equal positive and negative charge, we must have*

$$3\ Cr^{2+} = 3 \times (+2) = +6$$

$$2\ AsO_3^{3-} = 2 \times (-3) = -6$$

The formula is thus

$$Cr_3(AsO_3)_2$$

PROBLEM 2–17   Name the compounds represented by the following molecular formulas: $RbBr$, BeF_2, AlN, Cr_2O_3, $(NH_4)_2CO_3$, HBr, $K_2Cr_2O_7$, $Cr(HSO_4)_2$, $Fe(CN)_3$.

PROBLEM 2–18   Write molecular formulas for the following compounds: rubidium hypobromite, calcium phosphide, iron (III) phosphate, sodium hydrogen phosphate, calcium carbide, iron (III) oxalate, hydrogen selenide.

BINARY COVALENT COMPOUNDS OF NONMETALLIC ELEMENTS

A binary compound consists of only two elements. Certain elements, classified as nonmetals, can combine to form binary covalent compounds. These covalent compounds exist as molecules. There is a definite number of atoms of each element in a molecule of any given binary covalent compound. However, the same elements can combine in different proportions to form different compounds. For example, nitrogen and oxygen can form the compounds: N_2O, NO, NO_2, N_2O_4, N_2O_5. For this reason the name of the compound must specify the number of atoms of each type in the molecule. This is accomplished using prefixes:

NUMBER	PREFIX	NUMBER	PREFIX
1	mono-	5	penta-
2	di-	6	hexa-
3	tri-	7	hepta-
4	tetra-	8	octa-

The use of the prefix *mono-* is optional. It is most commonly used to emphasize the *mono-* compound when *di-* and *tri-* compounds also exist for the same elements. For example, CO—carbon *mono*xide, and CO_2—carbon *di*oxide.

The order of writing the names of the two elements is based on the concept of *electronegativity*. Electronegativity is defined qualitatively as the relative ability of one atom versus another to attract the shared pair of electrons in a covalent bond. There are several electronegativity scales which have been devised to define

TABLE 2–5 Electronegativity of several elements

ELEMENT	ELECTRONEGATIVITY	ELEMENT	ELECTRONEGATIVITY
F	4.0	Be	1.5
O	3.5	Al	1.5
N	3.0	Mg	1.2
Cl	3.0	Li	1.0
Br	2.8	Ca	1.0
C	2.5	Sr	1.0
I	2.5	Na	0.9
H	2.1	Ba	0.9
P	2.1	K	0.8
B	2.0	Rb	0.8
Si	1.8	Cs	0.7

electronegativity quantitatively. These will be discussed quantitatively and in detail in Chapter 6. For convenience a list of electronegativities of the more common elements is given in Table 2–5.

You will also be able to get an ordering of electronegativities from the periodic table (see Figure 6–12). The electronegativity generally increases as one goes from left to right in the periodic table (with the exception of the right hand column of inert gases in group VIIIA) and from the bottom to the top. Thus, the most electronegative element is fluorine.

Elements with low electronegativity are often referred to as *electropositive*. The order of naming the elements in the binary covalent compound is the more electropositive element first and the more electronegative element second. In addition, the electronegative element is given the characteristic *-ide* ending. Therefore, in naming a binary covalent compound one goes through the following mental steps: 1. choose the more electropositive element and order the elements accordingly, 2. select the proper prefixes designating the number of atoms of each type, 3. add the *-ide* suffix to the more electronegative element. Several examples of naming binary covalent compounds are given in the following list. Alternative common names are given in parenthesis for some compounds.

MOLECULAR FORMULA	NAME
SO_2	sulfur dioxide
SO_3	sulfur trioxide
N_2O	dinitrogen oxide (nitrous oxide)
NO	nitrogen oxide (nitric oxide)
NO_2	nitrogen dioxide
N_2O_4	dinitrogen tetroxide
N_2O_5	dinitrogen pentoxide

PROBLEM 2-19   Name the compounds represented by the following molecular formulas: TeF_4, SO_2, PCl_3, SF_6, ClO_2, Cl_2O_7, OF_2.

PROBLEM 2-20   Write molecular formulas for the following compounds: carbon tetrachloride, phosphorus pentachloride, bromine pentafluoride, diphosphorus pentoxide (frequently called phosphorus pentoxide), boron trifluoride, silicon tetrafluoride.

In some cases common names for compounds have been adopted and used for many years. Since these common names are not systematized in any way, it is unfortunately necessary to memorize or look up these names in order to associate them with the proper molecular formulas. These common names for compounds will be given as the compound is introduced in the text. These common names are analogous to slang expressions in a language, and they cause great difficulty in using the language.

2-D. MOLE CONCEPT

The atomic weight scale, as we have stated, is a relative scale of atomic masses based upon the isotope ^{12}C, which is assigned the value 12.0000 amu. Any molecular weight that we calculate is the mass of a typical molecule of the substance in amu. The mass of a molecule is found by taking a weighted average of the masses of molecules in a typical sample of that substance. For example, on the average

1 atom of Ar has a mass of 39.948 amu

1 atom of ^{12}C has a mass of 12.000 amu

1 molecule of H_2O has a mass of 18.015 amu

1 molecule of $NaCl$ has a mass of 58.443 amu

Chemists seldom work with single atoms or molecules in the laboratory, so a relative weight scale in amu is not practical. Generally, we work with samples whose masses can be given in milligrams or grams. So that we can give the relative masses of substances in units of grams, we define *gram-atomic weight* as the atomic weight expressed in grams and *gram-molecular weight* as the molecular weight expressed in grams. The terms *gram-atom* and *mole* can then be defined as

2-1   1 gram-atom (g-atom) = quantity of an element whose mass is the
gram-atomic weight (g-at wt)

1 mole = quantity of a compound whose mass is the gram-molecular
weight (g-mol wt)

The relative weights of atoms and molecules can now be expressed in terms of the quantities of a gram-atom and a mole. Using the examples used previously, we can state that

1 g-atom of Ar has a mass of 39.948 g

1 g-atom of ^{12}C has a mass of 12.000 g

1 mole of H_2O has a mass of 18.015 g

1 mole of $NaCl$ has a mass of 58.443 g

Note in the above examples that the masses of g-atoms and moles are in the same proportion as are the masses of atoms and molecules. The g-atom and mole have the same mass relationship as the atom and molecule except that the g-atom and the mole each contain a large number of atoms and molecules. The number of atoms in a g-atom or of molecules in a mole has been determined experimentally and is known as Avogadro's number, N.*

$$N = \text{Avogadro's number} = 6.023 \times 10^{23}$$

If the substance is an ionic substance, such as solid $NaCl$, Avogadro's number represents the number of formula units. A formula unit of $NaCl$ would be a sodium and a chloride ion. In the case of $BaCl_2$, the formula unit would consist of one Ba^{2+} and two Cl^-. Using Avogadro's number we can write convenient expressions showing the relationship between g-atoms and atoms, and between moles, molecules, or formula units.

2–2a   1 g-atom = 6.023×10^{23} atoms

2–2b   1 mole = 6.023×10^{23} molecules

2–2c   1 mole = 6.023×10^{23} formula units (ionic compound)

The concepts of mole, gram-molecular weight, Avogadro's number and the parallel concept of g-atom, gram-atomic weight, and Avogadro's number are essential to chemistry. You must be able to use them in calculations as well as understand their significance.

EXAMPLE 2–8   *Calculate the number of moles and the number of molecules of H_2O in a 25.0 g sample.*

The molecular weight of H_2O is 18.015, hence from equation 2–1

1 mole H_2O has a mass of 18.015 g

*For a review of exponentials, see Appendix C.

The 25 g of H_2O can thus be converted to moles by multiplying by the factor (1 mole/18.015 g).* Therefore

$$moles\ H_2O = 25.0\ g\ \left(\frac{1\ mole}{18.015\ g}\right) = 1.39\ moles$$

According to equation 2–2b

$$molecules\ H_2O = (1.39\ moles\ H_2O)\left(\frac{6.023 \times 10^{23}\ molecules}{mole}\right)$$

Therefore

$$molecules\ of\ H_2O = 8.36 \times 10^{23}\ molecules$$

Note the cancellation of grams in the first equation and moles in the second equation. It is essential that the units always be written down with each term and cancelled when appropriate. This procedure will insure that factors are used properly and that equations are dimensionally correct.

EXAMPLE 2–9   Calculate the mass and number of molecules in a sample of 2.5 moles of benzene which has a molecular formula of C_6H_6.
The molecular weight of benzene is found by

$$MW = 6(atomic\ weight\ C) + 6(atomic\ weight\ H)$$

$$MW = 6(12.01) + 6(1.008)$$

$$MW = 78.11$$

The gram-molecular weight for benzene is thus 78.11 g. The mass of the 2.5 moles of benzene sample is found using the expression

$$1\ mole\ benzene\ has\ a\ mass\ of\ 78.11\ g$$

$$mass\ benzene = (2.5\ moles)\left(\frac{78.11\ g}{1\ mole}\right)$$

$$mass\ benzene = 195.3\ g$$

Using Avogadro's number from equation 2–2

$$Molecules\ of\ benzene = (2.5\ moles\ benzene)\left(6.023 \times 10^{23}\ \frac{molecules}{mole}\right)$$

$$Molecules\ of\ benzene = 15.06 \times 10^{23}\ molecules$$

*For a discussion of the factor method for the conversion of units, see Appendix A.

PROBLEM 2–21   Calculate the number of g-atoms and atoms in a 5 g sample of oxygen whose molecular formula is O_2.

PROBLEM 2–22   Calculate the number of moles and molecules in a 5 g sample of oxygen recalling that oxygen exists as a diatomic molecule O_2.

PROBLEM 2–23   Calculate the number of Ca^{2+} and Cl^- ions in 3 moles of $CaCl_2$.

2–E. STATES OF MATTER

Thus far in the discussion of matter we have seen that a substance is composed of molecules which in turn are composed of atoms. Of course, molecules and atoms are far too small to be seen with the naked eye. Only large aggregates of the molecules in the form of material substances are visible. These aggregates of molecules can be solid, liquid, or gas. *Solid*, *liquid*, and *gas* are the three basic states of matter. Pure substances, as we see them in nature, exist in one of these three *states* or *phases*.

SOLID STATE

The solid state can be distinguished from the liquid and gas states by its ability to retain its shape and size. The solid consists of a three-dimensional network of molecules, atoms, or ions which are bonded or attracted more strongly to one another than are liquid or gas molecules. (See Figure 2–2 for the structure of solid NaCl.) The molecules, atoms, or ions occupy specific locations relative to one another in this three-dimensional structure. For this reason it is said that solids have long range order (order over long distances) in contrast to liquids in which there is ordering of the molecules, atoms, or ions but order is only short-range. Note in the structure of NaCl in Figure 2–2(a) that a Cl^- occupies each corner and each face of the cube. The Na^+ occupy each edge of this cube. This three-dimensional arrangement of Na^+ and Cl^- is repeated many times over distances (or ranges) through the crystalline solid.

The molecules, atoms, or ions in a solid are generally closer to one another than are liquid or gas particles, and the solid state is thus more dense than the corresponding liquid or gas state. Water is an exception to this rule. The density of liquid water, H_2O, is greater than that of ice, solid H_2O. As we know, ice floats on water. In mixtures of most solid and liquid substances, the solid will sink in the liquid since the density of the solid is greater than that of the liquid. We shall discuss these points further in Chapter 9–L.

A detailed discussion of the bonds or attractive forces between the molecules, atoms, or ions must be delayed until we have a deeper understanding of molecular structure and molecular properties. We can estimate the relative magnitude of the forces holding the solid together by considering the melting point and heat of

TABLE 2-6 Normal melting points and heats of fusion of some solid substances

SUBSTANCE	FORMULA	$t_{mp}(°C)$	$\Delta H_{fus}(cal/mole)$
hydrogen	H_2	−259.1	28.
oxygen	O_2	−218.4	106.3
hydrogen chloride	HCl	−114.8	476.0
water	H_2O	0.0	1436.
sodium chloride	NaCl	801.	7220.
silicon dioxide (quartz)	SiO_2	1710.	3400.

fusion of the solid. The *normal melting point* (t_{mp}) is defined as the temperature at which the solid at one atmosphere pressure is transformed to give the liquid state. The heat required to make this transformation once the solid is at the correct temperature is called the molar *heat of fusion* (ΔH_{fus}). The molar heat of fusion is generally expressed in units of calories of heat required to melt one mole of the substance. One calorie is defined as the amount of heat required to raise the temperature of one gram of water one °C.

$$solid \xrightarrow{\text{heat}} liquid$$

where $P = 1$ atmosphere, t = normal melting point = t_{mp}, and heat = molar heat of fusion = ΔH_{fus}

The liquid state is a more disordered state in which the intermolecular and interionic forces are weaker than in the solid. The heat of fusion is an indication of the decrease in intermolecular forces between the solid and liquid phases. The normal melting points and heats of fusion can vary over a very large range as shown in Table 2-6.

There are also solids, called *amorphous solids*, which retain a rigid shape but do not have an ordered array or network of molecules or ions. A common example of an amorphous solid is *glass*, which consists of calcium and sodium silicates. The silicon atoms are held in the solid phase by sequential Si—O—Si bonds of two types.

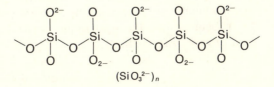

$(SiO_3^{2-})_n$

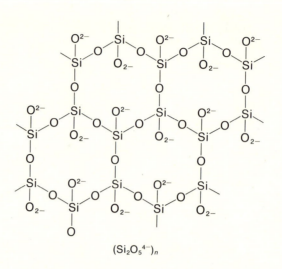

$$(Si_2O_5{}^{4-})_n$$

Consequently, the solid is rather rigid. Since these chains are not regular, however, the solid does not have a definite array or network of Ca, Na, Si, and O atoms or ions. In contrast to a crystalline solid the structure of glass is short range and variable throughout. Furthermore, there is *not* a discrete temperature at which an amorphous solid undergoes transition to the liquid phase. The amorphous solid softens and continues to soften as the temperature is raised. Eventually, the amorphous solid will become fluid but the change occurs over a large temperature range. Plastics, such as polyethylene, show properties characteristic of amorphous solids.

LIQUID STATE

The liquid state differs from the solid and gas states in its ability to retain its volume but not its shape. There are significant intermolecular, interatomic, or ionic forces in the liquid and these forces assist in holding the atoms, molecules, or ions together at short distances. However, the additional heat which the liquid state has absorbed allows the molecules, atoms, or ions to move more freely than the particles of a solid, which are held together rigidly. As a result the liquid state is fluid, but able to retain its volume. There is still some limited structure in the liquid state. The structure of ice and of water will be discussed in detail in Section 9–L.

When the liquid state is heated, a temperature is reached at which the molecules of the liquid can escape into the surrounding environment as a gas. If the liquid is at one atmosphere pressure, this temperature is called the *normal boiling point* of the substance and the heat required for the transformation is called the *molar heat of vaporization* (Table 2–7).

$$liquid \xrightarrow{heat} gas$$

TABLE 2–7 Normal boiling point and heat of vaporization of some liquid substances

SUBSTANCE	FORMULA	t_{bp}(°C)	ΔH_{vap}(cal/mole)
hydrogen	H_2	−252.5	216
oxygen	O_2	−183.0	1,630
hydrogen chloride	HCl	−84.9	4,680
water	H_2O	100.0	9,720
sodium chloride	NaCl	1,413.0	43,400

where P = one atmosphere, t = normal boiling point = t_{bp}, and heat = molar heat of vaporization = ΔH_{vap}.

When the liquid becomes a gas at one atmosphere pressure, the intermolecular, atomic, or ionic forces are greatly reduced. Consequently, the magnitudes of the normal boiling point and heat of vaporization are indications of the strengths of the intermolecular, atomic, and ionic forces in the liquid. Again a very large range of normal boiling points and heats of vaporization are encountered.

GASEOUS STATE

The gaseous phase consists of molecules or atoms which are widely spaced and free to move within a container. Ions are never produced by vaporization. Substances which are ionic or essentially ionic in the solid and liquid states generally form neutral molecules upon vaporization. Because of the freedom of the molecules or atoms in the gas phase, a gas has neither fixed volume nor shape.

In contrast to the forces characterizing solid and liquid states, the intermolecular or interatomic forces are almost negligible in the gas phase. This is especially true at low pressures, on the order of one atmosphere or less, because the distance between molecules is quite large at these low pressures. In fact, we usually ignore the intermolecular or interatomic forces altogether when we are discussing gases at low pressures. This assumption is part of the model for the ideal gas as discussed in Section 2–G.

2–F. SOLUTIONS—CONCENTRATION UNITS

A solution is a homogeneous mixture of two or more substances which has a unique set of properties characteristic of that specific solution. The homogeneity in the case of a true solution is at the molecular level. For example, when sugar, $C_{12}H_{22}O_{11}$, is dissolved in water, the $C_{12}H_{22}O_{11}$ molecules and H_2O molecules are intimately mixed. The $C_{12}H_{22}O_{11}$ molecules are completely dispersed throughout the H_2O and the solution will not have any properties characteristic of the pure solid sugar or pure water. A solution differs from a compound in that the proportions of the species in a solution can vary. Different solutions of sugar in water can be prepared by dissolving different amounts of sugar in a definite volume of water. In a compound, the atoms combine only in definite and unique proportions.

Solutions can exist in all three phases: gas, liquid, and solid. The most common solutions are in the liquid phase, but gas and solid solutions can also exist. Gases always mix to give a homogeneous gas mixture, such as air. Solid solutions are less common, but certain solid substances can mix homogeneously at the molecular or ionic level. Examples of solid solutions are metal alloys and solder. Since solid compounds have unique crystal networks of molecules or ions, the requirements are rather stringent in order for two solids to form a solid solution. A liquid solution can consist of the mixture of two liquids or a solid or gas dissolved in a liquid substance.

Most covalent compounds (such as sugar in water) dissolve as molecules, i.e., molecules of sugar and water are homogeneously distributed throughout the solution. However, ionic compounds, and some covalent compounds, will dissociate into ions in solution. For example, solid NaCl ($NaCl_{(s)}$) dissolves as Na^+ and Cl^- ions in water. The Na^+ and Cl^- ions will be homogeneously distributed throughout the solution. The Na^+ and Cl^- are completely surrounded by water molecules and we designate this condition with a subscript (*aq*) representing an *aqueous* solution. We can show the dissolution process in terms of an equation.

$$NaCl_{(s)} + \text{water} \longrightarrow Na^+_{(aq)} + Cl^-_{(aq)}$$

$$C_{12}H_{22}O_{11} + \text{water} \longrightarrow C_{12}H_{22}O_{11(aq)}$$

When the composition of a solution is varied, the properties of the solution will also vary correspondingly. For example, a solution containing 1 mole of sugar in 10 moles of water is quite different from a solution containing only 0.1 mole of sugar in 10 moles of water. For this reason we must state the composition of a solution in order to describe the solution completely. The composition of a solution can be expressed in various types of concentration units. Different concentration units are useful in different contexts.

The components of a solution are frequently identified and referred to as *solvent* and *solute*. The *solvent* is generally taken as the component in the solution in the highest concentration (major component). On the other hand, a *solute* (minor component) in solution is any component at a lower concentration than the solvent. Solids or gases which dissolve in a liquid to form a liquid solution are generally designated as the solutes. There can be any number of solutes in a solution and the concentration of each solute should be specified. In the following discussion of concentration units, only a two-component solution will be considered and where appropriate the solvent will be designated by A and the solute by B. This nomenclature will be retained throughout the text when we refer to solutions.

A simple way to express concentration is *percentage by mass*. Since the term "weight" is frequently misused in place of mass, this is also commonly referred to as *percentage by weight*. Letting *m* represent mass, the percentage by mass for components A and B can be calculated according to equation 2–3.

2–3a
$$\% \text{ A by mass} = \frac{m_A}{m_A + m_B} \times 100$$

2–3b
$$\% \text{ B by mass} = \frac{m_B}{m_A + m_B} \times 100$$

In an analogous manner the concentration can be expressed in terms of *percentage by volume.* Letting *V* represent volume, expressions for the calculation of percentage by volume are

2–4a
$$\% \text{ A by volume} = \frac{V_A}{V_A + V_B} \times 100$$

2–4b
$$\% \text{ B by volume} = \frac{V_B}{V_A + V_B} \times 100$$

In order to convert from percentage by mass to percentage by volume, the *densities* of the substances must be known. The density, ρ, is defined as the mass per unit volume and can be calculated from the known mass and volume of the substance

2–5
$$\rho = \frac{m}{V}$$

The concentration units parts per million (ppm) and parts per billion (ppb) are frequently used to express very small amounts of one substance (solute) in the presence of almost 100 % of another (solvent). Frequently, impurities are in small concentrations and are expressed as ppm or ppb. The units ppm and ppb are defined on the basis of *mass* of solute and solvent.

$$1 \text{ ppm} = 1 \text{ part solute}/10^6 \text{ parts solvent}$$

$$1 \text{ ppb} = 1 \text{ part solute}/10^9 \text{ parts solvent}$$

The concentration in ppm and ppb can be calculated from

2–6a
$$\text{ppm} = \frac{\text{mass}_{\text{solute}}}{\text{mass}_{\text{solvent}}} \times 10^6$$

2–6b
$$\text{ppb} = \frac{\text{mass}_{\text{solute}}}{\text{mass}_{\text{solvent}}} \times 10^9$$

For very low solute concentrations the mass of the solution is approximately the mass of solvent and the unit ppm is related to % by mass by

$$\text{ppm} = (\text{\% by mass}) \times 10^4$$

A concentration unit which is very commonly used in chemistry is the *mole fraction*. Mole fraction is simply the fraction of the moles of each species in the solution. Letting *n* represent moles, the mole fraction (X) can be calculated according to the equations:

2–7a

$$X_A = \frac{n_A}{n_A + n_B}$$

2–7b

$$X_B = \frac{n_B}{n_A + n_B}$$

The concentration unit *molality, m,* is defined as the number of moles of solute dissolved in 1 kilogram (1,000 g) of solvent. If the number of moles of solute is represented by n_B, the molality can be calculated from

2–8

$$m = \frac{n_B}{\text{kg solvent}}$$

The calculation of n_B from the mass of solute was discussed in a previous section using the equivalence expression in equation 2–1

$$n_B = [\text{mass}_{\text{solute}}\ (\text{grams})]\left(\frac{1\ \text{mole}}{\text{g-mol wt}}\right) = \frac{\text{mass (grams)}}{\text{MW (grams)}}\ \text{moles}$$

The *molarity, M,* of a solution is defined as the moles of solute dissolved in 1 liter of solution

2–9

$$M = \frac{n_B}{\text{liter solution}}$$

Again n_B can be calculated from the mass of solute as given above.

EXAMPLE 2–10   *Calculate the % by weight, % by volume, mole fraction, molality, and molarity of a solution containing 10 grams of liquid ethylene glycol, $OH(CH_2)_2OH$, in 500 grams of water. The density of ethylene glycol is 1.1088 g/ml and density of water is 1.00 g/ml. The density of the final solution is 1.004 g/ml.*

Using equations 2–3 through 2–9, we can calculate the concentration as:

% BY MASS

$$\% \ OH(CH_2)_2OH \ by \ mass = \frac{10 \ g}{500 \ g + 10 \ g} \ 100 = 1.96 \ \%$$

$$\% \ H_2O \ by \ mass = \frac{500 \ g}{500 \ g + 10 \ g} \ 100 = 98.04 \ \%$$

% BY VOLUME

From the density of $OH(CH_2)_2OH$ as 1.1088 g/ml we can write the equivalence expression

$$1.1088 \ g = 1 \ ml$$

The volume of 10.0 g of $OH(CH_2)_2OH$ is then

$$V_{OH(CH_2)_2OH} = (10.0 \ g)\left(\frac{1 \ ml}{1.1088 \ g}\right) = 9.03 \ ml$$

Similarly,

$$V_{H_2O} = (500 \ g)\left(\frac{1 \ ml}{1 \ g}\right) = 500 \ ml$$

The % by volume are then calculated:

$$\% \ OH(CH_2)_2OH \ by \ volume = \frac{9.03 \ ml}{500 \ ml + 9.03 \ ml} \ 100 = 1.77 \ \%$$

$$\% \ H_2O \ by \ volume = \frac{500 \ ml}{500 \ ml + 9.03 \ ml} \ 100 = 98.23 \ \%$$

PARTS PER MILLION

$$ppm \ OH(CH_2)_2OH = \frac{10 \ g}{500 \ g} \times 10^6 = 2 \times 10^4 = 20,000 \ ppm$$

MOLE FRACTION

The molecular weight of $OH(CH_2)_2OH$ is 62, then from equation 2–1

$$1 \ mole \ OH(CH_2)_2OH \ has \ a \ mass \ of \ 62 \ g$$

and the moles of $OH(CH_2)_2OH$ in 10.0 g is

$$n_{OH(CH_2)_2OH} = (10.0 \text{ g})\left(\frac{1 \text{ mole}}{62 \text{ g}}\right) = 0.161 \text{ moles}$$

Similarly the molecular weight of H_2O is 18 and

$$1 \text{ mole } H_2O \text{ contains } 18 \text{ g}$$

$$n_{H_2O} = (500 \text{ g})\left(\frac{1 \text{ mole}}{18 \text{ g}}\right) = 37.8 \text{ moles}$$

$$X_{OH(CH_2)_2OH} = \frac{0.161 \text{ moles}}{37.8 \text{ mole} + 0.161 \text{ moles}} = 0.00425$$

$$X_{H_2O} = \frac{37.8 \text{ moles}}{37.8 \text{ moles} + 0.161 \text{ moles}} = 0.99575$$

MOLALITY

Since $500 \text{ g } H_2O = 0.500 \text{ kg } H_2O$

$$m = \frac{n_{OH(CH_2)_2OH}}{\text{kg } H_2O} = \frac{0.161 \text{ moles}}{0.500 \text{ kg}}$$

$$m = 0.322 \text{ moles/kg}$$

MOLARITY

The density of the solution is 1.004 g/ml hence

$$1.004 \text{ g occupies } 1 \text{ ml}$$

and the volume of $(500 \text{ g} + 10 \text{ g}) = 510 \text{ g}$ of solution is

$$V_{soln} = (510 \text{ g})\left(\frac{1 \text{ ml}}{1.004 \text{ g}}\right) = 508 \text{ ml} = 0.508 \text{ liters}$$

The molarity is then calculated

$$M = \frac{n_{OH(CH_2)_2OH}}{\text{liters solution}} = \frac{0.161 \text{ moles}}{0.508 \text{ liter}}$$

$$M = 0.317 \text{ moles/liter}$$

PROBLEM 2–24   Calculate the mole fractions of ethanol (C_2H_5OH) and H_2O in a solution containing 95 % ethanol by mass.

PROBLEM 2–25   Calculate the molarity of a solution prepared by dissolving 1.5 grams of naphthalene, $C_{10}H_8$, in sufficient benzene, C_6H_6, to make the final volume 250 ml.

PROBLEM 2–26   0,30 mg of magnesium nitrate were dissolved in 1 ml of water. Calculate the molality of the Mg^{2+} ion and the NO_3^- in solution. Recall that the density of water is 1 g/ml.

2–G. IDEAL GAS LAW

Gases consist of molecules or atoms which are in continual motion. The pressure of the gas arises from the collisions of the gas molecules with the walls of the container. At ordinary pressures (1 atmosphere) and temperatures (> 0 °C) of a gas, the distance between the molecules is large compared to the size of the molecules. Under these conditions one may make the following approximations: 1. the size of the molecules can be neglected, i.e., the molecule has its mass at a point, thus occupying no volume, 2. the attractive force between molecules can be ignored at the large internuclear distances involved.

Under these conditions, the gas molecules collide with the walls of the container without influence from the other gas molecules in the sample. It can be shown mathematically that given these two assumptions, the pressure of a gas in a closed container is dependent on the number of moles of gas, the volume and temperature, but is independent of the nature of the molecular species. For example, one mole of N_2 gas and one mole of CH_4 gas would each exert the same pressure on the walls of a container if the volume and temperature were the same, even though N_2 and CH_4 have different molecular weights, sizes, and shapes. We say that these different gases, in behaving similarly, exhibit the properties of an *ideal gas*. Bear in mind that we are ignoring molecular volume and intermolecular forces when we attribute identical behavior to different gases. In the strictest sense, no gas is truly an *ideal* gas. However, at low pressure and high temperature, the pressure, volume, and temperature behavior of real gases are satisfactorily approximated by those of an ideal gas.

ABSOLUTE TEMPERATURE—CHARLES' LAW

There are three laws which describe the behavior of an ideal gas—Boyle's law, Charles' law, and Avogadro's law. These laws describe the relationship between the variables pressure, volume, temperature, and number of moles. According to Boyle's law, the pressure of a gas *varies inversely* with volume (temperature held

FIGURE 2-3 Volume of an ideal gas as a function of temperature.

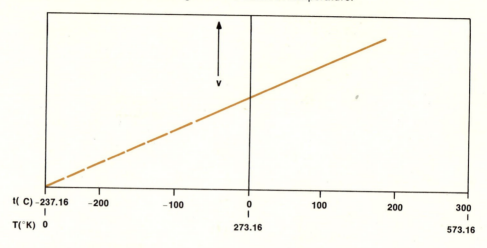

constant). Charles' law describes the relationship between volume and temperature and, furthermore, serves as a basis for the *absolute temperature scale*.
Charles' law states that the volume of a gas *varies linearly* with the temperature
when the pressure and number of moles of the gas are held constant.

If the experimental volume measurements are plotted against the experimental
temperature measurements, a straight line should be obtained. (See Appendix F
for further discussion of the linear function.) This line is described by the equation

2-10 $$V = at + b \quad (P, n = \text{constants})$$

where a = slope, b = intercept, t = temperature (°C). Graphically we can represent
the linear relationship as shown in Figure 2-3. If one extrapolates the linear function back to $V = 0$, the intercept is −273.16.* This extrapolation permits us to
define a new temperature scale called the *absolute temperature* which is assigned
a value of zero at −273.16 °C. We designate this absolute temperature scale,
which measures temperature in degrees Kelvin (°K), by the variable T.

The relationship between °C and °K is

$$T(°K) = t(°C) + 273.16$$

which corresponds to a simple displacement of the origin on the t(°C) axis. The
relationship between the t(°C) and T(°K) scales can be readily visualized by comparing the two thermometers in Figure 2-4.

*In order to obtain a number with this precision, one must also extrapolate very precise data back to
$P = 0$.

FIGURE 2-4   Comparison of the centigrade (Celsius) and absolute (Kelvin) temperature scales. The freezing point and boiling point of water are used as the reference points. Note that the scales differ only in the displacement by 273.16 degrees.

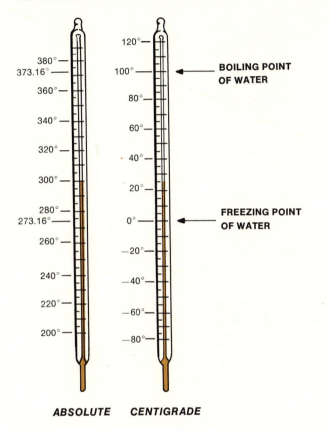

ABSOLUTE   CENTIGRADE

The displacement of zero makes V directly proportional to T, which is represented mathematically as $V \propto T$ or

2-11
$$V = aT$$

where a = slope of the line defined by equation 2-10.

At this time the only purpose for the extrapolation to $V = 0$ and for the definition of the absolute temperature scale is to make volume *directly proportional* to temperature. This absolute temperature scale simplifies the relationship between volume and temperature and leads to a rather simple general equation, called the ideal gas equation, which relates the variables P, V, n, and T.

GRAM-MOLECULAR VOLUME—AVOGADRO'S LAW

Avogadro's law states that a mole of any gas at a given temperature and pressure occupies the same volume. It has been determined experimentally that one mole of gas at 1 atmosphere and T = 273.16 °K (0 °C) occupies a volume of 22.4 liters which is defined as the *gram-molecular volume* (g-mol vol). The fact that this law is satisfied by different gases makes it possible to ignore variation in the molecular structure of gases. This law is an integral part of the *ideal gas equation*. The conditions of P = 1 atm and t = 0 °C are referred to as standard temperature-pressure (STP) conditions. Thus,

2–12   1 gram-molecular volume = 22.414 liters/mole at STP

IDEAL GAS EQUATION

Only three variables must be specified to completely describe the state or condition of an ideal gaseous substance. If we specify the pressure, temperature, and number of moles, the state of the gas is explicitly defined. The remaining variable, volume, is determined by the other three, and it is not necessary to specify its value. Alternatively, the variables P, T, and V could be given, and the number of moles, n, would then be automatically defined. It is only necessary to specify three of the variables P, V, T, and n to describe the state of a gas.

An *equation of state* is simply an equation which relates the variables that are used to describe the state of a system. The *ideal gas equation* is

2–13
$$PV = nRT = \frac{m}{MW} RT$$

where R = constant (called the molar gas constant), m = mass of the gas in grams, and MW = molecular weight of the gas in grams/mole.

If only three of the variables P, V, n, and T are known, this equation of state can be used to calculate the value of the fourth variable (if the gas is an ideal gas). There are several equations of state which relate the P, V, n, and T variables for gases that do not behave ideally; however, there is only one equation of state for ideal gases, and it is independent of the molecular species. Other equations of state contain constants whose specific values depend upon the nature of the molecular species described by the equation.

The value of the molar gas constant R can be found by considering one gram-molecular-volume, 22.414 liters at STP (standard temperature and pressure).

$$PV = nRT$$

$$P = 1 \text{ atm}$$
$$V = 22.414 \text{ liters}$$
$$T = 273.16 \text{ deg}$$
$$n = 1 \text{ mole}$$

Evaluating R

$$(1 \text{ atm})(22.414 \text{ liters}) = (1 \text{ mole})(R)(273.16 \text{ deg})$$

$$R = \frac{(1 \text{ atm})(22.414 \text{ liters})}{(1 \text{ mole})(273.16 \text{ deg})} = 0.08205 \frac{\text{liter-atm}}{\text{deg-mole}}$$

Note the units for R. It is necessary to specify units whenever R is written so that there is no chance for an incorrect dimensional analysis. R can have several values depending on the units associated with it:

$$R = 0.08205 \frac{\text{liter-atm}}{\text{deg-mole}} = 82.05 \frac{\text{ml-atm}}{\text{deg-mole}} = 1.987 \frac{\text{cal}}{\text{deg-mole}}$$

$$= 1.987 \times 10^{-3} \frac{\text{kcal}}{\text{deg-mole}}$$

Since pressure is frequently measured with a mercury barometer or manometer, the unit of pressure commonly used is mm Hg. More recently the unit mm Hg has been replaced with the unit torr.

$$1 \text{ mm Hg} = 1 \text{ torr}$$

Since one atmosphere pressure supports 760 mm Hg, the relationship between these pressure units is

$$1 \text{ atm} = 760 \text{ mm Hg} = 760 \text{ torr}$$

EXAMPLE 2–11   *Calculate the volume of a sample of N_2 containing 25 grams of N_2 at a pressure of 2.00 atmospheres and 27 °C, assuming the ideal gas law.*

We may use equation 2–13 expressed in terms of the mass and molecular weight of the gas. The absolute temperature is $T = 27 + 273.16 = 300$ °K.

$$PV = \frac{m}{MW} RT$$

$$(2.00 \text{ atm})(V) = \frac{(25 \text{ g})}{(28 \text{ g/mole})} \left(0.08205 \frac{\text{liter-atm}}{\text{deg-mole}}\right) (300 \text{ deg})$$

$$V = \frac{(25)(0.08205)(300)\ liters}{(2.0)(28)}$$

$$V = 11.0\ liters$$

There are many problems involving changes in the conditions (*P, V, T,* or *n*) of a gas that can be solved using the ideal gas equation of state. The general procedure involves writing down the ideal gas equation using first the initial conditions and then the final conditions. This will give two equations which can be solved simultaneously for the desired quantity. Unnecessary variables (such as nR in Example 2–12 and (2 moles) RT in Example 2–13) can be eliminated in solving the two equations.

EXAMPLE 2–12 *Suppose we had a sample of CH_4 at a pressure of 1 atm, a volume of 3 liters, and a temperature of 300 °K. If the gas was compressed to 1 liter and the temperature increased to 550 °K, what would be the new pressure of this gas?*
Substituting the initial information into equation 2–13

$$PV = nRT$$

$$(1\ atm)(3\ liters) = nR(300\ deg)$$

Substituting the final information into equation 2–13

$$P_f(1\ liter) = nR(550\ deg)$$

We can solve for the final pressure, P_f, by solving these two equations simultaneously. To do this, we solve each of these equations for nR and set them equal

$$nR = \frac{(1\ atm)(3\ liters)}{(300\ deg)} = \frac{P_f(1\ liter)}{(550\ deg)}$$

$$P_f = \frac{(1\ atm)(3\ liters)(550\ deg)}{(300\ deg)\ (1\ liter)}$$

$$P_f = 5.5\ atm$$

EXAMPLE 2–13 *Suppose we had a 2 mole sample of CO_2 at 3 atm pressure and a volume of 2 liters. If the sample pressure were decreased to 2 atm pressure, calculate the new volume if the temperature is held constant.*
Substituting the initial information into equation 2–13

$$PV = nRT$$

$$(3\ atm)(2\ liters) = (2\ moles)\ RT$$

Substituting the final information into equation 2–13

$$(2 \text{ atm})(V_f) = (2 \text{ moles}) \, RT$$

We can calculate the final volume, V_f, by solving these two equations simultaneously. Since the (2 moles)RT is constant in both equations

$$(2 \text{ moles}) \, RT = (3 \text{ atm})(2 \text{ liters}) = (2 \text{ atm})(V_f)$$

$$V_f = \frac{(3 \text{ atm})(2 \text{ liters})}{(2 \text{ atm})}$$

$$V_f = 3 \text{ liters}$$

EXAMPLE 2–14   Suppose we wanted to know the molecular weight of a given weight of unknown gas. If we measure the pressure, volume, and temperature of a sample of this gas, we can calculate the molecular weight using the ideal gas equation. The measurements on the gas are

$$P = 252 \text{ torr} \qquad T = 515 \,°K$$

$$V = 1.1 \text{ liters} \qquad mass = 0.50 \text{ grams}$$

According to equation 2–13

$$PV = \frac{m}{MW} \, RT$$

$$(252 \text{ torr})(1.1 \text{ liters}) = \frac{0.50 \text{ g}}{MW} \left(0.08205 \, \frac{\text{liter-atm}}{\text{deg-mole}}\right)(515 \text{ deg})$$

Note that the pressure units will not cancel. The pressure in mm Hg must be converted to atm. Since 1 atm = 760 torr,

$$\text{then} \left(\frac{1 \text{ atm}}{760 \text{ torr}}\right)(252 \text{ torr})(1.1) = \frac{0.50 \text{ g}}{MW} \left(0.08205 \, \frac{\text{atm}}{\text{mole}}\right)(515)$$

Solving for MW,

$$MW = \frac{(0.50 \text{ g})\left(0.08205 \, \frac{1}{\text{mole}}\right)(515)(760)}{(252)(1.1)}$$

$$MW = 58 \text{ g/mole}$$

The molecular weight cannot be used to identify the unknown substance since more than one compound can have this molecular weight. This unknown gas could be any of several compounds. Three of many possible compounds with a molecular weight of 58 are

$$
CH_3-CH_2-CH_2-CH_3 \qquad CH_3-CH_2-\overset{\overset{\displaystyle O}{\|}}{C}-H \qquad CH_2=C=CHF
$$

butane propionaldehyde 1-fluoroallene

PROBLEM 2–27   We have a gas at 3 atm pressure and a temperature of 300 °K. If we keep the volume constant and increase the temperature to 615 °K, calculate the pressure of the gas.

PROBLEM 2–28   What weight of benzene (C_6H_6) must be taken in order that the benzene gas at 500 °K will have a pressure of 5 torr in a volume of 30 liters? Assume benzene gas obeys the ideal gas equation under these conditions.

DALTON'S LAW OF PARTIAL PRESSURES

In a mixture of ideal gases each gas will make a contribution to the total pressure P of the gas. We define the *partial pressure*, P_i, of each component in the gas mixture as the contribution which each component makes to the total pressure. The total pressure is then the sum of the partial pressures of the components

2–14 $$P = P_1 + P_2 + \cdots\cdots$$

where P_1 is the partial pressure for component 1, P_2 is the partial pressure of component 2, etc.

   If we assume that the attractive force between molecules of two different types is also negligible, the entire gas sample behaves as an ideal gas

2–13 $$PV = n\,RT = (n_1 + n_2 + \cdots\cdot)\,RT$$

where n is the total number of moles, n_1 is the moles of component 1, etc. Since each gas is an ideal gas

2–15a $$P_1 V = n_1\,RT$$

2–15b $$P_2 V = n_2\,RT$$

. .

. .

. .

Dividing the equations 2–15 by equation 2–13 we obtain

$$\frac{P_1}{P} = \frac{n_1}{n} = X_1$$

$$\frac{P_2}{P} = \frac{n_2}{n} = X_2$$

$$\cdot \quad \cdot$$
$$\cdot \quad \cdot$$
$$\cdot \quad \cdot$$

where X is the mole fraction. These equations can be rearranged in the form of *Dalton's law of partial pressures*

2–16a $\qquad\qquad\qquad P_1 = X_1 P$

2–16b $\qquad\qquad\qquad P_2 = X_2 P$

$$\cdot \quad \cdot$$
$$\cdot \quad \cdot$$
$$\cdot \quad \cdot$$

Note that the partial pressure of each component is simply the mole fraction or fractional number of moles of that component times the total pressure.

PROBLEM 2–29   Air consists of approximately 78 % N_2 and 21 % O_2 by volume. The remaining 1 % consists of Ar, CO_2, H_2 and other inert gases. If the atmospheric pressure is 740 mm Hg, calculate the partial pressures of N_2 and O_2 in torr. Assume that all gases obey the ideal gas law and Dalton's law of partial pressures applies to the gas mixture. [Hint: In mixtures of ideal gases the volume percentage is the same as mole percentage.]

EXERCISES

1. Complete the following table

ELEMENT	ATOMIC NUMBER	NO. OF PROTONS	NO. OF ELECTRONS	NO. OF NEUTRONS	MASS NUMBER
magnesium	12	12	12	12	24
chlorine	17	17	17	18	35
tin	50	50	50	69	119
Ba^{2+}	56	56	54	81	137
S^{2-}	16	16	18	19	35

2. Naturally occurring boron consists of 80.20 % ^{11}B (mass = 11.009 amu) and 19.80 % ^{10}B (mass = 10.01). Calculate the atomic weight of boron.

3. a. Calculate the simplest formula of a compound which is 30.5 % nitrogen and 69.5 % oxygen by weight.
   b. Calculate the true molecular formula if 4.107 g of this gaseous compound occupies a volume of 0.50 liter at $P = 2$ atm and $t = 0$ °C.

4. When isopropyl alcohol (rubbing alcohol) is rubbed on the skin, the skin is cooled very noticeably. The boiling point of isopropyl alcohol is 82 °C.
   a. Explain the phenomenon which causes the sensation of cooling.
   b. If a child is running a high fever, the body temperature can be lowered quickly by swabbing the skin with isopropyl alcohol. Would you expect water to be as effective as isopropyl alcohol for this purpose? Would cooking oil be effective for this purpose? Explain.

5. Name the following compounds:
   a. CaH_2
   b. $Zn(C_2H_3O_2)_2$
   c. SrC_2O_4
   d. SnO_2
   e. $Ag_2S_2O_3$
   f. I_2O_5
   g. BN
   h. $CsIO$
   i. $CrBr_2$
   j. $SrCr_2O_7$

6. Write formulas for the following compounds:
   a. chromium (II) iodide
   b. sodium peroxide
   c. magnesium hydride
   d. gold (III) fluoride
   e. antimony (III) permanganate
   f. sodium silicate
   g. iron (II) permanganate
   h. calcium silicate
   i. sodium selenide
   j. calcium dihydrogen phosphate

7. Ammonium nitrate and ammonia (NH_3) gas are both used as fertilizers. Which of these two compounds contains the greatest percentage by weight nitrogen? Which would be the most economical to transport for the same amount of nitrogen it can supply as fertilizer?

8. The insecticide DDT has the structural formula

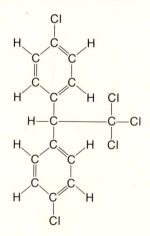

Write the empirical formula and calculate the percentage of Cl in DDT. Write the molecular formula designating the two

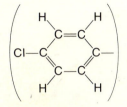

and Cl groups as shown in Problem 2–7.

9. Compounds with characteristic odors can generally be detected by the human senses at concentrations on the odor of ppm. The compound 1-octene-3-ol is the principal species in the volatile substances from mushrooms. This compound has the structural formula

$$H_2C\!=\!CH\!-\!\underset{\underset{\displaystyle H}{|}}{\overset{\overset{\displaystyle OH}{|}}{C}}\!-\!CH_2\!-\!CH_2\!-\!CH_2\!-\!CH_2\!-\!CH_3$$

a. If 1-octene-3-ol could be detected at a concentration of 1 ppm, calculate the amount of 1-octene-3-ol in grams that is required to give this concentration in a room (15 ft × 20 ft × 8 ft). Assume the molecular weight of air is 29 and $t = 25\,°C$, $P = 1$ atm.

b. Write the empirical molecular formula for 1-octene-3-ol.

c. Calculate the % by weight of C, H, and O.

10. a. Describe the gross structure of the isotope of potassium, $^{41}_{19}K$, giving the number of protons and neutrons in the nucleus and the number of electrons.

b. The atomic weight of K is 39.102. Account for the fact that the atomic weight is about 39 whereas the isotope of K in part (a.) had an isotopic mass about 41.

11. A compound is found to consist of 34.8 % sodium, 16.7 % boron, and 48.5 % oxygen. Determine its simplest formula.

12. The formula of a hypothetical compound of tellurium (at wt = 127) is M_2Te. Its composition is 80.0 % Te. Calculate the approximate atomic weight of the element M.

13. A gaseous compound contains 21.8 % hydrogen and 78.2 % boron. A sample of this gas weighing 19.0 mg at 100 torr and $t = 77\,°C$ occupied a volume of 150 ml.

a. Calculate the simplest formula from this data.

b. Calculate the molecular weight of the gas.

c. Determine the molecular formula for this compound.

14. Hemoglobin reacts with O_2 to form a complex which contains 4 moles of oxygen per mole of hemoglobin.

    a. Calculate the number of hemoglobin molecules necessary to complex 0.5 liters of O_2 gas at 25 °C and 1 atm.

    b. Calculate the volume of O_2 complexed in 10 ml of plasma at 25 °C. Assume the concentration of hemoglobin in blood is 16 g/100 ml and the molecular weight of hemoglobin is 70,000.

15. a. A procedure called for 10 grams of Na_2HPO_4 to be dissolved in distilled water to make 500 ml of solution. If only $Na_2HPO_4 \cdot 12H_2O$ were available, how much $Na_2HPO_4 \cdot 12H_2O$ must be used to supply 10 grams of Na_2HPO_4 in 500 ml.

    b. Calculate the molarity of the solution.

16. Nitrous oxide (N_2O) gas is mixed with oxygen for use in anesthesia. Calculate the partial pressure of each of these gases in a mixture containing 80 % oxygen and 20 % N_2O by volume. Assume the total pressure is one atmosphere and the gases obey the ideal gas equation.

17. Sodium exists as Na^+ in serum. One way in which to analyze for the Na^+ in serum is to precipitate the Na^+ as

$$NaZn(UO_2)_3(CH_3COO)_9 \cdot 9H_2O$$

The molecular weight of this precipitate is 1538. The Na^+ in a 10 ml sample of serum gave a precipitate of 1.5 g. Calculate the concentration of Na^+ in the serum expressed as

    a. gram-atoms of Na^+/liter serum

    b. percentage by weight, assuming the density of the serum is 1.00 g/ml.

    c. What weight of $NaCl$ would be required to produce this amount of Na^+ in 1 liter of serum?

18. The mineral kaliophilite has the molecular formula $K_2O \cdot Al_2O_3 \cdot 2SiO_2$ with a molecular weight of 316. Calculate the % SiO_2 in kaliophilite.

19. Magnetite (Fe_3O_4) and pyrite (Fe_2S) both contain iron. Which mineral would give the largest amount of iron on the basis of a unit weight of mineral?

20. When acid reacts with calcite ($CaCO_3$ or $CaO \cdot CO_2$), the CO_2 is liberated as gas. What weight of calcite is required to produce 3 ml CO_2 at 60 °C and 2 atm pressure.

21. A compound given the common name tabun is a very dangerous nerve gas. It has the structural formula

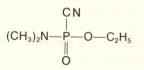

Assuming tabun is lethal at a concentration of one ppm:
a. Calculate the weight of tabun necessary to poison a lake that is used as a water supply. The lake is 25 miles long, 10 miles wide, and 20 ft deep (40.2 km × 16.1 km × 6.1 m). Assume the compound will not react with water.
b. Calculate the molar concentration of the one ppm tabun in water. Assume the solution has the same density as water (1 g/ml) at this low concentration.
c. Write the molecular formula for tabun.
d. What is the percentage of N in tabun?

22. The gases from a volcano on the island of Hawaii were analyzed. The composition, given as volume percent, is as follows:

GAS	VOLUME PERCENT
CO_2	20.93
SO_2	11.42
N_2	4.13
H_2O	61.56

There were also trace quantities ($< 1\%$) of CO, H_2, Ar, S_2, and SO_3.
a. What are the mole fractions of each of the major components?
b. If CO_2, SO_2, and H_2O were produced by the combustion of substances containing carbon, sulfur, and hydrogen, calculate the mole fraction and weight percent of C, S, N, and H in the original uncombusted material.

23. The partial pressure of O_2 in air is 158 mm Hg. The air which is expired after breathing has a partial pressure of O_2 of 115 mm Hg.
a. How many moles of O_2 are absorbed by the lungs per liter of air intake?
b. It requires two moles of O_2 to oxidize CH_3COOH to CO_2 and H_2O in the Krebs cycle. Calculate the volume of air in liters that must be inhaled in order to oxidize one mole of CH_3COOH.

24. A beta particle (electron) can be emitted from a nucleus by the transformation of a neutron into a proton

$$n \longrightarrow p^+ + e^-$$

An isotope of argon, $^{40}_{18}Ar$, is radioactive, giving off a β particle (e^-) in the disintegration process. What isotope remains after the $^{40}_{18}Ar$ nucleus gives off a β particle?

25. Calculate the amount of gold required to plate an object of area 100 in^2 with a thickness of 0.001 in. The density of gold is 19.282 g/cc. Express the amount of gold in (a) grams (b) g-atoms. (c) How many electrons would be required if the gold was electroplated according to the reaction

$$e^- + Au^+_{(aq)} \longrightarrow Au_{(s)}$$

26. Assume the density of a nucleus is 10^8 tons/cm^3 (9.06×10^{10} kg/cm^3). Calculate the diameter of an atom of gold, Au, (in miles) if the nucleus were the size of a ping pong ball (diameter approximately 2.5 cm or 1 in). The density of gold is 19.282 grams/cm^3. [Hint: Assume the radius is proportional to $V^{1/3}$.]

27. A new element X is discovered. Analysis of four of its compounds showed the following number of grams of X in one mole of each compound.

Compound:	I	II	III
Grams of X:	160	240	320

    What is the most probable atomic weight of X? How might your answer be incorrect?

28. Perchloric acid, $HClO_4$, is generally marketed as an aqueous solution containing 60 % by weight $HClO_4$. The density of the 60 % solution is 1.54 g/ml. Describe how you would prepare 500 ml of 0.5 M $HClO_4$.

29. Cannizzaro's method (1858) for the determination of atomic and molecular weights is based upon Avogadro's principle that two gases at the same T and P contain the same number of molecules. Therefore, the densities of the gases at the same T and P are in the same proportion as the molecular weights of the gases. One liter of an unknown gas weighs 0.305 g whereas 1 liter of oxygen weighs 0.125 g, both gases are at the same T and P. Calculate the molecular weight of the unknown gas, assuming oxygen has a molecular weight of 32.

30. Hemoglobin contains 0.34 % by weight iron. Calculate the minimum molecular weight of hemoglobin. [Hint: Each molecule of hemoglobin must contain at least one Fe atom or ion.]

CHAPTER THREE

CHEMICAL REACTIONS

In Chapter 2 we stated that a chemical symbol is a shorthand way to write the name of an element and a chemical formula is a shorthand way to describe a compound and to give the relative number of atoms of each type in the compound. Similarly, a chemical equation is a shorthand way to write a chemical reaction. A chemical reaction is a process whereby one or more substances react chemically to give new substances. The chemical equation tells the relative *numbers* of molecules, atoms, or ions which are involved in the chemical reaction as reactants or as products. An analogy can be made between this shorthand method of representation in chemistry and the letters of the alphabet, the words, and the sentences in a language.

For example, consider the reaction between hydrogen and bromine to produce hydrogen bromide.

$$H_2 + Br_2 \longrightarrow 2\,HBr$$

The plus sign should be read as ''reacts with'' and the arrow* means ''yields'' or ''produces.'' The chemical equation thus states: ''One molecule of H_2 reacts with 1 molecule of Br_2 to produce 2 molecules of HBr. Notice that the coefficient placed

*An equal sign (=) can also be used.

53

before HBr signifies 2 molecules of HBr. It is not necessary to place a "1" in front of H_2 and Br_2 since it is understood that a missing number is "1".

Another example of a chemical equation represents the reaction between ethyl alcohol (C_2H_5OH) and oxygen to give carbon dioxide and water

3–1
$$C_2H_5OH + 3O_2 \longrightarrow 2CO_2 + 3H_2O$$

This equation states that one molecule of C_2H_5OH and 3 molecules of O_2 produce 2 molecules of CO_2 and 3 molecules of H_2O.

The coefficients appearing before the formulas of reactants and products are the lowest numbers which can represent the *relative* number of molecules of reactants and products involved in the reaction. For example, not only does one molecule of C_2H_5OH react with 3 molecules of O_2 but also 2 molecules of C_2H_5OH react with 6 molecules of O_2, with the relative number of product molecules also doubling in this case.

$$2C_2H_5OH + 6O_2 \longrightarrow 4CO_2 + 6H_2O$$

Notice in the new equation that the proportion of reactant and product molecules remains the same as in equation 3–1. Equation 3–1 can also represent the reaction of much larger numbers of molecules, for example, 100 or even 10^{20} times as many molecules. In fact there is no reason why the chemical equation cannot be expressed in multiples of 6.023×10^{23} molecules. Since 6.023×10^{23} molecules is equivalent to one mole (equation 2–2b), the chemical equation also represents the multiple of moles of reactants and products.

$$1(6.023 \times 10^{23})C_2H_5OH + 3(6.023 \times 10^{23})O_2 \longrightarrow$$

$$2(6.023 \times 10^{23})CO_2 + 3(6.023 \times 10^{23})H_2O$$

$$1 \text{ mole } C_2H_5OH + 3 \text{ moles } O_2 \longrightarrow 2 \text{ moles } CO_2 + 3 \text{ moles } H_2O$$

The *relative* numbers of reactant and product involved in a reaction may be individual molecules or units of molecules such as moles.

PROBLEM 3–1   Silicon tetrachloride reacts with water according to the chemical equation

$$SiCl_4 + 4H_2O \longrightarrow H_4SiO_4 + 4HCl$$

If 6×10^{23} molecules of $SiCl_4$ react, how many molecules of HCl will be produced?

BALANCING EQUATIONS BY INSPECTION

In the previous examples of chemical equations, coefficients were placed in front of the molecular formulas in order to balance the equation. A balanced equation is one in which the same numbers of atoms of each type appear on both sides of the equation. The numbers must remain the same since matter is neither created nor destroyed in a normal chemical reaction. By normal chemical reactions, we exclude nuclear transformations, in which elements can be transformed into other elements and mass can be converted into energy. In nuclear transformations mass is not conserved; that is, mass does not remain constant.

Generally, the smallest possible set of whole numbers is used as coefficients in the balanced equation. The procedure we will use for balancing equations at this time is known as the method of inspection. It is applicable to all equations except oxidation-reduction reactions. Balancing oxidation-reduction equations will be deferred to Section D of this chapter. There is no set of fixed rules you must follow in order to balance equations by inspection. The most efficient approach can best be learned from examples and practice. In general, you should examine the elements on both sides of the equation, one element at a time. Place coefficients in front of the molecules to keep the number of atoms of each element constant. If coefficients are added to balance one element, you must check the other elements to see if the previously balanced atoms remain balanced. If any atoms are now unbalanced, they must be rebalanced and the procedure repeated. In some reactions a polyatomic ion remains intact on both sides of the equation. Such a polyatomic ion can be balanced as one unit. Some examples will illustrate this procedure.

EXAMPLE 3–1   *Balance this equation* $ZnS + O_2 \longrightarrow ZnO + SO_2$

Examining zinc first, we notice that Zn is in ZnS on the left side and ZnO on the right. Since each compound contains one atom, the equation is balanced with respect to the element zinc. Examining sulfur next we find S in ZnS and SO_2. Since each of these compounds contains only one S, the equation is balanced with respect to sulfur. The final element to balance is oxygen. Oxygen is in O_2 on the left side and in both $ZnO + SO_2$ on the right. The right side has 3 atoms of oxygen, the left 2 atoms. Since Zn and S are already balanced by the coefficients on ZnO and SO_2, the O should be balanced by adding a coefficient for O_2. A coefficient of 3/2 in front of O_2 will balance the oxygen atoms since this gives 3 atoms of oxygen on the left side.

$$ZnS + 3/2\ O_2 \longrightarrow ZnO + SO_2$$

We must now go back and examine each element again to see if they remain balanced. Examining Zn and S again, we conclude that there is no change in the number and the equation is balanced. Generally, whole numbers are used for coefficients in a chemical equation. To remove the 3/2 fraction, we multiply the coefficients for the entire equation by 2. The balanced equation is thus

$$2ZnS + 3O_2 \longrightarrow 2ZnO + 2SO_2$$

EXAMPLE 3–2   *Balance the equation* $Ca(OH)_2 + Ca(HCO_3)_2 \longrightarrow CaCO_3 + H_2O$

Balancing first the calcium, we notice that a coefficient of 2 must be placed in front of $CaCO_3$.

$$Ca(OH)_2 + Ca(HCO_3)_2 \longrightarrow 2CaCO_3 + H_2O$$

Balancing the carbon atoms, we note that there are two in $Ca(HCO_3)_2$ and only one in $CaCO_3$. However, the coefficient of 2 before $CaCO_3$ balances the carbon atoms. Since carbon is combined with 3 oxygen atoms in both HCO_3 and CO_3, we can consider CO_3 as a unit. In this case, the CO_3 group is also balanced since there are 2 CO_3 groups on both sides of the equation. The oxygen atoms apart from CO_3 must now be balanced. We note there are 2 oxygen atoms in $Ca(OH)_2$ and one in H_2O. This requires a coefficient of 2 before H_2O to balance the remaining O atoms.

$$Ca(OH)_2 + Ca(HCO_3)_2 \longrightarrow 2CaCO_3 + 2H_2O$$

Finally examining H, we note there are 2 atoms in $Ca(OH)_2$, 2 atoms in $Ca(HCO_3)_2$ and 4 atoms in $2H_2O$. Hence, the hydrogens are balanced with no change in coefficients required. Rechecking the balancing of Ca, we see no change. Thus, the equation is balanced.

EXAMPLE 3–3   *Balance the chemical equation* $Al_2(SO_4)_3 + KOH \longrightarrow Al(OH)_3 + K_2SO_4$

Balancing the Al: $Al_2(SO_4)_3 + KOH \longrightarrow 2Al(OH)_3 + K_2SO_4$

Balancing the SO_4: $Al_2(SO_4)_3 + KOH \longrightarrow 2Al(OH)_3 + 3K_2SO_4$

Balancing the K: $Al_2(SO_4)_3 + 6KOH \longrightarrow 2Al(OH)_3 + 3K_2SO_4$

Balancing the OH requires no further change.
Checking Al and SO_4, we see that the equation is balanced.

PROBLEM 3–2   Balance the following chemical equations
a. $Al + O_2 \longrightarrow Al_2O_3$
b. $PCl_3 + H_2O \longrightarrow H_3PO_3 + HCl$
c. $Ca_3P_2 + H_2O \longrightarrow Ca(OH)_2 + PH_3$
d. $Ca(OH)_2 + H_3PO_4 \longrightarrow Ca_3(PO_4)_2 + H_2O$

3–B. GENERAL CLASSIFICATION OF CHEMICAL REACTIONS

In studying chemistry, it is helpful to group chemical reactions into various categories or types. The classification in this section is quite general and includes essentially all inorganic reactions. Knowing the patterns into which chemical reactions fall helps you anticipate how two substances might react.

In later sections we introduce the additional classifications of acid-base and oxidation-reduction reactions. As will become apparent later, acid-base and oxidation-reduction classifications make it possible to order compounds or elements according to acid strength or oxidizing power. This allows us to predict whether a reaction of this type will occur. This is a very powerful tool for the chemist to use and for this reason it will be stressed throughout the book. The classifications in the present section, however, do not furnish any basis for predicting whether a chemical reaction will occur.

DIRECT COMBINATION

In this type of reaction two elements or compounds combine directly to give another compound (Figure 3–1). A general equation can be written showing the direct combination of two elements or of two compounds A and B.

$$A + B \longrightarrow AB$$

If A and B represent elements, the elements will be in their natural state and are not necessarily single atoms. A and B may exist as molecules. Furthermore, A and B need not combine in a 1:1 ratio. Examples of direct combination of elements are

$$Fe + S \longrightarrow FeS$$

$$2Mg + O_2 \longrightarrow 2MgO$$

$$2H_2 + O_2 \longrightarrow 2H_2O$$

Examples of direct combination of compounds are

$$NaO + CO_2 \longrightarrow NaCO_3$$

$$CH_2{=}CH_2 + H_2 \longrightarrow CH_3CH_3$$

$$CH_2{=}CH_2 + Cl_2 \longrightarrow CH_2Cl{-}CH_2Cl$$

FIGURE 3–1   Direct combination

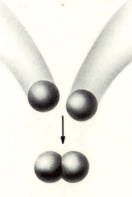

DECOMPOSITION

This type of reaction is the reverse of direct combination (Figure 3–2). Decomposition reactions involve the dissociation of a compound into its elements or into simpler compounds. The general equation for the decomposition of AB is

$$AB \longrightarrow A + B$$

Remember that if A and B represent elements they are in their natural state and are not necessarily single atoms. Examples of decomposition reactions are

$$2HgO \longrightarrow 2Hg + O_2$$

$$2KClO_3 \longrightarrow 2KCl + 3O_2$$

$$CaCO_3 \longrightarrow CaO + CO_2$$

$$CH_3-CH_2OH \longrightarrow CH_2{=}CH_2 + H_2O$$

DISPLACEMENT

In this type of reaction one of the elements in a compound is displaced by another element (Figure 3–3). The general reaction can be given as

$$AB + C \longrightarrow CB + A$$

Here A and B can be polyatomic ions as well as simple ions or atoms. Some examples are

$$Zn + 2HCl \longrightarrow ZnCl_2 + H_2$$

$$Fe + Cu(NO_3)_2 \longrightarrow Fe(NO_3)_2 + Cu$$

$$Zn + H_2SO_4 \longrightarrow ZnSO_4 + H_2$$

$$NH_4^+ + HCl \longrightarrow NH_4Cl + H^+$$

FIGURE 3–2 Decomposition

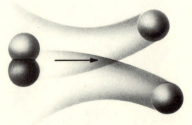

FIGURE 3–3 Displacement

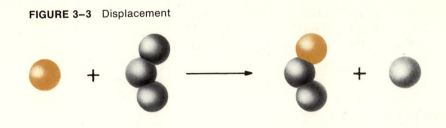

DOUBLE DECOMPOSITION (METATHETICAL)

Metathetical reactions generally involve compounds which have replaceable ions. In this reaction two compounds, each consisting of two simple or polyatomic ions, exchange ions forming two new compounds (Figure 3–4). The general equation, representing the reaction between AB and CD, is

$$AB + CD \longrightarrow AD + CB$$

Again, it should be pointed out that AB, CD, AD, and CB need not be combined in a 1:1 ratio. Examples of metathetical reactions are

$$Al_2(SO_4)_3 + 6KOH \longrightarrow 2Al(OH)_3 + 3K_2SO_4$$

$$2AgNO_3 + BaCl_2 \longrightarrow 2AgCl\downarrow + Ba(NO_3)_2$$

$$NaOH + HCl \longrightarrow NaCl + HOH$$

PROBLEM 3–3 Identify the type of reaction which occurs in the following:
a. $2P + 5Cl_2 \longrightarrow 2PCl_5$
b. $Pb + 2AgNO_3 \longrightarrow Pb(NO_3)_2 + 2Ag$
c. $Ca(OH)_2 + H_2CO_3 \longrightarrow CaCO_3 + 2H_2O$

FIGURE 3–4 Double decomposition (other mechanisms possible)

d. $Zn + 2HCl \longrightarrow ZnCl_2 + H_2$
e. $2Al + Fe_2O_3 \longrightarrow Al_2O_3 + 2Fe$
f. $2Al(OH)_3 \longrightarrow Al_2O_3 + 3H_2O$

3–C. CHEMICAL EQUILIBRIUM

In the chemical equations which we have written so far it has been assumed that reactions proceed from reactants to products (left to right in the direction of the arrow). However, many chemical reactions can also proceed from the products of the reaction back to the reactants, i.e., right to left. Reactions that can occur both in the forward and the reverse directions are appropriately called *reversible reactions*. We designate reversible reactions with two arrows to represent the simultaneous reaction in both the forward and reverse directions. For example, if we had the reaction between A and B to produce the products C and D, the reversible reaction would be represented by

$$A + B \rightleftharpoons C + D$$

The *rate of reaction* is a measure of the speed at which the reactants can combine to give the products of the reaction. The rate of reaction can be expressed, for example, in terms of (moles/liter) per minute.

If we start out with only reactants A and B in our reaction vessel, the reaction can proceed initially only from left to right to give products C and D.

$$A + B \longrightarrow C + D$$

If A and B are substances in solution, the reaction will occur at a rate which generally depends upon the concentration of these reactants. As A and B react, the concentrations of A and B will necessarily decrease and the rate of the forward reaction will correspondingly decrease. As soon as some products C and D are produced, the back reaction will immediately occur.

$$A + B \longleftarrow C + D$$

As more C and D are produced during the course of the reaction, the concentrations of C and D will increase and the rate of the reverse reaction will correspondingly increase. The change in concentration of reactants and products with time is shown in Figure 3–5.

As time passes, the rate of the forward reaction (left to right) will decrease and the rate of the reverse reaction (right to left) will increase. Eventually the rate of the forward reaction will become equal to the rate of the reverse reaction. At this time the rate of formation of products will equal the rate of production of reactants, and

FIGURE 3–5 Concentration of reactants and products during reaction. Brackets [ ] designate molar concentration.

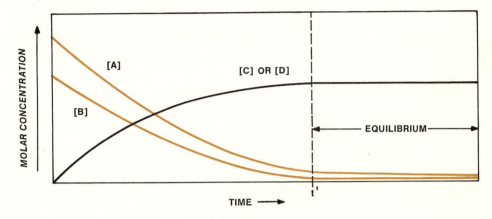

the concentrations of both products and reactants will remain constant. When there is no change in concentration with time, we say that *chemical equilibrium* has been established.

$$\text{rate}_{\text{forward}} = \text{rate}_{\text{reverse}}$$

$$A + B \rightleftharpoons C + D \qquad \text{(equilibrium)}$$

Note in Figure 3–5 that the concentrations of A, B, C, and D have become constant at time *t'*. From this point on the reaction is at chemical equilibrium.

In chemistry, equilibrium is a condition in which there is no change in the concentrations of the species. However, it is important to point out that at chemical equilibrium both the forward and reverse reactions still occur, and new reactants and products are continually being formed. For this reason we use the term *dynamic equilibrium*. This is in contrast to the ideal case of a static equilibrium, where no change of any type is occurring.

Notice in Figure 3–5 that a large concentration of products C and D were produced at equilibrium and the concentrations of A and B became correspondingly quite small. We can express the relative amount of product produced at equilibrium in terms of the percentage or fraction of reactant reacted. In the example shown in Figure 3–5 there is a large percentage reacted to give products and we would say that the *yield* of the reaction was high. In Chapter 13 the concentrations of reactants and products at equilibrium will be related to each other precisely: we will introduce the concept of an equilibrium constant, a numerical constant which will permit us to calculate the percentage or fraction reacted.

The percentage or fraction reacted to give products can vary from essentially no reaction (0 %) to complete reaction (100 %). Any intermediate value is possible.

FIGURE 3–6 Equilibrium concentrations at different yields

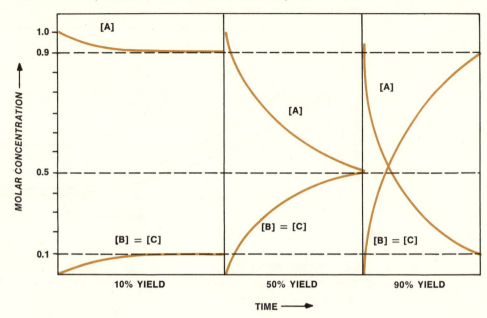

To illustrate variations in yield that can be encountered let us take the simple dissociation of A into products B and C.

$$A \rightleftharpoons B + C$$

It is assumed that there is no B or C present initially. Since an equal number of moles of B and C are produced, [B] must equal [C]. In Figure 3–6 are shown three possible situations where the percentage reacted to give products may be 10 %, 50 %, 90 %. Note in Figure 3–6A that the molar concentration of the reactant decreases from 1.0 to 0.9 whereas the product concentration reaches 0.1 at equilibrium. In Figure 3–6B the reactant concentration decreases from 1.0 to 0.5 whereas in Figure 3–6C the reactant concentration decreases from 1.0 to 0.1.

Throughout the rest of the book the double arrow will be used only whenever the reversibility of a reaction is to be emphasized or if it is pertinent to the discussion at that point (for example in calculating equilibrium concentrations). At all other times reactions will be designated by a single arrow, regardless of whether or not they are reversible.

3–D. OXIDATION-REDUCTION REACTIONS

Oxidation-reduction reactions are those in which there is an exchange of electrons between reactants. *Oxidation* is the loss of electrons whereas *reduction* is the gain of electrons. In a reaction of this type, one substance must be oxidized (loses e⁻)

and one substance must be reduced (gains e⁻). Since electrons cannot be created or destroyed, the number of electrons lost by one substance must be gained by the other substance. Since the substance being oxidized is causing reduction in the other substance, it is called the *reducing agent*. Likewise the substance being reduced is oxidizing the other substance and is called the *oxidizing agent*. Many biochemical reactions are oxidation-reduction reactions. Some examples of these will be discussed in Chapter 18. In order to identify the oxidizing agent and the reducing agent in an oxidation-reduction reaction we must first understand *oxidation numbers*.

OXIDATION NUMBERS

The concept of oxidation numbers has been used for many years in chemistry. It is most useful in balancing oxidation-reduction equations. However, the concept does not have any fundamental significance in chemistry except as an important tool in certain calculations. The value of this concept will become apparent as the student progresses through the course. Essentially, it helps us account for all of the electrons in a compound.

The *oxidation number* is the *apparent* positive or negative charge on an atom when the electrons in the molecule or ion are assigned to the atoms according to the following rules:

1. The atoms of all free elements (uncombined with any other element) are assigned a value zero. The free element is the state in which the element occurs in nature regardless of the complexity of the bonding of the atoms. The atoms in the following formulas would have an oxidation number of zero: H_2, O_2, N_2, S_8, Fe, Zn, Na, Pb, Cl_2, Ca.

2. The oxidation number for the H atom is always +1 and for the O atom is always −2 with the following four exceptions:
   a. the atoms in the free state of the elements have zero oxidation number as discussed in (1) previously
   b. metal hydrides—the H atom is assigned an oxidation number −1, e.g., NaH (sodium hydride)
   c. peroxides—the O atom is assigned an oxidation number −1, e.g., H_2O_2 (hydrogen peroxide), Na_2O_2 (sodium peroxide)
   d. OF_2—the oxygen atom is assigned a value of +2.
   Examples where the oxidation number of H is +1 and O is −2 are the following:

$$H_2O, \quad H_2SO_4, \quad Na_3PO_4, \quad Na_2O, \quad CH_4, \quad CH_3\overset{\displaystyle O}{\overset{\displaystyle \|}{C}}\!-\!OH$$

3. The charge of monatomic ions is also the oxidation number of this ion.
   Fe^{2+} oxidation number $= +2$
   Na^+ oxidation number $= +1$

Cl^- oxidation number $= -1$
Mg^{2+} oxidation number $= +2$
Fe^{3+} oxidation number $= +3$

4. When one or more pairs of electrons are shared between two atoms in a covalent bond, the pair (or pairs) of electrons are assigned to the atom with the greatest electronegativity. We have already said that a qualitative definition of electronegativity is the relative ability to attract a shared pair of electrons. A list of electronegativities was given in Table 2–5.

5. The sum of the oxidation numbers of all atoms in a molecule must be zero except in a polyatomic ion where the sum must equal the charge on the ion.

EXAMPLE 3–4   *The oxidation number of sulfur in H_2SO_4 can be calculated using the rules we have given. The oxidation number for hydrogen is $+1$ and for oxygen, -2. Let x represent the unknown oxidation number for sulfur. According to our rules, the sum of oxidation numbers of all atoms must be zero.*

$$\overset{(+1)\ (-2)}{H_2SO_4}$$

$$2(+1) + x + 4(-2) = 0$$

Oxidation number S $= x = +6$

EXAMPLE 3–5   *Similarly the oxidation number for sulfur in thiosulfate ion $S_2O_3{}^{2-}$ is calculated in the following manner. Let x $=$ oxidation number of sulfur.*

$$\overset{(-2)}{(S_2O_3)^{2-}}$$

$$2x + 3(-2) = -2$$

$$2x = +4$$

Oxidation number S $= x = +2$

EXAMPLE 3–6   *The oxidation number can also turn out to be fractional as in the case of sulfur in $Na_2S_3O_6$. Since sodium is a $+1$ ion (Na^+), the oxidation number is also $+1$. The procedure for calculating the oxidation number of S is as before.*

$$\overset{(+1)\ (-2)}{Na_2S_3O_6}$$

$$2(+1) + 3x + 6(-2) = 0$$

Oxidation number S $= x = 3\tfrac{1}{3}$

After you have had practice in determining oxidation numbers, you will be able to carry out the simple mathematical steps mentally. The determination of the oxidation numbers will then be done with ease and rapidity.

PROBLEM 3–4   Determine the oxidation numbers of the atoms in the following compounds and ions: HNO_3, HCl, $(PO_4)^{3-}$, $Ca_3(PO_4)_2$, Fe^{3+}.

EXAMPLES OF OXIDATION-REDUCTION

As the first example of an oxidation reduction reaction, let us take a simple reaction which is carried out in a water solution

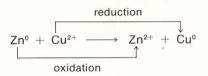

$$\text{Zn}^0 + \text{Cu}^{2+} \longrightarrow \text{Zn}^{2+} + \text{Cu}^0$$

The Cu^{2+} could come from a copper salt such as $Cu(NO_3)_2$, which in a water solution gives Cu^{2+} and NO_3^- ions. Since the reaction involves only Cu^{2+}, there is no need to write the NO_3^- ions. Cu^{2+} gains $2e^-$ in going to Cu^0, so copper is reduced. The Cu^{2+} is defined, therefore, as the *oxidizing agent*; it *oxidizes* Zn to Zn^{2+}. In this reaction metallic copper is deposited whereas the zinc metal goes into the solution as Zn^{2+} ions. Since the reaction proceeds from left to right we can say that Zn^0 is a better reducing agent than Cu^0.

Another simple oxidation-reduction reaction is

$$\overset{\text{oxidation}}{\text{Cu}^0 + 2\text{Ag}^+ \longrightarrow \text{Cu}^{2+} + 2\text{Ag}^0}$$
reduction

This reaction is also carried out in aqueous solution. Ag^+ is the oxidizing agent and Cu is oxidized to Cu^{2+}. The reaction proceeds from left to right (Cu goes to Cu^{2+}) because Ag^+ is a better oxidizing agent than Cu^{2+}. From these two reactions we know that the order of oxidizing power is:

$$
\begin{array}{l}
\text{Ag}^+ + e^- \longrightarrow \text{Ag}^0 \\
\text{Cu}^{2+} + 2e^- \longrightarrow \text{Cu}^0 \\
\text{Zn}^{2+} + 2e^- \longrightarrow \text{Zn}^0
\end{array}
$$

(oxidizing power ↑)

Numerous substances can be investigated experimentally and listed in the order of their oxidizing power. Once such a table is constructed, it can be used to predict whether a certain substance is capable of reducing or oxidizing another species. The details of this predictability are presented in Chapter 14, electrochemistry, where a quantitative measure of oxidizing power is described.

A final example of an oxidation-reduction reaction is

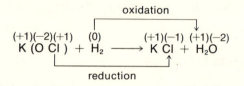

which is also carried out in solution. The oxidation numbers are determined and shown above the symbols, of the elements. Since Cl goes from +1 to −1, it is reduced by gaining 2e⁻/atom. H goes from 0 oxidation state to +1, hence, it is oxidized by losing 1e⁻/atom.

PROBLEM 3–5   Identify the species oxidized and reduced and state the total number of electrons required for each of the oxidation and reduction processes.

a. $2Au + 3F_2 + 6HCl \longrightarrow 2AuCl_3 + 6HF$

b. $3H_2S + Cr_2O_7^{2-} + 8H^+ \longrightarrow 3S + 2Cr^{3+} + 7H_2O$

c. $Cu + Cl_2 \longrightarrow CuCl_2$

d. $2Fe^{2+} + Br_2 \longrightarrow 2Br^- + 2Fe^{3+}$

PROBLEM 3–6 H^+ will oxidize Zn metal in solution; however, H^+ is incapable of oxidizing Cu.

$$Zn^0 + 2H^+ \longrightarrow Zn^{2+} + H_{2(g)}$$

$$Cu^0 + 2H^+ \longrightarrow \text{no reaction}$$

Where would you put the oxidizing power of H^+ in the table of oxidizing power above?

$$2H^+ + 2e^- \longrightarrow H_{2(g)}$$

Assume that the oxidizing power of H^+ is not equal to that of either Zn^{2+} or Cu^{2+}.

BALANCING OXIDATION-REDUCTION EQUATIONS

The balancing of oxidation-reduction equations differs from other reactions. When you balance an oxidation-reduction equation, you must make sure that the number of electrons transferred from the oxidized species is equal to the number of electrons accepted by the species being reduced. When this electron transfer is balanced, the net charge on each side of the equation resulting from the ionic species in the reaction is automatically balanced. In other words charge is conserved; it is neither created nor destroyed in the chemical reaction. In balancing oxidation-reduction equations we use the criterion of conservation of charge as a final check on the balanced equation.

We will give the procedure for balancing oxidation reduction equations as a series of steps, and we will work an example at the same time to illustrate the procedure. Consider the oxidation-reduction reaction

$$Au^0 + F_2 + H^+ \longrightarrow Au^{3+} + HF$$

1. Identify oxidation-reduction steps and determine the number of electrons transferred in each step. In our example we first evaluate the oxidation numbers of each element and write them directly over the symbol.

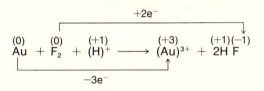

Since Au^0 goes to Au^{3+}, the Au^0 is oxidized by losing $3e^-$. The F_2^0 goes to a -1 oxidation state in HF. Since there are 2 fluorine atoms in F_2 we should place a 2 before HF and there will then be $2e^-$ gained in this reduction process.

2. Balance the electrons lost with the electrons gained, by putting appropriate coefficients in front of the reactants and products oxidized and reduced. In the above reaction the balance of electrons can be accomplished with coefficients 2 and 3.

$$2(3e^- \text{ lost/Au}) = 3(2e^- \text{ gained/}F_2)$$

In the oxidation-reduction equation we place a coefficient of 2 before Au and Au^{3+} and 3 before F_2 and 2HF.

$$2Au^0 + 3F_2 + H^+ \longrightarrow 2Au^{3+} + 3(2HF)$$

3. Balance the remaining elements, which are not oxidized or reduced, without changing the coefficients already established for the oxidized and reduced species.
   In the above reaction the only other element is H^+ and it can be balanced with a coefficient 6 since H only appears as $3(2HF) = 6HF$. The balanced equation is

$$2Au^0 + 3F_2 + 6H^+ \longrightarrow 2Au^{3+} + 6HF$$

4. Check the balanced oxidation-reduction equation by examining the balance of charge; the total charge on the left must equal the total charge on the right side of the equation.

$$\text{charge on left side} = 2(0) + 3(0) + 6(+1) = +6$$

$$\text{charge on right side} = 2(+3) + 6(0) = +6$$

Since the charge balanced and we have already seen that there are equal numbers of atoms on each side of the equation, the equation is balanced.

EXAMPLE 3–7   *Balance the following oxidation-reduction equation.*

$$H_2S + Cr_2O_7^{2-} + H^+ \longrightarrow S + Cr^{3+} + H_2O$$

First, evaluate the oxidation number of each element and determine which species is oxidized and which is reduced.

The oxidation number for Cr in $Cr_2O_7^{2-}$ is +6 and in Cr^{3+} it is +3. Since we have 2 Cr changing, the electron gain is $+6e^-$. Since one S^{2-} loses only two electrons, we give H_2S and S a coefficient of 3.

$$3H_2S + Cr_2O_7^{2-} + H^+ \longrightarrow 3S + 2Cr^{3+} + H_2O$$

The O atoms are balanced with a coefficient 7 for H_2O. Finally the H atoms are balanced with $8H^+$ on the left side.

$$3H_2S + Cr_2O_7^{2-} + 8H^+ \longrightarrow 3S + 2Cr^{3+} + 7H_2O$$

PROBLEM 3–7   Balance the following oxidation-reduction equations:
a. $AuBr_{4(aq)}^- + Hg_{(l)} \longrightarrow Au_{(s)} + Hg_2Br_{2(s)} + Br_{(aq)}^-$
b. $Fe(CN)_{6(aq)}^{3-} + Re_{(s)} + OH_{(aq)}^- \longrightarrow Fe(CN)_{6(aq)}^{4-} + ReO_{4(aq)}^- + H_2O$
c. $Zn_{(s)} + MnO_{4(aq)}^- + OH_{(aq)}^- \longrightarrow ZnO_{2(aq)}^{2-} + MnO_{2(s)} + H_2O$

EQUIVALENT WEIGHT FOR OXIDIZING AND REDUCING AGENTS

Equivalents and *equivalent weight* are useful quantities in calculating the amounts of reactants which enter into an oxidation-reduction reaction or into an acid-base reaction. For an oxidizing or reducing agent the *gram equivalent weight (g-eq wt)* is the mass of the substance which can accept or donate 6.023×10^{23} electrons = Avogadro's number of electrons in a particular reaction. One gram-molecular weight (g-mol wt) of the substance contains 6.023×10^{23} molecules and can gain or lose $6.023 \times 10^{23} \times$ (e$^-$ gained or lost/molecule). Therefore

3–2
$$\text{g-eq wt} = \frac{\text{g-mol wt}}{\text{e}^- \text{ gained or lost/molecule}}$$

One *equivalent*, eq, (also called gram-equivalent) is that quantity of an oxidizing or reducing agent which contains one gram-equivalent weight. Therefore, we can calculate the number of equivalents in a sample of mass *m*, by

3–3
$$\text{eq} = \frac{m}{\text{g-eq wt}}$$

The relationship between equivalents and mass in equation 3–3 can also be expressed in terms of moles by substituting equation 3–2 into equation 3–3.

$$\text{eq} = \frac{m}{\text{g-eq wt}} = \frac{m}{\dfrac{\text{g-mol wt}}{\text{e}^- \text{ gained or lost/molecule}}} = \frac{m}{\text{g-mol wt}} (\text{e}^- \text{ gained or lost/molecule})$$

Since m/g-mol wt is equal to moles, we have the relationship

3–4
$$\text{eq} = (\text{moles})(\text{e}^- \text{ gained or lost/molecule})$$

EXAMPLE 3–8

An example of an oxidation-reduction was given previously

$$KOCl + H_2 \longrightarrow KCl + H_2O$$

If we had 75 g of $KOCl$, how many gram-equivalents of $KOCl$ would this represent? The molecular weight of $KOCl$ is 90.55. The gram-equivalent weight of $KOCl$ is thus found from equation 3–2

$$\text{g-eq wt} = \frac{90.55 \text{ g}}{2} = 45.275 \text{ g}$$

Using equation 3–3, we can convert the 75 grams to equivalents

$$eq = \frac{m}{\text{g-eq wt}} = \frac{75 \text{ g}}{45.275 \text{ g}} = 1.658$$

PROBLEM 3–8

According to the reaction

$$2Au + 3F_2 + 3HCl = 2AuCl_3 + 6HF$$

a. Calculate the equivalent weight of F_2.
b. Calculate the number of moles and the number of equivalents of F_2 in a 50 gram sample. Use equation 3–3 to calculate the equivalents.
c. Show that equation 3–4 is in agreement with the results in b.
d. How many g-atoms of Au would react with this 50 g F_2?
e. What mass of Au would react with 50 g F_2?
f. How many equivalents of Au would react with this 50 g F_2?
g. Compare the equivalents of Au (in part f) with the equivalents of F_2 in 50 g (part b). Explain why they should be equal.

3–E. ACID-BASE REACTIONS

There are several different ways to define acids and bases. Initially we will restrict our definition of acids and bases to water, the most common solvent system. Other more general acid-base concepts are discussed at the end of this section.

Water (H_2O) dissociates slightly into the hydronium and hydroxide ions and is always in equilibrium with these ions. Since water ionizes itself, the process is called *autoionization*.

$$H_2O \rightleftharpoons H^+_{(aq)} + OH^-_{(aq)}$$

The longer arrow representing the reverse reaction emphasizes that the equilibrium is displaced towards the H_2O, i.e., the concentration of products is small compared to the concentration of water. (The concentration of $H^+_{(aq)}$ is only 10^{-7} M in pure water; the concentration of water is 55.5 M.) Generally the arrows representing reversibility are made of equal length regardless of the displacement of the equilibrium.

The H^+ and OH^- are not isolated ions in water. The ions have a very close association with H_2O molecules which drastically affects their properties. As in Chapter 2, this association with H_2O molecules in solution is designated by the subscript (aq) for aqueous. $H^+_{(aq)}$ is often written H_3O^+.

It is the above equilibrium that provides the basis for the classification of acids and bases which we will emphasize. At this point we will define acids and bases and give some examples of acid-base reactions. The equilibrium involving acid-base dissociation is most important in understanding acid and base strengths and will be discussed in Chapter 13. It is this ability to order acid and base strengths that makes the separate treatment of acid-base reactions useful.

ACID

An *acid* is any substance which, upon dissolving in water, increases the concentration of $H^+_{(aq)}$. The acidic compound usually contains a hydrogen atom which will dissociate in water to give $H^+_{(aq)}$. For example, HCl dissociates in water to give

$$HCl \xrightarrow{aq} H^+_{(aq)} + Cl^-_{(aq)}$$

This dissociation increases the concentration of $H^+_{(aq)}$. Similarly, HNO_3 is an acid where the dissociation is

$$HNO_3 \xrightarrow{aq} H^+_{(aq)} + NO^-_{3(aq)}$$

Acids such as HCl and HNO_3 are considered *strong acids* since they completely dissociate into ions in aqueous solution. Other common strong acids are HBr, HI, H_2SO_4, and $HClO_4$.

In addition there are also weak acids which do not completely dissociate into ions. Frequently, the dissociation of weak acids is quite small (0.001–5 %). Some acids are moderately strong; dissociation is intermediate (5–50 %). An example of a weak acid is HCN, which dissociates only 0.001 % at a concentration of 0.2 M. The equilibrium concentrations are written below the formulas in the equation

$$HCN_{(aq)} \rightleftharpoons H^+_{(aq)} + CN^-_{(aq)}$$
$$\phantom{HCN_{(aq)}} \text{0.2 M} \qquad \text{0.00001 M} \quad \text{0.00001 M}$$

Another example of a weak acid is acetic acid (CH_3COOH) which contains four hydrogen atoms. However, in each acetic acid molecule, only the one hydrogen bonded to the oxygen atom is capable of dissociation in water to give H^+. Accordingly,

$$CH_3COOH_{(aq)} \rightleftharpoons CH_3COO^-_{(aq)} + H^+_{(aq)}$$

Some substances are acidic although they do not contain any hydrogen atoms. These substances increase the concentration of $H^+_{(aq)}$ by reaction with the solvent. For example, aluminum chloride ($AlCl_3$) dissolves in water to produce Al^{3+} and Cl^-. The Al^{3+} can react with water to give H^+ ions.

$$AlCl_3 \xrightarrow{aq} Al^{3+}_{(aq)} + 3Cl^-_{(aq)}$$
$$+H_2O \updownarrow$$
$$Al(OH)^{2+}_{(aq)} + H^+_{(aq)}$$

The nomenclature of acids is systematic in relationship to the anion of the acid. If the anion has an *-ate* or *-ide* ending, the acid has an *-ic* ending. If the anion has an *-ite* ending, the acid has the *-ous* ending. For binary acids (acids of monatomic ions) the prefix *hydro-* is attached to the name of the acid. For acids containing polyatomic ions, the hydrogen is dropped from the name and the anion is named first, according to the rules we have just specified, followed by the word acid. Examples (see table below) should help clarify the procedure for naming acids.

FORMULA OF ACID	NAME OF ION	NAME OF ACID
HCl	chlor*ide*	*hydro*chlor*ic* acid
H_2SO_4	sulf*ate*	sulfur*ic* acid
H_2SO_3	sulf*ite*	sulfur*ous* acid
HNO_3	nitr*ate*	nitr*ic* acid
HBr	brom*ide*	*hydro*brom*ic* acid
$HBrO$	hypobrom*ite*	hypobrom*ous* acid
$HBrO_2$	brom*ite*	brom*ous* acid
$HBrO_3$	brom*ate*	brom*ic* acid
$HBrO_4$	perbrom*ate*	perbrom*ic* acid
$HC_2H_3O_2$	acet*ate*	acet*ic* acid
HCN	cyan*ide*	*hydro*cyan*ic* acid

PROBLEM 3–9   Name the following compounds: $KClO_3$, $HClO_3$, HCN, HF, NaH, KIO_4, HIO_4, $Ca(ClO_2)_2$, HNO_2, $HCOOH$ [Hint: $HCOO^-$ is the formate ion.]

There are some polyatomic anions which contain hydrogen atoms that can dissociate in solution to give H^+. These anions are also acids. Examples are HCO_3^- and $H_2PO_4^-$

$$HCO_{3(aq)}^- \rightleftharpoons H_{(aq)}^+ + CO_{3(aq)}^{2-}$$

$$H_2PO_{4(aq)}^- \rightleftharpoons H_{(aq)}^+ + HPO_{4(aq)}^{2-}$$

These anions are not named as acids despite the fact that they are acidic. They are referred to by the anion names given in Section 2–C.

BASE

A base is any substance which, upon dissolving in water, increases the amount of OH^- in solution. The base may or may not actually contain a hydroxide ion which is liberated in solution. The base may, however, react with H^+ to increase the amount of OH^-. An example of a substance containing OH^- is any metal hydroxide, such as sodium hydroxide, $NaOH$, or barium hydroxide, $Ba(OH)_2$.

$$NaOH_{(s)} \xrightarrow{aq} Na_{(aq)}^+ + OH_{(aq)}^-$$

$$Ba(OH)_{2(s)} \xrightarrow{aq} Ba_{(aq)}^{2+} + 2OH_{(aq)}^-$$

Most metal hydroxides which are soluble in water are *strong bases* and dissociation into ions is essentially complete. Other examples of strong bases are LiOH, KOH, and $Sr(OH)_2$.

An example of a substance which does *not* contain an OH^- but produces an increase in $(OH^-)_{aq}$ when it is dissolved in water is ammonia, NH_3. The reaction is

$$NH_{3(aq)} + H_2O \rightleftharpoons NH_{4(aq)}^+ + OH_{(aq)}^-$$
$$\text{0.1 } M \qquad\qquad\qquad \text{0.0013 } M \quad \text{0.0013 } M$$

Frequently, the reaction with water does not go to completion and NH_3 is therefore considered a *weak base*. When the concentration of $NH_{3(aq)}$ is 0.1 *M*, the reaction with water is only 1.3 %.

The nomenclature of bases containing the hydroxide ion is identical to the nomenclature of inorganic compounds based on ionic constituents. Examples are:

NaOH   sodium hydroxide

$Ca(OH)_2$ calcium hydroxide

$Al(OH)_3$ aluminum hydroxide

CuOH   copper(I) hydroxide
                (cuprous hydroxide)

$Cu(OH)_2$ copper(II) hydroxide
                (cupric hydroxide)

NEUTRALIZATION BY ACID-BASE REACTION

The reaction of an acid with a base is called *neutralization*. The reaction involves the combination of H^+ and OH^- to form H_2O, and of the remaining anion of the acid and the cation of the base to form a salt, which is generally soluble in the water solvent.

$$\text{acid} + \text{base} \longrightarrow \text{salt} + \text{water}$$

If the acid contains more than one ionizable H^+ the neutralization may involve the reaction of only one H^+, two H^+, or more, depending upon the conditions. Examples of neutralizations are:

$$NaOH + HCl \longrightarrow NaCl + H_2O$$

$$\text{or}$$

$$Na_{(aq)}^+ + OH_{(aq)}^- + H_{(aq)}^+ + Cl_{(aq)}^- \longrightarrow Na_{(aq)}^+ + Cl_{(aq)}^- + H_2O$$

$$Ca(OH)_2 + H_2SO_4 \longrightarrow CaSO_4 + 2H_2O$$

$$NaOH + H_3PO_4 \longrightarrow NaH_2PO_4 + H_2O$$

$$NaOH + NaH_2PO_4 \longrightarrow Na_2HPO_4 + H_2O$$

$$NaOH + Na_2HPO_4 \longrightarrow Na_3PO_4 + H_2O$$

PROBLEM 3–10 Complete and balance the following reactions: (assume all H^+ and OH^- react except where designated otherwise)

a. $NaOH + HC_2H_3O_2 \longrightarrow NaC_2H_3O_2$

b. $Ca(OH)_2 + HCOOH \longrightarrow Ca(HCOO)_2$

c. $HCl + NH_4OH \longrightarrow$

d. $H_2SO_4 + Al(OH)_3 \longrightarrow$

e. $H_2S + NaOH \longrightarrow NaHS +$

f. $NaOH + H_3PO_4 \longrightarrow$

EQUIVALENT WEIGHTS FOR ACIDS AND BASES

The equivalent weights for acids and bases are defined in a manner similar to that for oxidizing and reducing agents (equations 3–2 to 3–4). In this case, however, the important aspect is the number of protons or hydroxide ions that are furnished by the acid or base in a reaction. The definition of gram-equivalent-weight for an acid or a base is

3–5 $$\text{g-eq wt} = \frac{\text{g-mol wt}}{\text{number of } H^+ \text{ or } OH^- \text{ which react/molecule}}$$

The relationship between the number of equivalents and number of moles is:

3–6 $$\text{eq} = (\text{moles})(H^+ \text{ or } OH^- \text{ which react/molecule})$$

To determine the number of equivalents of acid or base taking part in a reaction, you must know the number of protons from the acid or hydroxide ions from the base which react. The same substance can have different equivalent weights if it reacts differently in two reactions. For example H_3PO_4 potentially has three hydrogens that can ionize or react with OH^-. However, in most reactions with a base only one or two of these protons is used as shown in the chemical equations

a. $NaOH + H_3PO_4 \longrightarrow NaH_2PO_4 + H_2O$

b. $2NaOH + H_3PO_4 \longrightarrow Na_2HPO_4 + 2H_2O$

In these reactions the H_3PO_4 would have the following equivalent weights.

for a. $$\text{eq wt } H_3PO_4 = \frac{\text{MW } H_3PO_4}{1}$$

for b. $$\text{eq wt } H_3PO_4 = \frac{\text{MW } H_3PO_4}{2}$$

PROBLEM 3–11 a. Calculate the equivalent weights of H_2SO_4 and $NaOH$ for the following neutralization reaction

$$H_2SO_4 + 2NaOH \longrightarrow Na_2SO_4 + 2H_2O$$

b. How many equivalents of H_2SO_4 would there be in a 100 g sample containing 25 % H_2SO_4 by mass?

c. How many equivalents of H_2SO_4 would there be in a 100 gram sample containing 25 % H_2SO_4 by mass if the neutralization reaction were

$$H_2SO_4 + NaOH \longrightarrow NaHSO_4 + H_2O$$

SOLVENT CONCEPT OF ACIDS AND BASES

The previous definition of acids and bases was restricted to water as the solvent. Chemical reactions can also be carried out in nonaqueous solvents. A system of acids and bases can be defined in nonaqueous solvents based upon the ionization of the solvent. For this reason this definition of acids and bases is called the *solvent concept*. For example, *liquid* ammonia dissociates slightly into NH_4^+ and NH_2^- ions. As in the case of water, these ions are completely surrounded by solvent NH_3 molecules and there is considerable interaction between the ions and the solvent molecules. The extent of dissociation into ions is again quite small. Comparing the ammonia dissociation (autoionization) with that for water

$$2NH_{3(l)} \rightleftharpoons NH_{4(ammonia)}^+ + NH_{2(ammonia)}^-$$
(pure liquid) $\sim 10^{-15}\ M$ $\sim 10^{-15}\ M$

$$H_2O_{(l)} \rightleftharpoons H_{(aq)}^+ + OH_{(aq)}^-$$
(pure liquid) $10^{-7}\ M$ $10^{-7}\ M$

An acid in liquid ammonia is any substance which increases the concentration of NH_4^+ and a base is any substance which increases the concentration of NH_2^-. The salt ammonium chloride, NH_4Cl, is a strong acid in liquid ammonia, analogous to HCl in aqueous solutions.

$$NH_4Cl \xrightarrow{NH_3} NH_{4(ammonia)}^+ + Cl_{(ammonia)}^-$$
acid

$$HCl \xrightarrow{aq} H_{(aq)}^+ + Cl_{(aq)}^-$$
acid

Substances which are acids in aqueous solutions are also acids in liquid ammonia. NH_4^+ is formed by the reaction

$$HCl + NH_3 \longrightarrow NH_{4(ammonia)}^+ + Cl_{(ammonia)}^-$$

A strong base in liquid ammonia is sodium amide, $NaNH_2$, analogous to $NaOH$ in water.

$$NaNH_{2(s)} \xrightarrow{NH_3} Na^+_{(ammonia)} + NH^-_{2(ammonia)}$$
$$\text{base}$$

$$NaOH_{(s)} \xrightarrow{aq} Na^+_{(aq)} + OH^-_{(aq)}$$
$$\text{base}$$

Neutralization between acids and bases in liquid ammonia is again analogous to neutralization in water.

$$NH_4Cl + NaNH_2 \longrightarrow 2NH_{3(l)} + Na^+_{(ammonia)} + Cl^-_{(ammonia)}$$
$$\text{acid} \qquad \text{base}$$

$$HCl + NaOH \longrightarrow H_2O_{(l)} + Na^+_{(aq)} + Cl^-_{(aq)}$$
$$\text{acid} \qquad \text{base}$$

Other nonaqueous solvents which illustrate the solvent concept of acids and bases are given in Table 3–1. The autoionization step is shown in each case and the ions characteristic of acids and bases are indicated. For each solvent an example of an acid and a base is shown. Again water is shown for comparison. The subscript (*sol*) designates that the ion is *solvated* (completely surrounded by solvent molecules) in that solvent.

It is interesting to note that there are numerous solvents available to a chemist which are quite different from water, yet in terms of ionization and acid-base properties are quite analogous to water. However, water is the most prevalent solvent on earth and is the fundamental solvent for all living matter here on earth. The unique properties of water enter into many of the discussions in this book. It might be appropriate to begin a chemistry course with the question: what is unique about water. It will take various aspects of many sections in this book to define fully this uniqueness.

TABLE 3–1 Non-aqueous solvent acid-base systems

SOLVENT	IONIZATION	ACID	BASE
water	$H_2O \rightleftharpoons H^+_{(aq)} + OH^-_{(aq)}$ $\qquad\quad$ acid $\quad$ base	HCl	$NaOH$
hydrogen fluoride	$HF \rightleftharpoons H^+_{(sol)} + F^-_{(sol)}$ $\qquad\quad$ acid $\quad$ base	$HClO_4$	KF
sulfur dioxide	$2SO_2 \rightleftharpoons SO^{2+}_{(sol)} + SO^{2-}_{3(sol)}$ $\qquad\qquad$ acid $\qquad$ base	$SOCl_2$	Na_2SO_3
hydrogen cyanide	$HCN \rightleftharpoons H^+_{(sol)} + CN^-_{(sol)}$ $\qquad\qquad$ acid $\qquad$ base	HCl	KCN

PROBLEM 3–12 Acetic acid, HAc, ionizes according to the equation

$$HAc \rightleftharpoons H^+_{(sol)} + Ac^-_{(sol)}$$

a. Classify the following compounds as acids, bases, or salts in the solvent HAc: NaAc, NH_3, HCl, NaCl, H_2SO_4, $Mg(Ac)_2$, BaI_2, H_2O. Consider the possibility that the substance could react with the solvent HAc, and that this reaction may produce acidic or basic conditions.
b. Write the reaction between $HClO_4$ and NH_4Ac.

PROBLEM 3–13 Dinitrogen tetroxide ionizes according to the equation

$$N_2O_4 \rightleftharpoons NO^+ + NO_3^-$$

Classify the following as acids, bases or salts in this solvent: NOCl, $NaNO_3$, KBr, AgCl, $AgNO_3$, NOBr.

BRÖNSTED-LOWRY CONCEPT OF ACIDS AND BASES

Another concept of acids and bases which is quite functional is the Brönsted-Lowry theory which is based on the relative ability of a substance to donate or accept a proton. According to this frequently used concept, an *acid* is defined as any substance which can *donate a proton* (H^+) and a *base* is any substance which can *accept a proton*. The strength of a substance as an acid or base is based upon its relative ability to donate or accept a proton. On this basis the acid strength of substances can be easily compared and evaluated, even if the substances function as both acids and bases in different solvents.

In this concept an acid, upon donating a proton, always produces a base. And a base, upon accepting a proton, becomes an acid. For example, HCN is an acid which upon donating a proton forms the base CN^-.

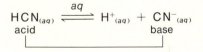

The base which is formed is called the *conjugate base* of the acid and the pair is referred to as a *conjugate acid-base pair*.

In this system an acid-base reaction consists of a simple proton transfer from an acid to a base. For example, reaction between HCl and NH_3 is

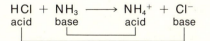

Similarly HCl and CN^- can react as an acid and a base, respectively

Each proton-donating species of a polyprotic acid (i.e., an acid containing more than one hydrogen atom) is also associated with a conjugate base. For example, H_2SO_4 dissociates as

$$H_2SO_4 \underset{}{\overset{aq}{\rightleftharpoons}} H^+_{(aq)} + HSO_4^-{}_{(aq)}$$

and also as

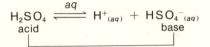

HSO_4^- can act as both an acid and a base in the Brönsted-Lowry concept.

The relative strength of an acid depends on its ability to donate a proton to another base. If an acid is a strong acid, its conjugate base must necessarily be a poor proton acceptor or, in other words, a weak base. The reaction will tend to proceed in the direction of donation of a proton by the strong acid. Conversely, a strong base has great proton affinity and must necessarily be associated with a weak conjugate acid. For example, HCl is a strong acid (good proton donor) and Cl^- is a very weak base (poor proton acceptor). On the other hand, HCN is a weak acid (poor proton donor) and CN^- is a very strong base (good proton acceptor).

By carrying out reactions between different acids and bases the relative strength of acid-base pairs can be established. A list of some acid-base pairs is given in Table 18 of Appendix T. This table can be used to predict the course of acid-base reactions. Any acid higher in this table (strong acid) will transfer a proton to any base lower in the table (strong base). The reaction will necessarily produce a weaker acid and a weaker base. The reactions of HSO_4^- with CH_3COO^-, NH_3, and CN^- illustrate this.

$$HSO_4^- + CH_3COO^- \longrightarrow CH_3COOH + SO_4^{2-}$$

$$HSO_4^- + NH_3 \longrightarrow NH_4^+ + SO_4^{2-}$$

$$HSO_4^- + CN^- \longrightarrow HCN + SO_4^{2-}$$

In most of these acid-base reactions there is an equilibrium between the reactants and products, and all species are present. However, when a strong acid reacts with a strong base, equilibrium is achieved with a high yield of products. For this reason when we say that an acid-base reaction will occur, we mean that it will produce a high yield of products. The greater the difference between the acid strength of the reacting acid and acid strength of the conjugate acid of the reacting base, the greater the yield for the reaction.

PROBLEM 3–14   Predict whether the following acid-base reactions can occur. If the reaction will proceed, complete the equation by giving the proper products. If the reaction will not occur write *N.R.* for no reaction. All reactions are carried out in water.

a. $HNO_3 + NH_3 \longrightarrow$

b. $HCN + Cl^- \longrightarrow$

c. $NH_4^+ + OH^- \longrightarrow$

d. $C_2H_5OH + C_6H_5O^- \longrightarrow$

e. $HClO_4 + SO_4^{2-} \longrightarrow$

3–F. RELATIONSHIP BETWEEN MOLES, MOLECULES, MASS, GRAM-MOLECULAR VOLUME, AND EQUIVALENT WEIGHT

You may have noticed that the term *mole* has appeared several times within this chapter and Chapter 2. We have already related moles to several other quantities.

1. Equation 2–2 relates number of moles to number of molecules through Avogadro's number.

2. Equation 2–1 relates number of moles to mass.

3. Equation 3–4 relates number of moles to equivalents for an oxidizing or reducing agent.

4. Equation 3–6 relates number of moles to equivalents for an acid or base.

5. Equation 2–12 relates number of moles to liters through the g-mol vol.

To emphasize the interrelationship between these quantities, a wheel is shown with moles at the hub in Figure 3–7. The term equivalents is shown only once to represent either equivalents of oxidizing-reducing agent or of acid-base. This wheel arrangement, in addition to illustrating the importance of the mole concept, can be used as a mnemonic device for recalling these relationships. A few examples will be given to illustrate the use of these relationships.

EXAMPLE 3–9   *Suppose we wish to calculate the number of molecules in three equivalents of $H_2C_2O_4$ where the acid-base reaction was*

$$H_2C_2O_4 + NaOH \longrightarrow NaHC_2O_4 + H_2O$$

FIGURE 3–7 The central role of the mole concept

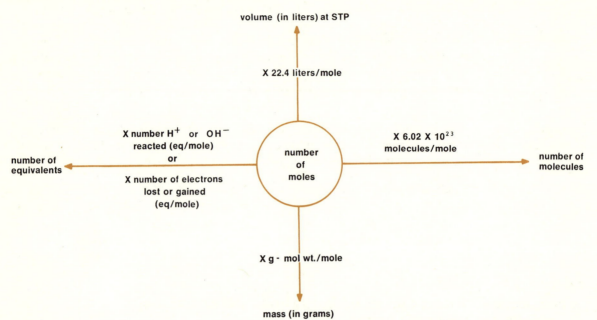

Since there is one H^+ which reacts per molecule of $H_2C_2O_4$

$$moles\ H_2C_2O_4 = (3\ eq)\left(\frac{1\ mole}{1\ eq}\right) = 3\ moles\ H_2C_2O_4$$

$$molecules\ H_2C_2O_4 = (3\ moles)\left(\frac{6.023 \times 10^{23}\ molecules}{mole}\right) = 18.069 \times 10^{23}\ molecules$$

EXAMPLE 3–10   *Calculate the number of moles of $H_2C_2O_4$ in 3 eq $H_2C_2O_4$ which are involved in the following oxidation-reduction reaction*

$$5\overset{(+3)}{H_2C_2O_4} + 2MnO_4^- + 6H^+ = 10\overset{(+4)}{C}O_2 + 2Mn^{2+} + 8H_2O$$

The oxidation number of C in $H_2C_2O_4$ is +3, and in CO_2, +4. Each carbon loses one e^-, and for 2C in $H_2C_2O_4$ there is a loss of $2e^-$. Therefore, 1 mole = 2 equivalents, and

$$moles\ H_2C_2O_4 = (3\ eq)\left(\frac{1\ mole}{2\ eq}\right) = 3/2\ moles\ H_2C_2O_4$$

Note that the relationship between moles and equivalents for $H_2C_2O_4$ is different in Examples 3–9 and 3–10. As emphasized earlier, the determination of equivalent weight and equivalents depends upon the specific reaction in which the compound is involved. $H_2C_2O_4$ can act as a reducing agent with permanganate and also as an acid with $NaOH$.

EXAMPLE 3–11

Calculate the number of molecules in 50 liters of O_2 at STP.
  Since 1 mole $= 22.414$ liters (STP)

$$\text{the number of moles in 50 liters} = (50 \, \cancel{liters})\left(\frac{1 \, mole}{22.414 \, \cancel{liters}}\right) = \frac{50 \, moles}{22.414}$$

$$\text{molecules in 50 liters} = \left(\frac{50}{22.414} \, \cancel{moles}\right)\left(\frac{6.023 \times 10^{23} \, molecules}{1 \, \cancel{mole}}\right)$$

$$= 13.4 \times 10^{23} \, molecules$$

Alternatively we could calculate the number of molecules using the ideal gas equation of state. The number of moles of gas can be found from

$$PV = n \, RT$$

$$n = \frac{PV}{RT} = \frac{(1 \, \cancel{atm})(50 \, \cancel{liters})}{\left(0.08205 \, \frac{\cancel{liter\text{-}atm}}{deg\text{-}mole}\right)(273 \, \cancel{deg})}$$

$$n = \frac{50 \, moles}{(0.08205)(273)}$$

and as before

$$\text{molecules in 50 liters} = \frac{50 \, \cancel{moles}}{(0.08205)(273)}\left(\frac{6.023 \times 10^{23} \, molecules}{\cancel{mole}}\right)$$

$$= 13.4 \times 10^{23} \, molecules$$

PROBLEM 3–15

Calculate the volume of 5 g N_2 at STP.

PROBLEM 3–16

Calculate the number of molecules in 4 eq $KMnO_4$ which reacts according to the equation in Example 3–10.

PROBLEM 3–17

Calculate the volume occupied by 1×10^{20} molecules of an ideal gas under STP conditions. Express the volume in milliliters, ml.

3–G. STOICHIOMETRY

The subject of *stoichiometry* deals with the amounts of reactants and products which are involved in a given chemical reaction. In other words, how much product will be produced from a certain amount of reactant, or how much reactant is required to produce a certain amount of product. The amount of product or reactant can be expressed in terms of weight, volume of a gas, or moles. Calculations involving any of these terms are very similar.

We began this chapter with a discussion of chemical equations and their significance. Reiterating very briefly, the coefficients in the chemical equation can designate the number of moles, g-atoms, or g-formulas of a substance which enter into the reaction. Consider the following reaction:

3–7 $$2ZnS + 3O_2 \longrightarrow 2ZnO + 2SO_2$$

Equation 3–7 tells us that 2 moles ZnS react with 3 moles O_2 to give 2 moles ZnO and 2 moles SO_2. From this we can write the following expressions of equivalence:

2 moles ZnS reacts with 3 moles O_2.
2 moles ZnS yields 2 moles ZnO
2 moles ZnS yields 2 moles SO_2
3 moles O_2 yields 2 moles ZnO
3 moles O_2 yields 2 moles SO_2
2 moles ZnO is produced with 2 moles SO_2

Any one of these expressions can be used to write a factor for converting units. For example, to calculate the moles of ZnS required for complete reaction of 9 moles of O_2

$$(9 \text{ moles } O_2) \frac{2 \text{ moles } ZnS}{3 \text{ moles } O_2} = 6 \text{ moles } ZnS$$

Since we also know the relationship between moles and (1) weight, (2) volume, (3) molecules, we are in a position to calculate the amounts of reactants and products in terms of any of these three quantities. In the following examples we will illustrate typical problems that are frequently encountered. These should serve as a guide to the logic involved so that the student can solve any new type of stoichiometric problem that may arise.

EXAMPLE 3–12 *Let us use the reaction in equation 3–7 to illustrate how to calculate the weight of reactant required to give a certain weight of product. Suppose we wish to prepare 100 g of ZnO. How many grams of ZnS must be reacted with oxygen? Since the chemical equation expresses a relationship between moles, we must always convert from a weight or volume to moles. We use the following factors of equivalence to make the conversion. The molecular weight of ZnO is 81.4 and that of ZnS is 97.4.*

$$(100 \text{ g } ZnO)\left(\frac{1 \text{ mole } ZnO}{81.4 \text{ g } ZnO}\right)\left(\frac{2 \text{ moles } ZnS}{2 \text{ moles } ZnO}\right)\left(\frac{97.4 \text{ g } ZnS}{1 \text{ mole } ZnS}\right) = 120 \text{ g } ZnS$$

UMBC BLOOD BANK PROGRAM

FOR STUDENTS
FACULTY AND STAFF

SOME

THINGS

TO

CONSIDER

IN

MAKING

YOUR

DECISION

ABOUT

PARTICIPATION

"WHAT IF I DON'T GIVE ONE PINT OF BLOOD?"

Blood when you need it?

Last year 82,000 pints of blood were needed for
the Baltimore area! The Red Cross collected
only 45,000 pints! Sufficient blood for your
need won't be available unless it is on hand.
When blood is not available, consider the time
consuming trouble of recruiting special donors.
How many of your friends would give to help you?
How many could give?

AND THE COST? $50 to $100 a pint!

Individuals undergoing major surgery may use up
to 15 pints of blood a day - it's staggering the
number of people your family would have to find
to meet this kind of blood need. The only other
recourse would be commercial blood banks involving businesses and donors trying to make a
profit. The need to pay the donors, technicians
and owners to obtain the blood increases its
cost phenomenally before it reaches you! If
you need 15 pints a day for two weeks to keep
you alive, how would you possibly afford it?

UNKNOWN, UNHEALTHY DONORS = UNWHOLESOME BLOOD.

Your failing to donate blood through the UMBC
Blood Bank Program increase the need for hos-

GENERAL INFORMATION

UMBC has established a group blood bank program.
As a member of the UMBC Blood Bank Program,
including all student, faculty and staff, you
will be entitled to the following benefits:

> If you are single - all the blood needs of
> yourself, your parents and your dependent
> brothers and sisters will be met.
>
> If you are married - all the blood needs of
> yourself, your spouse, your unmarried child-
> ren, your parents and the parents of your
> spouse will be met.

In order for the UMBC community to continue to
receive these benefits over 1000 donors will be
needed during the academic year 1974-75.

A professional staff from the Red Cross will be
on campus collecting blood from volunteer donors
on November 11, 12, 13th. While the actual giving
of blood takes just a few minutes, the whole process of registration, relaxation and refreshment
takes approximately 50 minutes.

You may safely donate blood up to five times per
year without parental permission if you are at
least 18 years of age and in normal good health.

A single donation of blood will cover for your
entire stay at UMBC. If you are a senior you
will be covered for one year from the date you
give blood.

EXAMPLE 3–13

If a problem similar to Example 3–12 calls for the volume of a gaseous reactant or product, we first convert the moles into volume. To illustrate, consider equation 3–7 again and answer the question "How many grams of ZnS are required to produce 50 liters SO_2 at a pressure of three atmosphere and 500 °K?"

We start by calculating the moles of SO_2 in the 50 liters at P = 3 atm and T = 500 °K from the ideal gas equation (equation 2–13).

$$PV = n\,RT$$

$$(3\ atm)(50\ liters) = n\left(0.08205\ \frac{liter\text{-}atm}{deg\text{-}mole}\right)(500\ deg)$$

$$n = \frac{(3)(50)}{(0.08205)(500)}\ moles\ SO_2$$

$$n = 3.67\ moles\ SO_2$$

$$(3.67\ \cancel{moles\ SO_2})\left(\frac{2\ \cancel{moles\ ZnS}}{2\ \cancel{moles\ SO_2}}\right)\left(\frac{97.4\ g\ ZnS}{1\ \cancel{mole\ ZnS}}\right) = 358\ g\ ZnS$$

PROBLEM 3–18

Calculate the weight of SO_2 produced by reacting 5 g O_2 with an excess of ZnS (reaction occurs according to equation 3–7).

PROBLEM 3–19

Calculate the number of molecules of O_2 that must be reacted with an excess of ZnS to give 10 moles SO_2 (reaction given by equation 3–7).

PROBLEM 3–20

Ethane burns in air according to the equation

$$C_2H_6 + \tfrac{7}{2}O_2 \longrightarrow 2CO_2 + 3H_2O$$

How many liters of O_2 at 1 atm and 700 °K are required to completely burn 1 g C_2H_6?

A slightly different type of stoichiometric problem arises when the masses of two or three reactants are given and the calculation involves finding the amount of product produced in the reaction. Often one reactant is in excess and another reactant limits the amount of product formed. In this situation we first calculate the amount of product formed on the basis of each separate reactant. The reactant which gives the smallest amount of product is the one which limits the reaction and determines the amount of product actually formed. An example will illustrate the necessary calculations.

EXAMPLE 3–14

Consider again the reaction in equation 3–7. Suppose we have 10 g of ZnS and 15 g of O_2. How much ZnO will be formed from the complete reaction of these materials.

The amount of ZnO produced from 10 g ZnS can be calculated as described previously.

$$(10 \text{ g } ZnS)\left(\frac{1 \text{ mole } ZnS}{97.4 \text{ g } ZnS}\right)\left(\frac{2 \text{ moles } ZnO}{2 \text{ moles } ZnS}\right)\left(\frac{81.4 \text{ g } ZnO}{1 \text{ mole } ZnO}\right) = 8.35 \text{ g } ZnO$$

The amount of ZnO produced from 15 g O_2 is calculated in a similar manner.

$$(15 \text{ g } O_2)\left(\frac{1 \text{ mole } O_2}{32 \text{ g } O_2}\right)\left(\frac{2 \text{ moles } ZnO}{3 \text{ moles } O_2}\right)\left(\frac{81.4 \text{ g } ZnO}{1 \text{ mole } ZnO}\right) = 25.5 \text{ g } ZnO$$

Since 8.35 g is less than 25.5 g, the amount of product ZnO actually produced is 8.35 g. All of the O_2 is not reacted and some O_2 is left as excess.

PROBLEM 3–21 A 20 g sample of ethane is burned with 5 g O_2.

$$C_2H_6 + \tfrac{7}{2}O_2 = 2CO_2 + 2H_2O$$

How many grams of CO_2 are produced?

In the following example quantities of two reactants are given, and you must determine which reactant is in excess.

EXAMPLE 3–15 *Suppose 1.0 g ZnS is reacted with 0.5 g O_2 according to equation 3–7. Which reactant is in excess and by what amount?*
  We may start by calculating the amount of ZnS necessary to react with 0.5 g O_2.

$$(0.5 \text{ g } O_2)\left(\frac{1 \text{ mole } O_2}{32 \text{ g } O_2}\right)\left(\frac{2 \text{ moles } ZnS}{3 \text{ moles } O_2}\right)\left(\frac{97.4 \text{ g } ZnS}{1 \text{ mole } ZnS}\right) = 1.015 \text{ g } ZnS$$

Since 1.015 g ZnS is greater than the 1.0 g ZnS actually present, then the O_2 must be in excess. We can now calculate the amount of O_2 that will be required to react with the 1.0 g ZnS.

$$(1.0 \text{ g } ZnS)\left(\frac{1 \text{ mole } ZnS}{97.4 \text{ g } ZnS}\right)\left(\frac{3 \text{ moles } O_2}{2 \text{ moles } ZnS}\right)\left(\frac{32 \text{ g } O_2}{1 \text{ mole } O_2}\right) = 0.483 \text{ g } O_2$$

The amount of O_2 remaining after the reaction is completed is thus

$$(0.5 \text{ g } - 0.483 \text{ g}) = 0.017 \text{ g } O_2 \text{ remaining}$$

PROBLEM 3–22 Suppose we ignite a mixture containing 5 g C_2H_6 and 3 g O_2. Will C_2H_6 or O_2 remain after the reaction is complete? Calculate the excess amount.

Another type of stoichiometric calculation involves reactions carried out in solution. In these calculations, amounts of reactants can be expressed in terms of volume or concentration of solution. The concentration of the reactant in solution may or may not be known. This type of stoichiometric calculation generally involves either acid-base or oxidation-reduction reactions.

Before going directly into these stoichiometric calculations, we must first define a new concentration unit, *normality*. Normality, N, is the number of equivalents per liter of solution.

3–8

$$N = \frac{eq}{V(\text{in liters})}$$

The calculation of equivalents from mass of acid or base or of oxidizing or reducing agent was presented earlier.

EXAMPLE 3–16   *Calculate the normality of oxalic acid, $H_2C_2O_4$, in a solution containing 2.000 g in 250 ml. The solution reacts with permanganate ion according to the reaction*

$$2MnO_4^- + 5H_2C_2O_4 + 6H^+ \longrightarrow 2Mn^{2+} + 10CO_2 + 8H_2O$$

It is not actually necessary to have the balanced equation to calculate the normality of the $H_2C_2O_4$ solution. In order to calculate the equivalent weight and hence the number of equivalents, we must know the number of electrons transferred/molecule of $H_2C_2O_4$. The oxidation numbers for carbon are evaluated as in section 3–D. Since there are two carbon atoms in $H_2C_2O_4$, we can calculate the change of e^- to produce $2CO_2$.

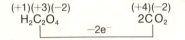

The oxidation number of carbon changes from +3 to +4 and there are two carbon atoms involved. There then must be a loss of $2e^-$/molecule of $H_2C_2O_4$.

$$\text{eqwt } H_2C_2O_4 = \frac{MW\ H_2C_2O_4}{2} = \frac{90.0}{2} = 45.0$$

Using equation 3–3

$$\text{eq } H_2C_2O_4 = \frac{m}{\text{g-eq wt}} = \frac{2.000\ g}{45.0\ g/eq} = 0.0444\ eq$$

By equation 3–8

$$\text{normality of } H_2C_2O_4 = \frac{0.0444\ eq}{0.250\ l} = 0.1776\ eq/l$$

PROBLEM 3–23   Calculate the normality of a solution containing 0.100 g of $H_2C_2O_4$ in 25 ml, where the $H_2C_2O_4$ reacts with $NaOH$ according to the equation

$$H_2C_2O_4 + 2NaOH \longrightarrow Na_2C_2O_4 + 2H_2O$$

Stoichiometric calculations involving acid-base and oxidation-reduction reactions are conveniently carried out using the concept of equivalents. In an oxidation-reduction equation the electrons lost must equal the electrons gained.

3–9 $$e^- \text{ lost} = e^- \text{ gained}$$

Since an equivalent of a species being oxidized loses 6.023×10^{23} e^-, the number of electrons lost in an oxidation-reduction reaction is 6.023×10^{23} times the number of equivalents of the oxidized species. Similarly, the number of electrons gained is 6.023×10^{23} times the number of equivalents of the reduced species. Substituting into equation 3–9

$$(6.023 \times 10^{23}) \times (\text{eq. oxidized species}) = 6.023 \times 10^{23} \times (\text{eq. of reduced species})$$

and cancelling the 6.023×10^{23}

3–10 $$\text{eq. oxidized species} = \text{eq. reduced species}$$

Similarly, in an acid-base reaction the equivalents of H^+ reacted must equal the equivalents of OH^- reacted.

3–11 $$\text{eq. acid reacted} = \text{eq. base reacted}$$

The number of equivalents on either side of the equation can be found by equation 3–3 or by equation 3–12 (equation 3–8 rearranged).

3–3 $$\text{eq.} = \frac{m}{\text{g-eq wt}}$$

3–12 $$\text{eq} = \text{volume} \times \text{normality}$$

EXAMPLE 3–17   *How many grams of $KMnO_4$ will react with 25 ml 0.1775 N $H_2C_2O_4$ according to the equation*

$$2MnO_4^- + 5H_2C_2O_4 + 6H^+ \longrightarrow 2Mn^{2+} + 10CO_2 + 8H_2O$$

We can let x = grams $KMnO_4$. The MnO_4^- changes from +7 oxidation state to +2 and must therefore gain $5e^-$

$$\underset{MnO_4^-}{\overset{(+7)}{}} \quad \underset{Mn^{2+}}{\overset{(+2)}{}}$$
$$\underset{+5e^-}{\underline{\qquad\qquad}}$$

$$eq\ wt\ KMnO_4 = \frac{MW\ KMnO_4}{5} = \frac{158}{5} = 31.6$$

The equivalents of $KMnO_4$ in x grams is thus

$$eq\ KMnO_4 = \frac{x}{g\text{-}eq\ wt\ KMnO_4}$$

The equivalents of $H_2C_2O_4$ is given by equation 3–12:

$$eq\ H_2C_2O_4 = V_{H_2C_2O_4}N_{H_2C_2O_4}$$

Substituting these expressions for equivalence into eq $KMnO_4$ = eq $H_2C_2O_4$:

$$\frac{x}{g\text{-}eq\ wt\ KMnO_4} = V_{H_2C_2O_4}N_{H_2C_2O_4}$$

$$\frac{x}{31.6\ g/eq} = (0.025\ liter)(0.1775\ eq/liter)$$

$$x = (31.6)(0.025)(0.1775)\ g$$

$$x = 0.140\ g\ KMnO_4$$

EXAMPLE 3–18   *What volume of an 0.250 N $H_2C_2O_4$ solution is required to react with 30 ml of 0.100 N $KMnO_4$?*

Since the quantity of both reactants can be expressed in terms of volume of solution, equation 3–12 can be used for both $H_2C_2O_4$ and $KMnO_4$.

$$eq\ H_2C_2O_4 = V_{H_2C_2O_4}\ N_{H_2C_2O_4}$$

$$eq\ KMnO_4 = V_{KMnO_4}\ N_{KMnO_4}$$

Since eq $H_2C_2O_4$ = eq $KMnO_4$

$$V_{H_2C_2O_4}\ N_{H_2C_2O_4} = V_{KMnO_4}\ N_{KMnO_4}$$

$$V_{H_2C_2O_4}(0.250\ eq/liter) = (0.030\ liter)(0.100\ eq/liter)$$

$$V_{H_2C_2O_4} = \frac{(0.030)(0.100)}{(0.250)} = 0.012\ liter = 12\ ml$$

PROBLEM 3-24   What volume of 0.200 N $KMnO_4$ solution is required to react with 0.175 g $H_2C_2O_4$, according to the reaction in example 3–17?

PROBLEM 3-25   What weight of $K_2Cr_2O_7$ is required to react with 25 ml of 0.05 N H_2S solution? The equation for the reaction is

$$3H_2S + Cr_2O_7{}^{2-} + 8H^+ \longrightarrow 3S + 2Cr^{3+} + 7H_2O$$

PROBLEM 3-26   What volume of 0.100 N NaOH solution is required to react with 0.250 g $H_2C_2O_4$? The equation for the reaction is

$$2NaOH + H_2C_2O_4 \longrightarrow Na_2C_2O_4 + 2H_2O$$

EXERCISES

1. Balance the following equations. For those reactions which are of the oxidation-reduction type, identify the oxidizing agent and the reducing agent.
   a. $H_3PO_4 + Mg(OH)_2 \longrightarrow Mg_3(PO_4)_2 + H_2O$
   b. $SO_3{}^{2-} + MnO_4{}^- + H^+ \longrightarrow SO_4{}^{2-} + Mn^{2+} + H_2O$
   c. $N_2O_5 + H_2O \longrightarrow HNO_3$
   d. $P_4O_{10} + H_2O \longrightarrow H_3PO_4$
   e. $PbCl_2 + CH_3COOH \longrightarrow Pb(CH_3COO)_2 + HCl$
   f. $Zn + H_2SO_4 \longrightarrow ZnSO_4 + H_2$
   g. $N_2O \longrightarrow N_2 + O_2$
   h. $PBr_3 + H_2O \longrightarrow H_3PO_3 + HBr$
   i. $PbCO_3 + HCl \longrightarrow PbCl_2 + CO_2 + H_2O$
   j. $Fe + HNO_3 \longrightarrow Fe(NO_3)_3 + NO + H_2O$
   k. $HNO_3 + HCl \longrightarrow NO + Cl_2 + H_2O$
   l. $Al + NaOH + H_2O \longrightarrow NaAlO_2 + H_2$

2. What volume of Cl_2 at 26 °C and 741 mm of pressure is produced by complete reaction of 52.68 g $KMnO_4$ according to the reaction

$$KMnO_4 + HCl \longrightarrow KCl + MnCl_2 + Cl_2 + H_2O \quad \text{(unbalanced)}$$

3. A portable hydrogen generator utilizes the reaction

$$CaH_2 + H_2O \longrightarrow Ca(OH)_2 + H_2 \quad \text{(unbalanced)}$$

   Calculate   a. the moles of H_2, and
   b. the volume of H_2 at 25 °C and 1.5 atm which can be produced from 50.0 g of calcium hydride.

4. Acetylene, C_2H_2, for use in welding, can be produced from the reaction of calcium carbide with water

$$CaC_{2(s)} + H_2O_{(l)} \longrightarrow C_2H_{2(g)} + Ca(OH)_{2(aq)} \qquad \text{(unbalanced)}$$

    a. How many moles of acetylene can be produced from 100 lbs CaC_2?
    b. If the acetylene produced were burned in a torch at the rate of 200 ml/minute (measured at 1 atm and 25 °C), how long could the torch be used for 100 lbs CaC_2?

5. You are given 20 ml 15 M H_2SO_4. If you add to this 100 ml 3 M H_2SO_4
    a. What is the concentration of the solution assuming the total volume is 120 ml?
    b. How many moles of H_2SO_4 are there in the 120 ml solution?
    c. How many milliliters of 6 M $Ca(OH)_2$ must be added to neutralize the acid in (b) to $CaSO_4$.

6. What weight of $SnCl_2$ (in grams) would be required to reduce 100 ml 0.1 M $K_2Cr_2O_7$? ($Cr_2O_7{}^{2-}$ is reduced to Cr^{3+}; Sn^{2+} is oxidized to Sn^{4+}).

7. 90.0 ml 0.43 M solution of H_2O_2 oxidized 64 ml of a solution of NaI to I_2. What is the molar concentration of the NaI solution? (The H_2O_2 is reduced to H_2O.)

8. In the following equations list the conjugate acid-base pairs.
    a. $HSO_3{}^- + H_2O \rightleftharpoons H_3O^+ + SO_3{}^{2-}$
    b. $NH_4{}^+ + H_2O \rightleftharpoons NH_3 + H_3O^+$ (slight reaction)
    c. $H_2O + H_2O \rightleftharpoons H_3O^+ + OH^-$ (slight reaction)
    d. $H_2O + CO_3{}^{2-} \rightleftharpoons HCO_3{}^- + OH^-$

9. The following reactions take place in aqueous solution
    a. $A + BH^+ \rightleftharpoons AH^+ + B$
    b. $C + H_3O^+ \rightleftharpoons CH^+ + H_2O$
    c. $BH^+ + H_2O \rightleftharpoons H_3O^+ + B$
    d. $AH^+ + H_2O \longrightarrow$ no reaction
    List the acid-base pairs AH^+–A, BH^+–B, CH^+–C, H_3O^+–H_2O in order of increasing acid strength as shown in Table 18, Appendix T.

10. Hydrochloric acid can be prepared using the hydrogen isotope of mass 2. This isotope of hydrogen is called deuterium, D, and the formula for the hydrochloric acid containing deuterium is DCl. In aqueous solution DCl has essentially the same properties as HCl. If you were to mix DCl with water, you would find upon analysis that some DHO and D_2O would be formed. Explain how this might occur in light of our discussion of Sections 3–C and 3–E.

11. Baking soda contains $NaHCO_3$. Upon heating, baking soda decomposes to give CO_2 gas.

$$NaHCO_3 \xrightarrow{\text{heat}} Na_2CO_3 + H_2O + CO_{2(g)} \qquad \text{(unbalanced)}$$

a. Balance the equation.
b. Calculate the volume of CO_2 given off from a handful of baking soda (approx. 50 g). Assume $t = 200\,°C$ and $P = 1$ atm.
c. Explain how baking soda is a good fire extinguisher, especially convenient for grease fires as would occur in the kitchen. [Hint: The density of CO_2 is greater than the density of air.]

12. Ammonia is produced by the Haber process from N_2 and H_2 gases.

$$N_2 + 3H_2 \longrightarrow 2NH_3$$

The reaction is run at high pressures for reasons which will be discussed in Chapter 13. A reaction vessel of 100 liters contained 25 mole % N_2 and 75 mole % H_2 initially, at 475 °C and 1,000 atm. If the product yield is 40 %, how many moles of NH_3 will be produced when equilibrium is established? Assume that all gases obey the ideal gas law, that the final pressure is 800 atm, and that the temperature remains 475 °C. Compute the partial pressures of N_2 and H_2 in the final mixture.

13. In liquid ammonia the following metathetical reactions occur. Complete and balance the equations.
a. $AgNO_3 + KNH_2 \longrightarrow$
[Hint: KNO_3 is insoluble in liquid ammonia.]
b. $Pb(NO_3)_2 + KNH_2 \longrightarrow PbNH_{(s)} + NH_3 +$
Write balanced equations for similar reactions in water solvent where KNH_2 is replaced by KOH.

14. The corrosion (oxidation) of iron occurs when it is in contact with *both* air and water. (A more detailed discussion of corrosion is given in Chapter 14). The overall reaction is

$$Fe_{(s)} + O_{2(g)} + H_2O_{(l)} \longrightarrow Fe_2O_3 \cdot H_2O_{(s)} \quad \text{(unbalanced)}$$

The $\cdot H_2O$ in the product designates that a molecule of H_2O is attached to each Fe_2O_3 unit. Suppose you have a sealed metal drum (30 cm radius and 1 m tall) which contains air at 1 atm and 25 °C and enough moisture to cause corrosion.
a. Calculate the extent of corrosion in terms of g-atoms of iron that would be converted to $Fe_2O_3 \cdot H_2O$ according to the above reaction if all the oxygen in the metal drum were used. Assume the air contains 20 % O_2 by volume.
b. If the corrosion occurred uniformly over the inside surface of the drum, what thickness of the metal drum will be corroded? The density of Fe is 7.86 g/cm³.

15. In Table 18, Appendix T the relative acid-base strengths were given according to the Brönsted-Lowry concept. The strength of an acidic species in a certain solvent depends upon how far it is displaced in this series above the ionization of the solvent. For example, HCl is far above

$$H_2O \rightleftharpoons H^+ + OH^-$$

in Table 18 and is a strong acid in water. However, HCN is only slightly above H_2O and is a weak acid. Similar considerations relative to the ionization of CH_3COOH and NH_3 can be used to classify relative acid and base strengths in the solvents CH_3COOH and NH_3.

a. In which solvent (CH_3COOH, NH_3, or H_2O) would HCN be the strongest acid?

b. Write the acid-base reaction between HCN and a base in each of these solvents.

c. In which solvent would KCN be the strongest base?

16. Acrylonitrile ($CH_2\!\!=\!\!CH\!\!-\!\!CN$) is used in making polymers which are eventually made into synthetic fibers for cloth. Recently acrylonitrile has been produced from propylene ($CH_2\!\!=\!\!CH\!\!-\!\!CH_3$), ammonia ($NH_3$) and oxygen from the air. The reaction is

$$CH_2\!\!=\!\!CH\!\!-\!\!CH_3 + NH_3 + O_2 \xrightarrow[\text{catalyst}]{\text{heat}} CH_2\!\!=\!\!CH\!\!-\!\!CN + H_2O \quad \text{(unbalanced)}$$

a. Balance the equation

b. If propylene at $t = 25\ °C$ and 5 atm pressure is fed into the reactor at the rate of 10 m³/hr, calculate the maximum possible amount of acrylonitrile (as kilograms) that could be produced in one day.

17. Oxygen can be prepared in the laboratory by the decomposition of potassium chlorate upon heating.

$$KClO_{3(s)} \xrightarrow[\text{catalyst}]{\text{heat}} KCl_{(s)} + O_{2(g)} \quad \text{(unbalanced)}$$

How many milliliters of oxygen gas would be produced at 25 °C and 745 mm pressure when 25.0 g of potassium chlorate are decomposed?

18. Gasoline consists primarily of octane, C_8H_{18}, which burns (combustion) according to the equation

$$C_8H_{18} + O_2 \longrightarrow CO_2 + H_2O \quad \text{(unbalanced)}$$

However, in the automobile the combustion is not always complete and small amounts of CO are also produced. CO complexes with the heme portion of hemoglobin in the blood and prevents the adsorption of O_2 necessary in the respiration process. Breathing CO at a concentration of 0.30 mole percent for 30 min is fatal.

Suppose a car is enclosed in a garage (20 ft × 20 ft × 8 ft) and is running on idle, consuming gasoline (density = 0.8 g/ml) at the rate of 0.5 gal/hr. For this particular car the exhaust gases contain 4 mole percent CO. How long would it take the air in the enclosed garage to acquire enough CO to reach the lethal concentration? Assume complete mixing of the gasoline exhaust with the air in the garage. Assume the air is at 0 °C and P = 1 atm.

19. One method for producing metallic zinc from zinc oxide consists of mixing the oxide with powdered coke and heating in clay retorts. The metallic zinc is vaporized and collected in an air-cooled clay condenser. The reaction is

$$ZnO_{(s)} + C_{(s)} = CO_{(g)} + Zn_{(s)}$$

The melting point of Zn is 693 °K and the boiling point is 1180 °K.
a. Calculate the amount of ZnO and C required to produce 100 kg Zn.
b. What temperatures are required in the reactor and condenser in order to remove the Zn from the reactor?
c. For every 100 kg of Zn produced, calculate the volume of CO gas which will be formed at 1200 °K and 1 atm pressure.
d. If the CO were vented to the atmosphere at 25 °C, calculate the volume of CO produced from 100 kg at this temperature. Assume P = 1 atm.
e. What problems might be incurred if CO were vented to the atmosphere? (See exercise 18 above.)

20. Sulfuric acid is a very important chemical commercially since it is used in numerous processes. Sulfuric acid can be produced by what is called the "contact process" where SO_3 is produced by having SO_2 and O_2 in *contact* with finely divided platinum.

$$SO_2 + O_2 \longrightarrow SO_3 \quad \text{(unbalanced)}$$

The yield is 96 % at 445 °C. The sulfur trioxide is then reacted with water to give sulfuric acid

$$SO_3 + H_2O \longrightarrow H_2SO_4$$

a. If the SO_2 is produced by burning elemental sulfur, how much H_2SO_4 (moles) can be produced from 100 kg sulfur?
b. What is the volume of the H_2SO_4 produced in part (a) if the density of pure H_2SO_4 is 1.84 g/ml?
c. If the SO_3 were bubbled into 100 liters of water to produce H_2SO_4, what molal concentration of H_2SO_4 would be produced starting with the 100 kg of sulfur as described in a.?

21. The fermentation of glucose to ethyl alcohol proceeds according to the following equation:

$$C_6H_{12}O_6 \longrightarrow C_2H_5OH + CO_2 \quad \text{(unbalanced)}$$

What volume of CO_2 will be produced at 25 °C and P = 1.5 atm from 5 lbs glucose? Assume a yield of 25 %.

CHAPTER FOUR

ATOMIC STRUCTURE

So far we have considered the structure of the atom only in terms of its gross features—a very dense nucleus, whose mass consists principally of protons and neutrons, and extranuclear electrons surrounding the nucleus at a greater distance (10^4 times the diameter of the nucleus). In this chapter we will look at the details of atomic structure. The electrons are found in areas around the nucleus called *orbitals*, which have various energies and various shapes. The orbitals, that the electrons in any given atom occupy, influence the properties of the atom. We are especially interested in the influence of orbitals on bonding between atoms.

4–A. CLASSICAL MECHANICS VERSUS QUANTUM MECHANICS

The motion of macroscopic bodies that we observe in everyday life (automobiles, missiles, planets) is governed by the laws of classical mechanics. Newton first presented these laws and they are as valid today as when they were first proposed. A detailed discussion of these laws is better undertaken by physicists, but it is important in chemistry to point out that classical mechanics places no restriction on the energy or velocity which a macroscopic body can acquire. For example, one can drive an automobile 20 mph, 25 mph, 30 mph or at any speed in between. Similarly, a planet or a satellite can travel at 25,000 ft/sec, 25,002 ft/sec, 25,001.1 ft/sec, or any other speed. The only limit is that the satellite's trajectory must be elliptical. Within reason there are *no restrictions* on the velocities, and hence energies, which these macroscopic bodies may attain.

The explanation of the behavior of macroscopic objects in terms of classical mechanics seems quite reasonable. However, the motion in systems containing small particles—electrons, neutrons, etc.—does not obey classical mechanics. Take as an example the hydrogen atom ($_1^1$H) consisting of a single proton for a nucleus, and an electron e⁻. The energy of the electron in this simple atom arises from its position and motion about the positive nucleus, but it can *not* take on just any value. The values of the energy of the electron in the $_1^1$H atom are *restricted*. This restriction *never* occurs in classical mechanics and a new system must be devised to describe atomic systems.

4–B. THE DUAL NATURE OF LIGHT

We must first discuss the properties of light, before we can consider electrons as small particles. Light is an *electromagnetic radiation* which consists of bundles of energy called *photons*. In order to explain adequately all of the experimental facts about light, it is necessary to postulate a dual nature of light—a particle nature and a wave nature. The energy of a photon is related to the frequency of that radiation through Planck's relationship

4–1
$$\epsilon = h\nu$$

where ν = frequency in sec⁻¹, h = Planck's constant = 6.626×10^{-27} erg-sec, and ϵ = energy of the photon in ergs. Through equation 4–1, *frequency uniquely defines energy*. Frequency is also related to wavelength λ and wave number, $\bar{\nu}$, by

4–2
$$\lambda = \frac{c}{\nu} = \frac{1}{\bar{\nu}}$$

FIGURE 4–1   A wave of $\lambda = 1 \times 10^{10}$ cm = 1×10^{18} Å, $\nu = 3$ cycles/sec

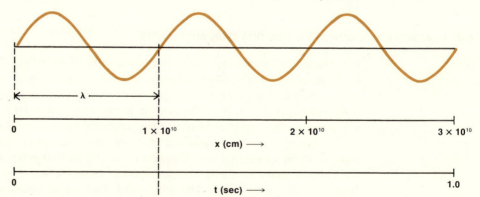

where c = velocity of light in cm/sec (and λ is then expressed in terms of cm). The wavenumber equals the number of cycles or oscillations that occur in 1 cm. The relationship between wavelength and frequency is illustrated in Figure 4–1.

As the wave in Figure 4–1 illustrates, the wavelength λ is the distance traveled in one complete cycle or oscillation. The frequency (ν) is the number of oscillations in one second (in Figure 4–1 this is 3 cycles/sec). The wave number $\bar{\nu}$ for the wave in Figure 4–1 can be calculated

$$\bar{\nu} = \frac{1}{\lambda} = \frac{1}{1 \times 10^{10} \text{ cm}} = 1 \times 10^{-10} \text{ cm}^{-1}$$

The wavelength of light is so small that it is convenient to define a smaller unit of length called the angstrom unit (Å)

$$1 \text{ Å} = 10^{-8} \text{ cm}$$

The size of atoms and molecules is also on the order of 10^{-8} cm, and it is, therefore, convenient to express atomic and molecular dimensions in terms of angstrom units.

The relationship between energy and wavelength is derived from equations 4–1 and 4–2.

4–3
$$\epsilon = \frac{hc}{\lambda}$$

According to equation 4–3, as the wavelength increases, the energy of the photon decreases.

M. Planck introduced the concept of light consisting of photons as early as 1901. At that time Planck was trying to explain the unique energy distribution given off by blackbody radiators. It was necessary in the derivation to postulate that light was composed of a fundamental unit called a light quantum. Today the term photon is more commonly used than light quantum.

The importance of this photon or light quantum concept became especially apparent when it was used by Einstein (1905) to explain the ejection of electrons from a metal surface when exposed to light. This phenomenon is called the photoelectric effect. The electrons ejected from the metal surface can be collected and measured as a current (Figure 4–2A). The energy of the photon striking the metal surface must exceed some critical energy to cause an electron to be ejected. In Figure 4–2B the current (arising from the ejected electrons) is plotted against the energy of the incident photon, $h\nu$. Notice that no current, I, is observed until the energy of the photon exceeds the critical energy, $h\nu_c$. Each metal has a unique critical energy. According to equation 4–1, if one photon is to eject one electron, the frequency of the light ν must exceed a critical frequency, ν_c. If light energy were not contained in a single particle, then ejection of an electron would depend

FIGURE 4–2A Photoelectric effect on a metal surface

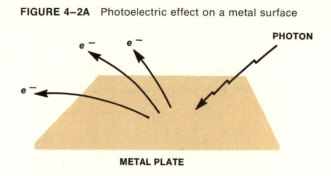

METAL PLATE

FIGURE 4–2B Photoelectric effect—dependence on photon energy

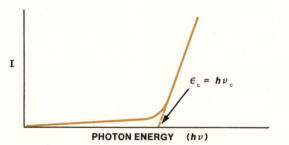

$\epsilon_c = h\nu_c$

upon the intensity of the light source (rate at which energy is supplied to the metal surface) rather than upon the frequency. Since we observe experimentally that the photoelectric effect depends on frequency and not intensity, we conclude that the energy of light is contained in photons. This conclusion supports the particle concept of light. In addition to the photon or particle model of light it is also necessary to utilize the wave model in order to explain other properties of light, such as diffraction.

EXAMPLE 4–1   *Calculate the energy of a photon for light of wavelength 6,000 Å.*

$$\epsilon = \frac{hc}{\lambda} = \frac{(6.62 \times 10^{-27} \text{ erg-sec})(3 \times 10^{10} \text{ cm/sec})}{(6,000 \text{ Å})\left(\frac{1 \text{ cm}}{10^8 \text{ Å}}\right)}$$

$$\epsilon = 3.31 \times 10^{-12} \text{ erg/photon}$$

If we wished to express this energy in kcal/mole we would first multiply by Avogadro's number and then use the conversion factor (Appendix A)

$$1 \text{ erg} = 2.390 \times 10^{-11} \text{ kcal}$$

In this case Avogadro's number represents the number of photons in one mole of photons

$$\epsilon = \left(3.31 \times 10^{-12} \; \frac{erg}{photon}\right)\left(6.023 \times 10^{23} \; \frac{photons}{mole}\right)\left(\frac{2.390 \times 10^{-11} \; kcal}{erg}\right)$$

$$= 47.6 \; kcal/mole$$

PROBLEM 4–1 It requires 6.73×10^{-12} ergs to remove an electron from the surface of Cr metal. Is light of wavelength 2500 Å of sufficient energy to eject electrons from the surface of Cr?

PROBLEM 4–2 Would you expect the photoelectric effect to occur on Cr metal if light of 3500 Å were used?

4–C. ATOMIC SPECTRA

When an atom absorbs energy, it moves from one energy level to another; this is called *excitation*. When an atom is excited it can emit energy of a specific frequency by falling down to a lower energy level. Each element, when excited, produces a unique *atomic spectrum* consisting of a set of frequencies. The spectrum can then be used to identify the elements in an unknown sample. There are several ways in which the atom can be excited to higher energy and the spectrum depends, to some extent, on the mode of excitation. The more common methods of excitation are absorption of high energy (short wavelength) light, flame, spark discharge, and direct current arc.

The hydrogen atom, consisting of a single proton and electron, is the simplest of all atoms and likewise has the simplest spectrum (as shown in Figure 4–3A). Only the spectrum in the visible region (the Balmer spectral series) is shown. The progression of lines becoming more closely spaced at shorter wavelengths is known as a spectral series. The origin of these spectral lines will be discussed in Section 4–D. The hydrogen atom also has a series of spectral lines in the vacuum ultraviolet called the Lyman series, and three similar series in the infrared region. The spectrum for the sodium atom in Figure 4–3B is also comparatively simple, but it is completely different from that of the hydrogen atom. The lines appear narrow except for the lines identified by D_1 and D_2. These are intense lines which give the yellow color characteristic of sodium.

The iron atom spectrum, in contrast to hydrogen and sodium, is quite complex. The spectrum from the ultraviolet through the infrared region consists of some 40,000 lines. Only the spectrum over a small region 3700–4000 Å is shown in Figure 4–3C, but this includes a large number of lines.

Note in Figure 4–3 that all of the spectra consist of very sharp lines corresponding to discrete frequencies of the light. A spectral line occurs because the atom has gone from a discrete higher energy level to a discrete lower energy level and

FIGURE 4–3 Emission spectra for the A. hydrogen atom in the visible region (Balmer series—the spectral lines H_α, H_β, etc.), B. sodium atom in the visible and ultraviolet regions, and C. iron atom in the region 3700–4000 Å.

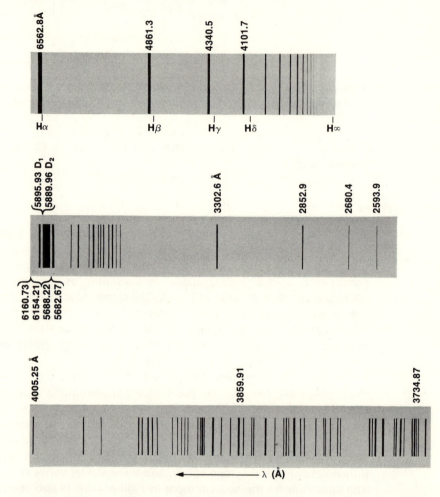

has given off the energy difference in terms of light of a specific frequency. There are numerous higher and lower energy levels in the atom from which transitions can occur since there are numerous discrete lines observed in the spectra. The various energy levels to which the atom can be excited are indicated by the horizontal lines in Figure 4–4, an energy level diagram. The vertical arrows indicate the transition of an excited atom from a higher energy level to a lower energy level, giving off a photon with a specific energy $h\nu_i$.

From this observation of atomic spectra, we conclude that the atom can have only *specific* energies and that the specific frequencies in the spectra arise from the difference in these specific energies. This is a quite different phenomenon

FIGURE 4–4 Typical energy level diagram

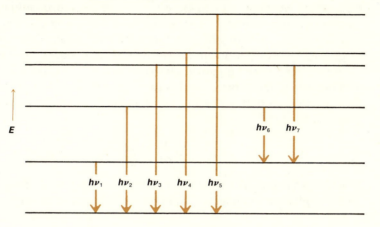

than that observed with macroscopic bodies. In systems involving submicroscopic bodies such as atoms there are restrictions on the energies; not all energies are allowed. If all energies of the atom were allowed, an atomic spectrum would be continuous and blurred, in contrast to the sharp lines seen in Figure 4–3. Consequently, classical mechanics cannot account for the sharp line-like atomic spectra and a new type of mechanics must be found to describe the motion of atomic and subatomic particles.

4–D. BOHR THEORY

Niels Bohr first attempted to explain the phenomenon of atomic spectra in 1913. He successfully derived an equation with which the energies of the hydrogen atom could be calculated very accurately. Using this energy formula the spectrum for hydrogen can be calculated and generally agrees with the experimental spectrum to an accuracy of six or seven significant figures. This was an amazing result since seldom does theory agree with experiment to that great precision.

In order to derive an equation for the electronic energy of the hydrogen atom, Bohr had to use concepts which were different from classical mechanics. Bohr accomplished the restriction to specific electronic energies by introducing a concept called *quantization*. In his derivation he assumed that the angular momentum of the electron, in its motion about the positive nucleus, was quantized and allowed to take on only multiples of $h/2\pi$. Mathematically this is expressed as

4–4 $$mvr = n(h/2\pi) \qquad n = 1, 2, 3, 4 \ldots$$

where mvr = angular momentum, h = Planck's constant, and n is any whole number and is called the *quantum number*. The theory was thus called quantum theory. Since there has been a new development of the theory in the form of wave mechanics, we now refer to Bohr's work as the old quantum theory. The new quantum theory refers to the development of wave mechanics. It is quantization that is the most distinguishing feature between classical mechanics and quantum mechanics. In every problem solved by quantum mechanics the energies will be restricted to so-called quantized levels.

In order to obtain a better understanding of the concept of quantization, we will give (without proof) the equation for the electronic energy of the hydrogen atom obtained by Bohr*

4-5
$$E_n = -C \frac{1}{n^2} \qquad n = 1, 2, 3, 4 \ldots$$

where E is the energy of a single hydrogen atom, C is a constant for an element, and n = quantum number as defined in equation 4–4. C is given by

$$\left(\frac{2\pi^2 Z^2 e^4 m}{h^2}\right) \frac{M}{M + m}$$

where Z = charge on the nucleus, e = unit of charge = 4.80×10^{-10} electrostatic units, m = mass of the electron, M = mass of the nucleus. Note that for every value of n there is a specific value of E designated as E_n, and there can be no other values of E. The atom thus has discrete electronic energies.

The excitation of the hydrogen atom involves the promotion of the single electron from a lower electronic energy level to a higher electronic energy level as given by equation 4–5. To remove the electron farther away from the positive nucleus requires energy. As n becomes larger, the electron is moved further away from the nucleus. If n goes to infinity (complete removal of the electron) then E_n becomes zero according to equation 4–5. When the electron is in any level $n < \infty$, the E_n is negative which corresponds to a stable state for the electron. The values of the electron energy in the hydrogen atom have been plotted in Figure 4–5 for various values of n.

The atomic spectrum for the hydrogen atom as shown in Figure 4–3A arises from transitions between the energy levels. When the energy of the electron changes from a higher electronic energy level to a lower electronic energy level, a photon of light is emitted (Figure 4–4). The emission spectral lines designated by H_α, H_β, H_γ, H_δ, in Figure 4–3A correspond to the transitions from energy levels $n = 3, 4, 5, 6$ to the energy level $n = 2$ as shown in Figure 4–5.

The energy of the photon, given by equation 4–3, equals the difference in the energy levels involved in the transition. We will let n_U represent the upper value of n and n_L the lower value of n. The energy of the photon is given by

4-6
$$\frac{hc}{\lambda} = E_U - E_L$$

*If you wish to work out the mathematical development of equation 4–5, see exercise 11 at the end of this chapter.

FIGURE 4-5 Energy levels in the hydrogen atom

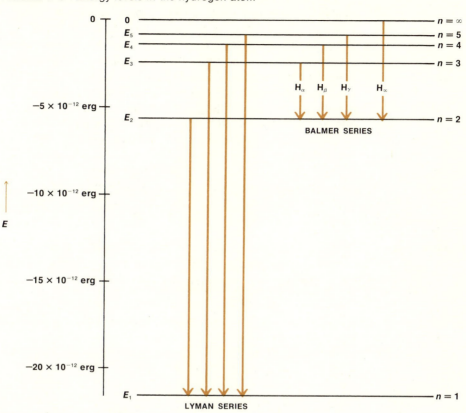

Substituting the expression for E_n from equation 4-5 into equation 4-6 we obtain

4-7

$$\frac{hc}{\lambda} = -C\left(\frac{1}{n_u^2} - \frac{1}{n_L^2}\right)$$

Since spectral lines are frequently expressed in terms of wavelength (λ), we will solve equation 4-7 for $\bar{\nu} = (1/\lambda)$ and remove the negative sign.

4-8

$$\bar{\nu} = \frac{1}{\lambda} = \frac{C}{hc}\left(\frac{1}{(n_L)^2} - \frac{1}{(n_u)^2}\right)$$

We can simplify equation 4-8 by defining the constant $R = \dfrac{C}{hc}$.

4-9

$$\bar{\nu} = \frac{1}{\lambda} = R\left(\frac{1}{(n_L)^2} - \frac{1}{(n_u)^2}\right)$$

For hydrogen $R = 109,677$ cm^{-1}. Equation 4-9 can be used to calculate the wavelength or wave number for the transition of the electron between any two energy levels.

The equations 4–5 to 4–9 have been discussed with respect to the hydrogen atom where $Z = 1$. Actually, the equations are also valid for monatomic ions which consist of a single electron in addition to the nucleus, e.g., He^+, Li^{2+}, Be^{3+}. In using the equations for these monatomic ions the proper Z and M should be used.

We will not discuss the Bohr Theory in any further detail since all of the accomplishments of this theory are also obtained from the newer concept of wave mechanics without many of the shortcomings of the Bohr theory. Two of the most noteworthy shortcomings of the Bohr theory are   1. that the concept of quantization was introduced arbitrarily without any more fundamental bases than that quantized energies are required in order to explain the line-like spectra,   2. that the position of the electron in the atom was assumed to be in a definite path travelling at a definite velocity about the nucleus. The latter of these is in disagreement with a principle, known as the Heisenberg uncertainty principle, which does not allow the exact position and velocity of the electron to be known simultaneously. This is discussed in Section 4–G.

EXAMPLE 4–2   *The constant R (Equation 4–9) for He^+ is 438,889.08 cm^{-1}. Calculate the frequency of the light absorbed when the electron undergoes the transition from the $n = 1$ to $n = 4$ energy level. Use equation 4–9 to calculate the wavelength.*

$$\bar{\nu} = \frac{1}{\lambda} = -(438{,}889.08 \ cm^{-1})\left(\frac{1}{4^2} - \frac{1}{1^2}\right)$$

$$\bar{\nu} = \frac{1}{\lambda} = 411{,}458.52 \ cm^{-1}$$

The frequency (ν) can be calculated using equation 4–1: $\nu = \dfrac{c}{\lambda}$

$$\nu = (411{,}458.52 \ cm^{-1})(2.997925 \times 10^{10} \ cm/sec) = 12.3351 \times 10^{15} \ sec^{-1}$$

PROBLEM 4–3   Calculate the frequency and the wavelength in angstrom units for the transition from $n = 5$ to $n = 1$ in the hydrogen atom. In what series would this spectral line belong? (See Figure 4–5.)

PROBLEM 4–4   The Balmer series for the hydrogen atom (Figures 4–3 and 4–5) arises from transitions from higher energies terminating at the $n = 2$ level. Using Figure 4–5 explain why the lines in the series become closer and closer together as they reach what is called the series limit H_∞ (lowest wavelength line in the series). Calculate the wavelength for H_∞.

PROBLEM 4–5   Calculate the energy required to remove the electron from a hydrogen atom when the hydrogen atom has the electron in its lowest energy level ($n = 1$).

$$H \longrightarrow H^+ + e^-$$

This energy is called the ionization potential for hydrogen.

PROBLEM 4–6   Comment on the statement "During the experiment the light from the tungsten lamp (a continuum source) could not be seen since the hydrogen atoms in the vessel absorbed all the light."

4–E. DE BROGLIE'S HYPOTHESIS

The next major breakthrough in the development of quantum or wave mechanics and our present concept of atomic structure was a hypothesis presented by DeBroglie in 1924. In Section B we discussed the concept that light has to be viewed both as a particle (photon) and as a wave in order to describe all of its properties (dual nature of light). DeBroglie, as a young scientist working towards his doctorate, made the daring proposal that all moving objects which have a mass should possess wave character just as does light. In other words, all species should have this dual character of both a particle and a wave.

For a photon we can obtain a relationship between mass and wavelength by using Einstein's mass-energy equivalence

4–10 $$E = mc^2$$

and Planck's relationship to frequency

4–1 $$E = h\nu$$

Setting equations 4–10 and 4–1 equal we have

$$mc^2 = h\nu$$

Since $\nu = c/\lambda$ from equation 4–2,

$$mc^2 = \frac{hc}{\lambda}$$

$$mc = \frac{h}{\lambda}$$

DeBroglie proposed an equation analogous to this for particles with mass m travelling at a velocity v

4–11a $$mv = \frac{h}{\lambda}$$

The wavelength of the particle is thus

4–11b $$\lambda = \frac{h}{mv}$$

DeBroglie's hypothesis was not readily accepted by scientists at that time. However, shortly thereafter in 1927, Davison and Germer showed experimentally that electrons could be diffracted in a manner analogous to light diffraction and therefore that particles do display a wave character. *Diffraction* is the phenomenon whereby light of multiple wavelengths is separated into the component wavelengths when reflected off a ruled grating. An understanding of the diffraction phenomenon depends upon the wave character of light. The results of the Davison and Germer experiment furthermore confirmed the exact relationship between m, v, and λ as given in DeBroglie's equation (4–11). DeBroglie's hypothesis has also been supported by experiments in which neutrons and protons have also been diffracted.

EXAMPLE 4–3 *Calculate the wavelength of a baseball travelling at 100 miles per hour. Assume the mass of a baseball is approximately 1/4 lb. Using equation 4–11b*

$$\lambda = \frac{h}{mv} = \frac{6.62 \times 10^{-27} \text{ erg-sec}}{(1/4 \text{ lb})(100 \text{ miles/hour})}$$

In order to make the units cancel, we must convert pounds to grams and miles/hour to cm/sec.

$$(1/4 \text{ lb})(453.6 \text{ g/lb}) = 113.4 \text{ g}$$

$$(100 \text{ mi/hr})(5{,}280 \text{ ft/mi})(12 \text{ in/ft})(2.54 \text{ cm/in})\left(\frac{1 \text{ hr}}{60 \text{ min}}\right)\left(\frac{1 \text{ min}}{60 \text{ sec}}\right)$$

$$= 4.47 \times 10^3 \text{ cm/sec.}$$

And, since 1 erg = 1 g-cm²/sec²

$$\lambda = \frac{6.62 \times 10^{-27} \text{ (g-cm}^2\text{/sec}^2\text{)(sec)}}{(113.4 \text{ g})(4.47 \times 10^3 \text{ cm/sec})} = 1.3 \times 10^{-32} \text{ cm}$$

This value is very small; obviously calculations can be made for macroscopic particles using wave mechanics but the results do not significantly affect our classical mechanical model for macroscopic objects.

PROBLEM 4–7 Calculate the wavelength of a 10 ev electron. (1 ev = 1.602×10^{-12} erg, mass of electron = 9.11×10^{-28} g). [Hint: Calculate the velocity v from the kinetic energy expression $1/2 \, mv^2$.]

PROBLEM 4 ·8 Calculate the wavelength of a 10 lb satellite travelling at 25,000 ft/sec. What effect would this wavelength have on its motion according to classical mechanics? In other words, would this wavelength affect the motion of the satellite to any significant degree?

4–F. THE DUAL NATURE OF ELECTRONS—WAVE MECHANICS

Just as we must consider two different models for photons in order to describe the nature of light, according to DeBroglie's hypothesis we also must consider both the particle concept and the wave nature to describe electrons. To consider electrons as particles is natural, since the electron has mass. As a particle we might consider the motion of its mass according to classical mechanics. The charge of the electron is also taken into account whenever there is a field present which can affect its acceleration. For many problems this approach is quite satisfactory.

However, there are also other problems where it is necessary to consider the wave nature of the electron. For example in the diffraction experiment of Davison and Germer just mentioned, electrons which are reflected off a crystal come off at preferred angles just as photons are diffracted off a grating. The passage of electrons through thin films of metal also resulted in the diffraction of the electrons (Thomson, 1927). In both of these experiments the results can only be explained by the electrons having wave properties, with the metals in both experiments acting as diffraction gratings.

Of utmost importance to chemistry is the nature of electrons in atoms. In this case again, the wave nature of electrons is important. A new type of mechanics appropriate for electrons in atoms, called *wave mechanics* or quantum mechanics, was developed by E. Schrödinger in 1926 based upon the wave nature of the electron. Since electrons displayed wave character, he assumed they should obey equations appropriate to wave phenomena. The famous Schrödinger wave equation was developed; it describes the behavior of electrons in an atom or molecule. Actually the Schrödinger wave equation is not restricted to only this problem but is a general equation applicable to any sub-atomic problem. Quantum mechanics applies to the motion of particles with small mass, while Newton's classical mechanics applies to macroscopic objects.

One big difference between the quantum mechanical problem and the classical mechanical problem is in the nature of the solution. In classical mechanics when we solve the equations of motion, the solution describes the exact position and velocity of the object. In quantum mechanics the solution of a wave equation is in the form of a wave function. The wave function does *not* describe the exact position and velocity of the particle. Instead, the *square of the wave function* represents the *probability* of finding the particle at some location. For electrons in atoms and molecules, therefore, the square of the wave function represents the probability of finding the electron at various locations in the atom or molecule.

The quantum mechanical solution also differs from classical mechanics in that unique solutions (wave functions) exist. For a given quantum mechanical problem each wave function is characterized by a set of quantum numbers and only the wave functions with these quantum numbers are allowed. This is analogous to the restrictions which Bohr introduced arbitrarily. However, in this case the restriction arises naturally from the Schrödinger wave equation and from certain restrictions which one must place on the wave functions to make them physically meaningful. Each unique wave function dictates a unique energy and only these energies are allowed for the system. Thus, the quantized or restricted values of the electronic energies, which are required to explain atomic spectra, also arise naturally from the solution of the Schrödinger equation. We shall make use of the properties associated with wave functions for the hydrogen atom solutions in Sections 4–H and 4–I in order to help clarify the nature of the quantum mechanical solution.

4–G. HEISENBERG UNCERTAINTY PRINCIPLE

As a result of the wave nature of the electron, and of small particles in general, it is impossible to know precisely the position and momentum of a small particle. The relationship between this uncertainty in position and momentum was developed by Heisenberg and is known as Heisenberg's uncertainty principle. If $p = mv$ is the momentum and x is position, Heisenberg's uncertainty principle states*

4–12 $$\Delta p \, \Delta x \geq h/2\pi$$

where Δp and Δx are the uncertainties in p and x. According to equation 4–12 the product of the uncertainties must exceed or at best equal $h/2\pi$. If one quantity, say Δx, is made very small by accurately determining the position, then the uncertainty in the momentum Δp must necessarily increase. An analogous uncertainty expression can be given in terms of the energy, E, and time, t

$$\Delta E \, \Delta t \geq h/2\pi$$

Other relationships can be derived from these using the relationship with ΔE or Δp. For example, from equation 4–1, $\Delta E = h\Delta \nu$. Several examples of physical problems in which the variables p and x are being measured simultaneously are discussed in a text by Heisenberg.**

4–H. QUANTUM MECHANICAL RESULTS FOR THE HYDROGEN ATOM

QUANTUM NUMBERS ALLOWED

The quantum or wave mechanical solution of the hydrogen atom involves the solution of a rather complicated mathematical equation called the Schrödinger wave equation. This wave equation is a second order partial differential equation. The solution of this equation is beyond the scope of this book, but the actual

---

*Analysis of some problems gives $h/2\pi$, and of others $h/4\pi$ or h. In general the term on the right is "on the order of h" and the exact number is not important. For simplicity we will use only $h/2\pi$.

**W. Heisenberg, *The Physical Principles of The Quantum Theory,* Dover Publications, Inc., 1950.

FIGURE 4–6   Representation of the position of the electron by polar coordinates

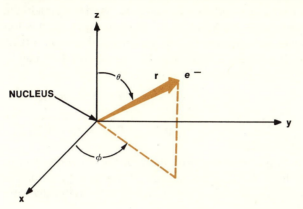

solutions can be understood and related to atomic structure without a knowledge of differential equations or calculus.

In the quantum mechanical solution of the hydrogen atom it is necessary to describe the position of the moving electron in relationship to the position of the nucleus. This could be done using the Cartesian coordinates x, y, z, but it is more convenient in this problem to represent the position of the electron with polar coordinates r, θ, ϕ. Figure 4–6 shows how these coordinates represent the position of the electron with respect to the nucleus at the origin.

The solutions of the wave equation are called wave functions, ψ. ψ is a function of the coordinates r,θ,ϕ. For the hydrogen atom, solving the wave equation is facilitated by the fact that the wave functions can be separated into a product of functions containing each of the three variables. Mathematically this is stated as

$$\psi(r,\theta,\phi) = R(r)\ \Theta(\theta)\ \Phi(\phi)$$

where $R(r)$ represents some function of the variable r, $\Theta(\theta)$ is a function of θ only, $\Phi(\phi)$ is a function of ϕ only. Using this form of the wave function, the solution of the wave equation gives wave functions which are described by three quantum numbers n, l, m. These quantum numbers must be integers. The *wave functions* for the hydrogen atom problem are also called *orbitals* and the electron can be in any one of these orbitals. The function $R(r)$ describes the dependence of the wave function, Ψ, on the radial distance, r, from the nucleus and, therefore, describes the size of the orbital. The functions $\Theta(\theta)$ and $\Phi(\phi)$ describe the dependence of the wave function on the angles θ and ϕ and, therefore, describe the orientation of the orbital. The quantum number n is called the *principal* quantum number and arises from only the $R(r)$ function. The $R(r)$ function depends on both n and l, but the quantum number n is the principal term and is the dominant factor determining the *size of the orbital*. On the other hand, the quantum numbers l and m are involved in the $\Theta(\theta)$ and $\Phi(\phi)$ functions, and their values govern the *shape and orientation of the orbital* in space. The influence of *n*, *l*, and *m* on the size and shapes of orbitals will be discussed in detail in the following section.

An example of a wave function for the hydrogen atom is the following solution which involves a unique set of quantum numbers. $a_0 = 0.529$ Å = Bohr radius = constant. The separate functions $R(r)$, $\Theta(\theta)$, and $\Phi(\phi)$ can be readily identified.

$$\psi = \underbrace{\frac{1}{4\sqrt{2\pi}}\left(\frac{1}{a_0}\right)^{3/2}}_{\text{constant}} \underbrace{\frac{r}{a_0}\,e^{-r/2a_0}}_{R(r)} \underbrace{\sin\theta}_{\Theta(\theta)} \underbrace{\cos\phi}_{\Phi(\phi)}$$

In the solution of the wave equation for the hydrogen atom, certain restrictions arise which limit the possible values of the quantum numbers *n*, *l*, and *m*. The allowed values for the quantum numbers arise naturally from the mathematical solution.

$$n = 1, 2, 3, \ldots$$

$$l = 0, 1, 2, \ldots n - 1$$

$$m = -l \to 0 \to +l$$

Quantum number *m* may take any integral value from $-l$ to $+l$, including zero. Each allowed combination of these three quantum numbers (*n*, *l*, *m*) represents a wave function for an e$^-$ in an atomic orbital of hydrogen. The orbitals which have the same *n* value are said to be in the same *shell*. Orbitals which have the same *n* and *l* values are in the same *subshell*. The orbitals in a subshell are commonly designated in the following way: the value of *n* is given first followed by a letter which designates the value of *l*. The letters used to designate *l* are as follows:

l	LETTER DESIGNATION
0	*s*
1	*p*
2	*d*
3	*f*
4	*g*
5	*h*

Table 4–1 has been constructed to show the more common orbitals. Note that for $n = 1$ only $l = (n - 1) = 0$ is permitted, and therefore *m* must also be zero. The subshell designation 1*s* contains "1" for the principal quantum number and "*s*" for $l = 0$. For $n = 2$, note that two values of *l* (0,1) are allowed. For $l = 0$, *m* can only be zero. However, for $l = 1$, *m* can have three possible values (−1,0,1). The

TABLE 4-1 Atomic orbitals

n	l	m	NUMBER OF ORBITALS PER SUBSHELL ($2l + 1$)	SUBSHELL DESIGNATION	NUMBER OF ORBITALS/SHELL (n^2)
1	0	0	1	1s	1
2	0	0	1	2s	
2	1	−1			4
2	1	0	3	2p	
2	1	1			
3	0	0	1	3s	
3	1	−1			
3	1	0	3	3p	
3	1	1			9
3	2	−2			
3	2	−1			
3	2	0	5	3d	
3	2	1			
3	2	2			

designation 2s arises from $n = 2$ and $l = 0$. The designation 2p arises from $n = 2$, and $l = 1$. Note that there are three orbitals in the 2p subshell. The number of orbitals in any subshell is $2l + 1$ and these are designated in the previous table. In general the number of orbitals in any shell is n^2. As stated in Table 4–1 the number of orbitals for the shell $n = 1$ is $n^2 = 1$, for the shell $n = 2$ the number of orbitals is $n^2 = 4$, etc.

PROBLEM 4–9 Show all allowed combinations of n, l, m and subshell designation for all orbitals with $n = 4$. How many orbitals are there in this shell?

PROBLEM 4–10 Is the orbital designated by the following quantum numbers allowed? Explain $n = 5$, $l = 2$, $m = −3$.

ORBITAL ENERGIES

A unique energy is associated with each wave function or orbital in the solution of the wave equation. For the hydrogen atom problem each orbital has associated with it an energy which is dependent only upon the quantum number n. The orbital energy is identical to that from the Bohr theory given in equation 4–5.

4–5
$$E_n = -C\left(\frac{1}{n^2}\right)$$

Since the energy of the orbitals depends only on the principal quantum number, n, the orbitals in the hydrogen atom with the same n value have the same energy.

For example, the orbitals characterized by the quantum numbers (2,0,0),(2,1,−1), (2,1,0) and (2,1,1) all have the same energy. Orbitals with the same energy are said to be *degenerate*. In this example there are four orbitals with the same energy, so the orbitals are said to be *fourfold* degenerate. Since the energy of an orbital in the hydrogen atom depends only on *n*, the orbital degeneracy of any shell is equal to the number of orbitals in that shell (n^2). (This same degeneracy is not present in other atoms.)

The various orbital energies for the hydrogen atom are shown in Figure 4–7. The orbitals are identified by their subshell designation and by the quantum numbers (*n, l, m*) below the line.

The energy level diagram in Figure 4–7 is the same as that shown in Figure 4–5 except that each energy level is now composed of several degenerate orbitals (n^2). The line spectrum for hydrogen that was explained using Figure 4–5 and the Bohr theory can likewise be explained from the wave mechanical solution and the orbitals in Figure 4–7. In fact the wave mechanical solution is superior to the Bohr result in that the intensities of the spectral lines can be explained by the wave mechanical result. The Bohr theory was unable to do this.

The lowest energy (ground) state for the hydrogen atom has the single electron in the 1*s* orbital; that is, the electronic configuration for the ground state for the H atom is $1s^1$. The hydrogen atom can also be in higher energy (excited) electronic states by elevating the electron to energy orbitals such as the 2*p*. The electronic configuration for this state would be $2p^1$. Many other excited states exist for H such as $3p^1$, $3d^1$, $4f^1$. The spectrum described previously in Figure 4–3 arises from transitions between these various electronic energy states of the H atom.

4–I. DESCRIPTION OF THE HYDROGEN ATOMIC ORBITALS

In the previous section we identified the wave functions associated with the hydrogen atomic orbitals in terms of three quantum numbers and gave the energy formula (equation 4–5) with which the energy of the atomic orbital can be calcu-

FIGURE 4–7   Diagram of hydrogen orbital energies

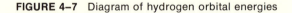

3*s* ____   3*p* ____ ____ ____   3*d* ____ ____ ____ ____ ____
   (3,0,0)   (3,1,−1)(3,1,0)(3,1,1)   (3,2,−2)(3,2,−1)(3,2,0)(3,2,1)(3,2,2)

2*s* ____   2*p* ____ ____ ____
   (2,0,0)   (2,1,−1)(2,1,0)(2,1,1)

E

1*s* ____
   (1,0,0)

lated. In this section we will describe the shapes of the hydrogen atomic orbitals. By the shape of an atomic orbital we mean the various positions that the electron can occupy in any atomic orbital of hydrogen.

The information about orbital shape comes from the wave function of that orbital. Though the wave function for an atomic orbital does not define the exact position in space, the square of the wave function, ψ^2, describes the probability of finding the electron in any region of space. Since the description of orbital shapes is in the form of a probability function, we must first understand how to interpret probability functions. We will use both contours and distributions to show orbital shape. These concepts are important since they occur several times in chemistry and often in biology, physics, and geology.

CONTOUR GRAPHS

The best example of a contour graph is an elevation map which describes the terrain in a given locality. The elevation map consists of a series of contours or curves, with each line corresponding to a given elevation, as in Figure 4–8. Intervals of 500 ft are shown in Figure 4–8, but the interval used is arbitrary. The contour map shows the nature of the terrain. The map in Figure 4–8 shows two mountains with elevations between 3,000–3,500 ft and 4,500–5,000 ft. A very close spacing of contours indicates a very rapid rise or change of elevation, so the approach to peak B is steeper than to peak A. The valley between the mountains lies between 1,500–2,000 ft.

Contour maps are used to describe a three-dimensional surface on two-dimensional paper. Besides describing terrain, contour diagrams can show the probability of finding the electron at various locations in an atom or molecule. The contours in this kind of diagram represent various levels of probability

FIGURE 4–8 Contour map of elevations

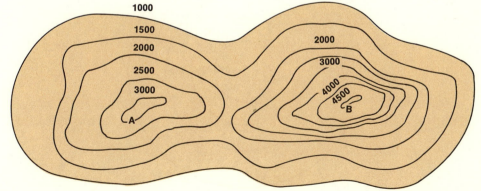

density. Though the contours of an atom or molecule are three-dimensional surfaces, it is conventional to graph the cross-section of the contours in a plane passing through the atom or molecule. Generally cross-sections in one or two planes will satisfactorily illustrate the shape of the three-dimensional contour.

TWO-DIMENSIONAL DISTRIBUTIONS

You are familiar with one-dimensional distributions that can be conveniently shown on a conventional two-dimensional graph (for example, probability of an exam grade is plotted versus the exam grade). The representation of a two-dimensional distribution on a graph presents more of a problem. In order to illustrate two-dimensional distributions we shall consider the distribution of bullet holes in a target. Since the object is to hit the bullseye, the bullet holes should be more concentrated about the center of the target (Figure 4–9).

Providing there is no misalignment of the gun sights, the bullet holes should be distributed equally in all directions from the center—that is, symmetrically about the target center. The probability of finding a bullet hole in some area on the target is represented by the concentration of holes in that area. We define the probability per unit area as the *probability density*. (The density of a three-dimensional object is the mass *per unit volume*.) Figure 4–9, which represents probability density by the density of bullet holes, is called a dot diagram. We will simply represent the area on the target where there is the most likely chance to find 20 % of the bullet holes by the circle which just encompasses 20 of the 100 bullet holes. This circle is designated 20 % in Figure 4–9. A circle at a specified distance from

FIGURE 4–9 Distribution of 100 bullet holes in a target. Probability level contours are shown at 20 %, 40 %, 60 %, and 80 % levels.

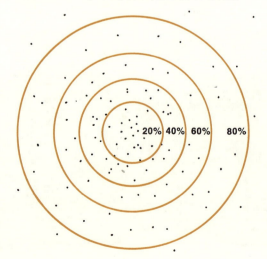

the center represents a *contour* of constant probability density or simply a probability level contour. Since the area defined by a contour delimits a 20 % probability of finding a bullet hole, we call this specific contour the 20 % probability level contour. We have defined the 40 %, 60 %, and 80 % contours but could have established any other probability level contour.

It is awkward to identify each contour in a diagram. In order to make the probability level contours easier to see, we can give the different areas different shadings, as in Figure 4–10. The black area represents the 20 % probability level contour and the 40, 60, 80 % levels are designated by successively lighter shades. These shadings are quite distinct and can be associated readily with each probability level contour. This representation of probability distributions in two-dimensions (1) gives the shapes of the contours, (2) gives a quantitative measure of the probability, and (3) is easy to interpret and understand.

HYDROGEN ATOMIC ORBITALS

As mentioned previously, each orbital of hydrogen is associated with a wave function, ψ. The square of this wave function, ψ^2, is a measure of the probability of an electron being at a location in space. We must use the probability function ψ^2 to describe the geometrical size, shape, and orientation of the orbitals of hydrogen. Since the wave function associated with each orbital is different, we can expect the geometrical configuration of each orbital to differ. We will limit our discussion in this chapter to the s and p orbitals. The discussion of d orbitals is deferred to Chapter 16 on transition metals.

Representation of electron probability functions in three dimensions is generally a rather difficult task. However, the use of probability level contours greatly

FIGURE 4–10 Probability level contours for the distribution of bullet holes in a target

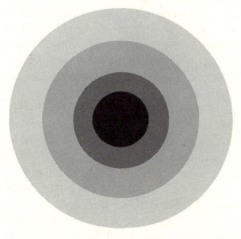

simplifies the representation. In this case the probability level contours are three-dimensional surfaces since ψ^2 is a three-dimensional probability distribution. ψ^2 is a measure of electron probability *per unit volume* or *electron probability density* (analogous to our previous use of probability density in two dimensions to represent bullet holes on a target).

The three quantum numbers n, l, m, associated with a wave function ψ govern the size, shape, and orientation of the atomic orbitals. The principal property of an atomic orbital determined by each quantum number is

n — size

l — shape

m — orientation in space

This will become apparent as we represent the different orbitals.

In order to show how we can go from a wave function ψ to a probability density distribution, we will consider the 1s orbital in detail. The wave function for the 1s orbital of the hydrogen atom is given by the equation

$$\psi_{1s} = \frac{1}{\sqrt{\pi}}\left(\frac{1}{a_0}\right)^{3/2} e^{-r/a_0}$$

Note that ψ_{1s} is a function only of the variable r and does not depend on the angles θ and ϕ. Consequently ψ_{1s} has the same value in all directions from the nucleus and we say that ψ_{1s} is *spherically symmetrical*. Since the square of the wave function is needed to give the probability density, we must calculate the function for ψ_{1s}^2:

$$\psi_{1s}^2 = \frac{1}{\pi}\frac{1}{a_0^3} e^{-2r/a_0}$$

A graph of this function is shown in Figure 4–11A. Note in Figure 4–11A that ψ_{1s}^2 is the greatest at the nucleus ($r = 0$) and decreases exponentially as r increases. Correspondingly, in Figure 4–11B the greatest probability density (density of dots) occurs at the nucleus and decreases as r is increased. Note that the distribution in B is the same in all directions from the nucleus. The contours (constant ψ_{1s}^2) are shown in Figure 4–11B for the 20 %, 40 %, 60 %, and 80 % probability levels. These are also shown in three dimensions in Figure 4–11C. These three-dimensional probability level contours represent the volume (sphere in this case) in which the electron can be found a certain percentage of the time. Note that the two-dimensional cross sections of the probability level contours for the 1s orbital in B are similar in appearance to those in the target diagrams in Figure 4–9. However, one must keep in mind that Figure 4–11B is only a cross section and does not represent the true three-dimensional distribution.

All s orbitals have $l = 0$ and they all have an electron probability density distribution which is spherically symmetrical about the nucleus. A spherically symmetrical

FIGURE 4–11   Representation of probability density for the 1s orbital   A. ψ_{1s}^2 as a function of r B. cross section of 1s probability density program   C. three-dimensional probability level contour diagram.

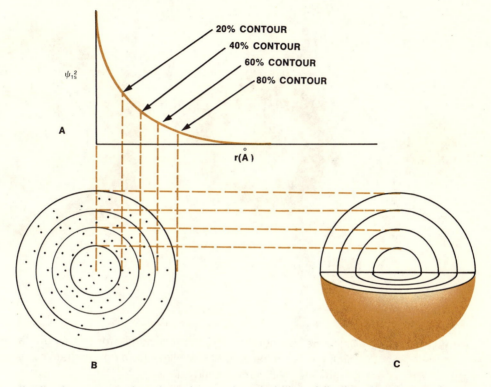

distribution means that there is equal probability of finding the electron in any direction at a given distance from the nucleus. Since the s orbitals are spherically symmetrical, the probability density graph can be used for any plane containing the nucleus. In Figure 4–12 the electron probability density distributions for the 1s, 2s, and 3s orbitals are represented by two-dimensional dot diagrams in a plane containing the nucleus.

Three things should be noted in Figure 4–12. The first is the increasing size as the principal quantum number n goes from 1 to 3. The second thing to be noted is the similarity in shape. All s-orbitals are spherically symmetrical. Recall that the shape is determined primarily by the value of l, and for s orbitals the l are all equal to zero. The third point to note is that there are surfaces where there is zero probability of finding the electron. These are called *nodal surfaces*. The dashed lines in the dot diagrams in Figure 4–12 are nodal surfaces and are spherical in the case of s orbitals. The number of spherical nodes is given by the formula

$$\text{spherical nodes} = n - l - 1$$

For the s orbitals $l = 0$, and the numbers of spherical nodes $= 0$, 1, and 2 for the atomic orbitals 1s, 2s, and 3s, respectively.

FIGURE 4–12 Electron probability density dot diagrams for *s* orbitals—1*s*, 2*s*, 3*s*—in order of increasing size. Nodal surfaces are shown by dashed lines.

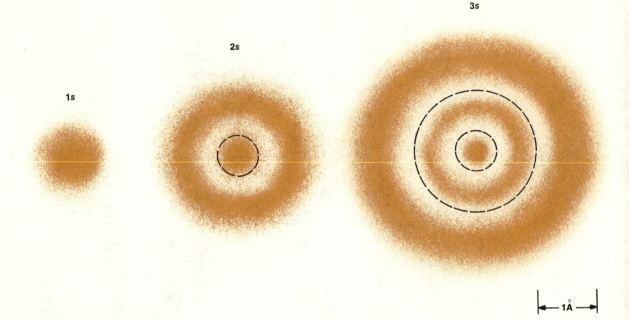

PROBLEM 4–11 Calculate the number of spherical nodes for the 4*s* orbital and show what you think the 4*s* orbital should look like. Use a dot diagram but simply shade the areas to indicate relative probability density.

The *s* orbital electron probability density distributions can also be represented by probability level contours, as in Figure 4–13. The same 20, 40, 60, 80 % levels are shown for the 1*s*, 2*s*, and 3*s* orbitals. For the 2*s* and 3*s* orbitals the 20 % probability level contour exists in more than one region. This is because the electron probability density ψ^2 is high in more than one region and a contour of constant probability density encompasses more than one region of space. Such a contour is also shown for the 2*p* orbital in Figure 4–16. (This is analogous to the elevation contours above 2,000 ft being in two regions in Figure 4–8.) The probability-level contours for the 1*s*, 2*s*, and 3*s* orbitals are also shown as in three-dimensional quarter cutaways in Figure 4–14. The spherical shape of all *s* orbitals is obvious from these models.

As was mentioned in Section 4–H, there are three 2*p* orbitals and three 3*p* orbitals. In each case these *p* orbitals can be represented by three orbitals with the same electron probability distribution, but oriented differently in space. The three 2*p* orbitals are designated $2p_x$, $2p_y$, and $2p_z$ and their principal axes are oriented along the *x*, *y*, and *z* axes. This will be shown later. For the present we will

FIGURE 4–13 Probability level contour for *s*-orbitals: 1*s* orbital, 2*s* orbital, 3*s* orbital. 20 %, 40 %, 60 %, and 80 % levels appear in order of lighter shading.

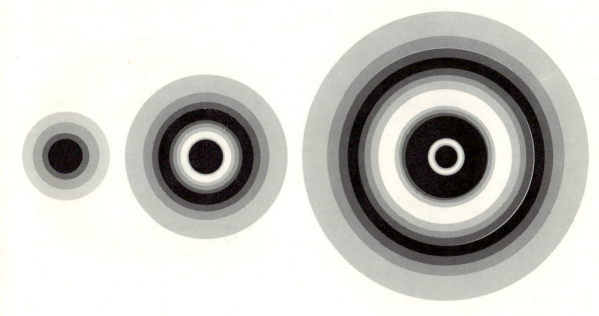

FIGURE 4–14 Three-dimensional probability level contours for *s*-orbitals shown with a quarter cut away. The 20 %, 40 %, 60 %, and 80 % probability level contours (surfaces) are shown.

consider only the shapes of the *p*-orbitals. For a *p*-orbital ($l = 1$), the electron probability density distribution is symmetrical about the axis. The dot-diagram in Figure 4–15 represents the electron probability density for the 2*p* and 3*p* orbitals in a plane containing this axis. Again the increase in size in going from a principal quantum number of $n = 2$ in 2*p* to $n = 3$ in 3*p* should be noted. The dashed lines also represent nodal surfaces in this diagram. The dashed line perpendicular to the principal axis is a *nodal plane* through the nucleus—a plane at which there is zero probability of finding the electron. There is a nodal plane perpendicular to the principal axis for both the 2*p* and 3*p* orbitals. The number of nodal planes passing through the nucleus for any orbital is given by the quantum number *l*. For the *p* orbitals $l = 1$ and there is only one nodal plane for all *p* orbitals. Also, note in Figure 4–15 that the spherical node on the 3*p* diagram is shown by the circular dashed line. For the 3*p* orbital $n = 3$, $l = 1$ and the number of spherical nodes equals $n - l - 1 = 3 - 1 - 1 = 1$.

FIGURE 4–15 Electron probability density dot diagram for *p*-orbitals, 2*p* and 3*p*, in order of increasing size. The distribution is symmetrical about the axis shown as a solid straight line. The dashed line represents a nodal plane perpendicular to the principal axis.

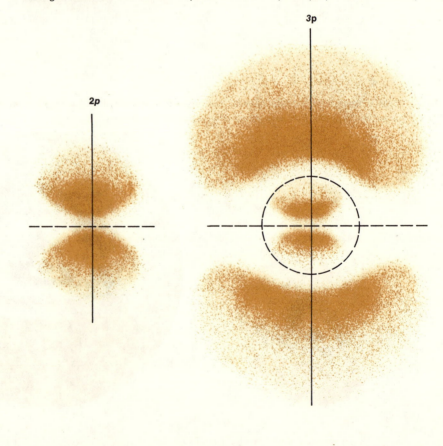

The electron probability level contour diagrams for the 2*p* and 3*p* orbitals are shown in Figure 4–16. The relative sizes of the orbitals is better seen using probability level diagrams than using other types of representation. Comparing the *p* orbitals with the *s* orbitals in Figure 4–13, it is clear that the 2*s* and 2*p* orbitals are about the same size. Also the 3*s* and 3*p* orbitals are approximately equivalent in size. This is expected since the quantum number *n* principally governs the size of the orbital. If the 2*s* and 2*p* are carefully compared, the 2*s* orbital has greater probability density near the nucleus compared to the 2*p* orbital. Similarly, the 3*s* orbital has greater probability near the nucleus than the 3*p*. This will have an important consequence in the ordering of the atomic orbitals by energy for multi-electron elements.

Since the *p*-orbitals are symmetrical about the principal axis (solid straight line in Figure 4–16), the three-dimensional probability level contours can be generated by rotating the two-dimensional probability level contour diagrams in Figure 4–16 about this principal axis. This will generate a three-dimensional surface which is the actual probability level contour for the 2*p* and 3*p* orbitals. A three-dimensional quarter cutaway is shown in Figure 4–17.

As we discussed earlier, there are three 2*p* orbitals, all of which have the same shape but have different orientations in space. The principal axes (solid straight line) of the 2*p* orbital are directed along the *x*, *y*, and *z* axes in Figure 4–18 for the p_x, p_y, and p_z orbitals. In order to show this orientation on paper we will use a single three-dimensional probability-level-contour for the $2p_x$, $2p_y$, and $2p_z$ orbitals. We have arbitrarily elected to use the 80 % probability level contour surfaces for the p_x, p_y, p_z orbitals in Figure 4–18. Similar contours could be constructed for the $3p_x$, $3p_y$, and $3p_z$ orbitals. In fact, the *p* orbitals for any *n* value consist of three orbitals which can be oriented along the *x*, *y*, *z* directions.

The representations with which we have described the atomic orbitals of hydrogen have given very precise information about the electron probability density. A good understanding of an accurate description of atomic orbitals is important

FIGURE 4–16 Probability level contour of *p*-orbitals. 20 %, 40 %, 60 %, and 80 % levels appear in order of lighter shading.

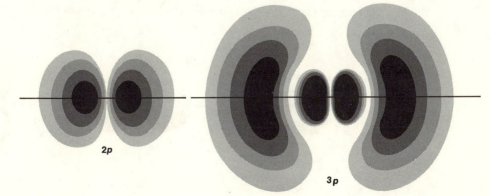

2*p*

3*p*

FIGURE 4–17 Three-dimensional probability level contours for *p*-orbitals (showing the 20 %, 40 %, 60 %, and 80 % contours)

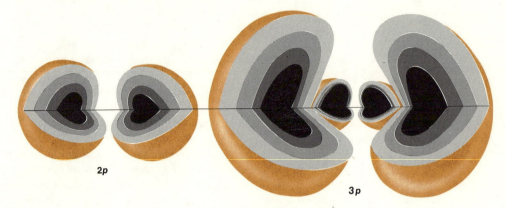

2*p*

3*p*

in the understanding of molecular bonding (Chapters 6 and 7). Although the representations of the atomic orbitals serve the purposes of this chapter quite well, they are too awkward for continued use. For simplicity we will therefore describe an atomic orbital by the cross section of the 80 % probability level contour in a plane containing the nucleus. For *p*-orbitals this plane will also contain the principal axis. The 80 % probability-level contours in two dimensions for the *s*- and *p*-orbitals are given together in Figure 4–19 for comparison.

PROBLEM 4–12   Calculate the number of nodal planes and spherical nodes for the 3*d* and 4*p* orbitals. Sketch the shape of a 4*p* orbital by shading appropriate areas in the two-dimensional cross section for the distribution.

FIGURE 4–18 The 80 % probability level contour surfaces for p_x, p_y, and p_z orbitals

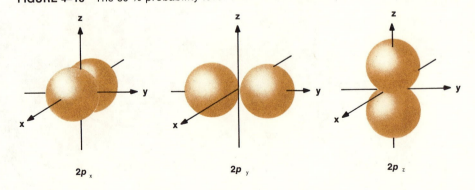

2*p*$_x$

2*p*$_y$

2*p*$_z$

FIGURE 4–19   80 % probability level contours in two dimensions for *s* and *p* orbitals

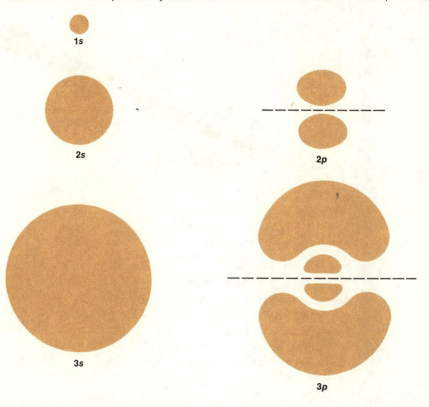

4–J. SPIN QUANTUM NUMBER

The three quantum numbers *n*, *l*, and *m* arise naturally from the hydrogen atom problem, and they are substantiated by experiments primarily in atomic spectroscopy. In addition, there is a fourth quantum number which arises from experiment. The experiment was performed by Stern and Gerlach (1921) and showed that a beam of silver atoms is split into two components when passed through a magnetic field. The simplified diagram in Figure 4–20 shows the splitting of the silver beam in an external magnetic field. This result was attributed to the fact that the electron has a spin and the interaction of a spinning charged particle with the magnetic field separates the beam of atoms. Since the beam separated into only two components, the spin of the electron must be restricted to only one of two orientations in space. We represent these two possible orientations with an arrow pointing in either of two directions.

↑   or   ↓

FIGURE 4–20 The Stern-Gerlach experiment with a beam of silver atoms in a magnetic field

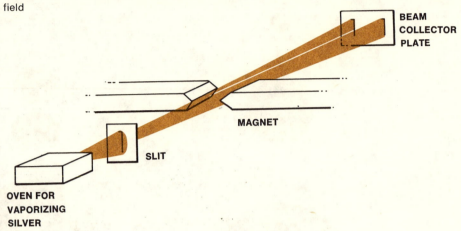

The arrow pointing up (↑) signifies the axis about which the electron spins which is parallel to the magnetic field and we arbitrarily assign a value of $+\frac{1}{2}$ for the spin quantum number with this orientation. Conversely, an electron with its spin axis opposed to the magnetic field (↓) is assigned a quantum number $-\frac{1}{2}$. Therefore, an electron in an atom has a fourth quantum number, m_s, with values of either $+\frac{1}{2}$ or $-\frac{1}{2}$. The concept of spin has also been derived theoretically by Dirac using quantum mechanics. However, the treatment is much more complex than that used in obtaining the solution of the hydrogen atom in terms of n, l, and m and we shall not consider it.

We now have four quantum numbers—n, l, m, m_s—which characterize an electron in an atom. All four quantum numbers will be important in describing the electron configuration in multielectron atoms.

4–K. CONCEPTS DETERMINING ELECTRONIC CONFIGURATIONS

When an atom or ion consists of two or more electrons, the quantum or wave mechanical solution becomes exceedingly difficult. For this reason we must use an approximation to the problem and must assume that the orbitals for the hydrogen atom are basically the correct form for multielectron atoms. However, due to the presence of the other electrons the hydrogen orbitals must be modified and in particular the relative orbital energies are changed. Consequently, some of the degeneracies of the hydrogen orbitals (as shown in Figure 4–7) will be removed. *Each electron in the atom has a specific energy.* This concept is justified by the fact that atomic spectra, as shown in Figure 4–3, contain discrete lines corresponding to transitions between unique energy states. It is also supported by experiments which can measure the *ionization potential*, the energy required to remove each of the electrons in the atom. Each of the measured ionization potentials show discrete energies for the electrons in an atom. More concerning ionization potential will be given in Chapter 5.

As the first approximation to the orbitals for multielectron atoms, each of the electrons is assumed to occupy one of the hydrogen atomic orbitals. We must now consider which of the atomic orbitals the electrons in an atom will occupy. The term *electronic configuration* is used to designate the pattern of occupation of the atomic orbitals.

PAULI EXCLUSION PRINCIPLE

The Pauli principle states that no two electrons in an atom can have all four quantum numbers the same. This restriction on the possible values of the quantum numbers limits the number of electrons per atomic orbital to two. Since an electron in any atomic orbital has n, l, and m quantum numbers specified, only the fourth quantum number m_s varies within that orbital. The values of m_s are restricted to $+\frac{1}{2}$ and $-\frac{1}{2}$, hence the restriction of *only two electrons per atomic orbital*.

HUND'S RULE

Two electrons which have the same spin quantum number are said to have *parallel* spins since their spin axes are aligned (are in the same direction). If two electrons have different spin quantum numbers, their spin axes are in different directions and their spins are said to be *antiparallel* or *paired*. Hund's rule is concerned with electrons occupying degenerate orbitals of the same type, such as the three $2p$ orbitals. It states that when electrons occupy degenerate atomic orbitals, they will occupy individual orbitals with parallel spins until all orbitals are singly occupied. When the number of electrons exceeds the number of degenerate orbitals, only then will two electrons occupy any one orbital with opposed spins. We can illustrate Hund's rule in the following diagram (the spin of the electron is designated by the arrows ↑ or ↓). The total number of electrons occupying these p-orbitals is shown in the left hand column.

NUMBER OF ELECTRONS	p_x	p_y	p_z
1	↑		
2	↑	↑	
3	↑	↑	↑
4	↑↓	↑	↑
5	↑↓	↑↓	↑
6	↑↓	↑↓	↑↓

Note that the electrons occupy separate orbitals with parallel spins until the number of electrons exceeds three. The consequences of this occupation of orbitals will be felt in both the physical and chemical properties of the elements (Chapters 5 to 7).

We conclude this section with some relationships which give the maximum number of electrons that can occupy the orbitals in a subshell (specified by l) and a shell (specified by n). Recall from Section I that the number of orbitals in a subshell is $2l + 1$ and the number of orbitals in a shell is n^2. Therefore, using the Pauli exclusion principle we obtain the following relationships

$$\text{number electrons/orbital} = 2$$

$$\text{number electrons/subshell} = 2(2l + 1)$$

$$\text{number electrons/shell} = 2n^2$$

PROBLEM 4–13   Show that the maximum number of electrons in the $n = 2$ shell is 8. Write the four quantum numbers for each electron in this shell with $n = 2$ (assume that the shell contains the maximum number of electrons).

PROBLEM 4–14   What is the maximum number of electrons that can be put into the 5f subshell?

4–L. ORDER OF ORBITAL ENERGIES

In general an atom will want to be in the most stable state, which is equivalent to the lowest energy or ground state. In order to make the energy of the atom minimal, the electrons must be put in the lowest energy orbitals such that the total energy of the electrons will be at a minimum. If we can establish an *order* of orbital energies, then the electrons can be put in the most stable arrangement among the atomic orbitals according to the rules we have established, and the electronic configuration for the ground state of the atom will be determined.

The order of orbital energies follows to a first approximation the order of the hydrogen atomic orbitals given in Figure 4–7. Recall that the energy of an electron in a hydrogen atomic orbital depended only on the principal quantum number n. The 1s orbital was lowest in energy, the 2s = 2p next, the 3s = 3p = 3d next in energy, etc. The most stable state for hydrogen (lowest energy) is obtained by putting the electron in the lowest energy 1s orbital. For helium, He, we have two electrons which can occupy the 1s orbital to give the lowest energy configuration.

The same considerations are taken in determining the electron arrangement for lithium, Li, which contains three electrons. Two electrons are placed in the 1s orbital and the remaining electron will go into the $n = 2$ shell. However, there are both 2s and 2p subshells in the $n = 2$ shell, so there is a question as to which orbital the remaining electron should occupy. In hydrogen these orbitals are of the same energy and it would be immaterial into which the electron was put. However, in atoms containing more than one electron, the energies of the 2s and 2p orbitals are not equal.

In the case of Li we have two electrons in the 1s orbital, and these will be quite close to the nucleus compared to an electron in the 2s or 2p orbitals (see Figures 4–13 and 4–16). The energy of an electron in either the 2s or 2p orbitals depends upon the attraction of the positive nucleus for this electron, but in the case of Li this attractive force is partially offset by the repulsion of the third electron by the two electrons in the 1s orbital. The average position of the electrons in a 1s orbital can be viewed diagrammatically as a cloud. The repulsion can thus be thought to arise between the 2s or 2p electron and the 1s charge cloud (Figure 4–21). The energy of the electron in a 2s or 2p orbital is thus dependent upon the net effect of the attraction and the repulsion for this electron. The resulting net attraction toward the nucleus (attraction to +3 nucleus vs. repulsion by two negative electrons) can be considered an attraction of the electron to a nucleus whose effective nuclear charge, $+Z_{eff}$, is different from the actual nuclear charge. The order of energies of a 2s electron or 2p electron is thus resolved to the question "In which orbital does the electron experience the greatest $+Z_{eff}$?"

As we can see from the electron distribution graphs in Figures 4–12 through 4–17 the electron in a 2s or 2p orbital does not spend all of its time outside the region of the 1s orbital but has a certain probability of being in the vicinity of the electrons in the 1s orbital and even a certain probability of being near the nucleus. In other words we can say that a 2s or 2p electron can *penetrate* the 1s charge cloud. This is shown in Figure 4–22 where the 2p orbital has been superimposed on the 2s orbital. The dashed line represents the 80 % probability level contour for the 1s orbital and one can see that the 2s orbital has more probability density falling within the 1s contour than has the 2p orbital. An electron in a 2s orbital is said to be more penetrating than a 2p electron. For this reason, an electron in a 2s orbital experiences a greater Z_{eff} than an electron in a 2p orbital, and an electron in a 2s orbital is of lower energy than an electron in a 2p orbital. Consequently, the third electron in lithium goes into the 2s atomic orbital rather than the 2p orbital. The spectrum of Li and its chemical properties are in agreement with this.

FIGURE 4–21   The attractive and repulsive forces from a 1s charge cloud on an outer electron

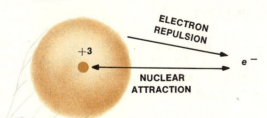

FIGURE 4–22 A comparison of the penetration of 2*s* and 2*p* orbitals. The 2*s* orbital is shown by black dots, the 2*p* orbital by colored dots, and the 1*s* orbital (the 80 % probability contour) by the dashed line

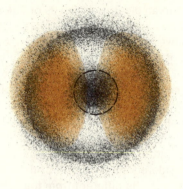

A similar analysis of *d* and *f* orbitals and an examination of the spectra and chemical properties of other atoms leads us to conclude that the order of penetration of orbitals is

$$s > p > d > f$$

This penetrating ability greatly affects the energy of an electron occupying one of these orbitals, and the order is quite different from the hydrogen atomic orbitals where all orbital energies are equal within a given shell. As we consider adding more electrons to atomic orbitals in building higher atomic number elements, the order of orbital energies is as follows:

$$1s < 2s < 2p < 3s < 3p < 4s < 3d < 4p < 5s < 4d < 5p < 6s < 4f < 5d$$

$$< 6p < 7s < 5f < 6d$$

The order of orbitals in increasing energy is also referred to as the *filling* order since this is the order in which one "fills in" electrons into orbitals in building up the electron configurations for the elements. A convenient method for ordering by correlation with the structure of the periodic table is shown in the following section.

4–M. ELECTRONIC CONFIGURATIONS OF THE ATOMS

Having established the necessary rules for filling and ordering orbitals, we can now write the electronic configurations for the elements in the periodic table. The electronic configuration for an atom is designated by the subshell with a post-superscript to indicate the number of electrons in that subshell, for example:

$$H \quad 1s^1$$
$$He \quad 1s^2$$

$$Li \quad 1s^2\ 2s^1$$
$$Be \quad 1s^2\ 2s^2$$
$$B \quad 2s^2\ 2s^2\ 2p^1$$
$$C \quad 1s^2\ 2s^2\ 2p^2$$
$$N \quad 1s^2\ 2s^2\ 2p^3$$

The electronic configurations for the other elements can be written in a similar manner if one uses the order of the atomic orbitals given in the previous section. However, an easier way to do this is to refer to the structure of the periodic table which associates into various groups the subshells being filled with electrons. In Figure 4–23, the outline of the periodic table is divided into the areas in which specific subshells are filled. A row in the periodic table is called a period for reasons that will be apparent from Chapter 5. We will designate the row or period number by r. The principal quantum number n of a subshell being filled is given by the period number r. In the region labeled rs electrons are added to s orbitals with a principal quantum number equal to r. In the $(r-1)d$ region, the electrons are added to orbitals in the d subshells with a principal quantum number $(r-1)$. In the rp region, electrons are added to orbitals in the p subshell with a principal quantum number equal to the period number r. At the point of the asterisks (* or **), orbitals in the $(r-2)f$ subshells are being filled.

FIGURE 4–23 The principal quantum number n of the subshell being filled as derived from the period number r of the periodic table

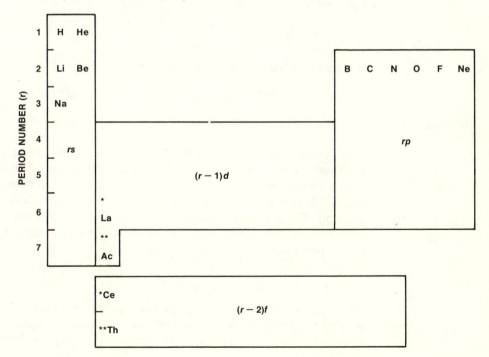

The electronic configuration for an element is obtained by starting at the left with the first period and sequentially going through the periods left to right adding the maximum electrons to the appropriate subshells until the element of interest is reached in the periodic table. As a first example, consider sodium, Na, with atomic number 11. We move across the first row filling the $1s$ orbital and across the second period filling the $2s$ and $2p$ orbitals. We now have 10 electrons added to orbitals but there are 11 electrons in sodium, so we proceed to the third period and add one electron to the $3s$ orbital. The electronic configuration for Na is

$$1s^2\ 2s^2\ 2p^6\ 3s^1$$

A more complex example is europium, Eu, with atomic number 63. The portion of the configuration and the number of electrons accounted for by each period is shown in the following table:

		NUMBER OF ELECTRONS
first period	$1s^2$	2
second period	$2s^2\ 2p^6$	8
third period	$3s^2\ 3p^6$	8
fourth period	$4s^2\ 3d^{10}\ 4p^6$	18
fifth period	$5s^2\ 4d^{10}\ 5p^6$	18
sixth period	$6s^2\ 4f^7$	9
		63

The complete electronic configuration for Eu is then

$$1s^2\ 2s^2\ 2p^6\ 3s^2\ 3p^6\ 4s^2\ 3d^{10}\ 4p^6\ 5s^2\ 4d^{10}\ 5p^6\ 6s^2\ 4f^7$$

PROBLEM 4–15   Write electronic configurations for chlorine and iron.

The use of the filling order of orbitals for finding the electronic configuration of the elements is very convenient and generally reliable. However, some of the atomic orbitals are very close in energy and occasionally a more stable configuration is attained by placing an electron in the orbital which is one higher in the ordering. The electronic configurations for all of the elements are given in Appendix H and some exceptions can be noted. As an example of an exception we should point out the electronic configuration for copper:

$$1s^2\ 2s^2\ 2p^6\ 3s^2\ 3p^6\ 4s^1\ 3d^{10}$$

Note that the $4s$ orbital has only one electron whereas the five $3d$ orbitals are filled with 10 electrons. By the procedure just discussed we would arrive at the configuration . . . $3p^6\ 4s^2\ 3d^9$ but this is not correct for copper. There are only some 9 exceptions to the filling order we have given and only a few have any significant consequences in terms of the chemical properties of the elements.

It should be emphasized again that the order of orbitals which we have just used is only the order in which the successive subshells are filled to obtain the electronic configurations of the elements. This procedure is generally satisfactory for giving the proper electron occupation of the atomic orbitals, but this filling sequence is not necessarily the order of the orbital energies after the electrons occupy these orbitals. The filling order is based on the orbital which has the lowest energy at the time the next electron is being added. However, in an element with a higher atomic number many electrons have been added and the nuclear charge has increased tremendously. Consequently, the energies of the electrons in the orbitals already occupied and, likewise, the order of orbital energies also change tremendously. For example, in Eu (63 electrons) an electron in the 2s or 2p orbital is in a much different charge environment than a 2s or 2p electron in Li (3 electrons).

For the transition elements, Groups IIIB to IIB in the periodic table (see inside back cover), the actual order of orbital energies differs from the filling order. This has an important effect on the ionization of these elements. For example, the *filling order* for Ni (atomic number 28) is

$$1s^2\ 2s^2\ 2p^6\ 3s^2\ 3p^6\ 4s^2\ 3d^8$$

However, the actual order of *orbital energy* within the nickel atom is

$$1s^2\ 2s^2\ 2p^6\ 3s^2\ 3p^6\ 3d^8\ 4s^2$$

and ionization of the nickel atom should remove the 4s electrons before the 3d electrons. There is good evidence that it is the $4s^2$ electrons which are removed in forming Ni^{2+}. The electronic configuration for Ni^{2+} is

$$1s^2\ 2s^2\ 2p^6\ 3s^2\ 3p^6\ 3d^8$$

The electronic configuration of the ion is important in understanding the chemical properties of the element.

PROBLEM 4–16   Write the electronic configurations for Mn, Mn^{2+}, Mn^{3+}.

EXERCISES

1. a. Write the electronic configuration for rubidium (at no. 37)
   b. Give the set of four possible quantum numbers for the electrons in each orbital, e.g., for $2s^2$, $n = 2$, $l = 0$, $m = 0$, $m_s = \pm 1/2$

2. What is the difference between the classical and the wave or quantum mechanical description of an electron?

3. A neutral atom of an element has 2 electrons with $n = 1$, 8 electrons with $n = 2$, 8 electrons with $n = 3$, 1 electron with $n = 4$. Supply as many of the following quantities as possible from this information.
a. atomic number
b. atomic mass (approximately)
c. total number of s electrons
d. total number of p electrons
e. total number of d electrons
f. number of protons in the nucleus
g. number of neutrons in the nucleus

4. Silver has the electronic configuration

$$1s^2 \ 2s^2 \ 2p^6 \ 3s^2 \ 3p^6 \ 3d^{10} \ 4s^2 \ 4p^6 \ 4d^{10} \ 5s^1$$

Note that the 5s orbital has a single electron whereas the 4d orbital is filled with 10 electrons. This is an exception to our filling order since normally we completely fill the 5s orbital before filling the 4d.
a. Why does Ag have the above electronic configuration rather than . . . $5s^2 \ 4d^9$?
b. What should be the most stable silver ion? Write the electronic configuration for that ion.

5. Should it require more or less energy to remove a 3d electron from Fe^{2+} than from Co^{3+}?

6. a. Calculate the number of spherical and planar nodes for the 4p and 5p orbitals.
b. Sketch the contour diagrams for the 4p and 5p orbitals as done for the 2p and 3p in Figure 4–16. How do the sizes of the p orbitals compare?
c. Taking into account the relative sizes of the 2p, 3p, 4p, and 5p orbitals, compare the charge densities in the outer region for these p orbitals. As we will see in Chapters 6 and 7, the sizes and charge densities of the orbitals play an important factor in determining bond strengths in molecules.

7. Sketch 80 % contour diagrams for the 4s and 5s orbitals. How do the shapes of the s and p orbitals compare as the principal quantum number increases to 4 and 5 in contrast to the 2s and 2p orbitals?

8. From your knowledge of the size and shape of atomic orbitals and electronic configurations, would you expect Li, Na, K, Rb, and Cs to follow any consistent trend of increase or decrease in size?

9. Mercury floodlights have a slightly blue and green color. These colors arise from radiation at the following wavelengths

$$\lambda = 4348 \ \text{Å (blue)}$$

$$\lambda = 5461 \ \text{Å (green)}$$

Calculate the energy in ergs/photon and kcal/mole for the radiation at each of these wavelengths. (A mole of photons $= 6.023 \times 10^{23}$ photons.)

10. The temperature of stars can be estimated by observing spectra of atoms and ions whose electrons are excited by the high temperatures in stars. Consider two empty atomic orbitals whose energies are ϵ_1 and ϵ_2 above a filled level. By adding energy to the atom or ion an electron can be promoted to either of these levels. When the electron returns to its original orbital, energy will be released in the form of electromagnetic radiation. The energy of the higher atomic orbital, ϵ_1 or ϵ_2, will be characterized by the observed wavelength λ_1 or λ_2 of the electromagnetic radiation. The intensities of these lines, λ_1 and λ_2 are proportional to the number of excited atoms which have electrons in that specific atomic orbital

$$I_1 = C_1 N_1$$

$$I_2 = C_2 N_2$$

where I = intensity, N = number of excited atoms, and C = constant. The relative values of C_1 and C_2 can be determined from experiments in the laboratory.

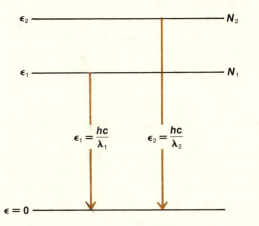

The temperature of the star governs the number of atoms that will be excited to any specific level. The relationship is given by the equation

$$N_1 = C \exp\left(-\frac{\epsilon_1}{kT}\right)$$

where C is a constant which is approximately the same for all energy levels, k = Boltzman constant = 1.38×10^{-16} erg/molecule. There is a similar expression for N_2.

a. Substitute N_1 and N_2 into the expressions for I_1 and I_2 and derive

$$\frac{I_1}{I_2} = \frac{C_1}{C_2} e^{-(\epsilon_1 - \epsilon_2)/kT}$$

and

$$\ln \frac{I_1}{I_2} = \ln \frac{C_1}{C_2} - \frac{(\epsilon_1 - \epsilon_2)}{kT}$$

or

$$\log \frac{I_1}{I_2} = \log \frac{C_1}{C_2} - \frac{(\epsilon_1 - \epsilon_2)}{2.303 \, kT}$$

b. If the intensities of two lines were measured with the following results,

$$\frac{I_1}{I_2} = 202$$

$$\frac{C_1}{C_2} = 5.63$$

$$\lambda_1 = 5,000 \text{ Å}$$

$$\lambda_2 = 4,000 \text{ Å}$$

what is the temperature of the star?

c. If the atom being excited was hydrogen, what is the ratio I_1/I_2 for the atomic orbitals $n_1 = 2$ and $n_2 = 3$ at 3,000 °K and 4,000 °K? Assume $C_1/C_2 = 1$.

11. Derive the equation

$$E_n = -C \, \frac{1}{n^2} = \frac{-2\pi^2 Z^2 e^4 m}{h^2} \, \frac{1}{n^2}$$

where Z = nuclear charge, e = unit of charge, m = mass of the electron. This equation is similar to equation 4–5 except that the term $M/(M + m)$ has been deleted. $M/(M + m)$ is a small correction factor which accounts for the slight motion of the nucleus. In the derivation you will need equation 4–4 which restricts the allowed values of the angular momentum. In addition you will need an equation for the total energy, E,

$$E = \text{kinetic energy} + \text{potential energy}$$

$$E = \frac{1}{2} mv^2 - \frac{Ze^2}{r}$$

Also, you will need the relationship resulting from equalizing the attractive force between the electron and the positive nucleus (Ze^2/r^2) and the centrifugal force (mv^2/r) on the electron.

$$\frac{Ze^2}{r^2} = \frac{mv^2}{r}$$

[Hint: Solve this last equation and equation 4–4 simultaneously to obtain expressions for r and v^2. Substitution into the expression for E should give the desired equation.]

12. Carry out a dimensional analysis on R in equation 4–9

$$R = \frac{2\pi^2 Z^2 e^4 m}{h^3 c} \frac{M}{M + m}$$

to show that R can have the units of cm^{-1}.
[Hint: The unit of charge, esu, can be expressed in cgs units. 1 esu = (1 erg cm)$^{1/2}$]

13. Plot the probability density (ψ^2) for the 2s orbital as a function of r. The wave function for the 2s orbital for the hydrogen atom is

$$\psi_{2s} = \frac{1}{4\sqrt{2\pi}} \left(\frac{1}{a_0}\right)^{3/2} \left(2 - \frac{r}{a_0}\right) e^{-r/2a_0}$$

$a_0 = 0.529$ Å. Explain mathematically why the 2s orbital has one radial node. [Hint: Consider values of r between 0 and 5 Å.]

CHAPTER FIVE

PERIODIC TABLE – CORRELATION OF ATOMIC PARAMETERS

In this chapter, we will be concerned with certain general properties of the elements which can be correlated with the periodic table – metallic properties, atomic and ionic size, ionization potential, electron affinity, and electrical conductivity. These properties show trends going from top to bottom and left to right in the periodic table. These properties and the periodic trends have a direct relationship to the physical and chemical behavior of the elements. Chapter 9 deals with the reactions and compounds of the elements as they are correlated to the periodic table.

5–A. THE PERIODIC TABLE

The periodic table was formulated chronologically well in advance of any knowledge of atomic structure and electronic configurations or even of atomic number. In fact, quantum mechanics was not at all known at the time Dmitri Mendeleev and Lothar Meyer (1869) independently observed periodicity in the elements and used it as a basis for constructing the periodic table. The table we use today is slightly different from the earlier one of Mendeleev and Meyer in that it is expanded and shows the transition elements in terms of groups. However, the form of their periodic table clearly showed the periodicity of the physical and chemical properties of the elements and their compounds.

It may be recalled from Chapter 4 that many elements have similar outer electronic configurations. For example, all the elements of group IIA have an outer shell configuration of ns^2:

Be $\quad$ $1s^2\,2s^2$
Mg $\quad$ $1s^2\,2s^2\,2p^6\,3s^2$
Ca $\quad$ $1s^2\,2s^2\,2p^6\,3s^2\,3p^6\,4s^2$
Sr $\quad$ $1s^2\,2s^2\,2p^6\,3s^2\,3p^6\,4s^2\,3d^{10}\,4p^6\,\,5s^2$
Ba $\quad$ $1s^2\,2s^2\,2p^6\,3s^2\,3p^6\,4s^2\,4p^6\,\,4d^{10}\,5s^2\,5p^6\,6s^2$

The arrangement of the periodic table is based on this periodic recurrence of an outer shell electronic configuration.

As you will see shortly, many chemical and physical properties of the elements depend upon the outer shell configuration. Since elements in a given group have the same electron configuration, the physical and chemical properties also display a *periodic* nature. It is the *periodicity* in the physical and chemical properties of elements and their compounds that is the basis for the periodic table.

All elements having the same type of outer orbital electronic configuration (but different n value) belong to a *group*. Groups are represented as vertical columns in the periodic table. *Periods* are horizontal rows of elements. Elements in a given period usually have the same n value, but different outer orbital configurations. For example, in the second period—elements lithium through neon—electrons are added to the second or $n = 2$ shell. Since the $n = 2$ shell can contain a maximum of 8 electrons ($2n^2$), eight elements, each one having a different number of electrons, can be in this period. Since the third period elements have electrons added to the third shell ($n = 3$), we would expect that it would contain 18 elements. However, as seen in Chapter 4, the energy of a higher n orbital, namely $4s$, is lower than that of $3d$, an orbital with a lower n value. Consequently the third period is complete after the $3p$ orbital is filled and contains only eight elements—sodium through argon. The fourth period begins with the filling of the 4s orbital in potassium, $4s^1$. Since the $3d$ level is filled prior to the $4p$, the fourth period is complete after filling the $4s$, $3d$, and $4p$ orbitals, giving a total of eighteen elements. The order of filling energy levels was discussed in Section 4–M (Figure 4–23); the result is the periodic table in its most general form. It is important to remember that the periodic table is simply the most useful classification system found to date. Should a more convenient classification be proposed, it might replace the periodic table.

The periodic table can be said to consist of four classes of elements:

1. *Representative elements* are the elements of the A groups IA–VIIA. These elements have an outer orbital configuration of ns, or an ns and np combination as

GROUP	ELEMENT	
IA	Li	$2s^1$
IIA	Be	$2s^2$
IIIA	B	$2s^2\,2p^1$
IVA	C	$2s^2\,2p^2$
VA	N	$2s^2\,2p^3$
VIA	O	$2s^2\,2p^4$
VIIA	F	$2s^2\,2p^5$

The outer shell is called the *valence shell*. The outer or valence shell electrons are those electrons generally involved in chemical reactions. A wide variety of reactions will be discussed in Chapter 9.

2. The *rare gases* are the elements of group VIIIA. Because of the limited reactivity of these elements, they are placed in a separate class. We will still consider them to be A-group elements which have an *ns,* or *ns* and *np,* outer orbital configuration. However, the rare gases *always* have a filled outer orbital configuration, such as $1s^2$ or ns^2p^6, as:

$$\text{Ne}\quad 1s^2\,2s^2\,2p^6$$

A filled valence shell of eight electrons represents a particularly stable electronic configuration; this is the reason for the limited chemical reactivity of rare gases.

3. *Transition elements* make up the B-groups IB–VIIIB. In the B-groups, the *d* subshell is partially filled as for iron, for example:

$$1s^2\,2s^2\,2p^6\,3s^2\,3p^6\,3d^6\,4s^2$$

4. *Lanthanides and actinides* appear at the bottom of the chart for convenience, although they follow group IIIB according to their atomic numbers. These elements contain partially filled *f* orbitals. The lanthanides, elements 57–71, contain unfilled 4*f* orbitals while the actinides, elements 89–103, contain unfilled 5*f* orbitals.

PROBLEM 5–1   List the first element of each period and its valence shell (outer orbital electronic) configuration.

PROBLEM 5–2   What factor is periodic in your answer to Problem 5–1?

5–B. GENERAL FEATURES AND TRENDS OF PROPERTIES

In Chapter 2, there was a brief discussion of ionic charge, ionic and covalent bonds, and electronegativity, and in Chapter 3 oxidation number was also briefly discussed. We know that

1. Ionic charge is the *net* positive or negative charge on an ion.
2. An ionic compound is one in which the bond is the result of interaction between ions.
3. A covalent compound is one in which the bond is the result of sharing of electrons between atoms.
4. Oxidation number or state is the *apparent* positive or negative charge on an atom within a covalently bonded compound or a complex ion such as SO_4^{2-}. In the case of monatomic ions the ionic charge on the monatomic ion is also the oxidation number or state.
5. Electronegativity is a measure of the relative electron attracting power of an atom in a bond.

In addition, we need to define here several other terms—ionization potential, electron affinity, atomic size, and ionic size—although these terms are considered in more detail in Sections 5–D, 5–E, 5–G, and 5–H.

The *ionization potential*, *IP*, is the energy required to *remove* an electron from a gaseous atom. For the present, we are concerned only with the lowest energy required for the reaction

$$A_{(g)} \longrightarrow A_{(g)}^+ + e_{(g)}^- \qquad \Delta E \text{(energy required)} = IP$$

The *electron affinity*, *EA*, is a measure of the *energy generally released* when an electron is *added* to an atom in the reaction

$$B_{(g)} + e^- \longrightarrow B_{(g)}^- \qquad \Delta E \text{(energy released)} = EA$$

The *atomic size* is considered to be the covalent radius. This radius is such that the sum of the two radii for any two atoms bound by a single electron pair is equal to the internuclear distance between the atoms (Figure 5–1).

The *ionic radius* is a measure of the size of the outer electron orbital of the ion. The sum of two ionic radii is equal to the internuclear distance in a crystal containing the two ions (Figure 5–1).

We are now in a position to state some general features and trends involving the properties we have just described.

TRENDS SEEN IN THE PERIODIC TABLE

1. Most of the elements to the left of group VA of the periodic table are metals. This includes the lanthanides and actinides. Notable exceptions are hydrogen, boron, and carbon, which are not metals.

FIGURE 5–1 Schematic of covalent and ionic radii

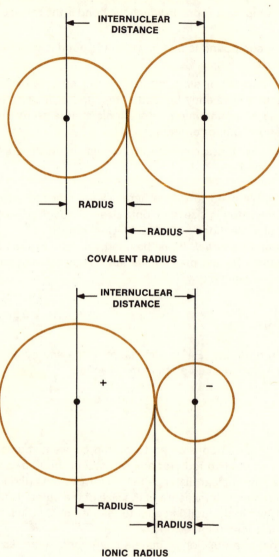

COVALENT RADIUS

IONIC RADIUS

2. Those elements whose symbols are in color in the periodic table inside the back cover of this book are nonmetals. All of the elements of groups VIIA and VIIIA are included.

3. As you proceed from the top of any one group to the bottom, the elements have more metallic character. The elements have less metallic character as you proceed from left to right within a period.

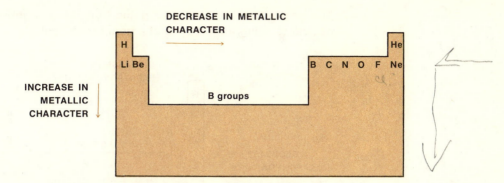

4. The change from metallic to nonmetallic elements is not sharply defined. The group of elements called *metalloids* have characteristics of both metals and nonmetals. These occupy a position adjacent to a diagonal region of the chart extending from Al to Sb as shown in the shaded area of the periodic table inside the back cover of this book.

5. The ionization potential, *IP*, decreases when proceeding from the top of a group to the bottom and increases going from left to right within a period

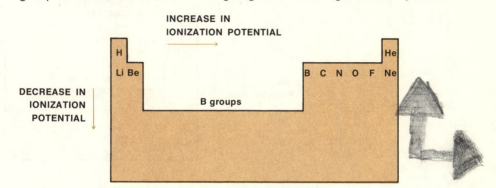

6. The electron affinity, *EA*, generally increases going from left to right within a period (not including group VIIIA). There is generally no clear trend within most groups.

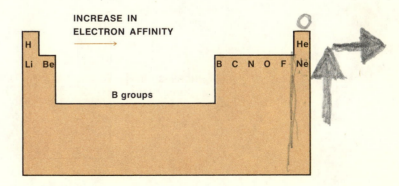

7. The electronegativity increases going from left to right within a period and decreases going from the top of a group to the bottom.

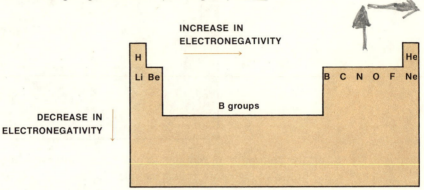

8. The atomic size decreases going from left to right within a period, but increases going from the top of a group to the bottom.

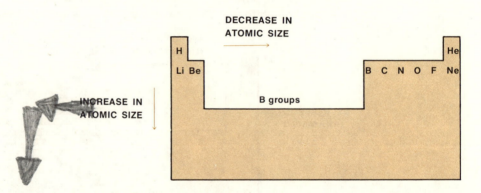

9. The ionic size increases going from the top of a group to the bottom. There is no trend within a period since the charge on the ion changes from positive to negative part way through the period.

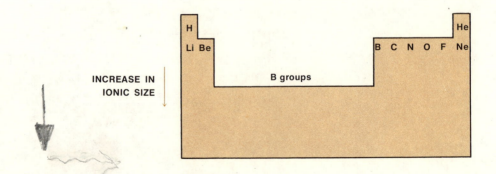

10. The elements of groups IA, IIA, and IIIA have 1, 2, and 3 electrons respectively in their valence shell. These elements lose electrons readily and attain a positive ionic charge (except Be and B) and an oxidation number equal to their group number.

11. The elements of groups VA, VIA, and VIIA have 5, 6, and 7 electrons, respectively, in their valence shell. These elements tend to gain a number of electrons equal to 8 minus the group number. Therefore, these elements will commonly have a negative oxidation number of 8 minus the group number, except when reacted with an element of greater electronegativity. When reacted with an element of greater electronegativity, the maximum possible positive oxidation number is equal to the group number.

12. The elements in group IVA have 4 electrons in their valence shell and tend to share electrons with other elements. The oxidation number can be either positive or negative.

13. The B groups, the transition elements, are metallic in nature. They tend to give up or share electrons. The group number commonly does *not* indicate the number of electrons in the valence shell. Their most common ionic charge is +2. Their oxidation number is positive and variable with a *maximum* value equal to the group number (see Section 5–I and Chapter 16).

14. The elements of group VIIIA are generally unreactive (Section 5–I; see also Chapter 9).

15. The lanthanides and actinides are metallic, and they give up electrons equal to their group number, IIIB. Therefore, the common ionic charge is +3 (Section 5–I).

5–C. TREND IN METALLIC CHARACTER

The elements increase in metallic character when going from the top to the bottom of the group. Criteria for *increasing* metallic character are:

1. decreasing ionization potential of the atoms
2. decreasing electron affinity of the atoms
3. decreasing electronegativity of the atoms
4. increasing tendency toward only positive oxidation states of the atom
5. increasing electrical conductivity of the solid

For some groups not all of the criteria are equally applicable. For example, in group IA all of the elements except hydrogen are considered to be metals, to have positive ionic or oxidation states, and to be excellent electrical conductors. Also, in group VIIA, all elements are considered to be nonmetals. Nonetheless, the ionization potential, electron affinity, and electronegativity decrease going from the top to the bottom of the groups IA and VIIA. For other groups, most of the criteria are applicable. For example in group VA, nitrogen at the top of the group is a gas and has a high ionization potential (14.5 ev) and electronegativity (3.0). It is a nonconductor and forms compounds in which the oxidation state of nitrogen can be positive or negative. However, bismuth at the bottom of the group is a metallic solid and has a relatively low ionization potential (8.0 ev), and electro-negativity (1.9), and forms only positive oxidation states and ions. The trends are general, and as with every generality, exceptions occur. In those cases where there is a nonmetal at the top of the group and a metal at the bottom, there is a transition between them via the metalloids.

PROBLEM 5–3   Utilizing the periodic table inside the back cover of this book, which of the ele-ments in each of the following groupings will have the most metallic character
a. Cl, Ca, As   b. S, Te, Se   c. Al, Si, P   d. Ga, Tl, B

PROBLEM 5–4   Using the groupings in Problem 5–3, answer the following questions.
a. Which element in Problem 5–3a will most likely have a negative oxidation state in its compounds?
b. In Problem 5–3c, which element will most likely have the lowest ionization potential?
c. Note that in Problem 5–3b, the elements are members of group VIA. Which member has the highest electron affinity and electronegativity?
d. Which element in Problem 5–3d has the largest atomic size?
e. Which element in Problem 5–3c is *not* a metalloid?

5–D. IONIZATION POTENTIAL

As previously described, the ionization potential is the energy *required* to *remove* an electron completely from a gaseous atom (or molecule) without giving any additional energy (kinetic) to the ejected electron or to the ion. This is equivalent to increasing the principal quantum number to infinity. Representing the atom by A, the ionization potential, *IP*, is the energy change, ΔE, for the process

5–1 $$A_{(g)} \longrightarrow A^+_{(g)} + e^-_{(g)} \qquad \text{energy required } (\Delta E) = IP$$

The A^+ is the charged species or ion resulting from electron loss. There are several ionization potentials for each element, corresponding to successive re-moval of electrons. The first ionization potential corresponds to the removal of one electron (equation 5–1), the second ionization potential corresponds to the removal of the second electron, and so forth. The energy required to remove one electron (first ionization potential) is always less than that for any subsequent electron (higher ionization potential)

$$Na \longrightarrow Na^+ + e^- \qquad IP = 5.1 \text{ ev}$$

$$Na^+ \longrightarrow Na^{2+} + e^- \qquad IP = 47.3 \text{ ev}$$

The energy is commonly measured in kcal or electron volts where 1 ev/molecule = 23.06 kcal/mole.

TRENDS OF IONIZATION POTENTIAL

There is a periodic relationship involving the ionization potential of the elements. Figure 5–2 shows the first ionization potential as a function of the atomic number. Note the cyclic or periodic pattern of the ionization potential. For each period it is lowest for elements of group IA and highest for group VIIIA and group VIIA of the representative elements. For example, Na has a value of 5.1 ev while that for Cl is 13.0 ev. This occurs because the effective nuclear charge, Z_{eff}, increases in going from left to right in a period. The Z_{eff} increases because the atomic number is regularly increasing across the period while the electrons are going into the same valence shell. This results in decreased screening of the nuclear charge by the electrons and thereby an increase in Z_{eff}.

Figure 5–3 is a plot of ionization potential versus group number. Note the decrease going from the top of a group to the bottom. For example, in group IA, the ionization potential for lithium is 5.4 ev whereas that for cesium is 3.9 ev. This occurs because the energy of the outermost electron increases with an increase in n, and the energy required to remove the electron to $n = \infty$ therefore decreases. In Chapter 4 it was pointed out that an electron in an atomic orbital has a negative energy, indicating a stable state for the electron. As the magnitude of n increases, the E increases to a less negative value and in fact at $n = \infty$, $E = 0$. Figure 5–4

FIGURE 5–2 First ionization potential versus atomic number

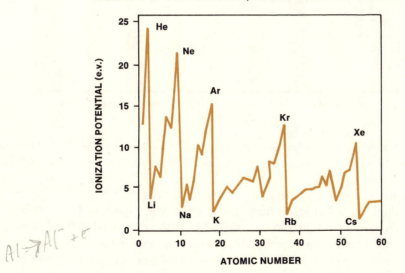

FIGURE 5–3 First ionization potential versus group number. Lines connect elements of the same period.

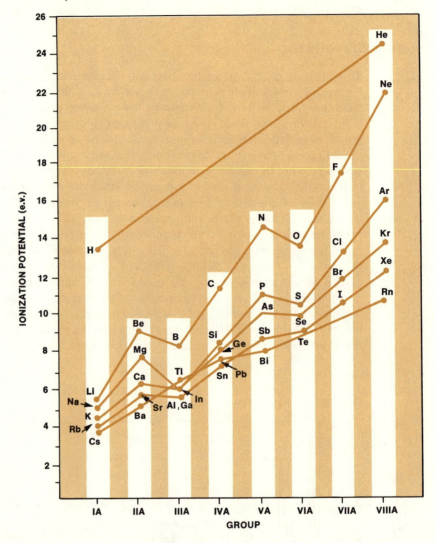

shows a schematic energy level diagram for the *ns* and *np* orbitals. The energy required to remove an electron, the ionization potential, becomes less as the value of *n* becomes greater.

Note the exception to the rule of increasing ionization potential in going from left to right, as we proceed from group VA to VIA, Figure 5–3. The elements of group VA have a configuration $\cdots np^3$, while those of group VIA have a configuration $\cdots np^4$. In the case of group VIA, one electron exists in a *p* orbital already containing another electron. This results in increased electron repulsion energy which

FIGURE 5–4 Energy level diagram for *ns*, *np* and *nd* orbitals. Note that at *n* = ∞. *E* = 0.

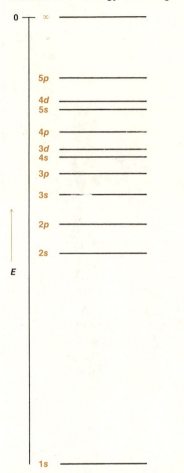

raises the total energy of the *p* electron and decreases the energy required for its ionization. Figure 5–5 gives the ionization potentials for all of the A-group elements shown in the form of a periodic chart.

IONIC CHARGE AND IONIZATION POTENTIAL

Figure 5–6 is a plot of the energy required to ionize an atom to a positive charge equal to its group number versus the charge on the ion which is equal to its group number. Note, for example, that the energy required to make the +3 ions of the elements of group IIIA is the sum of the first, second, and third ionization potentials. It can be seen from Figure 5–6 that the energy required to make ions of

FIGURE 5–5 First ionization potential of the A-group elements (in electron volts)

INCREASE IN IONIZATION POTENTIAL

IA	IIA		IIIA	IVA	VA	VIA	VIIA	VIIIA
Li 5.4	Be 9.3		B 8.3	C 11.3	N 14.5	O 13.6	F 17.4	Ne 21.6
Na 5.1	Mg 7.6		Al 6.0	Si 8.2	P 11.0	S 10.4	Cl 13.0	Ar 15.8
K 4.3	Ca 6.1	TRANSITION ELEMENTS	Ga 6.0	Ge 8.1	As 9.8	Se 9.8	Br 11.8	Kr 14.0
Rb 4.2	Sr 5.7		In 5.8	Sn 7.3	Sb 8.6	Te 9.0	I 10.5	Xe 12.1
Cs 3.9	Ba 5.2		Tl 6.1	Pb 7.4	Bi 7.3	Po 8.4	At 9.5	Rn 10.7

DECREASE IN IONIZATION POTENTIAL

higher and higher positive values within a period (from group IA as Na^+ to group VIIA as Cl^{7+}) increases very rapidly. This has several important consequences. Only the elements of the first three groups (IA, IIA, IIIA) commonly exist as positive ions (as Li^+, K^+, Ca^{2+}, Al^{3+}). This occurs because the energy required is too great to create highly charged positive ions such as Cl^{7+}. In addition, the ionization potential is the most important factor in determining the *reducing power* of an atom. Recall that this is the ability of an atom to give up an electron or transfer an electron to an acceptor. Since the ionization potential of the first three groups is relatively low (except boron and beryllium), these elements are reducing elements.

STABLE ELECTRON CONFIGURATIONS

It is necessary to be able to predict which of several electron configurations is most stable in order to understand reactivities of the elements. The electron configuration for positive ions results from removal of an electron from the highest filled orbital and succeeding ones from the same or next highest filled orbital as in equations 5–2 to 5–5.

5–2 $_{19}K(\cdot\,\cdot\,\cdot\,3s^23p^64s^1)$ + energy (4.3 ev) $\longrightarrow$ $_{19}K^+(\cdot\,\cdot\,\cdot\,3s^23p^6)$ + e^-

5–3 $_{19}K^+(\cdot\,\cdot\,\cdot\,3s^23p^6)$ + energy (31.8 ev) $\longrightarrow$ $_{19}K^{2+}(\cdot\,\cdot\,\cdot\,3s^23p^5)$ + e^-

5–4 $_{20}Ca(\cdot\,\cdot\,\cdot\,3s^23p^64s^2)$ + energy (6.1 ev) $\longrightarrow$ $_{20}Ca^+(\cdot\,\cdot\,\cdot\,3s^23p^64s^1)$ + e^-

5–5 $_{20}Ca^+(\cdot\,\cdot\,\cdot\,3s^23p^64s^1)$ + energy (11.9 ev) $\longrightarrow$ $_{20}Ca^{2+}(\cdot\,\cdot\,\cdot\,3s^23p^6)$ + e^-

It is worth noting that although the first ionization potential results from an electron being removed from the 4s level for both potassium (equation 5–2), and

FIGURE 5–6 The energy required to ionize an atom versus the charge on the ion (equal to its group number). Lines connect elements of the same period.

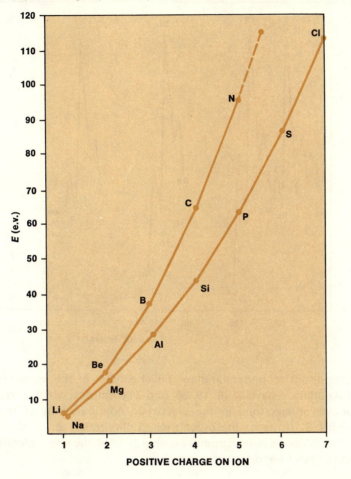

calcium (equation 5–4), the energy requirement for potassium is some 1.8 ev *less* than that for calcium. This happens because the number of protons or the charge on the potassium nucleus is one less than for calcium, but the electrons are in the same valence shell in both cases. This is equivalent to saying that the effective nuclear charge of potassium is less than that of calcium. However, in the second ionization, an electron is removed from the 3p level of potassium and from the 4s level of calcium (equations 5–3 and 5–5). Despite the fact that the number of protons or nuclear charge of K^+,19, is one less than that of Ca^+,20, the energy required for moving the second electron from K^+ is nearly 20 ev *greater* than that from Ca^+. It is necessary to break up a highly stable p^6 or $s^2p^6 = 8e^-$ configuration for K^+ (equation 5–3), whereas for Ca^+, this is not necessary.

The stability of ns^2np^6 electron configurations, which we have just mentioned,

FIGURE 5–7 Electron affinity for the elements versus atomic number

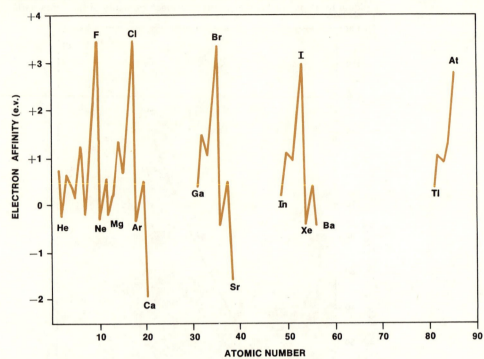

is important to understanding the chemistry of the elements. Stable electron configurations have 2, 10, 18, 36, and 54 electrons, which correspond to the total number of electrons in the electronic configurations of the rare gases. The number 8 recurs as the outer shell configuration for all the rare gases except helium (since its atomic number is only 2). Of all the stable electron configurations, 8 is the most important.

PROBLEM 5–5 Why is the energy required for the second ionization potential considerably higher than that of the first ionization potential for a given atom?

PROBLEM 5–6 Arrange the following atoms in the order of increasing ionization potential (lowest first): F, Cs, S, and Ca.

5–E. ELECTRON AFFINITY

The electron affinity is a measure of the *energy released* when an electron is *added* to an empty orbital of a gaseous atom (or molecule) as

5–6 $$Cl_{(g)} + e^- \longrightarrow Cl_{(g)} \qquad \text{energy released } (\Delta E) = EA$$

The nonmetallic elements such as fluorine, chlorine, and oxygen have relatively high electron affinities. On the other hand, the rare gases and some of the metallic elements have zero or negative electron affinities (E required). Also, since eight electrons (or two, in some cases) represent a highly stable electron configuration, the nonmetallic elements of groups VIA and VIIA become more stable as negative ions. Further, since the atomic number, within a period, is increasing while the electrons are added to a shell with the same principal quantum number, the shielding of the positive nuclear charge is less and less, meaning Z_{eff} increases within a period. Consequently, the nuclear attraction for the negatively charged electron increases and the electron affinity, therefore, generally increases.

TRENDS OF ELECTRON AFFINITY

The periodic relationship of electron affinities is shown in Figure 5–7 as a function of the atomic number, while Figure 5–8 correlates electron affinity with group number. Because of the general lack of experimental values, the data in these figures are a mixture of experimental and theoretical values. In Figure 5–7 note the

FIGURE 5–8   Electron affinity versus group number. Lines connect elements of the same period.

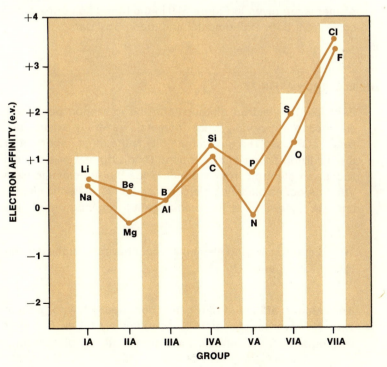

FIGURE 5–9 Electron affinities of the A-group elements (in electron volts)

INCREASE IN ELECTRON AFFINITY

IA	IIA				IIIA	IVA	VA	VIA	VIIA	VIIIA
Li 0.77	Be 0.38				B 0.18	C 1.29	N −0.21	O 1.46	F 3.50	Ne −0.22
Na 0.54	Mg −0.22				Al 0.20	Si 1.36	P 0.71	S 2.04	Cl 3.62	A −0.37
K 0.47	Ca −1.93		TRANSITION ELEMENTS		Ga 0.37	Ge 1.44	As 1.07	Se 2.12	Br 3.36	Kr −0.42
Rb 0.42	Sr −1.51				In 0.20	Sn 1.03	Sb 0.94	Te 1.96	I 3.06	Xe −0.45
Cs 0.39	Ba −0.48				Tl 0.32	Pb 1.03	Bi 0.95	Po 1.32	At 2.80	Rn

periodic recurrence of a high value for the electron affinity of the group VIIA elements. Note in Figure 5–8 that the electron affinity generally increases going from left to right within a period. This occurs because Z_{eff} is increasing in the same direction, so that the attraction for the added electron increases. Trends within a group are often erratic. Nonetheless, most commonly the electron affinity at the bottom of the group is less than at the top. The electron affinities of the A group elements are given in Figure 5–9 in the form of a periodic table. Again, the values come from both experimental and theoretical data.

IONIC CHARGE AND ELECTRON AFFINITY

Because the electron affinities and the ionization potentials of the nonmetallic elements are generally high, nonmetals generally exist as negative ions—F^-, Cl^-, Br^-, I^-, O^{2-}, S^{2-}. The electron affinity is the most important property determining the oxidizing power of an element. Oxidizing power is the ability of an element to accept an electron from a donor. Thus the nonmetallic elements of groups VA, VIA, VIIA act as oxidizing elements or agents while the elements of groups IA, IIA, and IIIA generally are not oxidizing but reducing agents (see Section 5–I).

PROBLEM 5–7   Which of the following elements would have the highest electron affinity and why: He, Cl, Ca, K, and B?

PROBLEM 5–8   The electron affinity of nitrogen is abnormally low for a nonmetal, approximately zero. However for carbon it is approximately 1.3 ev. Based on your knowledge of configurations, why should the electron affinity of nitrogen be abnormally low?

5–F. ELECTRONEGATIVITY

Relatively few atomic electron affinity values are accurately known, but another measure of the attractive power of atoms in a compound for electrons exists. The electronegativity of an element is the relative ability of that element to attract electrons when it is bonded. A low ionization potential indicates a greater relative tendency to give up electrons while a low electron affinity indicates a smaller relative tendency for attraction of electrons. In metals, both of these factors result in a relatively small electron attraction and, therefore, a relatively low electronegativity. Nonmetallic atoms tend to have high ionization potentials and electron affinities. These result in a relatively high electron attraction and, therefore, a relatively high electronegativity for nonmetals in compounds.

TRENDS IN ELECTRONEGATIVITY

There are trends in the electronegativities of elements both by period and group. The electronegativity increases going from left to right within a period, because

FIGURE 5–10   Electronegativity versus atomic number

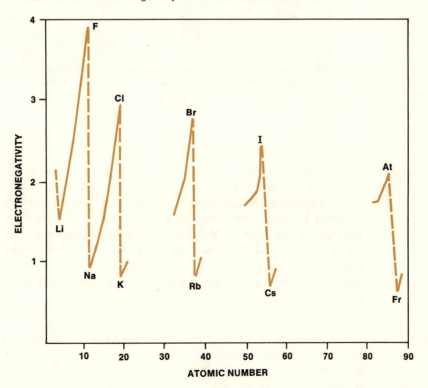

FIGURE 5–11 Electronegativity versus group number. Lines connect elements of the same period.

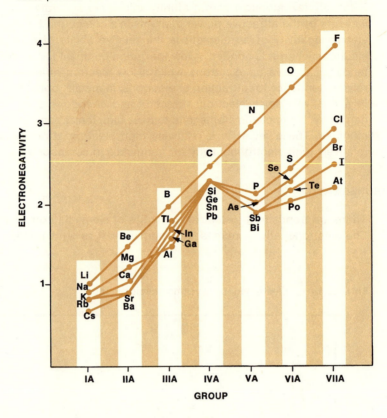

there is a transformation from metallic to nonmetallic character. This trend is shown in Figures 5–10 and 5–11. For example, there is a significant increase in electronegativities proceeding from lithium to fluorine within the first period.

Within a group, electronegativity decreases going from the top of the group to the bottom. For example, the electronegativity progressively decreases between F and I in group VIIA (Figures 5–10 and 5–11), because the elements become more metallic and the ionization potential decreases going down the group. Also, the electron affinity decreases or stays approximately constant going from top to bottom within a group.

PROBLEM 5–9 Which element of each of the following pairs would be expected to have the greater electronegativity?

a. C, Cl

b. S, O

c. Mg, Se

d. Sb, F

5–G. ATOMIC SIZE

The diffuse nature of the electronic charge distribution in an atom makes it impossible to define the volume occupied by an atom. That is, no sphere with a meaningful radius encompasses all of the electron charge cloud.

1. The *single bond covalent radius* is defined such that the sum of two radii is equal to the internuclear distance between the two atoms bound by a single pair of electrons (Figure 5–1). See Chapters 6 and 7 for a detailed discussion of the covalent bond.

2. The *metallic radius* is one-half the internuclear distance between nearest neighbor atoms in a metallic crystal.

3. The *van der Waals radius* is one-half of the internuclear distance between non-bonded atoms in the solid state.

Although there are minor differences for metals between metallic radii and covalent radii, the differences are small enough that we shall consider these two measures to be the same. With this in mind, Figure 5–12 plots atomic size in terms of single bond covalent radius as a function of atomic number. Note the periodic recurrences. For example, the maximum radius per period occurs for the members of group IA. Figure 5–13 is a plot of the single bond covalent radii as a function of the group number.

FIGURE 5–12   Covalent radius versus atomic number

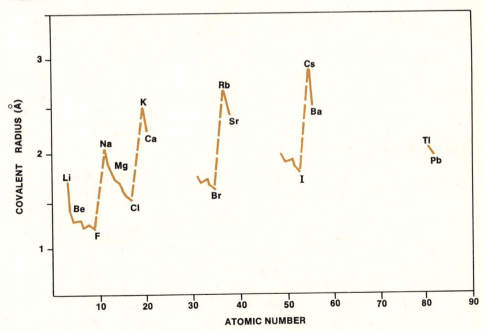

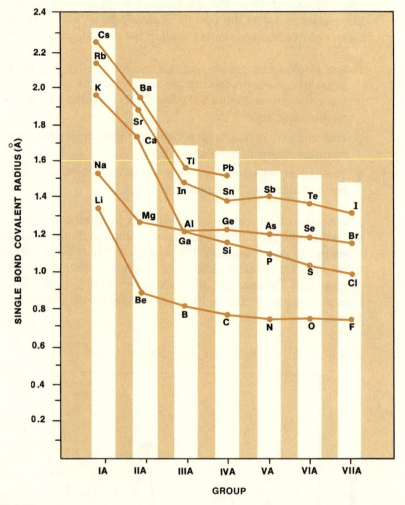

FIGURE 5–13 Single bond covalent radius versus group number. Lines connect elements of the same period.

TRENDS IN ATOMIC SIZE

A plot of van der Waals radii versus group number is given in Figure 5–14. It can be seen that for those atoms for which single bond covalent and van der Waals radii exist, the latter is always the larger. There are general trends for both the covalent and van der Waals radii. The size of the atom always increases when proceeding down a group (Figures 5–13 and 5–14) because the outermost (valence shell) electrons are going into shells with progressively higher n values and correspondingly higher radius values. Within a period, the size decreases when proceeding from left to right (particularly noticeable in Figure 5–13), since the effective nuclear charge (Z_{eff}) is undergoing a progressive increase. Figure 5–15 schematically portrays the covalent radii in the form of a periodic table.

FIGURE 5–14 Van der Waals radius versus group number. Lines connect elements of the same period.

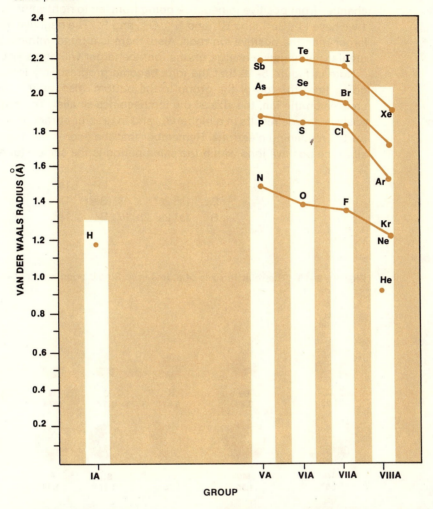

PROBLEM 5–10 The electronic configurations for several elements are as follows: (a) $1s^2$ (b) $1s^2\,2s^1$ (c) $1s^2\,2s^2\,2p^1$ (d) $1s^2\,2s^2$. Place the elements in the order of increasing size and justify your ordering.

5–H. IONIC SIZE

The appropriate measure of size for ions is the ionic radius. The ionic radius is defined such that the sum of two ionic radii is equal to the internuclear distance in a crystal containing the ions (Figure 5–1).

$$r_{\text{(interionic)}} = r_+ + r_-$$

where r_+ and r_- are the ionic radii of the positive and negative ions. Figure 5–16 is a plot of ionic radii versus group number. It should be noted that ionic charge changes from positive to negative going from left to right within a period. In period 1 we have Li^+, Be^{2+}, and B^{3+} and N^{3-}, O^{2-}, and F^-. The negative ion radii are always larger than the positive ion radii. Also, there is a trend within a period where the greater the negative charge on the ion, the larger will be its size.

Note in Figure 5–16 that the ions become progressively larger within a group going from the top of the group to the bottom. Also, as the positive charge increases on the ion, the size of the ion becomes smaller. The progressively higher positive charge results in a higher Z_{eff} and in greater attraction toward the nucleus for the remaining electrons. Remember that the "core" electron configuration of all of the positive ions within the same period is the same—for example

Li	$1s^2\ 2s^1$	Li^+	$1s^2$
Be	$1s^2\ 2s^2$	Be^{2+}	$1s^2$
B	$1s^2\ 2s^2\ 2p^1$	B^{3+}	$1s^2$

FIGURE 5–15   The relative sizes of atoms in Å. Single bond covalent radii are used.

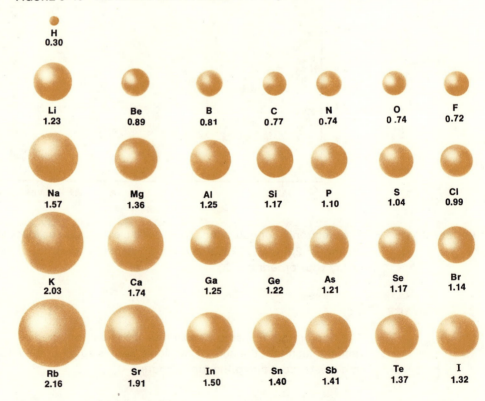

H 0.30						
Li 1.23	Be 0.89	B 0.81	C 0.77	N 0.74	O 0.74	F 0.72
Na 1.57	Mg 1.36	Al 1.25	Si 1.17	P 1.10	S 1.04	Cl 0.99
K 2.03	Ca 1.74	Ga 1.25	Ge 1.22	As 1.21	Se 1.17	Br 1.14
Rb 2.16	Sr 1.91	In 1.50	Sn 1.40	Sb 1.41	Te 1.37	I 1.32

FIGURE 5-16 Ionic radius versus the group number. Lines connect elements of the same period.

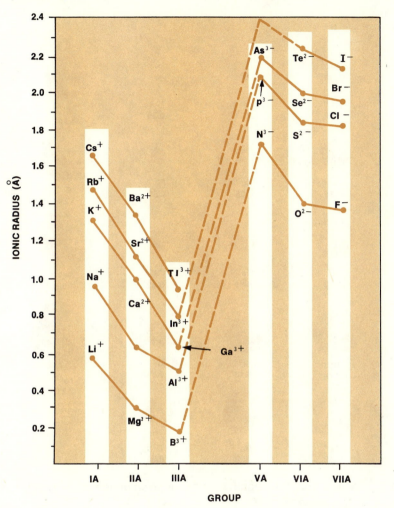

Figure 5-17 gives a good picture of the relative sizes of some common ions laid out in the form of the periodic table. This figure shows the trend of increasing size going from the top to the bottom of a group.

Isoelectronic species are those containing the same number of electrons. For example, $_{16}S^{2-}$, $_{17}Cl^-$, $_{18}Ar$, $_{19}K^+$, $_{20}Ca^{2+}$ all contain 18 electrons and therefore constitute an isoelectronic series containing both atoms and ions.

PROBLEM 5-11   Why should ions become larger going from the top of a group to the bottom?

FIGURE 5–17   The relative sizes of some ions. The ionic radius is given in Å in parentheses.

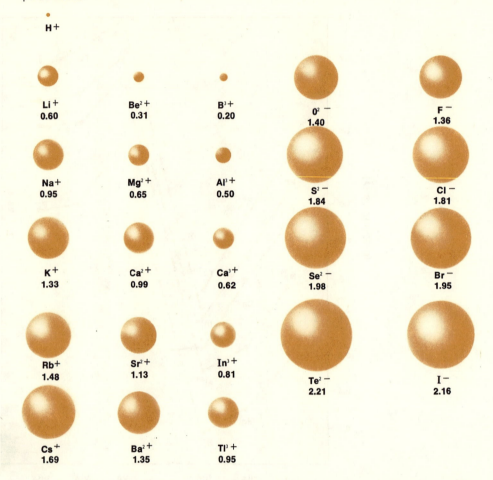

PROBLEM 5–12   Why should the negative ions of a given period be larger than positive ions of the same period?

5–I. METALS, METALLOIDS, AND NONMETALS

METALS

More than three-fourths of the known elements are metals. Two properties of metals are tensile strength and elasticity. *Tensile strength* is a measure of the force required to break a sample of a metal of a given cross section. The metals of the transition series have the highest tensile strength whereas metals such as tin (Sn) and lead (Pb) have low tensile strengths. *Elasticity* is a measure of the ability of an object to return to its original shape once it has been deformed (for example, when it is bent). A metal such as iron has high elasticity whereas metals such as sodium or lead have relatively low elasticity.

The general ability of metals to be made into thin wires or sheets depends upon two additional properties—ductility and malleability. *Ductility* is a measure of the ability of a sample to be permanently lengthened or stretched and still possess strength. *Malleability* is related to the ability of a material to be deformed, as by pounding, and yet possess strength. Gold (Au) shows a high degree of both ductility and malleability while tungsten shows a lower degree of both properties.

Metals are good conductors of electricity and heat. Usually these two go in parallel—the better the conduction of electricity, the better the conduction of heat. Silver (Ag) and copper (Cu) are among the best conductors of electricity while bismuth (Bi) is among the poorest (as are the other metals and metalloids of group VA). Almost all electrical wire is made of copper; silver is far too expensive to use.

The properties of tensile strength, elasticity, malleability, and ductility suggest that a rather special kind of bonding must be present in metals. The bonding must be sufficiently strong to account for the relatively high tensile strength and elasticity; yet it must be flexible enough to permit malleability and ductility. Moreover, once lengthening or pounding of the metal has occurred, bonding must still be present.

METALLIC BONDING

The metallic solid is made up of atoms, in contrast to ions and molecules in ionic and covalent solids. In addition. the atoms are closely packed. Because of this closeness, the valence shell electrons of the individual atoms are not localized on any one atom center but are free to move throughout the metal solid. We can arrive at a model where positively charged ions are in fixed positions imbedded in a sea of mobile electrons. The electrons are responsible for the overall binding of the positive ions to one another in the solid. Because all atom centers are involved in the bonding, residual bonding exist even after deformation; that is, there are not specific bonds which are solely responsible for holding atoms together in a metal.

Because the electrons in metallic structures are mobile, they move throughout the solid readily, which accounts for electrical conductivity (Figure 5–18). In

FIGURE 5–18 Schematic diagram of electrical conduction in a solid. Note that conduction involves progressive displacement of electrons.

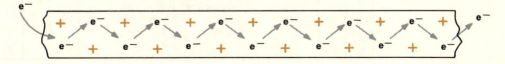

metals, an increase in temperature results in a decrease of conductivity. This decreased conductivity can be considered to be the result of an increase in the number of collisions of electrons with the positive nuclear centers which oscillate more as the temperature increases.

ATOMIC PROPERTIES AND METALLIC CHARACTER

It is possible to associate certain properties of the atom with the metallic character of the element. Atoms of metals generally have low ionization potentials and low electron affinities, and the electronegativities of metals are also low. Metal ions, as a result, are positively charged and have a positive oxidation state. Furthermore, many metallic compounds are ionic for the same reason. The low ionization potential of metals is one of the principal reasons metals are considered to be good reducing agents.

The metals of the representative elements of groups IA—IIIA exist as positive ions or have positive oxidation states equal to their group number. This is expected since a stable $ns^2\, np^6$ electron configuration exists when the charge is equal to the group number.

The B-group elements are all metals showing varying magnitudes of positive oxidation states. Recall that in these elements, the $3d$, $4d$, and $5d$ levels are being filled. The ground electronic state configuration is such that an ns level commonly is the valence shell (Chapters 4 and 16, Section 5–A). Thus the oxidation number and ionic charge are positive. Commonly, B-group ions have a $+2$ charge. The oxidation number can be as high as the group number. For example, osmium in group VIII-B has an oxide OsO_4 in which the oxidation number is $+8$.

FIGURE 5–19 The common oxidation states for the representative elements

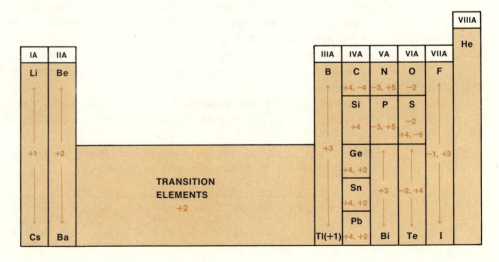

The lanthanides and actinides are silvery white reactive metals. In these elements, 4*f* and 5*f* sub-shells are being filled. The valence shell for the lanthanides is 6*s*. It is likely that the valence shell for the actinides is 7*s*. However, it is quite common for electrons to be lost from the inner *f* and *d* sub-shells as well. Consequently, the common ionic charge on these ions is +3, with +2 also existing in some instances. An oxidation state of +4 also exists for several of these elements.

The common oxidation states are shown in the form of a periodic table, Figure 5–19. In many instances, additional oxidation states are possible.

Much of the foregoing discussion dealt with the physical properties of the solid —the macrostructure. However, when discussing chemical properties we are dealing with individual atoms, molecules or ions—the microstructure. This difference can result in difficulty in classifying an element as a metal or metalloid. For example, even though solid tin (Sn) is commonly considered to be a metal, some of its compounds have chemical properties typical of a metalloid. The chemical properties of elements and compounds are discussed in Chapter 9.

METALLOIDS

These elements are usually considered to include silicon, germanium, arsenic, antimony, and tellurium. Boron also is often included. Certain compounds of tin, aluminum, and lead have properties which sometimes cause these elements to be included in this group. More concerning this point is given at the end of this section. Metalloids have physical properties which are intermediate between those of metals and nonmetals. They are commonly brittle. The ionization potentials of metalloids generally lie between those of metals and nonmetals. All metalloids can have positive oxidation states, and some, such as Sb and As, exist as positively charged ions in solution. One element, tellurium, can also exist in a negative oxidation state as in H_2Te.

ELECTRICAL CONDUCTION AND SEMICONDUCTORS

Generally the electrical conductivity of metalloids is low but nonetheless exists. Some metalloids are *semiconductors*. In semiconductors, the electrons are not as mobile as in metals, and the energy required to move the electrons is in between that of a conductor and that of an insulator. Consequently, any additional input of energy (such as in the form of heat) will increase electron mobility and thereby conductivity; that is, increase of temperature increases the conductivity of metalloids, which is opposite to the temperature behavior of metals.

Elements such as silicon and germanium can be made to become better conductors by the addition, called *doping*, of other elements such as phosphorus which contains one more electron in its valence shell than do either silicon or germanium. Consequently, there is one more electron in the doped semiconductor than is required for bonding, and this extra electron can move through the solid. The doping element is added in very low concentration (parts per million) but can increase the conductivity by a factor of one thousand or greater.

NONMETALS

Nonmetals have many properties inverse to those of metals. Most nonmetals occur in nature as diatomic molecules. Examples of gaseous nonmetals are N_2, O_2, and the halogens. Solid nonmetals such as sulfur and phosphorus occur as multiatomic molecules, S_8 and P_4. The rare gases are monoatomic and the interaction in the liquid and solid phases is weak. The nonmetals which are solids have low tensile strength, elasticity, ductility, and malleability. Obviously, these latter properties cannot apply to the gaseous nonmetals. Furthermore, nonmetals—for example, nitrogen, oxygen, and sulfur—are considered nonconductors of electricity and poor conductors of heat. The energy required to get the electrons to move is so great that such nonmetals may be used as *insulators*.

ATOMIC PROPERTIES OF NONMETALS

Generally, the ionization potential and electron affinity of the nonmetals are relatively high. The electronegativities are also relatively high and the nonmetal monatomic ions are negatively charged. The charge on the monatomic ion will be equal to 8 minus the group number. Chlorine is in group VII, therefore it will have a charge of 8 minus 7, or 1. The atom adds electrons until it has a stable $ns^2\,np^6$ electron configuration or 8 outer electrons. If nonmetals react with metals, the resulting product is often ionic (for example, KF). This happens because of the large differences between the electronegativities of metals and nonmetals. If nonmetals are reacted with elements of similar electronegativity, the resulting compound will be covalent—ICl, HI, and CCl_4.

The nonmetallic elements of groups VA-VIIA generally have a negative oxidation state equal to 8 minus the group number. The principal reason for this is the same as that given for the charge on the monatomic ions. However, the exact oxidation state depends upon whether the element is combined with an element of greater or lesser electronegativity. For example, consider sulfur in H_2S and SO_4^{2-}. In H_2S, sulfur has a -2 oxidation number and in SO_4^{2-}, it is $+6$. Sulfur is more electronegative than hydrogen but less electronegative than oxygen. Fluorine is the most electronegative element and it will always have an oxidation number of -1. Oxygen commonly has an oxidation number of -2, except in OF_2 (O^{2+}), peroxides (O^-), and superoxides (O_2^-).

The relatively high electron affinity of the nonmetals causes them to be oxidizing agents. Consequently, the oxidation number of a nonmetal is changed to a more negative value during oxidation-reduction reactions.

The elements near the bottom of group VA are more metallic than those at the top. Therefore, these commonly exist in positive oxidation states, and in some cases exist as positive ions, as Bi^{3+}.

The elements of group IV commonly have variable oxidation states. This is particularly true for carbon which can range from -4 to $+4$. The elements near the bottom of the group usually only have oxidation states of $+2$ and $+4$. Compounds formed from elements of group IVA with an oxidation state of $+4$ are covalent, such as CCl_4 and $SnCl_4$. Because the energy required to lose four electrons is so high (see Figure 5–5), it is energetically more favorable to share electrons with other atoms. Common oxidation states are given in the form of a periodic table, Figure 5–19, but other values exist.

The electron configuration of group VIIIA, the rare gases, is such that the *s* or *s,p* subshells are filled. Consequently, these elements are generally considered to be unreactive. However, it has been found recently that the rare gases with relatively low ionization potentials such as xenon and krypton can form compounds with fluorine and oxygen (which have high electron affinities). All of the rare gas compounds and complex ions are covalent.

EXERCISES

1. Fill in the table by putting checks in the appropriate places.

	Physical Character		Electric Conduction		Ionization Potential		Electro-negativity		Ionic Charge	
	metal	nonmetal	good	poor	high	low	high	low	+	−
Calcium (Ca)										
Oxygen (O)										
Rubidium (Rb)										
Bromine (Br)										

2. Explain why there is a trend within a period of an increase in the size of the negative ion as the negative charge increases.

3. Plot the first ionization potential versus the period number for the group A elements of periods 2, 3 and 4. What kinds of trends can be observed?

4. Plot the electron affinity versus the period number for the group A elements of periods 2, 3 and 4. What kinds of trends can be observed?

5. Would you expect the rare gases of group VIIIA to have high or low ionization potential and electron affinity? Why?

6. The ionization potentials for carbon, nitrogen, and oxygen are 11.3, 14.5 and 13.6 ev, respectively. Note that nitrogen is abnormally high and then a decrease occurs for oxygen. Explain this result.

7. Would you expect the second ionization potential of sodium to be more or less than that of magnesium? Why?

8. a. In general, do metals or nonmetals have the higher ionization potentials?
   b. Why would these elements tend to have high ionization potentials? [Hint: Think in terms of Z_{eff}].

9. The covalent radius takes a large jump when proceeding from a group VIIA element of one period to a group IA element of the next period. Explain.

10. Predict the ordering of the internuclear or bond distances for the following molecules:
    a. HF, HCl, HBr, HI
    b. Li_2, K_2, Na_2
    c. F_2, Cl_2, Br_2, I_2

11. Predict the interionic or bond distance in a crystal for the following: CsI, LiF, RbBr, CaO, MgS. [Hint: Consider the data in Figure 5–17].

12. Would you expect O^{2-} ion to be larger or smaller than Ne and why?

13. Would you expect Na^+ to be larger or smaller than Ne and why?

14. Which ions or atom of the following would you expect to be the smallest and largest and why?
    a. O^{2-} b. F^- c. Na^+ d. Mg^{2+} e. Ne

15. Note that the difference in the ionization potential between $_{20}Ca$ and $_{31}Ga$, as well as between $_{38}Sr$ and $_{49}In$, is very small despite a large increase in the atomic number. However, there is a considerably larger increase in ionization potential between $_{12}Mg$ and $_{13}Al$, despite the fact that Mg, Ca, and Sr are in the same group (IIA), and Al, Ga, and In are in the same group (IIIA). Explain.

16. Discuss and explain any relationship you can determine that exists between atomic size for the A elements and a. ionization potential   b. electron affinity.

17. What group of elements has the highest relative attraction for electrons in a bond (electronegativity)? Why?

18. Assume the electronegativity (χ) of an element can be calculated from the following formula:

$$\chi = \frac{IP + EA}{2}$$

where *IP* and *EA* are ionization potential and electron affinity (ev) and the signs are as given in Figures 5–5 and 5–9.
a. In general, what characteristic would cause an element to have the largest electronegativity?
b. Without doing specific calculations, what group of elements as a whole would have the lowest χ (except group VIIIA) and why?
c. Calculate the χ values for each element of group VIIA.

19. Justify the statement that electronegativity decreases from the top of a group to the bottom of a group. Use the formula of Exercise 18.

20. Answer the following questions about the elements with the following electronic configurations 1. · · · · $3p^6\, 4s^2$ 2. · · · · $3p^6\, 3d^{10}\, 4s^2\, 4p^5$.
    a. Are the elements metals or nonmetals?
    b. Do the elements have high or low *IP, EA*, and χ?
    c. Which element has atoms of the larger size?
    d. Which $+2$ ion is the smaller in size?
    e. Predict the oxidation or ionic charge number of each element.

21. The following atoms and ions constitute an isoelectronic series. The relative sizes are shown in Figures 5–15 and 5–17. Provide an explanation for the trend in size. $_{16}S^{2-}$, $_{17}Cl^-$, $_{18}A$, $_{19}K^+$, $_{20}Ca^{2+}$.

22. Place the following in order of increasing ionization potential and give the reasons why.

    F (group VIIA), Na (group IIA), Cs (group IA), Ne (group VIIIA)

23. Place the following in order of increasing size and give the reasons why.

    Ar, S^{2-}, Na^+, Cl^-, Li^+

24. Which of the following species is largest? Why?

    K^+, Ar, Cl, Cl^-

25. It is possible to cause electron ejection from the elements and to use the electrons for some purpose (e.g., photoelectric cell). Assume you wanted to use light to cause the electron ejection. Consider the average energy of such light to be about 5 ev. Assuming the energy required to eject the electrons from the element was the same as the ionization potential, what group of elements would you use?

CHAPTER SIX

MOLECULAR STRUCTURE AND BONDING IN DIATOMIC MOLECULES

In atoms, the electrons interact with the nucleus and with other electrons in the atom. However in molecules, there are nuclear-nuclear, electron-electron, and nuclear-electron interactions. The first two are repulsion forces while the last one is attractive. When two atoms are brought together, the compound or bond formed can ideally be ionic or covalent. The distinction was made earlier between ionic and covalent bonding: electrons are transferred in the former and shared in the latter. In this chapter, we shall pursue this and other aspects of chemical bonding in more detail. Further, we shall explore two theories—valence bond and molecular orbital—that can be used to describe covalent bonding in molecules.

6–A. IONIC BOND

An *ionic bond* is one in which essentially complete electron transfer occurs between atoms, and individual positive and negative ions are formed. The ions are held together principally by the coulombic attraction between the unlike charges. The words "essentially" and "principally" are used to indicate that no compound is purely ionic in nature.

The process of forming an ionic compound or ion-pair molecule in the gas phase can be viewed as occurring in steps beginning with the gaseous atoms. In the case of NaCl,

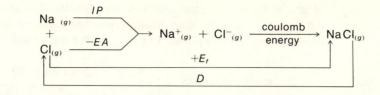

6–1

where IP is the ionization potential, EA is the electron affinity, D is the energy of bond dissociation, and E_f is the energy of formation of $NaCl_{(g)}$. Also, the coulomb potential energy term

6–2
$$\text{coulomb energy (ergs/mole)} = -N\frac{(n_+e)(n_-e)}{r}$$

is the energy of attraction between the ions (Appendix I) where N is Avogadro's number, e is the absolute value of the charge on the electron (4.80×10^{-10} esu), n_+ and n_- are the number of positive and negative charges on the ions, respectively, and r is the interionic or bond distance in centimeters (1 Å $= 10^{-8}$ cm) between the ions in the ion-pair molecule.

It can be seen from equation 6–1 that any *one* step can be determined if all the rest are known. This is true because of the law of conservation of energy. Since the initial state $[Na_{(g)} + Cl_{(g)}]$ and the final state $[Na_{(g)} + Cl_{(g)}]$ are the same, the energies involving the ionization potentials, electron affinities, energy of formation or energy of bond dissociation, and coulomb energy, with their appropriate signs, must sum to zero. Thus assigning a plus sign when energy is required and a minus sign when energy is released we will have

6–3
$$+IP - EA - \frac{N(n_+e)(n_-e)}{r} - E_f = 0$$

6–4
$$+IP - EA - \frac{N(n_+e)(n_-e)}{r} + D = 0$$

From equation 6–4

$$D = +EA + \frac{N(n_+e)(n_-e)}{r} - IP$$

The kind of cyclic process represented in equation 6–1 is known as a *Born–Haber cycle*.

Before continuing with the calculation of a bond dissociation energy, we shall examine just the coulomb term, equation 6–2, for a specific case, NaCl. The interionic or bond distance in NaCl in the gas phase is 2.36 Å while in the crystalline state it is $r = r_+ + r_-$. From Chapter 5, Figure 5–17, $r_{Na+} = 0.95$ Å and $r_{Cl-} = 1.81$ Å, so $r = 2.76$ Å. Recall from Section 2–B that ionic compounds such as NaCl consist of ions in the solid crystal rather than molecules. The spacing between the ions in the crystal is greater than the internuclear distance between the atoms in the molecules in the gas phase. For NaCl in the gas phase, $r = 2.36$ Å and

$$\text{coulomb energy} = -\frac{(6.02 \times 10^{23})(4.80 \times 10^{-10})(4.80 \times 10^{-10})}{2.36 \times 10^{-8}}$$

$$= -58.7 \times 10^{11} \text{ ergs/mole}$$

To convert to kilocalories, 1 erg $= 2.39 \times 10^{-8}$ cal

$$\text{coulomb energy} = -58.7 \times 10^{11} \text{ ergs} \times 2.39 \times 10^{-8} \text{ cal/erg} \times \frac{1 \text{ kcal}}{10^3 \text{ cal}}$$

$$= -140 \text{ kcal/mole}$$

With this in mind we can calculate the energy of formation and dissociation energy for NaCl. In the following equations we have changed the units of the ionization potential and electron affinity to kcal from ev since energies of formation and dissociation are commonly in kcal.

step 1. $\quad$ $Na_{(g)} \longrightarrow Na^+_{(g)} + e^-$ (ionization potential) $\qquad IP = +118$ kcal

step 2. $\quad$ $Cl_{(g)} + e^- \longrightarrow Cl^-$ (electron affinity) $\qquad EA = -83$ kcal

step 3. $\quad$ $Na^+_{(g)} + Cl^-_{(g)} \longrightarrow NaCl_{(g)}$ (coulomb energy) $\qquad -\dfrac{N(n_+e)(n_-e)}{r} = -140$ kcal

---

$Na_{(g)} + Cl_{(g)} \longrightarrow NaCl_{(g)}$ (energy of formation) $\qquad E_f = -105$ kcal

As can be seen from equations 6–1, 6–3, and 6–4, the energy of dissociation, D, is equal in magnitude but opposite in sign to that of the energy of formation, E_f, therefore $D = +105$ kcal. The energy required in step 1 is more than that released in step 2. Thus, it is by step 3 that the coulomb energy of attraction makes ionic bond formation possible, since there is a large amount of energy released. This release of energy is more than sufficient to make up for the energy deficiency from the summation of steps 1 and 2, as shown in the overall summation. Actually, the experimental value for the bond dissociation of $NaCl_{(g)}$ to atoms is +97 kcal/mole. The computed value, +105 kcal/mole, was based on a purely ionic model. This assumption is not completely correct; nonetheless, the close agreement between the calculated result and the experimental result clearly indicates that the ionic model is a good approximation for molecules such as NaCl.

A plot of the potential energy of the system $Na_{(g)}$, $Cl_{(g)}$, and $NaCl_{(g)}$ is shown in Figure 6–1. At large distances the potential energy between $Na_{(g)}$ and $Cl_{(g)}$ is zero; that is, there is no attraction or repulsion. As the distance decreases, the potential energy decreases, because electron exchange or transfer occurs; in addition, there is attraction between the oppositely charged Na^+ and Cl^- ions. This decrease

FIGURE 6–1 Potential energy curve for the formation of $NaCl_{(g)}$ molecules

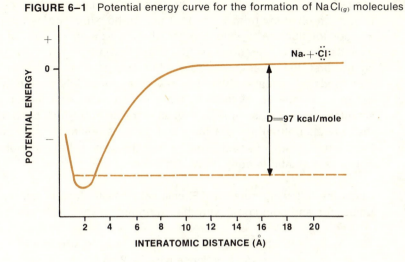

continues, reaching a minimum near 2 Å. The potential energy then begins to increase as the distance further decreases because of nuclear-nuclear and electron-electron interaction. The interaction of particles with the same charge is repulsive. The minimum near 2 Å corresponds to the bond distance between Na and Cl in the $NaCl_{(g)}$ molecule. The energy difference noted as *D* in Figure 6–1 is the experimental bond dissociation energy.

Table 6–1 gives some bond dissociation energies for various alkali metal halides. Note that energy is required for dissociation (+*D*) and that dissociation leads to *atoms*, not ions.

6–5
$$MX_{(g)} \longrightarrow M_{(g)} + X_{(g)} \qquad \Delta E = D_{MX}$$

The variation in the D_{MX} in Table 6–1 can be accounted for in terms of the parameters *EA*, *r*, and *IP*, as shown in equation 6–4. Several trends can be

TABLE 6–1 Experimental bond dissociation energies of alkali metal halide molecules

METAL	HALIDE			
	F	Cl	Br	I
Li	137	112	100	85
Na	114	97	87	73
K	118	101	91	77
Rb	116	101	90	77
Cs	120	106	96	82

seen. For a given alkali metal, the bond dissociation energy decreases as the halide ion changes from F^- to I^-. This can be explained by the decrease in *EA* (except for fluorine to chlorine) and the increase in r_-, both of which cause a decrease in D. On the other hand, for any given halogen, the bond dissociation energy first decreases going down the group of alkali metals in Table 6–1, and then tends to increase with only a few exceptions. This seemingly unusual trend can be accounted for by the increase in r_+ which causes a decrease in *D*, while the simultaneous decrease in *IP* causes an increase in *D* which becomes dominant by the time the bottom of the alkali metal group is reached. Apparently the increase in r_+ is the dominant effect in going from the small Li^+ to Na^+, but this increase in r_+ is less dramatic in going to K^+, Rb^+, and Cs^+.

PROBLEM 6–1 Given the following data for KF, calculate a value for the ionization potential of potassium and compare it with the experimental value (Chapter 5).

electron affinity of fluorine = 81 kcal/g atom
interionic distance $KF_{(g)}$ = 2.55 Å
bond dissociation energy $KF_{(g)}$ = 118 kcal/mole

PROBLEM 6–2 Note in Table 6–1 that the energy of dissociation for LiF is some 23 kcal/mole greater than that for NaF. Obviously, the same anion F^- is involved. However, the ionization potential for Na is less than that of Li. Qualitatively account for the ordering of the bond dissociation energies of these two molecules.

6–B. COVALENT BOND

Most molecules do not have ionic bonds. Instead, bonding electrons are *shared* between the atoms of the molecule and the bond is called *covalent*. In covalent bonds, certain measurable properties associated with the electrons in the molecule, such as the ionization potential, are different in magnitude from those associated with the electrons of a single atom.

Consider the formation of a hydrogen molecule from hydrogen atoms. Figure 6–2 is a schematic plot of the potential energy as a function of distance of the system of two hydrogen atoms. At large distances, the potential energy of the system is constant (equals zero) and unchanging. As the distance between the two atoms decreases, the potential energy decreases, finally reaching a minimum at that distance corresponding to the equilibrium distance between the atoms in the real molecule. This distance is called the *bond length* and is 0.74 Å in the hydrogen molecule. As the distance continues to decrease, the potential energy begins to increase very rapidly. The initial decrease in potential energy results principally from the simultaneous attraction of the electrons in H_2 by two positive nuclei rather than by one as in the hydrogen atom. The subsequent increase in potential energy results primarily from the nuclear-nuclear repulsion interaction. The bond

FIGURE 6–2   Potential energy curve for the formation of H_2

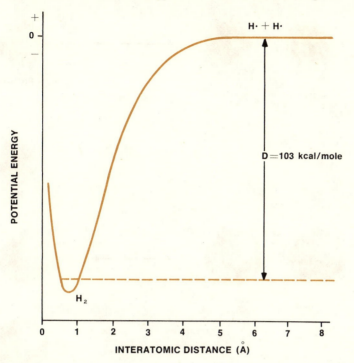

dissociation energy, D, is 103 kcal and is shown in Figure 6–2, where the reaction is

$$H_2 \longrightarrow 2H$$

It is possible to make quantum mechanical calculations on the charged species H_2^+ as well as on H_2. These are the simplest molecules on which such calculations can be performed. The bond dissociation energy, bond length charge distribution, and the shape of the potential energy curve can be quite accurately determined. In the case of individual hydrogen atoms, the electronic charge is spherically symmetrical (Chapter 4). When a hydrogen molecule, H_2, or hydrogen molecule ion, H_2^+, are formed, the electronic charge is no longer spherically distributed. In the process of bonding, the simultaneous attraction of each electron by two nuclei causes some of the electronic charge to move off the atomic centers and into the region between the two atomic centers. This can be illustrated in several ways, as shown in Figure 6–3; the charge distribution in the H_2 molecule is symmetrically distributed and no nodal planes exist. This can also be presented in an oversimplified manner as

$$\cdot H + H \cdot \longrightarrow H{:}H$$

FIGURE 6–3 The charge distribution in the H₂ molecule

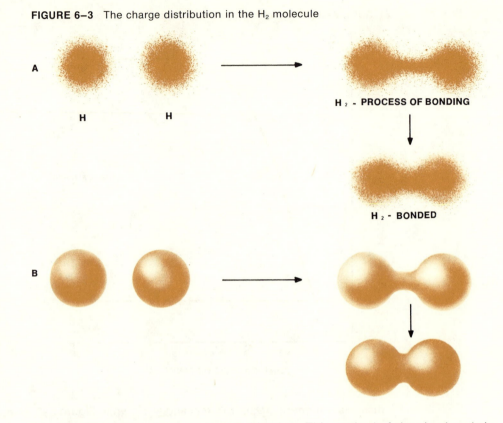

in which the electrons are represented by dots. This method of showing bonds in molecules is known as a Lewis electronic formula. When *one pair* of electrons is involved, the bond is a covalent *single bond*. We will see later that one, two and three electron pairs can be shared between atoms, and so single, double and triple bonds exist. Single, double and triple bonds also can be represented by single, double and triple lines between the atoms: H—H, O=O, N≡N. In covalent bonds, it is possible for one atom to donate both electrons to form the covalent single bond, as in the ammonium ion. This latter type of bond is often referred to as a *coordinate covalent bond*

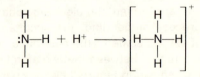

The representation of two electrons between the two hydrogen atoms is an oversimplification. In fact, the equivalent of only about 1/20 of an electron moves off each of the atomic centers to the region between the atoms, with about 1/10 of an electron responsible for the charge density between the atoms. Thus, the Lewis electronic formulation or the line description of a covalent bond is more convenient than accurate as a representation of a covalent bond, emphasizing the sharing of electrons.

Electrons cannot be distinguished from one another. For example, consider the formation of hydrogen fluoride from a hydrogen atom and a fluorine atom.

$$H\cdot + \cdot \ddot{\underset{\cdot\cdot}{F}}: \longrightarrow H:\ddot{\underset{\cdot\cdot}{F}}:$$

This description of bond formation may lead one to believe that the covalent bond in HF is made up of distinguishable H and F electrons, but the electrons are not identifiable as red, black, green, hydrogen, fluorine, sodium, etc., electrons. In other words, all electrons are the same and indistinguishable from one another, just as one covalent bond cannot be distinguished from another (as in the NH_4^+ ion).

6–C. ELECTRONIC FORMULAS (LEWIS FORMULAS)

It is very useful to use the elemental symbol with its valence shell configuration to develop criteria for the bonding in many molecules. Recall that two and eight electrons in a valence shell represent particularly stable electronic configurations for an atom. This is true not only for isolated atoms but for atoms in molecules. Also, remember it is generally the unpaired electrons in the valence shell of the atom that participate in bonding.

With these premises, Lewis electronic formulas can be easily developed for most molecules. In writing Lewis formulas remember that in all cases except hydrogen, eight electrons around each atom constitutes a particularly stable electron configuration.

EXAMPLE 6–1

What is the Lewis formula for the molecule H Cl? The chlorine atom has a valence shell configuration of 7 electrons (. . . $3s^2\,3p^5$) while hydrogen has 1 electron ($1s^1$). The one unpaired p electron in Cl combines with the unpaired s electron in H to give HCl

$$H\cdot + \cdot \ddot{\underset{\cdot\cdot}{Cl}}: \longrightarrow H:\ddot{\underset{\cdot\cdot}{Cl}}:$$

If we count the number of electrons around each atom we find there are two around H and eight around Cl. The electrons participating in the bond are counted for each atom.

Further examples are

$$H\cdot + \cdot\ddot{\underset{\cdot\cdot}{F}}: \longrightarrow H:\ddot{\underset{\cdot\cdot}{F}}: \text{ or } H{-}F$$

$$:\ddot{\underset{\cdot\cdot}{Cl}}\cdot + \cdot\ddot{\underset{\cdot\cdot}{Cl}}: \longrightarrow :\ddot{\underset{\cdot\cdot}{Cl}}:\ddot{\underset{\cdot\cdot}{Cl}}: \text{ or } Cl{-}Cl$$

$$:\ddot{\text{O}}\cdot \ + \ :\ddot{\text{O}}\cdot \ \longrightarrow \ :\ddot{\text{O}}::\ddot{\text{O}}: \ \text{ or } \ \text{O}{=}\text{O}$$

$$\text{K}\cdot \ + \ \cdot\ddot{\text{B}}\text{r}: \ \longrightarrow \ \text{K}^+:\ddot{\text{B}}\text{r}:^-$$

$$\cdot\dot{\text{C}}\cdot \ + \ 2\cdot\ddot{\text{O}}\cdot \ \longrightarrow \ :\ddot{\text{O}}::\text{C}::\ddot{\text{O}}: \ \text{ or } \ \text{O}{=}\text{C}{=}\text{O}$$

Because eight electrons around each atom is the criterion for a stable configuration, carbon dioxide is written $:\ddot{\text{O}}::\text{C}::\ddot{\text{O}}:$ and *not* $:\ddot{\text{O}}:\text{C}:\ddot{\text{O}}:$

Thus a given atom will tend to bond with other atoms such that the type and number of such atoms will result in the stable configuration of $8e^-$ surrounding each atom or $2e^-$ for hydrogen.

PROBLEM 6–3   Write the Lewis electron formulas for N_2, HBr, ICl, CO, CS_2 and LiF (ionic).

6–D. VALENCE BOND THEORY OF BONDING

The method we have chosen thus far to describe covalent bonding is known as the *valence bond theory* of bonding. According to this theory, bonding in a molecule is the result of bringing together individual atoms with their electrons. Furthermore, a pair of electrons constitutes a bond, and this electron pair is considered to belong to the particular pair of atoms that are bonded together. This theory provides a very useful description of the bonding in simple and more complex molecules, including the geometrical arrangement of the atoms. Furthermore, this description corresponds closely to the classical or conventional chemical picture of a reaction; that is, atoms (elements) react to form molecules (compounds). The valence bond theory was developed in 1927, just one year after the introduction of the wave description for electrons by Schrödinger.

The bonding between two atoms in valence bond theory occurs when a one-electron orbital of an atom overlaps a one-electron orbital of another atom. It is the valence shell electrons that are involved in the bonding. *An important criterion for bonding in this theory is that the orbitals involved have maximum overlap.* This important statement permits a qualitative prediction of the geometry of polyatomic

FIGURE 6–4 Formation of an F_2 molecule by head-to-head overlap of the *p*-orbitals of fluorine atoms

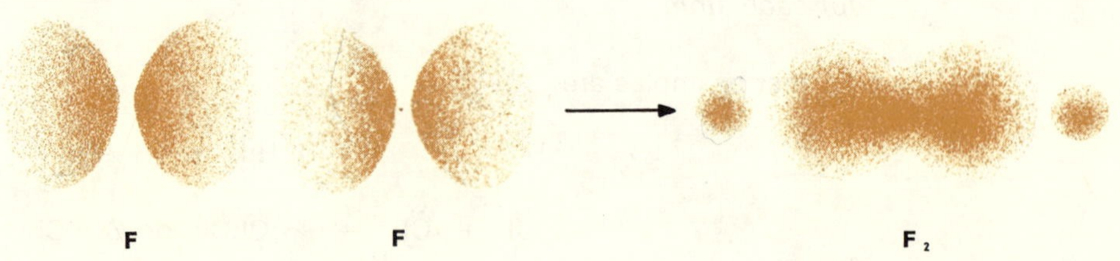

    F   F F_2

FIGURE 6–5 Formation of O_2 molecule by head-to-head and side-to-side overlap of the *p*-orbitals of oxygen atoms

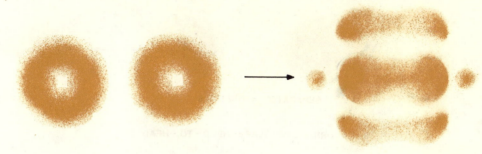

molecules to be made. Since only two atoms are involved in any diatomic molecule, a diatomic molecule must be linear.

In certain cases, bonding involving *p* orbitals occurs by a head-to-head overlap, as in F_2 in Figure 6–4. In the molecule O_2, one pair of *p* orbitals come together head-to-head and a second pair come together side-to-side, as shown in Figure 6–5.

The bonds formed by different types of electrons must be distinguished. When

1. *s*-orbitals overlap
2. *p*-orbitals overlap in a head-to-head fashion
3. *s*- and *p*-orbitals overlap

the resulting bond in all three cases is denoted as a sigma, σ, bond (Figure 6–6). Note that σ bonds do not have any nodal planes containing the internuclear axis (the bond).

When *p* orbitals overlap in a side-to-side fashion, the resulting bond is denoted pi, π. All π bonds have a nodal plane containing the nuclear axis or bond (Figure 6–6). It should be noted that in either case, σ or π, the charge distribution becomes concentrated between the atoms to form the bond. Table 6–2 summarizes this information.

TABLE 6–2 Types of molecular bonds

TYPES OF BOND	TYPE OF ATOMIC ORBITALS OR ELECTRONS INVOLVED
σ	$s + s$
σ	$s + p$
σ	$p + p$ (head-to-head)
π	$p + p$ (side-to-side)

FIGURE 6–6 Formation of σ and π bonding orbitals

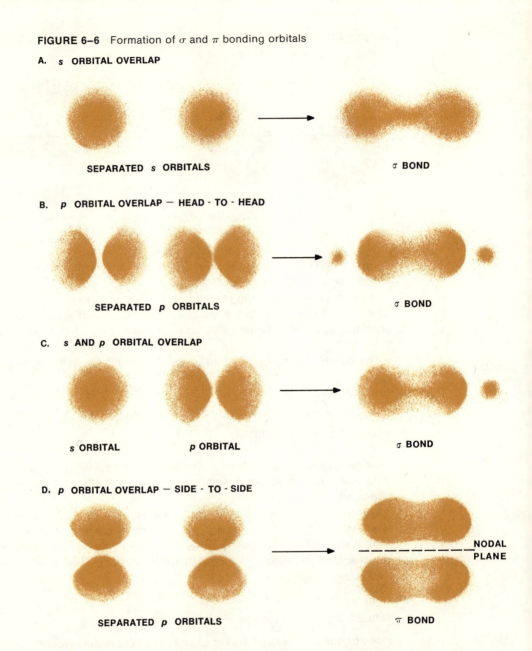

A. **s ORBITAL OVERLAP**

SEPARATED *s* ORBITALS σ BOND

B. **p ORBITAL OVERLAP — HEAD - TO - HEAD**

SEPARATED *p* ORBITALS σ BOND

C. **s AND p ORBITAL OVERLAP**

s ORBITAL   *p* ORBITAL σ BOND

D. **p ORBITAL OVERLAP — SIDE - TO - SIDE**

NODAL PLANE

SEPARATED *p* ORBITALS π BOND

One further point should be made relative to the question of how to bring *p* orbitals together. For F_2 (Figure 6–4), why were the *p* orbitals brought together head-to-head (σ), rather than side-to-side (π)? The answer to this question is that σ bonds are stronger than π bonds, and consequently, where a choice exists, a σ bond is made. If a second set of *p* orbitals can overlap, as in the case of O_2 (Figure

6–5), then because of the geometry of the remaining *p* orbitals, they must come together side-to-side to give a π bond.

Where more than one bond is formed between two atoms, this is known as a *multiple bond*. In the case of O_2, there are two bonds, a σ and a π, and this is known as a *double bond*. There are three bonds in N_2, one σ and two π, and therefore, a *triple bond* exists in N_2.

It is worth again noting the tendency to achieve a stable configuration of valence electrons around each atom in a diatomic molecule. That is, in all hydrogen containing molecules there are two electrons around hydrogen and commonly eight electrons around the other atom as

$$H:\ddot{F}: \qquad H:\ddot{Br}: \qquad H:H$$

In diatomic molecules not containing hydrogen, commonly eight valence electrons surround each atom as

$$:\ddot{Br}:\ddot{Br}: \qquad :\ddot{O}::\ddot{O}: \qquad :C:::O:$$

We can anticipate that this tendency is also true for polyatomic molecules,

$$\begin{array}{ccc} H & H & \\ H:\overset{\cdot\cdot}{\underset{H}{C}}:H & H:\overset{\cdot\cdot}{\underset{H}{N}}:H & H\overset{\cdot\cdot}{\underset{\cdot\cdot}{\overset{:\ddot{O}:}{}}}H \end{array}$$

methane   ammonia   water

Multiple bonds are stronger than single bonds. This might be expected since in order to dissociate a molecule with a double bond for example, two bonds must be broken. However, π bonds are weaker than σ bonds made from *p* orbitals. Nonetheless, the bond dissociation energies reflect the multiple character of the bond as

$$Li_2 \text{ (single bond)} \qquad D = 26 \text{ kcal/mole}$$
$$O_2 \text{ (double bond)} \qquad D = 117 \text{ kcal/mole}$$
$$N_2 \text{ (triple bond)} \qquad D = 170 \text{ kcal/mole}$$

The important point concerning valence bond theory is that molecules are made by bringing together *whole atoms* and bonds are formed by overlap of the atomic orbitals.

PROBLEM 6–4   Identify the bond as single, double, or triple for each of the molecules except LiF in Problem 6–3.

PROBLEM 6–5   What type of bonds, σ or π, will be formed when the following pairs of atoms form a molecule?
a. Na + Na   b. H + Br   c. I + Cl   d. N + N   e. I + I

PROBLEM 6-6   Why in Cl_2 are the *p*-orbitals of each chlorine atom brought together head-to-head to give a σ bond, rather than side-to-side to give a π bond?

6-E. MOLECULAR ORBITAL THEORY OF BONDING

The valence bond theory of bonding or formation of a molecule involves bringing together complete atoms which then interact to form a bond. The *molecular orbital theory* of bonding does not have the advantage of the analogy to the classical picture of a reaction between whole atoms to make a molecule. Instead, nuclei (or nuclei plus deep inner shell electrons) are brought to the position they will occupy in the final molecule and electrons are fed into polycentric molecular orbitals with definite energy. For example, in the H_2^+ molecule, the two nuclei are brought together at a distance appropriate to the experimental bond distance. Using quantum mechanics, the energy of the single electron in this potential field is calculated. The electron thus resides in a molecular orbital of a defined energy. For molecules other than H_2^+, we usually assume the solutions to be similar to those found for H_2^+, particularly regarding the shape of the molecular orbitals. Of course, the energies must be different in fact since with more than one electron, electron repulsion terms become important factors affecting the molecular orbital energy.

In this theory, the electrons are molecular electrons belonging to the molecule as a whole and residing in orbitals that encompass all of the nuclei of the molecule. It can be seen that the pictorial concept of specific electron pairs making up a bond, as in valence bond theory, is lost. However, both theories assert that the two electrons in the diatomic molecule H_2 do belong to both nuclei. A similar consideration is valid for other diatomic molecules as well in that the bonding electrons belong to both nuclei. In the case of polyatomic molecules the bonding electrons do not belong to individual atoms but are in polycentric orbitals which encompass the entire molecule.

Why use molecular orbital theory when valence bond theory provides a more attractive intuitive appeal related to the classical ideas regarding a chemical reaction to produce a molecule? The principal reason is that in the molecular orbital theory we can differentiate the various electrons by the specific, identifiable molecular orbitals they occupy. This is directly analogous to atoms where we identified the electrons in terms of the atomic orbitals which they occupied. Atomic orbitals and molecular orbitals are similar conceptual ideas. In atoms, electrons exist in orbitals of a definite energy around a *single* nucleus. In molecules, electrons exist in orbitals of definite energy around *several* nuclei which together constitute a potential energy field core.

One of the best ways to picture the generation of molecular orbitals is to combine atomic orbitals. We shall take the simplest of all molecular systems, H_2^+, which consists of two protons and one electron. The nuclei are placed at their internuclear distance in H_2^+ and the shape and energy of the molecular orbitals can be calculated by quantum mechanics. Two hydrogen atomic orbitals are com-

FIGURE 6–7 Formation of molecular orbitals from 1s atomic orbitals. The dotted line indicates a nodal plane (a plane with zero charge density).

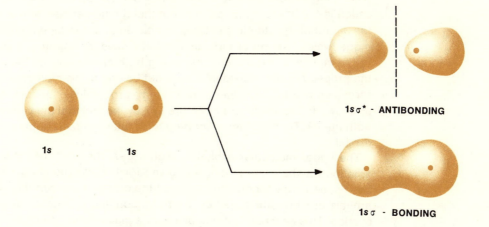

1$s\sigma^*$ - ANTIBONDING

1$s\sigma$ - BONDING

1s 1s

bined to form two molecular orbitals. After generating these molecular orbitals we can generate molecular electronic configurations for H_2^+. In addition, molecular orbital configurations can be generated for other diatomic molecules by adding a maximum of two electrons to each molecular orbital. This is analogous to the procedure used in electronic configurations for atoms where two electrons were added to each of the atomic orbitals. Again, the strongest bond is formed when the overlap between the atomic orbitals is at its maximum.

Consider further the case of H_2^+. Recall from Chapter 4 that the description of an electron in a particular atomic orbital can be given as a probability contour plot. This involves a spherical shape for s orbitals, dumbbell shape for p orbitals, etc. Molecular orbitals can be closely approximated by combining atomic orbitals in the probability contour picture. The molecular orbitals resulting from the combination of 1s atomic orbitals are shown in Figure 6–7. Note that from combining two 1s atomic orbitals two molecular orbitals are generated: 1$s\sigma$ and 1$s\sigma^*$. It is always true that combining two atomic orbitals will result in two molecular orbitals. We must now discuss how the shapes shown in Figure 6–7 arise, the source of the molecular orbital identification as 1$s\sigma$ and 1$s\sigma^*$, and the meaning of the terms bonding and antibonding.

Electrons have a wave nature; particular atomic orbital electrons have particular probability contour plots associated with them (e.g., 1s orbital electrons in Figure 6–7). The lower-energy molecular orbital is generated when there is overlap between the 1s atomic orbitals. This could be viewed in a sequence as

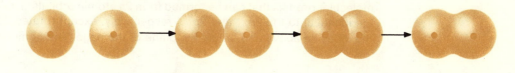

Just as for atoms, there are contours of constant charge density or probability. Since the molecular orbitals are made from combining $1s$ orbitals, and since combining s orbitals make σ orbitals (Section 6–D), the molecular orbital identification is $1s\sigma$. It should be pointed out that the nomenclature σ and π is a result of the *nature* of the atomic electrons interacted and is independent of the theoretical model, valence bond or molecular orbital, used for bonding. Also, overlap results in the molecular orbital encompassing both nuclei, so that the electrons are not the property of the individual atoms but belong to the molecule as a whole. Bond formation results in the electrons having a probability of being between the atoms, as shown in the lower part of Figure 6–7 by the fact that the contour encompasses both nuclei. Thus, the term *bonding* molecular orbital is applied to this molecular orbital.

The upper molecular orbital in Figure 6–7, the $1s\sigma^*$, has the $1s\sigma$ part of the notation for the same reasons as given for the $1s\sigma$ (bonding) molecular orbital. The * superscript denotes the fact that the orbital is *antibonding*. When the two $1s$ orbitals are brought together in an antibonding molecular orbital, they do *not* overlap. This results in a nodal plane, or a place of zero electron density, between the atoms. The resulting $1s\sigma^*$ orbital has a region between the atoms where the contours are not continuous, meaning that no contour encompasses both nuclei. This orbital has a nodal plane; as a matter of fact, the electrons in the antibonding orbital are repelled away from the region between the nuclei. This is pictorially represented by the egg shape with a greater probability of finding the electron toward the backside of each of the nuclei (Figure 6–7). The lack of overlap results in a situation which is *against* bonding or is antibonding.

A schematic energy level diagram relating the energy of the $1s$ atomic and the $1s\sigma$ and $1s\sigma^*$ molecular orbitals is shown in Figure 6–8. It should be noted that the $1s\sigma^*$ (antibonding) orbital is shifted ΔE_a above the energy of the $1s$ atomic orbital energies and the $1s\sigma$ (bonding) orbital is ΔE_b below the $1s$ atomic orbital energies. The absolute value of ΔE_a is greater than that of ΔE_b. This means that a single electron in the $1s\sigma^*$ orbital destabilizes a molecule more than a single electron in the $1s\sigma$ orbital stabilizes the molecule. Therefore, if we tried to make a hydrogen molecule, H_2, by putting one of the two molecular electrons in the $1s\sigma$ molecular orbital and one in the $1s\sigma^*$ molecular orbital, the total energy would be greater (less negative) than the energy of the atoms from which the hydrogen molecule was made. The resultant would give an overall instability and H_2 would immediately separate to make hydrogen atoms. The antibonding orbital is more destabilizing (repulsive) to bonding than the bonding orbital is stabilizing (attractive) to bonding (see Figure 6–8). The relative stabilizing and destabilizing property of the bonding and antibonding orbitals determines which diatomic molecules exist.

Other molecular orbitals of interest to us are those formed by the interaction of p orbitals, in particular the $2p$ orbitals. Figure 6–9 shows the two different types of molecular orbitals that can be formed from $2p$ atomic orbitals. The $2p\pi_z$ and $2p\pi_z^*$ are identical to the $2p\pi_y$ and $2p\pi_y^*$ respectively, except rotated 90°, and therefore the lobes lie above and below the plane of the page. The nomenclature arises as

FIGURE 6–8   Relative energies of 1s, 1sσ and 1sσ* orbitals

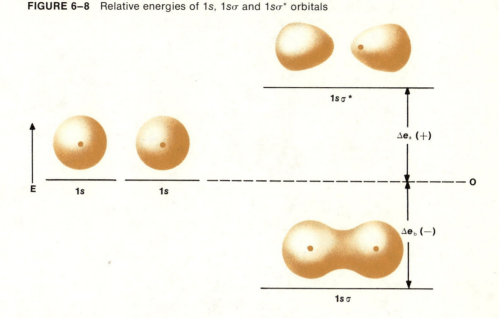

follows. The $2p\pi$ and $2p\pi^*$ molecular orbitals arise from $2p$ atomic orbitals which when brought together side-to-side give π orbitals (Section 6–D). The * superscript indicates that the orbital is antibonding. The bonding π orbital encompasses the space between the atoms, whereas there is a nodal plane or a region of no charge density between the atoms for the antibonding orbital. In the antibonding orbital there is a bulge in the lobes away from the atomic centers (and away from the bond axis), indicating a repulsion away from the potentially bonding region. This is a parallel situation to that seen for the $1s\sigma^*$ orbital.

In the case of the σ molecular orbitals made from $2p$ atomic orbitals, the $2p\sigma$ nomenclature arises because $2p$ atomic orbitals were brought together in a head-to-head fashion to give σ orbitals. The * again means antibonding. Once again, note in Figure 6–9 that in the $2p\sigma$ molecular orbital, charge density exists between the centers. In the case of the $2p\sigma^*$ molecular orbital in Figure 6–9, there is a nodal plane or region of no charge density between the centers. Further, the repulsive nature is shown by the fact that the lobes between the nuclei are small whereas those on the outside of the nuclei are larger and bulge away from the nuclei.

Figure 6–10 shows the relative energies of the molecular orbitals arising from the three $2p$ atomic orbitals. There is no one correct energy order of atomic orbital energies appropriate for all atoms. The same is true for the order of molecular orbital energies for diatomic molecules. In particular, the $2p\sigma$ and two $2p\pi$ orbitals can be exchanged in energy order depending upon the internuclear distance. Note also that the $2p\pi_y$ and $2p\pi_z$ orbitals are degenerate (as are the

FIGURE 6–9 Formation of molecular orbitals from $2p$ atomic orbitals. The dashed lines indicate nodal planes (planes of no charge density).

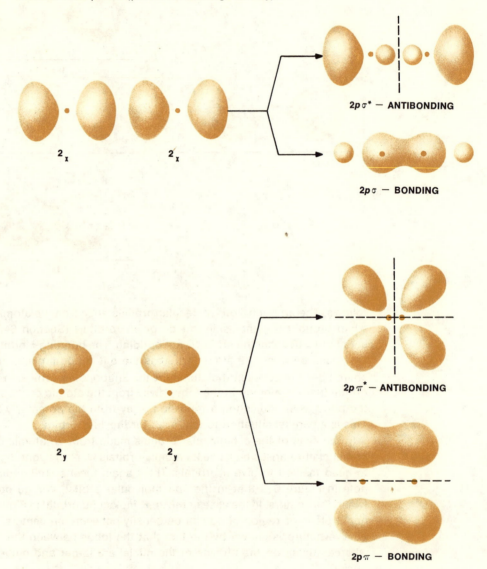

$2p\pi_y{}^*$ and $2p\pi_z{}^*$ orbitals). Degeneracy was encountered in Chapter 4 for the $2p$ atomic orbitals. There is no energy difference between the $2p_y$ and $2p_z$ orbitals because of the spatial geometry; therefore, no energy difference exists between the $2p\pi$ molecular orbitals generated from them. (The same is true for the $2p\pi^*$ molecular orbitals.)

It is possible to generate other molecular orbitals in a similar fashion using different atomic orbital contour representations, such as 2s. In this instance we would generate $2s\sigma$ and $2s\sigma^*$ molecular orbitals. We can then put all of the molecular orbitals on the same energy diagram and represent the order of molecular orbital energies as shown below and also in Figure 6–11.

$$1s\sigma < 1s\sigma^* < 2s\sigma < 2s\sigma^* < 2p\pi_y = 2p\pi_z < 2p\sigma < 2p\pi_y^* = 2p\pi_z^* < 2p\sigma^*$$

This order is not invariant and does depend on nuclear charge, electronic interactions, and internuclear distance. It is possible for the $2p\pi_y$ and $2p\pi_z$ orbitals to be above the $2p\sigma$ orbital. Based on the above order of molecular orbitals, it is possible to deduce molecular electron configurations. We simply count the number of electrons that exist in the molecule. Next we add these electrons to the individual molecular orbitals according to their energy—filling the lowest energy molecular orbital first. According to the Pauli principle, we can have two electrons

FIGURE 6–10   Relative energies of $2p$, $2p\sigma$ and $2p\pi$ orbitals

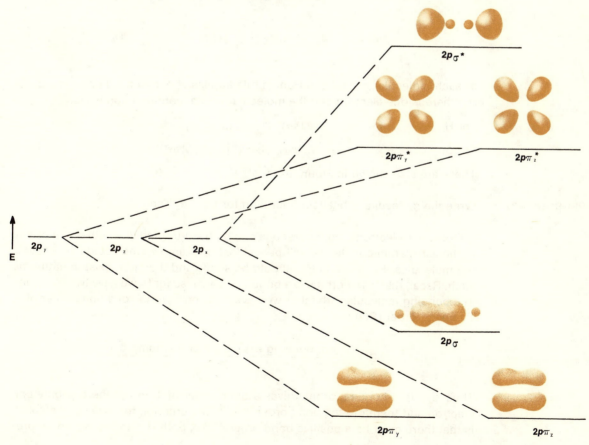

FIGURE 6–11 Molecular orbital energy order with a few molecular electronic configurations

$2p\sigma^*$

$2p\pi_y^*$ $2p\pi_z^*$

$2p\sigma$

$2p\pi_y$ $2p\pi_z$

E $2s\sigma^*$ ⇅

$2s\sigma$ ⇅

$1s\sigma^*$ ⇅

$1s\sigma$ ↑   ⇅   ⇅
                             H_2^+ H_2 Be_2

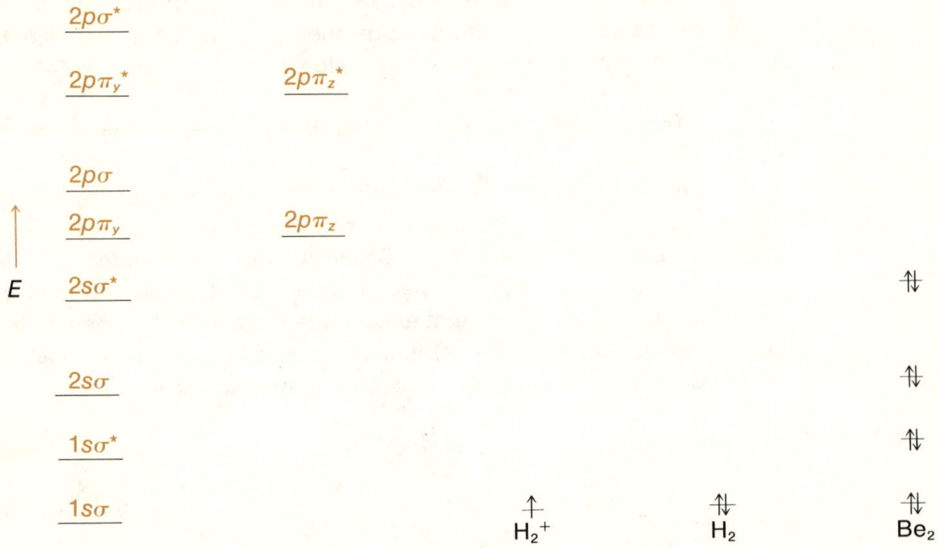

in each molecular orbital and Hund's rule applies as for atoms. For example, for H_2^+ there is one electron and the molecular orbital configuration is $(1s\sigma)^1$

For H_2 $(1s\sigma)^1 (1s\sigma)^1$ or $(1s\sigma)^2$

For Be_2 $(1s\sigma)^2 (1s\sigma^*)^2 (2s\sigma)^2 (2s\sigma^*)^2$

These are also shown in Figure 6–11.

PROBLEM 6–7   Write the molecular orbital configuration for Li_2, He_2^+, and B_2.

Molecular electronic configurations can be used to predict whether or not any particular diatomic molecule will be stable. For example, one could ask whether the molecules, H_2^+, H_2, and He_2 should be stable and therefore exist in nature as such. Recall that a $1s\sigma$ orbital is a bonding orbital, so for H_2 we have two electrons in a bonding molecular orbital or we have 1 "bond's worth" or a bond order of 1. The bond order (*BO*) is, therefore,

6–6 $$BO = \frac{(\text{no. bonding } e^-) - (\text{no. antibonding } e^-)}{2}$$

Thus, $2e^-$ in a bonding orbital gives a bond order of 1. In H_2^+ there is only one electron and therefore the bond order is 1/2. The criterion for molecular stability is that there must be a positive bond order. Thus, both H_2^+ and H_2 would be predicted to be stable with bond orders of 1/2 and 1, respectively. This is in agreement with experiment.

In the case of He$_2$ we have one bond from two electrons in the bonding molecular orbital, $1s\sigma$, and one antibond from two electrons in the antibonding molecular orbital, $1s\sigma^*$. The bond is cancelled by the antibond with an apparent zero resultant (or bond order = 0), and no bonding results. In fact, an antibond destabilizes more than a bond stabilizes. Thus, the destabilization energy is even greater than expected from the simple addition giving zero bond order. It would be predicted, therefore, that He$_2$ molecules should not exist at ordinary temperatures and that helium should exist in the atomic state as He. This is in agreement with experiment. Table 6–3 summarizes the situation for H$_2^+$, H$_2$, and He$_2$.

The valence bond theory made it obvious that H$_2$ could be formed by bringing two complete H atoms together to form a covalent bond

$$H\cdot + H\cdot \longrightarrow H:H$$

It was not so obvious that you could not do the same thing for helium

$$He: + He: \longrightarrow He::He$$

However, the explanation for the non-existence of He$_2$ arises naturally and easily from a consideration of the molecular orbital theory of bonding. It is important to note that in molecular orbital theory we do not insist that a bond be made up of an electron pair. Thus, H$_2^+$ is a stable molecule containing only one electron in a bonding molecular orbital ($BO = 1/2$). Recall that valence bond theory in its simplest, most straightforward form requires that an electron *pair* constitutes a bond.

The molecule O$_2$ has 16 electrons, and taking into account Hund's Rule (Section 4–L) it has the following predicted electronic configuration:

$$(1s\sigma)^2 (1s\sigma^*)^2 (2s\sigma)^2 (2s\sigma^*)^2 (2p\sigma)^2 (2p\pi_y)^2 (2p\pi_z)^2 (2p\pi_y^*)^1 (2p\pi_z^*)^1$$

Note that because of Hund's Rule, there is one electron in each of the $2p\pi_y^*$ and $2p\pi_z^*$ orbitals with parallel spins. Any substance that has molecules or ions that

TABLE 6–3 Molecular electron configurations and stability

MOLECULE	MOLECULAR ELECTRON CONFIGURATION	NUMBER BONDING e⁻	NUMBER ANTIBONDING e⁻	BOND ORDER (*BO*)	PREDICTED STABILITY
H$_2^+$	$(1s)^1$	1	0	1/2	stable
H$_2$	$(1s)^2$	2	0	1	stable
He$_2$	$(1s)^2(1s^*)^2$	2	2	0	unstable

contain a single electron or two or more electrons with parallel spins shows the property of *paramagnetism*. For our purposes, this property means that the substance will be attracted into a magnetic field. Thus, according to our predicted electronic configuration, O_2 should be paramagnetic. This is in agreement with experimental studies. This prediction was one of the early triumphs of molecular orbital theory. It is not possible to make such a prediction using valence bond theory, where a bond is represented by an electron pair.

PROBLEM 6–8 Write the molecular electronic configuration for the following molecules and predict whether they will be stable or unstable and by how much (in terms of the bond order): He_2^+, Li_2, Be_2, and Be_2^+.

We have said that it might be helpful to identify electrons according to the molecular orbitals they occupy. The concepts of bonding and antibonding molecular orbitals permit us to predict, as well as understand, the basis for the existence or non-existence of certain diatomic molecules. Furthermore, if we know the assignment of electrons to molecular orbitals, we can determine what happens when individual electrons are excited from one molecular orbital energy level to another molecular orbital energy level. For example, in H_2 the orbital configuration was determined to be

$$H_2 \text{ (ground state)} \quad (1s\sigma)^2$$

By the use of an external energy source, one of these electrons could be excited to a higher energy molecular orbital, for example

$$H_2 \text{ (excited)} \quad (1s\sigma)^1 (1s\sigma^*)^1$$

If the external energy source is light, absorption takes place and we have a molecular absorption spectrum. Of course, an electron can be excited to any of the other higher energy molecular orbitals as well:

$$H_2 \text{ (excited)} \quad (1s\sigma)^1 (2s\sigma)^1$$

This process is entirely analogous to that producing atomic spectra. In the latter process, an electron is excited to a higher energy atomic orbital and an atomic absorption spectrum is produced.

Earlier we discussed the result of trying to make H_2 by putting one electron in a $1s\sigma$ orbital and the other in a $1s\sigma^*$ orbital—it would immediately fly apart. Actually, of course, the way to "make" such H_2 is to begin with real H_2 and put in energy to excite an electron from $1s\sigma \longrightarrow 1s\sigma^*$. Such a condition is an excited state for the H_2 molecule and would be expected to result in dissociation of H_2 into H atoms[‡].

[‡]It is known that dissociation does occur when the electrons have parallel spins

$$H_2 \text{ (excited)} \quad (\uparrow) \ (\uparrow)$$
$$1s\sigma \ \ 1s\sigma^*$$

When light absorption results in electronic excitation to a stable excited state (resistant to dissociation), after a finite time period the electron returns to its original lower energy orbital with light emission. This process results in an emission spectrum, parallel to the situation possible in atoms.

The fact that molecules can undergo electronic excitation by light absorption is important in several respects. It is possible to learn about the structure of a molecule through studies of the absorption and emission spectra (Chapter 19). The fundamental prerequisite of photochemistry is that the molecule absorb light and be raised to an electronic excited state. While in this excited state it is possible for a bond to break in a diatomic molecule. In polyatomic molecules, (1) a bond(s) can break and an atom(s) be ejected, (2) the molecule can rearrange to a different geometrical shape, and (3) bonds can be formed such that an entirely different type molecule is made. These reactions can be represented as follows:

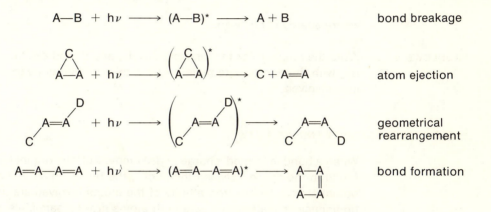

$A—B + h\nu \longrightarrow (A—B)^* \longrightarrow A + B$	bond breakage
	atom ejection
	geometrical rearrangement
	bond formation

Also, the absorbed energy can be transferred to another molecule which undergoes reaction. Many molecular processes, such as photosynthesis and certain aspects of smog production (Chapter 19), require light absorption. Deterioration of many plastics occurs through absorption of ultraviolet light followed by photochemical reactions (Chapter 19). One critically important photochemical reaction is

$$3O_2 \xrightleftharpoons{\text{light}} 2O_3$$

which occurs high above the earth's surface. This reaction requires high energy ultraviolet light, and therefore, its occurrence acts as a screen to prevent high energy ultraviolet light from reaching the earth's surface. If it were not for this reaction, it is likely that life on earth could not exist, since this ultraviolet light would cause highly destructive photochemical reactions in all biological systems. There has been considerable discussion concerning potential modification of this reaction if high flying supersonic jets were to exist in any substantial number.

Finally, the concepts of ionization potential and electron affinity can be applied to molecules as they were to atoms. Thus, the ionization potential (first) is the energy required to remove an electron from the highest filled molecular orbital. For example,

$$H_2 \ (1s\sigma)^2 \longrightarrow H_2^+ \ (1s\sigma)^1 + e^-$$

where 356 kcal/mole is required for the ionization. Also, the electron affinity is the energy released when an electron is added to a molecule (commonly considered to be located in the lowest-energy empty molecular orbital). For example,

$$F_2 \ (1s\sigma)^2 \ (1s\sigma^*)^2 \ (2s\sigma)^2 \ (2s\sigma^*)^2 \ (2p\pi_y)^2 \ (2p\pi_z)^2 \ (2p\sigma)^2 \ (2p\pi_y^*)^2 \ (2p\pi_z^*)^2 + e^- \longrightarrow$$

$$F_2^- \ (1s\sigma)^2 \ (1s\sigma^*)^2 \ (2s\sigma)^2 \ (2s\sigma^*)^2 \ (2p\pi_y)^2 \ (2p\pi_z)^2 \ (2p\sigma)^2 \ (2p\pi_y^*)^2 \ (2p\pi_z^*)^2 \ (2p\sigma^*)^1$$

where energy is released.

PROBLEM 6–9 Write the reaction for the ionization of He_2 and B_2 and electron addition to H_2 and Be_2 with the appropriate molecular orbital configurations for the molecules and ions involved.

6–F. ELECTRONEGATIVITY AND POLARITY

When a bond is formed, charge density moves off the original atom centers and is found in the space between the nuclei, the bond region. Because the ionization potential and the electron affinity of the atoms involved are different, except for homonuclear diatomics where both atoms are the same, electron density is not uniformly distributed in the bonding region. The relative ability of a bonded atom to attract the charge cloud or electron pair in a covalent bond was defined earlier as electronegativity, χ.

DEFINITION OF ELECTRONEGATIVITY—MULLIKEN VS. PAULING

The simplest, clearest quantitative definition of the electronegativity of atoms is that given by R. Mulliken

6–7
$$\chi = \frac{\text{ionization potential} + \text{electron affinity}}{2} = \frac{IP + EA}{2}$$

Although ionization potential data is generally available for almost all elements, the electron affinities of only a few elements are known. A quite complete table of electron affinities was given in Chapter 5; however, some of these are from theoretical or extrapolated data. L. Pauling first suggested on empirical grounds that the difference between the electronegativity of two atoms A and B could be satisfactorily derived from a consideration of the bond energy of the molecule A—B (see Section 6–H). The concept of electronegativity, then, has an experimental basis in terms of *IP* and *EA* (Mulliken) or bond energy (Pauling). Furthermore, the Pauling values of electronegativities of atoms can be related to those of the more fundamental Mulliken definition, with only a difference in absolute magnitude. Figure 6–12 presents electronegativity values based on the Pauling definition, because only a few values are available based on the Mulliken definition. We can deduce several important general features from Figure 6–12.

1. The nonmetals, particularly nitrogen, oxygen, fluorine, sulfur, chlorine, bromine, and iodine, have the highest electronegativity, while the metals have the lowest electronegativity.
2. The electronegativity increases going from left to right within a period; for example, in the second period: lithium (1.0), beryllium (1.5), carbon (2.5), nitrogen (3.0), oxygen (3.5), fluorine (4.0).
3. The electronegativity decreases going from the top of a group to the bottom of a group, as in group VII A: fluorine (4.0), chlorine (3.0), bromine (2.8), iodine (2.5).

FIGURE 6–12 Electronegativities of the elements in their common oxidation states[a]

IA	IIA	IIIB	IVB	VB	VIB	VIIB	VIIIB			IB	IIB	IIIA	IVA	VA	VIA	VIIA	O
1 H 2.1																	2 He —
3 Li 1.0	4 Be 1.5											5 B 2.0	6 C 2.5	7 N 3.0	8 O 3.5	9 F 4.0	10 Ne —
11 Na 0.9	12 Mg 1.2											13 Al 1.5	14 Si 1.8	15 P 2.1	16 S 2.5	17 Cl 3.0	18 Ar —
19 K 0.8	20 Ca 1.0	21 Sc 1.3	22 Ti 1.5	23 V 1.6	24 Cr 1.6	25 Mn 1.5	26 Fe 1.8	27 Co 1.8	28 Ni 1.8	29 Cu 1.9	30 Zn 1.6	31 Ga 1.6	32 Ge 1.8	33 As 2.0	34 Se 2.4	35 Br 2.8	36 Kr —
37 Rb 0.8	38 Sr 1.0	39 Y 1.2	40 Zr 1.4	41 Nb 1.6	42 Mo 1.8	43 Tc 1.9	44 Ru 2.2	45 Rh 2.2	46 Pd 2.2	47 Ag 1.9	48 Cd 1.7	49 In 1.7	50 Sn 1.8	51 Sb 1.9	52 Te 2.1	53 I 2.5	54 Xe —
55 Cs 0.7	56 Ba 0.9	57–71 La–Lu 1.1–1.2	72 Hf 1.3	73 Ta 1.5	74 W 1.7	75 Re 1.9	76 Os 2.2	77 Ir 2.2	78 Pt 2.2	79 Au 2.4	80 Hg 1.9	81 Tl 1.8	82 Pb 1.8	83 Bi 1.9	84 Po 2.0	85 At 2.2	86 Rn —
87 Fr 0.7	88 Ra 0.9	89– Ac– 1.1–1.7															

[a]Values from L. Pauling, *Nature of the Chemical Bond*, third edition, Cornell University Press, Ithaca, N.Y., 1960.

The trends in electronegativity are based on the expected values and trends for ionization potential and electron affinity of metals and nonmetals (see Sections 5-G and 5-H).

In covalent homonuclear diatomic molecules, such as H_2, O_2, N_2, and Cl_2, the electronic charge is symmetrically distributed in the bond. Since both atoms are the same, the electronegativity of each is the same and there is no tendency for charge to be displaced towards one atom or the other. However, in covalent heteronuclear diatomics such as HCl and HI, the difference in electronegativity between the atoms results in charge displacement and therefore charge asymmetry. Of course, the charge displacement will be towards the atom with the higher electronegativity. This charge asymmetry results in a dipole moment, μ, which means that one end of the molecule is relatively negative compared to the other. This can be shown as

$$\overset{\delta^+ \quad \delta^-}{\text{H}\!-\!\text{Cl}} \quad \text{or} \quad \overset{\longleftarrow}{\text{H}\!-\!\text{Cl}}$$

The Greek letters δ^+ and δ^- indicate a small relatively positive and relatively negative charge resulting from the charge asymmetry because $\chi_H < \chi_{Cl}$. The symbol $\leftrightarrow$ represents a vector with the arrow pointing to the negative end. The dipole moment is equal to the charge (q) in electrostatic units times the distance (d) in centimeters between the positive and negative centers.

6-8
$$\mu = q \times d$$

The unit commonly used is the *debye*, D, where 1D is 10^{-18} esu cm. For example, if approximately 0.2 of an electron ($0.2 \times 4.8 \times 10^{-10}$ esu) is placed 1 Å (10^{-8} cm) from a positive charge of the same magnitude, then the dipole moment is 1 D.

The relative electronegativity of atoms in a diatomic molecule determines the charge asymmetry and is responsible for the dipole moment. The difference in electronegativity of two atoms A and B should be nearly equal to the dipole moment of the molecule AB (μ_{AB}). Table 6–4 shows that this relation holds remarkably well for the hydrogen halides.

The dipole moment is of particular importance to us because of the effect which the interaction of molecular dipoles has upon such properties as solubility, boiling point and melting point (Chapter 8). Polyatomic molecules may or may not have

TABLE 6–4 Dipole moments and electronegativity differences of the hydrogen halides.

	HF	HCl	HBr	HI
$\chi_A - \chi_B$	1·9	0·9	0·7	0·4
μ (Debyes)	1·82	1·09	0·79	0·38

dipole moments even if different atoms are involved. The geometry of the poly-atomic molecule is the critical factor determining whether a dipole moment exists (see Chapter 7).

PROBLEM 6–10   Predict whether the following molecules will have a dipole moment and in which direction.
a. N_2 b. ICl   c. F_2 d. BrF

6–G. COVALENT VERSUS IONIC BONDING

In the section on electronegativity, reference was made to covalent molecules and the associated dipole moments; the larger the difference in χ_A and χ_B, the greater the dipole moment. If we react two atoms having very different electronegativities, then instead of electron sharing (covalent bond) we shall obtain essential electron transfer (ionic bond). Thus, we should be able to predict whether a molecule will be covalent of ionic. Furthermore, if the molecule is covalent, we can predict the direction of the dipole and roughly its magnitude. Generally, if the absolute difference in electronegativity, $|\chi_A - \chi_B|$, is greater than 2, the compound will be considered ionic. Table 6–5 gives examples of the correlation of the difference between χ_A and χ_B, some dipole moment values, and whether the compound is considered to be ionic or covalent.

TABLE 6–5   Correlation of electronegativity, dipole moment and bond type.

MOLECULE	$\chi_A - \chi_B$	DIPOLE MOMENT	DOMINANT BOND TYPE
H_2	0	0	covalent
Cl_2	0	0	covalent
N_2	0	0	covalent
HF	1.9	1.8	covalent
HCl	0.9	1.09	covalent
HBr	0.7	0.79	covalent
HI	0.4	0.38	covalent
KCl	2.2	8.0	ionic
CsCl	2.3	10.5	ionic
KF	3.2	8.6	ionic
BaO	2.6	—	ionic
CsF	3.3	—	ionic
ClF	1.0	0.83	covalent
BrF	1.2	1.29	covalent
ICl	0.5	—	covalent

As noted earlier, oxidation number refers to the apparent value of the positive or negative charge on an atom in a covalent compound. Ionic charge refers to the actual value of the positive or negative charge of a monatomic ion in an ionic compound or in solution. The assignment of oxidation number depends on the relative electronegativity of the atoms involved; that is, the electrons making up the bond are assigned to the more electronegative element. The number of electrons in excess of that normally present in the neutral atom constitutes the negative oxidation number for that atom. Of course, the other atom attains a positive oxidation number equal to the number of electrons lost by assignment to the other atom. For example, in HCl, chlorine is the more electronegative so both electrons in the bond are assigned to chlorine, making a total of 8. This is one electron in excess of those present in the free atom so chlorine has an oxidation number of -1. Conversely, hydrogen has an oxidation number of $+1$.

PROBLEM 6–11   Predict whether the following compounds will be ionic or covalent and what will be the ion charge or oxidation number of each of the elements: CaO, ClF, NO, CO, KCl, HI, SrO, NaH.

The application of the definitions of ionic charge and oxidation number is not restricted to diatomic species. Table 6–6 shows several polyatomic compounds with the ionic charge or oxidation number assigned to various atoms.

6–H. ELECTRONEGATIVITY AND BOND ENERGY

In general, we find that for a heteronuclear diatomic molecule A—B the bond dissociation energy ($D_{A—B}$) is greater than that expected from the mean of the two bond dissociation energies for A—A and B—B:

6–9
$$D_{A—B} > \frac{D_{A—A} + D_{B—B}}{2}$$

We might at first expect that the bond energy would be equal to the mean if the electron distribution were symmetrical (electron pair in the center). However, we know that when there are different atoms involved, the charge distribution is not symmetrical (or the electron pair is not in the center), because of the difference between χ_A and χ_B.

It has been observed that the difference in the bond energy from the mean of $D_{A—A}$ and $D_{B—B}$ increases as the difference in electronegativity of A and B increases. Equation 6–10 is the empirically derived relationship.

6–10
$$\left(D_{A—B} - \frac{D_{A—A} + D_{B—B}}{2} \right) = (|\chi_A - \chi_B|)^2$$

where $|\chi_A - \chi_B|$ indicates the absolute value of the difference in electronegativity,

TABLE 6–6   Ionic charge and oxidation number of the atoms in several compounds

COMPOUND	BOND TYPE	MOST ELECTRO-NEGATIVE ATOM	IONIC CHARGE		OXIDATION NUMBER FOR COVALENT COMPOUNDS	
Li_2O	ionic	O	Li^+	+1	—	
			O	−2	—	
BeO	ionic	O	Be	−2	—	
			O	−2	—	
CO_2	covalent	O	—		C	+4
			—		O	−2
H_2S	covalent	S	—		H	+1
			—		S	−2
CCl_4	covalent	Cl	—		C	+4
			—		Cl	−1
NaF	ionic	F	Na	+1	—	
			F	−1	—	
SO_2	covalent	O	—		S	+4
			—		O	−2
BF_3	covalent	F	—		B	+3
			—		F	−1
OF_2	covalent	F	—		O	+2
			—		F	−1
NH_3	covalent	N	—		H	+1
			—		N	−3
H_2O	covalent	O	—		H	+1
			—		O	−2
PbF_2	ionic	F	Pb	+2	—	
			F	−1	—	

and D_{A-B}, D_{A-A}, and D_{B-B} are in electron volts. This is the basis Pauling used to establish the χ values for the elements. For example, consider H—F. The bond dissociation energy calculated from the mean, $\bar{D}_{H-F}$, is

$$\bar{D}_{H-F} = \frac{D_{H-H} + D_{F-F}}{2}$$

$$= \frac{104.2 + 36.6}{2}$$

$$= 70.4 \text{ kcal/mole}$$

Actually, the bond dissociation energy for H—F is 134.6 kcal/mole, some 64 kcal

TABLE 6-7   Bond dissociation energies of some diatomic molecules[a]

MOLECULE	BOND ENERGY (D_{A-B}) (kcal/mole)	$\dfrac{D_{A-A} + D_{B-B}}{2}$ (kcal/mole)	Δ (kcal/mole)	$\lvert \chi_A - \chi_B \rvert$
H—H	104.2			
F—F	36.6			
Cl—Cl	58.0			
Br—Br	46.1			
I—I	36.1			
H—F	134.6	70.4	64.2	1.9
H—Cl	103.2	81.1	22.1	0.9
H—Br	87.5	75.2	12.3	0.7
Cl—F	60.6	47.3	13.3	1.0
I—Cl	50.3	47.1	3.2	0.5

[a]Note that equation 6-11 is not uniformly applicable to all compounds since a geometric mean often is required to calculate Δ instead of an arithmetic mean.

greater than expected from the simple mean. Table 6-7 gives other examples where a difference, Δ, exists.

6-11
$$\Delta = \left(D_{A-B} - \frac{D_{A-A} + D_{B-B}}{2} \right)$$

6-I. RESONANCE

The formulation A:B denotes that there is a symmetrical charge distribution. As we have seen, this is actually not always true; a difference between χ_A and χ_B creates an unsymmetrical distribution. We are able to improve our description of the electron charge distribution most clearly within the valence bond approximation by employing the concept of *resonance* or *resonance structures*. The individual resonance structures represent alternative ways to describe the bonding or electronic structure of the molecule. As an example, consider HCl. Possible descriptions of the electron distribution in the bond can be:

H:Cl   or   (H)$^+$(:Cl)$^-$ or   (H:)$^-$ (Cl)$^+$

I   II   III

Based on earlier considerations, we would expect HCl to be purely covalent which corresponds to the structure H:Cl. However, the important question is whether this description is accurate. The answer to this is no—the simple description is too naive. On the other hand, the purely ionic structures $(H)^+(Cl)^-$ and $(H)^-(Cl)^+$ are also inaccurate and, by themselves, are incorrect; HCl is not an ionic compound. The idea of resonance is to combine structures I, II and III, none of which individually describe accurately the bonding, but which come close to the correct description when superimposed in the proper proportion. The final resulting structure is often called the *resonance hybrid*. The best description of the bonding in HCl within the valence bond approximation involves a superposition of the electron distributions of structures I and II and III and is written as

$$H\!-\!Cl \longleftrightarrow H^+Cl^- \longleftrightarrow H^-Cl^+$$

Each of these is known as a resonance structure. It is important to point out that there are *not* three different kinds of HCl molecules, two ionic and one covalent, each existing for some finite time. There is only one kind of HCl and the above structures when combined provide a closer description of the true electron distribution or bonding than does any one structure alone. Resonance is symbolized by the double headed arrow ($\longleftrightarrow$). The final description, the resonance hybrid, is arrived at through a quantum mechanical calculation which determines the fraction of each of I, II and III that gives the lowest energy and thereby gives the best approximation to the electronic distribution in the bond.

We know HCl has a dipole moment of 1.07 D. Furthermore, because of the fact that the electronegativity of chlorine is greater than that of hydrogen, we would predict the existence of a dipole. Therefore, even though both the ionic resonance structures H^+Cl^- and H^-Cl^+ are required in principle, the fractional contribution of H^+Cl^- must be considerably greater than that of H^-Cl^+. This is generally true in all cases where there is a substantial difference in the electronegativity of the atoms involved; the chance of finding both electrons on the less electronegative atom is very small. Because $\chi_{Cl} \gg \chi_H$, the resonance structure description of HCl would be essentially

$$\underset{\text{I}}{H\!-\!Cl} \longleftrightarrow \underset{\text{II}}{H^+Cl^-}$$

The fractional contribution of the resonance structure II has been calculated to be 0.17. This means that the HCl bond has 17% ionic character. Knowing this, we would predict that HCl has a dipole moment with the negative end toward Cl. This prediction is in agreement with experiment. The larger is $|\chi_A - \chi_B|$ for a molecule AB, the greater will be the relative contribution of the ionic structure and the greater the percentage ionic character.

We can represent that the resonance hybrid of $H\!-\!Cl$ and H^+Cl^- as

$$\overset{\delta^+ \quad \delta^-}{H\!-\!Cl}$$

FIGURE 6–13 Energy relationship between the resonance structures and resonance hybrid of HCl. The energy of the resonance hybrid is always lower than that of any individual resonance structure.

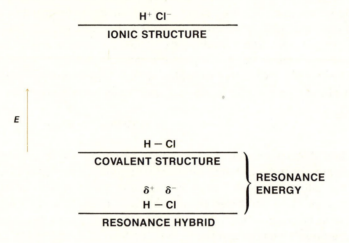

where the bond represents the covalent bonding (from H—Cl) and the δ^+ and δ^- indicate that charge assymmetry exists (from H$^+$Cl$^-$). Note that in Figure 6–13 the calculated energy of the resonance hybrid is lower than that of either of the individual resonance structures alone. We would expect this since neither one of the resonance structures alone provides the correct description of the bonding. The difference in energy between the singular resonance structure of lowest energy and that of the resonance hybrid is known as the *resonance energy*. It may be looked at as the energy lowering obtained by "mixing" the ionic resonance structure with the covalent resonance structure. The description of HCl is applicable to any other molecular system. In particular

1. any individual resonance structure will always be higher in energy than the final resonance hybrid and
2. the resonance hybrid consists of some fractional mixture of each resonance structure such that a minimum of energy is obtained.

Earlier in this section we noted that the greater the difference in electronegativity between the atoms in a diatomic molecule, the greater will be the relative contribution of the ionic structure to the resonance hybrid. In view of this we might expect that there would be a relationship between the difference in electronegativity of the atoms and the percentage ionic character. Such a relationship does exist as shown in Figure 6–14 where there is plotted the percentage ionic character versus the absolute value of the electronegativity difference of the atom, $|\chi_A - \chi_B|$, for a number of diatomic compounds. Table 6–8 gives data for some

FIGURE 6–14 Percentage ionic character versus the electronegativity difference of the atoms in some diatomic molecules. Experimental points based on dipole moment data are shown as dots. The solid line is the theoretically predicted relationship.

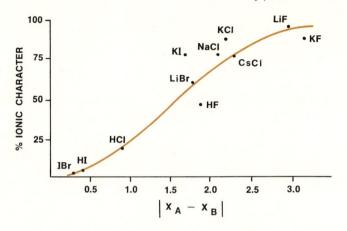

diatomic compounds pertinent to the percentage ionic character and the difference in electronegativity of the atoms.

It is not possible to accurately describe the electron distribution in a bond by writing one structure such as H—Cl. Realizing this, we have attempted to improve our description within the valence bond approximation by writing resonance structures which do not individually exist as real descriptions of the bonding in a molecule but which when combined give a close approximation to the actual situation. This is represented by double headed arrows between the structures. It is not possible to make a pictorial representation of this situation by molecular orbital theory. Ultimately, however, the results calculated by molecular orbital and

TABLE 6–8 Percentage ionic character of some diatomic molecules

COMPOUND	PERCENTAGE IONIC CHARACTER	$\lvert \chi_A - \chi_B \rvert$
H_2	0	0
Cl_2	0	0
HCl	17	0.9
HF	45	1.9
HI	4	0.4
IBr	~3	0.3
LiI	55	1.5
KI	75	1.7
KF	83	3.2

valence bond theories regarding the electron distribution in the bond are the same. The valence bond approximation does provide a satisfying pictorial mechanism to represent the nature of the bonding that exists via the use of resonance structures. The use of resonance structures is *not* limited to diatomic molecules. We will consider polyatomic molecules in the next chapter.

PROBLEM 6–12 Referring to Figure 6–14 and based upon your knowledge of the relationship between electronegativity and percentage ionic character answer the following:

a. Do you expect the percentage ionic character of ICl to be greater or lesser than that of IBr?
b. Do you expect the percentage ionic character of HBr to be greater or lesser than that of HI and HCl, respectively? Make an estimate of the percentage ionic character in HBr.
c. Will the percentage ionic character of $LiCl$ be greater or lesser than $NaCl$ and why?

EXERCISES

1. What is the principal difference in approach between the valence bond theory and molecular orbital theory of bonding?

2. What is meant by saying that a molecule is polar?

3. The energy released in the reaction $F + F \longrightarrow F_2$ is 37 kcal/mole. What is the bond dissociation energy of F_2?

4. Given that the internuclear distance in $LiF_{(g)}$ is 1.51 Å, calculate the energy of formation of $LiF_{(g)}$ from the ions.

5. For $KCl_{(g)}$ the internuclear distance is 2.6666 Å. Also, the ionization potential of potassium is 100 kcal/g-atom and the electron affinity of chlorine is 83.5 kcal/g-atom. Calculate
   a. the energy of formation gaseous KCl molecule from the ions.
   b. the bond dissociation of $KCl_{(g)}$.

6. The ionization potential of lithium is 5.39 ev/atom. The bond dissociation energy of LiF is 180 kcal/mole. With this information and the results of Exercise 4, calculate the electron affinity of fluorine.

7. The energy of formation of CaO from the ions is approximately 730 kcal/mole. However, the internuclear distance is similar to that in the alkali halides and is equal to 1.8 Å. Give a qualitative argument why the energy of formation should be significantly different than for the alkali halides. Also, verify quantitatively that this is the correct value.

8. Write Lewis electronic formulas for NaI, BaS, H_2O, OF_2, CF_4, MgF_2, and Na_2O.

9. What will be the valence or oxidation state of all of the elements in each of the following: HI, LiF, O_2, CS_2, SO_3, KBr, $CaCl_2$, Al_2O_3, NiO, $ZnCl_2$, Ag_2O, H_3PO_4, CO_3^{2-}, SO_4^{2-}, ClO_2^-, and NH_4^+?

10. a. Would you expect the following compounds to have a dipole moment?
    b. Which end will be negative?
    c. Will the dipole moment be relatively small or large?
    Cl_2, HI, CsBr, LiF, IBr, O_2, MgO, and CuF.

11. Predict whether the compounds of Exercise 10 will be ionic or covalent.

12. What would be the electronic configuration for the lowest energy excited state for the following molecules: N_2, Li_2, and C_2?

13. Utilizing the molecular orbital theory predict whether the following diatomic molecules will be stable or unstable: He_2^+, C_2, Ne_2, F_2, F_2^+.

14. What would be the resonance structures contributing to the following compounds? Which would you expect to be dominant and why? HI, NaF, IBr, RbCl, and BrCl.

15. Which molecule(s) of the following would be predicted to show the greatest deviation of its bond energy from that obtained by taking the arithmetic mean of A—A and B—B, and why? IBr, BrF, HBr, Cl_2, and NO.

16. Write the reaction for the first ionization of, and for electron addition to the following molecules, utilizing the appropriate molecular orbitals for the molecules and ions involved.
    a. B_2 b. C_2 c. O_2

17. a. Determine the bond order for the molecules and ions of Exercise 16.
    b. Compare O_2 and O_2^+. Do you expect the bond dissociation energy to be greater for O_2 or O_2^+, and why?
    c. Under what circumstances would the bond dissociation energy of an ion be greater than that of the neutral molecule?

18. He_2 formed from ground state He atoms is predicted to be unstable (see Section 6–E). Predict whether He_2 will be stable if formed from a ground state He atom and an excited state He atom with a $1s^1 2s^1$ configuration.

19. List four factors that affect the energy of an essentially ionic bond.

20. List three factors that affect the energy of a bond in a diatomic molecule which is essentially covalent.

21. The bond distances in the sodium halides are

MX	r
NaCl	2.3606 Å
NaBr	2.5020 Å
NaI	2.7115 Å

Calculate the coulomb energy for each compound and graph the coulomb energy and EA_x versus period number of the halide. Explain why the bond energies for NaCl, NaBr, and NaI vary as they do in Table 6–1.

22. The bond distance in the alkali chlorides are

MX	r
NaCl	2.3606 Å
KCl	2.6666 Å
RbCl	2.7867 Å

Calculate the coulomb energy and graph the coulomb energy and IP_M versus period number of the alkali metals. Explain why the bond energies for NaCl, KCl, RbCl, vary as they do in Table 6–1.

23. Do you expect the percentage ionic character of LiI to be greater than or less than the following: LiBr, NaBr, and NaCl? Why?

24. The percentage ionic character of HCl and HI are 17 and 4, respectively. Which of these will have a larger dipole moment and why?

CHAPTER SEVEN

MOLECULAR STRUCTURE AND BONDING IN POLYATOMIC MOLECULES

In this chapter we shall apply the concepts of bonding developed in Chapter 6 to polyatomic molecules. We shall be using these concepts to develop the capability of predicting certain new features such as the geometry of a molecule. Molecular orbital theory can be applied equally well to both polyatomic molecules and diatomic molecules. However, valence bond theory generally offers an easier approach to the understanding of bonding and geometry in polyatomic systems. Bonding in many polyatomic molecules cannot be explained using the concepts developed thus far—the new concept of hybridization is required.

In this chapter we will examine classes of compounds containing carbon. These compounds have classically been the focus of interest in the field of organic chemistry. We will be primarily concerned with the structural aspects of organic molecules in this chapter.

7–A. SINGLE BONDS IN POLYATOMICS

As noted in Chapter 6, the strongest bond is formed when the individual atomic orbitals overlap to the maximum extent. This is called the *criterion of maximum overlap*. The electronic configuration of fluorine is

$$1s^2\, 2s^2\, 2p_z^2\, 2p_y^2\, 2p_x^1$$

201

FIGURE 7–1 Formation of F_2 by overlap of p orbitals

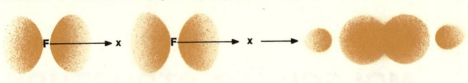

It is immaterial whether the one unpaired electron is in the p_x, p_y, or p_z orbital. The important point is that whatever orbital it is in, the formation of F_2 proceeds by overlapping along the coordinate axis containing the same subscript. In our case, F_2 is formed by the two fluorine atoms approaching along the x-axis in order to provide maximum overlap in a σ bond, as shown in Figure 7–1. In this and the following example we shall use a charge cloud representation to show the electronic distribution.

In a polyatomic molecule such as H_2O, the criterion of maximum overlap still applies. The principal approach to bond formation is the same, except that more than one bond is formed. The electronic configuration of oxygen is

$$1s \quad 2s \quad 2p_x \quad 2p_y \quad 2p_z$$
$$\uparrow\downarrow \quad \uparrow\downarrow \quad \uparrow \quad \uparrow \quad \uparrow\downarrow$$

The unpaired electrons in the $2p_x$ and $2p_y$ orbitals can each form a bond with a hydrogen atom which contains a single $1s$ electron (Figure 7–2). In this case, the $1s$ orbital of each hydrogen atom overlaps with the p_x or p_y orbital of oxygen by approaching along the x or y axis. This procedure provides for maximum over-

FIGURE 7–2 Formation of H_2O showing orbital overlap

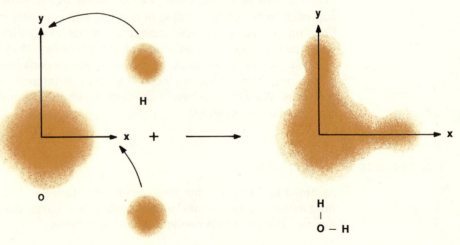

lap of orbitals. Since bonding occurs along axes (p_x and p_y) that are mutually perpendicular, we would predict the H—O—H angle in water to be 90°. In fact, it is somewhat different from this, ~105°, but the principle of maximum overlap gives us the qualitatively correct picture that the water molecule is clearly bent and not linear. We shall return shortly to a reconsideration of the H_2O molecule.

PROBLEM 7–1   What would you expect the bond angle to be in (a) H_2S, (b) OF_2, (c) SCl_2, and (d) PCl_3?

7–B. HYBRIDIZATION INVOLVING SINGLE BONDS

If we try to apply the valence bond approximation to carbon compounds, we find that problems arise. Carbon has the electron configuration $1s^2 2s^2 2p_x^1 2p_y^1$. Since a covalent bond arises from pairing valence electrons between atoms, the simplest stable compound of carbon and hydrogen would be expected to be CH_2 made by bringing the hydrogen atoms along the *x* and *y* directions. In this case the 1s orbitals of the hydrogen atoms would overlap with the p_x and p_y orbitals of carbon and the H—C—H angle would be approximately 90°. However, this is not the case; the simplest stable compound of carbon and hydrogen is CH_4, with an H—C—H angle of 109.5°. The concept of *hybridization* is introduced to explain such a circumstance.

Hybridization is a useful model for bonding in molecules where the obvious approach does not agree with experiment. In the hybridization model, two or three atomic orbitals within an atom can be thought of as being mixed together in certain proportions to generate new hybridized atomic orbitals having different shapes and directional properties than the orbitals from which they were made. Following this, the other atoms in the molecule are brought up along the direction of the hybridized orbitals to produce maximum overlap. Table 7–1 and Figure 7–3 show the hybrid orbitals commonly formed, the shape or geometry they define, and the angles between the hybrid orbitals. The angles are determined by theoretical calculations. Note that the number of hybrid orbitals generated is equal to the number of pure atomic orbitals (*s*, *p*, and *d*) mixed together. Also, all

TABLE 7–1   Constitution and properties of hybrid orbitals

PARTS s	PARTS p	PARTS d	HYBRID ORBITAL	NO. HYBRID ORBITALS	SHAPE OF HYBRID	ANGLE BETWEEN HYBRIDS	EXAMPLES
1	1	0	sp	2	linear	180°	$HgCl_2$
1	2	0	sp^2	3	triangle	120°	BF_3
1	3	0	sp^3	4	tetrahedron	109.5°	CH_4, NH_4^+
1	3	1	sp^3d	5	trigonal bipyramid	90°, 120°	PCl_5
1	3	2	sp^3d^2	6	octahedron	90°	SF_6

FIGURE 7–3 Geometrical properties of hybrid orbitals

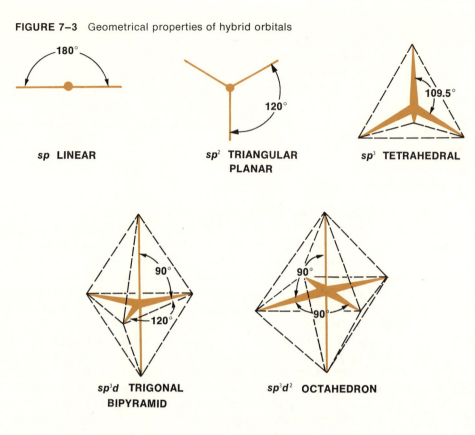

sp **LINEAR**

sp² **TRIANGULAR PLANAR**

sp³ **TETRAHEDRAL**

sp³d **TRIGONAL BIPYRAMID**

sp³d² **OCTAHEDRON**

hybrid orbitals of one type are equivalent. Heretofore we have used the word atomic preceding the term hybrid orbitals to emphasize their atomic nature. Hereafter, we shall drop this usage for convenience.

We shall first consider the tetrahedral hybridized orbitals. The formation of such orbitals can be conceptualized as shown in Figure 7–4. In step 1, the carbon atom is electronically excited from a $1s^2 2s^2 2p^1 2p^1$ configuration to a $1s^2 2s^1 2p^1 2p^1 2p^1$ configuration; hybrid orbitals do not yet exist at this stage. With further input of energy (step 2), the one *s* and three *p* atomic orbitals mix or combine to make the four hybrid orbitals. This step is a nonobservable excitation to a valence state where the hybridized orbitals exist. This is another way of saying that hybridization is a model which can explain experimental data but which in itself is not a physically observable phenomenon. Note again that the number of hybrid orbitals produced is the same as the number of pure atomic orbitals that are mixed, and all are equivalent (see Table 7–1). Steps 1 and 2 of Figure 7–4 require energy. This energy is compensated for when the hybridized carbon atom undergoes bonding with other atoms (step 3). In other words, the energy released in compound formation (step 3) is more than that required to form the four sp^3 orbitals on carbon (steps 1 and 2).

FIGURE 7–4 Diagram of energy levels for the formation of *sp³* orbitals. Step (1) corresponds to an electronic excitation. Step (2) is a non-observable excitation to the hybridized valence state. Step (3) represents the bonding of the hybridized carbon atom to other atoms.

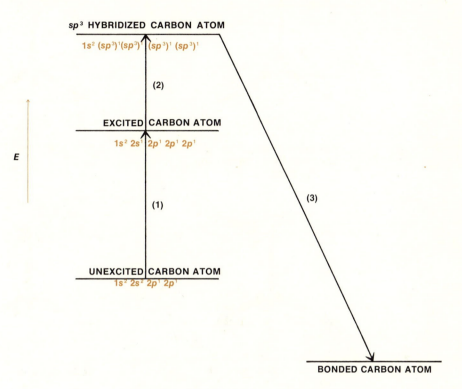

In methane, CH_4, we have experimentally determined the H—C—H bond angle to be 109.5°. We can understand why this is true if we say that the carbon atom in CH_4 is *sp³* hybridized (see Table 7–1 and Figure 7–4). Each *sp³* hybrid orbital overlaps with a 1*s* orbital of hydrogen to produce CH_4 (or there is a pairing between an *sp³* electron of carbon and a 1*s* electron of hydrogen). It is not necessary that carbon combine only with hydrogen. As a matter of fact, carbon combines with many other elements (Figure 7–5). For example, carbon can combine with chlorine atoms by overlap of *sp³* (C) and *p* (Cl) orbitals to produce carbon tetrachloride which has the expected Cl—C—Cl bond angle of 109.5°. In certain organic molecules known as alkanes or saturated aliphatic hydrocarbons (Section 7–H), only the elements carbon and hydrogen occur. In these compounds both C—C and C—H bonds exist. The *sp³* orbital of carbon can overlap with another *sp³* orbital of another carbon as shown in Figure 7–6. Two tetrahedra are brought together at one apex to form a C—C bond, and hydrogen atoms are placed at the other apices to form the C—H bonds.

FIGURE 7–5 Formation of tetrahedral carbon molecules

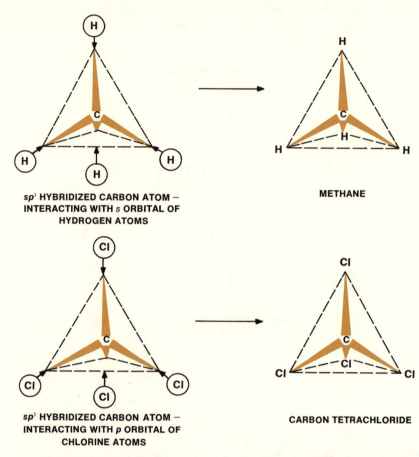

*sp*³ HYBRIDIZED CARBON ATOM –
INTERACTING WITH *s* ORBITAL OF
HYDROGEN ATOMS

METHANE

*sp*³ HYBRIDIZED CARBON ATOM –
INTERACTING WITH *p* ORBITAL OF
CHLORINE ATOMS

CARBON TETRACHLORIDE

A parallel situation exists for the formation of *sp* and *sp*² hybrid orbitals in carbon. However, these cases are somewhat more complicated because *p*-orbital electrons remain unused after the formation of the hybrid orbitals but are still available for bonding. We shall postpone these cases involving multiple bonds until Section 7–D. Other examples of *sp* and *sp*² hybridization exist for atoms besides carbon which are singly bonded to other atoms. We shall first consider the *sp* case. The molecules $BeCl_2$ and $HgCl_2$ are experimentally known to be linear.

$$Cl—Be—Cl$$

$$Cl—Hg—Cl$$

The configurations for beryllium and mercury are

$$Be \quad 1s^2 \, 2s^2$$

$$Hg \, \ldots \ldots \ldots \, 5d^{10} \, 6s^2$$

FIGURE 7–6 Formation of ethane by sp^3 orbital overlap

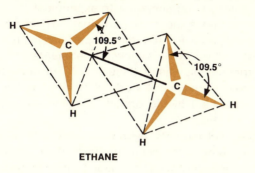

ETHANE

We can excite one of the ns electrons in each case to give

$$Be \quad 1s^2\,2s^1\,2p^1$$

$$Hg \quad \ldots \ldots \ldots \ldots 5d^{10}\,6s^1\,6p^1$$

and then hybridize the one s and one of the p orbitals to give two sp hybrid orbitals in each case. We can once again write electronic configurations for the atoms emphasizing the hybridized orbitals.

$$Be \quad 1s^2\,(sp)^1\,(sp)^1$$

$$Hg \quad \ldots \ldots \ldots \ldots 5d^{10}\,(sp)^1\,(sp)^1$$

Using Table 7–1 and Figure 7–3 we have said that sp hybrid orbitals are linear. The chlorine atoms are brought up along the direction of the sp hybrid orbitals, and the p-orbital of each chlorine overlaps an sp orbital to form a bond. The molecules are predicted to be linear, which is in agreement with experiment.

The molecule BF_3 is experimentally known to be planar with F—B—F angles of 120°. The electronic configuration of boron is

$$B \quad 1s^2\,2s^2\,2p^1$$

If we used just the one unpaired electron in boron to bond with fluorine, we would expect the molecule BF. However, we know from experiment that BF_3 is correct. If we excite an s electron to a p orbital and hybridize one s and two p orbitals, as

$$B \quad 1s^2 2s^2 2p^1 \xrightarrow{\text{excite}} B \quad 1s^2 2s^1 2p^1 2p^1 \xrightarrow{\text{hybridize}} B \quad 1s^2 (sp^2)^1 (sp^2)^1 (sp^2)^1$$

we will obtain three equivalent sp^2 hybridized orbitals. The three hybrid orbitals are in a plane with all angles equal to 120° (Figure 7–3). Consequently, three F atoms can bond along the directions of the sp^2 hybrid orbitals of B giving BF_3 with a predicted geometry in agreement with experiment.

Atomic orbitals other than s and p can participate in hybridization. For example, s, p, and d orbitals can hybridize in different proportions to produce hybrid orbitals with varying shape and direction. It is known experimentally that PCl_5 has the geometric form of a trigonal bipyramid (Figure 7–7). In phosphorus, sp^3d hybrid orbitals can be formed by mixing one s-orbital, three p-orbitals and one d-orbital. The resulting five equivalent hybrid orbitals point to the corners of a trigonal bipyramid (Figures 7–3 and 7–7).

A specific example involving sp^3d^2 hybrid orbitals would be sulfur in SF_6. SF_6 is known experimentally to be in the geometric form of an octahedron (Figure 7–8). In sulfur, the sp^3d^2 hybrid orbitals can be made by mixing the three orbitals in the ratio one part s: three parts p: two parts d. The resulting six equivalent hybrid orbitals point to the corners of a regular octahedron (Figure 7–3).

FIGURE 7–7   Structure of PCl_5

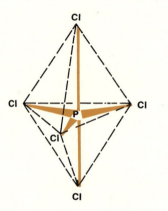

FIGURE 7–8   Structure of SF_6

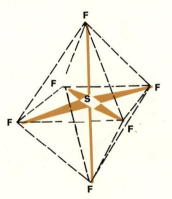

The hybridized orbitals have a different shape than the original *s*, *p*, and *d* orbitals from which they are made. The hybrid orbitals are all directed differently (Figure 7–3), and have a greater charge density along the direction in which they lie. The difference in charge density and shape of an sp^3 hybrid compared to a *p* orbital is shown in Figure 7–9. Note the significantly greater charge density and extension into space of one of the lobes of the sp^3 hybrid orbital compared to that of the *p* orbital. The same is generally true for all of the other hybrid orbitals, particularly compared to an *s* orbital which has a spherical shape, and is thus not directed at all.

Based on the criterion of maximum overlap and comparative charge density and space extension, we would expect a stronger bond between atoms if overlap occurs between hybridized orbitals compared to unhybridized orbitals. Let us consider the three molecules Li_2, F_2, and C_2H_6. In Li_2, the bond is made from overlap of *s* orbitals; in F_2 overlap of *p* orbitals occurs; in C_2H_6, the C—C bonds come from overlap of sp^3 hybrid orbitals of each carbon atom (Figure 7–10). The bond

FIGURE 7–9   Comparison of the shape and charge density of sp^3 and *p* orbitals

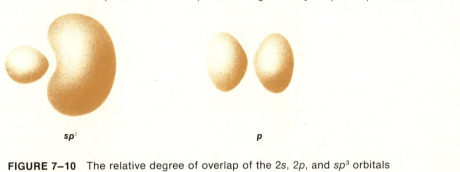

sp^3 *p*

FIGURE 7–10   The relative degree of overlap of the *2s*, *2p*, and sp^3 orbitals

2s   *2p*

sp^3

energies have been determined to be 26, 37, and 80 kcal/mole, respectively. Thus s-orbitals with no directionality overlap the least to give the weakest bond. The p-orbitals which have directionality overlap more, giving the next strongest bond. Finally, sp^3 orbitals, which have the most charge density and space extension along a direction, overlap to give the strongest bond of all. Figure 7–10 shows a schematic diagram of the relative degree of overlap of $2s$, $2p$, and sp^3 hybrid orbitals (made from $2s$ and $2p$ atomic orbitals). Not only does the concept of hybridization provide orbitals having the maximum possible overlap (therefore giving the strongest bond) but it permits us to understand the bond angles found experimentally and to predict those that have not yet been studied.

When we discussed bonding in H_2O, we suggested that bonding of hydrogen along the p_x and p_y axes of oxygen at 90 ° angles would provide maximum bond strength. Our simple approach to the problem involving overlap of pure p-orbitals (O) and s-orbitals (H) gave us a qualitatively correct but quantitatively incorrect geometry. We predicted an H—O—H angle of 90 ° instead of the observed angle ~105 °. The bonding in H_2O cannot be described as involving pure p- and s-orbital overlap. Instead, oxygen must have hybridization involving sp^3 hybrid orbitals. Two hydrogen $1s$ orbitals overlap two sp^3 orbitals and the other four oxygen orbital electrons remain as unbonded lone pairs of electrons in each of the two remaining sp^3 orbitals as shown in Figure 7–11. If no hydrogens atoms were bonded, the angle between the sp^3 orbitals would be 109.5 ° (tetrahedral oxygen); however, the presence of the covalently bonded hydrogen atoms reduces the electron repulsion between the sp^3 orbitals, allowing the angle to close up to ~105 °.

The situation with ammonia, NH_3, is interesting. We would predict that the H—N—H angle would be ~90 ° because each hydrogen atom $1s$ orbital overlaps with the p_x, p_y, p_z orbitals of nitrogen which are mutually perpendicular. Actually the H—N—H angle is ~108°, which is very nearly the tetrahedral angle. Therefore, we presume that the original s and p orbitals of nitrogen hybridize to give essentially sp^3 orbitals. The three hydrogen atoms and a lone electron pair are approximately directed to the corners of a tetrahedron as shown in Figure 7–11. If no hybridization had occurred, the lone electron pair would be in a nondirectional s

FIGURE 7–11 Structures of NH_3 and H_2O

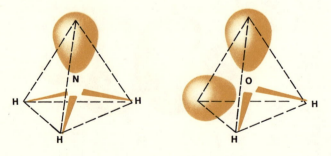

orbital. The fact that this lone electron pair is highly directed has an important consequence for the physical and chemical properties of ammonia. The same considerations are valid for H_2O regarding the important consequences of the directed lone electron pairs.

PROBLEM 7–2 $CdCl_2$ is essentially linear. Explain this result.

PROBLEM 7–3   Near room temperature $SiCl_4$ is a liquid with Cl—Si—Cl angles of ~109°.
a. Explain whether you would consider $SiCl_4$ to have ionic or covalent bonds. Why?
b. Account for the Cl—Si—Cl bond angle.

7–C. DIPOLE MOMENTS OF SINGLY BONDED MOLECULES

For diatomic molecules there is no difficulty determining if a dipole moment exists and if so, its direction. In *homo*nuclear diatomics, both atoms are identical, the electronegativity of both atoms must therefore be the same, and no charge asymmetry exists. In *hetero*nuclear diatomics, charge asymmetry must exist, because no two different atoms have the same electronegativity. The negative end of the dipole is always toward the atom with the greater electronegativity.

$$\overset{\longrightarrow}{H—Cl} \qquad \overset{\longrightarrow}{I—Cl}$$

For polyatomic molecules, even though there may be a dipole for an individual bond between two different atoms, the molecule as a whole may or may not have a dipole moment. For example, consider the linear molecule $BeCl_2$:

$$\overset{\longleftarrow}{Cl}—Be—\overset{\longrightarrow}{Cl}$$

Since beryllium and chlorine have different electronegativities, there is an individual bond dipole in the direction of each chlorine. Remember that the symbol ↔ represents a vector and therefore has both magnitude and direction. Since the vectors representing these bond dipoles are opposite in direction but equal in magnitude, there is a zero resultant or net cancellation, and no *molecular* dipole exists. This is parallel to two people pulling on a rope in opposite directions but with equal force—there is no net movement in either direction. This situation will always occur in linear molecules where there is a symmetrical distribution of atoms around a central atom as in $HgBr_2$ where mercury is the central atom.

$$\overset{\longleftarrow}{Br}—Hg—\overset{\longrightarrow}{Br}$$

If a molecule is not linear, adding the contributing vectors will show whether a dipole exists and in what direction. (Remember that a vector shows both direction and magnitude. The length of the arrow is proportional to the magnitude of the charge asymmetry, and the direction it points denotes the direction of the dipole.) Consider the molecule BF_3

where the molecule is symmetrically planar with all angles equal to 120°. Because of the opposed symmetrical distribution of the B—F bond dipoles, vector cancellation occurs and no molecular dipole should exist. This is in agreement with experimental evidence.

The bond dipoles in H_2O are not symmetrically opposed so a resultant molecular dipole does exist, as indicated by the dashed arrow

$$\overset{\uparrow}{\underset{H\;\;\;\;\;\;H}{O}}$$

The molecule PCl_5

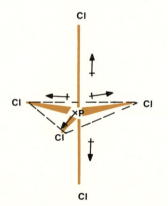

is a more complex case. The three bond dipoles in the triangle cancel as do the two above and below the plane of the triangle. Thus the resultant dipole moment is zero and PCl_5 has no dipole. Similarly in CH_4 since all angles are equal, the resultant of the three bond dipoles to the corners of the triangle base of the tetrahedron will exactly cancel the remaining bond dipole to the fourth hydrogen above the plane as shown in Figure 7–12. Thus CH_4 will have no dipole moment. In all cases where there is a *symmetrical distribution* of bond dipoles of equal magnitude, cancellation occurs and no molecular dipole exists.

PROBLEM 7–4   Predict whether the following molecules will have a dipole moment, and in so far as is possible, in what direction (even if only qualitatively).

(a) F_2 (b) HF (c) H_2S (d) NH_3 (e) H_2CCl_2

FIGURE 7–12 Cancellation of bond dipoles in CH_4 resulting in no dipole moment. The dashed dipole is the resultant of three dipoles from hydrogen to carbon.

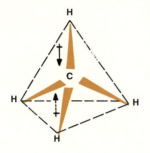

7–D. MULTIPLE BONDS

We have already discussed multiple bonding in diatomic molecules (Section 6–D). However, some discussion of polyatomic multiple bonding is necessary. Multiple bonding always involves a σ bond plus one or more π bonds as we saw for the case of O_2 and N_2.

In the case of sp^2 and sp hybridized carbon atoms there are one or two un-bonded p electrons per atom, respectively. Because the sp^2 orbitals are in a plane, the remaining p orbital must be perpendicular to the plane as

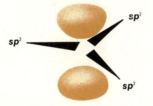

If we bring together two sp^2 hybridized carbon atoms that have hydrogen atoms on two of the three sp^2 orbitals, then we can overlap the third sp^2 orbital of each carbon atom to make a σ bond. Also, each of the p orbitals can overlap side-to-side to make a π bond. The combination of a σ and π bond gives us a double bond. This situation is exactly what happens in the ethylene molecule, $CH_2{=}CH_2$ (Figure 7–13). We expect the molecule to be planar since the sp^2 hybridized carbon orbitals are in a plane (Figure 7–3, Table 7–1), and the H—C—H angle to be 120°. This is in agreement with experimental data.

Another situation occurs when we have two sp hybridized carbon atoms with hydrogen atoms bonded to one of the sp orbitals. The remaining sp orbital of each carbon atom can overlap to form a σ bond, and the two p-orbitals can overlap side-to-side to form two π bonds. Acetylene is an example of a molecule contain-ing a triple bond between two carbons, $HC{\equiv}CH$ (Figure 7–14). We would expect the molecule to be linear since the original sp hybridized carbon orbitals were linear (Figure 7–3, Table 7–1). Again, this is in agreement with experimental data.

FIGURE 7–13 Formation of the σ and π bonds of ethylene

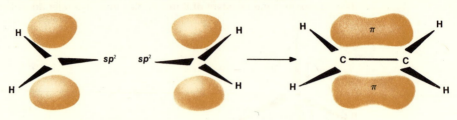

FIGURE 7–14 Bonding in acetylene—one σ and two π bonds. Note that overlap of the π-lobes results in a cylindrical charge distribution.

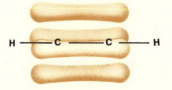

The pattern of multiple bond formation can be stated to be as follows

single bond $-\sigma$ bond

double bond $-\sigma + \pi$ bonds

triple bond $-\sigma + 2\pi$ bonds

The σ bond can arise from

$$s + s \longrightarrow \sigma$$

$$p + p \text{ (head-to-head)} \longrightarrow \sigma$$

$$\left.\begin{array}{l} sp + sp \\ sp^2 + sp^2 \\ sp^3 + sp^3 \end{array}\right\} \text{head-to-head} \longrightarrow \sigma$$

$$\left.\begin{array}{l} s + p \\ + sp \\ + sp^2 \\ + sp^3 \end{array}\right\} \longrightarrow \sigma$$

The π bond (for present consideration) can arise only from side-to-side overlap of two p orbitals:

$$p + p \text{ (side-to-side)} \longrightarrow \pi$$

It is not necessary that the atoms involved in single, double, or triple bonds be the same. For example, for single bonds:

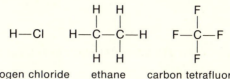

hydrogen chloride   ethane   carbon tetrafluoride

For double bonds:

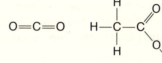

carbon dioxide   acetic acid

For triple bonds:

$$C\equiv O \qquad H-C\equiv C-H \qquad H-C\equiv N$$

carbon monoxide   acetylene   hydrogen cyanide

It might be anticipated that the presence of an extra one or two bonds should increase the bond dissociation energy relative to a single bond. This is indeed correct, as it is for diatomic molecules (Section 6–D, E). The carbon-carbon bond dissociation energies of the following single, double and triple bond systems is

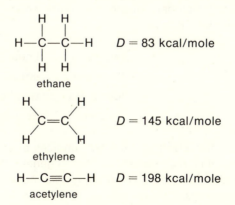

It should be noted that the bond energies of the double and triple bonds are not 2 and 3 times greater than that of the single bond. The bonding energy of side-to-side overlap of the p orbitals to make π bonds is of lower magnitude than the

bonding energy of head-to-head overlap of the sp^2 or sp orbitals to make σ bonds. Thus, a π bond is weaker than a σ bond. The same is generally true for diatomic molecules.

7–E. MOLECULAR STRUCTURE UTILIZING VALENCE SHELL ELECTRON PAIR REPULSION

The discussion so far has dealt with the development of molecular structure based on the concept of maximum orbital overlap and on a hybridization model approach. In this section, we will discuss another model whereby the geometry around an atom is largely determined by the repulsive interactions between the pairs of electrons in the valence shell of the central atom. It is important to emphasize that more than one approach, or model, can often be used to explain any result.

The electrons in a valence shell occupy localized atomic orbitals that are oriented around the remaining atom core such that the distance between such orbitals is maximized. This is a result of the repulsive interaction of like charges. When neutral molecules are formed, essentially localized electron pairs (both bonding and non-bonding) around the central atom adopt a geometry maximizing the average distance between them. Again this occurs because of the repulsive potential between the negatively charged electron pairs. The most important factor determining the geometry is the relative repulsion between the *bonding* pairs and the *nonbonding* pairs of electrons. The repulsive interaction is small when the overlap between the orbitals is small and large when the overlap is large. Moreover, the repulsion increases very rapidly as the overlap increases.

From elementary geometric considerations, the most probable location of different numbers of electron pairs which would minimize repulsions can be determined to be:

two pairs	linear (180 °)
three pairs	equilateral triangle (120 °)
four pairs	tetrahedral (109.5 °)
six pairs	octahedral (90 °)

For five electron pairs two choices exist: a trigonal bipyramid and a square bipyramid. It can be shown that in general the trigonal bipyramid is slightly favored. We will not be able to apply this model to transition metals with partially filled *d*-orbitals.

For a central atom in a molecule, where only single bonds exist, a general formula

$$A L_b N_n$$

could be written where A is the central atom, b is the number of bonding ligands L, and n is the number of nonbonding (or lone) pairs of electrons N. A *ligand* is an atom or group of atoms bonded to the central atom A. Ligands we have already discussed are Cl, F and NH_3. The geometry around A will be determined by the spatial arrangement of the bonding and nonbonding pairs of electrons (see Table 7–2).

Some of the geometric details in Table 7–2 require further consideration. First, nonbonding electron pairs repel any adjacent electron pairs more strongly than do bonding pairs. The nonbonding pair occupies more space than a bonding pair. Thus a nonbonding pair will overlap any adjacent pair more strongly than will a bonding pair and will give greater repulsion. The ordering of the repulsion is

lone pair–lone pair $>$ lone pair–bonding pair $>$ bonding pair–bonding pair

Consider the molecules CH_4, NH_3, and H_2O. All three will have a general tetrahedral shape because the central atom is surrounded by a total of four electron pairs (see Table 7–2). However, the number of bonding and nonbonding pairs is different

CH_4 4 bonding pairs, 0 nonbonding pairs

NH_3 3 bonding pairs, 1 nonbonding pair

OH_2 2 bonding pairs, 2 nonbonding pairs

Based on the relative magnitudes of repulsion, we would expect that

H—C—H angle $>$ H—N—H angle $>$ H—O—H angle

This is in agreement with experiment which gives

H—C—H angle	H—N—H angle	H—O—H angle
109.5 °	107.3 °	104.5 °

This is also true, in general, when the hydrogen atoms of NH_3 or H_2O are replaced by carbon groups (alkyl groups, Section 7–G), nitrogen, oxygen, or fluorine; the angle in the nitrogen and oxygen compounds is less than the tetrahedral angle.

Second, the bonding pair–bonding pair repulsion decreases with increasing electronegativity of the ligand. This is because the greater the electronegativity of the ligand, the greater is the contraction of the bonding orbital and the less the repulsion between them. The following bond angle ordering results

$OF_2(103°) < H_2O(104.5°)$

$NF_3(102°) < NH_3(107°)$

$PCl_3(100°) < PI_3(102°)$

TABLE 7–2 General shapes of molecules as a function of numbers of valence shell pairs

NUMBER OF VALENCE SHELL PAIRS AROUND CENTRAL ATOM A	GEOMETRY	SCHEMATIC OF GEOMETRY	EXAMPLES
2	linear	AL₂	$HgCl_2$, $CdCl_2$, $Zn(CH_3)_2$
3	3 bonding triangular	AL₃	BF_3, GaI_3
	2 bonding 1 nonbonding v-shaped	AL₂N₁	$SnCl_2$, $PbBr_2$
4	4 bonding tetrahedral	AL₄	CH_4, NH_4^+, $(BeCl_4)^{2-}$
	3 bonding 1 nonbonding trigonal pyramid	AL₃N₁	NH_3, H_3O^+, $AsCl_3$
	2 bonding 2 nonbonding v-shaped	AL₂N₂	H_2O, SCl_2, H_2S

TABLE 7–2 General shapes of molecules as a function of numbers of valence shell pairs

NUMBER OF VALENCE SHELL PAIRS AROUND CENTRAL ATOM A	GEOMETRY	SCHEMATIC OF GEOMETRY	EXAMPLES
5^a 5 bonding	trigonal bipyramid	AL_5	PCl_5, $NbCl_5$
6 6 bonding	octahedral	AL_6	SF_6, MoF_6, $[SnCl_6]^{2-}$ $[AlF_6]^{3-}$

^{a}Other geometries exist but are less common. For details consult article by R. J. Gillespie, *J. Chem. Ed. 40*, 295 (1963).

Also, differences can be noted when the central atom is not from the second period (Li to F) but is from the third (Na to Cl). In the latter case, appreciable bonding pair–bonding pair interaction does not occur until the L—A—L angle is near 90 °. In comparing H_2S to H_2O, the two nonbonding pairs of sulfur (of the third period) repel each other to the point where appreciable bonding pair interaction begins to occur (near 90 °). This results in the angle of H_2S being 92 ° whereas in H_2O it is 104.5 °. This electron pair repulsion model is a simple and useful model for understanding molecular geometry.

PROBLEM 7–5 Predict the general geometry of the following molecules.
(a) BCl_3 (b) CF_4 (c) $SbCl_5$ (d) $[Ag(CN)_2]^-$ (e) PF_6 (f) ZnI_2
(g) $PbCl_2$

PROBLEM 7–6 Predict which compound of the following pairs will have the larger bond angle and why.
(a) PBr_3 versus PI_3
(b) $SbCl_3$ versus SbI_3

7–F. RESONANCE

Earlier (Section 6–I), we discussed resonance as it pertained to diatomic molecules such as HCl. Within the valence bond approximation a bond is represented by an electron pair, and if we draw a single structure for a molecule, we assume this to be a reasonably accurate description of the bonding or charge density. However, there are certain cases where one structure is insufficient to give even a reasonably accurate description of the bonding. Consider the molecule SO_3. Sulfur and oxygen each have 6 valence shell electrons, so SO_3 should have 24 bonding and nonbonding electrons. Remembering to have eight electrons around each atom involved in bonding, we can write SO_3 in any of three ways

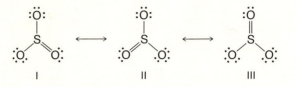

I   II   III

Experiment tells us that all three sulfur-oxygen bonds in SO_3 are equivalent (that is, have the same bond length). Each of I, II, and III indicate one different S—O bond, but if we superimpose the electronic configurations of I, II, and III, all S—O bonds will be equal. We say that I, II, and III are *resonance forms or structures* of SO_3. It must be re-emphasized that there is only one kind of SO_3 and that we use resonance structures to give the best possible description of the bonding within the limitations of the valence bond approximation. The limitation which interests us is that a bond is represented by an electron pair.

The same situation arises in describing many other molecules. Let us first consider the case of an ion, carbonate (CO_3^{2-})

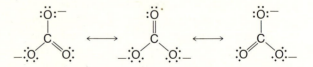

This presents a parallel situation to the SO_3 molecule. For the acetate ion, $C_2H_3O_2^-$ or (CH_3COO^-), we have

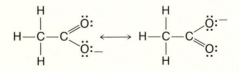

The classical use of the concept of resonance in organic chemistry is for the structure of benzene, C_6H_6. Experiment shows that the carbon atoms lie at the vertices of a regular planar hexagon (meaning all bond lengths as equal), with the six hydrogen atoms in the same plane, directed outward; all bond angles are 120°. All of the carbon atoms are sp^2 hybridized and the remaining p-orbital electron on each carbon atom forms a π bond with one of its neighboring carbon atoms, resulting in the apparent structure

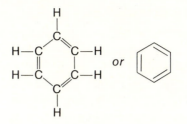

However, we can ask why one carbon neighbor should be preferred over the other for the location of the double bond. The structure

is equally valid. In fact, each represents a resonance form or structure such that we write

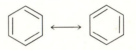

These are called Kekulé structures named after the chemist who first proposed them. (Actually, there are other possible resonance structures but these contribute much less to the actual bonding picture.) Superposition of the electron configurations of the Kekulé structures would result in all bonds being of equal length in benzene. The structure of benzene can be understood utilizing both the hybridization model and the resonance model. The molecular orbital theory model is also useful if we consider that the six p electrons of carbon are delocalized equally over the whole molecule. We will return to a discussion of the molecular orbital theory model when we discuss the structure of organic molecules.

Any resonance structure is an attempt to describe the nature of the bonding in a molecule. Where there is more than one resonance structure possible, the best description of the bonding involves contributions by all of the resonance structures. The bonding scheme of any one of the resonance structures alone results in higher energy than that arising from hybridization of all structures. Consequently, we can define the resonance energy as the difference between the energy of the one contributing structure with the lowest energy and the energy of the resonance hybrid (Section 6–I). The important physical and chemical properties which are influenced by resonance energies, such as the ionization of organic acids, are discussed in Chapters 10 and 13.

PROBLEM 7–7 Show the possible resonance forms for triangular shaped SO_2 assuming a maximum of 8 electrons surrounds sulfur.

PROBLEM 7–8 In planar NO_3^- all bonds are equivalent. Show the resonance forms to account for this. How does NO_3^- differ from CO_3^{2-}?

7-G. STRUCTURE OF ORGANIC MOLECULES

Carbon is a unique element since carbon-carbon bonds are strong enough to permit large numbers of carbon atoms to be bonded in long chains and large cyclic molecules. In addition, multiple bonds can exist between carbon atoms (Section 7-D). Carbon can bond not only to itself but also to other atoms, principally hydrogen, oxygen, nitrogen, sulfur, and the halogens. In addition, multiple bonds can exist between carbon and oxygen, nitrogen, and sulfur. No other atom has such a multifaceted character, and because of this, a chemistry based on carbon exists and is known as organic chemistry.

Organic chemistry is an extremely important branch of chemistry. Life itself — plant and animal — depends upon reactions involving organic molecules. Also, most of the synthetic products produced by man involve organic compounds. At this stage we shall discuss the classes of organic compounds that exist, some characteristics of each class, and we shall identify the nature of the bonding as well as special structural features of such compounds.

7-H. HYDROCARBONS

ALKANES

The first class we shall consider is the *alkanes* (or *saturated aliphatic hydrocarbons*). These compounds contain only carbon and hydrogen, with the general formula of C_nH_{2n+2}, and have a H—C—H angle of 109.5 °. This is consistent with the idea that the carbon atoms are tetrahedrally hybridized. Thus all C atoms in alkanes can be considered to be sp^3 hybridized. The parent molecule of this class is methane, CH_4, and the others are generated by the addition of CH_2 units except where the carbon is to be in a terminal position and then CH_3 is added.

$$H_3C—CH_2—CH_3 \qquad CH_3—CH_2—CH_2—CH_2—CH_3$$
propane   pentane

The *-ane* ending indicates that only carbon and hydrogen are involved and that only single bonds are present in the molecule. The fact that there are no double bonds in the molecule could be anticipated theoretically since it is not possible for sp^3 hybridized carbon to be involved in double bonds.

Both branched and unbranched chains exist.

$$H_3C—CH_2—CH_2—CH_3 \qquad CH_3—CH—CH_3$$
$$\qquad\qquad\qquad\qquad\qquad\qquad\qquad\quad |$$
$$\qquad\qquad\qquad\qquad\qquad\qquad\qquad CH_3$$

n-butane   isobutane (*i*-butane)
unbranched   branched

FIGURE 7–15 Structure of pentane

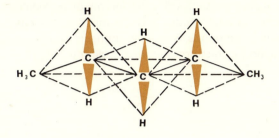

The letter *n* designates that structure as *normal*—no branching. When the name is used without a prefix, it is assumed that no branching of the chain occurs. Butane and isobutane are *structural isomers. Structural isomers have the same formula but a different arrangement of the atoms.*

Although a structure for an alkane is drawn as if it were a linear straight chain, it is in fact zig-zag. This is explained by the tetrahedral nature of carbon (sp^3 hybrids); the C—C bond results from tetrahedra joined at apices. A more accurate representation of the five carbon alkane, pentane, is shown in Figure 7–15.

As shown in Table 7–3, the names of all saturated hydrocarbons end in *-ane* and the length of the carbon chain is designated by the prefix: *meth*—one, *eth*—two, *prop*—three, *but*—four. After the first four members of the series the prefix is the Greek word for the number of carbon atoms in that chain. Note that the boiling point increases progressively as the number of carbon atoms increase. This increase will be explained in Section 8–E in terms of the increased attractive force between molecules of higher molecular weight.

TABLE 7–3 Boiling and melting points for some unbranched alkanes

MOLECULAR FORMULA	NAME	BOILING POINT (° C)	MELTING POINT (° C)
CH_4	methane	−161°	−184°
C_2H_6	ethane	− 88°	−183°
C_3H_8	propane	− 45°	−190°
C_4H_{10}	*n*-butane	+ 0.6°	−138°
C_5H_{12}	*n*-pentane	36°	−148°
C_6H_{14}	*n*-hexane	69°	− 94°
C_7H_{16}	*n*-heptane	98°	− 91°
C_8H_{18}	*n*-octane	126°	− 98°
C_9H_{20}	*n*-nonane	150°	− 51°
$C_{10}H_{22}$	*n*-decane	174°	− 32°

If hydrogen is removed from an alkane, the remainder (called an alkyl group) of the molecule is able to react with other atoms or molecules to form new compounds. The names of alkyl groups provide much of the basis for naming many classes of organic compounds. The alkyl group is named from the alkane from which it is produced by changing the *-ane* ending to *-yl*.

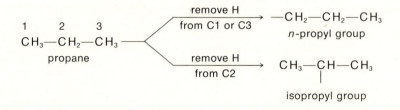

The line from the terminal or middle carbon atom of the *n*-propyl or isopropyl group respectively represents an electron capable of forming a bond to another atom or group of atoms. In contrast to propane, removal of any hydrogen from methane and ethane results in the same alkyl group being produced. The removal of a hydrogen atom from either *terminal* carbon of an unbranched alkane always gives the same alkyl group.

$$CH_3—CH_2—CH_2—CH_3 \xrightarrow[\text{from C1 or C4}]{\text{remove H}} CH_3—CH_2—CH_2—CH_2—$$

$$\begin{array}{cccc} 1 & 2 & 3 & 4 \end{array}$$

n-butane   *n*-butyl group

Table 7–4 contains the names and formulas for some common alkyl groups.

TABLE 7–4 Names and formulas for some common alkyl groups

NAME	STRUCTURAL FORMULA
methyl	$H_3C—$
ethyl	$H_3C—CH_2—$
n-propyl	$H_3C—CH_2—CH_2—$
i-propyl	$H_3C—CH—CH_3$ 	
n-butyl	$CH_3—CH_2—CH_2—CH_2—$

One of the important structural aspects of alkanes and alkyl groups is the possibility of twisting around C—C single bonds. Recall that a σ bond (a single bond) between the two carbons of ethane arises by head-to-head overlap of sp^3 hybrid orbitals to give a cylinder-like charge cloud.

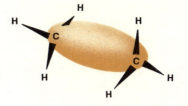

If we rotate one CH_3 group relative to the other around the C—C bond, we do not change the overlap of the atomic orbitals involved because of the cylindrical symmetry of the charge density in the bond. Consequently, we expect little or no energy change by this rotation. In fact, about 3 kcal are required to go from the staggered to the eclipsed form, because of the interaction through space of the electrons in the C—H bonds.

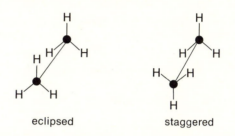

eclipsed   staggered

At room temperature there is sufficient energy for these forms to interconvert quite readily. Therefore, it is often said (though actually incorrectly), that there is free rotation about any single bond—meaning in effect no energy is required to do the rotation. We shall return to freedom of rotation when we consider twisting around a double bond.

One of the important uses of the alkanes is as a fuel, particularly in internal combustion engines. Gasoline is a mixture of straight and branched alkanes with an average composition of C_7H_{16}. High-compression engines often knock with many types of gasoline. The presence of branched chains and of unsaturated hydrocarbons (see the next section) can decrease or eliminate the amount of knocking. Also, the knocking can be reduced or avoided by the addition of a small amount of lead tetraethyl $Pb(C_2H_5)_4$, from which the name "ethyl gasoline" is derived. An area of current research interest is the improvement of the antiknock properties of gasoline and the elimination of the lead tetraethyl. The concern stems from so-called lead pollution resulting from the combustion or breakdown of $Pb(C_2H_5)_4$ to other components including lead itself.

PROBLEM 7–9   a. Do all the C—C bonds in propane have cyclindrical charge symmetry? Why?
b. What kinds of bonds, in terms of σ and π, are the C—H bonds of propane?

PROBLEM 7–10   Write the three possible structural isomers for C_5H_{12}.

CYCLIC ALKANES

The formulas for cyclic saturated hydrocarbons is C_nH_{2n} and differs from the general formula for alkanes, C_nH_{2n+2}. The more common ring sizes are those with 3, 4, 5, and 6 carbon atoms. The names of these are the same as those of the straight chain alkanes containing the same number of carbon atoms but are preceded by the prefix *cyclo* as

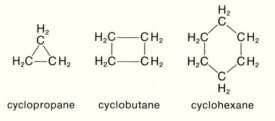

cyclopropane   cyclobutane   cyclohexane

Much of the time, the hydrogen atoms are omitted, as is the symbol C in the ring:

Cyclobutane is somewhat non-planar and cyclohexane is definitely non-planar.

ALKENES

Alkenes are compounds which contain only carbon and hydrogen but also have a double bond. All of the potential bonds of the carbon atoms are *not* saturated with hydrogen, leading to the name sometimes given to these compounds — *unsaturated aliphatic hydrocarbons*. The general formula for this class is C_nH_{2n} and the *-ene* ending is used to indicate unsaturation (or the presence of a double bond). In alkenes, the H—C—H and H—C—C angles are 120°. This is consistent with the idea that the carbon atoms are sp^2 hybridized.

The parent member of this class is ethylene, $H_2C{=}CH_2$. It is planar, consistent with the sp^2 hybridization of the carbon atoms (see also Section 7–D). The next member is obtained by substitution of a methyl (H_3C—) group for a hydrogen to give propene or propylene, H_3C—$CH{=}CH_2$. When there are four or more carbon atoms, the situation becomes more complicated, since the location of the double bond is not predetermined. An example is the two unbranched molecules

$$\overset{4}{C}H_3-\overset{3}{C}H_2-\overset{2}{C}H{=}\overset{1}{C}H_2 \qquad \overset{4}{C}H_3-\overset{3}{C}H{=}\overset{2}{C}H-\overset{1}{C}H_3$$

1-butene or butylene   2-butene or *i*-butylene

In propylene and longer alkenes, the molecule contains both sp^2 carbon atoms

$\left({>}C{=}C{<} \text{ portion}\right)$ and sp^3 carbon atoms (the alkyl portion).

The unsaturated compounds which contain more than one double bond are known as *polyenes*. The first of this series is butadiene, $H_2C{=}CH{-}CH{=}CH_2$. The double pair of unsaturated carbon atoms is indicated by the ending *-diene*. Of course, there are molecules with three double bonds (e.g., hexa*triene*), and so forth. In general, all of the polyenes are planar, consistent with the carbon atoms being sp^2 hybridized.

Recall that ethylene is a planar molecule containing a double bond (one σ and one π bond). Consider the possibility of rotation around this double bond compared with rotation around a single bond. Of course, when all atoms attached to carbon are the same (hydrogen) we can not tell if a rotation of 180° has occurred. However, if we substitute one —CH_3 group for one H on each carbon to make 2-butene (a substituted ethylene), then it is possible to have two distinct geometric arrangements:

trans-2-butene	*cis*-2-butene
methyl groups on opposite sides	methyl groups on same side

Is it possible to interconvert between the *cis* and the *trans* form? Earlier we found that rotation around a single bond (σ bond) cost little energy (about 3 kcal); therefore, essentially free rotation existed. Thus, the σ part of the double bond should present no barrier to twisting. The π-bond, however, arises from a side-to-side overlap of the *p*-orbitals, and as we twist one *p* orbital with respect to the other around the single bond we decrease the overlap of the *p*-orbitals, until at 90° there is no overlap (Figure 7–16). The most stable bond or the situation of lowest energy occurs when the maximum overlap is attained. Now, if maximum overlap occurs at 0° and by twisting to 90° there is no overlap, rotation must require energy. Furthermore, a second twist by 90° in the same direction increases the overlap once again and re-establishes the existence of the π-bond and gives the *trans* form. A schematic energy diagram of this is given in Figure 7–17. The

FIGURE 7–16 Schematic diagram showing the effect of twisting upon *p*-orbital overlap

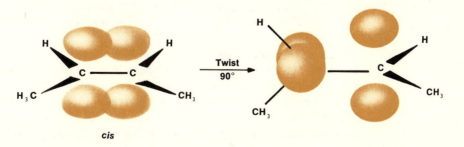

FIGURE 7–17 Energy diagram for rotation around a double bond. Note that the same energy is required to twist 90° beginning with either the *cis* or *trans* form.

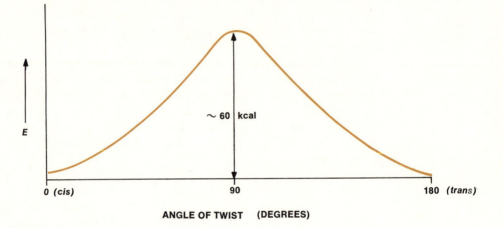

interconversion between the *cis* and *trans* forms, or isomers, requires approximately 60 kcal of energy, which hinders rotation about a double bond (in contrast to free rotation about a single bond). Because of this barrier to rotation, it is possible to isolate *cis* and *trans* isomers which have different physical properties. For example, the *cis* isomer of 2-butene boils at +1 °C. while the *trans* isomer boils at +4 °C.

If we were to measure the dipole moment of the *cis* and *trans* butene isomers, we would find they are different. For the sake of argument—and better available data—let us consider the case where we replace the —CH₃ groups with chlorine. There are significant differences in the dipole moments of the two dichloroethylenes as well as in other properties.

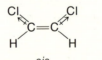

cis	*trans*
dipole moment = 1.85 D	dipole moment = 0.0 D
boiling point = 60°C	boiling point = 48°C
melting point = −80°C	melting point = −50°C

It can be seen that the vectors representing the dipole moment cancel for the *trans* isomer (equal in magnitude, opposite in direction). For the *cis* isomer, however, a resultant vector exists, giving a finite value for the dipole moment.

Alkenes are often used to produce complex large molecules called polymers. Polymers include plastics such as polyethylene or polypropylene, and rubber (polybutadiene). Not all polymers are plastics. Synthetic rubber is a polymer containing butadiene and another alkene linked together. Proteins are polymeric molecules that are discussed in Chapter 18. Polymers are discussed further in Section 7–M.

PROBLEM 7–11   Is it possible to have distinguishable *cis* and *trans* isomers of 1-butene (see text for formula)? Briefly defend your answer.

PROBLEM 7–12   Note in the potential energy diagram of Figure 7–17 that when proceeding in either direction from 90° the energy decreases. Briefly explain this result.

ALKYNES

The most highly unsaturated hydrocarbons contain a triple bond and constitute a class known as the *alkynes*. These have the general formula C_nH_{2n-2}. The parent and most important member of this class is *acetylene*, $HC{\equiv}CH$. Acetylene is a linear molecule, consistent with the two carbon atoms being *sp* hybridized. The hydrogen atoms can be replaced by alkyl groups to make methyl acetylene, for example:

$$CH_3{-}C{\equiv}CH$$

Acetylene is important as one of the reactants in the preparation of special synthetic rubbers such as neoprene. Also, it is used as a fuel in acetylene torches. Acetylenes are generally quite reactive compounds and as such are useful as reactants to synthesize more complex molecules.

7–I. ALIPHATIC ORGANIC COMPOUNDS CONTAINING OXYGEN

ALCOHOLS AND ETHERS

After carbon and hydrogen, oxygen is the next most common element found in organic molecules. The compounds of oxygen containing carbon in its lowest oxidation state are *alcohols* and *ethers*. The general formulas of these are

$$R{-}OH \qquad R_1{-}O{-}R_2$$
$$\text{alcohol} \qquad \text{ether}$$

where R, R_1, and R_2 represent any alkyl group. For example, $H_3C{-}OH$ is methanol or methyl alcohol, and $H_3C{-}O{-}CH_3$ is (di)methyl ether. Alcohols can be considered to be derived from alkanes by replacing one hydrogen with a hydroxyl group ($-OH$). Also, it is possible to picture this as one hydrogen in water being replaced by an alkyl group. A group of atoms that have distinct chemical and physical properties is known as a *functional* group. *The functional group of an alcohol is an —OH group.* In the case of the ether, R_1—O—R_2, the easiest way to see the derivation of the compound is to consider the two hydrogen atoms of water as being replaced by the same or different alkyl groups. *The functional group of an ether is the C—O—C linkage.* All carbon atoms of saturated alcohols and ethers are *sp*³ hybridized. The C—O—H and C—O—C bond angles are generally on the order of the tetrahedral angle of 109°. However, this bond angle can vary considerably depending upon the size of the R groups attached to the oxygen atom. For large R groups the angle can increase to 125°.

The common name of simple alcohols consists of the name of the R group which is bonded to the OH group, followed by the word alcohol. The more formal name of simple alcohols consists of the name of the parent alkane, with the terminal -e replaced by the suffix -ol. For example,

$$CH_3—OH \qquad CH_3CH_2—OH \qquad CH_3\overset{\overset{\displaystyle OH}{|}}{C}HCH_3$$

| methyl alcohol | ethyl alcohol | isopropyl alcohol |
| (methanol) | (ethanol) | (isopropanol) |

Ethers are named by using the names of the alkyl groups attached to the oxygen atom and adding the word ether. For example

$$CH_3—O—C_2H_5$$
methyl ethyl ether

If the alkyl groups are identical the prefix *di-* is used to indicate two groups. For example

$$CH_3—O—CH_3$$
dimethyl ether

Sometimes the prefix *di* is dropped and it is then understood that R_1 equals R_2. (Dimethyl ether is also called methyl ether.)

The most familiar alcohol is probably ethanol, which is the major component of wines, beers, and liquors and is responsible for the physiological reaction caused by these beverages. Isopropanol is commonly used in rubbing alcohol. Ethanol is one of the few alcohols that is not toxic to the human body—in small doses, of course. Alcohols are important reactants in many industrial processes. The most familiar ether is probably (di)ethyl ether, which acts as an anesthetic.

If a molecule has more than one hydroxyl group, it is called a polyol such as diol (two-OH groups) and triol (three-OH groups). Sugars are polyols that have other functional groups as well (Chapter 18). The simplest diol has the common name ethylene glycol

$$\overset{\displaystyle 1 \qquad 2}{\underset{\displaystyle \overset{|}{OH}\;\overset{|}{OH}}{H_2C—CH_2}}$$

and the simplest triol has the common name glycerine

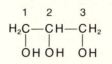

These can also be named as alkane derivatives. Ethylene glycol would be 1,2-ethanediol and glycerine would be 1,2,3-propanetriol. The glycol is the principal ingredient in anti-freeze. Glycerine is used in lotions and is a principal ingredient in the production of nitroglycerine.

The molecular formula for *both* ethanol and (di)methyl ether is C_2H_6O. Given this molecular formula, it would not be possible to know if it referred to the ether or the alcohol. On the other hand, the structural formulas distinguish the two: ethanol is C_2H_5OH or more correctly CH_3CH_2OH, while (di)methyl ether is $CH_3—O—CH_3$.

PROBLEM 7–13   Write the formula for the following:
(a) diethyl ether   (d) butyl alcohol
(b) methyl propyl ether   (e) isopropyl alcohol
(c) propanol

ALDEHYDES AND KETONES

The organic compounds containing oxygen doubly bonded to carbon are the *aldehydes*

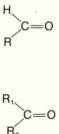

and *ketones*

Here again, R, R_1, and R_2 designate different alkyls group. The C=O portion of the molecule is commonly known as the *carbonyl group*. This portion of the molecule is planar, with bond angles of 120°, consistent with this carbon atom being sp^2 hybridized. The remaining carbons of the alkyl groups are sp^3 hybridized as expected. The simplest aldehyde is formaldehyde, $H_2C=O$, but the simplest aldehyde with an alkyl group is acetaldehyde, $CH_3—CH=O$. The simplest ketone is acetone (or dimethyl ketone), $CH_3—CO—CH_3$. The ketones have common names based on the types of alkyl groups they contain. For example, $CH_3—CO—C_2H_5$ is methyl ethyl ketone. In the case of dimethyl ketone, the common name acetone is customarily used.

Certain aldehydes and ketones occur naturally and are responsible for the smell of many substances, including the pleasant odor of newly mown hay. Aldehydes, especially formaldehyde, are made into polymeric plastics and resins. Aldehydes and ketones have medicinal properties, are used in perfumes, and have the ability to dissolve many organic compounds and therefore are used as solvents, as for example in paints.

PROBLEM 7–14   Write the formula for diethyl ketone.

CARBOXYLIC ACIDS AND ESTERS

The class of organic compounds with a hydroxyl group attached to a carbonyl group is the organic acids or *carboxylic acids*

$$R-C\underset{OH}{\overset{O}{<}} \quad \text{or} \quad RCOOH$$

where R again designates an alkyl group. The functional group —COOH is known as the *carboxyl group*. The $-C\underset{O}{\overset{O}{<}}$ portion of the molecule is planar with angles of 120°. Again, this is consistent with the carbon of the carboxyl group being sp^2 hybridized. The carbons of the alkyl group are sp^3 hybridized. The parent member is formic acid, HCOOH, and the first one containing an alkyl group is commonly called acetic acid, CH_3COOH. These compounds are acids, because in water a small fraction of the molecules give off protons (H^+) from the carboxyl group into the solution. The carbon atom of the carboxyl group is counted in determining the name of the acid (e.g., CH_3CH_2COOH is propionic or propanoic acid).

Esters are derivatives of acids in which the hydrogen of the —OH group is replaced by an alkyl group

These are commonly named as alkyl derivatives (R′) of the acid from which they are made by changing the *-ic* ending to *-ate*. Thus,

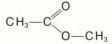

is called methyl acetate.

Acetic acid is probably the most familiar organic acid and is responsible for the acid properties (and sour taste) of vinegar. Esters are commonly found in nature and have a pleasant smell. Also, esters of glycerine constitute many fats and oils.

PROBLEM 7–15   Describe the structure, giving approximate bond angles, for butanoic acid, $CH_3CH_2CH_2COOH$.

OXIDATION STATES OF CARBON IN ORGANIC MOLECULES

The meaning of oxidation state in organic molecules is an important question. Remember at the beginning of this section we stated that the alcohols and ethers are the oxygen-containing organic molecules with carbon in its lowest oxidation state. The relative oxidation state of carbon can be surmised by evaluating the number of hydrogen atoms removed and oxygen atoms added. The carbon in the alkane is taken as the reference point and is in the lowest oxidation state. Consider the following sequence of oxidations of methane:

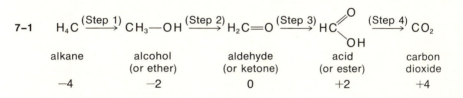

7–1

alkane	alcohol (or ether)	aldehyde (or ketone)	acid (or ester)	carbon dioxide
-4	-2	0	$+2$	$+4$

The oxidation state for carbon can be determined using the rules presented in Chapter 3. In organic compounds hydrogen has an oxidation number of $+1$ and oxygen has an oxidation number of -2. The oxidation number for carbon can then be evaluated as -4 in CH_4 and -2 in CH_3OH. The oxidation states for carbon in the sequential oxidation of methane are given underneath each compound in equation 7–1.

7–J. ALIPHATIC ORGANIC COMPOUNDS CONTAINING NITROGEN

We shall only be concerned about one kind of organic molecules containing nitrogen; these are the *amines*, R—NH_2.

The amines can be theoretically derived by replacing one hydrogen of ammonia (or two or three hydrogens) by an alkyl group.

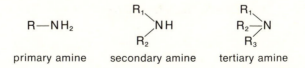

primary amine	secondary amine	tertiary amine

Also, of course, the amines could be considered to be derived from an alkane by replacing a hydrogen with —NH_2. The functional group —NH_2 (or $>N$—H or $\equiv N$) is known as the amino group. The names of amines are based on the alkyl group they contain (e.g., methylamine, CH_3—NH_2, and ethylamine, C_2H_5—NH_2). If there is more than one alkyl group of the same kind then the amine is called a di- or tri-alkyl amine (e.g., trimethylamine [$(CH_3)_3N$]).

All of the amines are bases, as is ammonia from which they may be thought to be derived. Thus, they increase the hydroxyl ion concentration when in a water solution

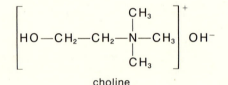

The carbon atoms of the alkyl groups are sp^3 hybridized, as might be expected. Based on the experimental R—N—H or R—N—R angles, the nitrogen atom must be approximately sp^3 hybridized.

Amines commonly occur in nature and can exist with additional substitution on the alkyl chain. For example, choline

$$\left[HO-CH_2-CH_2-\underset{\underset{\displaystyle CH_3}{|}}{\overset{\overset{\displaystyle CH_3}{|}}{N}}-CH_3 \right]^{+} OH^{-}$$

choline

is important in the utilization of fats in animals (including humans). A derivative of choline is an extremely powerful muscle contractant. Amines are also important as initiators of polymerization reactions to produce resins (for example, epoxy resins). Some of the most important biochemical molecules are the amino acids which are the building units for all proteins. Amino acids are molecules having an amino group at one end and a carboxylic acid group at the other.

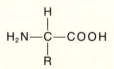

Atoms in addition to carbon may be included in the R portion of the molecule. These interesting compounds are discussed in Chapter 18.

PROBLEM 7–16 Write the formula for methylethylamine.

7–K. ALIPHATIC HALOGEN COMPOUNDS

Another class of compound of interest is the alkyl halides, with the general formula R—X, where X is any halogen. The H—C—H and X—C—H bond angles are near 109.5° (consistent with all carbon atoms being sp^3 hybridized). A typical compound is chloromethane, CH_3Cl, which is also commonly called methyl chloride. It is always possible to substitute for more than one hydrogen of an alkane to make polyhaloalkanes: dichloromethane, H_2CCl_2, or tetrachloromethane (often called carbon tetrachloride), CCl_4. One of the most important uses of the mixed haloalkanes, such as $CHCl_2F$ and CCl_2F—CCl_2F, is as refrigerants. The fluorinated compounds are inert and are particularly useful as lubricants and heat transfer agents. The completely fluorinated polymer —$(CF_2$—$CF_2)_n$— is known as Teflon (see Section 7–M).

PROBLEM 7–17   Name the following:
(a) CH_3CH_2Br (b) CH_3I

7–L. AROMATIC HYDROCARBONS AND DERIVATIVES

This class of hydrocarbons was not discussed with the others principally because of the unique nature of the bonding. In our discussion of resonance (Section 7–F), we noted that by both the valence bond approximation and molecular orbital theory, the double bonds of benzene (actually the π-bond portion) were not localized between pairs of atoms. Experimentally, it is known that all C—C bond distances are equal in benzene, and that the C—C—C and H—C—C bond angles are 120 °. This is consistent with the carbon atoms being sp^2 hybridized. There is a skeletal framework consisting of relatively localized σ bonds (from sp^2 orbital overlap). The remaining six p electrons (one from each carbon) overlap to form a π molecular orbital delocalized over the entire molecular ring:

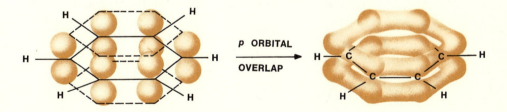

We have used the circle inside the hexagon to demonstrate that the bonds are equivalent and the π electrons are delocalized over the entire molecule.

Benzene is the parent member of the entire series of aromatic hydrocarbons. All of the aromatic hydrocarbons contain benzene rings fused together. Based on bond angles, the carbon atoms are *sp²* hybridized. All aromatic hydrocarbons are essentially planar. Some examples are

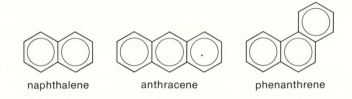

naphthalene   anthracene   phenanthrene

A vast number of substituted aromatic hydrocarbons exist since the hydrogen atoms can be replaced by many atoms or groups of atoms. These include the halogens, alkyls, $-NH_2$, $-OH$, $-COOH$, $-OCH_3$, $-CH=O$, and combinations of these. For example, some benzene derivatives are

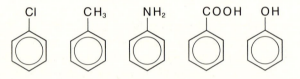

(In addition, the alkyl substituents can be further substituted with any of the groups noted above). The names of the derivatives are often derived from the nature of the substituent and the aromatic hydrocarbon. For example, in the above benzene derivatives, the chloro and methyl compounds are chlorobenzene and methylbenzene (or toluene).

In disubstituted benzene, the carbon atoms are identified by numbers going clockwise *or* by letters as *o-* (ortho), *m-* (meta) and *p-* (para)

where *S* is a substituent. For example

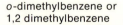

o-dimethylbenzene or
1,2 dimethylbenzene

1-methyl-3-chlorobenzene, *not*
m-chloro-1-methyl benzene

The simplest alcohol containing an aromatic ring is benzyl alcohol

When the hydroxyl group is attached directly to the ring it is known as phenol

In fact, the properties of this compound differ markedly from other alcohols, and phenol is not considered to be an alcohol. The phenols are weakly acidic whereas alcohols are not.

The aromatic hydrocarbons and their substituted derivatives are very important compounds. For example, aspirin is a substituted benzene derivative, as is oil of wintergreen,

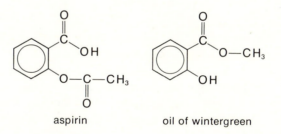

aspirin   oil of wintergreen

The old-fashioned mothballs were naphthalene while the new ones contain *p*-dichlorobenzene. Dopa, a compound used in the treatment of Parkinson's disease, is a benzene derivative.

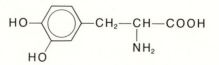

Thousands of compounds, important in medicine, plastics, resins, solvents, flavors, smells, and in biological systems, are substituted aromatic hydrocarbons.

7–M. POLYMERS

Polymers are formed by the linking together of a number of smaller molecules called monomers. Polymers are therefore often called macromolecules. An example of the formation of a polymer would be

$$H_2C{=}CH_2 + H_2C{=}CH_2 \longrightarrow \sim\!\!\sim CH_2{-}CH_2{-}CH_2{-}CH_2\sim\!\!\sim$$

where the greater the number of ethylene monomers, the longer is the polymer (polyethylene) produced. The process of reacting monomers to give polymers is called polymerization. Both synthetic and natural polymers exist. Table 7–6 gives some examples of synthetic polymers.

Many other polymers differ from those in Table 7–6 by substitutions and additional substituents. For example, if —CN is substituted for —Cl in polyvinylchloride then Orlon is the polymer. Also, if instead of polyvinylchloride

$$\left(CH_2{-}\underset{\underset{\textstyle Cl}{|}}{\overset{\overset{\textstyle H}{|}}{C}}\right)_n$$

we have

$$\left(CH_2{-}\underset{\underset{\textstyle Cl}{|}}{\overset{\overset{\textstyle Cl}{|}}{C}}\right)_n$$

then the polymer is Saran. Complete substitution of all hydrogens of polyethylene, Table 7–6, by fluorine gives Teflon: $(CF_2{-}CF_2)_n$.

Naturally occurring polymers include proteins, and polynucleotides like DNA. The proteins are polymers of amino acids which can be represented as

$$H_2N{-}\underset{R}{\overset{}{C}H}{-}\underset{O}{\overset{}{C}}\left(NH{-}\underset{R}{\overset{}{C}H}{-}\underset{O}{\overset{}{C}}\right)_n NH{-}\underset{R}{\overset{}{C}H}{-}COOH$$

where the repeating monomer unit is

$$\left(NH{-}\underset{R}{\overset{}{C}H}{-}\underset{O}{\overset{}{C}}\right)$$

Proteins can be of many different shapes. One important protein is hemoglobin, the oxygen-carrying protein in blood. More discussion of naturally occurring polymers will be given in Chapter 18.

TABLE 7–6 Some synthetic polymers

POLYMER	GENERAL FORMULAS

polyethylene $\left(CH_2-CH_2\right)_n$ or $\sim\sim\sim CH_2-CH_2-CH_2-CH_2 \sim\sim\sim$

polyvinylchloride $\left(\begin{array}{c} CH_2-CH \\ | \\ Cl \end{array}\right)_n$ or $\sim\sim\sim CH_2-\underset{|}{\underset{Cl}{CH}}-CH_2-\underset{|}{\underset{Cl}{CH}}\sim\sim\sim$

polypropylene $\left(\begin{array}{c} CH_2-CH \\ | \\ CH_3 \end{array}\right)_n$

polyester (Dacron) $\left(\begin{array}{c} C-\bigcirc-C-OCH_2CH_2O \\ \| \quad\quad\quad \| \\ O \quad\quad\quad O \end{array}\right)_n$

polymethylmethacrylate (Lucite, Plexiglas) $\left(\begin{array}{c} \quad\quad CH_3 \\ \quad\quad | \\ CH_2-C- \\ \quad\quad | \\ \quad\quad COOCH_3 \end{array}\right)_n$

polystyrene $\left(\begin{array}{c} CH_2-CH \\ \quad\quad | \\ \quad\quad \bigcirc \end{array}\right)_n$

Many synthetic polymers are well known and have many important uses. For example, Dacron is used in clothing, and polyvinylchloride is used as a plastic sheet to protect the color of meat (since O_2 diffuses through the plastic only very slowly). Polypropylene is used in indoor-outdoor carpet, and polyethylene and polypropylene in plastic items of all sorts. Lucite has been used as an unbreakable "glass." These are but a few examples of the myriad of polymers with an equal myriad of uses.

EXERCISES

1. Assuming no hybridization exists, what specific periodic property would account for the bond angle in H_2S and H_2O being 90°?

2. Could your answer to Exercise 1 be extended to other cases such as $(CH_3)_2S$ and $(CH_3)_2O$? Explain briefly.

3. Could your answer to Exercise 1 be generalized to other molecules with atoms different from sulfur and oxygen?

4. Assuming no hybridization, what would you expect the bond angles to be in
a. NH_3 b. H_2Se c. H_3Sb d. $AsCl_3$?

5. Is hybridization considered to occur for a single atom or does it occur between two or more atoms?

6. What does the criterion of overlap have to do with bonding and the structure of a molecule?

7. $PbCl_4$ is a low boiling liquid (105 °C). Explain why the boiling point is low and describe the nature of the bonding expected in this compound.

8. The molecule CH_3Cl has a dipole moment while the molecule CCl_4 does not. Discuss why this is true.

9. SnF_4 is a low boiling liquid. Do you expect SnF_4 to be ionic or covalent? Predict the geometry expected for this molecule.

10. The molecule 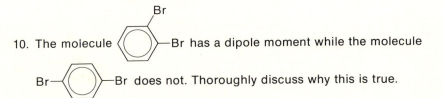 has a dipole moment while the molecule

Br—⟨ ⟩—Br does not. Thoroughly discuss why this is true.

11. Show whether the following have multiple bonds.
a. HCN (linear)   b. CS_2 (linear)   c. CO   d. OF_2 e. H_2CO.
Assume a maximum of eight electrons can surround any atom.

12. Predict the general geometry expected for the following:
a. CCl_4 f. PF_5
b. $[FeCl_4]^-$ g. $Sn(OH)_6^{2-}$
c. NF_3 h. $Au(CN)_2^-$
d. OF_2 i. WCl_6
e. $Cd(CH_3)_2$

13. Predict which compound of the following pairs will have the larger angle.
a. $AsCl_3$ vs. AsI_3
b. PF_3 vs. PBr_3
c. CF_4 vs. NF_3
d. NF_3 vs. OF_2

14. Show the possible resonance forms for the formate ion ($HCOO^-$).

15. Write the structural formula for cyclopentane. Do you expect the molecule to be planar? Explain.

16. Write the structural formula for
a. *n*-pentyl alcohol   b. methyl propyl ether   c. ethyl *n*-propyl ketone
d. diethylamine   e. *m*-diiodobenzene   f. 1-ethyl-2-bromobenzene.

17. Modern mothballs are *p*-dichlorobenzene. Write the structural formula for this.

18. In the molecule methanol, H_3COH, what would you expect the H—O—C bond angle to be? [Hint: Remember that this molecule can be considered as a derivative of water].

19. The simplest polyene is butadiene $H_2\overset{1}{C}=\overset{2}{C}H—\overset{3}{C}H=\overset{4}{C}H_2$. About which bonds should a substantial barrier exist to twisting by 180°?

20. Consider the molecule

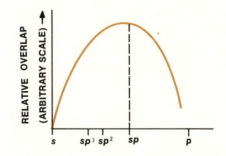

    a. Are there *cis* and *trans* isomers of this molecule? Briefly defend your answer.

    b. Will this molecule have a dipole moment? Briefly defend your answer.

21. In Chapters 4 and 5, we discussed effective nuclear charge. The effective nuclear charge of an *sp*³ hybridized carbon is less than that of an *sp*² hybridized carbon. Which carbon atom will be the more electronegative?

22. The following curve shows the relative overlap between two orbitals of the same type.

    a. Explain the results for the relative degree of overlap for each type orbital.

    b. Correlate this with the bond strength expected in each case.

23. A molecule such as $Ni(NH_3)_2Br_2$ can have either a tetrahedral structure (see Figure 7–3), with nickel at the center of the tetrahedron, or a square planar structure, as shown below

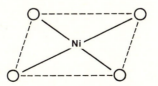

The circles represent either NH_3 or Br groups. If $Ni(NH_3)_2Br_2$ was found to have a zero dipole moment, what is the structure? [Hint: Can a tetrahedral structure of $Ni(NH_3)_2Br_2$ have a zero dipole moment?]

24. The compound $Co(NH_3)_4Cl_2$ has an octahedral structure (see Figure 7–3) with the cobalt at the center of the octahedron. The NH_3 and Cl occupy the corners of the octahedron.

a. Draw the two different structural isomers for his compound.

b. Predict which isomer will have a dipole moment.

CHAPTER EIGHT

MOLECULAR AND IONIC INTERACTIONS

In this chapter we shall be concerned about ion–ion, molecule–molecule, atom–atom and ion–molecule interaction. We will consider six types of interactions which affect one or more of the following properties: melting points, boiling points, solubility, and the overall or long-range structure of compounds. We will discuss these interactions beginning with the strongest and ending with the weakest. A rather special kind of molecule–molecule interaction, hydrogen bonding, exists, involving a hydrogen attached to an electronegative atom such as oxygen in an —OH group.

8–A. ION–ION

The interaction between positive and negative ions ranges in magnitude from 100 to 1000 kcal/mole. Ion–ion interactions were introduced in Chapter 6. When we are discussing ionic bonding, we must consider either the potential energy between the ions when a molecule in the gas phase is produced or the potential energy between the ions when a crystalline ionic solid is produced. Energy is released in both cases. The *lattice energy* is the energy released in the formation of one mole of the crystalline ionic substance. For example:

8–1a $\qquad Na^+_{(g)} + Cl^-_{(g)} \longrightarrow NaCl_{(g)} +$ energy (140 kcal/mole)

8–1b $\qquad Na^+_{(g)} + Cl^-_{(g)} \longrightarrow NaCl_{(s)} +$ energy (184 kcal/mole)

In the gas phase, the energy of interaction *per ion pair* is that noted in Chapter 6:

8–2

$$E = \frac{(n_+e)(n_-e)}{r}$$

where n_+ and n_- are the number of charges on the individual ions and r is the distance between the ions. If the right hand side of equation 8–2 is multiplied by Avogadro's number, N, then E will be the energy released *per mole* in equation 8–3a.

8–3a
$$M_{(g)}^+ + X_{(g)}^- \longrightarrow MX_{(g)} \qquad \text{energy released}$$

8–3b
$$M_{(g)}^+ + X_{(g)}^- \longrightarrow MX_{(s)} \qquad \text{energy released}$$

The energies involved in the formation of several different metal halides from atoms in the gas phase were given in Table 6–1. Table 8–1 gives the energies involved in the formation of some crystalline metal halides from ions (equation 8–3b). In a few cases in Table 8–1 the energy for formation of the gaseous molecules from ions is given in parenthesis for comparison. The lattice energy can also be determined by a Born-Haber cycle, Section 6–A, where $[NaCl]_s$ is substituted for $[NaCl]_g$ and therefore $E(NaCl)_s$ becomes identical to the lattice energy. Note that the trends in the lattice energy are parallel to the trends in energy for the formation of the gaseous molecule.

The calculation of lattice energy in the solid is not as simple as was the determination of energy of formation of the ion-pair molecule in the gas phase. In the gas phase only one positive and one negative ion are involved in the calculation of the coulomb energy—the energy of formation of the ion pair molecule. In the crystalline phase, recall that each positive and negative ion is surrounded by a number of ions of the opposite charge—six in the case of NaCl (Section 2–B). This means

TABLE 8–1   Energies of formation of some solid alkali halides ($MX_{(s)}$) from the ions[a] (kcal/mole released)

METAL ($M_{(g)}$)	HALIDE ($X_{(g)}^-$)			
	F	Cl	Br	I
Li	241 (220)	198		175 (139)
Na	216	184 (140)	175	165
K	192 (130)	167		151 (104)
Rb	184	163 (119)		146.5 (104)

[a]Numbers in parenthesis are the energies of formation of the gaseous molecule from the gaseous ions (equation 8–3a)

that the coulomb attractive energy is increased because of the simultaneous attraction of an ion by all its neighbors of opposite charge. Also, of course, the repulsive energy resulting from the interaction of the electron charge clouds is more significant than before. Nonetheless, it is possible to evaluate the lattice energies, and such calculations are in good agreement with experimental data.

Because of the strong interionic forces in an ionic crystal, the disruption of the lattice arrangement of the ions is difficult. This is reflected by the fact that ionic compounds have high melting points—NaCl (801 °C) and MgF_2 (1266 °C). High boiling points also reflect the strong interionic attractions. (The boiling point of NaCl is 1413 °C and for MgF_2, 2239 °C.) This is not the case for covalent crystals, where only relatively weak molecule–molecule interactions are involved (see Sections 8–C and 8–E).

The lattice energy is an important factor in determining the solubility of a salt in water. The larger the lattice energy, the more difficult it is to dissolve the salt, because the first step in dissolving a salt (equation 8–4) can be viewed as formation of the ions; this requires energy.

8–4
$$MX_{(s)} \longrightarrow M^+_{(g)} + X^-_{(g)} \quad \text{energy required}$$

The second step (equation 8–5) involves the interaction of the ions with the water molecule.

8–5
$$M^+_{(g)} + X^-_{(g)} \xrightarrow{aq} M^+_{(aq)} + X^-_{(aq)} \quad \text{energy released}$$

where *aq* stands for aqueous or for an undefined large amount of water. In this second step energy is released. The interaction between the water molecules and the solvated ions, $M^+_{(aq)}$, leads us naturally to a consideration of the ion-dipole type of interaction.

PROBLEM 8–1 Note in Table 8–1 that the difference in energy of formation of the potassium halides is only slightly greater than that for the same rubidium halides. However, the difference is significantly greater between the same potassium and sodium halides. Explain. [Hint: Consider what you learned in Chapter 5.]

8–B. ION–DIPOLE

Ion–dipole interactions involve less energy than do ion–ion interactions, but more energy than do the other interactions we will consider. The magnitude of the interaction is 10 to 160 kcal/mole. Although such interactions are not restricted to water as the polar molecule, water does represent the most common solvent. When water is involved the energy of interaction is known as the *hydration energy*.

Recall that the water molecule has a dipole moment pointing toward oxygen, resulting from the difference in electronegativity of hydrogen and oxygen. This can be pictorially represented as

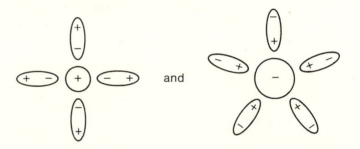

The + or − signs do *not* indicate ionic charges, but only the small charge difference (or partial charge separation) resulting from the electronegativity differences between the hydrogen atoms and oxygen atoms. This was written as δ^+ (H) and δ^- (O) in Chapter 6. The ion–dipole interaction between H_2O and the ions could then be schematically represented as

and

The energy of the interaction, hydration energy, depends on the number of water molecules surrounding the ion and on the charge density on the ion. The charge density (charge per unit volume) depends upon the ionic charge on the ion and its size (as deduced from the ionic radius).

The magnitude of the ion–dipole interaction has a significant effect on the solubility of salts. As noted in Section 8–A, energy is required to disrupt the lattice ion network. The solution process in water

8–6
$$MX_{(s)} \xrightarrow{aq} M^+_{(aq)} + X^-_{(aq)}$$

can be viewed as involving a first step, equation 8–4, which requires energy, and a second step involving an ion-dipole interaction, equation 8–5, which releases energy. The energy released in the ion–dipole interaction is critical in determining the tendency for an ionic compound to dissolve in a solvent.

We just noted that there is a dependency of the ion–dipole interaction, hydration, on the size and charge of the ion. Table 8–2 gives the energy of hydration for a number of ions in the reaction of equation 8–5. Several interesting points can be derived from Table 8–2. For ions with the same charge, the heat of hydration decreases as the size increases. A graph of hydration energy versus ionic radius is given in Figure 8–1. The hydration energy is inversely proportional to the ionic radius. Also, as the charge on a positive ion increases, the heat of hydration in-

TABLE 8-2 Heats of hydration of some ions

	ION	ENERGY RELEASED (kcal/g-atom)	IONIC RADIUS (Å), IN THE CRYSTAL[a]
Group IA	H^+	258	very small
	Li^+	121	0.67
	Na^+	95	0.97
	K^+	75	1.33
	Rb^+	69	1.48
	Cs^+	61	1.67
Group IIA	Be^{2+}	591	0.35
	Mg^{2+}	456	0.60
	Ca^{2+}	377	0.99
	Sr^{2+}	342	1.12
	Ba^{2+}	308	1.34
Group IIIA	Al^{3+}	1109	0.51
	In^{3+}	980	0.81
	Fe^{2+}	456	0.74
	Fe^{3+}	1041	0.64
Group VIIA	F^-	121	1.33
	Cl^-	90	1.81
	Br^-	82	1.96
	I	71	2.20
Group VIA	S^{2-}	330	1.84

[a]See Figure 5–17 of Chapter 5 where the relative sizes of some ions are shown.

creases. For example, compare Na^+, Ca^{2+} and In^{3+} (Table 8–2 and Figure 8–1), which have approximately the same ionic size. The same is true for negative ions; the heat of hydration of Cl^- is less than that of S^{2-}, which is in the same period and has approximately the same ionic size. The hydration energy varies approximately as the square of the charge on the ion. For example, the charge on the S^{2-} is -2 and the hydration energy is approximately $(2)^2 = 4$ times the hydration energy for Cl^-. The variation of the hydration energy with ionic charge and size can also be conveniently pictured in the periodic table shown in Figure 8–2.

Though the magnitude of the interaction does depend upon both the charge density and number of molecules surrounding the ion, the charge density is the more important factor. This can be seen in a couple of ways. The Ca^{2+} ion has a greater charge but is smaller than K^+, and therefore the charge density is greater for the Ca^{2+}. Of course, fewer water molecules are able to surround Ca^{2+} compared with K^+. Nonetheless, the heat of hydration of Ca^{2+} is considerably greater than that of K^+ (Table 8–2, and Figures 8–1 and 8–2). The same conclusion can be

FIGURE 8–1 Hydration energy versus ionic radius

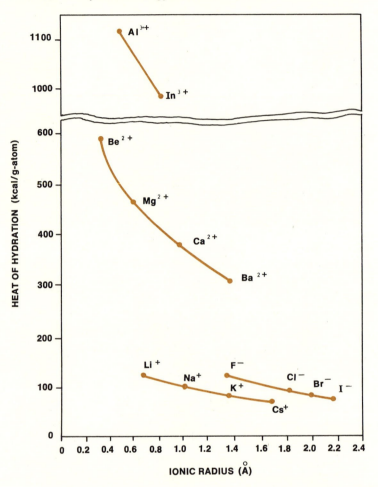

arrived at in the comparison of S^{2-} and Cl^- ions. These two ions are very nearly the same size (Figures 5–17 and 5–16). However, the heat of hydration of S^{2-} is nearly 4 times greater than that of Cl^- (Table 8–2, Figures 8–1 and 8–2).

Another very important consequence of the ion–dipole interaction involves the ionization in water of molecules which are normally considered to be covalent. Examples of this include HCl, HBr, HI, HNO_3 and H_2SO_4. Such a reaction can be represented as

$$HCl + H_2O \xrightarrow{aq} H_3O^+_{(aq)} + Cl^-_{(aq)}$$

Note that we have not written H^+ but H_3O^+, since it is impossible for a bare proton to exist in water. Thus, it is H_3O^+, the *hydronium* ion, that is hydrated in water.

FIGURE 8–2 Heats of hydration of some ions (kcal/mole)

IA									VIIIA
H⁺ 258	IIA		IIIA	IVA	VA	VIA	VIIA		He
Li⁺ 121	Be²⁺ 591		B	C	N	O	F⁻ 121		Ne
Na⁺ 95	Mg²⁺ 456	TRANSITION ELEMENTS	Al³⁺ 1109			S²⁻ 330	Cl⁻ 90		
K⁺ 75	Ca²⁺ 377	Fe²⁺ 456 Fe³⁺ 1041					Br⁻ 82		
Rb⁺ 69	Sr²⁺ 342		In³⁺ 980				I⁻ 71		
Cs⁺ 61	Ba²⁺ 308								

(H_3O^+ represents a hydrated proton but other species such as $H_5O_2^+$ and $H_9O_4^+$ also exist.) Note from Table 8–2 and Figures 8–1 and 8–2 that the heat of hydration of H^+ is unusually high (258 kcal/mole), especially for a monopositive charged ion. Thus, the energy of hydration of the H^+ (plus the Cl^- ion) is sufficient to result in the formation of ions from a molecule normally considered to be covalent. The same is true for the other examples given. This ionization reaction is principally responsible for the extremely high degree of solubility of such molecules in water.

PROBLEM 8–2 Assuming the Se^{2-} ion were stable in solution, would you expect the hydration energy to be greater or less than S^{2-}? Briefly explain why.

PROBLEM 8–3 Note that the heat of hydration of H^+ is unusually high even for a monovalent ion. Provide a brief explanation for why this is true.

PROBLEM 8–4 Using the hydration energies for K^+ and Ba^{2+}, show that the hydration energy depends upon the square of the ionic charge.

8–C. DIPOLE–DIPOLE

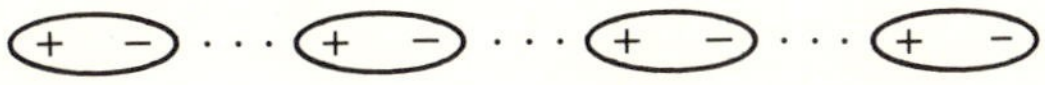

The interaction of the next greatest magnitude is that between covalent molecules that each have a dipole moment. The energy range of this interaction is approximately 1–6 kcal/mole. This is known as dipole–dipole interaction and can be represented as

or for HCl

$$\overset{\delta^+\delta^-}{HCl}\cdots\overset{\delta^+\delta^-}{HCl}\cdots\overset{\delta^+\delta^-}{HCl}$$

This is not meant to imply that it is necessary that there be a linear arrangement. In fact, except for diatomic molecules, a linear arrangement would not be the most probable one.

In the simplest case, where two linear molecules with dipoles, such as HCl, are interacting in the most favorable orientation, head to tail, the energy involved is

8-7
$$E = \frac{-\mu_1\,\mu_2}{r^3}$$

where μ_1 and μ_2 are the dipole moments of the two molecules and r is the distance between the dipoles:

Since the sign in equation 8–7 is negative, energy is released, and the interaction is an attractive one. A feeling for the energy involved can be obtained from the following consideration. If two linear molecules are aligned, each with a dipole moment of 1 D and 10 Å apart,

approximately 30 kcal/mole is required to separate them to an infinite distance. This presumes that a vacuum is present; the energy is less if they are imbedded in a solvent. Also, note that the energy depends upon $1/r^3$. In other words, the energy of attraction decreases quite rapidly with increasing distance between the dipoles.

In addition to the above considerations, there is a temperature dependence for dipole–dipole interactions. As the temperature increases, the thermal agitation prevents the ideal orientation from occurring; therefore the attraction decreases with an increase in temperature. For two dipoles in rapid thermal motion, the *average* potential energy of attraction is

8-8
$$E = \frac{-2}{3kT}\,\frac{\mu_1^2\,\mu_2^2}{r^6}$$

where k is the Boltzman constant (1.38×10^{-16} erg deg^{-1}) and T is the absolute temperature. It can be seen from equation 8–8 that at moderate temperatures the energy of interaction is proportional to $1/r^6$, instead of $1/r^3$.

The dipole–dipole interaction occurs both in the liquid and solid phases. Furthermore, the interaction is quite weak (considerably less than that of interionic attraction). Thus, the melting and boiling points of covalent polar molecules will be considerably less than for ionic compounds. For example, HCl has a melting point of $-115\,°C$ and a boiling point of $-85\,°C$, and for ClF these values are $-155\,°C$ and $-101\,°C$, respectively. Recall that for NaCl, an ionic compound, the melting point is $801\,°C$ and the boiling point is $1413\,°C$.

The dipole–dipole interaction is an important factor in the mutual solubility of one liquid in another. For example, the alkane hydrocarbon hexane is essentially insoluble in water whereas ethyl alcohol is essentially infinitely soluble. The principal reason for this is the existence of dipole–dipole interactions in the case of the $H_2O–C_2H_5OH$ system. This can be pictured as

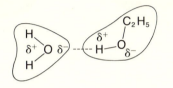

This interaction is not present in the $H_2O–C_5H_{12}$ system.

Even though a polyatomic molecule may have no *net* dipole, it may have individual bond dipoles. Therefore, there may be attractive interaction between individual parts of adjacent molecules. Dipole–dipole interactions can also account for the variation of other physical properties of solutions, as we will see in Chapter 12.

PROBLEM 8–5   Consider the case of HCl as a gas where molecular vibration is possible. Which equation would you use to calculate the dipole–dipole interaction energy, and briefly why?

PROBLEM 8–6   For Problem 8–5, will the dipole–dipole interaction energy depend on temperature, and which way will it change if the temperature is increased?

PROBLEM 8–7   Many organic molecules are not soluble in water. However, a common antifreeze, ethylene glycol, is obviously soluble in water. Account for this. [Hint: If you do not remember the formula for ethylene glycol, look in Chapter 7.]

8–D. DIPOLE–INDUCED DIPOLE

Thus far we have considered situations in which the species involved had a permanent charge (ion) or dipole (polar molecule). Some molecules, such as homonuclear diatomics, CH_4, $BeCl_2$ and others, do not have a permanent dipole (for reasons considered in Section 7–C). Also, recall that atoms such as helium or xenon do not have a dipole moment. Nonetheless, if a charge is brought close to such a molecule or atom, a charge asymmetry can be induced. This induced

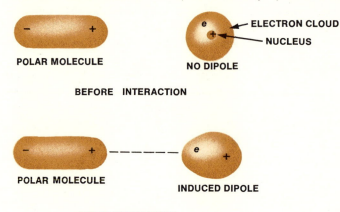

FIGURE 8–3 Inducement of a dipole moment by a polar molecule

charge asymmetry is equivalent to an induced dipole. This can be viewed as in Figure 8–3, where the charge which is inducing the dipole is one end of a polar molecule.

The amount of charge asymmetry that can be induced in a molecule or atom by a given charge depends upon the polarizability, α, of the molecule or atom. The magnitude of the polarizability depends upon how easily the electronic charge cloud can be distorted. Obviously, molecules with large and extensive charge clouds, such as I_2, can be distorted or polarized more easily than those with small, compact charge clouds such as F_2 or CH_4.

The magnitude of the induced dipole moment depends upon the polarizability of the charge cloud by a given charge. The energy involved in the dipole–induced dipole interaction is proportional to $1/r^6$. The exact formula for the potential energy of interaction is

8–9
$$E = \frac{-2\alpha\mu^2}{r^6}$$

where μ is the dipole moment of the polar molecule and α is the polarizability of the other nonpolar molecule or atom. The magnitude of the interaction energy is approximately 0.2–2 kcal/mole. Again, since the sign of the right hand side of equation 8–9 is negative, energy is released and the interaction is an attractive one.

The dipole–induced dipole interaction is in a large part responsible for the solubility of the inert gases in water. The water dipole is capable of sufficiently distorting the charge cloud of the inert gas atoms to induce a dipole. Also, crystalline hydrates of alkanes such as butane are known. The stability of this hydrate must largely have its origin from dipole (H_2O)–induced dipole (C_4H_{10}) interaction.

8—E. INDUCED DIPOLE—INDUCED DIPOLE

This type of interaction involves what are known as *London* or *dispersion forces*, and is perhaps the most difficult interaction to picture. However, the following should offer some insight. Assume that it were possible to take a photograph of a nonpolar molecule or an atom which shows the instantaneous location of electrons. A superposition of a large number of such photographs will show that there is no preference for any one distribution of electrons. However, individual photographs at various instants of time will show various distributions of electrons. Thus, we would have a large number of rapidly varying dipoles that in turn can polarize the charge clouds of neighboring molecules or atoms. These induced dipoles can then interact with the instantaneous dipoles which produced them (Figure 8—4).

Generally, this type of interaction will be weak, but it can be fairly strong if the molecular polarizability is high. The energy range is from 0.02–10 kcal/mole. The energy of interaction is proportional to $1/r^6$. The magnitude of the interaction

FIGURE 8—4 Induced dipole—induced dipole interaction. At any instant of time one system will have an asymmetric charge distribution and therefore a small dipole. This in turn can induce a dipole in another system. Of course, the roles could be reversed.

**TIME AVERAGED
CHARGE DISTRIBUTION**

**SOME INSTANT OF
TIME FOR ONE SYSTEM**

**SOME INSTANT OF
TIME FOR TWO SYSTEMS**

depends upon the polarizability of the molecules or atoms involved. For two species of the same kind, the induced dipole–induced dipole interaction is given by

8–10
$$E = \frac{-3}{4} \frac{(\alpha^2)(IP)}{r^6}$$

where α is the polarizability and IP is the ionization potential.

Despite the fact that this is a relatively weak interaction, it is responsible in large part for the liquefaction of the inert gases and other molecules such as CH_4 and H_2. Also, of course, because the interaction is generally weak in the foregoing liquids, the boiling points of such substances would be expected to be low. This is true for the boiling points of helium, −269 °C, of H_2, −253 °C, and of methane, −164 °C. The boiling points of the inert gases are shown in Table 8–3. We would expect helium to be much less polarizable than xenon, which has a large, extensive charge cloud. Since the magnitude of the interaction depends upon the polarizability of the species involved, we can note the progressive increase in the boiling point from helium to xenon. In the case of I_2, the polarizability is large and the induced dipole–induced dipole interaction is of sufficient magnitude that I_2 is a solid at room temperature.

Thus far our considerations have been largely limited to two interacting species, at least in terms of the energy equations. However, for long-chained compounds such as polyethylene the pair interactions between chains become additive. In this case we consider that the C—H bond dipole is essentially negligible. Even if this were not true, the induced dipole–induced dipole forces between pairs of CH_2— groups of adjacent chains makes a significant contribution to the interactions of polymer chains (Figure 8–5). Naturally occurring long-chain alkanes also interact via an induced dipole–induced dipole mechanism.

PROBLEM 8–8   In general, do you expect the polarizability of N_2 to be more or less than that of H_2? What evidence is there for your answer?

TABLE 8–3 Boiling points of the inert gases and some other molecules

MOLECULE	BOILING POINT
He	−269 °C
Ne	−249 °C
A	−189 °C
Kr	−157 °C
Xe	−107 °C
N_2	−196 °C
H_2	−253 °C
CH_4	−164 °C

FIGURE 8–5 Structure of polyethylene and a schematic view of the induced dipole–induced dipole interaction between chains

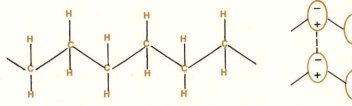

POLYETHYLENE POLYMER CHAIN   **INTERACTION OF TWO POLYMER CHAINS**

PROBLEM 8–9   a. In a general sense, what kind of electrons are polarizable in Kr and CH_4?
b. Compare the polarizability of Kr and CH_4.

PROBLEM 8–10   The boiling points of several alkanes are given in Table 7–3. Account for the variation in boiling points on the basis of intermolecular interactions.

8–F. ION–INDUCED DIPOLE

This interaction is the least common and the one for which we have the least data. In fact, the energies involved have not been well defined, so we shall only briefly mention one case. The energy of interaction between an ion and an induced dipole is proportional to approximately $1/r^4$. The formation of certain complex ions is caused at least in part by the energy released by this type of interaction, for example

$$I_2 + I^- \longrightarrow I_3^-$$

8–G. HYDROGEN BONDING

We are discussing hydrogen bonding separately since this type of interaction represents a rather special case of dipole interaction. It has been found that polar molecules which contain hydrogen attached to particularly nitrogen, oxygen, or fluorine have interactions with other molecules of a significantly greater magnitude than that which is expected from the magnitude of their dipole moments. The energy of a hydrogen bond is 2–10 kcal/mole.

A *hydrogen bond* is a relatively weak bond which is formed between a molecule containing a hydrogen covalently bonded to an electronegative atom (A) and a molecule (B) which is a proton acceptor; that is, B is a Brönsted base (Chapter 3). The hydrogen bond is commonly symbolized by a dotted line:

A—H···B

Note that the hydrogen is not a formal proton. However, because hydrogen is attached to an electronegative center it assumes a partial positive charge:

$$\overset{\delta^+}{H}\!-\!\overset{\delta^-}{A}$$

As the electronegativity of A increases, the partial positive charge on hydrogen increases. Because of the partial positive charge on the hydrogen atom, the hydrogen is called *protonic*.

It is found that as B becomes a stronger base (greater proton attraction), the resulting hydrogen bond (AH····B) becomes stronger. (This notation is not meant to imply that B is a single atom or ion, or that only diatomic molecules or singly bonded systems are involved.) Several examples of hydrogen bonded systems are:

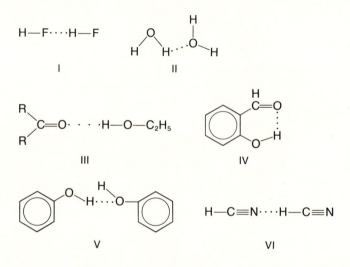

Note that in several cases A—H and B are identical—I, II, V, and VI—and it is an atom within the base B that acts as the proton acceptor. In compound IV, the hydrogen bonding occurs between groups within the same molecule. This is called *intramolecular* hydrogen bonding. If the interaction is between two molecules, it is called *intermolecular* hydrogen bonding. Hydrogen bonding, although a comparatively weak bond, is important in determining the long-range structures, or arrangement of molecules, of compounds in the liquid and solid states. For example, the long-range structure of liquid and solid water is determined in large part by hydrogen bonding (see Chapter 9). Hydrogen bonding is particularly important in helping extremely complex biochemical molecules maintain their shapes (see Chapter 18). Other examples of the importance of hydrogen bonding are given in the following pages.

Recall that the interaction of hydrogen bonding is greater than that expected from the dipole moment. This can be seen nicely by comparing the boiling point

TABLE 8-4 Comparison of the physical properties of C_2H_5OH and CH_3-O-CH_3

COMPOUND	DIPOLE MOMENT (D)	MELTING POINT (°C)	BOILING POINT (°C)
C_2H_5OH	1.7	−115	78
CH_3-O-CH_3	1.3	−141	−25

and melting point of ethyl alcohol (C_2H_5OH) with its isomer, dimethyl ether (CH_3-O-CH_3), as given in Table 8–4. Based on similarity of dipole moments, we would expect the interactions between the alcohol molecules to be comparable to those between ether molecules. However, because of the hydrogen bonding between alcohol molecules but not between ether molecules, the melting point and boiling point of the alcohol are 26 °C and 103 °C higher, respectively.

The hydrogen bonding for C_2H_5OH can be drawn as:

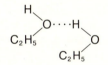

For CH_3-O-CH_3, the hydrogen atoms are not protonic, because of the relatively low electronegativity of carbon, and only a weaker dipole–dipole interaction exists:

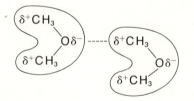

The easiest theoretical approach to understand hydrogen bonding is that employing the valence bond approximation. The principal resonance forms to be considered are

$$A-H \quad B \longleftrightarrow A^-H^+\cdots B \longleftrightarrow A^- \quad H-B^+$$
$$\quad\;\; I \qquad\qquad\qquad II \qquad\qquad\quad III$$

where I and II represent the covalent and ionic resonance forms of AH: $A-H \leftrightarrow A^-H^+$. The contributions of forms I and II account for the polar (dipole) character of the AH molecule; this in turn results in a dipole–dipole interaction with polar molecule B. The strength of the hydrogen bond cannot be satisfactorily accounted for solely on the basis of the dipole–dipole interaction resulting from combining forms I and II. The resonance form III makes some contribution to the strength of the hydrogen bond. Resonance form III involves a covalent bond between H and B⁺; the extent of its contribution is still a matter of question.

The strength of the hydrogen bond depends in part on the electronegativity of the atom bonded to hydrogen. Hydrogen bonds, depicted as A—H$\cdots$B, are strongest when A and B are fluorine, next when A and B are oxygen, and next when A and B are nitrogen. The hydrogen bond strength increases as the electronegativity of the atoms A and B increases.

In essentially all cases, the hydrogen bond, —H$\cdots$B, is longer than the covalent bond, A—H. However, in the case of the $(FHF)^-$ ion, there is only one H—F bond length of 2.26 Å, indicating that the hydrogen is at a point midway between the two fluorines. Usually hydrogen bonds have a bond dissociation energy of ~2–10 kcal/mole. However, in the case of $(FHF)^-$, the H—F bond energy is estimated to be approximately 58 kcal/mole.

Further indication of the influence of the electronegativity of the atom to which hydrogen is attached comes from boiling and melting point data. Figure 8–6 is a plot of the boiling points of a series of hydride (AH_n) molecules. The boiling points of H_2O, NH_3, and HF are abnormally high compared to the hydrides of the other elements belonging to their respective groups. For example, based on the trend of the boiling points of H_2Te, H_2Se, and H_2S, H_2O would be expected to have a considerably lower boiling point than it does. The deviation occurs because the electronegativity of oxygen is appreciably higher than that of tellurium, selenium, and sulfur. This produces a high degree of hydrogen bonding, which results in an abnormally high boiling point for H_2O. A similar argument explains the abnormally high boiling points of NH_3 and HF compared with the other molecules in their respective groups. However, note for the group SnH_4, GeH_4, SiH_4, and CH_4, that CH_4 has a lower boiling point than the other molecules. This happens because there is no hydrogen bonding in CH_4 to upset the expected trend.

As mentioned earlier, hydrogen bonding principally originates with a hydrogen atom covalently bonded to oxygen, nitrogen, or fluorine. However, it is possible for hydrogen bonding to involve other electronegative atoms as well. Consider the following examples:

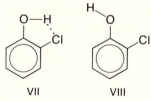

It is possible to experimentally observe the hydrogen bonding shown in formulation VII of *o*-chlorophenol. The tendency to form this hydrogen bond is so great that about 91 % of all such molecules exist as formulation VII and only about 9 % are not hydrogen bonded as in formulation VIII.

The existence of hydrogen bonding is responsible for many deviations from expected behavior and is responsible for structure or ordering of many organic

FIGURE 8–6 Boiling points for some hydride molecules versus period. Lines connect molecules containing atoms from the same group of the periodic table.

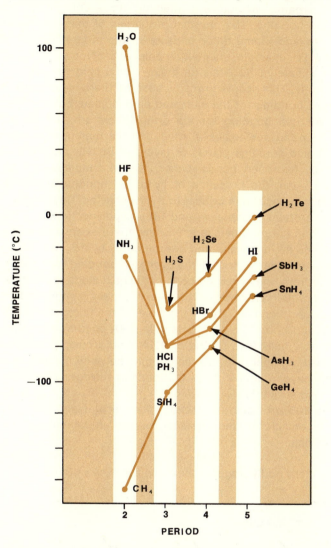

and inorganic compounds. Both acetic acid and formic acid form highly stable dimers resulting from hydrogen bonding. The structure of the carboxyl group, COOH, is such that two hydrogen bonds can form between two molecules. For example, the formic acid dimer has the structure:

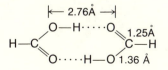

The hydrogen bond energy is 7.1 kcal/mole, which is quite high (for acetic acid it is 7.6 kcal/mole).

Most organic molecules containing an —OH group show intermolecular hydrogen bonding. The physical properties of many compounds having similar structures and atoms differ because of the difference in the kind of hydrogen bonding. For example, the melting point of o-chlorophenol [formulas VII and VIII] is approximately 0 °C, while for the m- and p-chlorophenols it is 29 °C and 41 °C, respectively. This occurs primarily because there is predominantly *inter*molecular hydrogen bonding for the meta and para isomers but predominantly *intra*molecular hydrogen bonding for the ortho. Therefore, the interaction *between* the o-chlorophenol molecules is small and the melting point correspondingly low.

Boric acid contains planes or layers of $B(OH)_3$ molecules which are held together by hydrogen bonds. Other inorganic molecules also exhibit hydrogen bonding. Also, the presence of hydrogen bonding between many biologically important molecules is critical to their behavior. This is true of natural polymers such as proteins (Chapters 7 and 18) and polynucleotides (Chapter 18).

In certain natural fibers, such as silk, there exists a fibrous protein whose protein chains lie in a plane and are united by hydrogen bonding (Figure 8–7). In a synthetic polymer, Nylon 66, the hydrogen bonding is of a similar type, except the distance from one hydrogen bond to the next along the chains is greater. In other cases, a protein chain takes the form of a helix (spiral), and the helix is held rigid by hydrogen bonding. This is the case for the fibrous proteins in hair, fingernail, muscle and horn. The polynucleotide DNA occurs in chromosomes in the form of a double helix (two chains of DNA wind around a common axis). The chains are joined together by hydrogen bonding between the hydrogens of amino groups and the oxygen atoms of C=O groups or nitrogen atoms (which are parts of a cyclic ring structure).

Table 8–5 provides a summary of molecular and ionic interactions, the factors responsible for the interactions, and an estimate of the magnitude of the interactions.

PROBLEM 8–11   Why is there no hydrogen bonding in the case of CH_4?

FIGURE 8–7 Silk fibers

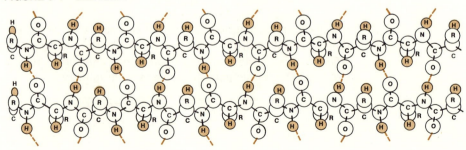

TABLE 8–5 Summary of molecular and ionic interactions

TYPE OF INTERACTION	PRINCIPAL FACTORS RESPONSIBLE FOR INTERACTION AND r DEPENDENCE	MAGNITUDE OF INTERACTION
ion–ion	charge on ions $$\frac{1}{r}$$	~100–1000 kcal/mole
ion–dipole	charge density ion, dipole moment	~10–160 kcal/mole
dipole–dipole	dipole moment $$\frac{1}{r^3} \text{ or } \frac{1}{r^6}$$	~1–6 kcal/mole
dipole–induced dipole	dipole moment, polarizability $$\frac{1}{r^6}$$	~0.2–2 kcal/mole
induced dipole–induced dipole	polarizability $$\frac{1}{r^6}$$	0.02–10 kcal/mole
ion–induced dipole	charge on ion, polarizability $$\frac{1}{r^4}$$	not well known
hydrogen bonding	dipole moment, electronegativity, and basicity	~2–10 kcal/mole

EXERCISES

1. Place the following compounds in order of increasing melting point and justify the ordering: LiF, RbI, RbF, CaF_2

2. The melting point of HCl and $LiCl$ are vastly different. Which do you expect to be higher and why?

3. Place the following in order of increasing boiling point and justify the ordering: $NaCl$, He, Br_2

4. Note in Table 8–2 and Figure 8–1 that the heats of hydration of ions at the beginning and end of a series have nearly the same value. Give some plausible explanation of how such similarities could exist.

5. The ionic radius of Tl^{3+} is approximately 0.95 Å. Assuming that Tl^{3+} ion could exist in water, would you expect it to have a greater or lesser heat of hydration than Al^{3+} and why?

6. Gasoline and water for all practical purposes do not mix—meaning there is no mutual solubility. On the other hand the antifreeze methyl alcohol and water obviously do mix. Give an explanation for these results. (See Chapter 7 if you have forgotten what these organic molecules are.)

7. Would you expect HI to dissolve in water to form hydriodic acid? Briefly explain why.

8. Crystalline hydrates of such ionic compounds as $CoSO_4$ and $CuSO_4$ are quite stable. However, as noted in this chapter, the crystalline hydrate of butane is relatively unstable. Offer an explanation for this observation.

9. The boiling point of H_2 is -253 °C while that for Cl_2 is -34 °C. Give a reason why this should be true. [Note: do not be concerned with the difference in molecular weights.]

10. The melting point and boiling point of acetic acid are 17 °C and 118 °C, respectively. The melting point and boiling point of the ethyl ester of acetic acid (Section 7–I) are -83.6 °C and 77 °C. Give an explanation for the differences between the molecules.

11. The melting point of *o*-hydroxybenzoic acid (Section 7–L) is 159 °C. The melting point of *p*-hydroxybenzoic acid (Section 7–L) is 213 °C. Offer a reason for this experimental result.

12. Note in Table 8–1 that most of the data for the bromides are missing.
a. Make a plot of the energies of formation of the lithium halides and the rubidium halides.
b. Predict by interpolation the energy of formation of $LiBr$ and $RbBr$.

13. In Section 8–C it was noted that the melting point of HCl was -115 °C and that of ClF was -155 °C. Offer an explanation for this difference.

14. Everyone knows ordinary salt ($NaCl$) is quite soluble in water. A friend of yours would like to know if and approximately how much salt in qualitative terms would dissolve in (a) octane and (b) ethyl alcohol. Provide an answer as well as an explanation to your friend.

15. Despite the fact that $NaCl$ is quite soluble in water, it is not as soluble as HCl (35.7 g/100 ml H_2O vs. 82.3 g/100 ml H_2O). Offer some possible qualitative explanation for this difference.

16. The melting points of two series of hydrides is as follows:

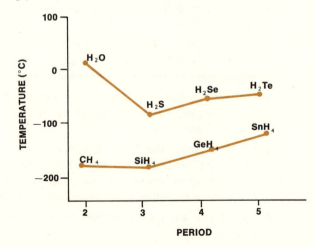

Offer an explanation for the relative positions of H_2O and CH_4 in each of the series.

17. Long-chained alcohols such as $CH_3-(CH_2)_n-OH$ where n is in the vicinity of 12 are often placed on the surface of large reservoirs to prevent water evaporation. Provide a physical explanation of how these compounds work. What is the nature of the interactions that could occur?

18. The interaction between ions is called a long-range interaction whereas induced dipole–induced dipole is considered a short-range interaction. Show why this is true by plotting E versus r for each of these interactions.

19. Place the following ions in order of increasing heats of hydration and give the reasons why:

$$Al^{3+}, Mg^{2+}, Na^+, Cs^+$$

CHAPTER NINE

CORRELATION OF MOLECULAR PROPERTIES AND REACTIONS WITH THE PERIODIC TABLE

We shall first be concerned in this chapter with general molecular properties and with the trends in these properties for some classes of compounds. The primary emphasis will be on physical and chemical properties displayed by all compounds of a given type. This approach discusses physical and chemical trends as a function of the relative location of the element concerned in the periodic chart. It will be much easier to understand this chapter if you refer frequently to the periodic table inside the back cover of the book as you read the chapter.

The chemical properties of the elements will then be discussed by group. Our primary focus will be on the reactions common to a group as a whole. Where important exceptions exist, these will be noted. In addition, the physical and chemical properties of some of the compounds produced will be considered. A table at the beginning of the discussion of each group will show properties of the elements and the acid-base properties of the oxides. In this chapter, you will begin to gain some insight into the common properties of the elements and compounds without being overwhelmed with a myriad of details to memorize. A table at the end of the discussion of each group will summarize the general reactions of the elements and some compounds.

9–A. GENERAL MOLECULAR PROPERTIES AND TRENDS

We shall first describe some general properties and trends of representative A-group elements. We will discuss the basis of these properties in more detail later on.

1. Reactions between elements of groups adjacent or nearly adjacent in the periodic table give compounds which are covalent. Compounds formed by reactions between groups far removed from one another are ionic.

2. The oxides and hydroxides of metals are bases, of metalloids amphoteric, and of nonmetals acids. An amphoteric oxide or hydroxide is one that can act as an acid or a base. Oxides and hydroxides become more acidic going from left to right in a period and less acidic going from top to bottom within a group.

3. The binary hydrogen compounds of the elements illustrate the combining power of the elements; for example, consider the third period.

IA	IIA	IIIA	IVA	VA	VIA	VIIA
NaH	MgH_2	AlH_3	SiH_4	PH_3	H_2S	HCl

The combining power of elements IA–IIIA is equal to their group number, and hydrogen is in a -1 oxidation state. For groups IVA–VIIA the combining power of the elements is equal to 8 minus the group number, and hydrogen has an oxidation state of $+1$ or -1 depending on the location of the element within the group. There is a marked change in the nature of the binary hydrogen compounds within a period. The hydrogen compounds of groups IA and IIA are ionic solids, while those of groups IVA–VIIA are generally gaseous and covalent. There are only minor changes in the nature of binary hydrogen compounds within a group; these are of no particular consequence in our considerations.

4. The binary compounds of chlorine can be used to illustrate the combining power of elements with the halogens; for example, consider the third period.

IA	IIA	IIIA	IVA	VA	VIA	VIIA
$NaCl$	$MgCl_2$	$AlCl_3$	$SiCl_4$	PCl_3	SCl_2	Cl_2

The combining power of the elements with halogens is in general
a. equal to the group number for groups IA–IIIA
b. equal to 8 minus the group number for groups IVA–VIIA
The chlorides can be considered to be generally representative of the halides. Within a period there is a change from ionic compounds at the left (groups IA and IIA) to covalent compounds on the right (groups IVA–VIIA). Also, there is a trend toward more ionic compounds going from the top of a group to the bottom of a group.

5. Alkyl compounds of the elements exist. In the third period, for example, with Me as methyl:

IA	IIA	IIIA	IVA	VA	VIA	VIIA
NaMe	$MgMe_2$	$AlMe_3$	$SiMe_4$	PMe_3	Me_2S	MeCl

Once again the combining power of the elements is readily seen to be
a. equal to the group number for groups IA–IIIA
b. equal to 8 minus the group number for groups IVA–VIIA
It is possible to consider the alkyl group as a single atom in order to establish the relative charge distribution in the molecules. For the group IA metal compounds, the alkyl group is negative relative to the metal atom. On the other hand, for the group VIIA compounds, the alkyl group is relatively positive. It can be seen that there is a trend within a period such that the alkyl groups are relatively negative on the left and positive on the right. This change affects both the physical and chemical properties of the compounds. There is a trend within the groups, particularly groups IVA–VIIA, for the alkyl group to become less positive (or become slightly negative) going from the top of the group to the bottom.

9–B. IONIC VERSUS COVALENT COMPOUNDS

We have seen that whether a compound is ionic or covalent may be considered a function of the groups from which the elements reacted come. Recall that electronegativity generally increases going from left to right within a period. Recall from Section 6–G that if the absolute value of the difference in the electronegativities of the elements is greater than 2, the compound will be ionic. The basis of our first generality (Section 9–A, 1) is that the reaction of adjacent groups involves elements of nearly the same electronegativity, leading to the formation of a covalent compound. The reaction of elements of groups quite far removed involves elements with significantly different electronegativities, to form an ionic compound. This can be nicely seen, for example, when we consider the melting points of the third period fluorides:

NaF	MgF_2	AlF_3	SiF_4	PF_3	SF_6	ClF
995 °C	1263 °C	1290 °C	−90 °C	−152 °C	−51 °C	−156 °C

The principal factors responsible for the marked change from left to right are the change from ionic to covalent bonding and the change in the atomic arrangement (see also Section 8–A). In NaF, Na^+ is surrounded by six F^- ions and each F^- by six Na^+. This structure is the same as that for NaCl given in Section 2–B. A giant network of Na^+ and F^- exists with strong bonds in many directions. On the other hand, in SiF_4 each Si is surrounded by four F and *molecules* of SiF_4 exist which are held together by relatively weak interactions (of the induced dipole–induced dipole type) as shown in Figure 9–1.

FIGURE 9–1   Arrangement of SiF_4 molecules in the solid

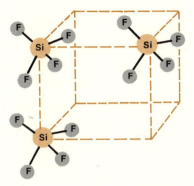

SUMMARY

We expect the compounds produced from reaction of group VIIA with groups IA and IIA to be essentially ionic and those from reaction of group VIIA with groups IVA, VA, and VIA to be essentially covalent; for example, KCl and $CaCl_2$ are essentially ionic while $SiCl_4$, PCl_3 and SCl_2 are essentially covalent. Also, we expect compounds made of elements from adjacent groups, such as nitrogen (group VA) and oxygen (group VIA), to be essentially covalent.

PROBLEM 9–1   Write the formulas for barium bromide and phosphorous (III) bromide. Do you expect these to be ionic or covalent? Why?

PROBLEM 9–2   Predict the nature of the bonding in K_2O and in NO. Justify your answer.

9–C. OXIDES AND HYDROXIDES OF THE ELEMENTS

BASES

The hydroxides of metals are hydroxyl ion donors and are, therefore, bases. The hydroxides of the nonmetals are hydrogen ion donors and are, therefore, acids. Metal hydroxides are most commonly written $M(OH)_n$; specific examples are NaOH and $Ca(OH)_2$. One of the important reasons why $M(OH)_n$ are OH^- donors is because of the low electronegativity of the metal and the high electronegativity of oxygen; the metal-oxygen bond is essentially ionic. Consequently, ionization in water occurs to give, for example:

$$NaOH \xrightarrow{aq} Na^+_{(aq)} + OH^-_{(aq)}$$

The hydroxides and oxides of metals are bases in that they react with acids and acidic oxides to produce salts:

$$\text{basic oxide} + \text{acid} \longrightarrow \text{salt} + H_2O$$

$$BaO + 2HNO_3 \longrightarrow Ba(NO_3)_2 + H_2O$$

$$\text{base} + \text{acidic oxide} \longrightarrow \text{salt} + H_2O$$

$$Ca(OH)_2 + SO_3 \longrightarrow CaSO_4 + H_2O$$

The hydroxides and oxides of groups IA and IIA (except beryllium) are all bases. In all of the other groups, only the hydroxides of indium and thallium (group IIIA) are solely bases.

ACIDS

The oxides and hydroxides of the nonmetals are acidic, with the general formula being $XO_m(OH)_n$, as in $SO_2(OH)_2$, or H_2SO_4, and $NO_2(OH)$, or HNO_3. The structural formulas for the two examples of H_2SO_4 and HNO_3 show that they are hydroxides.

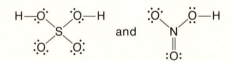

In these cases, the electronegativities of the nonmetallic central atoms nitrogen and sulfur are relatively high, comparable to that of oxygen. This results in a covalent bond between the central atom and oxygen and a highly polar O—H bond. In water, the ion-dipole interactions with water, as discussed in Section 8–B, result in H$^+$ ionization or donation. The oxides and hydroxides of nonmetals also are acids in that they react with bases and basic oxides to produce salts:

$$\text{acidic oxide} + \text{base} \longrightarrow \text{salt} + H_2O$$

$$SO_3 + 2KOH \longrightarrow K_2SO_4 + H_2O$$

$$\text{acidic oxide} + \text{basic oxide} \longrightarrow \text{salt}$$

$$CO_2 + CaO \longrightarrow CaCO_3$$

$$\text{acid} + \text{basic oxide} \longrightarrow \text{salt} + H_2O$$

$$H_2CO_3 + CaO \longrightarrow CaCO_3 + H_2O$$

The known hydroxides of the elements of groups VIA (except oxygen) and VIIA are all acids. Most of the hydroxides of group VA are also acidic. The hydroxides of the first element of group IIIA (boron) and group IVA (carbon) are the only ones in those groups that are solely acidic.

The oxides of group VIA (except oxygen) are acidic as are those of phosphorus and nitrogen (as NO_2) of group VA. The oxides of boron, carbon, and silicon are also acidic. The oxides of group VIIA are generally relatively unstable. Commonly, these do not react with bases directly.

AMPHOTERIC HYDROXIDES AND OXIDES

Those oxides and hydroxides which have a central atom which is a metalloid show both acidic and basic character. This property is called *amphoterism*. This means that the oxide or hydroxide can react with either a strong base or a strong acid:

$$Al(OH)_3 + 3HNO_3 \longrightarrow Al(NO_3)_3 + 3H_2O$$

and

$$Al(OH)_3 + NaOH \longrightarrow NaAlO_2 + 2H_2O$$

The actual product species that exist in solution are more complex than this formula suggests; they are hydrated as $Al(H_2O)_6^{3+}$ and $Al(OH)_4^-$. Oxides and hydroxides showing amphoteric behavior occur in those groups headed by a nonmetal and terminated at the bottom with a metal. This includes groups IIIA, IVA, and VA. Also, although beryllium is commonly considered to be a metal, its oxide and hydroxide are amphoteric.

SUMMARY

The trends in the acidic-basic character of the oxides are given in Figure 9–2. Those oxides within the heavy borders are amphoteric. Those oxides to the left are basic and those to the right are acidic. The trends in the hydroxides are given in Figure 9–3. Again, those hydroxides within the colored borders are amphoteric. Note that the formulas of the acidic hydroxides are written in the two ways discussed earlier in this section. There are some difficulties in considering the hydroxides since hydroxides of certain elements do not exist. Nonetheless, Figure 9–3 is a fair representation of what is known. There is a trend from ionic to covalent character going from left to right within a period for both the oxides and hydroxides.

PROBLEM 9–3   Complete and balance the following reactions, or note "no reaction" if none will occur.

FIGURE 9–2 The acid-base nature of oxides

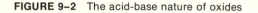
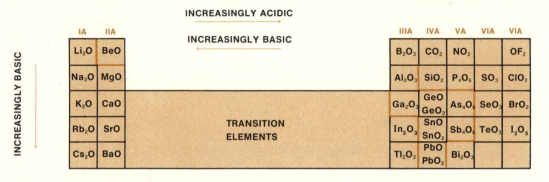
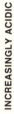

FIGURE 9–3 The acid-base nature of hydroxides

INCREASINGLY ACIDIC

INCREASINGLY BASIC

INCREASINGLY BASIC (left) — INCREASINGLY ACIDIC (right)

IA	IIA		IIA	IVA	VA	VIA	VIIA
LiOH	Be(OH)₂		B(OH)₃ / H₃BO₃	CO(OH)₂ / H₂CO₃	NO₂OH / HNO₃		FOH / HFO
NaOH	Mg(OH)₂		Al(OH)₃		PO(OH)₃ / H₃PO₄	SO₂(OH)₂ / H₂SO₄	ClO₂OH / HClO₃
KOH	Ca(OH)₂		Ga(OH)₃	hydrous GeO	As(OH)₃ / H₃AsO₃	SeO₂(OH)₂ / H₂SeO₄	BrO₂OH / HBrO₃
RbOH	Sr(OH)₂	TRANSITION ELEMENTS	In(OH)₃	hydrous SnO	hydrous Sb₂O₅	TeO(OH)₃ / H₃TeO₄	IO₂OH / HIO₃
CsOH	Ba(OH)₂		Tl(OH)₃	hydrous PbO			

a. $MgO + SO_3 \longrightarrow$

b. $Na_2O + \text{?} \longrightarrow Na_2CO_3$

c. $Ca(OH)_2 + HClO_3 \longrightarrow$

d. $LiOH + Mg(OH)_2 \longrightarrow$

PROBLEM 9–4 Consider the oxides Rb_2O and SO_3.

a. Do you expect these to be basic or acidic and why? How would you prove it?

b. Do you expect these to be ionic or covalent and why? How would you prove it?

9–D. BINARY COMPOUNDS OF HYDROGEN

The combining power of the elements with hydrogen is represented in Figure 9–4. The hydrogen compounds of groups IA and IIA in particular contain hydrogen in a −1 oxidation state. When an electric current is passed through a liquid melt of

FIGURE 9–4 The ionic-covalent nature of binary hydrogen compounds. Melting points are given in °C and d means "decomposes."

IA	IIA			IIIA	IVA	VA	VIA	VIIA
LiH 680	BeH$_2$ d			B$_2$H$_6$ −165.5	CH$_4$ −182.5	NH$_3$ −77.7	H$_2$O 0	HF −83.1
NaH d	MgH$_2$ d			not well defined	SiH$_4$ −184.7	Ph$_3$ −133.8	H$_2$S −85.5	HCl −114.2
KH d	CaH$_2$ d	TRANSITION ELEMENTS		Ga$_2$H$_6$ −21.4	GeH$_4$ −165.9	AsH$_3$ −116.3	H$_2$Se −65.7	HBr −86.9
RbH a	SrH$_2$ d			not well defined	SnH$_4$ −150	SbH$_3$ −88	H$_2$Te −51	H I −50.8
CsH d	BaH$_2$ 1200				PbH$_4$			

IONIC   COVALENT

→ INCREASINGLY COVALENT

such compounds, H$_2$ is liberated at the positive electrode (anode). This means that hydrogen exists as the negative hydride ion, H$^-$, in these compounds. The hydrogen compounds of groups IA and IIA are known as hydrides, are ionic and saltlike, and have high melting points. These hydrides react with water to produce H$_2$ and the corresponding hydroxide:

$$LiH + H_2O \longrightarrow H_{2(g)} + LiOH$$

$$BaH_2 + 2H_2O \longrightarrow H_{2(g)} + Ba(OH)_2$$

The binary hydrogen compounds of the elements in groups IIIA–VIIA are covalent and have low melting points. With groups VIA and VIIA, hydrogen has an oxidation state of +1. With groups IIIA–VA, except for the compounds with carbon and nitrogen (CH$_4$, NH$_3$), hydrogen generally has a −1 oxidation state. However, these compounds are still covalent and the pure liquid cannot be decomposed with electricity, as can be those of groups IA and IIA.

The binary hydrogen compounds of groups VIA and VIIA are acidic in water:

$$HCl \xrightarrow{aq} H^+_{(aq)} + Cl^-_{(aq)} \qquad \text{strong acid}$$

$$H_2S \rightleftharpoons H^+_{(aq)} + HS^-_{(aq)} \rightleftharpoons H^+_{(aq)} + S^{2-}_{(aq)} \qquad \text{very weak acid}$$

The group VIIA compounds are all strong acids except for HF. The compounds of group VIA are all very weak acids. The acid strength of the hydrogen compounds of group VIA increases proceeding from the top of the group to the bottom.

The binary hydrogen compound of nitrogen (in group VA) is basic in water solution: NH_3 reacts with water to produce NH_4^+ and OH^-. The remaining binary hydrogen compounds of group VA are neutral. Except for NH_3 they are all exceedingly poisonous, and SbH_3 is very unstable. The binary hydrogen compounds of group IVA are all neutral in water solution.

Note the exception in group IIIA of B_2H_6 (and Ga_2H_6) to the existence of simple binary hydrogen compounds. This compound is *electron deficient*; that is, if we attempt to draw a dot electronic formula:

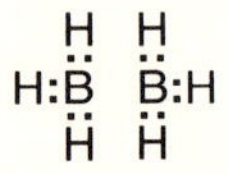

there are insufficient valence electrons to bond all of the atoms together. Based on experimental results, the structure of B_2H_6 is the one shown in Figure 9–5.

Note that in B_2H_6 no B—B bond exists and the bridging hydrogen atoms lie above and below a plane defined by the two BH_2 portions of the molecule. Thus the binding of the BH_2 portions occurs through two three-center molecular orbitals, each involving the two B atoms joined via an H bridging atom.

SUMMARY

The binary hydrogen compounds of the electropositive group IA and IIA metals are ionic saltlike compounds, whereas the compounds of elements of group IIIA–VIIA are covalent (Figure 9–4). The stability of the latter compounds decreases from the top of the group to the bottom. There is a trend to increasing acidic nature for the compounds going from left to right within a period. At the same time, the hydrogen atom changes oxidation state from -1 (groups IA and IIA) to $+1$ (groups VIA and VIIA). The group VIIA compounds, except for HF, are all strong acids in water solution.

PROBLEM 9–5

Complete and balance the following reactions:

a. $NaH_{(l)} \xrightarrow[\text{current}]{\text{electric}}$

b. $CaH_2 + H_2O \longrightarrow$

c. $HI + CaO \longrightarrow$

d. $NH_3 + HNO_3 \longrightarrow$

FIGURE 9–5 Structure of B_2H_6 showing bridging hydrogen atoms

H   H   H
 \   .   /
  B 97° B 122°
 /   \
H   H   H

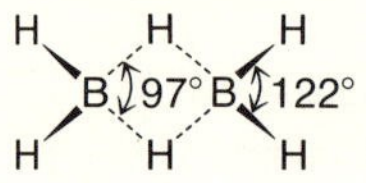

9–E. BINARY HALOGEN COMPOUNDS

The combining power of the elements with halogens is represented in Figure 9–6, using the chlorides as specific examples. Chlorine is in a −1 oxidation state in all of its binary compounds, except in Cl_2O and ClF, which are actually the oxide and fluoride of chlorine.

As might be expected from our discussion in Section 9–B, the halogen compounds exist in two major classes, ionic and covalent. The halides of groups IA and IIA are ionic, whereas those of groups IVA–VIIA are generally covalent. The latter is particularly true when an element of groups IVA–VIIA is in its highest oxidation state. For example, $SnCl_4$ and $PbCl_4$ are liquids with relatively low melting and boiling points (i.e., covalent compounds). However, $SnCl_2$ and $PbCl_2$ are high melting solids. Moreover, the molten salt of $PbCl_2$ is a moderately good conductor, indicating its ionic nature. As might be expected there is a trend toward increasing ionic nature going from the top of a group to the bottom. This is particularly noticeable in groups IIIA–VIA (Figure 9–6).

Certain compounds shown in the table of binary chlorides are relatively unstable, such as Cl_2O and SCl_2. The more stable compounds are S_2Cl_2 and Cl_2O_7. This occurs for other halides also—for example, the fluorides of highest stability for the elements of group VIA (except oxygen) are SF_6, SeF_6, and TeF_6. For group VIIA, BrF_5, and IF_5 are the most stable fluorides. It is not surprising that the elements combining with fluorine are in their highest oxidation state, since fluorine is the most electronegative element of all.

It is also not uncommon that certain halides do not exist. No stable iodide of sulfur, selenium, or nitrogen exists, and many of the interhalogen compounds (compounds formed between halogens) do not exist or have only low stability. It appears that only about thirteen interhalogen compounds are well defined. Note that there is always only one atom of the less electronegative element per molecule, for example: $BrCl$, ClF, BrF_3, IF_5, ICl.

FIGURE 9–6 The ionic-covalent nature of binary halogen compounds. Melting points are given in °C.

IA	IIA						IIIA	IVA	VA	VIA	VIIA	
	LiCl 610						BCl₃ −107	CCl₄ −22.9	NCl₃	Cl₂O	ClF −155.6	
	NaCl 808	MgCl₂ 714					Al₂Cl₆ 192	SiCl₄ −68	PCl₃ −92	SCl₂	Cl₂	
	KCl 772	CaCl₂ 782		TRANSITION ELEMENTS			Ga₂Cl₆ 77.5	GeCl₄ −49.5	AsCl₃ −16	SeCl₂	BrCl	
	RbCl 717	SrCl₂ 875					InCl₃ 586	SnCl₄ −33.3	SbCl₃ 73.2	TeCl₄ 224.1	ICl 27.3	
	CsCl 645	BaCl₂ 962					TlCl₃	PbCl₂ 498	BiCl₃ 232			

INCREASINGLY COVALENT →

← INCREASINGLY IONIC

INCREASINGLY IONIC

INCREASINGLY COVALENT

SUMMARY

Within a period the halides change from essentially ionic on the left to covalent on the right. The halides increase in ionic character going from the top of the group to the bottom.

PROBLEM 9–6 $AlBr_3$ is a nonconductor of electricity whereas AlF_3 is a conductor. Explain these observations.

PROBLEM 9–7   Write the formulas for the following. State whether they should be ionic or covalent.
a. cesium iodide
b. carbon tetrafluoride
c. magnesium fluoride
d. phosphorous (III) bromide

9–F. ALKYL COMPOUNDS OF THE ELEMENTS

In the cases where the element bonded to the alkyl group is a metal, compounds are referred to as *organometallics*. The combining power of elements with alkyl groups is shown in Figure 9–7. Note that the methyl compounds are similar to the hydrogen and halogen compounds shown previously with respect to the number of groups which combine with the element.

As we noted in Section 9–A, considering the alkyl group as a single atom permits an assignment of the relative charge distribution in the molecule. The alkyl groups are relatively negative when combined with elements of groups IA–IIIA. The names of the methyl derivatives of groups IA–IIIA are based on the negative methide ion, CH_3^-; for example,

$$NaCH_3 \qquad Ba(CH_3)_2$$

sodium methide   barium methide

FIGURE 9–7   The covalent nature of alkyl compounds. The methyl group (Me) is used as a representative alkyl group.

IA	IIA	INCREASINGLY COVALENT →			IIIA	IVA	VA	VIA	VIIA
LiMe	BeMe$_2$				BMe$_3$	Me$_4$C	Me$_3$N	Me$_2$O	MeF
NaMe	MgMe$_2$				Al$_2$(CH$_3$)$_6$	SiMe$_4$	Me$_3$P	Me$_2$S	MeCl
KMe					GaMe$_3$	GeMe$_4$	Me$_3$As	Me$_2$Se	MeBr
RbMe			TRANSITION ELEMENTS		InMe$_3$	SnMe$_4$	Me$_3$Sb	Me$_2$Te	MeI
					TlMe$_3$	PbMe$_4$	Me$_3$Bi		

When the alkyl group is relatively negative, the compounds undergo two reactions of interest to us. These occur particularly for all alkyl compounds of groups IA and IIA, as follows:

1. spontaneous inflammability in air
2. hydrolysis to produce the metal hydroxide and the alkane:

$$NaCH_3 + H_2O \longrightarrow NaOH + CH_4$$

The alkyl compounds of carbon are organic molecules (Sections 7–G and 7–H). For example, the following can be thought of as representative derivatives of methane:

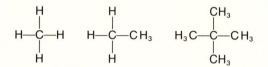

The alkyl compounds of nitrogen are amines (Section 7–J), and therefore are derivatives of ammonia:

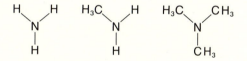

All of these are bases. The alkyl compounds of oxygen are ethers:

$$H_3C—O—CH_3$$

The alkyl compounds of the halogens are aliphatic halides. As will be seen in Section 9–H, these can be reacted with magnesium to make compounds which are important in the synthesis of organic compounds.

The methyl compound of aluminum is $Al_2(CH_3)_6$. This is another example of an electron deficient compound.

PROBLEM 9–8   Silicon is a member of group IVA. Write the molecular formula for the methyl and ethyl derivatives of silicon. Describe the structure of these molecules.

PROBLEM 9–9   The methyl derivative of lead is tetramethyl lead. What A-group do you expect lead to be in, based on the formula for tetramethyl lead? What structure do you expect for tetramethyl lead?

PROBLEM 9–10   Describe the structure of the methyl derivatives of the second period elements (i.e., LiMe, BeMe$_2$, . . . MeF).

*9–G. GROUP IA–ALKALI METALS

Some properties of the alkali metals are given in Table 9–1.

The alkali metals are never found in nature as such because they are extremely reactive. All are soft, silvery metals with low melting points and low boiling points. In fact they can be cut with a knife, and cesium in a degassed vial can be melted by body heat. In the gas phase, diatomic molecules exist, such as Na_2. All the alkali metals are considered to be good conductors of electricity and heat. Cables of sodium encased in a tough plastic are being tested as a replacement for copper wire, an important application, as we are beginning to approach exhaustion of the known copper reserves. Sodium can be obtained cheaply and in large quantities from sea water. In its liquid state sodium can be used as a heat transfer medium for nuclear reactors.

The elemental form of the alkali metals can be prepared by passage of an electric current through the molten salt, $MX_{(l)}$. The process is known as *electrolysis*, of which details are given in Chapter 14.

The alkali metals display the expected general trends, proceeding from the top of the group in the periodic table to the bottom, in atomic and ionic radius, ionization potential, and electronegativity. All of the alkali metals are good reducing agents. Because they show a strong tendency to give up electrons to an acceptor, almost all alkali metal compounds are ionic.

The alkali metals react violently with water to produce the hydroxides and H_2, and considerable heat is released:

$$2Na + 2H_2O \longrightarrow 2Na^+ + 2OH^- + H_{2(g)} + 80 \text{ kcal}$$

TABLE 9–1 Properties of elements of group IA

	Li	Na	K	Rb	Cs
atomic number	3	11	19	37	55
valence configuration	$2s^1$	$3s^1$	$4s^1$	$5s^1$	$6s^1$
common form	metal-atomic	metal-atomic	metal-atomic	metal-atomic	metal-atomic
ionization potential (kcal)	124	118	100	96	90
electronegativity	1.0	0.9	0.8	0.8	0.7
atomic radius (Å)	1.23	1.57	2.03	2.16	2.35
common oxidation state	+1	+1	+1	+1	+1
ionic size (Å)	0.60	0.95	1.33	1.48	1.69
melting point (°C)	179	97	64	39	28
boiling point (°C)	1336	880	760	700	670
heat of hydration (kcal/mole)	121	95	75	69	61
acid-base character of oxide and hydroxide	basic	basic	basic	basic	basic

This reaction is an example of the reducing ability of the alkali metals. Electrons are given to the hydrogen ion in water to produce elemental hydrogen.

The alkali metals all react directly with the halogens, hydrogen, sulfur, and phosphorus to produce the corresponding halides, hydrides, sulfides, and phosphides. Lithium is the only one of the group that reacts with O_2 to produce the oxide. The others react to produce compounds containing the O_2^{2-} ion and known as peroxides (Section 9–L), such as Na_2O_2, or containing the O_2^- ion and known as superoxides (Section 9–L), such as KO_2, RbO_2, CsO_2.

The reaction of the alkali metals with liquid ammonia is particularly interesting. The alkali metals first dissolve in liquid ammonia to produce blue solutions which are excellent electrical conductors, due to the fact that cations and ammonia solvated electrons are present.

$$M + (m + n)NH_3 \rightleftharpoons M^+(mNH_3) + e^-(nNH_3)$$

The solvated electrons also are responsible for the color. Upon standing, a metal amide and H_2 are produced.

$$2K + 2NH_3 \longrightarrow 2KNH_2 + H_2$$

Nearly all of the salts of the alkali metals are soluble in water, including

1. halides, MX
2. nitrates, MNO_3
3. sulfates, M_2SO_4
4. hydroxides and oxides, MOH and M_2O
5. carbonates, M_2CO_3
6. phosphates, M_3PO_4
7. acetates, CH_3COOM
8. chlorates, $MClO_3$
9. sulfides, M_2S

The heats of hydration of the alkali metal ions tend to be low (Table 9–1) because of the low charge/size ratio of the ions (Chapter 8).

Table 9–2 gives a summary of some of the more important reactions of the elements and compounds of this group.

PROBLEM 9–11   In the vapor phase, the alkali metals are known to occur as diatomic molecules.
a. Predict the bond length of Li_2 and Cs_2.
b. Which molecule do you expect will have the highest bond dissociation energy? Why?

TABLE 9–2 Some general reactions of compounds of the group IA elements

REACTION[a]	COMMENTS
1. $2MX_{(l)} \xrightarrow{\text{electric current}} 2M + X_2$	used in preparation of pure metal
2. $2M + 2H_2O \longrightarrow 2MOH + H_2$	violent reaction
3. $2M + H_2 \longrightarrow 2MH$	elevated temperatures required; produces ionic hydride
4. $2M + X_2 \longrightarrow 2MX$	
5. $2M + S, Se, Te \longrightarrow M_2S$	
6. $3M + P, As, Sb \longrightarrow M_3P$	
7. $2M + 2NH_3 \longrightarrow 2MNH_2 + H_2$	high temperature required with $NH_{3(g)}$; in liquid NH_3 gives solvated e^- followed by reaction
8. $MOH + acid \longrightarrow salt + H_2O$	oxide M_2O also reacts with acid
9. $MOH \xrightarrow{aq} M^+_{(aq)} + OH^-_{(aq)}$	oxide M_2O also reacts with water to give same ions
10. $MH + H_2O \longrightarrow MOH + H_2$	
11. salts $+ H_2O \longrightarrow$ aq. solution	nearly all salts are soluble in water

[a]M stands for group IA metal; X is any halogen (F,Cl,Br,I); $MX_{(l)}$ is liquid (molten) salt

PROBLEM 9–12 Given that NaBr and RbBr are both soluble in water, which will have the higher heat of hydration? Why?

PROBLEM 9–13 Write balanced equations for the following:
a. $Li_2O + H_2O \longrightarrow$
b. $CsOH + H_2SO_4$ (complete reaction) $\longrightarrow$
c. $KH + H_2O \longrightarrow$
d. $Na + S \longrightarrow$
e. $Na + NH_3 \longrightarrow$
f. $Cs + H_2 \longrightarrow$
g. $RbBr \xrightarrow{aq}$
h. $NaBr_{(l)} \xrightarrow{\text{electrical energy}}$
i. $Cs + P \longrightarrow$

PROBLEM 9–14 Identify the cation and anion of the following and the oxidation state of each.
a. Li_2O
b. Na_2SO_4
c. RbI
d. $KMnO_4$
e. $Na_2Cr_2O_7$
f. CsH
g. KOH

*9–H. GROUP IIA–ALKALINE EARTH METALS

Some properties of the alkaline earth metals are given in Table 9–3.

The alkaline earths are not found in nature as elements because they are so reactive. All are metals varying in hardness, with beryllium being very hard and barium quite soft. Beryllium is found in the silicate mineral beryl, $Be_3Al_2(SiO_3)_6$; beryllium and many of its compounds are considered to be quite toxic. Magne-

TABLE 9–3 Properties of the elements of group IIA

	Be	Mg	Ca	Sr	Ba	Ra
atomic number	4	12	20	38	56	88
valence configuration	$2s^2$	$3s^2$	$4s^2$	$5s^2$	$6s^2$	$7s^2$
common form	metal-atomic	metal-atomic	metal-atomic	metal-atomic	metal-atomic	metal-atomic
ionization potential (kcal)	214	175	140	132	120	121
electronegativity	1.5	1.2	1.0	1.0	.9	.9
atomic radius (Å)	0.89	1.36	1.74	1.91	1.98	
common oxidation state	+2	+2	+2	+2	+2	+2
ionic size (Å)	0.31	0.65	0.99	1.13	1.35	1.43
melting point (°C)	1278	651	842	769	725	
boiling point (°C)	2970	1107	1487	1384	1140	
heat of hydration (+2 ion) (kcal/mole)	591	456	377	342	308	
acid-base character of oxide and hydroxide	amphoteric	basic	basic	basic	basic	basic

sium also occurs in a wide variety of silicates and is found abundantly in the ocean. Calcium occurs as calcium carbonate ($CaCO_3$) in chalk, marble, and limestone. Both strontium and barium occur most commonly as sulfates. All isotopes of radium are radioactive.

All of the alkaline earth elements are considered to be good conductors of electricity. The elements are commonly prepared by electrolysis of the fused halide salts.

A fact which was not pointed out in the discussion of the group IA elements is that the chemical properties of lithium, as the first member of the group, are different in several respects from those of the other members of group IA. For example, lithium is the only member of the group which can react directly with nitrogen. There are several other chemical differences, including the ready reaction of lithium with carbon and the relative thermal instability of its carbonate. These differences arise primarily because lithium, as the first member of the group, is small, and the ion has a high charge density. The same is true for beryllium in group IIA and, in fact, is generally true for all first members of groups. Consequently, these first members often show different chemical properties than the remaining elements of the group.

The general trends (in the periodic table) of ionization potential, electronegativity, atomic and ionic size, and heat of hydration are followed for the group IIA elements. The atomic and ionic size of the elements of group IIA are less than those of group IA because of the greater effective nuclear charge. The hydration energies of the group IIA ions are considerably greater than those of the group IA ions because of the considerably higher charge density of the ion. All of the alkaline earth metals are good reducing agents, although not as good as the alkali metals.

The reaction of the group IIA metals with water varies. Calcium, strontium, and barium react readily with cold water, whereas hot water is required for a reaction with magnesium, and beryllium does not react at all. When there is a reaction, it is of the same type as with the group IA metals; that is, H_2 is evolved and the metal hydroxide is formed.

$$Ba + 2H_2O \longrightarrow Ba(OH)_2 + H_2$$

The group IIA metals all react directly with oxygen and all other members of group VIA (sulfur, selenium, and tellurium).

$$2Ca + O_2 \longrightarrow 2CaO$$

$$Ba + S \longrightarrow BaS$$

The group IIA elements also react with carbon to form carbides:

$$M + 2C \longrightarrow MC_2$$

An important carbide is CaC_2 which can be reacted with water to commercially produce acetylene:

$$CaC_2 + 2H_2O \longrightarrow H—C\equiv C—H + Ca(OH)_2$$

However, the most feasible method to make CaC_2 is not by direct combination of calcium and carbon, but rather by the reaction:

$$CaO + 3C \text{ (coke)} \longrightarrow CaC_2 + CO$$

The elements calcium, strontium, and barium dissolve in liquid ammonia to give colored solutions, as do all the metals of group IA. Likewise, after the solution stands for a time, the metal amide is formed—for example, $Ba(NH_2)_2$ from an ammonia solution of barium.

The principal distinction of beryllium is in regard to the covalent nature of its compounds and the amphoteric nature of its oxide and hydroxide. The oxide or hydroxide can react directly with both acids and bases. Many of the compounds of Be, including the halides and oxides, show very low ionic character and therefore are largely covalent. For example, the electrical conductivity of $BeCl_2$ is 1,000 times less than that of $NaCl$.

Magnesium is unique from one particular point of view. It is possible to react magnesium metal with an organic halide to make a magnesium organic halide known as a Grignard reagent.

$$Mg + R—X \longrightarrow R—Mg—X$$

Commonly X is Cl, Br, or I. These Grignard reagents play an important role in the synthesis of many organic compounds.

For beryllium, the halide, nitrate, sulfate, carbonate, acetate, chlorate, and chromate (CrO_4^{2-}) are all very soluble. In the case of compounds of the other group IIA elements in water:

1. Halides are quite soluble except for the fluorides, which are relatively insoluble.
2. The nitrates are all relatively soluble.
3. The acetates are all relatively soluble.
4. The chlorates are all relatively soluble.
5. The oxides are relatively insoluble and react to produce hydroxides, except for BeO, which shows no solubility nor reaction.
6. The hydroxides are relatively insoluble, especially $Be(OH)_2$; $Ba(OH)_2$ is the most soluble (4.2 g/100 g H_2O).
7. The sulfates are quite insoluble, except for magnesium.
8. The carbonates are all relatively insoluble.
9. The chromates are relatively insoluble, except for magnesium.

In conclusion, Table 9–4 gives a summary of some of the more important reactions of the elements and compounds of group II.

PROBLEM 9–15   a. Do you expect the first ionization potential of the group IIA elements to be higher or lower than those of IA? Why?

TABLE 9–4   Some general reactions of group IIA elements and compounds

REACTION[a]	COMMENTS
1. $MX_{2(l)} \xrightarrow{\text{electric current}} M + X_2$	used in preparation of pure metal
2. $M + 2H_2O \longrightarrow M(OH)_2 + H_2$	elevated temperature required for Be and Mg
3. $M + \frac{1}{2}O_2 \longrightarrow MO$	
4. $M + S, Se, Te \longrightarrow MS$	
5. $M + H_2 \longrightarrow MH_2$	generally occurs at elevated temperature; produces ionic hydrides
6. $M + X_2 \longrightarrow MX_2$	
7. $3M + N_2 \longrightarrow M_3N_2$	occurs at high temperature and also with high pressure
8. $M + 2NH_{3(l)} \longrightarrow M(NH_2)_2 + H_2$	elevated temperatures required with $NH_{3(g)}$; liquid NH_3 gives solvated e^- followed by reaction
9. $M(OH)_2 + acid \longrightarrow salt + H_2O$	oxide MO also reacts with acid; Be compounds are amphoteric
10. $M(OH)_2 + H_2O \longrightarrow M^{2+} + OH^-$	oxide MO reacts with water to give same ions; Be compounds are amphoteric
11. $MH_2 + H_2O \longrightarrow M(OH)_2 + H_2$	
12. salts + $H_2O \longrightarrow$ solution	halides, nitrates, acetates, chlorates

[a]M stands for group IIA elements; X is any halogen (F,Cl,Br,I).

b. The first ionization potential of barium (group IIA) is less than that of lithium and about equal to that of sodium. Explain.

PROBLEM 9–16   a. Calculate the total heat of hydration of the ions of $CaBr_2$ and compare the value with that of the ions of $NaBr$. [Hint: Use data given in Chapter 8.]
b. Note that Na^+ and Ca^{2+} are nearly identical in size and yet have considerably different heats of hydration—why?

PROBLEM 9–17   How many moles of electrons would it take to produce 1 mole (actually 1 g-at wt) of K and Ba from KCl and $BaCl_2$.

PROBLEM 9–18   Write balanced equations for the following, if a reaction occurs:
a. $BaO + HCl \longrightarrow$ f. $Ba + NH_3 \longrightarrow$
b. $Na_2O + CaO \longrightarrow$ g. $Be + H_2O \longrightarrow$
c. $Sr(OH)_2 + HBr \longrightarrow$ h. $Ca + Br_2 \longrightarrow$
d. $Mg + O_2 \longrightarrow$ i. $BaCl_2 + H_2O \longrightarrow$
e. $Ba(OH)_2 + RbOH \longrightarrow$ j. $Ca + N_2 \longrightarrow$

∗ 9–I. GROUP IIIA ELEMENTS

Some properties of the group IIIA elements are given in Table 9–5.
The group IIIA elements are not found in nature as free elements. Boron is commonly obtained from borax, $Na_2B_4O_7 \cdot 10H_2O$. (This formula, employing the symbol

TABLE 9–5 Properties of the elements of group IIIA

	B	Al	Ga	In	Tl
atomic number	5	13	31	49	81
valence configuration	$2s^22p^1$	$3s^23p^1$	$4s^24p^1$	$5s^25p^1$	$6s^26p^1$
common form	nonmetal-covalent	metal-atomic	metal-possibly Ga_2 molecules	metal-atomic	metal-atomic
ionization potential	191	138	138	133	141
electronegativity	2	1.5	1.6	1.7	1.8
atomic radius (Å)	0.80	1.25	1.24	1.50	1.55
common oxidation state	+3	+3	+3	+3	+3
ionic size (Å)	0.20	0.50	0.62	0.81	0.95
melting point (°C)	2200	660	30	157	304
boiling point (°C)	2550	2467	2403	2000	1457
heat of hydration (kcal/mole)		1109	1115	980	990
acid-base character of oxide and hydroxide	acidic	amphoteric	amphoteric	basic	basic

·10H$_2$O, implies that although molecules of water are involved, the precise nature of their location and bonding is generally unknown.) Aluminum principally occurs as bauxite, Al$_2$O$_3$·2H$_2$O, from which the pure metal is obtained by electrolysis. The other elements of this group commonly occur as compounds mixed with zinc blende (ZnS). These elements are commonly purified by passing an electric current through aqueous solutions of the salts. Fibers of pure elemental boron have extremely high tensile strength. Boron is essentially a nonconductor of electricity; aluminum and indium are good conductors, and the remainder are relatively poor conductors.

The atomic and ionic size of the group IIIA elements is less than that of the group IIA elements primarily because of the increase in the effective nuclear charge. The hydration energies of the group IIIA ions are greater than those of the group IIA ions because of the considerably higher charge density on the group IIIA ions.

The elements of group IIIA show essentially no reactivity with water. At least in the case of aluminum, this nonreactivity is the result of protection of the metal by a coating of Al$_2$O$_3$ (aluminum oxide), and if this coating is removed, aluminum can react with water to produce H$_2$. All of the elements will react with hydrochloric acid to produce +3 ions, except for thallium, which produces a +1 ion. The elements also react with oxidizing acids, such as HNO$_3$ and H$_2$SO$_4$, except for aluminum, which again does not react because of the formation of the surface layer of Al$_2$O$_3$.

All of the group IIIA elements react with O$_2$ and S to produce the corresponding oxide and sulfide, for example:

$$4In + 3O_2 \longrightarrow 2In_2O_3$$

Thallium takes on the +1 oxidation state to produce Tl$_2$O and Tl$_2$S. The reaction to produce Al$_2$O$_3$ is particularly interesting since it is a very highly exothermic reaction—approximately 400 kcal produced per g-mol/wt Al$_2$O$_3$. This means that Al$_2$O$_3$ is extremely stable, and aluminum will reduce nearly any other metallic oxide to the free metal. The reaction of aluminum with Fe$_2$O$_3$ is so highly exothermic that it produces molten iron useful for welding in what is known as the *thermite process*. The oxides and hydroxides of the group IIIA elements show acidic, amphoteric, and basic properties: the compounds of boron are acidic, of aluminum and gallium amphoteric, and of indium and thallium basic.

The elements aluminum and gallium will react directly with strong bases, but boron requires molten alkali metal hydroxides. Salts containing complex anions, such as AlO$_2^-$ and BO$_3^{3-}$, are formed. In solution, AlO$_2^-$ exists as the Al(OH)$_4^-$ or [Al(OH)$_4$(H$_2$O)$_2$]$^-$ ion.

Nitrides of all of these elements exist. However, only boron and aluminum react directly with N$_2$ at elevated temperatures.

$$2B + N_2 \longrightarrow 2BN$$

The usual form of boron nitride is interesting because it has a layered structure like that of graphite (Section 9–J). High temperature and pressure will cause this layer form to convert to a diamondlike structure which is even harder than diamond. The elements of group IIIA essentially do not dissolve in or react with liquid ammonia, in contrast to the elements of groups IA and IIA.

The chlorides, bromides, and iodides are covalent. The boron trihalides act as electron acceptor molecules and form a variety of complex compounds. Common electron donor molecules are those containing nitrogen or oxygen, for example:

$$F_3B \ + \ :NH_3 \ \longrightarrow \ F_3B\!-\!NH_3$$

$$F_3B \ + \ :O\!\!\begin{matrix} \diagup CH_3 \\ \diagdown CH_3 \end{matrix} \ \longrightarrow \ F_3B\!-\!O\!\!\begin{matrix} \diagup CH_3 \\ \diagdown CH_3 \end{matrix}$$

The halide (all exist), nitrate (all exist except for boron), and sulphate (all exist except for boron) salts of the metals in the +3 oxidation state are soluble in water. The acetates and carbonates exist for all except boron and are soluble in water, but a reaction occurs with water such that the corresponding ions do not exist in solution. This reaction with water is known as *hydrolysis* (see also Chapters 13 and 16), for example:

$$Al(C_2H_3O_2)_3 \ + \ 3H_2O \ \rightleftharpoons \ Al(OH)_3 \ + \ 3HC_2H_3O_2$$

In the case of the halides, nitrates, and sulphates, partial hydrolysis of the hydrated metal ion occurs to produce an acidic solution, as:

$$[M(H_2O)_6]^{3+} \ + \ H_2O \ \rightleftharpoons \ [M(H_2O)_5(OH)]^{2+} \ + \ H_3O^+$$

In the case of aluminum, this reaction occurs to an extent sufficient to produce an acidity nearly the same as that of acetic acid in water. The Tl^+ ion (which, as we have mentioned, is a common form of thallium) is stable in water solution, whereas Tl^{3+} undergoes extensive hydrolysis. Boron halides hydrolyze in water to give boric acid.

Table 9–6 gives a summary of some of the more important reactions of the elements and compounds of group III.

PROBLEM 9–19   Why should the bromides of group IIA be essentially ionic while those of group IIIA are essentially covalent?

PROBLEM 9–20   The ionic size of the +3 ions of group IIIA are significantly smaller than those of +2 ions of group IIA. Explain.

PROBLEM 9–21   Write balanced equations for the following if a reaction occurs:
a. $Ga \ + \ O_2 \ \longrightarrow$ b. $Al(OH)_3 \ + \ HCl \ \longrightarrow$

TABLE 9-6   Some general reactions of group IIIA elements and compounds

REACTION[a]	COMMENTS
1. $M + H_2O \longrightarrow$ no reaction	Al will react if Al_2O_3 coating removed
2. $4M + 3O_2 \longrightarrow 2M_2O_3$	requires elevated temperatures
3. $2M + 6HCl \longrightarrow 2MCl_3 + 3H_2$	except Tl gives Tl^+; B, Ga, In react with other acids in parallel manner
4. $2M + 3X_2 \longrightarrow 2MX_3$	also TlX
5. $B(OH)_3 + base \longrightarrow BO_3^{3-} + H_2O$	B only
6. $M(OH)_3 + acid \longrightarrow M^{3+} + water$	Al and Ga
7. $2M + 3S \longrightarrow M_2S_3$	requires elevated temperatures
8. $2M + N_2 \longrightarrow 2MN$	B and Al require elevated temperatures; others by reaction with NH_3
9. $M + H_2 \longrightarrow$ no reaction	hydrides must be prepared indirectly; complex B hydrides exist
10. $[M(H_2O)_6]^{3+} + H_2O \rightleftharpoons [M(H_2O)_5(OH)]^{2+} + H_3O^+$	hydrolysis of +3 ion in solution for all except B; Tl^+ stable, while Tl^{3+} is hydrolyzed

[a]M stands for group IIIA elements; X is any halogen (F,Cl,Br,I).

c. $In(OH)_3 + HCl \longrightarrow$

d. $In(OH)_3 + NaOH \longrightarrow$

e. $Tl + H_2 \longrightarrow$

f. $BF_3 + (CH_3)_3N \longrightarrow$

g. $Al_2O_3 + HCl \longrightarrow$

h. $In + I_2 \longrightarrow$

i. $Ga + H_2SO_4 \xrightarrow{\text{complete}}$

＊ 9-J. GROUP IVA ELEMENTS

Some properties of the Group IVA elements are given in Table 9-7.

Of the Group IVA elements, only carbon is found in both the free and combined states. In the free state, carbon principally exists as diamond and graphite, whereas in the combined state it exists as hydrocarbons such as natural gas (CH_4), petroleum, and coal. Silicon is the second most abundant element on earth and always occurs combined as silica (SiO_2) or in silicate minerals like granite. Germanium and lead commonly exist in nature as the sulfides (PbS). Tin is most common in nature as the oxide (SnO_2).

Carbon in the form of diamond is a nonconductor of electricity. However, graphite is a good conductor. Silicon and germanium have only a very low conductivity and are considered to be semiconductors. Tin and lead are generally considered to be good conductors, with tin the better of the two. Graphite, tin, and lead are also good conductors of heat, with the latter two being better than graphite.

TABLE 9–7 Properties of elements of group IVA

	C	Si	Ge	Sn	Pb
atomic number	6	14	32	50	82
valence configuration	$2s^2 2p^2$	$3s^2 3p^2$	$4s^2 4p^2$	$5s^2 5p^2$	$6s^2 6p^2$
common form	nonmetal-atomic-covalent	nonmetal-atomic-covalent	metal-atomic-covalent	metal-atomic	metal-atomic
ionization potential	260	188	182	169	171
electronegativity	2.5	1.8	1.8	1.8	1.8
atomic radius (Å)	0.77	1.17	1.22	1.41	1.54
common oxidation states	+4, −4	+4, −4	+4, +2	+4, +2	+4, +2
+2 ionic size (Å)			0.73	0.93	1.20
melting point (°C)	>3550	1410	937	232	328
boiling point (°C)	4827 (diamond)	2355	2830	2270	1744
heat of hydration of +2 ion (kcal/mole)					350
acid-base character of oxide and hydroxide	acidic	acidic	amphoteric	amphoteric	amphoteric

Because of the rather substantial differences between graphite and diamond, we shall briefly consider the structures of these *allotropes*. Allotropes are different forms of an element existing in the same physical state. Figure 9–8 shows the crystal structure of graphite and diamond. Each carbon atom in diamond is tetrahedrally, covalently bonded (sp^3 hybridized) to four others in a giant molecule. However, in graphite, each carbon atom is strongly covalently bonded to three others in a hexagonal ring arrangement, which gives an infinitely extensive molecule in two-dimensional layers. The fact that these layers are only weakly bonded to one another is in part responsible for the lubricant properties of graphite. A better way to describe the situation in graphite is that carbon is sp^2 hybridized, while the remaining p electrons of each carbon atom form a molecular orbital delocalized over the entire two-dimensional molecule. In the case of diamond, there are no delocalized relatively weakly bound electrons. In graphite, therefore, when an electrical potential is applied, the electrons in the delocalized molecular orbital can move, and this results in electrical conductivity. As would be expected, the bonding differences are also responsible for the differences in hardness between diamond and graphite.

Note that carbon is substantially different, with respect to ionization potential and electronegativity, from the other members of the group; there is a sharp break between carbon and silicon, and then substantially no change. The break can be partially attributed to the fact that silicon is a metalloid. In addition, other changes expected throughout the rest of the group, such as in atomic radius, are modified by the fact that inner d shells are being filled.

FIGURE 9–8 Structures of   A. graphite and   B. diamond. In B the dark spheres define one tetrahedron with carbon at all four corners.

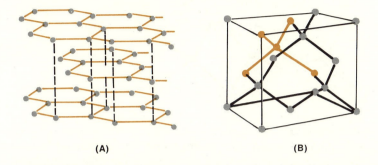

(A)   (B)

None of the elements of group IVA show any reactivity with water, although several of them react with acids. Silicon seems essentially unreactive in all acids whether dilute or concentrated, but tin and lead react with almost all acids and form the divalent metal ion and H_2 in dilute HCl. In concentrated oxidizing acids, lead, tin, and germanium react, but do not produce H_2.

All of the group IVA elements except carbon react with concentrated alkali metal hydroxide solutions or with the molten hydroxide. In all cases of reactions with the hydroxide solutions H_2 is liberated. Lead and tin exist in +2 oxidation states in hydrated complex ions:

$$Pb + 2NaOH + 2H_2O \longrightarrow Pb(OH)_4^{2-} + 2Na^+ + H_2$$

Silicon and germanium apparently exist in the +4 oxidation states, as for example, in GeO_3^{2-} or $Ge(OH)_6^{2-}$.

All of the group IVA elements react with fluorine, and except for carbon, all react with the other halogens to produce the corresponding halide. In all cases except tin (which has an oxidation state of +2), the most common oxidation state of the group IVA elements is +4. The tetrahalides are generally covalent. The Pb(II) fluoride, chloride, and bromide are ionic, a fact which points out the clearly metallic character of lead.

All of the group IVA elements react with O_2. All of the oxides produced contain these elements in the +4 oxidation state, except Pb(II). All of these elements can also form monoxides, although SiO is not common. Of the dioxides and monoxides, CO_2 and CO are the only ones gaseous at room temperature. There are a couple of points worth discussing concerning CO_2 and SiO_2. The two oxides are very different from each other in that individual molecules of CO_2 exist, whereas SiO_2 exists as a polymeric three-dimensional solid, *silica*, consisting of SiO_4^{4-} tetrahedra. Figure 9–9 shows a portion of the structure of silica. Note that each silicon atom is surrounded by four oxygen atoms, and each oxygen atom is common to two tetrahedra and therefore a neighbor to two silicon atoms. The reason for this difference between CO_2 and SiO_2 is considered to be the fact that

FIGURE 9–9 A portion of the structure of silica

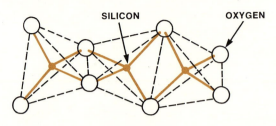

silicon does not form multiple bonds (so that only single σ type bonds can exist with oxygen, leading to the formation of condensed tetrahedra). Quartz is an example of pure silica. Because of the three-dimensional bonding and the Si—O bonding strength, quartz is a hard, high-melting substance.

CO_2 and SiO_2 are acidic, while the other dioxides are all amphoteric. CO_2 is fairly soluble in water and exists to a very small extent as the hydroxide, carbonic acid ($CO(OH)_2$), which is a weak acid (although it is not possible to isolate it). Ionization of carbonic acid gives very low concentrations of HCO_3^- and a still lower concentration of CO_3^{2-}, in addition to H^+ (or H_3O^+), according to the equilibrium:

$$H_2CO_3 \xrightleftharpoons{aq} HCO_{3(aq)}^- + H_{(aq)}^+ \rightleftharpoons CO_{3(aq)}^{-2} + H_{(aq)}^+$$

Carbonate equilibria are principally responsible for maintaining constant H^+ concentration in the blood and for the transportation to the lungs of CO_2 produced in metabolic processes. Carbonate salts react with acids to produce CO_2:

$$CaCO_3 + 2HCl \longrightarrow CO_2 + CaCl_2 + H_2O$$

Silica is essentially insoluble in water, and no well-defined hydroxide exists (hydrous SiO_2 does exist). Silica reacts with alkali metal hydroxides to form silicates. Many naturally occurring minerals such as feldspar, mica, and talc are silicates. Although many of the silicate minerals have a tetrahedral framework, as does quartz, they may contain other metal constituents besides Si. When Al, for example, replaces Si in some of the SiO_4^{4-} tetrahedra, another metal ion has to be present to maintain electrical neutrality. Often, this metal ion is from group IA, as in feldspar, $KAlSi_3O_8$, where some AlO_4^{5-} tetrahedra exist. (The formulas of silicates represent the empirical composition and not the molecular structure.)

Zeolites are sodium aluminosilicates, such as $Na_2Al_2Si_4O_{12}$, which are useful in chemistry. The tetrahedral framework and an open channeled structure give these materials very unusual properties. Zeolites are excellent for softening water. When water percolates through the channels, ions such as Fe^{3+} and Ca^{2+} exchange with the Na^+ and are removed, resulting in "soft" water. Zeolites are often called "molecular sieves" because of their ability to separate molecules of different sizes.

Calcium magnesium aluminosilicates occur as fibers and are widely used as asbestos. The semiprecious gem lapis is a sodium aluminosilicate containing S_x^{2-} anions, such as S_2^{2-} and S_3^{2-}, which are responsible for the blue color of the gemstone.

Ordinary glass, called *soft glass*, contains a mixture of silicates, about 84 % silicon and oxygen, and the remainder sodium, calcium, and aluminum. *Pyrex* glass is a boro-aluminosilicate glass containing about one-quarter the amount of sodium and calcium that is in soft glass. Pyrex glass is not as easily attacked by water or bases as soft glass is, and is resistant to breakage when subjected to rapid temperature changes. Unreacted cement is a mixture of calcium silicates and calcium aluminate ($Ca_3Al_2O_6$). When reacted with water, a series of complex reactions occur which ultimately produce highly interlocked calcium alumino-silicate crystals.

Carbon monoxide is toxic at very low concentrations because it competes with O_2 for binding to hemoglobin, the oxygen-carrying component of blood; hemoglobin has a higher affinity for CO than for O_2. Carbon dioxide is important in photosynthesis (Chapters 18 and 19).

Binary hydrogen compounds exist for all of these elements. The thermal stability of these compounds markedly decreases going from the top of the group to the bottom. As we discussed earlier, it is also possible for carbon atoms to be linked in long-chain carbon-hydrogen compounds of the alkane and alkene type. However, silicon compounds of the type Si_nH_{2n+2} (silanes), which are comparable to the alkanes (C_nH_{2n+2}), exist only up to $n = 6$; the germanes of formula Ge_nH_{2n+2} exist only up to $n = 3$; and only the simple hydrides of tin and lead exist (SnH_4 and PbH_4). There are no known compounds of silicon, germanium, tin, or lead with hydrogen where a double bond exists between the respective group IVA atoms. These facts, among others, point to the very special nature of carbon among all of the elements.

Recall that it is possible to make alkyl derivatives of the elements of this group. An important one is $Pb(C_2H_5)_4$, an antiknock additive for gasoline (Section 7–H). Silicones are polymers of (alkyl)$_2$SiO units, for example:

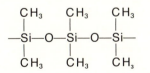

Such silicones have unusually good thermal stability and are chemically inert, making them valuable as lubricants. Other polymerized silicones exist—for example, in the form of silicone rubber.

All of the tetrahalides undergo hydrolysis in water except CCl_4, which essentially does not dissolve or react. In general no +4 ions exist in solution; but Sn^{2+} and Pb^{2+} can exist in solution, although Sn^{2+} undergoes hydrolysis. The nitrate and acetate are the only salts of Pb^{2+} that have any reasonable solubility in water.

TABLE 9-8 Some general reactions of group IVA elements and compounds

REACTION[a]	COMMENTS
1. $E + H_2O \longrightarrow$ no reaction	
2. $E + O_2 \longrightarrow EO_2$	requires elevated temperatures and Pb gives PbO; monoxides exist for all elements except Si
3. $3E + 4HNO_3 \longrightarrow 3EO_2 + 4NO + 2H_2O$	Ge and Sn only; Pb gives $Pb(NO_3)_2$
4. $E + 2HCl \longrightarrow ECl_2 + H_2$	Sn and Pb only
5. $E + OH^- \longrightarrow EO_2^{2-} + H_2$	Sn and Pb; Si and Ge give EO_3^{2-}
6. $E + 2F_2 \longrightarrow EF_4$	Pb gives PbF_2
7. $E + 2(Cl_2, Br_2, I_2) \longrightarrow EX_4$	except C; Pb gives PbX_2
8. $EO_2 + 2OH^- \longrightarrow EO_3^{2-} + H_2O$	C and Si; Ge, Sn, Pb exist as $E(OH)_6^{2-}$; dioxides of latter three elements are amphoteric
9. $EO + acid \longrightarrow E^{2+} + H_2O$	Ge, Sn, Pb only; also monoxides of these three elements are amphoteric
10. $E + H_2 \longrightarrow$ no reaction	hydrides must be prepared indirectly

[a]E stands for Group IVA element

Table 9-8 gives a summary of some of the more important reactions of the group IVA elements and compounds. Note that carbon and silicon, but especially carbon, participate in only a few of the reactions.

PROBLEM 9-22   Why would you expect graphite to be slippery and therefore be a lubricant whereas diamond is not?

PROBLEM 9-23   Tin and strontium are in the same period, but Sn (II) is significantly smaller than Sr (II). Explain.

PROBLEM 9-24   Write balanced reactions for the following, if a reaction occurs:
a. $CO(OH)_2 + HNO_3 \longrightarrow$ f. $SnO + LiOH \longrightarrow$
b. $Na_2CO_3 + HI \longrightarrow$ g. $SiO_2 + OH^- \longrightarrow$
c. $Pb + NaOH \longrightarrow$ h. $C + HNO_3 \longrightarrow$
d. $Si + H_2O \longrightarrow$ i. $C + H_2 \longrightarrow$
e. $SnO + HCl \longrightarrow$

✳ 9-K. GROUP VA ELEMENTS

Some properties of the group VA elements are given in Table 9-9.

The element nitrogen is found in the uncombined state as N_2. It also commonly exists in the combined state as the nitrate ($NaNO_3$). Phosphorus is found only

TABLE 9–9 Properties of elements of group VA

	N	P	As	Sb	Bi
atomic number	7	15	33	51	83
valence configuration	$2s^22p^3$	$3s^23p^3$	$4s^24p^3$	$5s^25p^3$	$6s^26p^3$
common form	gaseous-diatomic molecules	solid-non-metallic	solid-metallic-atomic	solid-metallic-atomic	solid-metallic-atomic
ionization potential (kcal)	335	254	226	199	168
electronegativity	3.0	2.1	2.0	1.9	1.9
atomic radius (Å)	0.74	1.10	1.21	1.41	1.52
common oxidation state	−3, +5	−3, +5	+3	+3	+3
+3 ionic size (Å)				0.76	0.96
melting point (°C)	−210	44[a]	814 (36 atm)	631	271
boiling point (°C)	−196	280[a]	633	1380	1560
acid-base character of oxide and hydroxide[b]	acidic N(V)	acidic P(V)	amphoteric As(III)	amphoteric Sb(III)	basic Bi(III)

[a] the white allotrope

[b] oxidation state of element is in parenthesis

in the combined state, chiefly as the phosphate—for example, $Ca_3(PO_4)_2$. The remaining members—arsenic, antimony, and bismuth—can be found in the uncombined state but most commonly exist as the sulfides (in the +3 oxidation state).

Nitrogen is a nonconductor of heat and electricity. Phosphorus exists in at least three allotropic forms: white, red, and black. The black form is the least common, but it is an electrical conductor whereas the others are not. Arsenic and antimony exist principally as two allotropes: yellow nonmetallic and gray metallic, the latter being the more stable. Bismuth exists only in a gray metallic form. Arsenic, antimony, and bismuth all have low electrical conductance in the metallic form.

Note that the ionization potentials of this group are unusually high (nitrogen has a greater ionization potential than oxygen). This can be explained by the p orbital being exactly half full, with one electron in each of the p_x, p_y, and p_z orbitals. The extra degree of stabilization which exists causes the ionization potentials to be higher than might be expected. There is a relatively sharp break between nitrogen and phosphorus in regard to ionization potential and atomic radius, followed by a modest rate of change for the remainder of the group. The filling of inner levels again becomes a prime factor modifying expected behavior beginning after phosphorus.

None of the group VA elements reacts with water or with acids such as HCl or H_3PO_4; however, all except nitrogen react with oxidizing acids such as HNO_3 and H_2SO_4. In the case of phosphorus and arsenic, the corresponding oxyacid is formed, H_3PO_4 and H_3AsO_4, whereas for antimony and bismuth, Sb_4O_6 and Bi^{3+} are formed.

All of the group VA elements except nitrogen can react with the halogens to produce the trihalides (except for phosphorus, which produces PF_5). Only a few pentahalides exist, for any of the group VA elements. Only four binary halides are known for nitrogen—N_2F_2, N_2F_4, NF_3, and NCl_3. All of the trihalides except NF_3 react with water to produce the corresponding hydrohalic acid, for example:

$$PBr_3 + 3H_2O \longrightarrow H_3PO_3 + 3HBr$$

$$BiCl_3 + H_2O \longrightarrow BiOCl + 2HCl$$

The NF_3 compound is unreactive with water, but NCl_3 is highly reactive, producing NH_3 and $HOCl$.

All of the group VA elements can react with O_2. In all cases except nitrogen, the element is in its +3 oxidation state. Nitrogen forms several oxides: N_2O, NO, NO_2, and N_2O_5. The oxide NO is produced in automobile engines, and in the air it is rapidly oxidized to NO_2. The compound NO_2 then participates in the initial step of smog production via a photochemical reaction (Chapter 19). NO_2 and oxides of phosphorus and arsenic, where the elements are in the +3 oxidation state, all react with water to produce acids. P_4O_{10} reacts with different amounts of water to produce several kinds of phosphoric acids—the one of interest to us is H_3PO_4. The oxides As_4O_6 and Sb_4O_6 are amphoteric, giving AsO_2^- and SbO_2^- ions with bases, and complex positive ions such as SbO^+ with acids.

The structure of the hydroxides of these elements varies significantly and therefore requires some discussion. A first point is that although the molecular formula for phosphorous acid, H_3PO_3, is parallel to that for arsenous acid, H_3AsO_3, different numbers of protons can ionize in the two acids. In the case of H_3PO_3 the structure is tetrahedral:

$$\overset{\displaystyle :\!\overset{\displaystyle ..}{O}\!:}{\underset{\displaystyle H}{\overset{\displaystyle |}{H\overset{\displaystyle ..}{\underset{\displaystyle ..}{O}}\!-\!\overset{\displaystyle |}{\underset{\displaystyle |}{P}}\!-\!\overset{\displaystyle ..}{\underset{\displaystyle ..}{O}}H}}$$

whereas for H_3AsO_3 it is

$$\underset{\displaystyle \overset{\displaystyle |}{OH}}{HO\!-\!As\!-\!OH}$$

Thus, the hydrogen atom attached directly to phosphorus *cannot* ionize to give H^+, whereas all three hydrogens can ionize in H_3AsO_3. However, H_3AsO_3 is a very weak acid, and the ionization of even one H^+ is very slight. The H_3PO_3 acid is considered a moderately strong acid, but its second H^+ ionizes only to a very small extent. The other acid of interest is (*ortho-*)phosphoric H_3PO_4 which has a tetrahedral structure

In water, only one H^+ ionizes to any significant extent,

$$H_3PO_{4(aq)} \rightleftharpoons H_2PO_{4(aq)}^- + H_{(aq)}^+$$

and H_3PO_4 is considered to be a moderately strong acid. However, all three H^+ ions can be replaced in steps by reaction with a strong base:

$$3NaOH_{(aq)} + H_3PO_{4(aq)} \longrightarrow Na_3PO_4 + 3H_2O$$

Phosphates commonly occur in washing detergents. However because they promote the growth of algae in streams and lakes, disrupting the natural ecology, the phosphates are being removed from detergents. Organic phosphate esters exist where one of the H of the —OH groups is replaced by an organic molecule. One such molecule of particular importance is called a nucleoside and consists of a sugar which is in turn attached to an organic base. Reaction or condensation of nucleoside-substituted phosphates, called *nucleotides*, through one of the free-OH groups of the phosphate produces the biologically important nucleic acid polymers, deoxyribonucleic acid (DNA) and ribonucleic acid (RNA). This reaction (discussed in more detail in Chapter 18) can be schematically represented as follows:

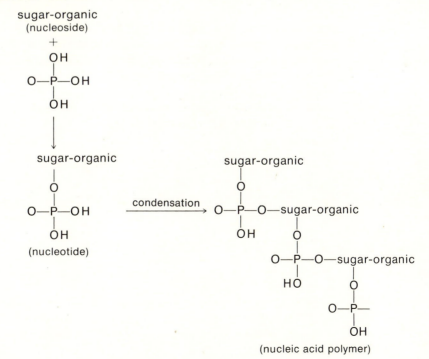

Because nitrogen is such an important element we shall pursue several additional aspects of its chemistry. Even though molecular N_2 is extremely inert at room temperature, nitrogen fixing bacteria are able to convert N_2 into organic nitrogen compounds. The sequence of reactions is not completely understood. However, we are obtaining some insight into this problem by studying transition metal complexes (Chapter 16).

The binary hydrogen compound of nitrogen, ammonia, is the only one of all those in group VA that is prepared directly from the elements. (General properties of the binary hydrogen compounds were considered in Section 9–D.) Recall that the melting and boiling points of NH_3 are abnormally high, compared with other hydrogen compounds of elements of group VA, because of hydrogen bonding. Ammonia, with a boiling point of $-33\,°C$, is a gas at room temperature.

The nitrogen atom in NH_3 can be thought of as approximately sp^3 hybridized, and the unbonded pair of electrons on nitrogen is highly spatially directed. Because of this NH_3 should be a good electron pair donor. (Recall from Section 9–I that NH_3 can react with BF_3 by essentially donating an electron pair for the bond.) NH_3 is highly soluble in water, mainly because of hydrogen bonding, reacting to produce some NH_4^+ and OH^-. Furthermore, because it is a base, it can react directly with hydrogen chloride or with hydrochloric acid as well as other acids to give the ammonium ion:

$$NH_{3(aq)} + HCl_{(aq)} \longrightarrow NH_{4(aq)}^+ + Cl_{(aq)}^-$$

Liquid ammonia is a good solvent. We mentioned earlier that the group IA and IIA metals dissolve in liquid ammonia followed by reaction to give the amide.

As noted previously, N_2 is quite unreactive, chiefly because of the high bond dissociation energy of N_2, 226 kcal/mole. This energy is so high because N_2 contains a triple bond: $:N\equiv N:$.

Table 9–10 gives a summary of some of the more important reactions of the group VA elements and compounds.

PROBLEM 9–25   Write the dot electronic formula for the nitrate ion.

PROBLEM 9–26   Why could the white allotrope of phosphorus have such a low melting point?

PROBLEM 9–27   Write balanced reactions for the following if they occur:

a. $NaOH + PO(OH)_3 \xrightarrow{complete}$ f. $K + NH_{3(l)} \longrightarrow$
b. $PO(OH)_3 + SO_2(OH)_2 \longrightarrow$ g. $P + Cl_2 \longrightarrow$
c. $NH_{3(aq)} + HNO_3 \longrightarrow$ h. $As + HCl \longrightarrow$
d. $Sb_4O_6 + HBr \longrightarrow$ i. $P_4O_6 + H_2O \longrightarrow$
e. $Sb_4O_6 + KOH \longrightarrow$ j. $As + O_2 \longrightarrow$

TABLE 9–10 Some general reactions of group VA elements and compounds

REACTION[a]	COMMENT
1. $E + H_2O \longrightarrow$ no reaction	
2. $4E + 3O_2 \longrightarrow E_4O_6$	P, As, Sb; Bi gives Bi_2O_3 and N gives NO
3. $E + HCl, H_3PO_4 \longrightarrow$ no reaction	
4. $E + HNO_3, H_2SO_4 \longrightarrow$ reaction	for all except N; see text
5. $E + NaOH \longrightarrow$ oxy-anion	except for N; oxy-anions are $(H_2PO_2^-)$, (AsO_2^-), (SbO_2^-)
6. $2E + 3F_2 \longrightarrow 2EF_3$	except for N; P gives PF_5
7. $2E + 3(Cl_2, Br_2, I_2) \longrightarrow 2EX_3$	except for N
8. $EX_3 + H_2O \longrightarrow HX +$ product	Except for N; product for P and As is H_3EO_3 and for Sb and Bi it is EOCl; NCl_3 also reacts
9. $E_4O_6 + 6H_2O \longrightarrow 4H_3EO_3$	· for P and As; NO_2 gives HNO_3 and P can give other acids
10. $E_4O_6 + acid \longrightarrow EO^+ + H_2O$	As, Sb, Bi (as Bi_2O_3); As and Sb are amphoteric; not N
11. $N_2 + 3H_2 \longrightarrow 2NH_3$	other hydrides prepared indirectly
12. $E + S \longrightarrow E_2S_3$	As, Sb, Bi; sulfides insoluble in H_2O

[a]E stands for a Group VA element

✳ 9–L. GROUP VIA ELEMENTS

Some properties of the group VIA elements are given in Table 9–11.

The group VIA elements are essentially nonmetals. We shall generally exclude polonium from our discussion because it is radioactive; note, however, that its chemistry is clearly more metallic in character than that of the remainder of the elements.

Oxygen, found both uncombined (O_2) and combined, is the most abundant element in nature. It constitutes about 50 % of the weight of the earth's crust. Sulfur is found in both the free state and in the combined state primarily as sulfides (S^{2-}) and sulfates (SO_4^{2-}). Selenium, tellurium, and polonium, which usually occur in the combined state, are very rare.

O_2 is a nonconductor of heat and electricity although it is paramagnetic (will be attracted into a magnetic field—Section 6–E). Sulfur is also a nonconductor of heat and electricity. Both sulfur and selenium exist in several allotropic forms. One of the forms of selenium is a poor conductor of electricity, but the conductivity increases when exposed to light. The metallic gray form exhibits a photovoltaic action—a reaction in which light is converted into electricity—which makes it useful in photo and solar cells. Polonium commonly exists as a low-melting metal with properties of conductivity quite typical of metals.

The chemistry of oxygen has already been discussed in relation to the chemistry of the elements of other groups. There are, however, a few things that can be added. Oxygen exists most commonly as O_2, but it also has another allotrope

TABLE 9–11   Properties of elements of group VIA

	O	S	Se	Te	Po
atomic number	8	16	34	52	84
valence configuration	$2s^2 2p^4$	$3s^2 3p^4$	$4s^2 4p^4$	$5s^2 5p^4$	$6s^2 6p^4$
common form	gaseous diatomic molecules	solid-non-metallic-S_8 molecules	solid-Se_n	solid-Te_n	solid-metallic-atomic
ionization potential (kcal)	314	239	225	208	194
electronegativity	3.5	2.5	2.4	2.1	2.0
atomic radius (Å)	0.74	1.04	1.17	1.37	1.64
common oxidation state(s)	−2	−2,+4,+6	−2,+4	−2,+4	+2,+4
−2 ionic size (Å)	1.40	1.84	1.98	2.21	
melting point (°C)	−218	113	217	452	254
boiling point (°C)	−183	445	685	1390	962
hydration energy (kcal/mole)		330(S^{-2})			
acid-base character of oxide and hydroxide		acidic	acidic	acidic	

called *ozone*, O_3 (boiling point of −112 °C). Ozone exists in significant quantities in the upper atmosphere, as well as in smog, and is produced via a photochemical reaction (Section 6–E; Chapter 19). Ozone is a bent molecule

with two principal resonance forms

$$O\diagdown_{O}\diagup^{O} \longleftrightarrow O\diagdown_{O}\diagup O$$

It is much more reactive than O_2 and is a strong oxidizing agent. For example, it can react with mercury at room temperature to produce HgO, whereas heating is required if O_2 is the reactant.

Oxygen reacts directly with all of the elements except those of group VIIA and VIIIA to produce the corresponding oxides, which will show acidic, amphoteric, or basic properties depending upon whether the element combining is a metal, metalloid, or non-metal. Oxygen can exist as the oxide ion, O^{2-}, superoxide ion, O_2^-, and the peroxide ion, O_2^{2-}.

Hydrogen peroxide, H_2O_2, is not planar. The two —OH groups are approximately perpendicular to one another (Figure 9–10). H_2O_2 is a liquid (boiling point of 150 °C, melting point of 0.9 °C) in which the molecules are strongly hydrogen

FIGURE 9–10   Structure of H_2O_2

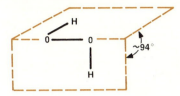

bonded. The compound is slightly acidic and is an oxidizing agent, but it is relatively unstable and decomposes:

$$2H_2O_2 \longrightarrow 2H_2O + O_2$$

Probably the most important compound of oxygen is H_2O. In Chapter 7 we discussed the structure of a water molecule in which the oxygen can be described as approximately tetrahedrally hybridized. In liquids, including water, there is a *long-range* structure, which means that there is an overall organized arrangement of the molecules within the liquid (Chapter 8). This is also true in solids including ice. The existence of hydrogen bonding in water is an important feature responsible for its abnormally high melting point (0 °C) and boiling point (100 °C) compared with the other binary hydrogen compounds of the group. This hydrogen bonding is also the principal factor producing long-range structure in the solid and liquid.

The structure of ice is such that each oxygen atom is covalently bonded to two hydrogen atoms and is hydrogen bonded to two other oxygens from adjacent H_2O molecules. Actually, an oxygen atom of a water molecule is surrounded tetrahedrally by the oxygen atoms of four adjacent water molecules, and the O—H bonds are directed along a line connecting any two oxygen atoms, as shown in Figure 9–11. Thus, solid water is a polymeric substance with hexagonal holes. This open structure is responsible for the fact that the density of ice is less than the density of liquid water.

The structure of liquid water is still an unsettled issue. However, all agree that there exist some water molecules that are not hydrogen bonded. Because of this the overall packing of the water molecules can become tighter relative to the solid, thereby causing the density to increase upon melting. It also appears that clusters of water molecules still exist that are tetrahedrally bound as in ice. In any event, upon melting there is both an overall decrease in the degree of hydrogen bonding and an overall collapse of the open structure present in ice. It appears that somewhere in the vicinity of 45 % of the hydrogen bonds present in ice are broken upon going to liquid H_2O. The energy required to break a mole of hydrogen bonds is 3.7 to 4.5 kcal.

The following discussion refers primarily to sulfur, selenium, and tellurium. Polonium will be excluded. None of these elements (including oxygen) reacts with water. Sulfur appears to be the only element in the group (including oxygen) that

FIGURE 9–11 Structure of ice. The large circles are oxygen atoms and the small circles are hydrogen atoms.

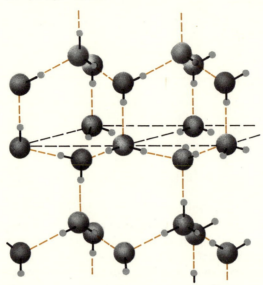

reacts with strong bases. However, all except oxygen react with concentrated HNO_3 and with O_2 to produce the corresponding dioxide; SO_2 is the only gaseous dioxide of the group. The dioxides react with water to produce hydroxides which are oxyacids, for example, sulfurous acid

$$SO_2 + H_2O \longrightarrow SO(OH)_2 \text{ or } H_2SO_3$$

Pure H_2SO_3 or H_2TeO_3 have not been made. Each of the acids is moderately strong and can dissociate two protons. However, dissociation is not complete, and the second H^+ ion comes off with considerably greater difficulty than the first. The dioxides and hydroxides are all acidic and show the typical reactions (Section 9–C). Sulfur trioxide, SO_3, is an acidic oxide which can react with water to produce the oxyacid sulfuric acid, H_2SO_4. In H_2SO_4, which is

$$H\ddot{O}-\overset{\displaystyle :\ddot{O}:}{\underset{\displaystyle :\ddot{O}:}{S}}-\ddot{O}H$$

two protons can dissociate in H_2O. It is considered a strong acid. Although removal of the second proton is not complete, it is nonetheless highly dissociated

$$H_2SO_{4(aq)} \xrightarrow{\ 100\ \%\ } H^+ + HSO_4^- \rightleftharpoons H^+ + SO_4^{2-}.$$

Obviously both SO_3 and H_2SO_4 react with bases or basic oxides. Peroxysulfuric acids exist, the most important of which is peroxydisulfuric acid (or persulfuric):

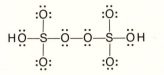

This acid can lose protons to produce the persulfate ion $S_2O_8^{2-}$, which is one of the strongest known oxidizing agents.

Metals react with oxygen, sulfur, selenium, and tellurium to produce the corresponding binary compound. The resulting alkali metal (group IA) and alkaline earth metal (group IIA) compounds are soluble in water, but the group VIA ion undergoes hydrolysis. For example:

$$S^{2-} + H_2O \longrightarrow HS^- + OH^-$$

Sulfides of the metals whose hydroxides are weak bases completely hydrolyze in water to give H_2S, for example:

$$Al_2S_3 + 6H_2O \longrightarrow 2Al(OH)_3 + 3H_2S$$

All of the group VIA elements except oxygen form binary hydrogen compounds which are all very weak acids in H_2O, for example:

$$H_2S \overset{aq}{\rightleftharpoons} HS^-_{(aq)} + H^+_{(aq)}$$

$$HS^-_{(aq)} \rightleftharpoons S^{2-}_{(aq)} + H^+_{(aq)}$$

All of the elements except oxygen react with F_2 to produce the corresponding hexafluoride. SF_6 is highly inert to almost any reaction even at elevated temperature, and is an excellent gaseous electric insulator. Other di- and tetrahalides exist.

Table 9–12 gives a summary of some of the more important reactions of the group VIA elements, excluding polonium.

PROBLEM 9–28   The -2 ionic sizes of the elements of group VIA are considerably larger than the $+2$ ionic sizes of the group IIA ions. Explain.

PROBLEM 9–29   Given that PbS is insoluble, write the reactions leading to the precipitation of PbS, beginning with $Pb(NO_3)_2$ and H_2S.

PROBLEM 9–30   Write balanced reactions for the following if they occur:
a. $SO_3 + K_2O \longrightarrow$ b. $H_2Se + H_2O \rightleftharpoons$

TABLE 9–12 Some general reactions of group VIA elements and compounds

REACTION[a]	COMMENT
1. $E + H_2O \longrightarrow$ no reaction	
2. $E + O_2 \longrightarrow EO_2$	SO_3 also produced
3. $yE + xM \longrightarrow M_xE_y$	sulfide, selenide, telluride
4. $3E + 4HNO_3 \longrightarrow 3EO_2 + 2H_2O + 4NO$	except O
5. $EO_2 + H_2O \longrightarrow H_2EO_3$	except O
6. $EO_3 + H_2O \longrightarrow H_2EO_4$	except O; Te gives H_4TeO_6
7. $\left.\begin{matrix}H_2EO_3\\ H_2EO_4\end{matrix}\right\} + OH^- \longrightarrow$ salt $+ H_2O$	except O
8. $ME + 2HCl \longrightarrow H_2E + MCl_2$	
9. $H_2 + E \longrightarrow H_2E$	except Te
10. $H_2E + H_2O \rightleftharpoons HE^- + H^+ \rightleftharpoons E^{2-} + H^+$	except O for the second step
11. $E + 3F_2 \longrightarrow EF_6$	except O
12. $2E + Cl_2 \longrightarrow E_2Cl_2$	except O; tetrachlorides and bromides also exist

[a]E is a Group VIA element except for Po; M is a metal.

c. $SeO_2(OH)_2 + HCl \longrightarrow$ g. $Se + F_2 \longrightarrow$

d. $SeO_2(OH)_2 + 2LiOH \longrightarrow$ h. $SO_2 + HCl \longrightarrow$

e. $SO_2 + H_2O \longrightarrow$ i. $Na_2S + HCl \longrightarrow$

f. $S + O_2 \longrightarrow$ j. $S + HNO_3 \longrightarrow$

✳ 9–M. GROUP VIIA—THE HALOGENS

Some properties of the group VIIA elements are given in Table 9–13; astatine is excluded because it is radioactive and detailed information regarding its chemical and physical properties is lacking. All of the halogens are nonmetals. The symbol X represents a halogen.

None of these elements exists in the free state; all occur in the combined state as halides, although an important source of iodine is $NaIO_3$. F_2 and Cl_2 are commonly obtained by electrolysis of molten salts, while Br_2 is obtained by oxidation of bromide salts obtained primarily from sea water:

$$2Br^- + Cl_2 \longrightarrow Br_2 + 2Cl^-$$

Iodine is obtained by reducing the iodate ion (IO_3^-). None of the halogens are conductors of electricity, and they are poor conductors of heat.

TABLE 9–13 Properties of elements of group VIIA

	F	Cl	Br	I
atomic number	9	17	35	53
valence configuration	$2s^2sp^5$	$3s^23p^5$	$4s^24p^5$	$5s^25p^5$
common form	gaseous-diatomic molecule	gaseous-diatomic molecule	liquid-diatomic molecule	solid-diatomic molecule
ionization potential (kcal)	402	300	273	241
electronegativity	4.0	3.0	2.8	2.5
atomic radius (Å)	0.72	0.99	1.14	1.33
common oxidation state	−1	−1	−1	−1
−1 ionic size (Å)	1.36	1.81	1.95	2.16
melting point (°C)	−218	−101	−7	+113
boiling point (°C)	−188	−35	+59	+183
heat of hydration of −1 ion (kcal/mole)	121	90	82	71
bond dissociation energy of X_2 (kcal/mole)	37	58	46	36

The halogens (except F_2) dissolve in and react with water; they vary in solubility. In addition to varied amounts of dissolved hydrated halogens, the following reaction occurs:

$$X_2 + H_2O \rightleftharpoons H^+ + X^- + HOX$$

This reaction is quite important for Cl_2. F_2 reacts completely to produce O_2 instead of HOF; thus no dissolved hydrated F_2 exists.

Generally, except for F_2, the halogens react with strong bases to produce either XO^- or XO_3^- and X^- depending upon the concentration and temperature of the base.

Oxides of all of the halogens exist, although except for I_2O_5 they are thermally unstable. They are all made indirectly. The halogens can exist in several possible oxidation states, for example: Cl_2O, ClO_2, and Cl_2O_7. The only important oxide(s) of fluorine is F_2O; of bromine, Br_2O and BrO_2; and of iodine, I_2O_5.

There are several oxyacids of the halogens (only one exists for fluorine). These have the general formula $(XO_n)OH$, where $n = 0$ to 3 depending upon the nature of X. Table 9–14 shows the most important ones. Of all of the oxyacids, only HIO_3, $HClO_4$, and HIO_4 can be isolated in the pure states. All of the acids in Table 9–14 are good oxidizing agents. Salts of all of these acids are known, most commonly with the alkali metal (group IA) and alkaline earth metal (group IIA) ions, for example: $Ca(OCl)_2$ and $KClO_3$.

TABLE 9-14 Some oxyacids of the halogens

OXIDATION NUMBER	F	Cl	Br	I	EXAMPLE OF NAME	ACID STRENGTH
+1	HOF	HClO	HBrO	HIO	hypobromous	very weak
+2		HClO$_2$			chlorous	moderate
+3		HClO$_3$	HBrO$_3$	HIO$_3$	iodic	strong
+4		HClO$_4$	HBrO$_4$	HIO$_4$	perchloric	strong

Binary hydrogen compounds of all of the halogens exist. All are covalent but soluble in water through ionization via ion-dipole interaction. All except HF are strong acids and undergo 100 % dissociation. Although HF dissociates in water, a subsequent reaction occurs to produce the very stable HF$_2^-$ ion:

$$HF + H_2O \rightleftharpoons H_3O^+ + F^-$$

$$F^- + HF \longrightarrow HF_2^-$$

The hydrogen halides can be made by the direct reaction of the elements, but are commonly made by reaction of a metal salt and an acid:

$$NaCl \ (or \ NaF) + H_2SO_4 \rightleftharpoons HCl \ (or \ HF) + NaHSO_4$$

HBr and HI must be made with a non-oxidizing acid since an oxidizing acid such as H$_2$SO$_4$ will produce I$_2$ or Br$_2$. The fluoride ion reacts with the surface of teeth to form fluoroapatite, Ca$_5$F(PO$_4$)$_3$, which substantially lowers the likelihood of dental cavities.

We saw in our earlier discussions that the halogens can react with any metal of the A groups. Those halides of groups IA and IIA are especially ionic. Halogens can also react with other halogens to produce interhalogen compounds. All XX' compounds except IF exist, and certain XX'$_n$ compounds also exist, as ClF$_3$.

Almost all chlorides, bromides, and iodides are soluble in water, except for those of Ag(I), Hg(I), Pb(I), and Tl(I). The fluorides of most metals are relatively insoluble, although except for lithium, the salts of the alkali metals are soluble. Most salts of the group IA and IIA halates (XO$_3^-$) and perhalates (XO$_4^-$) are soluble in water. The sodium and potassium hypohalites (XO$^-$) are also soluble in water. (See Section 2-B for the nomenclature of such polyatomic ions.)

Table 9-15 gives a summary of some of the more important reactions of the halogens and their compounds.

PROBLEM 9-31 The melting point and boiling point of the diatomic halogens progressively increase from F$_2$ to I$_2$. Propose a reason why this should be true.

TABLE 9–15 Some general reactions of group VIIA elements and compounds

REACTION[a]	COMMENT
1. $X_2 + H_2O \longrightarrow H^+ + OH^- + HOX$	except that F gives O_2 instead of HOF
2. $X_2 + O_2 \longrightarrow$ no reaction	oxides are prepared indirectly
3. $X_2 + 2OH^-_{(cold)} \longrightarrow XO^- + X^- + H_2O$	except F_2
4. $3X_2 + 6OH^-_{(hot)} \longrightarrow XO_3^- + 5X^- + 3H_2O$	except F_2
5. $X_2 + H_2 \longrightarrow 2HX$	
6. $NaX + H_2SO_4 \longrightarrow HX + NaHSO_4$	F and Cl; Br and I require non-oxidizing acid
7. $HX + H_2O \longrightarrow H_3O^+ + X^-$	except that HF is less dissociated and forms HF_2^-
8. $HX + OH^- \longrightarrow X^- + H_2O$	
9. $HX +$ basic oxide $\longrightarrow$ salt $+ H_2O$	
10. $X_2 + X_2' \longrightarrow XX'$	except for IF, and some XX'_n exist
11. $3X_2 + 2P \longrightarrow 2PX_3$	
12. $X_2 + 2S \longrightarrow S_2X_2$	for Cl and Br only; SF_6 with F_2
13. $X_2 + M \longrightarrow$ metal halide	any metal of A groups and some others as well
14. metal halides $\xrightarrow{aq}$ soluble	except F and Ag(I), Hg(I), Pb(I), and Tl(I) salts

[a]X stands for a halogen and M for metal

PROBLEM 9–32 The heat of hydration of the -1 ions is approximately the same as that for the $+1$ ions of group IA. Explain the factors which make this a reasonable result.

PROBLEM 9–33 Write balanced reactions for the following if they occur:

a. $HI \xrightarrow{aq}$

b. $HF + Cs_2O \longrightarrow$

c. $(ClO_3)OH + Ca(OH)_2 \longrightarrow$

d. $I_2 + H_2 \longrightarrow$

e. $HF + SiO_2 \longrightarrow SiF_4 +$

f. $Br_2 + OH^-_{(hot aq)} \longrightarrow$

g. $Cl_2 + H_2O \longrightarrow$

h. $P + Cl_2 \longrightarrow$

i. $Ca + Br_2 \longrightarrow$

j. $CsBr \xrightarrow{aq}$

* 9–N. GROUP VIIIA—RARE GASES

Some properties of the group VIIIA elements are given in Table 9–16.

The rare gases are found free in nature. They are commercially produced as by-products from the fractional distillation of liquid air. Helium may also be obtained from natural gas. The ionization potential of a rare gas is greater than that

TABLE 9–16 Properties of elements of group VIIIA

	He	Ne	Ar	Kr	Xe
atomic number	2	10	18	36	54
valence configuration	$1s^2$	$2s^22p^6$	$3s^23p^6$	$4s^24p^6$	$5s^25p^6$
common form	gas-monatomic	gas-monatomic	gas-monatomic	gas-monatomic	gas-monatomic
ionization potential	566	497	363	322	278
atomic radius (Å) (van der Waals)	0.93	1.12	1.54	1.69	1.90
melting point (°C)	−272	−249	−189	−157	−112
boiling point (°C)	−269	−246	−186	−153	−107

of any other element in its period. This is so because of the high stability produced by having 8 electrons in the outer shell (or 2 in the case of helium). Note that there is very little difference between the melting and boiling points of the individual rare gases. This indicates that the forces of attraction in the liquid and solid must be comparable. Only an induced dipole–induced dipole interaction can be operating.

Until recently, there were no known chemical reactions involving these elements. We know of hydrates of argon, krypton, and xenon which are unstable above the freezing point of water. It has also been recently discovered that the rare gases with relatively lower ionization potentials are reactive, but only with the most electronegative elements, fluorine and oxygen. The first compound found was $XePtF_6$, which is formed by the reaction of PtF_6 and Xe. It is also possible to directly react Xe and F_2 (at 400 °C) to produce XeF_4 (with a melting point of about 114 °C). The structure is square coplanar. In addition, it is possible to make XeF_2 (melting point about 150 °C), XeF_6 (melting point about 50 °C), and KrF_4.

The xenon fluorides can react with water to produce various compounds, including XeO_3, $Xe(OH)_4$, and $XeOF_4$. XeO_3 is apparently explosive when dry. A water solution of XeO_3 is weakly acidic and also is a strongly oxidizing medium. The ion XeO_6^{4+} has been obtained by reacting XeF_6 with NaOH.

EXERCISES

1. Ionic compounds typically have high melting points while covalent compounds have low melting points. Explain why this should be true.

2. Would you expect a liquid ammonia solution of potassium to be electrically conductive? Why? [Hint: Assume no KNH_2 exists.]

3. What is an electron deficient compound?

4. Why would you expect the rare gas elements to have generally a lower melting point than the halogens, even though both have no dipole moment?

5. What are allotropes? Give an example.

6. Water is unusual in that the density *increases* when ice melts to liquid water. Explain the role of hydrogen bonding in this unusual change.

7. $RbCl$ and PCl_5 are both white solids. Think of as many ways as possible which you could use to distinguish one from the other. (Exclude tasting since you would not want to take the chance of killing yourself.)

8. Assume a geologist found a diamond which was essentially completely encased in limestone ($CaCO_3$). How could he chemically free the diamond without harming it?

9. Explain why the boiling point of H_2Te is several hundred degrees below that of LiH.

10. Note that the heats of hydration of both positive and negative ions are exothermic. Does this mean that when a salt such as KBr is put in water the solution will necessarily heat up? Explain the basis for your answer.

11. Why should hydrogen in NaH have an oxidation state of -1?

12. The molecule NF_3 has bond angles of approximately 103° to 104°. Also, despite the large electronegativity difference in nitrogen and fluorine, the molecule has a very low dipole moment. Suggest a structure for NF_3 that would be compatible with this information. [Hint: Consider the structure and the existence of a lone pair of electrons.]

13. Suppose you wanted to make 11.2 liters of HBr by the following reaction:

$$2NaBr + H_3PO_4 \longrightarrow 2HBr + Na_2HPO_4$$

How many grams of $NaBr$ are required?

14. a. Calculate the weight of H_2 that would be produced by the complete reaction of 10.02 g calcium with water.
    b. What volume of H_2 would be produced in part (a) at STP?

15. How many electrons would be required to produce 21.9 g strontium from $SrBr_2$ in an electrolysis reaction?

16. Which of the following oxides will react with water to produce an acid?
    a. Bi_2O_3 d. P_2O_5
    b. CO_2 e. Al_2O_3
    c. CaO f. SeO

17. How many grams of O_2 would be necessary to completely react with 3.01×10^{21} atoms of indium?

18. Which of the following do you expect will have the highest dipole moment? Why?

$$HBr, HCl, HF, HI$$

19. Which of the following is the least basic?

$$LiOH, P(OH)_3, Sr(OH)_2, As(OH)_3$$

20. K(group IA), Ga(group IIIA), and Br(group VIIA) are elements belonging to the same period.
    a. Which will have the greater electronegativity, potassium or bromine?
    b. Which of the atoms is the smallest?
    c. Write the Lewis (dot) formula for the compound formed between gallium and bromine.
    d. Will the oxide of potassium be acidic, basic, or amphoteric?

21. Predict the formula for the compounds formed from each species in column 1 reacting with each species in column 2.

COLUMN 1	COLUMN 2
Cs	I
Sr	Se
In	O

22. Which of the following exhibits the largest electrical conductivity in the liquid state?

$$I_2, NH_3, HCl, N_2, Na$$

23. Which chloride should exhibit the most covalent type of bond?
    a. $NaCl$ b. KCl c. $CaCl_2$ d. $BaCl_2$ e. $BeCl_2$

24. Complete where necessary and balance the following reactions, if they occur.
    a. $Se + H_2O \longrightarrow$
    b. $GeO + HCl \longrightarrow$
    c. $LiNO_3 + aq \longrightarrow$
    d. $In + HBr \longrightarrow$
    e. $As + I_2 \longrightarrow$
    f. $2NaI + H_3PO_4 \longrightarrow$
    g. $P + O_2 \longrightarrow$
    h. $Rb_2O + HCl \longrightarrow$
    i. $NaH + H_2O \longrightarrow$
    j. $Ca + N_2 \longrightarrow$
    k. $Al + Fe_2O_3 \longrightarrow$
    l. $Al + KOH + H_2O \longrightarrow KAlO_2 + H_2$
    m. $Sn + HNO_3 \longrightarrow SnO_2 + NO + H_2O$
    n. $As + HNO_3 \longrightarrow H_3AsO_4 + NO_2 + H_2O$
    o. $SbCl_3 + H_2O \longrightarrow SbOCl +$
    p. $Se + HNO_3 \longrightarrow NO +$

CHAPTER TEN

CHEMICAL THERMODYNAMICS

In Chapters 4 through 9 we approached the study of chemistry through the *molecular structure and properties* of substances—microscopic properties. In this chapter a completely different approach will be taken—a *thermodynamics* approach—based upon observations and measurements of the *macroscopic properties* of a substance. A few such macroscopic properties are density, pressure, and temperature. Using measurements of these variables we can calculate other quantities, such as heat and work, which can then be used to formulate the three basic *laws of thermodynamics*. These laws give us a criterion for predicting the likelihood or feasibility of a reaction occurring.

Remember that a *law* is simply a generalized statement which summarizes a large number of experimental facts, whereas a *theory* is an explanation of these facts based on the derivation or development of a basic model (see Section 1–D). The study of molecular structure and consequent properties is based fundamentally on the *theory of quantum mechanics*, whereas thermodynamics is based upon three *laws*. The subjects of phase equilibria, solutions, chemical equilibrium, and electrochemistry, which will be discussed in Chapters 11–14, are based mainly upon thermodynamic principles. Wherever it is appropriate, both the quantum mechanical and thermodynamic approaches will be used to gain further insight into a phenomenon.

Thermodynamics is generally concerned with various *energy changes* which may occur in a process, and based upon the laws of thermodynamics we can place certain restrictions upon which energy changes may occur. There are various forms of energy that can be involved, such as heat and work. As can be seen from this description, thermodynamics is concerned with a very broad area involving engines of all types as well as physical and chemical processes. We shall restrict this chapter to *chemical thermodynamics*—the application of these principles to processes relating to chemistry. However, it should be pointed out that the same fundamental laws are the basis of thermodynamic concepts in general and are independent of the application.

10–A. OBJECTIVES AND LIMITATIONS

OBJECTIVES OF CHEMICAL THERMODYNAMICS

There are two basic objectives which we wish to achieve in our consideration of chemical thermodynamics:

1. to establish a criterion for determining whether a chemical reaction *may* occur spontaneously; that is, "Is the reaction possible?"
2. to calculate the maximum yield of products that can be obtained from a chemical reaction

In the first of these objectives the use of the word "may" should be emphasized. Chemical thermodynamics cannot guarantee that a reaction will take place in any reasonable time period. In other words, chemical thermodynamics does not establish the rate at which a reaction will occur. The use of the word "may" infers that the reaction is possible; however, you must wait the requisite amount of time, which may be exceedingly long. On the other hand, when chemical thermodynamics predicts that a reaction will not occur, we can definitely conclude that the reaction will indeed not occur. In this chapter we will show how this criterion can be established based on the first two laws of thermodynamics.

The second objective is calculation of the maximum quantity of product (yield) that can be expected from the reaction. The maximum yield will be obtained when the *reaction mixture* has come to equilibrium. This objective will be dealt with in Chapter 13, where a mathematical relationship between the equilibrium concentrations and thermodynamic quantities will be given. Using the equilibrium concentrations, the yield of the reaction can be calculated. We considered equilibrium in Chapter 3, but only qualitatively. With a thermodynamic background we can consider equilibrium quantitatively and, given the proper thermodynamic data, we can calculate the concentrations of the different species in a reaction mixture at equilibrium.

LIMITATIONS OF CHEMICAL THERMODYNAMICS

There are two basic limitations of chemical thermodynamics which restrict its use in predicting the probability of the occurrence of chemical reactions:

1. the inability to determine the *rate* of the reaction or the *time* required to attain the maximum (equilibrium) yield
2. the inability to give any information on the *mechanism* of the chemical reaction

The first of these two limitations was mentioned earlier in conjunction with the possibility of a reaction. The second of these limitations is concerned with the path the chemical reaction follows (i.e., what intermediate species are involved in going to the final reaction products). The rate and the mechanism of chemical reactions are the concerns of the subject of chemical *kinetics,* which is presented in Chapter 15. As can be seen from the objectives of thermodynamics and kinetics, the two subjects are complementary to one another.

To assist in understanding the objectives and limitations of chemical thermodynamics, we will consider the analogy of a mechanical system: the possible movement of a boulder located on the side of a mountain as illustrated in Figure 10–1. If our boulder is at rest at position A, its energy will be only the potential energy, U, which depends upon its height. At position A we designate its potential energy by U_A. We know from experience that an object such as the boulder can fall to a lower position such as position B. Associated with this change from position A to B will be a corresponding decrease in potential energy.

$$\Delta U_{AB} = U_B - U_A < 0$$

FIGURE 10–1   Potential energy change for a mechanical process

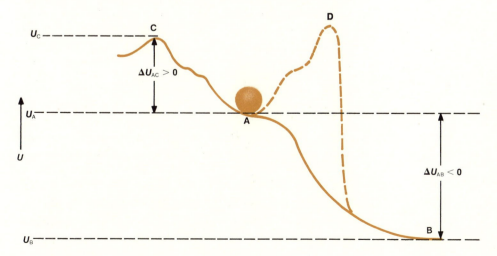

Thus, we have a criterion (for a decrease in potential energy, $\Delta U =$ a negative number) to determine whether the mechanical process *may* occur spontaneously. (Similarly, the first objective in thermodynamics is to find a criterion for whether a chemical reaction *may* occur spontaneously.)

Conversely, we know from experience that the elevation of a boulder at rest at position A will not spontaneously increase to the height of position C in Figure 10–1. Associated with this change in position from A to C will be an increase in potential energy. The change in potential energy, U_{AC}, is

$$\Delta U_{AC} = U_C - U_A > 0$$

Thus, we conclude that a mechanical process will *not* occur spontaneously if there is an increase in potential energy ($\Delta U =$ a positive number). (As we shall see at the conclusion of this chapter, our thermodynamic criterion can also state *positively* that a chemical reaction will *not* occur spontaneously.)

For this mechanical process illustrated in Figure 10–1 we have arrived at a criterion for whether a process *may* or may not be spontaneous. However, even though this criterion may be met, it *cannot* be guaranteed that the process *will* occur in any finite period of time. For example, there may be a mountain between positions A and B as illustrated by the dashed curve in Figure 10–1. In this case the boulder will not proceed on its own accord to position B. Additional energy must be given to the boulder at position A to raise it to position D from which it can go to B without any additional energy (indeed, with the release of energy). In other words, there may be a barrier such as D between the initial state A and the final state B that must be overcome before the process can actually occur. Similarly in chemical reactions there may be a barrier which prevents a reaction from taking place despite the fact that the thermodynamic criterion for spontaneity has been met. It is for this reason that thermodynamics is unable to determine the rate of the reaction or the time required for the reaction to occur. (In other words, though chemical thermodynamics can be used to establish a criterion that the reaction *may occur spontaneously*, it cannot be used to *guarantee* that the reaction will occur in any finite period of time.)

10–B. DEFINITIONS AND TERMINOLOGY

There are certain basic terms which are convenient to use in the discussion of thermodynamics. These will be defined and examples given to illustrate their significance.

1. *system* — that portion of the universe under investigation

2. *boundaries* — real or imaginary surfaces or lines which define the limits of the system.

3. *surroundings*—all objects outside the boundaries of the system that may interact with the system. These may constitute forces (for example, weights) against
which work can be done, or heat sources where heat may be stored or from
which heat may be transferred to the system.

As an illustration of these three terms, consider a gas held in a cylindrical container by a piston (Figure 10–2). The weights apply an external force on the gas
through the piston, and the temperature of the cylinder and gas is maintained
constant by immersing it in a large water bath. The *system* in this example would
consist of the gas, and the *boundaries* would consist of the walls of the cylinder
and the piston. The *surroundings* in this example consist of the weights, which
by raising and lowering the piston cause work to be done, and the water bath,
from which heat can be absorbed or to which heat can be released by the gas
sample in the cylinder. In this example, the boundaries are physically discrete
surfaces. However, another system consisting of an ideal gas could be defined
without specifying the container, since the nature of the container does not affect
the properties of the gas. In this case the *boundary* would be an *imaginary* surface
which contained the gas. Similarly, in liquid solutions the system could be simply
the liquid sample, and the boundary as defined by the container need not be
specified.

4. *equilibrium*—a condition in which there is no *net* change in the properties of
the system and surroundings.
5. *reversible process*—a process in which the system undergoes change at a
sufficiently slow rate that at any instant the system is in equilibrium with the
surroundings.

FIGURE 10–2 Example of a system (gas) and its surroundings

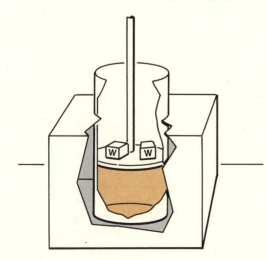

6. *property*—characteristic attribute which describes a system, such as color, density, or temperature.

7. *state*—a specific condition of a system that is completely described by its properties. For example, a gas at a specified temperature, pressure, and volume.

8. *state function* or *state variable*—a variable or function which is dependent only upon the state of the system. Therefore, when a system undergoes a change from one state to another, the change in the state function is independent of the path taken.

As an example of these last five terms, let us return to the example of the gas contained in the cylinder (Figure 10–3). The system is in *state A* when both weights contribute to the external pressure exerted on the gas. State A is defined by the *properties*, P_1, T_1, n_1, which completely describe the system. Knowing the values of these three variables allows anyone to reproduce the state of this system. State B is also a *state* of the system, although different from state A, and it is completely described by the *properties* P_2, T_1, n_1. In each of states A and B the piston remains stationary at a different height and the temperature remains at T_1. Since there are no changes in the system or surroundings while in states A and B, we say that they are each in a condition of *mechanical equilibrium*.

Now let us consider the process by which we can go from state A to state B. Since the pressure and volume change in going from state A to state B, we can describe the process on a P vs. V graph (Figure 10–4). The state A is at the coordinates (V_1, P_1). Similarly, state B is at (V_2, P_2). One process by which we can go

FIGURE 10–3 Example of a change in state of a system

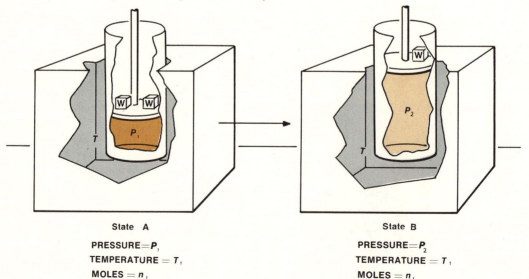

State A
PRESSURE = P_1
TEMPERATURE = T_1
MOLES = n_1

State B
PRESSURE = P_2
TEMPERATURE = T_1
MOLES = n_1

FIGURE 10–4 Two alternate paths from state A to state B

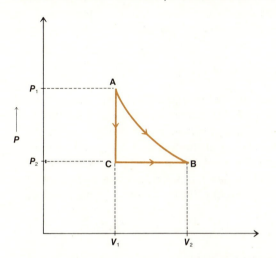

from state A to state B is by instantaneously removing one of the weights on the piston. The external pressure *immediately* drops from P_1 to P_2 without any change in volume or in temperature; this corresponds to the step from A to C on the graph. At point C the system is *not* in equilibrium with the surroundings and will immediately expand to state B. Since the system is not in equilibrium with the surrounding at all times during the expansion of the gas from C to B, the process is said to be *irreversible*. On the other hand, if the weight was removed in infinitesimally small pieces, then the external pressure (equivalent to the weights on the piston) would change at an infinitely slow rate and the external pressure would remain essentially equal to the pressure of the gas in the system; that is, the system and surroundings would remain in equilibrium with each other at each infinitesimal step in the expansion. The graph of such a process would follow the curved path connecting A directly to B. Since the system would be in equilibrium with the surroundings at all times, this process is called *reversible*.

The variables P and V are *state functions* since the magnitude of these variables depends only upon the state. For example, in state A, $V = V_1$ and in state B, $V = V_2$. The change in volume in going from A to B is

$$\Delta V = V_2 - V_1$$

and is *independent* of the path taken; that is, the change in volume occurring along the path ACB is the same as the change in volume along the reversible path AC.

$$\Delta V_{ACB} = \Delta V_{AB} = \Delta V = V_2 - V_1$$

The same is true for the pressure of the system, and hence P and V are both called *state functions*.

PROBLEM 10-1   Suppose you were interested in studying a transformation in a rock formation at the bottom of the ocean. Define what you call the system, surroundings, boundaries, properties of the system, and state of the system for this study.

10-C. WORK AND HEAT

Work is the product of a *force* times the displacement of the object upon which the force is acting. For example, *mechanical work* is expended when a person lifts a pile of books. The force acting on the moving books is the force of gravity, measured as the weight of the books. Energy can be defined as the capacity for doing work, and work involves the transfer of energy from one system to another. Work (W) necessarily has the same units as energy and mathematically can be expressed as the product of force (f) times displacement (d)

10-1
$$W = (f)(d)$$

The force can also have the form of *electrical potential*, $\mathcal{E}$. The displacement of charge, q, against this electrical potential constitutes *electrical work*.

10-2
$$W_{el} = (\mathcal{E})(q)$$

This will be important in Chapter 14 on electrochemistry.

For the present we are interested only in the form of mechanical work associated with the expansion or compression of substances such as gases. In this case "force" can be replaced by the external pressure (P_{ex}) on the system and "displacement" is replaced by a change in volume (ΔV) (see Appendix J). Provided P_{ex} remains constant, Equation 10-3 can be used to calculate the work for the irreversible process ACB described in Figure 10-4.

10-3
$$W = P_{ex}\Delta V \quad (P_{ex} = \text{constant})$$

In this irreversible process the $P_{ex} = P_2 =$ final pressure. Note that in equation 10-3 if P_{ex} is expressed in atmospheres and V in liters, the units will be (liter–atm). In order to convert to units of energy, a conversion factor must be used: 1(liter–atm) = 0.0242 kcal. For conversion to other energy units, consult Appendix A for the proper factors.

Equation 10-3 can also be used to calculate the work associated with slowly heating a substance at constant pressure. In this case the $P_{ex} = P =$ constant. Since the surroundings are in equilibrium with the system (same pressure), the *process is reversible*. The initial and final volumes, V_1 and V_2, must be known or can be calculated from the initial and final temperature (°K). If the substance is an ideal gas then

$$V_1 = \frac{nRT_1}{P}$$

$$V_2 = \frac{nRT_2}{P}$$

EXAMPLE 10–1   *As an example of the use of equation 10–3, let us consider the calculation of work in the expansion of an ideal gas described in the following graph*

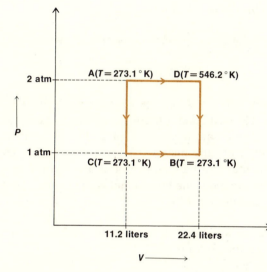

We will calculate the work for the two different paths ACB and ADB.

PATH ACB
This path is irreversible since P_{ex} is decreased instantaneously from 2 atm to 1 atm at constant temperature (AC) and the expansion (CB) is carried out against $P_{ex} = 1$ atm. Using equation 10–3

$$W_{ACB} = P_{ex}\Delta V = P_{ex}(V_2 - V_1)$$

$$= (1\ atm)(22.4\ liters{-}11.2\ liters)\left(0.0242\ \frac{kcal}{liter{-}atm}\right)$$

$$W_{ACB} = 0.271\ kcal$$

PATH ADB
This path actually consists of two paths:
step AD: constant pressure, $T = 273.1° \longrightarrow T = 546.2°$, reversible
step DB: constant volume, $T = 546.2° \longrightarrow T = 273.1°$, reversible
The work for step AD is given by equation 10–3

$$W_{AD} = P_{ex}(V_2 - V_1)$$

$$W_{AD} = (2\ atm)(22.4\ liters-11.2\ liters)\left(0.0242\ \frac{kcal}{liter-atm}\right)$$

$$W_{AD} = 0.542\ kcal$$

Since $\Delta V = 0$ and no displacement takes place in step DB, $W_{DB} = 0$
Therefore, $W_{ADB} = W_{AD} + W_{DB} = 0.542\ kcal$

Notice that the work done in the two processes in Example 10–1 is different:

$$W_{ACB} \neq W_{ADB}$$

We can visualize this difference in work by comparing the weights lifted for the two paths in Figure 10–5. At state A there are two equal weights on the piston of sufficient magnitude to create a pressure of 2 atm. In path ACB one weight is removed immediately from the piston in its original position and therefore only the second weight is lifted to the final height of the piston (state B). In path ADB both weights are lifted to the same final height as in state B, and therefore twice as much work is done along path ADB.

PROBLEM 10–2   Calculate the work for the path AEB in the following diagram. The system is slowly cooled to lower the pressure.

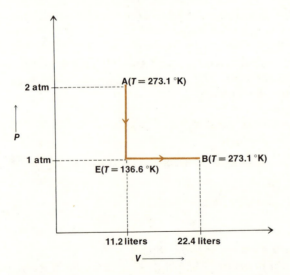

Compare W_{AEB} calculated in this problem with W_{ACB} and W_{ADB} from Example 10–1.

Another expression for work which we will find useful is that for reversible expansion at constant temperature (path AB in Figure 10–4 of Section 10–B).

FIGURE 10–5   Comparison of work done along two different paths

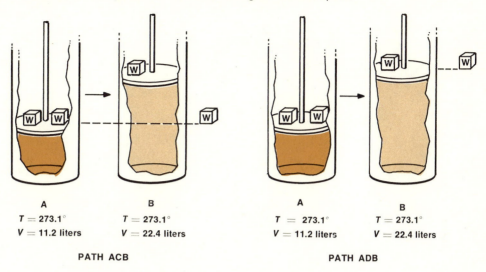

A	B	A	B
T = 273.1°	T = 273.1°	T = 273.1°	T = 273.1°
V = 11.2 liters	V = 22.4 liters	V = 11.2 liters	V = 22.4 liters

PATH ACB · PATH ADB

A process carried out at constant temperature is called an *isothermal* process. The expression for the work in an isothermal, reversible expansion is especially simple for an ideal gas (see Appendix K for derivation).

10–4
$$W = nRT \ln \frac{V_2}{V_1} = 2.303 \; nRT \log \frac{V_2}{V_1} \qquad \begin{cases} \text{isothermal, reversible} \\ \text{process; ideal gas} \end{cases}$$

where ln = natural logarithm and log = logarithm to the base 10 (see Appendix D for a review of logarithms).

EXAMPLE 10–2   *In Example 10–1 we considered two paths (ACB and ADB) in going from A to B. In the present example we will calculate the work for the reversible, isothermal expansion of one mole of an ideal gas directly from A to B (path AB in the graph on page 318).*

The work is given by equation 10–4

$$W = 2.303 \; nRT \log \frac{V_2}{V_1}$$

$$W_{AB} = (2.303)(1 \; mole)\left(1.987 \times 10^{-3} \; \frac{kcal}{deg\text{-}mole}\right)(273.1°) \log \frac{22.4 \; liters}{11.2 \; liters}$$

$$W_{AB} = 0.376 \; kcal$$

Note that W_{AB} is different from the W_{ACB} and W_{ADB} calculated in Example 10–1.

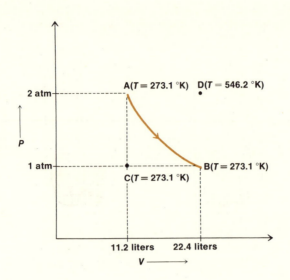

PROBLEM 10–3   From the results of Examples 10–1 and 10–2 and Problem 10–2, can you state whether or not work is a state function? Explain.

The heat absorbed in a process depends upon the nature of the process involved. The two processes with which we will be concerned are a change in temperature at (1) constant pressure or (2) constant volume. The *heat*, Q, absorbed by a substance can be calculated from

10–5
$$Q = mc(t_2 - t_1) = mc(T_2 - T_1)$$

where m = mass in grams, c = *specific heat* in cal/g-deg, and the temperature change can either be expressed on the centigrade scale as $(t_2 - t_1)$, or on the absolute scale as $(T_2 - T_1)$. Specific heat is the heat required to raise the temperature of one gram of a substance one degree centigrade (or one degree Kelvin, see page 41). Equation 10–5 is true only if c = constant. In general, c can vary with temperature. However, over a reasonably small temperature range, we can assume c is a constant. In thermodynamics, we define a term analogous to c, except that it is based on moles, *molar heat capacity*, C. The expression for Q is then

10–6
$$Q = nC(T_2 - T_1)$$

where n = number of moles or g-atoms of the substance, C is in cal/mole-deg and $(T_2 - T_1)$ = change in temperature.

Both c and C are dependent on pressure. In order to eliminate the dependence on this variable we define the molar heat capacity at constant pressure, C_P, or constant volume, C_V.

10–7
$$Q = nC_P(T_2 - T_1) \qquad \begin{cases} C_P = \text{constant} \\ P = \text{constant} \end{cases}$$

10–8
$$Q = nC_V(T_2 - T_1) \qquad \begin{cases} C_V = \text{constant} \\ V = \text{constant} \end{cases}$$

In using equation 10–7, the condition of constant pressure must be true from the initial state through to the final state. For example, the irreversible expansion described along path ACB in Example 10–1 does not occur at constant pressure, so that equation 10–7 would not be applicable. Similarly, equation 10–8 can be used only when the volume remains constant from the initial state to the final state. A table of heat capacities is given in Appendix T, Table 11.

EXAMPLE 10–3   *Calculate the heat required to raise the temperature of a 500 g block of iron from 300 °K to 800 °K. Assume the heat capacity is constant, $C_P = 6$ cal/g-atom-deg.*
  The atomic weight of iron is 55.85, hence the number of g-atoms of iron in the 500 g block is

$$n = (500 \text{ g})\left(\frac{1 \text{ g-atom}}{55.85 \text{ g}}\right) = \frac{500}{55.85} \text{ g-atom}$$

Using equation 10–7

$$Q = \left(\frac{500}{55.85} \text{ g-atom}\right)\left(6 \frac{cal}{g\text{-atom-deg}}\right)(800° - 300°)$$

$$Q = 26.9 \times 10^3 \text{ cal} = 26.9 \text{ kcal}$$

PROBLEM 10–4   In Example 10–1 we calculated the work for path ADB and in Problem 10–2 we calculated the work for path AEB.

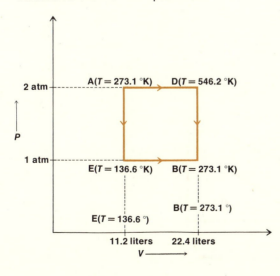

a. Calculate Q for paths ADB and AEB. Assume that $C_V = \frac{3}{2}R$ and $C_P = \frac{5}{2}R$.

b. Compare Q_{ADB} and Q_{AEB}. Are they equal? Is Q a state function?

10–D. FIRST LAW OF THERMODYNAMICS

The first law of thermodynamics is a statement of the conservation of energy in a process where there is a possibility of heat being exchanged and work being done by or on the surroundings. The first law can be expressed mathematically as equation 10–9.

10–9
$$\Delta E = Q - W$$

where Q and W are the heat and work as defined in Section 10–C and ΔE is the change in a new function called *internal energy*, E. As the name suggests, the change in internal energy is the increase or decrease of energy stored *within* the system. For example, if the system absorbs 10 kcal of heat and expends 6 kcal of work against the surroundings, the internal energy of the system will be increased by 4 kcal.

One very important property of E is the fact that it is a state function. Therefore, the change in internal energy between two states, ΔE, is independent of the path taken. This is the most amazing feature of the first law, since Q and W *are* dependent on the path taken. The independence of ΔE from the path taken makes the quantity E important in thermodynamics. E is commonly referred to as a *thermodynamic function* or *variable*. All of the thermodynamic functions or variables are also state functions.

The fact that ΔE is independent of the path taken can be illustrated by Figure 10–6, containing three paths between states A and B. The following general statements concerning Q, W, and ΔE in Figure 10–6 follow from the first law of thermodynamics:

$$Q_1 \neq Q_2 \neq Q_3$$

$$W_1 \neq W_2 \neq W_3$$

$$\Delta E_1 = \Delta E_2 = \Delta E_3$$

FIGURE 10–6   Alternate paths between states A and B

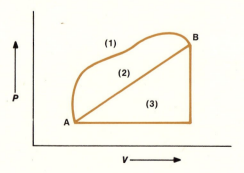

One or more of the Q's may, by chance, be equal to another; however, they need not be and generally are not equal. A similar situation can exist for the W's. On the other hand, ΔE_1, ΔE_2, and ΔE_3 must always equal one another.

A final comment should be made concerning the sign convention for Q and W. Notice in equation 10–3 that if ΔV is positive, W will be positive, corresponding to the work done *by* the system. Note that by equations 10–7 and 10–8, when heat is *absorbed* by the system ($T_2 > T_1$), then Q is positive. To help in recalling the general sign conventions, remember that:

Q positive—heat absorbed by the system

Q negative—heat liberated by the system

W positive—work done by the system

W negative—work done on the system

PROBLEM 10–5   In Problem 10–4 the heats for paths ADB and AEB were calculated. In Example 10–1 and Problem 10–2 the work for each of these processes was also calculated.
a. Using these results calculate ΔE_{ADB} and ΔE_{AEB}.
b. Compare ΔE_{ADB} and ΔE_{AEB}.
Do you expect them to be different?

PROBLEM 10–6   The vaporization of one mole of water at 100 °C and $P = 1$ atm requires 9.71 kcal of heat. The work associated with this vaporization arises from the change in volume between the water as a liquid (V_l) and as a gas (V_g).

$$\Delta V = V_g - V_l$$

Assume ideal gas behavior in calculating the V_g for water. The volume of liquid water can be calculated from its density, which is 1.00 g/ml. Calculate W and ΔE for the vaporization of water. Note that $V_g > V_l$.

10–E. ENTHALPY AND THERMOCHEMISTRY

We can define another thermodynamic function called *enthalpy*, H. This function is particularly convenient to use for processes carried out at constant pressure. Most chemical reactions are carried out in open vessels exposed to an approximately constant atmospheric pressure. Enthalpy is defined as

10–10
$$H = E + PV$$

where E is the internal energy, P is pressure, and V is volume. When a system undergoes a change, the change in enthalpy is given by

10–11
$$\Delta H = \Delta E + \Delta(PV)$$

(At this point we will only describe enthalpy by these mathematical equations. At the conclusion of this section, after you have seen the relationship between ΔH and bond energies, we can give a more physical meaning to H and ΔH in terms of molecular structure and bonding.) The internal energy E is a state function, as discussed in the previous section. The product PV is also a state function, since P and V are defined explicitly for any state. Therefore, it should be obvious from the definition of enthalpy in equation 10–10 that H is also a state function. As a consequence of this property, change in enthalpy, ΔH, is independent of the path taken in going between any two states.

EXAMPLE 10–4   *In Example 10–1 we considered two paths ACB and ADB in going from A to B*

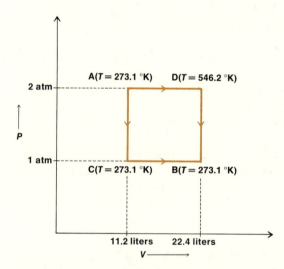

As an example of the calculation of the change in enthalpy let us evaluate ΔH for the path ADB. Using equation 10–11 we can evaluate ΔH from ΔE and $\Delta(PV)$ for this path.

$$\Delta H_{ADB} = \Delta E_{ADB} + \Delta(PV)$$

$$\Delta H_{ADB} = \Delta E_{ADB} + (P_B V_B - P_A V_A)$$

ΔE_{ADB} *was calculated in Problem 10–5 and found equal to zero. Substituting the initial and final P and V for the process, we obtain,*

$$\Delta H_{ADB} = 0 + (1\ atm)(22.4\ liters) - (2\ atm)(11.2\ liters)$$

$$\Delta H_{ADB} = 0$$

PROBLEM 10–7   Calculate ΔH for path AEB described in Problems 10–2 and 10–4.

PROBLEM 10–8   Calculate ΔH for the vaporization process described in Problem 10–6.

CHANGE IN ENTHALPY (ΔH) AT CONSTANT PRESSURE

For a process in which the pressure is maintained constant from the initial state through to the final state, the change in enthalpy is simply the heat change for the process. This can be shown as follows:

$$\Delta H = \Delta E + \Delta(PV)$$

$$\Delta H = Q - W + \Delta(PV)$$

Since the process is at constant pressure, $\Delta(PV)$ becomes $P\Delta V$

$$\Delta H = Q - W + P\Delta V$$

If we have only pressure-volume work (at constant pressure)

$$\Delta H = Q - P_{ex}\Delta V + P\Delta V$$

Since again the pressure remains constant,

$$P_{ex} = P$$

and

10–12 $$\Delta H = Q \quad \left\{ \begin{array}{l} \text{constant pressure} \\ \text{pressure-volume work only} \end{array} \right.$$

PROBLEM 10–9   Using equation 10–12, calculate ΔH for
a. steps AD and EB in Problem 10–4,
b. the vaporization process in Problem 10–6.

Note that equation 10–12 has been derived with only the two restrictions shown (i.e., constant pressure and pressure-volume work) and is thus applicable to any type of process satisfying these conditions. For example, in a chemical reaction (only pressure-volume work possible) exposed to a constant external pressure, the ΔH for the reaction would equal, according to equation 10–12, the heat of the reaction, Q_P. In fact the principal reason for defining H as we did in equation 10–10 was so that we could obtain the simple relationship given in equation 10–12.

HEAT OF REACTION

For a reaction carried out at constant pressure, the heat of reaction, Q, can be calculated from the difference in the enthalpies of the products and reactants for the reaction ($Q = \Delta H$). If we have the general chemical reaction

10–13 $$\underbrace{aA + bB}_{\text{initial state}} \longrightarrow \underbrace{cC + dD}_{\text{final state}}$$

the heat of reaction at constant pressure is

10–14
$$Q_P = \Delta H = \underbrace{(cH_C + dH_D)}_{\text{final } H} - \underbrace{(aH_A + bH_B)}_{\text{initial } H}$$

where H_A, H_B, H_C, H_D, are the enthalpies of the substances A, B, C, and D, and a, b, c, and d represent the number of moles of each species involved in the reaction. The calculation of ΔH is reasonably straightforward, provided the H_A, H_B, H_C, and H_D, are known. However, we encounter the problem that H is defined in equation 10–10 in terms of E, but we do not know how to obtain *absolute* values of internal energy. Only the *change* in internal energy, ΔE, can be evaluated from the first law. Consequently, we are unable to obtain absolute numerical values of H for various substances.

Because of this inability to obtain absolute H values, we establish a scale of *relative* enthalpy values. The zero on this scale is chosen to be the enthalpy of the elements in their *most stable form at 1 atm pressure*. For example, the most stable form for hydrogen at 1 atm pressure is a gas containing diatomic molecules H_2; accordingly we assign the value of zero enthalpy to hydrogen gas. As another example, the most stable form of carbon at 1 atm pressure is graphite to which we assign the value of zero enthalpy. The other elemental form of carbon is diamond, but since it is *not* the most stable form at one atmosphere, the enthalpy value for diamond is not zero. With this reference zero, the enthalpy of any substance can be evaluated, using *heats of reactions* which are measured by experiment. The enthalpy scale is illustrated in Figure 10–7. The enthalpy values of the substances on this scale are referred to as *heats of formation* (ΔH_f), because they correspond to the heat change for the formation of one mole of a substance from the elements in their zero reference state. For example, the ΔH_f for $H_2O_{(l)}$ at 25 °C corresponds to the heat of the reaction

$$H_{2(g)} + \tfrac{1}{2}O_{2(g)} = H_2O_{(l)} \qquad \Delta H = \Delta H_{f,H_2O_{(l)}} = -68.3 \text{ kcal}$$

Similarly, for benzene, $C_6H_{6(g)}$, the ΔH_f at 25 °C corresponds to the ΔH for the reaction

$$6C_{(graphite)} + 3H_{2(g)} = C_6H_{6(g)} \qquad \Delta H = \Delta H_{f,C_6H_6(g)} = +19.820 \text{ kcal}$$

Having devised this relative scale of heats of formation, we can now evaluate the ΔH of a chemical reaction without requiring absolute values of E. For the general reaction (equation 10–13), the heat of reaction is given by the sum of the heats of formation of the products minus the sum of the heats of formation of the reactants.

10–15
$$\Delta H = (c\Delta H_{f,C} + d\Delta H_{f,D}) - (a\Delta H_{f,A} + b\Delta H_{f,B})$$

For the enthalpy of a liquid, solid, or ideal gas we define the *standard conditions* as a *pressure of one atmosphere*. The standard condition is designated by adding

FIGURE 10–7   Relative scale of enthalpy—heats of formation at 25°C and 1 atm pressure.

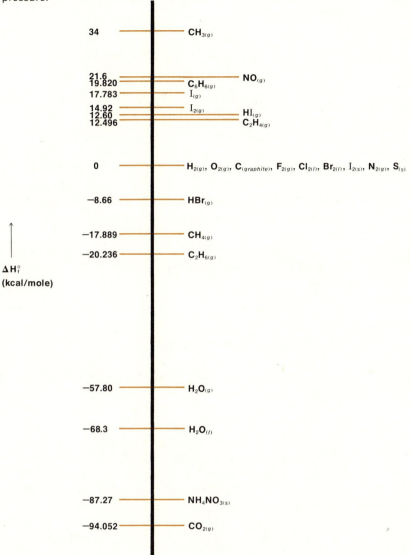

a post-superscript zero to the symbol. If standard conditions are employed, the standard heat of reaction for the general reaction 10–13 would be

10–16 $\Delta H° = (c\Delta H°_{f,C} + d\Delta H°_{f,D}) - (a\Delta H°_{f,A} + b\Delta H°_{f,B})$

Most tables give the value of $\Delta H°_f$ at 25 °C, as do Tables 3, 4, and 5 of Appendix T; however, some tables give thermodynamic data for different temperatures.

EXAMPLE 10–5   *To illustrate the calculation of the standard heat of reaction we will calculate the heat resulting from the burning of iso-octane (2,2,4-trimethylpentane), a component of gasoline:*

$$C_8H_{18(g)} + \frac{25}{2} O_{2(g)} \rightleftharpoons 8CO_{2(g)} + 9H_2O_{(g)}$$

$$\Delta H^\circ = 8\Delta H^\circ_{f,CO_{2(g)}} + 9\Delta H^\circ_{f,H_2O_{(g)}} - \Delta H^\circ_{f,C_8H_{18(g)}} - \frac{25}{2}\Delta H^\circ_{f,O_{2(g)}}$$

$$= 8(-94.052) + 9(-57.80) - (-53.57) - \frac{25}{2}(0)$$

$$\Delta H^\circ = -1,219.05 \ kcal$$

Note the negative sign on ΔH°, which designates that there is a decrease in enthalpy and heat is liberated in the reaction.

PROBLEM 10–10   Calculate the standard heat of reaction at 25 °C for the formation of glucose and oxygen from carbon dioxide and water as part of the photosynthetic process in plants

$$6CO_{2(g)} + 6H_2O_{(l)} \rightleftharpoons C_6H_{12}O_{6(aq)} + 6O_{2(g)}$$

Is heat absorbed or liberated in this process?

PROBLEM 10–11   Calculate the heat of vaporization of H_2O from the heats of formation of $H_2O_{(l)}$ and $H_2O_{(g)}$.

$$H_2O_{(l)} \rightleftharpoons H_2O_{(g)} \qquad \Delta H^\circ = \Delta H \text{ vaporization}$$

It is frequently necessary to consider an overall reaction in terms of a series of intermediate steps, the sum of which gives the overall reaction. If a reaction is broken down into two or more reaction steps, the ΔH for the overall reaction is simply the sum of the ΔH's for the reaction steps. To illustrate this, consider the oxidation of graphite to $CO_{2(g)}$ via the intermediate $CO_{(g)}$:

$$C_{(graphite)} + \tfrac{1}{2}O_{2(g)} \rightleftharpoons CO_{(g)} \qquad \Delta H_1$$

$$\underline{CO_{(g)} + \tfrac{1}{2}O_{2(g)} \rightleftharpoons CO_{2(g)} \qquad \Delta H_2}$$

$$C_{(graphite)} + O_{2(g)} \rightleftharpoons CO_{2(g)} \qquad \Delta H$$

$$\Delta H = \Delta H_1 + \Delta H_2$$

Since H is a state function, ΔH is independent of the path taken for the formation of $CO_{2(g)}$. We can illustrate the reaction as in Figure 10–8.

FIGURE 10–8 ΔH for a reaction as the sum of ΔH's for reaction steps

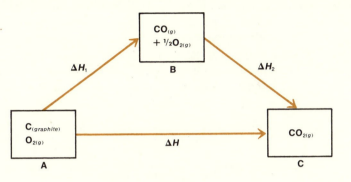

Since ΔH is independent of the path, ΔH for the overall reaction from A directly to C must equal $\Delta H_1 + \Delta H_2$ (for the path ABC).

This independence of path further emphasizes a point made at the beginning of this chapter. One of the limitations of thermodynamics is that it is "unable to give any information on the mechanism or pathway of the chemical reaction." The above example illustrates the fact that the magnitude or value of ΔH for the reaction depends only on the *initial state* $[C_{(graphite)} + O_{2(g)}]$ and the *final state* $[CO_{2(g)}]$ and is unaffected by the existence or non-existence of an intermediate such as $CO_{(g)}$.

PROBLEM 10–12 The following two reaction paths (mechanisms) have been proposed for the reaction between $H_{2(g)}$ and $I_{2(g)}$ to give $HI_{(g)}$.

PATH A	PATH B
$I_2 \longrightarrow 2I$	$H_2 \longrightarrow 2H$
$I + H_2 \longrightarrow HI + H$	$H + I_2 \longrightarrow HI + I$
$H + I \longrightarrow HI$	$I + H \longrightarrow HI$
$H_2 + I_2 \longrightarrow 2HI$	$H_2 + I_2 \longrightarrow 2HI$

(All of the species in these two mechanisms are in the gas phase.) Show that ΔH is the same for the two proposed mechanisms. You may obtain $\Delta H^{\circ}_{f,H}$ and $\Delta H^{\circ}_{f,I}$ from appendix T, Table 8.

BOND DISSOCIATION ENERGY

The *bond dissociation energy* is defined as the energy required to break a specific bond in a molecule in the gas phase. Recall that we defined this term in Chapter 6 and showed its relationship to bond formation in Figure 6–2. Bond dissociation energies were used in Chapters 6 and 7 in order to better understand the nature of bonding in molecules. From a rigorous standpoint, the gas should be at 0 °K; frequently, however, bond dissociation energies are given for 298 °K which is approximately room temperature. The symbol used to represent the bond dissociation energy is a capital D with a subscript formula of the molecule. The formula for the molecule is written with *a dash to designate the bond that is being broken.* For example, the bond dissociation energies of ethane are designated as follows:

$$C_2H_{6(g)} \longrightarrow CH_{3(g)} + CH_{3(g)} \qquad \Delta E^{\circ}_{0^{\circ}K} = \Delta H^{\circ}_{0^{\circ}K} = D_{H_3C-CH_3} = 88 \frac{kcal}{mole}$$

$$C_2H_{6(g)} \longrightarrow C_2H_{5(g)} + H_{(g)} \qquad \Delta E^{\circ}_{0^{\circ}K} = \Delta H^{\circ}_{0^{\circ}K} = D_{CH_3CH_2-H} = 98 \frac{kcal}{mole}$$

The bond dissociation energy is equal to both the change in internal energy (ΔE) and the change in enthalpy (ΔH) at 0 °K. The internal energy and enthalpy are equal at $T = 0$ °K, because the $\Delta(PV)$ term becomes zero. This can be shown using the equation of state of an ideal gas to represent the PV for each gas.

$$\Delta H = \Delta E + \Delta(PV)$$

$$\Delta H = \Delta E + \Delta(nRT)$$

At 0 °K
$$\Delta H_{0^{\circ}K} = \Delta E_{0^{\circ}K} + 0$$

A table of bond dissociation energies at 25 °C is given in Appendix T, Tables 6 and 7. A small correction (~1 kcal) would have to be made to convert these 25 °C values to true bond dissociation energies at 0 °K.

In the event that the bond dissociation energy, D, were not known, the heats of formation of the radicals and the compound itself could be used to calculate D. For example in the case of ethane:

$$C_2H_{6(g)} \longrightarrow C_2H_{5(g)} + H_{(g)}$$

$$D_{CH_3CH_2-H} = \Delta H^{\circ}_{f,C_2H_5(g)} + \Delta H^{\circ}_{f,H(g)} - \Delta H^{\circ}_{f,C_2H_6(g)}$$

(Since the ΔH_f of the radicals and the compound are given at 298° in Appendix T, Tables 5, 8, and 9, the $D_{CH_3CH_2-H}$ would be the bond dissociation energy at 298 °K rather than at 0 °K.)

$$D_{CH_3CH_2-H} = 25.7 + 52.10 - (-20.236) \text{ kcal} = 98 \text{ kcal}$$

PROBLEM 10–13   Calculate the C—H bond dissociation energy for

$$CH_3-H \qquad C_2H_5-H \qquad CH_3CH_2CH_2-H \qquad CH_3\overset{\overset{\displaystyle H}{|}}{C}HCH_3$$

using ΔH_f from the Appendices. Compare the bond dissociation energies. On the basis of your understanding of molecular structure, would you expect these bond dissociation energies to be similar?

PROBLEM 10-14 Express the ΔH of the reaction in Problem 10–12 in terms of the bond dissociation energies D_{I-I}, D_{H-H}, and D_{H-I}. Note that for the reaction step

$$I + H_2 \longrightarrow HI + H$$

the enthalpy change is the bond dissociation energy for H_2, D_{H-H}, (heat absorbed) minus the bond dissociation energy for HI, D_{H-I} (heat released). Calculate ΔH by summing up the ΔH for all three steps in each mechanism.

BOND ENERGY OR AVERAGE BOND ENERGY

The term *bond energy* or *average bond energy*, represented by the symbol ϵ, is very useful, especially when the bond dissociation energy of a specific bond in a specific molecule is not known. The bond energy of a given type of bond, such as C—H, is considered to be an average of all the C—H bond dissociation energies of typical molecules containing this bond. For example, in Problem 10–13 the C—H bond dissociation energies in CH_4, C_2H_6 and C_3H_8 were calculated and found to be 104, 98, 94.5 kcal, respectively. As one may have expected from the discussion of bonding in Chapter 7, where the C—H bond is described as arising from overlap of an sp^3 orbital of carbon and the $1s$ orbital of hydrogen, the C—H bond dissociation energies of each molecule should be similar. Given this information, the bond energy for C—H can be taken as the average of the following four C—H bond dissociation energies:

$$D_{CH_3-H} = 104.0 \text{ kcal}$$

$$D_{C_2H_5-H} = 98.0 \text{ kcal}$$

$$D_{H-nC_3H_7} = 98.0 \text{ kcal}$$

$$\underline{D_{H-iC_3H_7} = 94.5 \text{ kcal}}$$

$$\epsilon_{C-H} = \text{typical bond energy} = \frac{394.5}{4} = 99 \text{ kcal}$$

In general, large numbers of bond dissociation energies of specific molecules are considered in determining a bond energy. Since a bond energy is an average of bond dissociation energies, it is only an *estimate* of a bond dissociation energy in a molecule. Average bond energies should be used with caution, although in the absence of other data this may be the only information available. Use values for bond dissociation energies whenever you can. A table of average bond energies is given in Appendix T, Table 10.

PROBLEM 10-15   Calculate the bond energy for C—Br using the bond dissociation energies for the following compounds.

COMPOUND	D_{R-Br}(kcal)
CH_3—Br	70
C_2H_5—Br	69
i-C_3H_7—Br	68
t-C_4H_9—Br	63
C_6H_5—Br	80
$C_6H_5CH_2$—Br	51
CCl_3—Br	54
CF_3—Br	70

What is the largest deviation of the D_{R-Br} values from the mean?

ESTIMATION OF ΔH_f FROM BOND ENERGIES

In the investigation of a new or unknown compound, one may wish to evaluate a thermodynamic parameter, such as ΔH, for a reaction involving this compound. Since the heat of formation, ΔH_f, of the new or unknown compound will be unknown, one must first obtain an estimate of this quantity before calculating ΔH for the reaction. An estimate of ΔH_f can be obtained from average bond energies or bond dissociation energies. Remember that the accuracy of this estimate depends on how well the bond dissociation energies are approximated by average bond energies. The procedure used is illustrated in Example 10–6. The heats of formation used are taken from Table 8 in Appendix T.

EXAMPLE 10-6   *Suppose we were interested in the reaction*

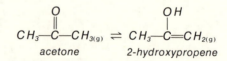

$$
\text{acetone} \qquad \text{2-hydroxypropene}
$$

We can determine experimentally the heat of formation for acetone, but it is not possible to determine the heat of formation of 2-hydroxypropene.
a. Estimate the heat of formation of 2-hydroxypropene.
b. Estimate the ΔH for the reaction using the result of (a) and the experimentally measured ΔH_f° for acetone.

a. Since the compound 2-hydroxypropene contains 3 carbon atoms, 6 hydrogen atoms, and one oxygen atom we first consider the formation of these atoms from the elements in their standard states, steps 1–3 in the following sequence of reactions. We then form the necessary bonds and estimate the ΔH for each step, using the average bond energy found in Table 10, Appendix T, as is shown in steps 4 through 8.

$$\Delta H \text{ (kcal)}$$

1. $\quad\quad\quad 3C_{(s)} \quad\quad\quad \longrightarrow 3C_{(g)} \quad\quad\quad 3\Delta H^{\circ}_{f,C(g)} = 3(170.9)$

2. $\quad\quad\quad 3H_{2(g)} \quad\quad\quad \longrightarrow 6H_{(g)} \quad\quad\quad 6\Delta H^{\circ}_{f,H(g)} = 6(52.10)$

3. $\quad\quad\quad \frac{1}{2}O_{2(g)} \quad\quad\quad \longrightarrow O_{(g)} \quad\quad\quad \Delta H^{\circ}_{f,O(g)} = (59.56)$

4. $\quad C_{(g)} + C_{(g)} \quad\quad \longrightarrow C{-}C_{(g)} \quad\quad -\epsilon_{C-C} = -(80)$

5. $C{-}C_{(g)} + C_{(g)} \quad\quad \longrightarrow C{-}C{=}C_{(g)} \quad\quad -\epsilon_{C=C} = -(146)$

6. $C{-}C{=}C_{(g)} + O_{(g)} \quad\quad \longrightarrow \overset{\displaystyle O}{\underset{\displaystyle |}{C}}{-}\overset{|}{C}{=}C_{(g)} \quad\quad -\epsilon_{C-O} = -(81)$

7. $\overset{\displaystyle O}{\underset{\displaystyle |}{C}}{-}\overset{|}{C}{=}C_{(g)} + 5H_{(g)} \quad\quad \longrightarrow CH_3{-}\overset{\displaystyle O}{\underset{\displaystyle |}{C}}{=}CH_{2(g)} \quad\quad -5\epsilon_{C-H} = -5(99)$

8. $CH_3{-}\overset{\displaystyle O}{\underset{\displaystyle |}{C}}{=}CH_{2(g)} + H_{(g)} \quad\quad \longrightarrow CH_3{-}\overset{\displaystyle OH}{\underset{\displaystyle |}{C}}{=}CH_2 \quad\quad -\epsilon_{O-H} = -(111)$

$$3C_{(s)} + 3H_{2(g)} + \tfrac{1}{2}O_{2(g)} \longrightarrow CH_3{-}\overset{\displaystyle OH}{\underset{\displaystyle |}{C}}{=}CH_{2(g)} \quad\quad \Delta H = -28 \text{ kcal}$$

The sum of steps 1 to 8 gives the final reaction which is the formation of 2-hydroxy-propene from the elements. Therefore, the ΔH for this reaction is the estimated heat of formation

$$\Delta H^{\circ}_{f, \; CH_3{-}\overset{OH}{\underset{|}{C}}=CH_{2(g)}} = \Delta H = -28 \text{ kcal}$$

b. The $\Delta H^{\circ}_{f, \; CH_3{-}\overset{O}{\overset{\|}{C}}{-}CH_3}$ is found from Table 4 in Appendix T to be -51.72 kcal/mole. The estimated ΔH for the reaction

$$CH_3{-}\overset{\displaystyle O}{\overset{\displaystyle \|}{C}}{-}CH_{3(g)} \longrightarrow CH_3{-}\overset{\displaystyle OH}{\underset{\displaystyle |}{C}}{=}CH_{2(g)}$$

is obtained using the $\Delta H^{\circ}_{f, \; CH_3{-}\overset{OH}{\underset{|}{C}}=CH_{2(g)}}$ from part a.

$$\Delta H = \Delta H^0{}_{f,\ CH_3-\overset{OH}{\underset{|}{C}}=CH_2} - \Delta H^0{}_{f,\ CH_3-\overset{O}{\underset{||}{C}}-CH_3}$$

$$\Delta H = -28\ kcal - (-51.72)$$

$$\Delta H = 24\ kcal$$

PROBLEM 10–16 Estimate ΔH for the reaction

$$CH_2{=}C{=}O \longrightarrow HC{\equiv}C{-}OH_{(g)}$$

The ΔH_f^0 for ketene ($CH_2{=}C{=}O$) is given in Table 4, Appendix T.

RELATIONSHIP BETWEEN ΔH OF REACTION AND BOND ENERGIES OF MOLECULES

In order to obtain a better understanding of enthalpy, H, and heat of reaction, ΔH, we will consider the relationship between ΔH and bond energies. In Example 10–6 we considered the reaction

$$CH_3-\overset{O}{\underset{||}{C}}-CH_{3(g)} \longrightarrow CH_3-\overset{OH}{\underset{|}{C}}=CH_{2(g)}$$

and estimated ΔH to be 24 kcal. We might now ask ourselves: What factors determine whether the ΔH of such a reaction will be positive or negative? If we tried to answer this question for all reactions, the answer would be very complex. However, if we consider only reactions in the gas phase and restrict the temperature of the reaction to 25 °C, the answer is rather simple. The dominant factor governing the value of ΔH in this case is the energies of the bonds that are formed to give the products. This can be seen readily with the reaction from Example 10–6. In order for the reaction to proceed we must (1) convert a C=O bond to a single bond C—O, (2) convert a C—C bond to a C=C bond, (3) break a C—H bond and form an O—H bond. These changes are outlined in the following sequence of steps and the relationship of ΔH to the bond energies is given:

$$\Delta H\ (kcal)$$

1. $CH_3-\overset{O}{\underset{||}{C}}-CH_3 \longrightarrow CH_3-\overset{O}{\underset{|}{C}}-CH_3 \qquad \epsilon_{C=O} - \epsilon_{C-O} = 177 - 81$

2. $CH_3-\overset{O}{\underset{|}{C}}-CH_3 \longrightarrow CH_3-\overset{OH}{\underset{|}{C}}-CH_2 \qquad \epsilon_{C-H} - \epsilon_{O-H} = 99 - 111$

3.

$$\frac{\begin{array}{c} \quad\quad OH \\ \quad\quad | \\ CH_3-C-CH_2 \longrightarrow CH_3-C=CH_2 \\ \\ \quad O \quad\quad\quad\quad\quad OH \\ \quad\quad \| \quad\quad\quad\quad\quad\quad | \\ CH_3-C-CH_3 \longrightarrow CH_3-C=CH_2 \end{array}}{}$$

$$\epsilon_{C\text{-}C} - \epsilon_{C=C} = 80 - 146$$

$$\Delta H = 18 \text{ kcal}$$

Note that this estimate differs from our previous estimate of $\Delta H = 24$ kcal. This difference of 6 kcal is expected and is the result of using average bond energies. We could obtain an accurate value if we could replace the average bond energies with the actual bond dissociation energies, but it is very seldom that we have all of these.

In the above example the ΔH is estimated as 18 kcal. Quite obviously ΔH will be negative (*exothermic* reaction) when the bonds formed are stronger than the bonds broken, and ΔH will be positive (*endothermic* reaction) when the reverse is true. With this example you should have a better understanding of what ΔH is in terms of molecular structure and bonding (see also Chapters 6 and 7).

PROBLEM 10–17   Consider the reaction in which hydrogen is added to ethylene to give ethane

$$CH_2=CH_{2(g)} + H_{2(g)} \longrightarrow CH_3-CH_{3(g)}$$

a. Describe which bonds must be broken and which bonds must be formed in order for this reaction to occur.
b. Use average bond energies to estimate the ΔH of reaction.
c. Calculate the ΔH of reaction using heats of formation from Table 5, Appendix T. Compare the results of b. and c.

10–F. ENTROPY–SECOND LAW OF THERMODYNAMICS

In discussing the first law of thermodynamics, we introduced two thermodynamic variables, E and H. As you will recall, the ability to predict the feasibility of a reaction is one of the prime objectives of this chapter. However, the ΔE or ΔH for reactions or physical processes that *can occur in nature* may increase, decrease, or remain the same. Quite obviously, then, the sign of ΔE or ΔH is not a satisfactory criterion of whether a reaction or physical process may *spontaneously* occur.

To illustrate this point we will describe some processes that occur in nature. We know that ice and water coexist at a pressure of one atmosphere and a temperature of 0 °C. (We can also say that ice is in equilibrium with water at 0 °C.) However, a sample of ice at +10 °C will completely melt to liquid water. For the melting of ice at 10 °C, $\Delta E \approx \Delta H$ has been determined experimentally and found to be positive:

$$H_2O_{(s)} \longrightarrow H_2O_{(l)} \text{ at } 10 \text{ °C} \quad\quad \Delta E \approx \Delta H = 1.53 \text{ kcal/mole}$$

Another process that occurs in nature is the freezing of water at −10 °C. In this case the ΔE and ΔH have been evaluated and found to be negative.

$$H_2O_{(l)} \longrightarrow H_2O_{(s)} \text{ at } -10 \text{ °C} \qquad \Delta E \approx \Delta H = -1.35 \text{ kcal/mole}$$

A third process that occurs in nature is the expansion of an ideal gas into a vacuum, as shown in Figure 10–9. Experimentally it is observed that there is no change in T upon expansion of an ideal gas into a vacuum; therefore, Q must be zero for the process. Since no work is performed by expansion into a vacuum, ΔE must be zero:

$$\Delta E = Q - W = 0 - 0 = 0$$

The term $\Delta(PV)$ is also zero, since $\Delta T = 0$, hence

$$\Delta H = \Delta E + \Delta(PV) = 0 + 0 = 0$$

Thus, we see that ΔH and ΔE for the processes of melting, freezing, and expansion have values which are positive, negative, and zero; nevertheless, all of these three processes occur readily in nature. As was stated earlier, the sign of ΔE or ΔH by itself cannot be used as a criterion of whether a reaction may occur. Something more than the first law of thermodynamics is needed to establish a criterion for the feasibility of a process. As we will see shortly, the information which we need to establish this criterion will be obtained from the second law of thermodynamics.

There are several ways to state the second law of thermodynamics, all of which can be shown to be equivalent. The form which we will use is most directly applicable to chemical problems.

The *second law of thermodynamics* can be written very succinctly:

10–17
$$\Delta S = \frac{Q_{rev}}{T} \qquad (T = \text{constant})$$

10–18
$$\Delta S > \frac{Q_{act}}{T} \qquad (T = \text{constant})$$

FIGURE 10–9 Expansion of an ideal gas into a vacuum

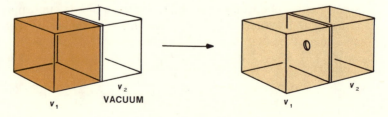

where S is a new thermodynamic function called *entropy*, T is the *absolute temperature*, Q_{rev} is the heat change along a reversible path, and Q_{act} is the actual heat change for any process that can occur (i.e., a process that is feasible). The function S is a state variable, which means that ΔS depends only on the initial and final states of a process. As we have seen earlier, Q_{rev} is generally dependent on the path; however, the factor of $1/T$ makes the ratio Q_{rev}/T independent of the path taken. Equation 10–17 can thus be used to calculate the ΔS for all reversible paths from the same initial state to the same final state. Equation 10–18 applies to any process which may occur spontaneously in nature. Processes that occur spontaneously in nature are always *irreversible* processes. Summarizing, the second law states that a process may proceed if the inequality of equation 10–18 is satisfied and that the process cannot proceed if the inequality is not satisfied. The second law can be stated in even more general terms in which it is not necessary to restrict temperature to a constant T. However, since we will be concerned primarily with chemical reactions at constant temperature, the second law as stated will suffice for our purposes.

In order to illustrate how equations 10–17 and 10–18 can be used to predict whether a process is feasible, let us consider the example of the expansion of the ideal gas into a vacuum (Figure 10–9). We must calculate ΔS and Q_{act}/T in order to see if the inequality in equation 10–18 is satisfied. The Q_{act}/T should be calculated for the irreversible expansion from V_1 to $(V_1 + V_2)$. In this case, the heat of the actual, irreversible expansion is zero,

$$\frac{Q_{act}}{T} = 0$$

To determine ΔS using equation 10–17 requires calculating Q_{rev}/T. Therefore, we must calculate the heat change for an isothermal, reversible expansion of the ideal gas from the same initial volume V_1 to the same final volume $(V_1 + V_2)$. Such a path can be visualized as one in which the expansion is carried out against a piston which is slowly moved until the gas expands from V_1 to $(V_1 + V_2)$ as shown in Figure 10–10. In the previous consideration of the isothermal expansion of the ideal gas (page 334) from V_1 to $(V_1 + V_2)$ we found that

FIGURE 10–10 Reversible expansion of an ideal gas

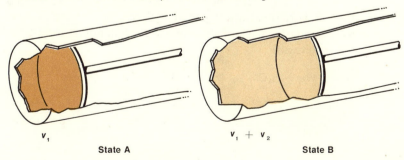

V_1

State A

$V_1 + V_2$

State B

$$\Delta E = 0$$

Since ΔE is independent of the path, ΔE must also be zero for the reversible, isothermal expansion from V_1 to $(V_1 + V_2)$. Consequently, from the first law, equation 10–9,

$$\Delta E = Q_{rev} - W_{rev} = 0$$

$$Q_{rev} = W_{rev}$$

We can calculate W_{rev} by equation 10–4; hence

$$Q_{rev} = W_{rev} = nRT \ln \frac{V_1 + V_2}{V_1}$$

Therefore, the change in entropy in going from A to B is

$$\Delta S = \frac{Q_{rev}}{T} = nR \ln \frac{V_1 + V_2}{V_1}$$

Now according to Equation 10–18 we are to compare ΔS with Q_{act}/T. If $\Delta S > Q_{act}/T$, the process may occur spontaneously. Since

$$\frac{Q_{act}}{T} = 0$$

$$\Delta S = nR \ln \frac{V_1 + V_2}{V_1}$$

and

$$(V_1 + V_2) > V_1$$

then $nR \ln (V_1 + V_2)/V_1 > 0$ and equation 10–18 is satisfied. Thus, the second law predicts that the ideal gas may expand spontaneously into a vacuum. Of course, we know in practice that a gas does expand spontaneously into a vacuum. However, the second law cannot guarantee that the process *will* occur. It can only state that it is possible that the expansion *may* occur.

In a similar manner, the second law of thermodynamics can predict that the melting of ice at 10 °C and the freezing of water at −10 °C may occur spontaneously. For such processes which occur at constant pressure, equation 10–12 stated that $Q_{act} = \Delta H$, and the ΔS therefore can be calculated from any reversible path constructed between the initial and final states (as was done above for the expansion of a gas into a vacuum). We shall not go through these calculations but the results are

$$H_2O_{(s)} \longrightarrow H_2O_{(l)} \qquad t = 10 \text{ °C}$$

$$\frac{Q_{act}}{T} = \frac{1.53 \text{ kcal/mole}}{283.1 \text{ deg}} = 5.37 \frac{\text{cal}}{\text{deg-mole}}$$

$$\Delta S = 5.56 \frac{\text{cal}}{\text{deg-mole}}$$

$$H_2O_{(l)} \longrightarrow H_2O_{(s)} \qquad t = -10\ °C$$

$$\frac{Q_{act}}{T} = \frac{-1.35\ \text{kcal/mole}}{263.1\ \text{deg}} = -5.13\ \frac{\text{cal}}{\text{deg-mole}}$$

$$\Delta S = -4.92\ \frac{\text{cal}}{\text{deg-mole}}$$

In both cases, $\Delta S > \dfrac{Q_{act}}{T}$

which is consistent with the observed fact that both processes do occur in nature.

It should be pointed out that in practice liquid water can exist temporarily at temperatures below 0 °C. Water below 0 °C is commonly called supercooled water. In light of our previous results where $\Delta S > (Q_{act}/T)$ for the freezing of water at -10 °C, the existence of supercooled water (i.e., liquid water at -10 °C) appears to contradict the second law of thermodynamics. However, this is not true, since the second law of thermodynamics *does not guarantee* that $H_2O_{(l)}$ at -10 °C will instantaneously freeze. It simply states that the freezing of water is possible (or allowed) at -10 °C but cannot state *when* the supercooled water will actually freeze.

On the contrary if

$$\Delta S < \frac{Q_{act}}{T}$$

the second law *guarantees* that the process will not occur. For example, consider the process of freezing water at $+10$ °C. Since the melting of ice at $+10$ °C is the reverse of the freezing of water at $+10$ °C, the Q_{act}/T and ΔS for freezing at $+10$ °C have the same magnitude but opposite signs as for the melting of ice at 10 °C.

$$H_2O_{(l)} \longrightarrow H_2O_{(s)} \qquad t = 10\ °C$$

$$\frac{Q_{act}}{T} = -5.37\ \text{cal/deg-mole}$$

$$\Delta S = -5.56\ \text{cal/deg-mole}$$

Thus in this case, $-5.56 < -5.37$, and $\Delta S < \dfrac{Q_{act}}{T}$ and the second law states that this process *cannot* occur spontaneously, no matter how much time elapses.

10–G. ENTROPY AND ORDER—THIRD LAW OF THERMODYNAMICS

Entropy can be viewed as the measure of the *disorder* or *randomness* of a system. When ΔS is positive for a process, then there is an increase in disorder or randomness of the system. For example, when the ideal gas in the previous section was expanded from the smaller volume V_1 to the larger volume ($V_1 + V_2$) the $\Delta S > 0$; that is, there was an increase in entropy—corresponding to more disorder in the larger volume $V_1 + V_2$. This makes sense since the molecules are free to move over a larger region of space.

For the solid state of a compound, which has a definite ordering of the molecules in the lattice, we expect the entropy to be low. The entropy of a liquid is higher, since the molecules are less ordered than in a solid and reasonably free to move throughout the liquid phase. When ice melts at 10 °C, $\Delta S = +5.27$ cal/deg-mole > 0; this means that there is an increase in entropy upon melting. Conversely, the freezing of water causes a decrease in entropy; ΔS at 0 °C$=-5.27$ cal/deg-mole < 0, corresponding to greater order or less disorder in the solid phase.

Another factor which contributes to disorder in a compound is the motion of the atoms within a molecule. Even in the solid state the atoms undergo a periodic motion about some average position in the solid. As the temperature is lowered this motion of the atoms decreases, and it is thought that at absolute zero, it is at a minimum. The motion is never completely stopped even at 0 °K; however, it does become a minimum at absolute zero, and the crystalline solid tends toward a state of minimum disorder or minimum entropy.

THIRD LAW OF THERMODYNAMICS

The term *perfect crystalline substance* refers to a solid which is not only pure but in which there is a perfect ordering of the atoms or molecules in the crystalline lattice. The third law of thermodynamics states that the entropies of perfect crystalline substances at 0 °K are equal and can be *assigned* the value of zero.

On the basis of the third law of thermodynamics *absolute values of entropy* can be calculated for both elements and compounds. A table of such entropy values at 25 °C is given in the last column in Appendix T, Tables 3, 4, 5.

ENTROPY CHANGE FOR A CHEMICAL REACTION

Since S is a state function, we can calculate the ΔS for a reaction in a manner similar to the way we calculated ΔH. However, we do not need to define a relative scale for S as we did for ΔH_f, since we can obtain absolute entropies using the third law of thermodynamics. For the general reaction (equation 10–13), the change in standard entropy is

10–19
$$\Delta S° = cS_C° + dS_D° - aS_A° - bS_B°$$

PROBLEM 10–18 Using Tables 3, 4, 5 in Appendix T, calculate $\Delta S°$ for the reaction

$$6\,CO_{2(g)} + 6\,H_2O_{(l)} = C_6H_{12}O_{6(aq)} + 6\,O_{2(g)}$$
$$\text{glucose}$$

10–H. FREE ENERGY—CRITERION FOR THE FEASIBILITY OF A REACTION

The entropy function, in conjunction with the second law of thermodynamics, can be used to predict the likelihood of a chemical reaction occurring. The procedure is identical to our previous calculation for the expansion of an ideal gas. We calculate ΔS and Q_{act}/T for the reaction and then compare these to see if the reaction is feasible (allowed).

However, most of our chemical reactions are carried out not only at constant T but also at constant P. Under these conditions a new criterion can be formulated to predict the likelihood of occurrence of a reaction. This involves the change in a new thermodynamic function called free energy which we will now introduce. The second law of thermodynamics, expressed in equations 10–17 and 10–18, can be written more concisely for both reversible and actual, irreversible processes:

$$\Delta S \geq \frac{Q}{T} \quad (T = \text{constant})$$

If the reaction is also carried out at constant P and only pressure-volume work is performed, $Q = \Delta H$, and the expression becomes

$$\Delta S \geq \frac{\Delta H}{T} \quad \begin{cases} T = \text{constant} \\ P = \text{constant} \end{cases}$$

Multiplying by T

$$T\Delta S \geq \Delta H$$

Transposing ΔH

$$T\Delta S - \Delta H \geq 0$$

If we reverse the signs, we must reverse the direction of the inequality.

10–20
$$\Delta H - T\Delta S \leq 0 \quad \begin{cases} T = \text{constant} \\ P = \text{constant} \end{cases}$$

The expression $\Delta H - T\Delta S$ can be combined into one thermodynamic function called *Gibbs free energy*, G, which is defined as

10–21
$$G = H - TS$$

At constant T

10–22
$$\Delta G = \Delta H - T\Delta S$$

The criterion for predicting whether a reaction is feasible can now be written succinctly as

10–23
$$\Delta G \leq 0 \qquad \begin{cases} T = \text{constant} \\ P = \text{constant} \end{cases}$$

In this equation, the $<$ sign applies to a reaction or process which may occur (is feasible), whereas the $=$ sign applies to a reaction or process at equilibrium. *In other words, if $\Delta G = 0$, the reactants and products of the reaction are in equilibrium.*

If the reaction is carried out with the reactants and products at standard conditions, the appropriate change in standard free energy is then given by

$$\Delta G^\circ = \Delta H^\circ - T\Delta S^\circ$$

The criterion for the feasibility of a reaction under standard conditions is

$$\Delta G^\circ \leq 0$$

It should be emphasized that the use of ΔG° for the prediction is restricted to the situation in which both reactants and products are at standard conditions.

The use of equation 10–23 to determine whether a reaction is possible is as straightforward as the calculation of ΔH or ΔS. Just as we did not know absolute values of enthalpy, we likewise do not know absolute values of free energy, since G is defined in terms of H in equation 10–21. We assign a ΔG_f° of zero to the elements in their standard states:

$$\Delta G_{f,\text{elements}}^\circ = 0$$

With this reference point a relative scale of free energies, called *free energies of formation*, is obtained and we can now calculate ΔG_f° for compounds (see Tables 3, 4, 5 of Appendix T).

The calculation of ΔG° from ΔG_f° of the products and reactants is done by the same equation we used to calculate ΔH° and ΔS°. For the general reaction

$$a\text{A} + b\text{B} \longrightarrow c\text{C} + d\text{D}$$

10–24
$$\Delta G^\circ = c\Delta G_{f,\text{C}}^\circ + d\Delta G_{f,\text{D}}^\circ - a\Delta G_{f,\text{A}}^\circ - b\Delta G_{f,\text{B}}^\circ$$

PROBLEM 10–19   Using Tables 3, 4, 5 in Appendix T, calculate ΔG° for the reaction

$$6\,CO_{2(g)} + 6\,H_2O_{(l)} = C_6H_{12}O_{6(aq)} + 6\,O_{2(g)}$$
$$\text{glucose}$$

PROBLEM 10–20   For the reaction in Problem 10–19, the ΔH° was calculated in Problem 10–10 and the ΔS° in Problem 10–18. At constant T

$$\Delta G = \Delta H - T\Delta S$$

Show that this calculation is in agreement with the result of Problem 10–19.

To aid in the understanding of the use of Gibbs free energy as a criterion for predicting the feasibility of a chemical reaction taking place, we shall refer to the analogy described first in Figure 10–1. We shall now let free energy, G, replace potential energy or height, as in Figure 10–11. Suppose we have reactants at point A (similar to the boulder in Figure 10–1), and ask whether it is possible for the reactants at A to go to products at B without any additional input of energy. The calculation of $\Delta G_{AB} = G_B - G_A$ is negative and we would say that the path of going from A to B is possible. However, by knowing only G_A and G_B we do not know whether or not there is an obstacle to overcome, such as the energy barrier at D (dashed line) to the right of A. If this energy barrier were present, it would prevent the reaction from A to B. However, if the energy barrier at D could be removed or lowered significantly, the reaction could then proceed to products at B.

On the other hand, if we were to ask whether the reactants at point A could give the products at C, we would calculate $\Delta G = G_C - G_A$. Since it is positive, we can say *with certainty* that this process will not proceed at this temperature and pressure (without the addition of free energy from an external source). Regardless of possible energy barriers between A and C, there is *no* way in which the reaction could possibly go from A to C *of its own accord*. With this diagram the student can readily see why thermodynamics can be used *to predict absolutely that a chemical reaction cannot occur*. On the other hand, it is limited to only *predicting the feasibility of* a reaction.

FIGURE 10–11   Free energy change for a chemical reaction

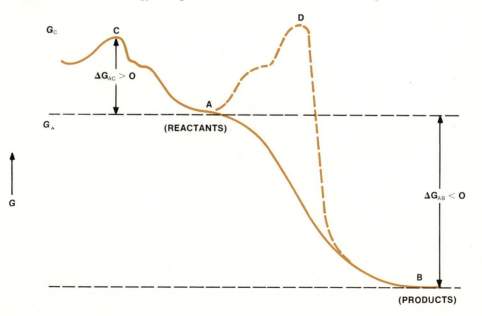

10–I. SIGNIFICANCE OF ΔG OF REACTION

Now that we have developed a criterion for feasibility of a reaction in terms of ΔG, we might ask: What in the reaction determines the value of ΔG? Why should some reactions have a negative ΔG whereas other reactions have a positive ΔG? These are not easy questions to answer. In fact if we had specific answers, we could *predict* whether a chemical reaction could occur without having to determine experimentally the ΔG_f for all the substances in the reaction. Unfortunately, we do not have explicit answers to these questions unless the reaction is exceedingly simple. However, we can identify the terms which influence ΔG and we do have at least a rough idea of how molecular properties influence these terms. In particular, we can make use of the theoretical information on atomic and molecular structure and its influence on molecular properties from Chapters 4 through 8.

Generally we carry out most reactions at constant temperature, so that ΔG for the reaction is given by equation 10–22. As you can see, equation 10–22 is composed of two terms: the ΔH term and the $T\Delta S$ term, which we will consider separately here.

ΔH TERM

As you have seen in Section 10–E, the enthalpy change for a chemical reaction depends principally on the strength of the chemical bonds which are broken and formed in the process of the reaction. An estimate of ΔH is obtained from the difference between the bond energies or bond dissociation energies (D_{AB}) of the bonds broken in the reactants and the bonds formed in the products. Chapters 6 and 7 gave a theoretical understanding of bond formation and of what factors influence the strength of bonds, for example, differences in electronegativity, orbital overlap, hybridization, resonance, multiple bonds. All of this information can be applied in understanding the ΔH for a reaction. Consider, for example, a very simple chemical reaction that occurs in the gas phase at very high temperatures.

$$Na_{(g)} + CH_3Cl_{(g)} \longrightarrow NaCl_{(g)} + CH_{3(g)}$$

The C—Cl bond in CH_3Cl is broken and the NaCl bond is formed in this reaction. The ΔH for the reaction is thus given by

$$\Delta H \approx D_{Na-Cl} - D_{H_3C-Cl}$$

We know that the Na—Cl bond is essentially ionic and hence very strong, whereas the H_3C—Cl bond is covalent and only moderately strong. In this reaction we would expect, then, a rather large negative ΔH. *A large negative ΔH favors the feasibility of a reaction, since it makes ΔG more negative.*

For reactions carried out in solution, the solvation energies of the reactants and products also make significant contributions to ΔH. This is especially true if ionic species are involved in the reaction (see Chapter 8). In addition, some reactions may be very complex, involving the breakage and formation of several bonds in the overall reaction. For example, a reaction we have considered previously,

$$6\,CO_{2(g)} + 6\,H_2O_{(l)} \longrightarrow C_6H_{12}O_{6(aq)} + 6\,O_{2(g)}$$
glucose

involves breaking the C=O bonds in CO_2 and O—H bonds in H_2O to form O=O bonds in O_2 and various types of bonds in glucose. There are several forms in which glucose can exist, one of which was represented in Chapter 2 and is shown again in the following diagram

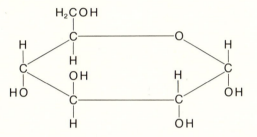

Analysis of the ΔH in terms of bond energies is much more difficult in such complex reactions.

(−TΔS) TERM

This term is more difficult to understand quantitatively on a molecular basis. We know from our previous discussion of the third law of thermodynamics that entropy is a measure of the disorder of a system. For a chemical reaction, the change in entropy, ΔS, reflects the change in order in going from reactants to products. There are two general factors which influence the orderliness of the molecular reactants or products: 1. number of species 2. freedom of motion of atoms within the molecules. The first is easy to evaluate since we can simply look at the reaction to see if there are more or fewer product molecules (or atoms or ions) than reactant molecules. If there is an increase in number of species, this means that there is an increase in disorder during the reaction, and this will tend to make ΔS positive. *Since there is a minus sign in the −TΔS term, a positive ΔS will make a negative contribution to ΔG and this will contribute towards making the reaction feasible.* Conversely, when there is a decrease in the number of molecules, ΔS will generally be negative and contribute towards making the reaction not feasible. This means that those chemical reactions are favored in which molecules break down to form more molecular species. For example, in the glucose-forming reaction considered above, there is a decrease in the number of molecules of products—7—compared to reactants—12—and a corresponding decrease in entropy ($\Delta S = -43.6$ cal/deg-mole).

The second factor contributing to disorder, freedom of motion of atoms within a molecule, is more difficult to evaluate. The atoms in molecules can vibrate*, and this motion contributes to the disorder of the molecule. However, in a more tightly bonded molecule this motion is more restricted and such a molecule has a lower entropy. For a reaction in which stronger bonds are formed, ΔS will tend to be negative, making $(-T\Delta S)$ a positive quantity. This positive contribution tends to offset somewhat the negative contribution from ΔH to ΔG which would be expected for a reaction forming stronger bonds.

EXERCISES

1. Suppose we go from state A to state B by two different paths, path no. 1 and path no. 2.

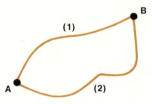

If this is all that is specified, which of the following equations are definitely applicable *without* further restrictions? If you cannot be absolutely certain that the equation applies, answer no.

a. $Q_1 = Q_2$

b. $\Delta H_1 = \Delta H_2$

c. $\Delta H_2 = \Delta E_1 + P_B V_B - P_A V_A$

d. $\Delta G_1 = \Delta H_1 - T\Delta S_1$

e. $W_2 = P_2(V_B - V_A)$

f. $W_1 = nRT_1 \ln \dfrac{V_2}{V_1}$

g. $\Delta S_1 = \dfrac{Q_1}{T_1}$

h. $\Delta S_1 = \Delta S_2$

i. $\Delta G_1 \leq 0$

j. $\Delta H_1 = \Delta E_1 + W$

*See Chapter 19 for a more detailed discussion of molecular motions and energies.

2. The standard heats of formation and entropies at 298 °K (based on the third law of thermodynamics) for diamond and graphite are as follows:

	ΔH_f° (kcal/mole)	S° (cal/deg-mole)
diamond	0.450	0.58
graphite	0	1.36

Consider the transformation

$$\text{diamond} \longrightarrow \text{graphite}$$

a. Can this transformation possibly occur at 298 °K and 1 atm?
b. Considering the results of a., why shouldn't women owning expensive diamond rings worry about their diamonds turning to graphite overnight?

3. Using thermodynamic data from Appendix T, Tables 3 and 4, predict the feasibility of the following reaction occurring:

$$CO_{(g)} + Cl_{2(g)} \longrightarrow COCl_{2(g)}$$

4. Consider the following three transformations of ice to water, all at $P = 1$ atm.
a. $H_2O_{(s, 50°C)} \longrightarrow H_2O_{(l, 50°C)}$
b. $H_2O_{(s, 0°C)} \longrightarrow H_2O_{(l, 0°C)}$
c. $H_2O_{(s, -50°C)} \longrightarrow H_2O_{(l, -50°C)}$
Realizing that the normal melting point of ice is 0 °C, state whether ΔG will be positive, negative, or zero for the three processes.

5. Criticize the statement: An investigation has been initiated to find a new reaction sequence so that the ΔE for the reaction could be lowered.

6. If methane and butane cost the same price per pound, which of the two would be the most economical fuel for heating? Assume complete combustion in both cases.

7. Calculate the heat of hydrogenation for the two isomers of 2-butene.

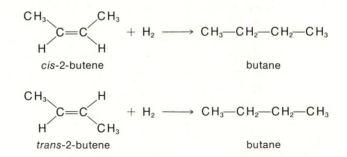

cis-2-butene   butane

trans-2-butene   butane

Compare these heats of hydrogenation. Can you rationalize why one value should be larger than the other on the basis of the relative stability of *cis* and *trans*-2-butene? Considering the molecular structure of these isomers would you expect one to be more stable than the other? Explain. [Hint: A methyl group is approximately the same size as the chlorine atom. Do the CH_3 groups get in the way of one another in these compounds?]

8. a. Calculate the heat of hydrogenation for

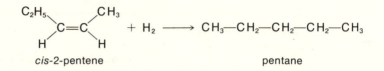

cis-2-pentene   pentane

Compare this with a similar value for *cis*-2-butene which is −28.45 kcal/mole. Note that they are very similar and in general one could conclude that

$$R\diagdown \atop H \diagup C=C \diagup R' \atop \diagdown H \quad + \quad H_2 \quad \longrightarrow \quad R-CH_2-CH_2-R'$$

has a heat of hydrogenation = 28.4 kcal

b. Using the estimate of the heat of hydrogenation for R—CH=CHR' in part a., estimate the heat of hydrogenation of benzene, *assuming* it has 3 alternate double and single bonds

c. Calculate the *actual* heat of hydrogenation of benzene to cyclohexane.

$$C_6H_{6(g)} + 3H_{2(g)} \longrightarrow C_6H_{12(g)}$$

using the actual heats of formation in the Appendices.

d. Account for the discrepancy between the estimated heat of hydrogenation in part (b.) and the actual heat of hydrogenation calculated in part (c.). [Hint: Recall the discussion of resonance in Section 7–F.]

9. In Problem 10–15 several bond dissociation energies were given for the C—Br bond. Account for the following variation in bond dissociation energy:

$$D_{CH_3-Br} = 70 \text{ kcal} > D_{CH_3CH_2-Br} = 69 \text{ kcal} > D_{(CH_3)_2CH-Br} = 68 \text{ kcal} >$$
$$D_{(CH_3)_3C-Br} = 63 \text{ kcal}$$

Assume the size of the CH_3 group is approximately the same as the covalent radius of Cl.

10. Calculate the amount of natural gas (methane) that must be burned in order to raise the temperature of the air in a house from 32 °F to 72 °F. Assume that there is complete combustion of the methane, and that the heat capacity of the air is $C_P = 7/2\,R$. Neglect the amount of air that would escape from the house upon heating at a constant pressure of 1 atm. Assume the house has an area of 2,000 ft^2 and 8 ft ceilings.

11. The temperature of a flame can be estimated by assuming the heat of the reaction is used in heating the products of the reaction. Estimate the flame temperature of an oxy-acetylene flame

$$CH{\equiv}CH_{(g)} + \tfrac{5}{2}O_{2(g)} \longrightarrow 2CO_{2(g)} + H_2O_{(g)}$$

and a hydrogen-oxygen flame

$$H_{2(g)} + \tfrac{1}{2}O_{2(g)} \longrightarrow H_2O_{(g)}$$

Assume the heat capacities are constant: $C_{P,CO_2} = 13$ cal/deg-mole, $C_{P,H_2O} = 10$ cal/deg-mole. [Hint: For simplicity in the calculations assume one mole of acetylene and one mole of H_2 have been burned.] Would you expect the temperature of a flame using air instead of oxygen to burn at a higher or lower temperature? Explain.

12. The reaction

$$Mg_2SiO_4 + SiO_2 \longrightarrow 2MgSiO_3$$
$$\text{fosterite} \quad \text{quartz} \qquad \text{enstatite}$$

cannot be carried out rapidly enough to determine ΔH directly from the heat of the reaction. For this reason ΔH must be determined indirectly. Use the following heats of solution in hydrogen fluoride to evaluate the ΔH for the above reaction.

	ΔH (kcal/mole)
$Mg_2SiO_4 + \infty HF \longrightarrow$ Solution I	−95.4
$SiO_2 + \infty HF \longrightarrow$ Solution II	−33.0
Solution I + Solution II $\longrightarrow$ Solution III	0
$MgSiO_3 + \infty HF \longrightarrow$ Solution III	−62.9

13. Calculate $\Delta H^\circ_{f, Mg_2SiO_4}$ from the following data

$$\Delta H^\circ_{f, SiO_2} = -205.4 \text{ kcal/mole}$$
$$\Delta H^\circ_{f, MgO} = -143.8 \text{ kcal/mole}$$
$$SiO_2 + 2MgO \longrightarrow Mg_2SiO_4 \qquad \Delta H = -15.1 \text{ kcal/mole}$$
$$\text{quartz} \quad \text{periclase} \qquad\quad \text{fosterite}$$

14. Calculate the change in enthalpy per mole necessary to cool steam at 105 °C to a liquid at 40 °C.

$$C_{P(steam)} = 8.7 \text{ cal/mole-deg}$$

$$C_{P(liquid)} = 18.0 \text{ cal/mole-deg}$$

$$\Delta H_{vap} = 9,850 \text{ cal/mole at } 100 \text{ °C and } 1 \text{ atm}$$

15. Calculate the standard heat of combustion of one mole of acetylene (C_2H_2) to form $CO_{2(g)}$ and $H_2O_{(l)}$.

16. Given the following

$$C_{(s)} + O_{2(g)} \longrightarrow CO_{2(g)} \qquad \Delta H° = -94.1 \text{ kcal}$$

$$Ca_{(s)} + \tfrac{1}{2}O_{2(g)} \longrightarrow CaO_{(s)} \qquad \Delta H° = -151.9 \text{ kcal}$$

$$Ca_{(s)} + C_{(s)} + \tfrac{3}{2}O_{2(g)} \longrightarrow CaCO_{3(s)} \qquad \Delta H° = -288.5 \text{ kcal}$$

Calculate the heat of reaction for

$$CaO_{(s)} + CO_{2(g)} \longrightarrow CaCO_{3(s)}$$

17. Very pure HCl is produced commercially by the direct combination of the pure elements in a flame. The elements H_2 and Cl_2 react in a flame similar to the burning of H_2 in the presence of O_2. Figure 10–12 describes the cooling procedure required to get liquid HCl. Note that the reaction vessel is jacketed with cooling water to remove most of the heat of reaction. The HCl vapor is then condensed to an aqueous solution by the addition of water and additional cooling. Suppose the reactants H_2 and Cl_2 enter the reactor at 25 °C and the HCl leaves the reactor at 100 °C.
    a. Calculate the heat of the reaction at 25 °C.
    b. Calculate the heat required to raise the temperature of HCl from 25 °C to 100 °C. Assume the C_P for HCl is 7 cal/deg-mole.
    c. From the results of (a) and (b) calculate the heat of reaction for:

$$H_{2(g, 25°C)} + Cl_{2(g, 25°C)} \longrightarrow 2HCl_{(g, 100°C)}$$

    d. If the reactants H_2 and Cl_2 are added at the rate of 20 liters/min., calculate the rate at which heat is liberated according to the reaction described in (c). Assume the reaction is complete and the H_2 and Cl_2 are at 2 atmospheres pressure.
    e. If the cooling water to the reactor is 25 °C and flows at the rate of 3.0 liters/min., calculate the temperature of the water as it leaves the reactor cooling jacket.

FIGURE 10–12 Hydrogen chloride is produced by the gas burner at the left. The gas is led over water in the silica pipes and dissolves to form concentrated hydrochloric acid. (Adapted from K. A. Kobe, *Inorganic Process Industries,* Macmillan, 1948.)

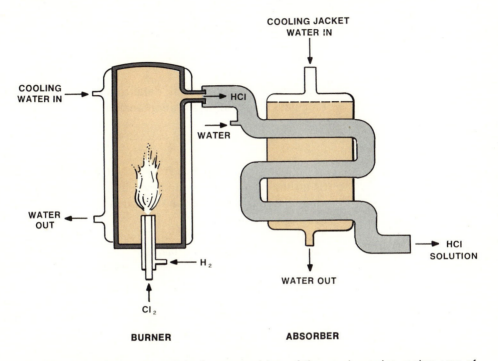

18. An alternative way to state the second law of thermodynamics makes use of the principle that heat can flow only from a hot body to a cold body. We know that this process does occur in nature; let us use the second law of thermodynamics to prove that this process may occur spontaneously. Let us take two large blocks of iron: block A at 150 °C and block B at 0 °C. Let us assume that 100 kcal of heat is transferred from block A to block B. Prove that this process may occur.

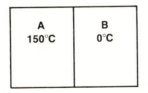

[Hint: Assume that the blocks are large enough to prevent a significant change in the temperature of blocks A and B. Calculate ΔS_A and ΔS_B. ΔS for the system will be

$$\Delta S = \Delta S_A + \Delta S_B$$

The heat change, Q_{act}, must be zero, since there is no exchange of heat with the surroundings. The heat given up by A is absorbed by B. For the calculation of ΔS_A and ΔS_B, you may assume the reversible process is the loss of 100 kcal reversibly from A and the gain of 100 kcal reversibly by B.]

19. Consider the reaction for the burning of hydrogen

$$H_{2(g)} + \tfrac{1}{2}O_{2(g)} \longrightarrow H_2O_{(g)}$$

   a. Calculate $\Delta H°$ and $\Delta S°$ using data from Table 3, Appendix T.
   b. Calculate the bond dissociation energy for

$$H_2O_{(g)} \longrightarrow OH_{(g)} + H_{(g)}$$

using thermodynamic data from Tables 3, 8, and 9, Appendix T.
   c. Using the result of (b) and known bond dissociation energies in Table 6, Appendix T, account for the fact that $\Delta H°$ is a negative quantity for the burning of hydrogen.
   d. Account for the fact that $\Delta S°$ for burning H_2 is negative in terms of the concept of the change in disorder.
   e. Which term, $\Delta H°$ or $T\Delta S°$, makes the dominant contribution to $\Delta G°$?

20. The heat of formation of $NaCl_{(s)}$ can be related to various fundamental quantities by expressing the reaction in terms of the following reaction steps:

$Na_{(s)} \longrightarrow Na_{(g)}$	ΔH_{vap}
$Cl_{2(g)} \longrightarrow 2Cl_{(g)}$	$D_{Cl\text{-}Cl}$
$Na_{(g)} \longrightarrow Na^+_{(g)} + e^-_{(g)}$	IP_{Na}
$Cl_{(g)} + e^-_{(g)} \longrightarrow Cl^-_{(g)}$	EA_{Cl}
$Cl^-_{(g)} + Na^+_{(g)} \longrightarrow NaCl_{(s)}$	U_{NaCl} (lattice energy)

This series of reaction steps can be arranged in a cycle (see equation 6–1, a Born Haber cycle).
   a. Show that the $\Delta H_{f,NaCl}$ is given by the following equation:

$$\Delta H_{f,NaCl} = \Delta H_{vap} + IP + \tfrac{1}{2}D_{Cl\text{-}Cl} - EA_{Cl} - U_{NaCl}$$

   b. Explain on the basis of these fundamental quantities why F_2 reacts more readily with the alkali metals than do the other halogen gases Cl_2, Br_2, and I_2. [Values of D_{x-x} are in Appendix T, Table 6; EA in Table T–17; U_{Nax} in Table 8–1.]

21. The synthesis of sucrose from glucose and fructose involves an increase in free energy:

$$\text{glucose} + \text{fructose} \longrightarrow \text{sucrose} + H_2O \qquad \Delta G = +5 \text{ kcal}$$

Since ΔG is positive, this reaction will not proceed by itself. However, there is another reaction involving the hydrolysis of ATP (adenosine triphosphate) to ADP (adenosine diphosphate)

$$ATP + H_2O \longrightarrow ADP + P \qquad \Delta G = -12 \text{ kcal}$$

which can promote the synthesis of sucrose. This latter reaction is said to couple with the synthesis step of the mechanism

$$ATP + glucose \longrightarrow glucose\text{-}P + ADP$$

$$glucose\text{-}P + fructose \longrightarrow sucrose + P$$

Calculate the ΔG for the coupled reaction. [Hint: ΔG is a state function and can be added as was shown for ΔH.]

CHAPTER ELEVEN

PHASE TRANSITIONS AND EQUILIBRIA

In the previous chapter we considered the basis for thermodynamics and how it could be used to predict whether a process or reaction may occur. In the following four chapters we will see how the criterion of $\Delta G < 0$ can be applied to chemical systems. In this chapter and the next, the thermodynamic principles are applied to phase transitions and to equilibria between the phases. This chapter is restricted to *single component* systems (involving only one chemical species). Phase equilibria involving two component solutions will be discussed in Chapter 12.

In the following discussions, we will refer to the molecules in a substance. However, the discussion is not restricted to substances composed of polyatomic molecules, but is equally applicable to monatomic elements and ionic compounds, where instead of molecules we have atoms (in all phases) or ions (in the liquid phase) respectively.

11–A. KINETIC MOLECULAR THEORY OF GASES

In Section 2–G we presented the ideal gas equation

2–13
$$PV = nRT$$

and showed how it could be used in making calculations for gases. The ideal gas equation is based upon experimental observations of pressure, volume, temperature, and moles of the gas and on analysis of the relationships between these variables. Generally equation 2–13 satisfactorily represents the experimental data at low pressures and high temperatures.

In this section we wish to present *the kinetic molecular theory* which represents a *molecular model* for gases. This model describes the behavior of gases in terms of the *motion* of gas molecules, hence the name *kinetic*. The model for an ideal gas consists of the following basic assumptions:

1. The molecules are in continual translational or straight-line motion throughout the volume of the container.
2. The volume of the molecules is negligible compared to the large distances between molecules.
3. As a consequence of the large distances between molecules, intermolecular forces (discussed in Chapter 8) can be neglected.
4. The molecules travel in straight-line paths unless they collide with the walls of the container. It is assumed that the magnitude of velocity of the molecules remains the same after collision with the wall. Such collisions are called *elastic* collisions. The *pressure* of the gas arises from the collisions of the molecules of mass (m) and velocity (v) with the walls of the container.
5. The average kinetic energy of the molecules ($\frac{1}{2}mv^2$) is proportional to the absolute temperature.

The equation of state of an ideal gas can be derived from the assumptions of the molecular model. The equation of state allows us to understand the behavior of an ideal gas in terms of the kinetic molecular theory. Furthermore, it dictates what conditions are necessary for real gases to behave as ideal gases. For example, assumptions 2 and 3 in our model require that the distance between molecules be large. This can only occur at low pressures and we may therefore expect real gases to obey the ideal gas law only at low pressures. This restriction to low pressure is even more stringent for polar molecules since larger intermolecular distances are required in order to be able to neglect the very strong intermolecular attractions between polar molecules.

MOLECULAR VELOCITY

According to assumption 5 we know that the average kinetic energy for an ideal gas is proportional to the absolute temperature

11–1
$$\tfrac{1}{2}mv^2 \; \alpha \; T$$

Molecules of gases at the same temperature have the same kinetic energy even if the gas is a mixture of different molecular weights. If we solve equation 11–1 explicitly for v,

11-2
$$v \; \alpha \; \left(\frac{T}{m}\right)^{1/2}$$

We can see that the velocity of molecules of a gas of given mass increases with T. The velocity of molecules of a gas at constant temperature depends inversely on the mass of the molecules. A heavier gas will necessarily have a lower velocity at a given temperature than a light gas, so that kinetic energy may remain constant.

DIFFUSION OF GASES

The diffusion of a gas is the dispersion or movement of a gas from a region of high concentration to a region of low concentration. Figure 11-1 shows the mixing of two gases which are initially separated by a removable wall. The rate of diffusion for either gas is determined by the rate at which the gas passes the plane originally occupied by the removable wall.

The rate at which a gas can diffuse is directly proportional to the average molecular velocity. The dependence of the average velocity on temperature and

FIGURE 11-1   Mixing of two gases

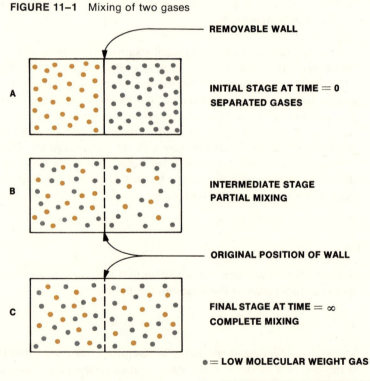

REMOVABLE WALL

A   INITIAL STAGE AT TIME = 0
SEPARATED GASES

B   INTERMEDIATE STAGE
PARTIAL MIXING

ORIGINAL POSITION OF WALL

C   FINAL STAGE AT TIME = ∞
COMPLETE MIXING

● = LOW MOLECULAR WEIGHT GAS

● = HIGH MOLECULAR WEIGHT GAS

mass was given in equation 11–2. Using this equation, the rate of diffusion (R) of a gas is related to temperature and mass

11–3
$$R \; \alpha \; \left(\frac{T}{m}\right)^{1/2}$$

From equation 11–3 we see that lighter molecules can diffuse more rapidly than heavier molecules.

The dependence of the rate of diffusion on molecular weight has been used to separate different gases, as illustrated in Figure 11–1. When mixing is partial the lighter gas (designated by the small gray dot) has diffused more rapidly past this plane into the other compartment than has the higher molecular weight gas (shown by the colored dot). Different isotopic species can be separated by utilizing this difference in rates of diffusion. Initially a mixture of two gases with different molecular weights is under pressure on one side of a porous plug (Figure 11–2). The gases can diffuse through small openings or channels in the porous plug. Since the low molecular weight gas can diffuse more rapidly than the high molecular weight gas, the gas mixture which initially diffuses through the porous plug is enriched in the lower molecular weight gas. In order to obtain a complete separation, this process is repeated many times.

Using equation 11–3, we can find an expression for the ratio (R_1/R_2) of the rates of diffusion of two gases with different mass.

$$R_1 \; \alpha \; \left(\frac{T}{m_1}\right)^{1/2}$$

$$R_2 \; \alpha \; \left(\frac{T}{m_2}\right)^{1/2}$$

If we assume that both gases are at the same temperature, the ratio is then given by the equation

FIGURE 11–2 Separation of gases by gaseous diffusion

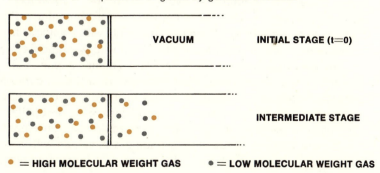

VACUUM   INITIAL STAGE (t=0)

INTERMEDIATE STAGE

● = HIGH MOLECULAR WEIGHT GAS   • = LOW MOLECULAR WEIGHT GAS

11–4a
$$\frac{R_1}{R_2} = \left(\frac{m_2}{m_1}\right)^{1/2}$$

commonly referred to as Graham's law. Since molecular weights reflect the relative masses of molecules, we can replace the ratio m_1/m_2 in equation 11–4a with the ratio of molecular weights

11–4b
$$\frac{R_1}{R_2} = \left(\frac{MW_2}{MW_1}\right)^{1/2}$$

EXAMPLE 11–1 *Calculate the ratio of the diffusion rates of H_2 and HD. D is deuterium, the $_1^2H$ isotope. Using equation 11–4b*

$$\frac{R_{H_2}}{R_{HD}} = \left(\frac{MW_{HD}}{MW_{H_2}}\right)^{1/2} = \left(\frac{3}{2}\right)^{1/2} = 1.22$$

PROBLEM 11–1 In order to make the first atomic bomb, the uranium isotope $_{92}^{235}U$ had to be separated from the $_{92}^{238}U$ isotope. This was done by using the relative rates of diffusion of the molecules of the gas UF_6 which contain either isotope. Calculate the ratio of the rates of diffusion of $_{92}^{235}UF_6$ to $_{92}^{238}UF_6$. Atomic masses are: $_{92}^{235}U$, 235.0 and $_{92}^{238}U$, 238.1.

DISTRIBUTION OF MOLECULAR VELOCITIES

In the kinetic molecular theory of gases the molecules in a gas are assumed to be in continual motion. However, all of the molecules do not travel with exactly the same velocity even though they do have the same mass and are at the same temperature. There is a *distribution* of velocities and these molecular velocities can range anywhere from zero to infinity. The mathematical expression representing the distribution of molecular velocities was first derived by Maxwell in 1860. A graphical representation of the distribution of molecular velocities is shown in Figure 11–3. The total area under the distribution curve corresponds to unit probability (100 %) of finding the velocity anywhere from $v = 0$ to $v = \infty$. The area under the distribution curve in Figure 11–3 from v_1 to v_2 then represents the probability from v_1 to v_2 as some fraction or percentage. For example, if there was a 40 % probability of finding a molecule with a velocity between v_1 and v_2, the area under the curve in Figure 11–3 from v_1 to v_2 would have a value of 0.40.

The distribution curve for molecular velocities depends upon the temperature. As the temperature is increased the entire distribution curve of molecular velocities is moved to the right, resulting in more molecules having high velocities (Figure 11–4). The shift in the point of maximum probability corresponds to an increase in *average velocity* of the molecules as the temperature is increased. Also note that the probability of finding a molecule with a velocity greater than any given velocity, such as v^*, increases with increasing temperature.

FIGURE 11–3 Distribution of molecular velocities at temperature *T*. The ordinate gives the probability of a molecule having a velocity between *v* and $(v + \Delta v)$. The shaded area is the probability of a molecular velocity in the region v_1 to v_2.

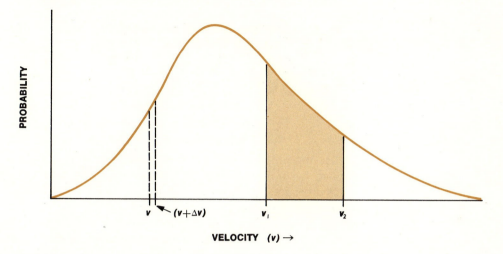

FIGURE 11–4 Distribution of molecular velocities at a range of temperatures

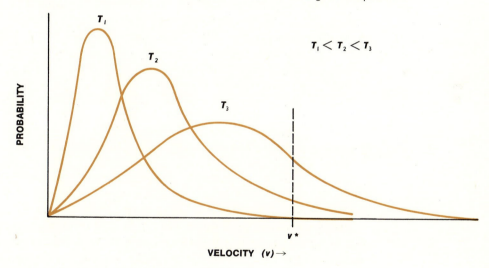

11–B. MOLECULAR DESCRIPTION OF PHASE TRANSITIONS

In the liquid phase (see also Section 2–E), molecules are closer together, and the intermolecular forces (Chapter 8) are more important. As you recall, the energy term resulting from each of these forces varies as $1/r^n$, where *n* may range from 1 to 6 depending on the nature of the specific force. These interactions greatly

FIGURE 11–5 Difference in forces acting upon a molecule at the surface and within the liquid phase

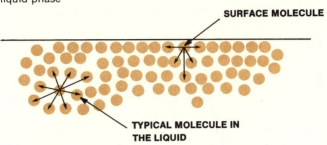

influence the properties of the liquid. Because molecules in the gas phase are separated by great distances, the intermolecular forces are less important. At low gas pressures, these forces can be neglected, and consequently, all gases at low pressures obey the ideal gas equation.

On the average a molecule within a liquid is attracted equally in all directions to its neighboring molecules. However, a molecule at the surface of the liquid will experience only intermolecular interactions directed towards the bulk of the liquid, as shown in Figure 11–5. As a consequence, the surface of a liquid has different properties from the main bulk of the liquid. Because of decreased interactions at the surface, the surface has an extra energy called *surface free energy*, in addition to the free energy of the liquid. As a result of this, there is a *surface tension* (γ), which is defined as the *increase* in surface free energy (ΔG) per unit area of surface produced (ΔA):

$$\gamma = \frac{\Delta G_{\text{surface}}}{\Delta A}$$

When energy is expressed in units of ergs, the units for γ are

$$\text{ergs/cm}^2 = \text{dyne-cm/cm}^2 = \text{dyne/cm}$$

Surface tension is thereby expressed as force/unit length.

One consequence of surface tension is that water droplets are spherical in shape. Since a sphere is the volume with the smallest surface area/unit weight of material, the surface free energy will be at a minimum for a sphere. As you will recall from Chapter 10, at constant temperature and pressure a system will tend to change in such a manner as to decrease the free energy. In other words, a favored reaction is one in which $\Delta G < 0$.

PROBLEM 11–2 Calculate by what factor the surface energy of a cube exceeds that of a sphere of the same volume. [Hint: Assume some volume of material such as 1 cm³ and calculate the surface area of the cube and sphere required to contain this volume.]

PROBLEM 11–3   If two drops of water are placed on a glass surface and brought together they will combine to form one drop. Why?

The molecules in a liquid are in continuous motion and are more or less free to move within the liquid. However, the motion in the liquid phase is *more restricted* than the motion of molecules in the gas phase. Because neighboring molecules are at much smaller intermolecular separations, greater intermolecular forces are present in a liquid. Nevertheless, the molecules in the liquid *are* in continuous motion, and their velocities increase as the temperature of the liquid is raised. The liquid molecules continually undergo collisions with one another in which they exchange energy. At any given time, molecules of a liquid will have a range of different energies. Figure 11–6 shows the distribution of molecular energies at each of two temperatures. Note that the shape of the distribution curve and the energy of maximum probability varies with temperature in a manner similar to the distribution of molecular velocities in the gas phase.

The area under the curve in any region of Figure 11–6 represents the fraction of the molecules with energies in that region. Note that the fraction of molecules with energy greater than a critical energy, E_c, increases with increasing temperature (curve T_2). That is, the probability of the molecules having an energy greater than E_c increases with temperature. Since kinetic energy is related to velocity by

$$E_k = \tfrac{1}{2}mv^2$$

there is a unique critical velocity, v_c, corresponding to the critical energy, E_c. The fraction of molecules with $v > v_c$ will also increase with increasing temperature.

FIGURE 11–6   Distribution of molecular energies in a liquid

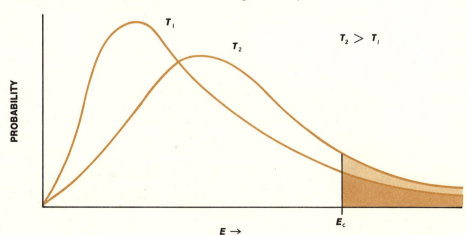

The process of *vaporization* occurs when surface molecules have velocities in the direction of the surface of the liquid that are sufficient to overcome the intermolecular forces at the surface. Not all surface molecules will escape into the vapor phase at any given time, since the molecules must have sufficient velocity and must be moving in the proper direction. Precisely speaking, the component of the velocity normal (perpendicular) to the surface, v_n, must exceed some critical velocity, v_c. Note in Figure 11–7 that some surface molecules have a larger velocity than others and that the direction of motion of some molecules is away from the bulk of the liquid; these are the molecules that can escape to the vapor phase.

The process of *vaporization* occurs at a given rate at a given temperature.

$$\text{liquid} \longrightarrow \text{gas} \qquad \text{(vaporization)}$$

In addition to vaporization we have also the process of *condensation* of gas molecules back to the liquid phase.

$$\text{liquid} \longleftarrow \text{gas} \qquad \text{(condensation)}$$

When the rate of vaporization equals the rate of condensation, there is no *net* change in the number of molecules in either of the phases. This means that the liquid phase is in equilibrium with the gas phase.

$$\text{liquid} \rightleftharpoons \text{gas}$$

It is important to point out that at equilibrium the processes of vaporization and condensation are still occurring and this is referred to as a *dynamic equilibrium*. This is in contrast to the ideal situation of a static equilibrium where no changes are taking place between the two phases.

The thermodynamic criterion for equilibrium at constant temperature and pressure is equation 10–23

$$\Delta G = 0$$

FIGURE 11–7 Molecules at a liquid–gas surface at a given instant. The length of each arrow is proportional to the magnitude of the velocity of that molecule at that instant. v_n illustrates a velocity component perpendicular to the surface of the liquid.

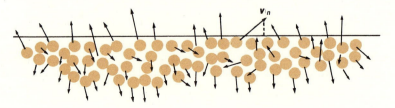

If ΔG for the liquid-gas equilibrium is $\Delta G_{f,g} - \Delta G_{f,l}$, then at equilibrium

$$\Delta G_{f,g} = \Delta G_{f,l}$$

Since pressure is constant in the vaporization process, the change in enthalpy (ΔH) is equal to the heat, Q (equation 10–12). The *heat of vaporization*, ΔH_{vap}, equals $\Delta H_{f,g} - \Delta H_{f,l}$. The ΔH_{vap} for H_2O at 100 °C is 540 cal/g or 9.72 kcal/mole.

The pressure of a gas which is in equilibrium with a liquid at some temperature is called the *vapor pressure* of the liquid. As the temperature of the liquid-gas system is increased, the rate of vaporization is increased, causing the vapor pressure to increase. The graph of vapor pressure versus temperature is shown in Figure 11–8 for water.

FIGURE 11–8   Water–water vapor equilibrium — vapor pressure versus temperature

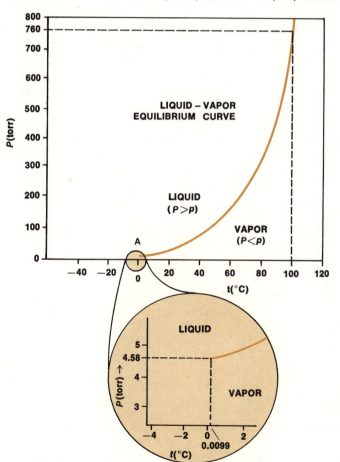

Any point on the curve in Figure 11–8 represents the pressure and temperature at which the phases are in equilibrium. For example at 60 °C the liquid and vapor are in equilibrium if the pressure is 150 torr. However, at any pressure (*P*) greater than the vapor pressure (*p*) (the region above the curve) only the liquid phase can be present. Similarly, at any pressure less than the vapor pressure (the region below the curve), only the vapor or gas phase can exist.

The *boiling point* of a liquid is the temperature at which the vapor pressure equals the atmospheric pressure. Therefore, the curve in Figure 11–8 also represents the graph of boiling point for water as a function of pressure. The *normal boiling point* is defined as the boiling point at one atmosphere pressure. For water the normal boiling point is 100 °C.

The vapor pressure curve in Figure 11–8 increases continuously with increasing temperature, indicating that the pressure required to maintain liquid and vapor in equilibrium increases as the temperature is increased. However, for each substance there is a unique point, called the *critical point*, at which this curve stops abruptly. At temperatures above this *critical temperature* (t_c) only the gas phase can exist and no matter how great a pressure is applied, the gas cannot be made to condense. The pressure at this critical point is called the *critical pressure* (p_c). The critical values for water are $t_c = 374.1$ °C and $p_c = 218.3$ atm. These values are not shown on the graph in Figure 11–8. However, if the graph is extended to higher pressures and temperatures to include the critical point, as in Figure 11–9, the vapor pressure curve shows a discontinuity at the critical point and becomes a vertical line at $t_c = 374.1$ °C.

The magnitudes of the critical temperature and pressure vary tremendously for different compounds. Table 11–1 gives some critical values. Note that the

FIGURE 11–9 Phase diagram for water in the vicinity of the critical temperature

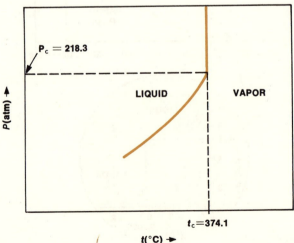

TABLE 11–1   Critical temperatures (t_c) and critical pressures (p_c)

COMPOUND	FORMULA	t_c (°C)	p_c (atm)
methane	CH_4	−82.1	45.8
ethane	C_2H_6	32.2	48.2
propane	C_3H_8	96.8	42.0
n-butane	C_4H_{10}	152.0	37.5
methyl alcohol	CH_3OH	240.0	78.5
acetone	$CH_3\overset{\displaystyle O}{\overset{\displaystyle \|}{C}}-CH_3$	235.5	47.0
acetaldehyde	$CH_3\overset{\displaystyle O}{\overset{\displaystyle \|}{C}}H$	187.8	54.7
methyl ether	CH_3OCH_3	127.0	52.6
acetic acid	$CH_3\overset{\displaystyle O}{\overset{\displaystyle \|}{C}}-OH$	321.6	57.1
ammonia	NH_3	132.5	112.5
water	H_2O	374.1	218.3
hydrogen fluoride	HF	188.0	64.0
hydrogen chloride	HCl	51.4	82.1
hydrogen bromide	HBr	90.0	84.5
hydrogen iodide	HI	150.0	81.9
carbon dioxide	CO_2	31.0	72.9

critical temperature increases as the intermolecular forces of attraction increase. For example, in the alkane series from CH_4 to C_4H_{10} there is a continual increase in t_c as the chain length is increased. As we have noted in Table 7–3, increasing the chain length increases the boiling point and the heat of vaporization. In Chapter 8, we attributed this to an increase in intermolecular interaction of the induced dipole-induced dipole type.

Also note the increase in t_c in the series HCl, HBr, and HI, which correlates with an increase in intermolecular interactions. The t_c for HF is higher than the t_c for the other hydrogen halides but this is expected due to the large intermolecular interaction that occurs in HF from hydrogen bonding HF⋯HF.

SOLID–LIQUID EQUILIBRIUM

The solid or crystal phase, in contrast to the liquid phase, has molecules, atoms, or ions at fixed locations in a three-dimensional crystal lattice structure (see Section 2–E). For example, you will recall the structure for ice shown in Figure

9–11, in which the oxygen atom of one water molecule is surrounded tetrahedrally by the oxygen atoms of four adjacent water molecules. The molecules in ice are held together principally through hydrogen bonds where the hydrogen atoms occupy sites along a line joining any two oxygen atoms. Similarly, in diamond (Figure 9–8) the carbon atoms occupy specific sites whereby a carbon atom is covalently bonded to four carbon atoms in a tetrahedral configuration. The crystal structure for NaCl is shown in Figure 2–2, where the Na^+ and Cl^- occupy fixed locations in a cubic arrangement. Each Na^+ is surrounded by six Cl^- in an octahedral arrangement, and each Cl^- is surrounded by six Na^+ in a similar octahedral arrangement.

The molecules, atoms, or ions in all three examples occupy definite locations in the crystal. However, this does not mean that the molecules, atoms, or ions in a solid are not free to move. On the contrary, these species undergo periodic motion about some equilibrium or average position, a motion referred to as *lattice vibration*. It is through such vibrations that a solid is capable of absorbing heat (*i.e.*, these vibrations determine the *heat capacity* of a solid).

As heat is added to a solid and the temperature of the solid is thereby raised, the lattice vibrations of the crystal increase. Eventually at some temperature the amplitude of the lattice vibrations is sufficient to break down the solid structure to the more randomly oriented liquid phase. This process is called *fusion* or *melting*,

$$\text{solid} \longrightarrow \text{liquid}$$

The reverse process is called *solidification* or *crystallization*

$$\text{solid} \longleftarrow \text{liquid}$$

A solid-liquid equilibrium is established when the rates of these two processes are equal:

$$\text{solid} \rightleftharpoons \text{liquid}$$

Since the solid-liquid system is at equilibrium at some constant temperature and pressure, the free energy change for the process is

10–23
$$\Delta G = 0$$

Since
$$\Delta G = \Delta G_{f,l} - \Delta G_{f,s}$$

then
$$\Delta G_{f,l} = \Delta G_{f,s} \quad \text{(equilibrium)}$$

The ΔH for the transition of solid to liquid is called *heat of fusion* (ΔH_{fus}).

$$\Delta H = \Delta H_{fus} = \Delta H_{f,l} - \Delta H_{f,s}$$

The free energies of the solid and liquid phases are not very sensitive to pressure changes. A large pressure change can occur without changing significantly the temperature of the solid–liquid equilibrium. The pressure versus temperature curve, at which solid and liquid are at equilibrium, has a very steep slope (Figure 11–10). The curve is generally very close to a straight line and for most substances the slope of the curve is positive. However, water is an exception in that the slope of the curve, as shown in Figure 11–10, is slightly negative. (It is very difficult to see that the curve has a negative slope since the deviation from the vertical is so small.)

At temperatures to the left of the solid–liquid equilibrium line in Figure 11–10, only the solid phase will exist. At temperatures to the right of the solid–liquid equilibrium line, only the liquid phase can exist.

FIGURE 11–10 Ice–water equilibrium — pressure versus temperature

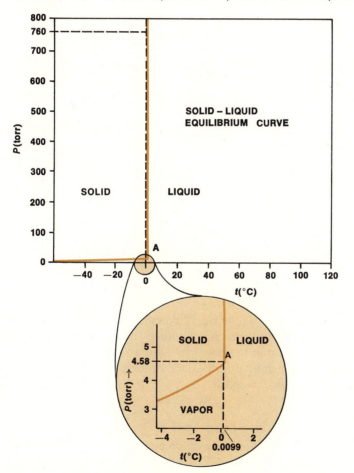

The melting point is the temperature at which the solid is in equilibrium with the liquid; hence, the P versus T curve shown in Figure 11–10 also represents the slight dependence of the melting point upon pressure. The normal melting point is defined as the melting point at one atmosphere pressure, which for H_2O is 0 °C.

SOLID-VAPOR EQUILIBRIUM

At a sufficiently low pressure, the liquid phase will not exist. For water this pressure is 4.58 mm (corresponding to point A in Figures 11–8 and 11–10). At all pressures below 4.58 mm, the solid phase will be transformed directly into the gas phase. The phase transition from solid to gas is called *sublimation*.

$$\text{solid} \longrightarrow \text{vapor}$$

The reverse process is *condensation*

$$\text{solid} \longleftarrow \text{vapor}$$

Equilibrium can be established when the rates of sublimation and condensation are equal.

$$\text{solid} \rightleftharpoons \text{vapor}$$

At constant temperature and pressure, the thermodynamic criterion for equilibrium is

10–23
$$\Delta G = 0$$
$$\Delta G = \Delta G_{f,g} - \Delta G_{f,s}$$
$$\Delta G_{f,g} = \Delta G_{f,s}$$

The ΔH for the process is called the *heat of sublimation*:

$$\Delta H = \Delta H_{sub} = \Delta H_{f,g} - \Delta H_{f,s}$$

The solid–vapor equilibrium depends on pressure and temperature in the same way as the liquid–vapor equilibrium. The curve for the solid–vapor equilibrium for H_2O is shown in the detail for Figure 11–10. At temperatures to the left of the curve, only the solid phase exists. At temperatures to the right of the curve, only the vapor phase can exist.

SOLID–SOLID EQUILIBRIUM

There are some substances which can exist in more than one *allotropic* form (see Section 9–J)—for example, carbon as graphite or diamond, sulfur as monoclinic or rhombic, and phosphorus as red or black. Just as we had equilibrium between the *condensed phases* of solid–liquid, we can also have equilibrium between two solid forms. We will designate the solid phases by α and β:

$$\alpha \rightleftharpoons \beta$$

Again, if the transition occurs at constant temperature and pressure

10–23 $$\Delta G = 0$$

$$\Delta G = \Delta G_{f,\beta} - \Delta G_{f,\alpha}$$

$$\Delta G_{f,\alpha} = \Delta G_{f,\beta}$$

The heat change for the process is called the *heat of transition*:

$$\Delta H_{trans} = \Delta H_{f,\beta} - \Delta H_{f,\alpha}$$

There are also pressure versus temperature curves for solid–solid equilibria. Examples of these will be given later.

11–C. PHASE DIAGRAMS FOR SINGLE COMPONENT SYSTEMS

A *phase diagram* is a graph which designates the phases that exist at specified conditions of the system. For a single component system (one chemical species), the conditions are specified by the pressure and temperature of the system. The phase diagram for a single component system can be represented just as we have represented equilibria between two phases in Section 11–B. For mixtures of substances, containing two or more components, the composition must be specified in addition to temperature and pressure. The phase diagrams of mixtures containing three or more components is more complicated. In this text we shall consider only one and two component systems.

PHASE DIAGRAM FOR H_2O

In Section 11–B we considered the phase equilibria between all three H_2O phases and represented them on P versus T graphs in Figures 11–8 and 11–10. The phase diagram for H_2O is a composite of these two graphs as shown in Figure 11–11. The dashed curve extending the liquid–vapor curve past point A to lower temperature and pressure corresponds to the vapor pressure curve of super-cooled water. As mentioned in Chapter 10, at temperatures below point A the solid phase is normally most stable; thus, supercooled water is in a metastable or temporary state which eventually will solidify to give ice.

Point A is called the triple point at which all three phases are in equilibrium simultaneously. There is a unique pressure and temperature for this point. For any point on a solid line in a phase diagram there are two phases present and only one variable, temperature or pressure, need be specified. For any point in the diagram where only one phase is present, both pressure and temperature must be specified in order to completely describe the state of the system.

PROBLEM 11-4   Suppose we had a sample of ice at −50 °C.

a. For a constant pressure of 2 atm, what phase changes occur and at what approximate temperature, as the sample is warmed to 200 °C?

b. For a pressure of 0.01 mm, describe the phase changes that occur as the sample is warmed to 200 °C. The vapor pressure of $H_2O_{(s)}$ at −50 °C is 0.2955 mm.

FIGURE 11-11   Phase diagram for H_2O

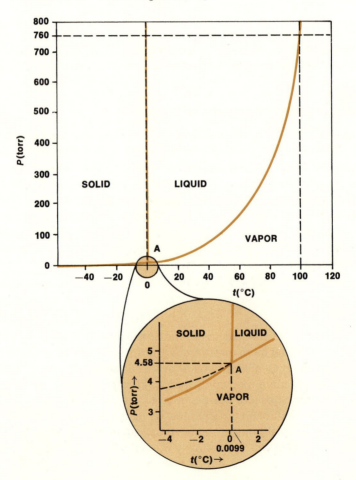

FIGURE 11–12   Phase diagram for CO_2

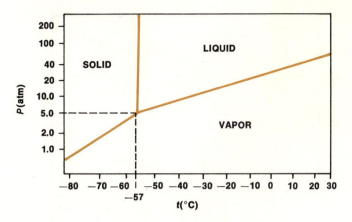

PROBLEM 11–5   If you had a piece of ice at 0 °C, could you cause the ice to melt by applying a high pressure with a sharp edge like a knife blade? Explain. [Hint: Recall that the solid–liquid equilibrium line has a negative slope.]

PROBLEM 11–6   Suppose an investigator reported that he had a sample of H_2O at 4.0 mm pressure with both the solid and gas phases present. What would be the temperature of his system?

PROBLEM 11–7   Estimate the boiling point of H_2O at 650 mm pressure. Why should eggs boiled at 650 mm pressure cook more slowly than at one atmosphere pressure?

PROBLEM 11–8   The phase diagram for CO_2 is given in Figure 11–12. Note that the scale of the t axis is not linear and the P axis is a logarithmic scale in order that a large range of pressures from 0.5 atm to 200 atm can be shown.
   a. Describe the phase transitions that occur as you warm a sample of $CO_{2(s)}$ from −85 °C to 100 °C at 1 atm pressure.
   b. Why does $CO_{2(s)}$ (dry ice) keep food colder than ice, $H_2O_{(s)}$?
   c. Under what conditions can we form liquid CO_2?

PHASE DIAGRAM FOR SULFUR

Sulfur can exist in either of two solid phases: monoclinic and rhombic. The phase diagram for sulfur is given in Figure 11–13, with the ordinate plotted in log P in order to show a large range of P values. Note in Figure 11–13 that there are *three* triple points (A, B, and C). In order to observe transitions from the liquid phase to the monoclinic form the liquid phase must be cooled quite slowly. Apparently the crystallization into the monoclinic form is a slow process and sufficient time must be allowed in order to attain equilibrium. If the liquid is cooled rapidly, rhombic crystals are formed directly.

FIGURE 11–13 Phase diagram for sulfur

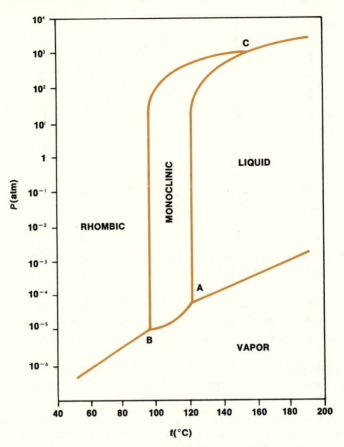

PROBLEM 11–9 Describe the phase transitions and approximate temperatures as liquid sulfur at 200 °C is slowly cooled. Assume the liquid sulfur is exposed to an atmospheric pressure of 1 atm.

PROBLEM 11–10 Recall from Section 9–J that carbon exists in two allotropic forms: graphite and diamond. The phases of carbon are shown in the following diagram. The pressure unit of kg/cm² is approximately equal to an atmosphere (1 atm = 1.033 kg/cm²).
a. What is the most stable form of carbon at room temperature and an atmospheric pressure of 1 atm?
b. Under what conditions of pressure and temperature can diamonds be produced from graphite?
c. Describe the processes that would occur if you heated graphite at 1 atm pressure.
d. Does the phase diagram give you any information as to the time required to convert diamond to graphite at room temperature and pressure of 1 atm? Explain.

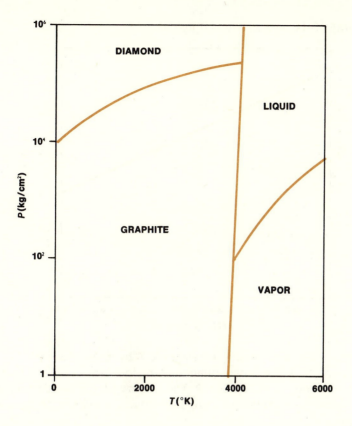

11–D. DEPENDENCE OF VAPOR PRESSURE ON TEMPERATURE

In the previous phase diagrams we showed graphically how the vapor pressure of a solid or a liquid phase varies with temperature. Mathematical equations can be derived from thermodynamic principles to describe accurately this relationship between vapor pressure (p) and the absolute temperature (T). The simplest of such equations, known as the Clausius-Clapeyron equation, is of the form

11–5
$$\ln p = -\frac{\Delta H}{RT} + \ln A$$

where $\ln p$ is the logarithm of the vapor pressure to the base e, $\Delta H =$ heat of vaporization or heat of sublimation, and A is a constant characteristic of the compound. We can also write the equation in terms of the logarithm to the base 10.

11–6
$$\log p = -\frac{\Delta H}{2.303\ RT} + b$$

FIGURE 11–14 Log p versus $1/T$ for $H_2O_{(l)}$

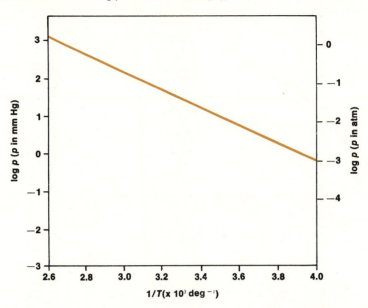

where b is a constant and is related to A by $\ln A = 2.303\,b$.

Equation 11–6 is a linear equation in the form $y = ax + b$, where $y = \log p$, $x = 1/T$, and $a = $ slope $= -\Delta H/2.303\,R$. The graph of $\log p$ versus $1/T$ for $H_2O_{(l)}$ is shown in Figure 11–14. Note in Figure 11–14 that the units of pressure do not affect the slope.

We can express vapor pressure, p, directly by taking the antilogarithm of equation 11–5.

11–7
$$p = Ae^{-\frac{\Delta H}{RT}}$$

The general relationship of pressure to temperature can be easily analyzed from equation 11–7. As the temperature increases the magnitude of $\Delta H/RT$ diminishes. However, since this is a negative exponent, the value of $e^{-\frac{\Delta H}{RT}}$ actually increases as T increases. Thus, we see that p will also increase with T. The graph of vapor pressure (p) versus $T = t\,(°C) + 273.1$ for liquid water is given in Figure 11–8. Note how rapidly the curve rises as the temperature is increased.

EXAMPLE 11–2

From the graph of log p versus $1/T$ in Figure 11–14, calculate the ΔH_{vap} for $H_2O_{(l)}$.

In order to calculate ΔH_{vap} we must evaluate the slope of the line. Let us assume that p is expressed in mm. We must read two points off the graph to calculate the slope. In order to have greater accuracy, the two points should be separated by a reasonably large spacing. Let us estimate the value of log p (mm) at $1/T = 2.7 \times 10^{-3}$ deg^{-1} and 3.9×10^{-3} deg^{-1}.

$$1/T_1 = 2.7 \times 10^{-3}\ deg^{-1} \qquad 1/T_2 = 3.9 \times 10^{-3}\ deg^{-1}$$

$$log\ p_1 = 2.83 \qquad\qquad log\ p_2 = 0.02$$

These 1/T values correspond to T = 373 °K and 256 °K, respectively. The slope is then calculated

$$slope = \frac{0.02 - 2.83}{3.9 \times 10^{-3}\ deg^{-1} - 2.7 \times 10^{-3}\ deg^{-1}} = \frac{-2.81}{1.2 \times 10^{-3}\ deg^{-1}}$$

$$= -2.34 \times 10^3\ deg.$$

Since

$$slope = -\frac{\Delta H_{vap}}{2.303\ R}$$

$$\Delta H_{vap} = -2.303\ R\ (slope)$$
$$\Delta H_{vap} = -2.303(1.987 \times 10^{-3}\ kcal/deg\text{-}mole)(-2.34 \times 10^3\ deg)$$
$$\Delta H_{vap} = 10.7\ kcal/mole$$

Note that the value of ΔH_{vap} calculated in Example 11–2 differs from the one given previously in Section 11–B for water at 100 °C (373 °K), $\Delta H_{vap} = 9.72$ kcal/mole. This is because ΔH_{vap} depends on T; the value given in Example 11–2 is an *average* value of ΔH_{vap} in the range −17 °C to 100 °C.

PROBLEM 11–11   The following are vapor pressures for methyl alcohol (CH_3OH) at various temperatures.

t (°C)	p (mm)
−44.0	1.0
−16.2	10.0
5.0	40.0
21.2	100.0
49.9	400.0
64.7	760.0

Prepare a graph of log p versus $1/T$ and evaluate ΔH_{vap} for methyl alcohol from the slope.

PROBLEM 11–12   A graph of log p versus $1/T$ for methane (CH_4) shows a slope of -0.445×10^3 deg. Evaluate ΔH_{vap} for methane and compare this with the ΔH_{vap} for CH_3OH evaluated in Problem 11–11. On the basis of the intermolecular interactions discussed in Chapter 8 account for the difference between the ΔH_{vap} of CH_4 and CH_3OH.

11–E. LE CHATELIER'S PRINCIPLE

Le Chatelier's principle is a general statement concerning the change which a system at equilibrium undergoes when variables such as pressure and temperature are changed. The principle can be applied to systems in physical equilibrium, such as the equilibrium between two phases, as well as to systems in chemical equilibrium. In this chapter we will apply the principle to physical equilibrium between phases of a one component system. In Chapter 12 the principle will be applied to physical equilibria between phases which may have two components (solutions) and in Chapter 13 to chemical equilibria.

Le Chatelier's principle states: *When a stress is applied to a system in equilibrium, the equilibrium will shift so as to relieve the stress.* The stress usually results from the changing of a variable, such as temperature or pressure. Although Le Chatelier's principle has been given as an independent principle, in fact it can be supported on a rigorous thermodynamic basis. A mathematical equation to express quantitatively the change in the equilibrium will be given in Chapters 12 and 13 for solutions and chemical equilibria.

EFFECT OF CHANGE IN PRESSURE ON EQUILIBRIUM

We will consider the equilibrium of reactants (A) and products (B) in which there is a difference between the volume of reactants and products:

$$A \rightleftharpoons B$$

For this reaction the change in volume, ΔV, is simply the volume of products, V_B, minus the volume of reactants, V_A.

$$\Delta V = V_B - V_A$$

The volumes V_A and V_B must be for the same mass of material. The volume for a substance in a given phase is usually expressed as per gram, where it is called *specific volume*, or as per mole, where it is called *molar volume*, $\bar{V}$. For example, liquid water has a density of approximately 1 g/ml. The specific volume is then

$$\text{specific volume} = \frac{1}{\rho} = \frac{1}{1 \text{ g/ml}} = 1 \text{ ml/g}$$

The molar volume of liquid water is

$$\bar{V}_{H_2O_{(l)}} = (1 \text{ ml/g})(18 \text{ g/mole})$$
$$\bar{V}_{H_2O_{(l)}} = 18 \text{ ml/mole}$$

The effect which a change in external pressure will have on this equilibrium system depends upon the sign of ΔV. According to Le Chatelier's principle, an increase in pressure (stress) can be relieved if the equilibrium shifts in the direction of decreasing volume. Therefore, if the volume decreases in going from left to right ($V_B < V_A$), equivalent to a negative ΔV, then an increase in pressure will shift the equilibrium to the right. The reverse is also true; that is, if the volume decreases in going from right to left ($V_A < V_B$), or ΔV positive, an increase in pressure will shift the equilibrium to the left. Of course, a decrease in pressure will cause the reverse effects. These results are valid for all equilibrium systems and are summarized as follows:

$$\text{Given } V_B < V_A \begin{cases} \text{an increase in } P \text{ shifts equilibrium to right} \\ \text{a decrease in } P \text{ shifts equilibrium to left} \end{cases}$$

$$\text{Given } V_A < V_B \begin{cases} \text{an increase in } P \text{ shifts equilibrium to left} \\ \text{a decrease in } P \text{ shifts equilibrium to right} \end{cases}$$

It is not necessary to memorize these results since they follow quite readily from Le Chatelier's principle.

EXAMPLE 11–3   *Consider the equilibrium between liquid water and water vapor at 50 °C. The vapor pressure of water at 50 °C is 92.5 torr. Therefore, the pressure on the liquid–vapor equilibrium system must be 92.5 torr in order for these two phases to remain in equilibrium at 50 °C*

$$H_2O_{(l)} \rightleftharpoons H_2O_{(g)}$$

a. If the pressure is increased to 150 torr, what is the effect on the equilibrium?
b. Show that the effect of increasing pressure is in agreement with the phase diagram for water in Figure 11–11. The molar volume for liquid water is approximately 18 ml/mole whereas the molar volume for H_2O vapor at 92.5 torr can be found from the ideal gas law

$$PV = nRT$$

$$P\left(\frac{V}{n}\right) = P\bar{V} = RT$$

$$\bar{V}_{H_2O_{(g)}} = \frac{RT}{P} = \frac{(0.08205 \text{ liter-atm/deg-mole})(323 \text{ deg})}{(92.5 \text{ torr})(1 \text{ atm}/700 \text{ torr})}$$

$$\bar{V}_{H_2O_{(g)}} = 286 \text{ liter/mole} = 286,000 \text{ ml/mole}$$

a. Obviously the volume of the vapor exceeds the volume of liquid

$$\bar{V}_{H_2O_{(g)}} > \bar{V}_{H_2O_{(l)}}$$

or $$\Delta V > 0$$

Therefore, according to Le Chatelier's principle, increasing the pressure will shift the equilibrium to the left, causing the vapor with a larger molar volume to condense to the liquid with a smaller molar volume.
b. Note in Figure 11–11 (p. 368) that at t = 50 °C increasing the pressure above 92.5 torr will cause the vapor to condense to liquid water. Thus the effect of increasing the pressure as interpreted from the phase diagram is consistent with Le Chatelier's principle.

PROBLEM 11–13   Consider the equilibrium between ice and water vapor at −3 °C. The vapor pressure of ice at −3 °C is 3.568 torr and the density of ice at −3 °C is 0.917 g/ml.
a. Using Le Chatelier's principle predict the effect which decreasing pressure (<3.568 torr) has on this equilibrium.
b. Interpret the effect of decreasing pressure using the phase diagram for water.

PROBLEM 11–14   a. Using Le Chatelier's principle, describe the effect of increasing pressure on the equilibrium between ice and liquid water at 0 °C and 1 atm. At 0 °C and 1 atm the density of water is 1.00 g/ml and the density of ice is 0.917 g/ml.
b. Interpret the effect of increasing pressure on the phase diagram for water.

PROBLEM 11–15   a. Using the results of Problem 11–14, explain how the pressure from the blade of an ice skate could melt ice. Apparently the ability to ice skate depends on this phenomenon.
b. Do you think skating would be effective on solid benzene? The densities of solid and liquid benzene are 1.014 g/ml and 0.895 g/ml.

EFFECT OF CHANGE IN TEMPERATURE ON EQUILIBRIUM

The effect which a change in temperature has on the general equilibrium between reactants A and products B

$$A \rightleftharpoons B$$

$$\Delta H = \Delta H_{f,B} - \Delta H_{f,A}$$

depends upon the sign of ΔH (see Section 10–E). If heat is absorbed for the reaction of A to B, ΔH is positive; if heat is released ΔH is negative for the reaction of A to B. According to Le Chatelier's principle, the stress of an increase in the temperature will be relieved if the equilibrium shifts in the direction which absorbs the heat (thereby lowering temperature). Therefore, if heat is absorbed (ΔH positive) in going from reactants to products, then an increase in temperature will shift the equilibrium to the right (towards the products). For a decrease in temperature the reverse effects are observed as summarized in the following:

If ΔH is positive $\begin{cases} \text{an increase in } T \text{ shifts the equilibrium to right} \\ \text{a decrease in } T \text{ shifts the equilibrium to left} \end{cases}$

If ΔH is negative $\begin{cases} \text{an increase in } T \text{ shifts the equilibrium to left} \\ \text{a decrease in } T \text{ shifts the equilibrium to right} \end{cases}$

PROBLEM 11–16   Consider the equilibrium

$$H_2O_{(s)} \rightleftharpoons H_2O_{(l)} \qquad \begin{array}{l} t = 0\,°C \\ P = 1\,atm \end{array}$$

a. Using Le Chatelier's principle predict the effect which increasing temperature has on this equilibrium.

b. Describe this effect of increasing temperature using the phase diagram for water and show how it is consistent with part a.

PROBLEM 11–17   Predict the effect of decreasing temperature on the equilibrium between ice and water vapor at $-3\,°C$ and 3.568 torr.

✳ 11–F. LIQUID CRYSTALS

In this chapter we have considered transformations and equilibria between the gas, liquid, and solid phases. Most compounds show only these three phases and the transition from one phase to another is very distinct. There are, however, some substances which do not undergo a complete and distinct transition between the solid and liquid states. Upon melting, these solids undergo a transition to one or more intermediate states in which the phase is fluid-like but has more highly ordered and localized molecular arrangement than in a normal liquid. For this reason, these intermediate phases are called *liquid crystals*, or mesomorphic or paracrystalline states. Some people prefer to call this rather unusual behavior the fourth state of matter. However, the liquid crystal state is exhibited by relatively few compounds; it is not general to all compounds. These intermediate liquid-crystal phases are isolated species, and in some cases each phase has a unique color. In some compounds there is more than one liquid-crystal phase and the transition from one liquid crystal phase to another occurs as the temperature changes. Compounds which show different colored liquid-crystal phases can thus be used to monitor temperatures on surfaces by the color of the liquid-crystal.

Liquid crystals usually occur in compounds whose molecules are unsymmetrical in shape. Unsymmetrical molecules can have a variety of intermolecular interactions involving different geometrical arrangements of the different functional groups in the molecule. In the liquid-crystal phases, different types of intermolecular interactions will be predominant. An example of a compound which forms liquid crystals is *p*-azoxyanisole which melts at 82 °C to form a liquid

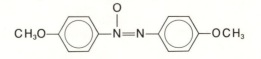

crystalline state which is stable to 150 °C. Note that intermolecular interactions can either occur through the O—CH$_3$ functional groups at the ends or by having more than one molecule lying parallel to each other.

There are two classes of liquid crystals: *thermotropic* and *lyotropic*. Thermotropic liquid crystals are formed simply by heating a pure solid substance, such as *p*-azoxyanisole, to the appropriate temperature. Lyotropic liquid crystals are formed by mixing two or more compounds, one of which is highly polar, such as water. Lyotropic liquid crystals have not been studied as extensively as thermotropic but they have recently been found in living systems, such as protein solutions. At this time we will restrict our discussion to thermotropic liquid crystals.

There are three types of thermotropic liquid crystal phases: *smectic*, *nematic*, and *cholesteric*. These three phases are differentiated on the basis of molecular arrangement. They are diagrammed in Figure 11–15 where they are contrasted with the ordered crystalline solid and the disordered liquid state. In compounds

FIGURE 11–15 Comparison of molecular arrangements in crystal, liquid–crystalline, and liquid phases

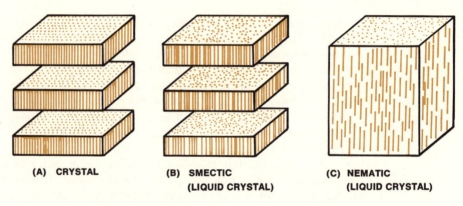

(A) CRYSTAL   **(B) SMECTIC (LIQUID CRYSTAL)**   **(C) NEMATIC (LIQUID CRYSTAL)**

(D) CHOLESTERIC (LIQUID CRYSTAL)

(E) LIQUID

which exhibit liquid crystal behavior the elongated molecules lie parallel to one another so that the long axis of each molecule is in the same direction. In the solid crystal phase such molecules should lie in layers in a highly ordered structure (Figure 11–15a). When the temperature of the solid is increased the additional kinetic energy of the molecules may be sufficient to break up the intermolecular interaction between the end functional groups of the molecules but not sufficient to affect the lateral interactions between the molecules. Thus the molecules remain parallel with their long axes pointing in the same direction.

In the smectic phase (Figure 11–15b) the molecules remain parallel to one another in a series of layers generally one molecule thick. The long axes of all the molecules in a given layer are perpendicular or almost perpendicular to the plane of the layer. The planes are relatively free to glide over one another and this probably explains why the properties of smectic liquid crystals resemble those of many soaps. In fact, the name smectic is derived from the Greek word *smector*, meaning soap-like. The smectic phase is usually turbid and viscous.

The structure of the *nematic* liquid crystal is shown in Figure 11–15c. In this phase the molecules remain parallel but are not constrained to different layers. There are differences of opinion as to the exact structure in the nematic phase. One hypothesis is that the molecules are arranged in swarms and within each swarm all the molecules are parallel to one another. The swarms themselves are randomly arranged. Another hypothesis assumes that *all* the molecules in this liquid crystal phase are parallel. Nematic phases are usually turbid but rather mobile.

The structure of the *cholesteric* phase is shown in Figure 11–15d. The name cholesteric was chosen because the majority of compounds in this type are derivatives of cholesterol. In the cholesteric liquid crystal the long axes of the molecules are parallel to the plane of the layers and, consequently, the layers are thin. Because of the molecular geometry the long axes of the molecules in each layer are slightly rotated, on the order of 15 minutes of arc relative to those in the preceding neighboring layers, giving a helical appearance to the crystal.

Compounds can exhibit one or more of these thermotropic liquid crystal phases; however, the transition from nematic to cholesteric has never been observed. Generally transitions occur between different smectic liquid crystal phases, between smectic and nematic, or between smectic and cholesteric. For example, two smectic and one nematic mesophases occur in ethyl *p*-(4-ethoxy-benzylideneamino)cinnamate (again an elongated molecule with functional groups at both ends)

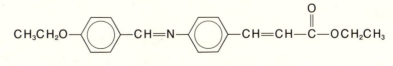

$$crystal \rightleftharpoons_{78\ °C} smetic\ I \rightleftharpoons_{110\ °C} smetic\ II \rightleftharpoons_{154\ °C} nematic \rightleftharpoons liquid$$

There are several important applications of liquid crystals and from recent research in the field it appears that there will be many more forthcoming. One use of liquid crystals is to map the temperature or difference in temperature over large surfaces. Substances which have cholesteric liquid crystal phases produce a series of bright colors when the material is illuminated with white light. The particular colors which arise depend upon the nature of the substance and its temperature, among other things. Small differences in temperature can produce strikingly different colors. By mixing such substances, any desired temperature-color combination can be obtained. These mixtures have been fabricated into flexible films and tapes and are used to test airplane wings, car springs, and transistor circuits for temperature differences resulting from friction or electrical resistance. Hot spots as small as one mil* in size can be detected in complicated electronic circuits, thus pinpointing high resistance junctions that could lead to an eventual breakdown. Thermal mapping of the human body is carried out by applying these cholesteric materials to the surface of the skin to locate veins and arteries and other internal structures that transmit heat at different rates from that of the surrounding tissue.

The color of cholesteric substances is also highly dependent upon foreign chemical vapors. The use of cholesteric materials in chemical plants for the detection of small amounts of toxic chemical vapors appears promising as a warning device of when a dangerous level of vapor contamination is reached. There are also some unique electrical and optical effects in nematic type liquid crystals which recent technology has applied to electronic display panels, electronically controlled windows and optical filters, simple numeric indicators, and all-electronic clocks.

PROBLEM 11–18   Which of the following molecules would you expect to exhibit a liquid crystal phase? On what basis did you make your choice?

a. $ClCH_2CH_2Cl$ b. $CH_3CH_2CH_2CH_2CH_2CH_2\overset{\overset{\displaystyle O}{\|}}{C}H$ c. CH_3O⟨◯⟩$\overset{\overset{\displaystyle H}{|}}{C}=\overset{\overset{\displaystyle H}{|}}{C}-\overset{\overset{\displaystyle O}{\diagup\!\!\diagdown}}{C}_{OH}$

[Hint: The structure of the $CH_3CH_2CH_2CH_2CH_2CH_2-$ in b. is not linear (see Section 7–H).]

EXERCISES

1. List the following compounds and elements in the order of increasing molecular velocity in the gas phase at the same temperature: CH_2Cl_2, Cl_2, Xe, $Pb(CH_3)_4$, XeF_4, HF

*1 mil = 1/1000 inch

2. Which of the compounds in the following pairs would you expect to have the highest ΔH_{vap} and why?
   a. He or Xe
   b. CH_4 or CH_3Cl
   c. NaCl or HCl
   d. CH_3OH or CH_3OCH_3
   e. CH_3CH_3 or CH_3CH_2OH
   f. H_2 or I_2

3. Which of the following pairs of compounds could be separated most readily by gaseous diffusion? Why?
   a. H_2 and HD
   b. $^{235}UF_6$ and $^{238}UF_6$
   c. CH_4 and CD_4
   The symbol D is used to represent the ^2_1H isotope, deuterium, which has a mass of 2.0140.

4. Which of the following gases would you expect to show the greatest deviation from the ideal gas law? Why? Assume all gases are at the same temperature and pressure.
   a. CH_4 b. CH_3Cl c. CH_3OH

5. a. List and briefly describe four types of solid where the classification is made according to the nature of the bonding and structure. Consult Chapter 9 for examples.
   b. Classify the following compounds according to these four types: diamond, iron, methane, benzene, cesium fluoride, carbon disulfide, copper, xenon, boron.

6. Explain the advantage of cooking vegetables and stews in a pressure cooker. Why would cooking with a pressure cooker be cheaper than normal boiling?

7. Explain why methane is sold as a gas at pressures as high as 140 atm whereas propane is sold as a liquid at a pressure of approximately 10 atm.

8. Ethyl alcohol (C_2H_5OH) and ethyl ether ($C_2H_5OC_2H_5$) both burn readily. However, the vapors coming from an open container of ethyl ether at room temperature are highly inflammable whereas the vapors of ethyl alcohol are much less inflammable. Explain why this is true. [Hint: Consider the relative concentrations of these substances in the gas phase.]

9. Refer to the phase diagram for carbon in Problem 11–10.
   a. Since graphite can be converted to diamond by increasing the pressure, which phase has the greatest density?
   b. Will solid graphite float on top of liquid carbon?

10. Sulfur often exists naturally in the pure elemental form. These deposits of elemental sulfur are found below the earth's surface and are recovered by the Frasch process. In this process a well is bored and concentric pipes containing superheated water under pressure and air are forced into the deposit to melt the sulfur as shown in the following figure.

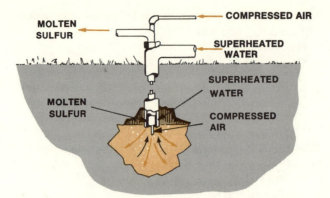

The molten sulfur is forced out the middle concentric pipe between the superheated water (outer pipe) and the compressed air (inner pipe) and thereby pumped to the surface. Refer to the phase diagram for sulfur (Figure 11–13) to answer the following questions.

a. Assuming that the elemental sulfur is in the rhombic form, what minimum temperature of the superheated water is required to melt the sulfur at 1 atm.

b. The phase diagram for H_2O, given in Figure 11–11, only goes to 1 atm pressure. The following data will allow you to extend the liquid-vapor curve above 1 atm and 100 °C.

TEMPERATURE (°C)	PRESSURE (atm)
110	1.458
120	1.956
130	2.67
140	3.56
150	4.70

What minimum pressure must the superheated water be under in order that the water can melt sulfur (part a)? What is the melting point of sulfur at this pressure?

c. The specific heat of fusion of sulfur is 9.2 cal/g. If one gram of superheated water is used to melt one gram of sulfur, what initial temperature must the superheated water have in order to melt the sulfur? What minimum pressure must the superheated water be under in order to attain this temperature?

11. In the following table are given the vapor pressures of solid and liquid methane as a function of temperature:

	t (°C)	p (mm)		t (°C)	p (mm)
$CH_{4(s)}$	−195.5	10.0	$CH_{4(l)}$	−181.4	100.0
	−187.7	40.0		−167.8	400.0

a. Graph ln p or log p versus $1/T$ for both the solid and liquid CH_4. Explain why the intersection of these two lines is the triple point for methane. Estimate the temperature and pressure for the triple point.

b. Sketch the phase diagram for methane, assuming only one solid phase. Identify the phases in each region.

12. Show that for sublimation

$$\Delta H_{sub} = \Delta H_{fus} + \Delta H_{vap}$$

13. Using the vapor pressure data for methane in Exercise 11, calculate ΔH_{sub} and ΔH_{vap} according to equations 11–5 or 11–6. Calculate ΔH_{fus} using the relationship given in Exercise 12.

14. Recall from Chapter 9 that SiO_2 can exist as a polymeric solid with a melting point of 1710 °C. Actually there are several solid phases of SiO_2 as shown in the following phase diagram.

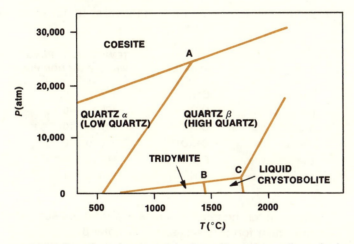

a. Describe the phases and phase transformations in heating quartz from 500 °C to 2100 °C at 1 atmosphere pressure.

b. The phase transition quartz $\alpha \rightleftharpoons$ quartz β occurs at t = 573 °C, $P = 1$ atm. The change in volume for one mole is 0.2 cm³/mole. The heat of transition from quartz α to quartz β is +150 cal/mole. Using Le Chatelier's principle, predict what effect increasing pressure and increasing temperature have upon this equilibrium system.

15. Identify the triple points for SiO_2 and state the phases which are in equilibrium (see Exercise 14 for phase diagram).

16. In the *Handbook of Chemistry and Physics* (Fiftieth edition. The Chemical Rubber Co., Cleveland, Ohio, 1969) is a table of vapor pressure variation with temperature. The vapor pressure dependence on temperature is given by the equation

$$\log_{10} p = -\frac{0.05223a}{T} + b$$

where p is in torr.

a. How is a in the above equation related to ΔH?

b. The values of a and b for some compounds are

COMPOUND	FORMULA	TEMPERATURE RANGE FOR WHICH EQUATION HOLDS (°C)	a	b
krypton	Kr	−169 to −150	9,377	6.92387
hydrobromic acid	HBr	−86 to −66	17,960	7.427
sodium chloride	NaCl	976 to 1155	180,300	8.3297

Account for the relative values of a for these substances in terms of the intermolecular and ionic forces discussed in Chapter 8.

17. a. The values of a and b (see Exercise 16) for some organic compounds are

COMPOUND	FORMULA	TEMPERATURE RANGE FOR WHICH EQUATION HOLDS (°C)	a	b
methane	CH_4	−174 to −163	8,517	6.8626
propane	C_3H_8	−136 to −40	19,037	7.217
n-propyl alcohol	C_3H_7OH	−45 to −10	47,274	9.5180
methyl ether	CH_3OCH_3	−70 to −20	23,025	7.720
acetaldehyde	CH_3CHO	−24.3 to 27.5	27,707	7.8206
propionic acid	C_2H_5COOH	20 to 140	46,150	8.715

Account for the relative values of a for these substances in terms of the intermolecular forces discussed in Chapter 8.

b. Derive the equation relating the normal boiling point with a and b. Use the equation relating the vapor pressure dependence on temperature given in Exercise 16. [Hint: Recall that the normal boiling point is the temperature when the vapor pressure, P, is 1 atm = 760 torr.] Does a high value of a correspond to a high boiling point? How is the boiling point affected by b? Which of the terms, a or b, is the dominant factor in relationship to the normal boiling point? Do you expect ΔH_{vap} and the normal boiling point to be related?

18. Account for the relative magnitudes of the critical temperature for the organic compounds given in Table 11–1. The order of increasing critical temperatures is:

$$CH_4 < C_2H_6 < C_3H_8 < CH_3OCH_3 < CH_3\overset{\displaystyle O}{\overset{\displaystyle \|}{C}}H < CH_3OH < CH_3\overset{\displaystyle O}{\overset{\displaystyle \|}{C}}{-}OH$$

19. Suppose we have a gas enclosed in a cylinder by a piston

a. As the pressure is increased on the piston, thereby compressing the gas, explain the various phases that appear. In one case assume the temperature is less than the critical temperature of the gas and in the other assume the temperature is greater than the critical temperature.

b. Explain why the following pressure vs. volume curves are obtained for the two different cases considered in a.

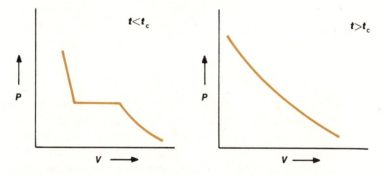

Describe the phase transitions that occur and identify the transitions on the *P* versus *V* curves.

CHAPTER TWELVE

SOLUTIONS — TWO-COMPONENT SYSTEMS

In Chapter 2 we introduced types of solutions and the units used to express the concentrations or relative amounts of substances in a solution. In this chapter we will consider solutions in much greater detail, based upon our knowledge of molecular structure, properties, and interactions (Chapters 6 through 9), and thermodynamics (Chapter 10). Furthermore, we will use the properties of solutions to explain such phenomena as freezing point depression and osmotic pressure.

12–A. MOLECULAR NATURE OF SOLUTIONS

GAS SOLUTIONS

Though we will be concerned primarily with liquid solutions in this chapter, it is useful to consider solutions of gases first. Let us consider two ideal gases A and B in adjacent compartments separated only by a single wall, as shown in Figure 12–1. If the wall separating the two compartments is ruptured, we know from experience that the two gases will mix. Since the gases are assumed to be ideal gases (intermolecular interactions are absent), there is no change in heat (Q) or temperature for this process of mixing the gases. That is, no heat is exchanged between the system (gases) and the surroundings of the container. The gases are

FIGURE 12–1 Mixing of two ideal gases

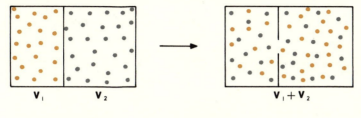

$\bullet$ = "A" MOLECULE $\bullet$ = "B" MOLECULE

assumed to be at the same pressure, which remains constant upon mixing since the gases are ideal. Therefore, the change in enthalpy as given by equation 10–12, is

12–1
$$\Delta H = Q = 0$$

Obviously, since $\Delta H = 0$ for mixing ideal gases, an enthalpy change does not influence the mixing.

The changes in entropy and in free energy for this mixing process can also be calculated quite easily. The entropy change is the sum of the entropy changes for the gases A and B (equation 10–19).

$$\Delta S = \Delta S_A + \Delta S_B$$

Since the mixing of A with B illustrated in Figure 12–1 involves work in the expansion of A from volume V_1 to $(V_1 + V_2)$, the change in entropy for gas A is given by equation 10–4.

$$\Delta S_A = n_A R \ln \frac{V_1 + V_2}{V_1}$$

Similarly, equation 10–4 can be used to calculate ΔS_B, for which the initial volume is V_2 and the final volume $(V_1 + V_2)$

$$\Delta S_B = n_B R \ln \frac{V_1 + V_2}{V_2}$$

The change in entropy of mixing is then

12–2
$$\Delta S = n_A R \ln \frac{V_1 + V_2}{V_1} + n_B R \ln \frac{V_1 + V_2}{V_2}$$

Since the temperature remains constant, the change in free energy is given by equation 10–22

$$\Delta G = \Delta H - T\Delta S$$

From equations 12–1 and 12–2 for ΔH and ΔS, ΔG of mixing is

$$\Delta G = 0 - T\left(n_A R \ln \frac{V_1 + V_2}{V_1} + n_B R \ln \frac{V_1 + V_2}{V_2}\right)$$

12-3
$$\Delta G = -n_A RT \ln \frac{V_1 + V_2}{V_1} - n_B RT \ln \frac{V_1 + V_2}{V_2}$$

Note that since

$$\frac{V_1 + V_2}{V_1} > 1 \quad \text{and} \quad \frac{V_1 + V_2}{V_2} > 1$$

then

$$\ln \frac{V_1 + V_2}{V_1} > 0 \quad \text{and} \quad \ln \frac{V_1 + V_2}{V_2} > 0$$

Consequently, since n_A, n_B, R, and T are positive quantities, the ΔS and ΔG for mixing ideal gases are

12-4
$$\Delta S > 0$$

12-5
$$\Delta G < 0$$

Analyzing these results, we can see that the mixing of gases arises from an increase in the entropy of the system. Upon mixing, the A and B molecules are free to move in a larger volume $(V_1 + V_2)$, and the system is thus more disordered. When ΔH is equal to zero as in this case, according to the second law of thermodynamics an increase in ΔS is expected; that is, *real processes tend to increase in entropy*. Since $\Delta H = 0$, an increase in ΔS corresponds to a decrease in free energy for the mixing of ideal gases. As you will recall from Section 10–H, the criterion for feasibility of a process occurring spontaneously at constant temperature and pressure is equation 10–23

$$\Delta G < 0$$

In summary it is important to emphasize that ideal gases do mix ($\Delta G < 0$) because entropy increases upon mixing, causing a more disordered system. Real gases, which differ from an ideal gas at higher pressure and temperature, also always mix to give homogeneous solutions; thus the ΔG of mixing must be less than zero. Even though real gases do not behave ideally and ΔH for mixing may not equal zero, the ΔS of mixing is positive and of sufficient magnitude to counteract any possible positive ΔH.

LIQUID SOLUTIONS

In contrast to gases, there are obviously very large intermolecular or ion-molecule interactions in liquids which can under no circumstances be neglected. These interactions play an important role in determining whether two liquids will mix to form a solution. The mixing of two liquids to form a liquid solution is illustrated in Figure 12–2. In this mixing process we keep the temperature and pressure constant. The change in free energy upon mixing is again given by equation 10–22

$$\Delta G = \Delta H - T\Delta S$$

Whether or not the liquids will mix to give a homogeneous solution will depend upon the sign of the free energy change. If ΔG is negative, a homogeneous solution will form.

The ΔS term is positive in the mixing of liquids for essentially the same reason as in the mixing of gases. In the solution the molecules of A and B are allowed to move over a larger volume than in the pure liquids, so the disorder of the system is increased. Consequently the entropy of the system increases upon mixing, and the contribution of entropy to the ΔG term is

$$(-T\Delta S) < 0$$

Since the $(-T\Delta S)$ is negative, this term predicts solution formation.

The value of ΔH plays an important role in determining whether ΔG of mixing will be positive or negative. The sign for ΔH of mixing can best be explained in terms of intermolecular interactions. (We will assume for the present discussion that the solute and solvent do not ionize in solution, thereby restricting our discussion to molecular interactions. Solutes which ionize in solution will be considered

FIGURE 12–2   Mixing of liquids A and B to form a homogeneous solution

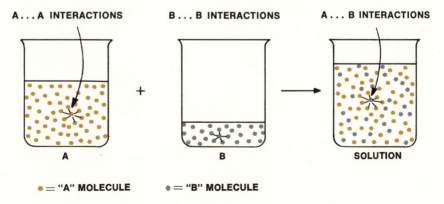

separately.) In pure liquid A there are obviously only interactions between molecules A (*i.e.*, A$\cdots$A interaction). Likewise in pure liquid B there are only B$\cdots$B intermolecular interactions. However, in a solution of A and B, there will be A$\cdots$B intermolecular interactions in addition to A$\cdots$A and B$\cdots$B interactions (Figure 12–2).

If the energy released in the A$\cdots$B intermolecular interaction exceeds the average of the energies released in the A$\cdots$A and B$\cdots$B interactions, then ΔH of mixing should be negative. This of course would help make ΔG more negative and in general one would expect such liquids to form a solution. This situation is summarized in the second line of Table 12–1, which gives the example of a solution of CH_3OH (methanol) and CH_3—O—CH_3 (dimethyl ether) in which hydrogen bonds are important. The corresponding terms for the mixing of ideal gases are shown in the first line of Table 12–1.

If the A$\cdots$B interaction is less than the average of the A$\cdots$A and B$\cdots$B interactions, then ΔH is positive. In this case whether ΔG is positive or negative depends upon the relative magnitude of ΔH and $(-T\Delta S)$. A homogeneous solution may or may not form. This situation is summarized in line 3 of Table 12–1. An example of a solution for which ΔG is apparently positive is the combination of water and C_6H_{14}, hexane. H_2O is very strongly hydrogen bonded to itself; whereas the $H_2O\cdots C_6H_{14}$ interaction is very small. In fact, there is a tendency for nonpolar molecules to interact with each other in preference to interacting with water. H_2O and C_6H_{14}, do *not* mix to give a homogeneous solution. We can generalize from this to say that a *highly polar substance will not mix with or dissolve in a nonpolar substance*. In the case of water and dioxane, even though $\Delta H > 0$ and water-dioxane interactions are less desirable than water-water or dioxane-dioxane, $\Delta G < 0$ and the two will form a solution.

Another situation that may occur (line 4 of Table 12–1), is that the A$\cdots$B interaction is on the same order as the average of A$\cdots$A and B$\cdots$B; $\Delta H \approx 0$ and ΔG is negative. This situation arises whenever the two substances being mixed are both nonpolar or both polar. In either case a solution should form. This phenomenon

TABLE 12–1 Thermodynamic quantities for various types of solutions

TYPE OF SOLUTION	INTERMOLECULAR INTERACTIONS	ΔH	$-T\Delta S$	ΔG	EXAMPLE
ideal gas	none	0	negative	negative	$O_{2(g)} + N_{2(g)}$
liquid	$A\cdots B > \dfrac{A\cdots A + B\cdots B}{2}$	negative	negative	negative	$CH_3OH_{(l)} + CH_3OCH_{3(l)}$
liquid	$A\cdots B \leqslant \dfrac{A\cdots A + B\cdots B}{2}$	positive	negative	positive or negative	$H_2O_{(l)} + C_6H_{14(l)}$ $H_2O_{(l)} + C_4H_8O_2$ (dioxane)
liquid	$A\cdots B \approx \dfrac{A\cdots A + B\cdots B}{2}$	0	negative	negative	$CCl_{4(l)} + C_6H_{14(l)}$
ideal solution	$A\cdots B \approx A\cdots A \approx B\cdots B$	0	negative	negative	$CCl_{4(l)} + CCl_3CCl_{3(l)}$

is frequently summarized as "like dissolves like" where "like" refers to the polarity or nonpolarity of the substance.

Finally, the situation may arise where the A$\cdots$B interaction is identical to the individual A$\cdots$A and B$\cdots$B interactions. In this case ΔH is zero and ΔG again negative (line 5 of Table 12–1). Such a solution is called an *ideal solution*. This type of solution frequently arises when the two substances are of the same type — when A and B are both aliphatic hydrocarbons or both alcohols or both ethers. Later we will define an ideal solution explicitly in terms of the vapor pressure behavior.

In the previous examples it was assumed that the solutions consisted of *molecules* of A and B. Another possibility is that the molecules of a compound may dissociate into *ions* in solution (*e.g.*, liquid HCl dissociates into $H^+_{(aq)}$ and $Cl^-_{(aq)}$ when dissolved in water). Such a solution is called an *electrolyte* since a solution containing charged ions is able to conduct an electrical current. Solutions which do *not* contain a sufficient concentration of ions to conduct an electrical current are called *nonelectrolytes*. In the case of electrolytes the sign of ΔH depends upon several factors. These are summarized in equation 12–6 which has been written as a Born-Haber cycle (see equation 6–1). The formula XY represents a molecule of A or B in the liquid phase.

12–6

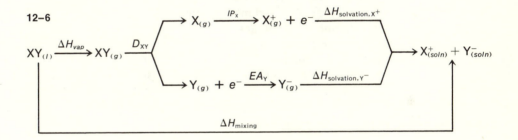

The ΔH for the entire solution process is the sum of the enthalpy change for each of the individual steps:

$$\Delta H = \Delta H_{vap} + D_{XY} + IP_X - EA_Y - \Delta H_{solvation,X^+} - \Delta H_{solvation,Y^-}$$

Obviously the sign of ΔH depends upon several factors. The first three terms are positive (energy absorbed) whereas the last three terms are negative (energy released).

Of particular importance are the $\Delta H_{solvation,X^+}$ and $\Delta H_{solvation,Y^-}$. Generally these terms are highly dependent on the nature of the solvent and are large when there are strong ion-molecule interactions. Solvents which are highly polar (have a high dipole moment) are especially good solvents for ionic compounds, since ion–dipole as well as ion–induced dipole interactions can exist. Water is a good example of this type of solvent.

The size and charge of the ion are also important factors in the magnitude of $\Delta H_{\text{solvation}}$, as discussed in Section 8–B. In particular, the $\Delta H_{\text{solvation}}$ for H^+ in H_2O is exceptionally large (258 kcal) for a singly charged ion. It is apparently this factor, combined with high EA_Y, which causes acidic compounds such as HCl (Cl being Y) to dissociate into ions in solution.

$$HCl \xrightleftharpoons{aq} H^+_{(aq)} + Cl^-_{(aq)}$$

Organic acids (see Section 7–H), such as acetic acid CH_3COOH, also dissociate into ions in solution, although in this case the dissociation into ions is not complete. For example, the dissociation for acetic acid at 1 M concentration in water is only 0.4 %.

PROBLEM 12–1   On the basis of possible intermolecular interactions discussed in Chapter 8, which of the following pairs of liquids would you expect to form (1) solutions (2) ideal solutions? Predict the sign of ΔS, ΔH, and ΔG of mixing to use as a basis for your answers.
a. hexane and heptane
b. water and octane
c. water and ethyl alcohol
d. methyl alcohol and methyl amine
e. CS_2 and nitromethane
f. CCl_4 and benzene
g. HBr and hexane
h. CH_2BrCH_2Br and CH_3—$CHBr$—CH_2Br

12–B. LIQUID SOLUTION–VAPOR EQUILIBRIUM

Just as a pure liquid can be in equilibrium with its vapor, a solution of liquids can also be in equilibrium with its vapor. In this case the vapor will be a mixture of the same volatile substances that are in the solution. The molecular arrangement at the surface of a solution is shown in Figure 12–3.

The escape of a molecule of A or B from the surface of the solution occurs in the same manner as does the vaporization of a pure substance (Section 11–B). For a molecule to escape, it must have both the necessary velocity and the proper direction. The A and B molecules each escape from the surface at independent rates. Likewise, the A and B molecules in the gas phase condense on the surface of the liquid with specific rates. Eventually the rates of vaporization and condensation become equal so that component A in the liquid solution is in equilibrium with A in the gas phase and B in the liquid solution is in equilibrium with B in the gas phase.

$$A_{(soln)} \rightleftharpoons A_{(g)}$$

$$B_{(soln)} \rightleftharpoons B_{(g)}$$

FIGURE 12–3   Molecular nature of the surface of a solution

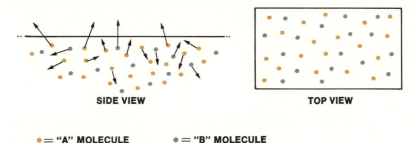

SIDE VIEW   **TOP VIEW**

●= "A" MOLECULE   ●= "B" MOLECULE

The pressures of A and B in the gas phase at equilibrium are the vapor pressures of these components of the liquid solution.

Since the thermodynamic criterion for equilibrium at constant temperature and pressure is $\Delta G = 0$, the change in free energy from the solution to the gas phase at equilibrium for the components A and B is

$$\Delta G_A = 0 = \Delta G_{f,A(g)} - \Delta G_{f,A(soln)}$$

$$\Delta G_B = 0 = \Delta G_{f,B(g)} - \Delta G_{f,B(soln)}$$

Therefore,

12–7a $$\Delta G_{f,A(soln)} = \Delta G_{f,A(g)}$$

12–7b $$\Delta G_{f,B(soln)} = \Delta G_{f,B(g)}$$

As the top view of Figure 12–3 shows, the surface of the liquid consists of both A and B molecules. The fraction of each of these species *at the surface* is represented by the mole fractions of A and B in the solution. Obviously then the vapor pressure of each of the components in the solution is dependent upon the concentration of that component.

RAOULT'S LAW

A particularly simple relationship between vapor pressure and concentration results when the intermolecular interaction between A and B is on the order of that of A···A and B···B (an ideal solution.) In this case, a surface molecule of A or B in the solution experiences the same intermolecular interactions as in the pure liquid A or B. The vapor pressure of either of the components in solution will then be the vapor pressure of the pure liquid multiplied by the fraction of molecules (mole fraction) of that species in the solution. This relationship between vapor pressure and concentration is known as *Raoult's law*. Mathematically Raoult's law is expressed as

12–8
$$p_A = X_A p_A^{\bullet}$$

$$p_B = X_B p_B^{\bullet}$$

where $p_A^{\bullet}$ and $p_B^{\bullet}$ represent the vapor pressure of pure liquid A and B, respectively.

When the vapor pressure of one component (either nonvolatile or at low concentration) is negligibly small, then Raoult's law is obeyed by the major component in the solution ($X_A > 0.95$), regardless of the intermolecular interactions or whether the solution is ideal. The major component of a solution is referred to as the *solvent*. Several examples of this will be given in Section 12–D.

HENRY'S LAW

The vapor pressures of A and B obey Raoult's law in an ideal solution where the A$\cdots$B, A$\cdots$A, and B$\cdots$B interactions are of the same magnitude. However, for other liquid solutions the A$\cdots$B interactions can be quite different from A$\cdots$A and B$\cdots$B interactions and the vapor pressures of A and B in these nonideal solutions do not follow Raoult's law. In this case, the dependence of the vapor pressures of A and B on mole fraction is quite complicated. However, in dilute solutions the vapor pressure of the solute, say component B, can be expressed in terms of a rather simple equation (12–9) known as Henry's law.

12–9
$$p_B = X_B K_B^H$$

where K_B^H is a constant called Henry's law constant. For any given solute-solvent combination there is a particular Henry's law constant.

In order to understand how Henry's law can arise, let us consider a dilute solution of a non-ionizing solute B in a solvent A, as illustrated in Figure 12–4. (We shall take into consideration the effect which ionization has on solutions later.) You should note that the B molecules are essentially completely surrounded by molecules of the solvent A. The B$\cdots$B interactions found in pure liquid B no longer exist in the dilute solution and therefore the ability of a B molecule to escape from the surface depends solely upon the A$\cdots$B interactions, and the term $p_B^{\bullet}$ in Raoult's law is replaced by K_B^H of equation 12–8. However, since the vapor pressure of B depends on the fraction of B molecules at the surface, the vapor pressure of B will still be proportional to the mole fraction of B just as it was in Raoult's law.

FIGURE 12–4   Dilute solution of solute B in solvent A

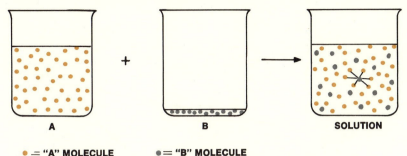

A   +   B   →   SOLUTION

• = "A" MOLECULE   ● = "B" MOLECULE

PROBLEM 12–2   The vapor pressures of pure hexane and heptane at 25 °C are

$$p_{\text{hexane}}^{\bullet} = 153 \text{ mm}$$

$$p_{\text{heptane}}^{\bullet} = 45.5 \text{ mm}$$

For a solution containing a 0.25 mole fraction of heptane, calculate the vapor pressure of hexane and heptane. Can you justify using Raoult's law for these calculations based upon your knowledge of intermolecular interactions?

PROBLEM 12–3   Graph p_{heptane} versus the mole fraction of heptane in hexane assuming heptane obeys Raoult's law. Use the values given for $p^{\bullet}$ in Problem 12–2.

PROBLEM 12–4   The vapor pressures of pure acetone (A) and chloroform (C) at 35 °C are

$$p_A^{\bullet} = 345 \text{ mm}$$

$$p_C^{\bullet} = 293 \text{ mm}$$

A solution consisting of 0.50 mole fraction $CHCl_3$ has the vapor pressures

$$p_A = 160 \text{ mm}$$

$$p_C = 100 \text{ mm}$$

a. Do these substances obey Raoult's law? Explain.
b. On the basis of your knowledge of intermolecular interactions account for these results. [Hint: $CHCl_3$ can form a hydrogen bond.]

12–C. IDEAL SOLUTION

An *ideal solution* is defined rigorously as a solution in which all components obey Raoult's law (equation 12–8) at all concentrations. A graphical representation of p_A and p_B versus mole fraction is shown in Figure 12–5. The abscissa in Figure

FIGURE 12–5 Vapor pressure curves for an ideal solution

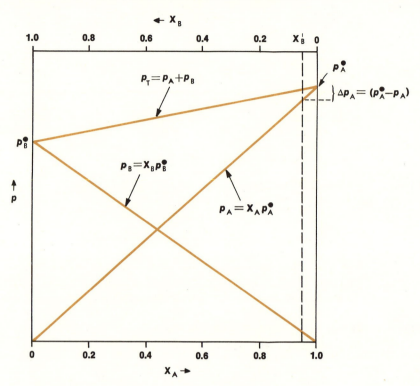

12–5 gives X_A running from left to right and X_B from right to left. Since the sum of the mole fractions must be one

12–10a
$$X_A + X_B = 1$$

12–10b
$$X_B = 1 - X_A$$

The equation $p_A = X_A p_A^\bullet$ should be a straight line. The value $X_A = 1$ corresponds to pure liquid A and the vapor pressure for pure liquid A is designated by $p_A^\bullet$. Similarly the equation $p_B = X_B p_B^\bullet$ is a straight line. Again at $X_B = 1$ (pure liquid B) the p_B is the vapor pressure of pure liquid B, $p_B^\bullet$. The total vapor pressure of the solution is the sum of the vapor pressures of A and B ($p_T = p_A + p_B$), where p_T is the total vapor pressure and also a straight line.

Figure 12–5 also shows that the vapor pressure of A is lower in the solution than in pure A, since $p_A = X_A p_A^\bullet$ and $X_A < 1.0$. The decrease in vapor pressure of component A (Δp_A) for any value of $X_A < 1$ is given by

12–11
$$\Delta p_A = (p_A^\bullet - p_A)$$

FIGURE 12–6   Vapor pressure curves for a $C_2H_4Br_2$–$C_3H_6Br_2$ solution

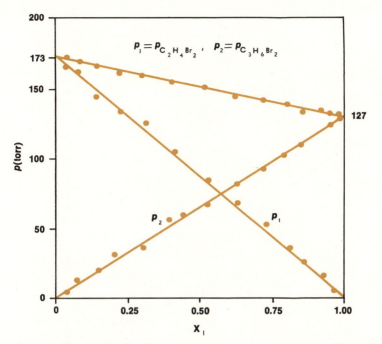

As we will see shortly, the decrease in vapor pressure affects the boiling point of a solution as compared to the boiling point of pure solvent.

In order to test experimentally whether a mixture of two liquids forms an ideal solution, vapor pressure measurements are made of the components in the solution. A graph similar to Figure 12–5 is then constructed to see if Raoult's law is obeyed by all constituents. A graph using the experimental vapor pressures for ethylene dibromide ($C_2H_4Br_2$)–propylene dibromide ($C_3H_6Br_2$) solutions is given in Figure 12–6.

PROBLEM 12–5   Calculate the decrease in vapor pressure of benzene when 5 grams of naphthalene are dissolved in 100 grams of benzene. Assume Raoult's law is obeyed and the temperature is 26.1 °C. The vapor pressure of benzene is 100 mm at 26.1 °C.

PROBLEM 12–6   The following vapor pressures were measured at 35 °C for solutions of acetone and carbon disulfide (CS_2).

X_{CS_2}	p_{CS_2} (mm)	$p_{acetone}$ (mm)
0.00	0	344
0.25	320	280
0.50	415	245
0.75	455	200
1.00	512	0

Do acetone and carbon disulfide form an ideal solution?

FIGURE 12-7 Vapor pressure curves for a nonideal solution—negative deviation. (---) ideal solution, (—) nonideal solution, (. . . .) Henry's law.

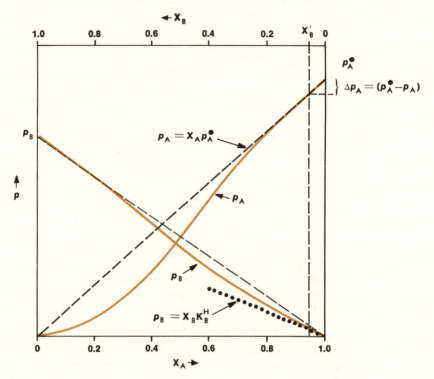

12-D. NONIDEAL SOLUTIONS

The vapor pressures of the components of nonideal solutions show marked deviations from Raoult's law, especially for a component which is at low concentration. We will again restrict the present discussion to components in the solution which do not ionize. Nonideal solutions can be classified into two categories based on how they deviate from the ideal solution. The first type of nonideal solution is one in which the A⋯B intermolecular attraction is greater than the A⋯A or B⋯B attraction. In this case the individual vapor pressure of either A or B is less than that expected from Raoult's law. The vapor pressure curves for such a situation are illustrated in Figure 12-7. Since the vapor pressures of A and B are lower than those expected from an ideal solution, this solution is said to show a *negative deviation*.

At low concentrations, the vapor pressure of a component of a nonideal solution will obey Henry's Law. Henry's Law was given earlier in equation 12-9:

$$p_B = X_B K_B{}^H$$

A graph of Henry's law is included in Figure 12-7; the vapor pressure calculated for solute B is given by the dotted line (⋯). Note that the actual vapor pressure of

FIGURE 12–8 Vapor pressure curves for a nonideal solution—positive deviation.
(---) ideal solution, (——) nonideal solution, (. . . .) Henry's law.

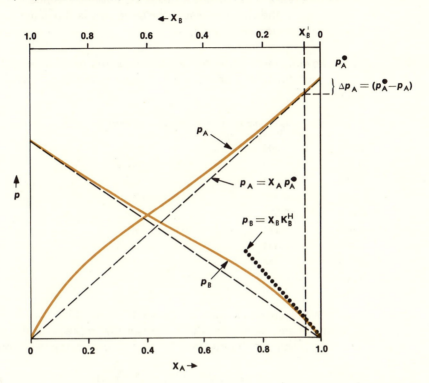

B (solid line) coincides with the dotted line ($\cdots$) at low solute concentration
($X_B < X_B'$). Henry's law is therefore obeyed by the solute B, provided that the
solute is at low concentration.

In the concentration region where the vapor pressure of solute B obeys Henry's
law, solvent A is at high concentration, and its vapor pressure obeys Raoult's law
(dashed line in Figure 12–7 in the region $X_B < X_B'$). Also recall that the vapor
pressure depression is given by equation 12–11

$$\Delta p_A = (p_A^{\bullet} - p_A)$$

Since the vapor pressure of A (p_A) obeys Raoult's law in the concentration region
$X_B < X_B'$, the value of Δp_A is identical to that for the ideal solution shown in
Figure 12–5.

For a nonideal solution in which the A$\cdots$B intermolecular attraction is *less* than
the A$\cdots$A or B$\cdots$B attractions, the actual vapor pressure curves will show *positive*
deviation from those expected in an ideal solution. This is illustrated in Figure
12–8. Note that the vapor pressure of B(p_B) calculated using Henry's law is valid

at low concentrations of $B(X_B < X_B')$. In this same concentration region ($X_B < X_B'$) the vapor pressure of solvent A obeys Raoult's law. Since p_A obeys Raoult's law in this region, the vapor pressure lowering Δp_A is the same as that for the ideal solution illustrated in Figure 12–5.

PROBLEM 12–7 The following table gives the vapor pressures of toluene and acetic acid as a function of concentration. Assume toluene is the solvent and acetic acid is the solute.

X_A (toluene)	X_B (acetic acid)	p_A (mm)	p_B (mm)
0.000	1.000	0.0	136
0.231	0.769	84.8	110.8
0.402	0.598	117.8	95.7
0.535	0.465	137.6	83.7
0.662	0.338	155.7	69.3
0.829	0.171	176.2	46.5
0.906	0.094	186.1	30.5
0.956	0.044	193.5	17.2
1.000	0.000	202.0	0.0

a. Graph this data as shown in Figures 12–7 and 12–8.
b. Estimate the region in which Henry's law represents the vapor pressure of solute (acetic acid) and Raoult's law represents the vapor pressure of solvent (toluene).
c. What is Henry's law constant for acetic acid?
d. Write Raoult's law for toluene.

12–E. COLLIGATIVE PROPERTIES — VAPOR PRESSURE LOWERING OF SOLVENT

In the previous discussion of both ideal and nonideal solutions we noted that the vapor pressure of the solvent (A) in a solution containing a non-ionizing solute (B) is lowered by Δp_A from that of the pure solvent. In the low solute concentration region ($X_B < X_B'$ in Figures 12–5, 12–7 and 12–8), the vapor pressure of a solvent obeys Raoult's law, $p_A = X_A p_A^\bullet$. Substituting Raoult's law into equation 12–11 for Δp_A:

$$\Delta p_A = p_A^\bullet - X_A p_A^\bullet$$

$$\Delta p_A = (1 - X_A) p_A^\bullet$$

Then according to equation 12–10b:

12–12 $$\Delta p_A = X_B p_A^\bullet$$

We see from equation 12–12 that the vapor pressure lowering of the solvent A is directly proportional to the mole fraction of the non-ionizing solute X_B, and to the vapor pressure of the pure solvent. In the concentration region $X_B < X_B'$ the Δp_A is independent of the nature of the solute. In other words, all non-ionizing solutes at the same concentration (X_B) will lower the vapor pressure of a solvent by the same amount. As we will see shortly, there are several properties of solutions (freezing point lowering, boiling point elevation, osmotic pressure) which depend only upon the nature of the solvent and are independent of the chemical properties of the solute. These properties are called *colligative properties*, derived from the word colligate: "to bind together by means of some suitable conception." In reference to solutions, we then define a *colligative property* as: a property of solutions whose magnitude depends only upon the solute concentration and the nature of the solvent, but is independent of the nature of the solute. Some of these colligative properties are a consequence of the vapor pressure lowering given in equation 12–12.

If the solute B is nonvolatile, the vapor pressure of the solution is equal to the vapor pressure of the solvent. The vapor pressure of a solution containing the nonvolatile solute then can be represented as a function of temperature (Figure 12–9) as we did in Figure 11–4 for a pure liquid. The vapor pressure of the pure solvent ($p_A^{\bullet}$) is also shown in Figure 12–9 (dashed line) for reference. The vapor pressure lowering, Δp_A, is the vertical distance between the two curves.

Vapor pressure lowering can also be calculated in terms of the molal concentration of solute B in the low concentration region $X_B < X_B'$. In order to show this mathematically, we must first express X_B in terms of moles of solvent A and solute B

$$X_B = \frac{n_B}{n_A + n_B}$$

At low solute concentrations the moles of solvent A are far in excess of moles of solute B

$$n_A \gg n_B$$

For this reason we can neglect n_B in the denominator ($n_A + n_B$) and our equation for X_B becomes

$$X_B = \frac{n_B}{n_A}$$

In order to convert to molality we will assume the mass of solvent A is 1000 grams; then the molality, m, will equal n_B. The moles of solvent A is then $1000/MW_A$ and our expression for X_B becomes

12–13
$$X_B = \frac{m}{\dfrac{1000}{MW_A}} = \frac{MW_A m}{1000}$$

FIGURE 12-9 Vapor pressure curve for water containing a nonvolatile solute. (——) vapor pressure of solution, (---) vapor pressure of pure solvent.

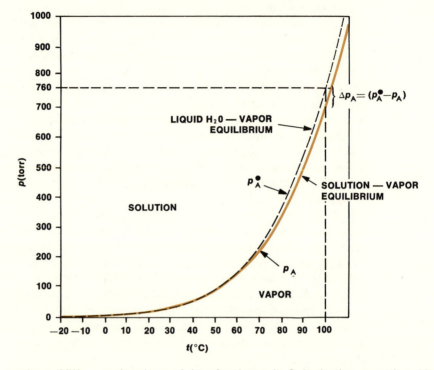

where MW_A = molecular weight of solvent A. Substituting equation 12–13 into equation 12–12 we obtain an expression for Δp_A in terms of molality, m.

$$12\text{–}14 \qquad \Delta p_A = \frac{MW_A p_A^\bullet}{1000} m$$

Equation 12–14 shows that Δp_A is also directly proportional to the molality, m, of the non-ionizing solute. As we will see shortly, the colligative properties are frequently expressed in terms of molality of solute.

PROBLEM 12-8   Calculate the vapor pressure lowering for a 0.5 molal solution of a nonvolatile solute in water at the normal boiling point. [Hint: What is $p_A^\bullet$ at the *normal* boiling point?]

12–F. COLLIGATIVE PROPERTIES OF NONELECTROLYTES

In the previous section we saw how a nonvolatile, non-ionizing solute lowers the vapor pressure curve of the pure solvent. Likewise, the presence of a nonvolatile solute B in the liquid phase (solution) also alters the equilibrium between solid A and the solution.

$$\text{solid solvent A} \rightleftharpoons \text{solvent A [solution of A + B]}$$

In this discussion, we will assume that the nonvolatile solute is present only in the liquid phase (solution) and is not present in the solid (*i.e.*, the solid phase is pure solid A). This assumption is not always valid but in the majority of cases it is a good approximation.

In order to understand this phenomenon, let us consider water as the solvent. As we know from Chapter 11, ice and water are in equilibrium at 0 °C and 1 atm:

$$H_2O_{(s)} \rightleftharpoons H_2O_{(l)} \qquad t = 0\text{ °C}, \Delta G = 0$$

The rate at which molecules in the solid phase go into the liquid phase (melt) depends only upon the nature of the solid phase; addition of a solute B to the liquid phase will not alter this rate.

$$H_2O_{(s)} \longrightarrow H_2O_{(\text{solution: H2O + solute})} \qquad t = 0°$$

Note that the solid phase is pure ice. The reverse process is solvent molecules in the solution going to the solid phase (freezing)

$$H_2O_{(s)} \longleftarrow H_2O_{(\text{solution: H2O + solute})} \qquad t = 0°$$

The rate of freezing is lowered, since the mole fraction of water (X_{H_2O}) in the solution is less than one, which decreases the ability of H_2O molecules to escape the liquid phase. (This is analogous to the decrease in the vapor pressure—the ability of molecules to escape into the gas phase—of solvent A in a solution, as a result of there being fewer A molecules on the surface.) The net result of the addition of a nonvolatile solute to the solution is the melting of $H_2O_{(s)}$ at 0 °C to dilute the solution

$$H_2O_{(s)} \longrightarrow H_2O_{(\text{solution: H2O + solute})} \qquad t = 0\text{ °C}, \Delta G < 0$$

From a thermodynamic viewpoint, the free energy of H_2O in the solution is less than that of pure $H_2O_{(l)}$ and the ΔG for melting is negative.

In order to establish equilibrium between the $H_2O_{(s)}$ and the H_2O in solution, the temperature must be lowered. If the solution contains the nonvolatile solute at a concentration of 2 molal, the temperature required to establish equilibrium is −3.72 °C. That is

$$H_2O_{(s)} \rightleftharpoons H_2O_{(\text{solution: H2O + 2m solute})} \qquad t = -3.72\text{ °C}, \Delta G = 0$$

The presence of the nonvolatile solute in H_2O at a concentration of 2 molal shifts the solid–liquid equilibrium curve to the left by 3.72 °C relative to the solid–liquid equilibrium curve for pure H_2O. The phase diagram for a liquid solution containing a nonvolatile, non-ionizing solute can now be constructed (Figure 12–10).

FIGURE 12–10 Phase diagram for an aqueous solution containing a nonvolatile, non-ionizing solute. (– – –) phase diagram for pure H_2O, (——) phase diagram for 2 molal solution.

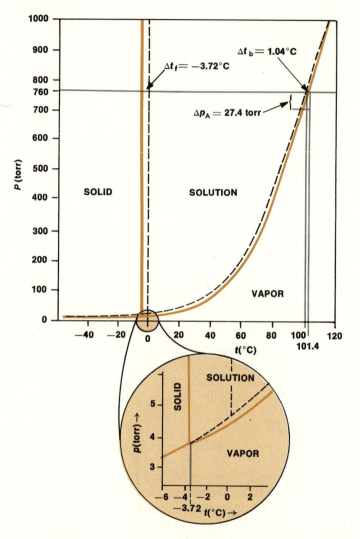

Remember that the solute is assumed to be present only in the liquid solution and that the solid and gas phases consist of only water. Notice in Figure 12–10 that the triple point has moved to -3.72 °C compared to $+0.0099$ °C for pure H_2O.

The consequence of these changes in the solid–solution and solution–gas equilibrium curves is apparent when the change at a pressure of 1 atm (760 torr) is noted. Since the normal melting point and normal boiling point are specified at one atmosphere pressure, the shift of the solid–solution curve to the left of the solid–liquid H_2O curve indicates the depression or lowering of the melting point (Δt_f) due to the nonvolatile solute. For a 2 molal solution this amounts to a shift of 3.72 °C. Similarly, the shift of the solution–gas curve to the right of the liquid H_2O–gas curve shows an elevation of the boiling point (Δt_b) due to the nonvolatile solute. For a 2 molal solution, this shift amounts to 1.04 °C. Just as the vapor pressure lowering was shown to be directly proportional to molality in equation 12–14, the freezing point depression (Δt_f) and the boiling point elevation (Δt_b) are also directly proportional to *m*. In the case of water, a nonvolatile, nonionizing solute at a concentration of 1 molal will result in $\Delta t_f = 1.86$ °C and $\Delta t_b = 0.52$ °C. (These values are different for other solvents.) Now that we have seen how the colligative properties arise and how they relate to the phase diagram for the solution (Figure 12–10), the mathematical equations relating the concentration dependence to each of the colligative properties will be presented. In addition, the dependence of the magnitude of the colligative properties upon the properties of the solvent will be explicitly shown.

VAPOR PRESSURE LOWERING (Δp_A)

The equation for vapor pressure lowering expressed in terms of molality (equation 12–14) may be rewritten as

12–15
$$\Delta p_A = K_{vp} m$$

where

12–16
$$K_{vp} = \frac{MW_A p_A^\bullet}{1000}$$

The term K_{vp} *depends only upon certain characteristics of the solvent*, namely the molecular weight and the vapor pressure of the pure solvent. The molal concentration, *m*, of the nonvolatile solute does not depend on the type of solvent.

BOILING POINT ELEVATION (Δt_b)

The relationship between the boiling point elevation, Δt_b, and the concentration of solute can be derived using thermodynamic functions. This result can be written as

12–17
$$\Delta t_b = K_b m$$

where K_b is a constant.

12–18
$$K_b = \frac{RT_b^2 MW_A}{1000 \, \Delta H_{vap}}$$

where T_b = normal boiling point of the solvent in °K, M_A = molecular weight of the solvent, ΔH_{vap} = molar heat of vaporization of the solvent (Section 11–B). The term K_b depends (as for Δp_A) only on the properties of the solvent, namely the heat of vaporization, molecular weight, and normal boiling point. The second term (*m*) in equation 12–17 is simply the molal concentration of the nonvolatile solute.

FREEZING POINT LOWERING (Δt_f)

The equation for the freezing point lowering, Δt_f, is very similar to that for Δt_b and also can be derived using thermodynamic parameters.

12–19
$$\Delta t_f = K_f m$$

12–20
$$K_f = \frac{RT_f^2 MW_A}{1000 \, \Delta H_{fus}}$$

where T_f is the freezing point of the solvent in °K and ΔH_{fus} is the molar heat of fusion of the solvent (Section 11–B). The term K_f is dependent on the properties of the solvent; the second term (*m*) in equation 12–19 is again simply the molality of the nonvolatile solute.

EXAMPLE 12–1   The ΔH_{fus} for H_2O is 1,436 cal/mole. Show that the K_f for water is 1.86 deg (kg solvent)/(mole solute).
From equation 12–20

$$K_f = \frac{RT_f^2 MW_A}{1000 \, \Delta H_{fus}}$$

The 1000 in the denominator has the units grams/kilogram and

$$K_f = \frac{(1.987 \text{ cal/deg-mole})(273.1 \text{ deg})^2(18 \text{ g/mole})}{(1000 \text{ g/kg})(1,436 \text{ cal/mole})}$$

$$K_f = 1.86 \text{ deg} \frac{kg \text{ solvent}}{mole \text{ solute}}$$

PROBLEM 12–9   Camphor ($C_{10}H_{16}O$) has an exceptionally high molal freezing point depression constant, $K_f = 37.7$ deg $\dfrac{\text{kg camphor}}{\text{mole solute}}$. The normal melting point of camphor is 178.4 °C.

Calculate the ΔH_{fus} for camphor from this data. What property of camphor accounts for its high K_f value compared to H_2O?

PROBLEM 12–10   Calculate the freezing point lowering for a 0.2 molal solution of a nonvolatile, non-ionizing solute in (a) H_2O (b) camphor.

PROBLEM 12–11   Calculate the boiling point elevation when 10 grams of a substance of molecular weight 5,000 are dissolved in 500 grams of water.

OSMOTIC PRESSURE

The concepts of osmosis and osmotic pressure are best understood through an illustration of how osmotic pressure is measured. Figure 12–11 shows two compartments, each containing a liquid, separated by a semipermeable membrane. The name semipermeable arises from the fact that only some chemical substances are able to pass through the membrane. The semipermeable membrane contains small holes or pores through which the molecules can pass. Only molecules of less than a certain size can cross the membrane. In Figure 12–11 the membrane used is permeable to solvent A but not to solute B.

The process of *osmosis* is the diffusion of a solvent through a semipermeable membrane from a solution in which the *solvent* is at a high mole fraction to a solution in which the solvent is at a lower mole fraction. In an extreme case the solution of higher mole fraction could be pure solvent A, as shown in Figure 12–11. Pure solvent A was placed in the left hand compartment and a solution of A and B in the right hand compartment. Since only A can pass through the membrane, the A molecules in the pure solvent (higher concentration) will diffuse or pass through the membrane to the solution with a lower mole fraction of A.

In order to understand the process of osmosis and the generation of osmotic pressure in more detail, we will consider the molecular and thermodynamic as-

FIGURE 12–11   Schematic diagram showing apparatus for the measurement of osmotic pressure

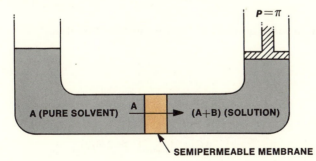

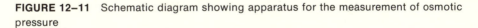

FIGURE 12–12 Diagram showing the permeability of A through a semipermeable membrane

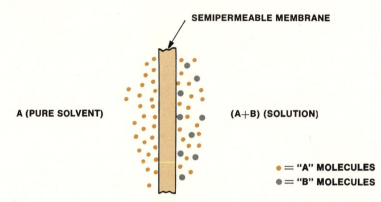

SEMIPERMEABLE MEMBRANE

A (PURE SOLVENT)   (A+B) (SOLUTION)

● = "A" MOLECULES
● = "B" MOLECULES

pects of the process. Figure 12–12 shows the molecular picture of the pure liquid and solution in contact with the semipermeable membrane. Molecules of A can pass through the membrane from the pure solvent into the solution and at the same time molecules of A can pass through the membrane in the opposite direction. Since the number of A molecules at the membrane surface in the pure solvent exceeds that in the solution, the rate of transfer from pure solvent A to the solution should exceed the rate in the opposite direction. Consequently, there should be a net transfer of solvent A molecules from the pure solvent to the solution, which is what we observe experimentally. If the concentration of solute B in the solution is increased, some A molecules at the surface of the membrane will be displaced and the fraction of A molecules at the surface decreases. Consequently the rate at which solvent A molecules pass from the solution to the pure solvent decreases and the net rate of transfer of solvent A (osmosis) from pure solvent to solution increases.

From a thermodynamic standpoint, the change in free energy for the net transfer of A by osmosis

$$A_{(pure\ solvent)} \longrightarrow A_{(solution)}$$

is

$$\Delta G = \Delta G_{f,A_{(solution)}} - \Delta G_{f,A_{(pure\ solvent)}}$$

Since the process actually occurs, it must be true that

$$\Delta G < 0$$

Therefore,

$$\Delta G_{f,A_{(solution)}} - \Delta G_{f,A_{(pure\ solvent)}} < 0$$

or

$$\Delta G_{f,A_{(pure\ solvent)}} > \Delta G_{f,A_{(solution)}}$$

In other words the free energy of A as pure solvent exceeds the free energy of A in solution and the transfer of A from pure solvent to solution is accompanied by a decrease in free energy. If the concentration of solute B in solution is increased, the free energy of A in solution is further lowered since the mole fraction of A is lowered.*

In order to establish an equilibrium between A in both compartments, pressure may be applied to the solution. This will increase the free energy of A in solution. Osmotic pressure (π) is the pressure that must be applied to the solution in order to establish this equilibrium between A in the two compartments. Figure 12–11 shows the osmotic pressure (π) as being applied to the solution by a piston. When equilibrium is established

$$\Delta G = \Delta G_{f,A_{(solution)}, \pi} - \Delta G_{f,A_{(pure\ solvent)}, P = 1\ atm} = 0$$

$$\Delta G_{f,A_{(pure\ solvent)}, P = 1\ atm} = \Delta G_{f,A_{(solution)}, \pi}$$

If the concentration of solute B in the solution is increased, the osmotic pressure (π) must necessarily be increased to maintain equilibrium.

Because the osmotic pressure is a function of the concentration of B in the solution, an equation relating the osmotic pressure to the molality of the solute, m, can be derived.

12–21 $$\pi = K_\pi m$$

where

12–22 $$K_\pi = RT\rho_A$$

where ρ_A is the density of pure solvent. Equation 12–21 is valid only in the concentration range where Raoult's law represents the vapor pressure of solvent A and Henry's law represents the vapor pressure of the solute B. Note that equation 12–22 has the same form as equations 12–16, 12–18, and 12–20 and that K_π depends only on the solvent. The second term in equation 12–21 is again simply the molal concentration of the solute.

The equations (12–15 through 12–22) which we have obtained for the colligative properties (Δp_A, Δt_b, Δt_f, π) can be used for different purposes, as illustrated in Example 12–2 and Problems 12–12 and 12–13. The use of osmotic pressure to determine the molecular weights of large polymeric molecules is particularly important. For convenience in working these problems a compilation of K_b, K_f, and K_π for various solvents is given in Table 12–2.

*$\Delta G_{f,A} = \Delta G_{f,A}^\circ + RT \ln X_A$, which is similar to the expression for $\Delta G_{f,B}$ as derived in Appendix M.

TABLE 12–2 Constants for colligative properties

SOLVENT	FORMULA	K_b(deg-kg/mole)	K_f(deg-kg/mole)	K_π(atm-kg/mole)
water	H_2O	0.52	1.86	24.4
ethanol	CH_3CH_2OH	1.23	—	19.3
acetic acid	CH_3COOH	2.93	3.90	25.5
benzene	C_6H_6	2.64	4.90	21.5
toluene	$C_6H_5CH_3$	3.37	—	21.2
naphthalene	$C_{10}H_8$	—	6.8	—
camphor	$C_{10}H_{16}O$	—	37.7	24.2

K_b is at the normal boiling point, K_f is at the normal melting point, K_π is at 25 °C.

EXAMPLE 12–2   *Calculate the molecular weight of a nonvolatile, non-ionizing solute for which 15 mg in 1 gram of camphor lowers the freezing point of camphor from 178.4 °C to 175.9 °C.*

From Equation 12–19, the freezing point lowering is

$$\Delta t_f = K_f m = K_f \frac{(mass\ of\ solute)/MW\ solute}{(mass\ of\ solvent\ in\ kg)}$$

$$\Delta t_f = 178.4 - 175.9 = 2.5\ °C$$

The mass of solvent (camphor) is 1 gram = 1×10^{-3} kg and the mass of solute is 15 mg = 15×10^{-3} g. Substituting this information into the above equation

$$2.5\ deg = 37.7\ deg\ \frac{kg}{mole}\ \frac{15 \times 10^{-3}\ g/MW}{1 \times 10^{-3}\ kg}$$

$$MW = \frac{(37.7)(15 \times 10^{-3})\ g}{(2.5)(1 \times 10^{-3})\ mole}$$

$$MW = 226$$

PROBLEM 12–12   Calculate the osmotic pressure between pure water and a solution obtained by dissolving 100 mg of a nonvolatile, non-ionizing solute in 100 grams of water. The solute is a polymer with a molecular weight of 5,000.

PROBLEM 12–13   Calculate the molecular weight of an unknown solute (non-ionizing) from the following measurement of osmotic pressure between pure water and a solution containing this unknown solute:

$$\pi = 1.22\ atm$$

mass of unknown solute in solution = 50 grams

mass of solvent in solution = 100 grams

12–G. COLLIGATIVE PROPERTIES OF ELECTROLYTES—ACTIVITY

Thus far in our discussion of colligative properties, the solute has been assumed to be non-ionizable as well as nonvolatile. The resulting solution is thus a non-electrolyte. If the solute dissociates into ions upon dissolving in the solution, the same equations can be used to calculate the colligative properties. However, in this case the molality is the total molality of all species, molecules plus ions, produced from the ionizable solute in solution. For example, if the solute were 0.1 molal NaCl, the total ion concentration would be 0.2 molal.

$$NaCl \xrightarrow{\ aq\ } Na^+_{(aq)} + Cl^-_{(aq)}$$

0.1 molal   0.1 molal 0.1 molal

0.2 molal

If the solute is K_2SO_4, the molality of K_2SO_4 must be multiplied by 3, since

$$K_2SO_4 \longrightarrow 2K^+ + SO_4^{2-}$$

m $2\,m$ m

$3\,m$

We will illustrate the effect of ionic compounds on colligative properties by the freezing point depression, Δt_f, observed for aqueous solutions of solutes which dissociate into ions. For solutes which dissociate into i number of ions the freezing point expression is

12–23 $$\Delta t_f = iK_f m$$

where m is the molality of the solute as added. The ratio $\Delta t_f/m$ should then be constant for a given solute

$$\frac{\Delta t_f}{m} = iK_f$$

The ratio $\Delta t_f/m$, as determined from experiment, is shown in Table 12–3 for various solutes at different concentrations.

Note that the ratio $\Delta t_f/m$ at 0.001 m is approximately the same for the solutes which dissociate into 2 ions and is larger for K_2SO_4 which dissociates into 3 ions. The last column gives iK_f which is the expected $\Delta t_f/m$ for each of these solutes at low concentrations where Henry's law is valid. Note that the experimental $\Delta t_f/m$ at 0.001 m are approximately equal to iK_f for each of these solutes.

You should also note in Table 12–3 that the ratio $\Delta t_f/m$ decreases as the molality of the ionizable solute is increased from 0.001 to 0.1 m. According to our equation for freezing point lowering, equation 12–19, the ratio $\Delta t_f/m$ should remain con-

TABLE 12-3 Freezing point lowering in aqueous solutions of ionized solutes

SOLUTE	IONS FORMED	i	0.1 m	0.01 m	0.001 m	iK_f
			EXPERIMENTAL $\Delta t_f/m$			
NaCl	$Na^+ + Cl^-$	2	3.48	3.60	3.66	3.72
KCl	$K^+ + Cl^-$	2	3.45	3.61	3.66	3.72
$MgSO_4$	$Mg^{2+} + SO_4^{2-}$	2	2.64	3.01	3.38	3.72
K_2SO_4	$2K^+ + SO_4^{2-}$	3	4.57	5.15	5.28	5.58
$La(NO_3)_3$	$La^{3+} + 3NO_3^-$	4	5.61	6.23	7.00	7.44
HCl	$H^+ + Cl^-$	2	3.52	3.60	3.69	3.72

stant. However, as we have pointed out earlier, the equations for all of the colligative properties, such as freezing point depression, are valid only in the concentration region where the solvent obeys Raoult's law and the solute obeys Henry's law (region $X_B < X_B'$ in Figures 12–5, 12–7, and 12–8). For solutions which contain ionizable solutes, Raoult's law is obeyed only at much lower solute concentrations than for solutions containing non-ionizable solutes. Ions in solution have strong ion–ion attractions and repulsions (see Section 8–A) and since these ion–ion interactions are strong they will have a strong influence even at reasonably low solute concentrations. Note that the $\Delta t_f/m$ data in Table 12–3 deviate significantly from the $iK_f = i(1.86)$ even at concentrations of 0.01 molal. At 0.1 m the deviation is quite large.

In order to account for this deviation, caused by the interaction of the ions or molecules of the solute, the concentration term for the ion is replaced by a term called *activity* (*a*). The activity can be thought of as the idealized concentration of the solute such that the solute–solute interactions can be neglected (as at infinite dilution). In order to relate the activity to the actual concentration, a factor called the *activity coefficient* (γ) is used to correct the concentration to give the activity.

12-24
$$a = \gamma m$$

Equations such as 12–24 are used to represent the activity of both ions or molecules of a solute. If the concentration of an ion is less than 10^{-3} m the activity coefficient is approximately one. For an uncharged solute molecule the concentration must be less than approximately 10^{-1} m in order for $\gamma = 1$. When γ is close to 1 the concentration, *m,* can be used in place of activity, *a*. In this text we will work only at low concentrations where we can assume $\gamma = 1$. In general you should remember that activities must be used when the solution is at high solute concentrations. If you neglect this you can introduce a significant error in your calculations.

In Chapter 13 we will express chemical equilibria in terms of activities, but there it will be convenient to relate the activity to the molar concentration instead of molality. Similarly, in Sections 12–H and 12–I it will be convenient to express the activity in terms of molarity.

PROBLEM 12–14   Calculate the boiling point elevation for water containing 0.05 molal lanthanum sulfate, $La_2(SO_4)_3$. [Hint: Lanthanum sulfate should completely ionize in water at this concentration. Assume no other reactions occur and $\gamma = 1$.]

12–H. SOLUBILITY OF SOLIDS

STANDARD FREE ENERGY CHANGE AND SOLUBILITY

In Section 12–A we discussed the mutual solubility or mixing of solvents in terms of ΔG for the process. We can also talk about the solubility of a solid in a solvent in terms of the free energy change. For the process of dissolving a solid solute B, the change in B can be represented by the equation

$$B_{(s)} \longrightarrow B_{(soln,[B])}$$

The free energy change for the *solute* is

12–25 $$\Delta G_{(solute)} = \Delta G_{f,B(soln)} - \Delta G_{f,B(solid)}$$

The free energy of B in solution depends upon the concentration of B dissolved in the solvent for which the following equation can be derived (see Appendix M).

12–26 $$\Delta G_{f,B(soln)} = \Delta G_{f,B(soln)}^\circ + RT \ln a_B$$

where a_B is the activity of solute B and is related to molar concentration by

$$a_B = \gamma_B[B]$$

and where γ_B is the activity coefficient. We will assume γ_B is one; then we can replace the activity a_B with the molar concentration, $[B]$. The activity coefficient becomes one in the low concentration region where Henry's law is obeyed. The term $\Delta G_{f,B(soln)}$ is the standard free energy for the solute where $[B] = 1$ M.

A graph of the free energy of B in solution versus $[B]$ is shown in Figure 12–13 in comparison with the free energy of solid B. Note that the free energy of B in solution increases as the molar concentration of B increases. At low concentrations the solution is *unsaturated* with respect to B ($[B] < S_B$) and

$$\Delta G_{f,B(soln)} < \Delta G_{f,B(solid)}$$

or ΔG of solution is negative, so that solid B can dissolve in a solution of concentration $[B]$.

Solid B will continue to dissolve in the solution and the ΔG will become less negative as the [B] in solution increases. Eventually the solution will become saturated ([B] = S_B) and at this point

$$\Delta G_{f,B_{(soln)}} = \Delta G_{f,B_{(solid)}}$$

The concentration of B in the saturated solution is called the *solubility* of B in the solvent A. In Figure 12–13 S_B is expressed in terms of moles of solute B/liter solution. (Solubility is also commonly expressed as grams of solute/100 g solvent.) The $\Delta G = 0$ corresponds to an equilibrium between solid B and B in solution at a concentration S_B.

At higher concentrations of B ([B] > S_B), the solution is *supersaturated*. Consequently

$$\Delta G_{f,B_{(soln)}} > \Delta G_{f,B_{(solid)}}$$

and ΔG of solution is positive. The supersaturated solution will eventually result in precipitation of solid B to produce a saturated solution where [B] = S_B.

TEMPERATURE DEPENDENCE

Using Le Chatelier's principle (Section 11–E), we can readily determine the effect which temperature has on the solubility of solids. You may recall that the effect which a temperature change has on an equilibrium system depends upon ΔH for

FIGURE 12–13 Free energy change as a function of the concentration of solid solute

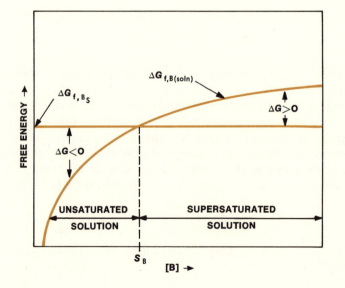

the process. In order to apply this principle to the solubility of a solid, we consider the equilibrium between solid solute and the saturated solution

$$B_{(s)} \rightleftharpoons B_{(soln,[B]=S_B)} \qquad \Delta H^\circ = \Delta H^\circ_{f,B_{(soln)}} - \Delta H^\circ_{f,B_{(solid)}}$$

The standard enthalpy change, ΔH°, can be determined directly from experiment or from experimentally determined heats of formation. If the solid dissociates into ions in solution, the standard enthalpy of formation of B in solution can be expressed in terms of the standard enthalpy of formation of the ions.

According to Le Chatelier's principle, if ΔH° is positive (endothermic) the solubility will increase (equilibrium shift to the right) when the temperature is increased. A graph of the solubility of various solutes as a function of temperature is given in Figure 12–14. Notice that most substances show an increase in solubility with an increase in temperature, corresponding to an increase of enthalpy. However, NaCl and PbBr$_2$ show very little change of solubility with increasing temperature and Ce$_2$(SO$_4$)$_3$ shows a decrease in solubility with increasing temperature. The effect that temperature has on solubility can also be derived from changes in free energy, ΔG°, independent of Le Chatelier's principle.

FIGURE 12–14   Solubility of various solutes

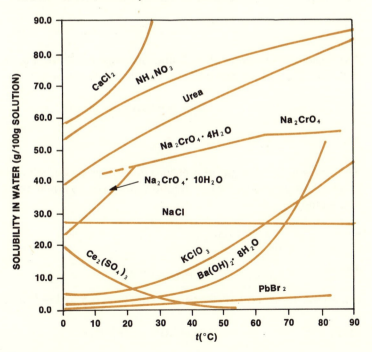

PROBLEM 12–15   The standard heats of formation for urea (NH_2CONH_2), solid and in aqueous solution are:

$$\Delta H_{f,NH_2CONH_{2(s)}} = -79.634 \text{ kcal}$$

$$\Delta H_{f,NH_2CONH_{2(aq)}} = -76.30 \text{ kcal}$$

Would you expect the solubility of urea to increase or decrease with an increase in temperature?

PROBLEM 12–16   Is the $\Delta H°$ of solution for $Ce_2(SO_4)_3$ positive or negative? (See Figure 12–14.)

12–I. SOLUBILITY OF GASES

STANDARD FREE ENERGY CHANGE AND SOLUBILITY

We can look at the solubility of a gas in a liquid in much the same way as we did the solubility of a solid in a liquid. The gas (B) will be at a pressure P_B in equilibrium with the liquid solution containing the gas at concentration [B]. The change in B for this process can be represented by the equation

$$B_{(g,P_B)} \rightleftharpoons B_{(soln,[B])}$$

The system is at temperature T and pressure P_B. When the solution is saturated with gas, the gas (B) and solution containing B are at equilibrium,

$$\Delta G = 0 = \Delta G_{f,B_{(soln)}} - \Delta G_{f,B_{(g)}}$$

12–7 $$\Delta G_{f,B_{(soln)}} = \Delta G_{f,B_{(g)}}$$

The free energy for a gas at pressure P_B is given by equation 12–27 (see Appendix L) and the free energy for B in solution was given by equation 12–26 (γ_B is assumed to be one).

12–27 $$\Delta G_{f,B_{(g)}} = \Delta G°_{f,B_{(g)}} + RT \ln P_B$$

12–26 $$\Delta G_{f,B_{(soln)}} = \Delta G°_{f,B_{(soln)}} + RT \ln [B]$$

Equations 12–26 and 12–27 can be substituted into 12–7 to obtain

12–28 $$\Delta G° = -RT \ln K$$

where the solubility constant of the gas is

12–29
$$K = \frac{[B]}{P_B}$$

and

$$\Delta G^\circ = \Delta G^\circ_{f,B_{(soln)}} - \Delta G^\circ_{f,B_{(g)}}$$

From equation 12–29 we see that the solubility of a gas B is directly proportional to the pressure P_B. The units of K are moles/liter-atm. Values of K depend upon the specific temperature T. Since equation 12–26 is restricted to low concentrations of solute B (Henry's Law region $X_B < X_{B'}$ in Figures 12–5, 7, and 8), equation 12–28 includes this restriction. The K values for various gases in water are shown in the third column of Table 12–4.

Nonpolar gases, such as O_2, N_2, H_2 and the inert gases, which cannot experience strong dipole–dipole interactions with water, generally show a very low solubility. Such molecules can interact with polar water molecules through dipole–induced dipole forces; these increase with increasing number of electrons/molecule or atom (as seen in Table 12–4). Polar molecules such as O_3, COS, and H_2S have solubilities on the order of 10–100 times that of the nonpolar molecules. Furthermore, gases such as NH_3, CO_2, and SO_2 react with water:

$$NH_{3(g)} + H_2O \rightleftharpoons NH_{3(aq)} \rightleftharpoons NH_{4(aq)}^+ + OH_{(aq)}^-$$

$$CO_{2(g)} + H_2O \rightleftharpoons H_2CO_{3(aq)} \rightleftharpoons H_3O_{(aq)}^+ + HCO_{3(aq)}^-$$

Because of these chemical reactions with water, the solubility of these compounds is much higher than expected and equation 12–29 cannot be used except at very low concentrations.

The constant K in equation 12–29 is a measure of the solubility of a gas at pressure P_B. In other words the greater the solubility, the greater the value of K. Previously (equation 12–9) we defined vapor pressure in terms of a Henry's law constant, K_B^H. Equation 12–9 can be written

12–30
$$\frac{1}{K_B^H} = \frac{X_B}{p_B}$$

It can be shown (Appendix M) that at low concentrations of B, X_B is directly proportional to the molar concentration of B, [B]. Therefore, the right hand side of equation 12–30 is proportional to $[B]/p_B$ and by comparison with equation 12–29 we see that

$$K \, \alpha \, \frac{1}{K_B^H}$$

TABLE 12-4 Solubility constants of gases in water

GAS	$t(°C)$	$K \times 10^3$ (mole/liter atm)	DIPOLE MOMENT (μ in Debye units)	No. e/molecule
He	25	0.41	0	2
H_2	25	0.78	0	2
N_2	25	0.65	0	14
CO	25	0.82	0.11	14
O_2	25	1.29	0	16
CH_4	25	1.34	0	10
C_2H_6	25	1.83	0	18
Ar	25	2.29	0	18
NO	25	3.00	0.15	15
Xe	25	4.87	0	54
N_2O	25	23.2	0.17	22
O_3	0	21.9	0.53	24
COS	20	22.5	0.71	30
H_2S	0	179	1.1	18
SO_2	0	3570.	1.63	32
NH_3	25	50,000	1.46	10
CO_2	25	32.9	0	22

That is, the solubility constant of a gas is inversely proportional to the Henry's law constant for that gas.

The solubility of a gas is sometimes expressed as that volume of the gas at 0 °C and 1 atm pressure (STP) which dissolves in 1 liter of water. From equation 12–29 we can calculate the solubility of B in moles/liter. The moles (n) of gas can be readily converted to volume of gas at STP using the ideal gas law

$$V = \frac{nRT}{P} = \frac{n(0.08205)(273)}{1} \text{ liters.}$$

Since gas solubility is generally low, the one liter volume of the solution can be assumed to be one liter of solvent.

PROBLEM 12-17 Calculate the solubility of O_2 and N_2 (from air) in water, at one atmosphere pressure and 25 °C. Assume the composition of air, given in mole percent, is 20 % O_2 and 80 % N_2.

a. Express solubility in terms of molar concentration.
b. Express solubility in terms of milliliters of gas at STP in one liter of water.

PRESSURE DEPENDENCE

The dependence of the solubility of a gas on the pressure is apparent from the expression for K in equation 12–29. Since K is constant, an increase in p_B must result in an increase in [B]. Equation 12–29 can be used to calculate the [B] at a specified pressure.

  The dependence of the solubility of gases on pressure can also be understood in qualitative terms from Le Chatelier's principle. Since the effect of pressure on gas solubility depends upon the change in volume we will consider ΔV for the process

$$B_{(g, p_B)} \rightleftharpoons B_{(soln, [B])} \qquad \Delta V = V_{B_{(soln)}} - V_{B(g, p_B)}$$

Obviously the volume of the gas is much larger than the volume which B contributes to the solution and for the dissolution of gas we always have a negative ΔV. From Le Chatelier's principle (Section 11–E) an increase in pressure will shift the equilibrium to the right (direction of smaller volume) and hence increase the solubility. This is in accordance with the previous conclusion based on the $K = [B]/p_B$ expression.

PROBLEM 12–18   Calculate the solubility of N_2 and He in water at 20 atm pressure and 25 °C. Why do you think deep sea divers use a helium-oxygen mixture rather than nitrogen-oxygen?

PROBLEM 12–19   Are the results of Problems 12–17 and 12–18 in accord with Le Chatelier's principle? When the partial pressure of N_2 is changed from 0.8 atm to 20 atm (as considered in problems 12–17 and 12–18), does K change? Explain.

TEMPERATURE DEPENDENCE

According to Le Chatelier's principle the temperature dependence of an equilibrium system depends upon the heat change for the process. For this reason we should consider $\Delta H°$ for the dissolution of a gas

$$B_{(g, p_B)} \rightleftharpoons B_{(soln, [B])} \qquad \Delta H° = \Delta H°_{f, B_{(soln)}} - \Delta H°_{f, B_{(g)}}$$

The $\Delta H°$ for this process is always negative (exothermic). This can be understood from the fact that the intermolecular interactions (B $\cdots$ B) in the gas phase are negligible (ideal gas approximation), whereas in solution the A $\cdots$ B interactions are quite significant. Since $\Delta H°$ is always negative for the process, the solubility of the gas will decrease (shift equilibrium to the left) when the temperature is increased. As mentioned in the previous section the effect that temperature has on the solubility can be derived from changes in free energy, $\Delta G°$.

  As we discussed previously, the solubility of a gas is greatly dependent upon the pressure of the gas; however, this effect does not occur through a change in K.

On the contrary K is a constant independent of the pressure applied to the gas. This can be readily understood from the relationship between K and $\Delta G°$ given in equation 12–28. The quantity $\Delta G°$ is calculated from the standard free energies of the solute in the gas phase and in solution and these are specified at 1 atm pressure. For this reason there will be no way in which $\Delta G°$ can vary by changing pressure. Consequently, the constant K is independent of pressure.

✳ 12–J. COLLOIDS

We have previously differentiated solutions from mixtures by the fact that true solutions are homogeneous at the molecular level whereas mixtures contain macroscopic particles of the individual compounds. Also a solution has properties which are different from the properties of the individual compounds whereas in the mixture the particles have the same properties as the pure compounds. Between solutions and mixtures there is also a classification called *colloids* which are mixtures of compounds where one of the compounds is dispersed into particles with diameters ranging anywhere from 10 Å to 10,000 Å. Because of the small size of the dispersed colloidal particles, the colloidal mixture in some ways resembles a true solution. For example, colloids will generally pass through ordinary filter paper as do true solutions. However, there are other properties of colloids which are markedly different from those of solutions and for this reason colloids occupy a unique classification of matter in themselves.

TYPES OF COLLOIDS

Most colloidal systems can be classified into two types: *lyophobic* or *lyophilic*. The term lyophobic has its origin from the Greek word phobos meaning fear or dislike. *Lyophobic colloids* are "solvent-hating;" that is, there is little or no attraction between the colloidal particles and the solvent. Consequently, lyophobic colloids are generally unstable from a thermodynamic standpoint and potentially the colloidal particles could combine to form larger noncolloidal particles which separate to form two phases. *Lyophilic colloids* (from the Greek philos, loving) contain either large molecules or aggregates (micelles) of smaller molecules which have an *attraction with the solvent*. Lyophilic colloids thus form stable systems similar to true solutions except that the solute particles have the dimensions of a colloid. Lyophilic colloids include many synthetic polymers and naturally occurring polymerlike substances such as proteins, nucleic acids and starches, and also soaps, detergents, and emulsifying agents. Lyophilic colloids consisting of aggregates or *micelles* of smaller molecules are also called *association colloids*. We will discuss their behavior in more detail later.

CLASSIFICATION OF LYOPHOBIC COLLOIDS

Colloids can be formed by dispersing any of the three phases — solid, liquid, or gas — in a dispersing agent composed of solid, liquid, or gas. All combinations of these phases can produce colloids except gas dispersed in gas, since this always produces a true gaseous solution. Lyophobic colloids can be classified according to the physical state of the dispersing medium.

1. *aerosol* (solid or liquid dispersed in a gas) The colloidal particles in an aerosol consist of small liquid or solid particles dispersed in a gas such as air. Common solid aerosols are smoke and dust whereas common liquid aerosols are clouds and fog. We also produce aerosols as sprays from pressurized containers to disperse insecticides, hair sprays, deodorants, and paints, to name a few. The use of aerosols has become a rather common practice in our everyday life.

2. *sol* (solid dispersed in a liquid) Sols are usually formed by breaking down a solid into small particles of colloid dimensions and dispersing these small particles in a liquid phase. The most common use of sols is in paints where the liquid dispersing agent eventually evaporates after spreading the paint out over a large surface area. Sols can also be produced by building up the colloidal particles from aggregates of molecules. For example, if HCl is added rapidly to a solution of silver nitrate, a milky solution containing colloidal AgCl will be produced. Initially the AgCl and water will be in a colloidal state, and only after aging and heating will a true precipitate of AgCl settle out from the solution.

3. *emulsion* (liquid dispersed in a liquid) A common example of this type of colloid is milk where globules of fat are dispersed in water.

4. *foam* (gas dispersed in a liquid or solid) Foams are usually produced by beating air throughout a liquid phase, as in soap suds and whipping cream. Foams can also be produced with pressurized containers as in the dispensing of shaving cream. Many plastics are solid foams such as polyurethane and polystyrene. Air is generally dispersed in the liquid state prior to solidification. Solid foams are excellent insulating materials.

LIGHT SCATTERING

A property that is common to colloidal dispersions is the ability to scatter light. When visible light is passed through a true solution, the light will either be transmitted by the solution or some of it will be absorbed. In either event an observer at right angles to the light beam would not be able to see the light as it passes into or through the solution.

On the other hand, when light passes into a colloidal suspension, it strikes the surface of the colloidal particles and is scattered in various directions. Consequently, an observer at right angles finds no difficulty in seeing the scattered light from the colloidal dispersion (Figure 12–15). This phenomenon is commonly

FIGURE 12–15 Scattering of light by a colloid (Tyndall effect)

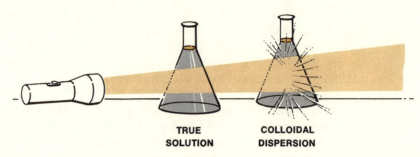

TRUE SOLUTION   **COLLOIDAL DISPERSION**

called the *Tyndall effect*. The Tyndall effect is observed every day in nature. For example, we observe clouds as a result of light scattered off the colloidal water particles, which prevents some of the sunlight from reaching the earth's surface. Also anyone who has had the experience of driving in the fog at night has observed the Tyndall effect when the bright light from an automobile headlight is scattered in all directions, thus reducing visibility.

ASSOCIATION COLLOIDS

Molecules which form association colloids in general have a dual character; one portion of the molecule is lyophobic whereas another portion is lyophilic. Examples of compounds which form association colloids in water are:

$$C_{17}H_{33}\overset{\displaystyle O}{\overset{\displaystyle \|}{C}}-O^- \ Na^+$$
sodium oleate

$$C_{12}H_{25}OSO_3^- \ Na^+$$
sodium dodecyl sulfate

$$(C_{16}H_{33})(CH_3)_3N^+ \ Br^-$$
cetyltrimethylammonium bromide

Note that in each molecule there is a portion of the molecule which is a long chain hydrocarbon (e.g., the $C_{17}H_{33}$ group in sodium oleate). Since there is little attraction between hydrocarbons and water, this long chain hydrocarbon portion of the molecule is lyophobic, or, since the dispersing agent is water, it is called *hydrophobic*. At the other end of each of the above molecules is an ionic group which is attracted strongly to the polar water molecules through ion-dipole interactions. This portion of the molecule is thus lyophilic, or, since the dispersing agent is water, we refer to it as *hydrophilic*.

The exact structure of the micelles is not known for certain, but the most likely structure has the hydrophobic hydrocarbon chains attracted to one another in the center of a spherical aggregate with the ionic groups at the periphery of the aggregate. This structure is illustrated in Figure 12–16 for sodium oleate. The structure of the micelle allows for the maximum number of interactions and thus gives the lowest energy or the most stable arrangement. The diameter of the micelle is approximately twice the extended length of the hydrocarbon chain, although the chains are not extended in straight lines. The hydrocarbon chains are flexible and able to entangle with one another in the central portion of the micelle. This central portion probably resembles a solution of long chain hydrocarbons. The ionic groups which extend into the water are ionized but a layer of Na^+ resides nearby. In the case of sodium dodecyl sulfate, experiments suggest that only 15 to 30 percent of the negative ions are not associated with Na^+ ions. Micelle formation for such compounds occurs only at concentrations higher than about 0.03 M concentration. At lower concentrations the salts appear to dissolve and behave in a manner similar to a salt such as sodium acetate

$$(Na^+ \; {}^-O-\overset{\overset{\textstyle O}{\|}}{C}-CH_3)$$

FIGURE 12–16 Aggregate or micelle of sodium oleate. The $C_{17}H_{33}$ group is represented by the zigzag line. Water molecules are not shown, but they would be present around the Na^+ and $-COO^-$ ions.

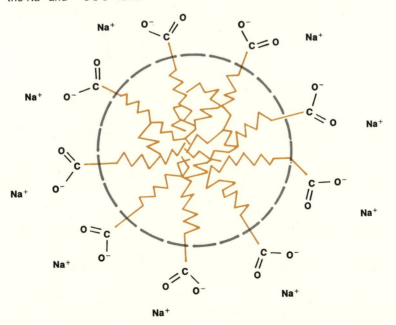

PROTECTIVE COLLOIDS — DETERGENCY

The compounds which can form association colloids, such as sodium oleate, also can act as *protective colloids*. This ability to act as a protective colloid is a most important property since this is the function of *soaps* and *detergents* in removing grease and dirt from fabrics. Lubricating oils and greases are long chain hydrocarbons and are generally insoluble in water. If a lubricating oil is dispersed in water by very rapid mixing, the result is a lyophobic colloid consisting of small colloidal oil droplets dispersed in the water. This is an unstable colloid and the oil droplets would eventually coalesce and form larger oil drops that would eventually rise to the surface and separate as an oil layer. For this reason water by itself is not able to remove oil and grease from fabrics.

Sodium oleate, which was just described as an association colloid, removes oil and grease from fabric by forming a protective colloid around the small oil and grease particles. The resulting colloid is lyophilic since the ionic group $—COO^-$ in the oleate ion has a strong affinity for the water molecules. A diagram of the resulting colloid is shown in Figure 12–17. Note that the long hydrocarbon chain in sodium oleate is soluble in the hydrocarbon oil or grease and the colloid is stabilized by the interaction of the $—COO^-$ ion with the water. The structure of this colloid is very similar to the structure of the association colloid shown in Figure 12–16; the only difference is the central oil drop in the soap–oil colloid. The ability to form an association colloid and to act as a protective colloid requires that the molecules have this dual lyophilic–lyophobic character. All of the common soaps and detergents have this basic structure and their detergent action is a result of this structure. Examples of other common soaps and detergents are given below:

$$CH_3—CH_2—CH_2—CH_2—CH_2—CH_2—CH_2—CH_2—CH_2—CH_2—\overset{\overset{\displaystyle O}{\|}}{C}—O^-Na^+$$

sodium laurate

$$CH_3—CH_2—CH_2—CH_2—CH_2—CH_2—(CH_2)_5—CH_2—CH_2—CH_2—CH_2—CH_2—CH_2—\overset{\overset{\displaystyle O}{\|}}{C}—O^-Na^+$$

sodium stearate

$$CH_3—CH_2—CH_2—CH_2—CH_2—CH_2—CH_2—CH_2—CH_2—CH_2—CH_2—CH_2—O\overset{\overset{\displaystyle O}{\|}}{\underset{\underset{\displaystyle O}{\|}}{S}}—O^-Na^+$$

sodium lauryl sulfate

$$CH_3—CH_2—CH_2—CH_2—CH_2—CH_2—CH_2—CH_2—CH_2—CH_2—CH_2—CH_2—\text{⟨benzene⟩}—\overset{\overset{\displaystyle O}{\|}}{\underset{\underset{\displaystyle O}{\|}}{S}}—O^-Na^+$$

sodium lauryl benzene p-sulfonate

FIGURE 12–17 Protective colloid formed by a soap (sodium oleate) around an oil or grease particle

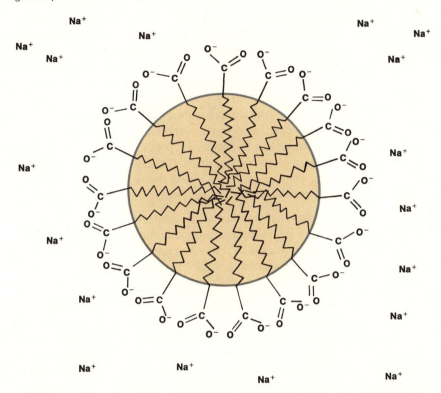

Note that all of the above soaps and detergents have the same basic molecular structure containing the hydrophobic hydrocarbon at one end and the hydrophilic anion group at the other end. They differ in the length and unsaturation in the hydrocarbon and in the nature of the anion. The compounds with the sulfate and sulfonate anions have the advantage that their Mg^{2+} and Ca^{2+} salts are soluble and for this reason these detergents can be used in hard water. (The hardness of water depends chiefly on its content of Ca^{2+} and Mg^{2+} salts.) On the other hand, the laurate, stearate, and oleate anions form insoluble salts with Mg^{2+} and Ca^{2+} and these so-called soft soaps can only be used in soft water where Mg^{2+} and Ca^{2+} have been removed.

EXERCISES

1. Sketch the phase diagram (P versus $t(°C)$) for an aqueous solution containing 1 molal $MgCl_2$. Assume that the equations for colligative properties as given in this chapter are valid at this concentration of $MgCl_2$. Show two graphs:

a. one for the range −10 °C to 110 °C

b. one showing a blowup of the region of the triple point (See Figure 11–11.)

2. An aqueous solution containing 100 grams of nonvolatile, non-ionizing solute in 2000 g of water, separated from pure water by a semipermeable membrane has an osmotic pressure of 0.068 atm at 25 °C.

a. Calculate the molecular weight of the solute using this data.

b. If the osmotic pressure measurement is limited to an accuracy of ±0.00005 atm, how accurate is your molecular weight calculated in part (a)? [Hint: Change the osmotic pressure by 0.00005 atm and see how much change there is in the calculated molecular weight.]

c. What would be the vapor pressure lowering for this solution? The vapor pressure of pure water at 25 °C is 23.8 mm. If the vapor pressure measurement is accurate to ±0.1 mm, could the vapor pressure lowering be used to determine the molecular weight of this solute?

3. A nonvolatile, non-ionic compound containing carbon, hydrogen, and oxygen has the following percentage composition: 40.0 % carbon, 6.67 % hydrogen, the remainder is oxygen. A solution of 1 g of this solute in 100 g of camphor lowers the melting point 2.1 °C. Calculate the molecular formula for this compound.

4. An aqueous solution containing 0.01 molal $NaCl$, and a quantity of pure water are placed in two compartments of the apparatus described in Figure 12–11. If the piston is removed from this apparatus and water were allowed to pass through the membrane, how high must the column of solution on the right be in order to establish equilibrium? Assume that the water being added to the solution does not change the concentration of the solution significantly. [The density of mercury is 13.6 times the density of water.]

5. When a certain solute is dissolved in water the resulting solution feels cool just after the solute is added. Does the solubility of this substance increase or decrease with an increase in temperature? Explain.

6. In making homemade ice cream, a temperature far below 0 °C is required. To attain this low temperature, table salt ($NaCl$) is added to ice at 0 °C in an insulated container. The resulting ice–salt solution is at a temperature well below 0 °C. Explain how this can occur. Assume the insulated container does not allow heat transfer to the atmosphere.

7. The solubility of PbF_2 is 2×10^{-3} M at 0 °C. Calculate the freezing point of an aqueous solution saturated with PbF_2.

8. Strong acids, such as HCl, completely dissociate into ions in aqueous solution. However, some solutes, such as weak acids and weak bases, do not completely dissociate into ions. As a result, the colligative property is between that expected for a molecular species which does not dissociate and one which dissociates completely into ions.

a. For trichloroacetic acid, the dissociation is:

$$Cl_3C—\overset{\overset{\displaystyle O}{\|}}{C}—OH_{(aq)} \rightleftharpoons H^+_{(aq)} + Cl_3C—\overset{\overset{\displaystyle O}{\|}}{C}—O^-_{(aq)}$$

If α is the fraction of trichloroacetic acid molecules which dissociate, there will remain $(1 - \alpha)$ fraction of trichloroacetic acid in solution and αH^+ and αCl_3CCOO^- will be formed. When the molality of trichloroacetic is 0.1 molal before dissociation, α is 0.13. Calculate the freezing point depression for 0.1 molal trichloroacetic acid in water.

b. A solution prepared by dissolving 0.100 g of HF per 50.0 g of H_2O shows a freezing point of -0.198 °C. What is the percent dissociation of HF?

9. In order to remove ice from streets, calcium chloride is sprinkled over the ice. Explain how the salt $CaCl_2$ can cause the ice to melt.

10. Red blood cells can be involved in osmosis, since water can pass through the cell membrane. If red blood cells are placed in pure water, there will be a net flow of water into the blood cell and the cell may burst (hemolyze). On the other hand, if the red blood cells are placed in a solution containing a solute, the solute concentration can be adjusted so that the osmotic pressure is zero. This solution is called an *isotonic medium.* A solution of 0.9 grams NaCl in 100 ml of water is isotonic with red blood cells. Calculate
a. the total solute concentration in red blood cells
b. the osmotic pressure of red blood cells in pure water.

11. The solubility constants for O_2 at different temperatures are given in the following table:

t(°C)	K (moles/liter-atm)
0	2.19×10^{-3}
10	1.72×10^{-3}
20	1.41×10^{-3}
30	1.195×10^{-3}
40	1.055×10^{-3}
50	0.952×10^{-3}

a. Graph K versus t and estimate the solubility constant for O_2 at 48.5 °C.
b. Calculate the solubility of O_2 from air which has a partial pressure of O_2 of 0.21 atm at 48.5 °C.
c. In Exercise 17 at the end of Chapter 10 we calculated that the water cooling a reaction vessel was 48.5 °C as it left the reactor. A certain aquatic species cannot survive over a long period if the O_2 concentration is less than 1.34×10^{-4} moles/liter. Using the results of b., can this aquatic species survive in

water which has its O_2 concentration diminished by previous heating to 48.5 °C? Assume the water has not absorbed any additional O_2 upon cooling down from the 48.5 °C temperature.

d. What is the maximum temperature to which this water could be previously heated in order that the oxygen concentration be sufficient for this specific aquatic species to survive?

12. The ocean contains $NaCl$ at a concentration of 0.6 M. Estimate the freezing point of the oceans assuming equation 12–23 holds even at this high concentration. [In fact considerable error is involved when using equation 12–23 for such high concentrations of any electrolyte. The actual freezing point of a 0.6 M $NaCl$ aqueous solution is −2.02 °C.]

13. It is not known whether a certain aqueous solution containing $Pb(ClO_4)_2$ is unsaturated, saturated, or supersaturated. Devise a simple test to ascertain which of these situations exists.

14. In an experiment in which a gas such as oxygen is produced, the gas is usually collected over water. The collection of the gas is illustrated in the following diagram.

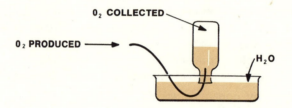

The volume of the gas is V_g and the volume of liquid water is V_l. Generally it is assumed that the solubility of the gas in the liquid water is negligible. Calculate the moles of O_2 that are dissolved in the water at 25 °C if $V_g = 100$ ml and $V_l = 200$ ml. Assume that the pressure of O_2 remains at 1 atm and the water does not contain any O_2 initially. What is the percent error that arises by neglecting the solubility of O_2 in water at 25 °C?

15. a. List three factors which determine whether a solute will increase or decrease the vapor pressure of the solution compared to the solvent.

b. Arrange the following aqueous solutions in the order of increasing vapor pressure: 1.0 M CH_3OH, 0.5 M $C_6H_{12}O_6$, 0.2 M Na_2SO_4, 0.2 M $CsBr$, 0.3 M CH_3COOH. The vapor pressures of pure CH_3OH and CH_3COOH are 120 mm and 17 mm, respectively, at 25 °C. The other compounds in the solutions have negligible vapor pressures except for water which is 23.7 mm at 25 °C.

16. An aqueous solution containing 0.25 g of a neutral polysaccharide in 100 ml exerts an osmotic pressure of 24.2 torr at 25 °C. The polysaccharide (a polymer made up of simple sugar molecules) has the formula $(C_6H_{10}O_5)_n$ where n is variable. From the above information calculate the average molecular weight and the average value of n for this polysaccharide.

17. When solutions of $Ba(NO_3)_2$ and Na_2SO_4 are mixed, a white milky-looking solution is first formed. After allowing this solution to sit for a few days, a white precipitate of $BaSO_4$ particles will eventually form. Explain the nature of the intermediate state where the solution has a milky appearance.

18. A protective colloid which stabilizes emulsions is called an emulsifying agent. Milk consists of an emulsion of fat particles in water which is stabilized by the emulsifying agent casein. Casein is a protein which contains the ionic groups $-COO^-$ and $-NH_3^+$. Describe how casein acts as an emulsifying agent in milk.

19. Referring to Figure 12–14, which hydrate of Na_2CrO_4 has the most endothermic ΔH of solution?

CHAPTER THIRTEEN

CHEMICAL EQUILIBRIUM

The two previous chapters have dealt with equilibria of chemical species in two different phases. In each case the equilibrium established was for a process involving a *physical* change; the chemical species were simply transferred from one phase to another with no new chemical species being produced. In this chapter we shall apply the same thermodynamic principles of equilibria to reactions in which new chemical species are formed.

The concept of chemical equilibrium applies to chemical reactions that occur in solution as well as in the gas phase, and to chemical reactions involving more than one phase, called heterogeneous reactions. Chemical equilibrium finds the greatest application in reactions in which both ionic and polar species are in solution. For this reason we shall emphasize in this chapter chemical equilibria in solution, and shall consider gas reactions only briefly near the end of the chapter.

13-A. EQUILIBRIUM CONSTANT—CHANGE IN STANDARD FREE ENERGY

In Chapter 10 two objectives of chemical thermodynamics were stated: the development of a criterion to determine whether a chemical reaction is possible (reaction is possible if $\Delta G < 0$) and the ability to calculate the equilibrium yield (ΔG° is related to the equilibrium concentrations).

For simplicity let us consider a reaction which involves only a single reactant, B, and a single product, C. The reaction involves b moles of B, and c moles of C, and we designate this equilibrium as

$$bB \rightleftharpoons cC$$

Just as in the case of phase equilibria, chemical equilibria are dynamic equilibria: b molecules of B continually react to give c molecules of C and c molecules of C continually react to give b molecules of B. When the two rates are equal, equilibrium is established. Equilibrium systems are generally kept at constant temperature and pressure. Since the thermodynamic criterion for equilibrium is $\Delta G = 0$ (equation 10–23) and

$$\Delta G = c\Delta G_{f,C} - b\Delta G_{f,B}$$

then

13–1
$$c\Delta G_{f,C} = b\Delta G_{f,B}$$

Thus far we have not specified the phases of B and C. The reactant B and product C can be in any phase: solid, liquid, gas, or solution. For each phase the same general approach is taken to derive the final equilibrium expression, but the expressions for the free energy will differ. The free energy of a solid or liquid does not change significantly with changes in pressure and we may assume the free energy of the solid or liquid to be the standard free energy (at one atmosphere pressure). Expressions for the free energy of a gas and an ideal solution have been derived in Appendices L and M. The expression for the free energy of formation of a pure solid or liquid at equilibrium is

$$\Delta G_{f,i} = \Delta G_{f,i}^{\circ}$$

for an ideal gas (equation 12–27)

$$\Delta G_{f,i} = \Delta G_{f,i}^{\circ} + RT \ln P_i$$

for a solute in solution (equation 12–26)

$$\Delta G_{f,i} = \Delta G_{f,i}^{\circ} + RT \ln a_i$$

where $a_i = \gamma_i [i]$. To derive the equilibrium expression for any phase, the appropriate expressions for $\Delta G_{f,i}$ are substituted into equation 13–1.

SOLUTION EQUILIBRIA

We will be concerned primarily with aqueous solutions of B and C and we will designate this by an (aq) subscript on the free energy terms for B and C.

13-2
$$\Delta G_{f,B_{(aq)}} = \Delta G^{\circ}_{f,B_{(aq)}} + RT \ln a_B$$

$$\Delta G_{f,C_{(aq)}} = \Delta G^{\circ}_{f,C_{(aq)}} + RT \ln a_C$$

Substituting equations 13-2 into equation 13-1 and simplifying we can derive equation 13-3 (see Appendix N for the derivation).

13-3
$$\Delta G^{\circ} = -RT \ln K = -2.303 \, RT \log K$$

where K is the equilibrium constant

13-4
$$K = \frac{(a_C)^c}{(a_B)^b} = \frac{(\gamma_C)^c}{(\gamma_B)^b} \frac{[C]^c}{[B]^b}$$

If B and C are solutes at low concentration (Henry's law region) or if the solution is ideal, the activity coefficients are equal to one and

13-5
$$K = \frac{[C]^c}{[B]^b} \qquad \text{(in the low concentration region)}$$

It should be noted that the change in *standard* free energy (ΔG° of equation 13-3) has only one value. Therefore only one equilibrium constant exists for a chemical reaction regardless of the concentrations of the species at the time the reaction is carried out. On the other hand, the change in free energy (ΔG) for the reaction does depend upon the concentrations of the reactants and products in the reaction mixture and will change as the concentrations change during the reaction. (The relationship between ΔG and ΔG° will be given in Section 13-C.) Equation 13-3 is useful in the calculation of K, provided that the standard free energies of formation of products and reactants are known. It is important to point out that the $\Delta G^{\circ}_{f,i}$ need only be determined once for any substance; subsequently, this value can be used numerous times to calculate K for any reaction involving this species. Equations 13-4 and 13-5 are necessary to calculate the concentrations of the reactants and products in the equilibrium mixture (see the procedures described in Sections 13-D through 13-G).

There are some equilibrium constants which are used quite frequently, such as acid and base dissociation constants and solubility product constants. In these special cases the equilibrium constants themselves are tabulated in addition to the free energies of formation of the individual species (Appendix T, Tables 12 to 15).

EXAMPLE 13-1

Calculate the equilibrium constant for the ionic dissociation of hydrocyanic acid in an aqueous solution at room temperature:

$$HCN_{(aq)} \rightleftharpoons H^{+}_{(aq)} + CN^{-}_{(aq)}$$

The standard free energy of formation of the cyanide ion and for hydrocyanic acid are given in Appendix T, Table 4.

$$\Delta G^\circ = \Delta G^\circ_{f,H^+_{(aq)}} + \Delta G^\circ_{f,CN^-_{(aq)}} - \Delta G^\circ_{f,HCN_{(aq)}}$$

$$\Delta G^\circ = 0 + 39.6 - 26.8$$

$$\Delta G^\circ = 12.8 \text{ kcal}$$

$$\Delta G^\circ = -2.303 \text{ RT } log \text{ K}$$

$$12.8 = -(2.303)(1.987 \times 10^{-3} \text{ kcal/deg-mole})(298 \text{ deg}) log \text{ K}$$

The unit of mole in the denominator can be ignored since it actually cancels as a result of ΔG° being calculated for reaction of moles of material.

$$log \text{ K} = -9.31 = -10 + .69$$

$$\text{K} = 4.9 \times 10^{-10}$$

EXAMPLE 13–2   Calculate the equilibrium constant at 298 °K for the reaction

$$Fe^{2+} + Hg^{2+} \rightleftharpoons Fe^{3+} + \tfrac{1}{2}Hg_2^{2+}$$

The standard free energies can be found in Appendix T, Table 3.

$$\Delta G^\circ = \Delta G^\circ_{f,Fe^{3+}} + \tfrac{1}{2}\Delta G^\circ_{f,Hg_2^{2+}} - \Delta G^\circ_{f,Fe^{2+}} - \Delta G^\circ_{f,Hg^{2+}}$$

$$\Delta G^\circ = -2.52 + \tfrac{1}{2}(36.79) - (-20.30) - (39.38) \text{ kcal}$$

$$\Delta G^\circ = -3.2 \text{ kcal}$$

$$\Delta G^\circ = -2.303 \text{ RT } log \text{ K}$$

$$-3.2 \text{ kcal} = -(2.303)(1.987 \times 10^{-3} \text{ kcal/deg-mole})(298 \text{ deg}) log \text{ K}$$

$$log \text{ K} = 2.34$$

$$\text{K} = 2.19 \times 10^2$$

Note that K (and K_P in the following section) is written without units; this may seem confusing because equations 13–4 and 13–8 imply that K should have units. If these equations were rigorously derived, you would see that the concentration or pressure terms do not represent actual pressures or concentrations, but are in fact ratios of the actual pressures or concentrations to the standard state ([B] = 1M; $P_B = 1$ atm). As ratios, these terms are unitless, and therefore K or K_P is unitless. When the equilibrium constants are used to calculate equilibrium concentrations, as we will do later in this chapter, the pressure or concentration terms must be expressed in atmospheres or moles/liter, in order to agree with the standard state for which K has been defined.

PROBLEM 13–1   Calculate the equilibrium constant at 298 °K for

$$NH_{3(aq)} + H_2O \rightleftharpoons NH_{4(aq)}^+ + OH_{(aq)}^-$$

The ΔG_{fi}° are given in Appendix T, Table 3.

PROBLEM 13–2   Calculate the equilibrium constant at 298 °K for

$$CN_{(aq)}^- + H_2O \rightleftharpoons HCN_{(aq)} + OH_{(aq)}^-$$

Use the standard free energies in Appendix T, Tables 3 and 4.

IDEAL GAS REACTIONS AND EQUILIBRIA

In deriving the equilibrium expression for chemical reactions carried out in the gas phase, equation 12–27 is used to express the free energy of formation of an ideal gas. If B and C are assumed to be ideal gases, the free energies of formation are given by

$$\Delta G_{f,B_{(g)}} = \Delta G_{f,B_{(g)}}^\circ + RT \ln P_B$$

13–6

$$\Delta G_{f,C_{(g)}} = \Delta G_{f,C_{(g)}}^\circ + RT \ln P_C$$

Substitution of equations 13–6 into 13–1 leads to equation 13–7

13–7 $$\Delta G^\circ = -RT \ln K_P$$

where

$$\Delta G^\circ = c\Delta G_{f,C_{(g)}}^\circ - b\Delta G_{f,B_{(g)}}^\circ$$

and

13–8 $$K_P = \frac{(P_C)^c}{(P_B)^b}$$

The subscript P on the equilibrium constant designates that the concentrations are expressed in partial pressures. Note the similarity between equations 13–4 and 13–8; the a_i is replaced by P_i. The $\Delta G°$ in equations 13–3 and 13–7 both represent the change in *standard* free energy, but the standard states appropriate to each equation are different. For solutions (equation 13–3), the standard state has an activity equal to one ($a_i = 1$), whereas the standard state for the ideal gas (equation 13–7) is one atmosphere ($P_i = 1$).

PROBLEM 13–3   As you will recall from Chapter 10, the standard free energy of the elements in their most stable form is assigned a value zero. Consequently, $\Delta G°_{f,N_{2(g)}} = \Delta G°_{f,O_{2(g)}}$ = 0. The standard free energies of formation of four nitrogen-oxygen binary compounds have been determined experimentally (all in kcal/mole).

$$\Delta G°_{f,N_2O_{(g)}} = 24.76, \ \Delta G°_{f,NO_{(g)}} = 20.72, \ \Delta G°_{f,NO_{2(g)}} = 12.39, \ \Delta G°_{f,N_2O_{4(g)}} = 23.49$$

a. Write the expression for the equilibrium constant for the reaction

$$N_2O_{(g)} + \tfrac{1}{2}O_{2(g)} \rightleftharpoons 2NO_{(g)}$$

Calculate $\Delta G°$ and K_P for this reaction
b. Repeat the work of (a) for the reaction

$$2N_2O_{(g)} + O_{2(g)} \rightleftharpoons 4NO_{(g)}$$

How does the K_P in b. compare to that in a.? Would the difference in K_P values affect the equilibrium concentrations of the gases? How does the $\Delta G°$ in b. compare to that in a.? Is the difference in $\Delta G°$ values in agreement with the difference in K_P values? [Hint: $2 \ln x = \ln x^2$.]
c. There are four $\Delta G°_{f,i_{(g)}}$ values given above for the nitrogen-oxygen compounds. How many *different* chemical reactions can you write involving any of these four compounds in addition to N_2 and O_2? (there are at least 10). For each reaction there is an equilibrium constant. If you had a set of 1,000 compounds containing just carbon, hydrogen, and oxygen which could react with one another and with these elements to produce other compounds, would it be more practical to make a table of $\Delta G°_{f,i}$ values or of equilibrium constants for each reaction?

HETEROGENEOUS EQUILIBRIA

The equilibrium expression for reactions which involve more than one phase can be derived in the same manner as the equilibrium expressions for the solution or gas phase reactions. If one of the substances is in a pure liquid or solid phase, then the free energy of formation of this substance is given by the standard free energy of formation

$$\Delta G_{f,i_{(l\,or\,s)}} = \Delta G°_{f,i_{(l\,or\,s)}}$$

The term $\Delta G_{f,i_{(l\,or\,s)}}$ *does not depend on any concentration terms and therefore the concentrations of liquids or solids will not appear in the equilibrium constant expression. However,* $\Delta G^\circ_{f,i_{(l\,or\,s)}}$ *will be included in the overall* ΔG° *term.*

In order to illustrate heterogeneous equilibria let us assume that C in our previous reaction is a liquid and B is a gas

$$bB_{(g)} \rightleftharpoons cC_{(l)}$$

The expressions for free energies of formation are

$$\Delta G_{f,B_{(g)}} = \Delta G^\circ_{f,B_{(g)}} + RT \ln P_B$$

$$\Delta G_{f,C_{(l)}} = \Delta G^\circ_{f,C_{(l)}}$$

Substituting these into equation 13–1 leads to equation 13–7

$$\Delta G^\circ = -RT \ln K_P$$

where

$$\Delta G^\circ = c\Delta G^\circ_{f,C_{(l)}} - b\Delta G^\circ_{f,B_{(g)}}$$

and

13–9
$$K_P = \frac{1}{(P_B)^b}$$

Note that equation 13–9 does *not* include a P_C term. However, the standard free energy of formation of liquid C must be included in the calculation of ΔG°. The equilibrium expression for more complex heterogeneous reactions can be written by inspection from the stoichiometric equations. Remember that the concentration of condensed phases does not appear in K_P, but the standard free energies of formation of all phases are included in the calculation of ΔG°.

PROBLEM 13–4   Calcium carbonate decomposes upon heating according to the reaction

$$CaCO_{3(s)} \rightleftharpoons CaO_{(s)} + CO_{2(g)}$$

Write
a. the expression for ΔG° in terms of the standard free energies of formation
b. the expression for K_P in terms of partial pressures.

As we have seen, the standard free energies of formation are very useful in the calculation of ΔG°'s and equilibrium constants, and the tabulation of values of standard free energy of formation is more practical than the tabulation of equilibrium constants. This does not mean to infer that equilibrium constants can only be determined from the thermodynamic parameters. Equilibrium constants are frequently determined directly from experimental measurements (next section) and can then be used to calculate $\Delta G°$. Many of the $\Delta G_f°$ values given in Tables 3, 4, and 5 of Appendix T have been determined in this manner.

13–B. EXPERIMENTAL DETERMINATION OF EQUILIBRIUM CONSTANTS

The equilibrium constant for a reaction can be determined experimentally by measuring the concentrations of one or more of the reactants or products after the reaction has reached equilibrium; the concentration of the species not measured directly can be determined from the stoichiometry of the reaction. Substitution of the concentration values into the equilibrium expression will give the equilibrium constant. This type of calculation is illustrated in the following example.

EXAMPLE 13–3
Calculate the equilibrium constant for the dissociation of chloroacetic acid in H_2O at 25 °C.

$$ClCH_2-\overset{\overset{O}{\|}}{C}-OH_{(aq)} \rightleftharpoons ClCH_2-\overset{\overset{O}{\|}}{C}-O^-_{(aq)} + H^+_{(aq)}$$

The initial concentration of chloroacetic acid is 0.01 M, and the $[H^+_{(aq)}]$ at equilibrium is determined experimentally at 25 °C to be 0.00311 M.

Since only the $[H^+]$ was determined at equilibrium, we must first determine the $[ClCH_2COOH]$ and $[ClCH_2COO^-]$. According to the stoichiometric equation, one $ClCH_2COO^-$ ion must be produced for every H^+ produced. Therefore

$$[ClCH_2-\overset{\overset{O}{\|}}{C}-O^-] = [H^+] = 0.00311 \text{ M}$$

Furthermore, since one molecule of $ClCH_2COOH$ must be lost for every H^+ ion produced, the initial molar concentration is reduced by 0.00311 M.

$$[ClCH_2-\overset{\overset{O}{\|}}{C}-OH] = (0.01 - 0.00311) \text{ M}$$

This can be summarized as follows:

	INITIAL CONDITIONS (M)	EQUILIBRIUM (M)
$[H^+]$	0	0.00311
$[Cl-CH_2-\overset{\overset{O}{\|}}{C}-O^-]$	0	0.00311
$[Cl-CH_2-\overset{\overset{O}{\|}}{C}-OH]$	0.01	(0.01–0.00311)

Substitution into the equilibrium expression gives

$$K = \frac{\left[\begin{array}{c} O \\ \parallel \\ ClCH_2-C-O^- \end{array}\right][H^+]}{\left[\begin{array}{c} O \\ \parallel \\ ClCH_2-C-OH \end{array}\right]} = \frac{(x)(x)}{(a-x)} = \frac{(3.11 \times 10^{-3})(3.11 \times 10^{-3})}{(0.010 - 0.00311)}$$

$$K = \frac{9.69 \times 10^{-6}}{0.00689} = 1.4 \times 10^{-3}$$

PROBLEM 13–5   Calculate the equilibrium constant at 25 °C for the reaction

$$Ag_{(s)} + Fe^{3+}_{(aq)} \rightleftharpoons Ag^+_{(aq)} + Fe^{2+}_{(aq)}$$

An initial concentration of 0.100 M $Fe^{3+}_{(aq)}$ produced 0.0785 M $Ag^+_{(aq)}$ at equilibrium. An excess of $Ag_{(s)}$ was present, so that it was never depleted.

13–C. ΔG, ΔG°, AND EQUILIBRIUM

In Chapter 10 we arrived at the following criteria which apply to any process, such as a chemical reaction, carried out at constant temperature and pressure:

$$\Delta G < 0 \qquad \text{spontaneous process}$$

$$\Delta G = 0 \qquad \text{equilibrium}$$

Now in this chapter we have shown that

$$\Delta G° = -RT \ln K$$

A question now arises rather frequently: When $\Delta G > 0$, the reaction should not proceed, yet we find from our calculation with $\Delta G° = -RT \ln K$ that there are always some products formed in the equilibrium state. How can we first predict that a reaction cannot proceed and shortly thereafter calculate the amount of product produced from this reaction? This seemingly paradoxical situation can be understood by closely examining what is meant by ΔG and $\Delta G°$ and to see how ΔG changes as equilibrium is approached.

In order to be more specific in our discussion, let us consider the reaction first used to describe equilibrium in Figure 3–6.

$$A \rightleftharpoons B + C$$

The Δ*G* for this reaction is the change in free energy from the reactants to the products calculated for the state in which each species exists at a given time. The prediction that the process is spontaneous, Δ*G* < 0, means that the reaction can proceed from left to right to produce additional products B and C. Contrarily, if Δ*G* > 0, the forward reaction is not spontaneous under the existing conditions and the reverse reaction may proceed to produce additional A. If Δ*G* = 0, the given mixture of A, B, and C is in equilibrium and no net change will take place.

Now Δ*G*° for the reaction corresponds, on the other hand, to the change in free energy when *all* species are present *in their standard states.** If Δ*G*° < 0, the reactant A in its standard state can produce more B + C in their respective standard states. If Δ*G*° > 0, the products B and C in their standard states can produce A in its standard state. In other words, the magnitude of Δ*G*° has the same significance as the magnitude of Δ*G* except that both reactants and products must be present in their standard states.

The relationship between Δ*G* and Δ*G*° for a chemical reaction can be derived from thermodynamic principles (see Appendix P). For the reaction of A to form products B and C, Δ*G* and Δ*G*° are related by equation 13-10

13-10
$$\Delta G = \Delta G° + RT \ln \frac{[B][C]}{[A]}$$

The terms [A], [B], and [C] are the concentrations of A, B, and C in the reaction mixture and Δ*G* is the free energy change for the reaction when A, B, and C are *at these concentrations*. The term [B][C]/[A] is of the same form as the equilibrium constant; however, it differs from the equilibrium constant expression in that the concentrations in equation 13-10 are not necessarily at equilibrium but are the concentrations of the reactants and products in the reacting mixture. The concentration terms in the equilibrium constant expression correspond to the equilibrium concentrations.

Now suppose we were to start out the reaction with pure reactant A and ask: Can we produce any product and, if so, how much? The answer to the first part of the question is definitely yes since there is no product present at the start of the reaction; that is, [C] = [B] = 0. Δ*G* can be calculated

$$\Delta G = \Delta G° + RT \ln \frac{(0)(0)}{[A]}$$

Since ln 0 = −∞ and we cannot do calculations with −∞, let us assume for the sake of calculation that [C] = [B] are very small (*close* to zero). Therefore ln([B][C]/[A]) is a *very large negative* number.

---

*In practice this is possible if A, B, and C are present in a large volume so that the additional one mole of B and one mole of C which is produced does not change the standard state (concentration 1 molar for solutions).

$$\Delta G = \Delta G^\circ + RT \, (- \text{ very large number})$$

$$\Delta G < 0$$

However, the fact that we *can* produce the product must be qualified. The amount of product that *will* be produced depends upon the value of ΔG° and thereby upon the value of the equilibrium constant. If ΔG° is sufficiently large and positive then the equilibrium constant will be very small and only a very small amount of product can be produced. In fact, the ΔG° could be so large that the amount of product formed would be too small to detect. On the other hand if ΔG° is very much less than zero, the equilibrium constant will be very large, the reaction will go almost to completion, and at equilibrium very little reactant A will be left.

In order to clarify this point further we have shown in Figure 13–1 three graphs of the free energies of A, B, and C as a function of concentration for the reaction

$$A \rightleftharpoons B + C$$

Three different ΔG° have been chosen so as to give yields of 10 %, 50 %, and 90 % as was shown for the same reaction in Figure 3–6. The ΔG for the reaction at any point in time is the difference between the two curves since

$$\Delta G = (\Delta G_{f,B} + \Delta G_{f,C}) - \Delta G_{f,A}$$

The change in ΔG can be calculated from equation 13–10. Notice that equilibrium is obtained in all three cases when the $\Delta G_{f,A}$ is equal to the sum $(\Delta G_{f,B} + \Delta G_{f,C})$. The curve for $\Delta G_{f,A}$ as a function of the concentration of A has the same structure in all three cases; this is also true for the $\Delta G_{f,B} + \Delta G_{f,C}$ curve. The only difference between the three cases is the displacement of one curve relative to the other on the y coordinate as governed by ΔG°. If $\Delta G^\circ = 2.67$ kcal, equilibrium is reached at only 10 % yield ($[B] = [C] = 0.1$ M). If ΔG° is instead 0.411 kcal, then the yield is increased to 50 %. Finally, if the ΔG° is -1.24 kcal, then the yield is increased to 90 %. If ΔG° had been $+10$ kcal the yield at equilibrium would have been only 0.02 % ($[B] = [C] = 0.00021$), whereas if ΔG° were -10 kcal the yield would be 99.999996 %, essentially a complete reaction.

13–D. ACID-BASE EQUILIBRIA

The general equilibrium expressions which were developed in Section 13–A can be applied to acid and base equilibria. Using these expressions, we can calculate the equilibrium concentrations of $H^+_{(aq)}$ and $OH^-_{(aq)}$ and thereby obtain a quantitative measure of the strength of an acid or base. We will restrict our consideration of acids and bases at the present to aqueous solutions, where an acid is a substance which increases the concentration of $H^+_{(aq)}$ and a base decreases the concentration of $H^+_{(aq)}$ (Section 3–E).

FIGURE 13–1 Change in ΔG during three reactions with yields of 10 %, 50 %, and 90 %

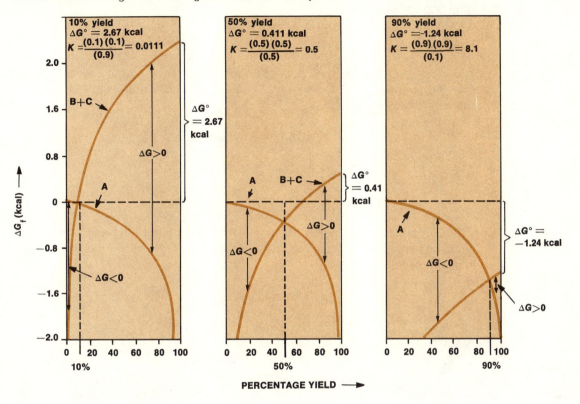

10% yield
$\Delta G° = 2.67$ kcal
$K = \dfrac{(0.1)(0.1)}{(0.9)} = 0.0111$

50% yield
$\Delta G° = 0.411$ kcal
$K = \dfrac{(0.5)(0.5)}{(0.5)} = 0.5$

90% yield
$\Delta G° = -1.24$ kcal
$K = \dfrac{(0.9)(0.9)}{(0.1)} = 8.1$

PERCENTAGE YIELD ⟶

AUTOIONIZATION OF WATER

We know that water ionizes into solvated protons, $H^+_{(aq)}$, and solvated hydroxide ions, $OH^-_{(aq)}$.

13–11
$$H_2O \rightleftharpoons H^+_{(aq)} + OH^-_{(aq)}$$

The solvated proton exists principally as the hydronium ion, H_3O^+, although other solvated species including $H_5O_2^+$ and $H_7O_3^+$ also exist to some extent in solution. We will designate the solvated proton as $H^+_{(aq)}$, realizing that it may exist as several species in aqueous solution.

The equilibrium expression for the autoionization process shown in equation 13–11 is

$$K = \frac{[H^+][OH^-]}{[H_2O]}$$

where the designation of aqueous solution has been dropped for simplicity. As we will see shortly, the extent of ionization is quite small, and for all practical purposes the $[H_2O]$ remains constant. Therefore, it is common to define a constant, K_w, which is

13–12
$$K_w = K[H_2O] = [H^+][OH^-]$$

At 25 °C $K_w = 1.0 \times 10^{-14}$. K_w depends upon temperature and increases with increasing temperature. The value of K_w at 25 °C is consistent with the $\Delta G°$ calculated from

$$\Delta G° = \Delta G°_{f,H^+_{(aq)}} + \Delta G°_{f,OH^-_{(aq)}} - \Delta G°_{f,H_2O_{(l)}}$$

$$\Delta G° = 0 + (-37.595) - (-56.69)$$

$$\Delta G° = +19.10 \text{ kcal}$$

The importance of equation 13–12 is that the product of $[H^+]$ and $[OH^-]$ must remain constant at all times regardless of any other ions or solutes in solution. In a neutral solution, the

$$[H^+] = [OH^-]$$

$$[H^+]^2 = K_w$$

$$[H^+] = [OH^-] = 10^{-7}$$

Obviously if the $[H^+]$ increases, the $[OH^-]$ must necessarily decrease. As a solution becomes more acidic the $[H^+]$ increases and the $[OH^-]$ decreases; as a solution becomes more basic the $[H^+]$ decreases and the $[OH^-]$ increases. A Brönsted base in aqueous solution is defined as a substance which is a proton acceptor; this is equivalent to the solvent concept definition of a base as a substance which increases the $[OH^-]$.

PROBLEM 13–6
A water sample containing some unknown solutes was analyzed and found to have a H^+ concentration of 5×10^{-3} M.
a. Calculate the OH^- concentration.
b. If you analyzed the other species in the solution and found that $[Na^+] = 0.01$ M, $[Br^-] = 0.0098$ M, $[CH_3COO^-] = 0.0002$ M, $[CH_3COOH] = 0.05$ M, would this information change your answer to part (a)? Explain.

pH AND pOH

$H^+_{(aq)}$ and $OH^-_{(aq)}$ ions are always present in aqueous solutions, in equilibrium with the solvent, H_2O. Furthermore, the $H^+_{(aq)}$ and $OH^-_{(aq)}$ can react with other ions, and the $[H^+]$ and $[OH^-]$ can influence the concentrations of these other ions in the solution. For this reason, we frequently have occasion to refer to the concentrations of the $H^+_{(aq)}$ and $OH^-_{(aq)}$. To assist in this regard, the terms called pH and pOH have been defined.

13–13a
$$pH = -\log [H^+]$$

13–13b
$$pOH = -\log [OH^-]$$

Generally pH and pOH values are positive quantities, since the $[H^+]$ and $[OH^-]$ usually range from 10^{-14} M to 1 M. For concentrations exceeding 1 M, the pH or pOH is negative. Similarly, the term pK is defined as

13–14
$$pK = -\log K$$

pK values can be positive or negative. Keep in mind that large K values correspond to low pK values and vice versa.

A convenient relationship between pH and pOH is obtained by taking the negative logarithm of equation 13–12:

$$-\log K_w = -\log [H^+] - \log [OH^-]$$

Using our definitions in equations 13–13 and 13–14

$$pK_w = pH + pOH$$

At 25 °C, p$K_w = 14$

13–15
$$14 = pH + pOH$$

The evaluation of pOH from a pH value and vice versa can be carried out readily using equation 13–15.

The scale of pH and pOH values in relationship to the $[H^+]$ and $[OH^-]$ is shown in Figure 13–2. Note that *low pH* values correspond to *high acidity* and *low pOH* values correspond to *high basicity*. Furthermore, note that a difference of one pH or pOH unit corresponds to a change in molar concentration by a factor of 10. Don't be misled into thinking that a pH of 4 is twice as acidic as a pH of 2.

EXAMPLE 13–4   *Calculate the pH and pOH of a solution where $[H^+] = 5.0 \times 10^{-8}$ M.*

$$pH = -\log [H^+] = -\log 5.0 \times 10^{-8} = -(-8 + \log 5.0)$$
$$pH = 8 - \log 5.0 = 8 - .70 = 7.30$$
$$pOH = 14.0 - pH = 14.0 - 7.30$$
$$pOH = 6.70$$

PROBLEM 13–7   Calculate the pH and pOH of a solution where $[OH^-] = 3.3 \times 10^{-3}$ M.

FIGURE 13–2 pH and pOH scales

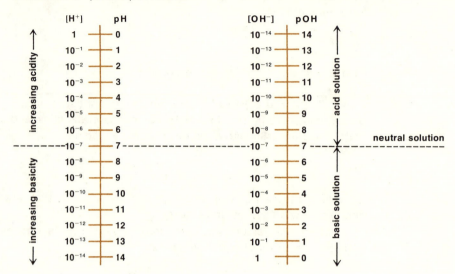

ACID DISSOCIATION

Some acids, such as the halogen acids, are considered to undergo complete dissociation into ions in water. We refer to these acids as *strong acids* because the complete dissociation of the ionizable hydrogens produces the maximum $[H^+_{(aq)}]$ for a given amount of acid. Since dissociation is assumed to be complete, the concentration of H^+ is equal to the concentration of the strong acid added to the solution. For example, suppose 0.3 moles of $HClO_4$ (a strong acid) were added to water to make 1 liter of solution. The concentration of ions would be calculated according to the dissociation

$$HClO_4 \xrightarrow{aq} H^+_{(aq)} + ClO^-_{4(aq)}$$

$$[H^+_{(aq)}] = [ClO^-_{4(aq)}] = 0.3 \text{ M}$$

On the other hand, there are also many acids for which dissociation is not complete. These are classified as *weak acids*; the strength of these acids, as measured by the extent of their dissociation, varies quite significantly. Since dissociation is not complete, an equilibrium exists between the weak acid (HA) and the dissociated ions.

13–16

$$HA_{(aq)} \underset{}{\overset{K_a}{\rightleftharpoons}} H^+_{(aq)} + A^-_{(aq)}$$

The expression for the equilibrium constant is

13–17
$$K_a = \frac{[H^+][A^-]}{[HA]}$$

where K_a is called the *acid dissociation constant*. The magnitude of K_a reflects the extent of dissociation of the acid and hence is a measure of the acid strength.

Let us represent the initial concentration of HA by a and the unknown $[H^+]$ at equilibrium by x. According to the stoichiometry in equation 13–16 we would have

$$[H^+] = [A^-] = x$$

The concentration of HA at equilibrium is then $(a-x)$. To summarize these changes in concentration:

	INITIAL CONDITIONS	EQUILIBRIUM
$[HA]$	a	$a - x$
$[H^+]$	0	x
$[A^-]$	0	x

Substituting these equilibrium concentrations into equation 13–17

13–18
$$K_a = \frac{(x)(x)}{(a - x)}$$

If the acid is relatively weak ($K_a < 10^{-5}$) and a reasonably large ($a > 0.01$ M), then x is small compared to a and

$$(a - x) \approx a$$

If we use this approximation in equation 13–18, we can readily solve for x:

$$K_a = \frac{x^2}{a}$$

13–19
$$x = (K_a a)^{1/2}$$

Obviously when K_a is large, x becomes large.

In order to illustrate the range of K_a values for weak acids, we will compare the following nitro substituted phenols. The structure for phenol is

COMPOUND	K_a	pK_a
phenol	1.28×10^{-10}	9.89
4-nitrophenol	7×10^{-8}	7.15
2,4-dinitrophenol	1.1×10^{-4}	3.96
2,4,6-trinitrophenol	4.2×10^{-1}	0.38

Note that nitro substitution on phenol increases the acidity dramatically, changing phenol, a weak acid, to the rather strong acid 2,4,6-trinitrophenol. Additional K_a values are given in Appendix T, Table 12.

When the K_a is large, as in 2,4,6-trinitrophenol, equation 13–19 cannot be used to calculate the amount of dissociation into ions. For stronger acids, equation 13–18 must be solved exactly by the quadratic formula.

$$K_a(a - x) = x^2$$

$$x^2 + K_ax - K_aa = 0$$

13–20
$$x = \frac{-K_a \pm \sqrt{K_a{}^2 + 4K_aa}}{2}$$

PROBLEM 13–8   Calculate the pH and $[OH^-]$ of a 0.02 M aqueous solution of HI (assume complete dissociation).

PROBLEM 13–9   Calculate the $[H^+]$ for a 0.1 M aqueous solution of phenol using
a. equation 13–19 and   b. equation 13–20. Compare the results of a. and b.

PROBLEM 13–10   Repeat the calculations of Problem 13–9 for a 0.1 M aqueous solution of 2,4,6-trinitrophenol.

PROBLEM 13–11   Repeat the calculations of Problem 13–9 for a 0.0001 M aqueous solution of phenol.

BASE DISSOCIATION

The dissociation of bases is analogous to acid dissociation. Some bases, such as the alkali metal hydroxides, undergo complete dissociation leading to a maximum yield of $OH^-_{(aq)}$. Because of this, these hydroxides are called strong bases. For example:

$$KOH_{(s)} \xrightarrow{aq} K^+_{(aq)} + OH^-_{(aq)}$$

As in the case of acids, there are also weak bases for which dissociation into ions is not complete. These weak bases may be metal hydroxides, such as $Ca(OH)_2$:

$$Ca(OH)_{2(aq)} \rightleftharpoons CaOH^+_{(aq)} + OH^-_{(aq)}$$

but weak bases can be any substance that reacts with water to produce OH^-:

$$NH_{3(aq)} + H_2O \rightleftharpoons NH_{4(aq)}^+ + OH_{(aq)}^-$$

The latter group includes numerous organic compounds which contain a basic nitrogen atom, such as CH_3NH_2

$$CH_3NH_{2(aq)} + H_2O \rightleftharpoons CH_3NH_{3(aq)}^+ + OH_{(aq)}^-$$
methylamine

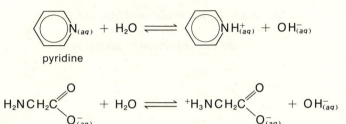

pyridine

$$H_2NCH_2C \overset{O}{\underset{O^-_{(aq)}}{\diagup}} + H_2O \rightleftharpoons {}^+H_3NCH_2C \overset{O}{\underset{O^-_{(aq)}}{\diagup}} + OH_{(aq)}^-$$

glycinate ion

The equilibrium expression for the base dissociation can be written down directly from the stoichiometric equation. If the weak base is a metal hydroxide, such as $Ca(OH)_2$:

$$Ca(OH)_{2(aq)} \rightleftharpoons CaOH_{(aq)}^+ + OH_{(aq)}^-$$

the base dissociation constant, K_b, is

$$K_b = \frac{[CaOH^+][OH^-]}{[Ca(OH)_2]}$$

The more common reaction of a weak base, B, is

13–21
$$B_{(aq)} + H_2O_{(l)} \rightleftharpoons BH_{(aq)}^+ + OH_{(aq)}^-$$

where the K_b is

13–22
$$K_b = \frac{[BH^+][OH^-]}{[B]}$$

and $\Delta G°$ is

13–23
$$\Delta G° = \Delta G°_{f,BH^+_{(aq)}} + \Delta G°_{f,OH^-_{(aq)}} - \Delta G°_{f,B_{(aq)}} - \Delta G°_{f,H_2O_{(l)}}$$

The $[H_2O]$ has been left out of the equilibrium constant since H_2O is the solvent and $[H_2O]$ does not change significantly at low solute concentrations. Note that $\Delta G°_{f,H_2O_{(l)}}$ must be included in the $\Delta G°$ expression of equation 13–23.

The calculation of the equilibrium concentrations of OH^- is analogous to the calculation of $[H^+]$ in acid dissociation. Let x represent the $[OH^-]$ at equilibrium. From the stoichiometric equation (equation 13–21)

$$x = [OH^-] = [BH^+]$$

If b represents the initial concentration of B before dissociation, then

$$[B] = (b - x)$$

Summarizing these concentration changes

	INITIAL CONDITION	EQUILIBRIUM
$[B]$	b	$b - x$
$[BH^+]$	0	x
$[OH^-]$	0	x

Our equilibrium expression becomes

13–24
$$K_b = \frac{(x)(x)}{(b - x)}$$

The solution of equation 13–24 is identical to the solution of equation 13–18, used for the calculation of acid dissociation. The same restrictions apply with respect to using the approximation $(b - x) \approx b$: small K_b ($< 10^{-5}$) or large pK_b (> 5) and large b (> 0.01 M).

Frequently, the acid dissociation constant K_a of the BH^+ ion is given for a base which is an organic compound. However, the base dissociation constant for B, K_b, can be calculated from the K_a value for BH^+. The acid dissociation step is

$$BH^+_{(aq)} \rightleftharpoons B_{(aq)} + H^+_{(aq)} \qquad K_a = \frac{[B][H^+]}{[BH^+]}$$

and

$$H_2O_{(l)} \rightleftharpoons H^+_{(aq)} + OH^-_{(aq)} \qquad K_w = [H^+][OH^-]$$

Dividing K_w by K_a, $[H^+]$ is eliminated to give equation 13–22

$$K_w/K_a = \frac{[H^+][OH^-]}{\dfrac{[B][H^+]}{[BH^+]}} = \frac{[BH^+][OH^-]}{[B]}$$

hence

$$K_w/K_a = K_b$$

As an example, the K_a for protonated aniline, $C_6H_5NH_3^+$, at 25 °C is

$$K_a = 2.34 \times 10^{-5}$$

The K_b for aniline, $C_6H_5NH_2$, is thus

$$K_b = \frac{K_w}{K_a} = \frac{10^{-14}}{2.34 \times 10^{-5}} = 0.427 \times 10^{-9}$$

A table of K_a values for the BH^+ of organic compounds is given in Appendix T, Table 14 and a short table of K_b values is given in Appendix T, Table 13.

PROBLEM 13–12   Calculate the $[OH^-]$ and pH for an aqueous solution at 25 °C containing 0.01 M *p*-toluidine (NH$_2$—⬡—CH$_3$). The K_a for $^+NH_3$—⬡—CH$_3$ at 25 °C is 8.32×10^{-6}.

COMMON ION EFFECT

In our previous calculations of weak acid and weak base dissociation, we assumed that the ionic species arose only from the dissociation of the acid or base. However, the equilibrium concentrations of the ions resulting from the dissociation can change significantly if another substance is present which contains ions like (common to) those arising from the dissociation of the acid or base. The effect of adding ions that are the same as those already in equilibrium is called the *common ion effect*.

For example, consider the dissociation of a weak acid HA:

$$HA_{(aq)} \rightleftharpoons H^+_{(aq)} + A^-_{(aq)}$$

The addition of a salt MA produces the anion $A^-_{(aq)}$ which is common to the anion of the acid.

$$MA \xrightarrow{aq} M^+_{(aq)} + A^-_{(aq)}$$

In calculating the equilibrium concentrations, both sources of $A^-_{(aq)}$ must be taken into account. A similar approach is taken when a salt, which has a cation common to that of a weak base, is added to the weak base.

EXAMPLE 13–5   *Hydrofluoric acid has a $K_a = 3.53 \times 10^{-4}$ at 25 °C. Calculate the equilibrium $[H^+]$ produced   a. from a 0.10 M HF solution and*
b. from a 0.10 M HF solution containing 0.20 M KF.

a. The acid dissociation is

$$HF_{(aq)} = H^+_{(aq)} + F^-_{(aq)}$$

The concentration changes are

	INITIAL CONCENTRATION	EQUILIBRIUM
$[HF]$	0.1	$(0.1 - x)$
$[H^+]$	0	x
$[F^-]$	0	x

$$K_a = \frac{[H^+][F^-]}{[HF]}$$

$$x = [H^+] = [F^-]$$

$$3.53 \times 10^{-4} = \frac{x^2}{0.1 - x} \approx \frac{x^2}{0.1}$$

$$x = [H^+] = 0.603 \times 10^{-2} M$$

b. The KF will dissociate completely

$$KF \xrightleftharpoons{aq} K^+_{(aq)} + F^-_{(aq)}$$

Since the concentration of KF is 0.2 M, this will produce an initial $[F^-] = 0.2$ M. *The expression for the equilibrium concentration of* $F^-_{(aq)}$ *must also contain this term. Letting x represent the* $[H^+]$, *we can summarize the concentration changes*

	INITIAL CONDITION	EQUILIBRIUM
$[HF]$	0.1	$0.1 - x$
$[H^+]$	0	x
$[F^-]$	0.2	$0.2 + x$

The equilibrium expression becomes

$$3.53 \times 10^{-4} = \frac{(x)(0.2 + x)}{(0.1 - x)}$$

Since x will be small ($\sim 10^{-4}$) compared to 0.1, we can assume

$$(0.2 + x) \approx 0.2$$

$$(0.1 - x) \approx 0.1$$

and

$$3.53 \times 10^{-4} = \frac{x(0.2)}{0.1}$$

$$[H^+] = x = 1.77 \times 10^{-4}$$

Note that the presence of 0.2 M KF decreased the $[H^+]$ by a factor of 34.

PROBLEM 13–13   Ammonia in water has a $K_b = 1.8 \times 10^{-5}$. Calculate $[OH^-]$ for a solution containing
a. 0.01 M NH_3 or   b. 0.01 M NH_3 plus 0.10 M NH_4Br.
Assume NH_4Br dissociates completely in aqueous solution.

13–E. POLYPROTIC ACIDS

In the previous section we restricted the discussion of acids to those in which only a single H^+ can be ionized. There are other acids which have more than one ionizable H^+ and are called *polyprotic* acids. These acids are also called *polybasic* since more than one proton per molecule is capable of neutralizing an OH^- ion. The more common polyprotic acids are: H_2SO_4, H_2S, H_2CO_3, $H_2C_2O_4$, H_3PO_4, and H_3AsO_4. Acids with two ionizable protons are called diprotic and those with three are called triprotic acids.

The ionization of a polyprotic acid generally occurs in steps, and there is an equilibrium expression associated with each ionization process, as illustrated with H_3PO_4.

$$H_3PO_{4(aq)} \rightleftharpoons H^+_{(aq)} + H_2PO^-_{4(aq)} \qquad K_1 = \frac{[H^+][H_2PO_4^-]}{[H_3PO_4]} = 7.52 \times 10^{-3}$$

$$H_2PO^-_{4(aq)} \rightleftharpoons H^+_{(aq)} + HPO^{2-}_{4(aq)} \qquad K_2 = \frac{[H^+][HPO_4^{2-}]}{[H_2PO_4^-]} = 6.23 \times 10^{-8}$$

$$HPO^{2-}_{4(aq)} \rightleftharpoons H^+_{(aq)} + PO^{3-}_{4(aq)} \qquad K_3 = \frac{[H^+][PO_4^{3-}]}{[HPO_4^{2-}]} = 2.2 \times 10^{-13}$$

The equilibrium constants are designated by subscripts corresponding to the successive dissociative steps. For each equilibrium step there is a·corresponding ΔG_1°, ΔG_2°, or ΔG_3°. These values of change in standard free energy are related in the conventional manner to the difference in the free energies of formation of products and reactants for each reaction step.

The successive ionization constants generally become smaller in magnitude. This is understandable since the charge of the negative ion that is formed in each dissociation step increases from -1 to -2 to -3. Separation of a H^+ ion from an anion with a larger negative charge requires more energy, and consequently the successive K_i's should become smaller as more protons are removed. Note the decrease in the K_i value for phosphoric acid, from approximately 10^{-3} to 10^{-8} to 10^{-13}. Acid dissociation constants for other polyprotic acids are given in Appendix T, Table 12.

The calculation of the ionic concentrations in a solution containing a polyprotic acid can be rather complex, since several equilibrium expressions must be solved simultaneously for several unknown concentrations. However, the calculations are simplified greatly if the K_i's are reasonably small. When the K_i's are small, each equilibrium expression can be solved independently, as will be illustrated in the following example for the dibasic acid, H_2S.

EXAMPLE 13–6   *Calculate the $[H^+], [HS^-]$, and $[S^{2-}]$ for an aqueous solution containing $[H_2S] = 0.1$ M.*

The equilibrium steps are:*

$$H_2S_{(aq)} \rightleftharpoons H^+_{(aq)} + HS^-_{(aq)} \qquad K_1 = \frac{[H^+][HS^-]}{[H_2S]} = 1.1 \times 10^{-7}$$

$$HS^-_{(aq)} \rightleftharpoons H^+_{(aq)} + S^{2-}_{(aq)} \qquad K_2 = \frac{[H^+][S^{2-}]}{[HS^-]} = 1 \times 10^{-14}$$

In the first equilibrium step, H^+ and HS^- are produced in equal amounts. In the second step, some of the HS^- produced in the first step further dissociates but only a very small amount more of H^+ ($K_2 \approx 10^{-14}$) will be produced compared to the H^+ produced in the first step. Therefore, we can make the assumption,

$$[H^+] = [HS^-]$$

Calculation from the first step gives:

$$1.1 \times 10^{-7} = \frac{[H^+][HS^-]}{[H_2S]} = \frac{[H^+]^2}{(0.1)}$$

$$[H^+] = [HS^-] = (1.1 \times 10^{-8})^{1/2} = 1.05 \times 10^{-4} M$$

From the second step:

$$1 \times 10^{-14} = \frac{[H^+][S^{2-}]}{[HS^-]} = \frac{(1.05 \times 10^{-4})[S^{2-}]}{(1.05 \times 10^{-4})}$$

$$[S^{2-}] = 1 \times 10^{-14}$$

Note that the $[S^{2-}]$ is simply equal to K_2, since $[H^+] = [HS^-]$. The $[S^{2-}]$ is independent of the concentration of $[H_2S]$, provided the latter is large enough so that $[H_2S] - [H^+] \approx [H_2S]$. The $[H^+]$, and consequently the pH, is dependent primarily on the first ionization step when both K_i's are small.

*K_2 is not known precisely and could be in error by as much as a factor of 10.

PROBLEM 13–14 Calculate $[H^+]$, $[HCO_3^-]$, and $[CO_3^{2-}]$ for an aqueous solution containing 0.01 M CO_2. The acid dissociation constants are $K_1 = 4.30 \times 10^{-7}$ and $K_2 = 5.61 \times 10^{-11}$.

PROBLEM 13–15 Calculate $[PO_4^{3-}]$ in an aqueous solution containing a. $[H_3PO_4] = 0.1$ M or b. $[H_3PO_4] = 0.01$ M.

PROBLEM 13–16 Calculate the $[HS^-]$ and $[S^{2-}]$ in a solution containing 0.1 M H_2S plus 0.01 M HCl. [Hint: The $[H^+]$ comes primarily from the HCl; the H^+ from H_2S can be neglected.]

13–F. BUFFER SOLUTIONS

When an acid or base is added to pure water, the concentration of H^+ and OH^- change quite dramatically, as we have seen from the examples and problems in Sections 13–D and 13–E. However, as we will see shortly, certain solutions, called buffer solutions, prevent a large change in H^+ or OH^- concentration when an acid or base is added. Buffer solutions have many applications and play an important role in controlling the $[H^+]$ or pH of solutions in various biological processes.

Buffer solutions generally consist of a weak acid plus a salt containing an anion common to that of the weak acid, or a weak base plus a salt containing a cation common to that of the base (see the discussion of the common ion effect in Section 13–D). The buffer action of a solution consisting of a weak acid plus salt of the acid arises from the fact that (a) added acid reacts with the common anion to produce the weak acid and (b) added base reacts with the weak acid to form more of the common anion. Calculation will show that these reactions do not produce a large change in pH of the solution. Similar reactions occur for the buffer solutions consisting of a weak base and salt containing a cation common to the weak base. For example, consider the weak acid benzoic acid (C_6H_5COOH) which has a $K_a = 6.46 \times 10^{-5}$. In aqueous solution we have the equilibrium

$$C_6H_5COOH_{(aq)} \rightleftharpoons C_6H_5COO^-_{(aq)} + H^+_{(aq)}$$

A buffer solution would be formed if we added the salt sodium benzoate (C_6H_5COONa), which produces $C_6H_5COO^-$ in solution.

$$C_6H_5COONa \xrightarrow{aq} Na^+_{(aq)} + C_6H_5COO^-_{(aq)}$$

Using the procedure discussed in the previous section for calculating the common-ion effect:

$$K_a = \frac{[C_6H_5COO^-][H^+]}{[C_6H_5COOH]}$$

Solving for $[H^+]$:

$$[H^+] = K_a \frac{[C_6H_5COOH]}{[C_6H_5COO^-]}$$

We can see that $[H^+]$ differs from K_a by a factor of the ratio $[C_6H_5COOH]/[C_6H_5COO^-]$. This ratio remains approximately constant in a buffer solution. Let us assume our buffer solution contains 0.1 M C_6H_5COOH and 0.1 M C_6H_5COONa. Then:

$$\frac{[C_6H_5COOH]}{[C_6H_5COO^-]} = 1$$

and for the buffer solution

$$[H^+] = K_a$$

Now let us see the effect of adding base or acid to this solution. Since C_6H_5COOH is a weak acid, addition of a strong acid to the buffer solution will produce C_6H_5COOH by the reaction of H^+ with the $C_6H_5COO^-$.

$$H^+_{(aq)} + C_6H_5COO^-_{(aq)} \longrightarrow C_6H_5COOH_{(aq)}$$

Suppose we add a few drops of concentrated HCl to make the concentration in solution 0.02 M HCl. Since we have added only a small volume of concentrated HCl, there will be very little dilution of the buffer solution. The 0.02 M HCl will react to give an additional 0.02 M C_6H_5COOH. The concentration changes are summarized as follows:

	INITIAL CONCENTRATION	ADDITION OF HCl	INTERMEDIATE CONCENTRATION
$[C_6H_5COOH]$	0.1 M		(0.1 + 0.02) M
$[C_6H_5COO^-]$	0.1 M		(0.1 − 0.02) M
$[Na^+]$	0.1 M		0.1 M
$[H^+]$	0.0000646 M	+0.02 M	x
$[Cl^-]$	0	+0.02 M	0.02 M

The $[H^+]$ for this solution is then calculated as before

$$[H^+] = K_a \frac{[C_6H_5COOH]}{[C_6H_5COO^-]} = K_a \frac{[0.12]}{[0.08]} = K_a \,(1.5)$$

The $[H^+]$ has increased by a factor of 1.5 or 50 % upon addition of HCl. A similar addition of concentrated HCl to pure H_2O to make a concentration of 0.02 M HCl would change the $[H^+]$ from 10^{-7} to 2×10^{-2} or by a factor of 2×10^5 (or 20,000,000 %).

PROBLEM 13–17   Calculate the $[H^+]$ for the addition of a concentrated NaOH solution to a buffer solution (0.1 M C_6H_5COOH and 0.1 M C_6H_5COONa) to make an initial concentration of 0.02 M NaOH.

PROBLEM 13–18   a. Calculate the $[H^+]$ of a buffer consisting of 0.05 M NH_3 and 0.05 M NH_4Cl.
b. Calculate the $[H^+]$ for this buffer solution when a few drops of concentrated HCl are added to make the initial concentration 0.01 M HCl.

Polyprotic acids and the salts formed by partially neutralizing one or more of the protons can also be used in buffer solutions. For example, equal quantities of the sodium salts NaH_2PO_4 and Na_2HPO_4 are used as a buffer in biological studies. The sodium salts completely dissociate in aqueous solution.

$$NaH_2PO_4 \xrightarrow{aq} Na^+ + H_2PO_4^-$$

$$Na_2HPO_4 \xrightarrow{aq} 2Na^+ + HPO_4^{2-}$$

In this case $H_2PO_4^-$ is a weak acid and HPO_4^{2-} is the anion common to this weak acid. The $[H^+]$ in the buffer can be calculated from the equilibrium

$$H_2PO_4^- \rightleftharpoons HPO_4^{2-} + H^+ \qquad K_2 = \frac{[H^+][HPO_4^{2-}]}{[H_2PO_4^-]}$$

Solving for $[H^+]$

$$[H^+] = K_2 \frac{[H_2PO_4^-]}{[HPO_4^{2-}]}$$

Since

$$[HPO_4^{2-}] = [H_2PO_4^-]$$

then

$$[H^+] = K_2$$

Addition of an acid or base simply changes the ratio $[H_2PO_4^-]/[HPO_4^{2-}]$, and the $[H^+]$ itself changes very little. The reactions upon addition of acid or base are

$$HPO_4^{2-} + H^+ \longrightarrow H_2PO_4^-$$

$$H_2PO_4^- + OH^- \longrightarrow HPO_4^{2-} + H_2O$$

PROBLEM 13–19   If a phosphate buffer containing 0.1 M NaH_2PO_4 and 0.1 M Na_2HPO_4 is prepared, calculate the $[H^+]$ for the addition of concentrated acid and base to make the initial concentrations   a. 0.01 M H^+ and   b. 0.01 M OH^-. Compare each answer with $[H^+] = K_2$.

PROBLEM 13–20   Describe how a mixture of H_3PO_4 and NaH_2PO_4 could act as a buffer. Calculate the pH of the buffer if equal amounts of H_3PO_4 and NaH_2PO_4 at 0.5 M were used.

13-G. HYDROLYSIS

Hydrolysis is a general term which refers to the reaction of a salt with water. The reaction can vary depending upon the nature of the cations and anions in the salt. Suppose the salt is one, such as $NaBr$ or $BaCl_2$, that results from the neutralization of a strong acid and a strong base:

$$HBr + NaOH \longrightarrow NaBr + H_2O$$

$$2HCl + Ba(OH)_2 \longrightarrow BaCl_2 + 2H_2O$$

These salts will simply dissociate into ions upon dissolving in water and there will be essentially no reaction with the solvent, water.

$$NaBr_{(s)} \xrightarrow{aq} Na^+_{(aq)} + Br^-_{(aq)}$$

$$BaCl_{2(s)} \xrightarrow{aq} Ba^{2+}_{(aq)} + 2Cl^-_{(aq)}$$

Since there is no reaction with water, the $NaBr$ and $BaCl_2$ solutions will be neutral with a pH $= 7$.

On the other hand, if we dissolve any salt which has been derived from the neutralization of a weak acid, a weak base, or both, the ions of the salt will undergo extensive reaction (hydrolysis) with the water. For example, if we dissolve a salt such as CH_3COONa, derived from a weak acid and a strong base, the ions Na^+ and CH_3COO^- will be formed initially. However, H^+ and OH^- are also present in solution from the dissociation of H_2O, and since CH_3COOH is a weak acid, H^+ will react with the CH_3COO^- to give acetic acid, CH_3COOH. The reactions can be summarized as follows:

$$CH_3COONa \xrightarrow{aq} Na^+_{(aq)} + \boxed{CH_3COO^-_{(aq)}}$$
$$H_2O \rightleftharpoons OH^-_{(aq)} + \boxed{H^+_{(aq)}}$$
$$\Downarrow$$
$$CH_3COOH_{(aq)}$$

The reaction of the CH_3COO^- ion with H^+ to give the weak acid CH_3COOH (to satisfy the acetic acid dissociation equilibrium expression) will necessarily result in a decrease in the H^+ concentration. Since the OH^- does not react with the Na^+ ($NaOH$ is a strong base), the final solution will have a higher concentration of OH^-, making the solution basic.

In summary, the hydrolysis of a salt derived from a weak acid and a strong base will result in the formation of the weak acid and an increase in the OH^- concentration (basic solution). On the other hand, the hydrolysis of a salt derived from a weak base and a strong acid will result in the formation of the weak base and an increase in the H^+ concentration (acidic solution). The hydrolysis of a salt derived from a weak acid and a weak base will result in the formation of both the weak acid and the weak base and the resulting solution will be approximately neutral *if* the dissociation constants for the acid and base are of approximately the same magnitude.

PROBLEM 13–21   Describe the hydrolysis of NH_4ClO_4 and predict whether the solution produced is acidic or basic.

* 13–H. ACID-BASE TITRATION

The neutralization reaction between an acid and a base has been discussed in Chapter 3. As you will recall, neutralization is accomplished when equal numbers of equivalents of acid and base are mixed. The experimental technique of *titrimetry* is used to determine how much acid is required to neutralize a certain amount of base or, vice versa, the amount of base required to neutralize a certain amount of acid. In titrimetry the amount of material required to bring about the neutralization is determined by measuring the volume of solution *titrated* (dispersed) from a buret. The buret is a device which can accurately measure the volume of solution dispensed. A common laboratory buret is shown in Figure 13–3.

The solution of the substance being titrated is placed in a flask and the solution to be added, called the *titrant*, is placed in the buret. As the titrant is added to the solution, the acid-base reaction occurs and the pH changes. For example, if the solution to be titrated is acidic, the pH of the solution will be low at the beginning of the titration. As base is added, the solution becomes less acidic and the pH increases. When the equivalents of base added approach the equivalents of the acid originally present, the pH increases very rapidly. At the *equivalence point*, where the equivalents of base equal the equivalents of acid, the pH changes at its most rapid rate. Figure 13–4 shows the titration of a strong base, such as NaOH, against a strong acid, such as HCl.

The equivalence point can be determined experimentally by detecting this rapid change in pH. One way to do this is to measure the pH of the solution directly with a pH meter. The pH meter is an instrument which generates a potential (from an electrochemical cell) which is directly proportional to the pH of the solution. (Electrochemical cells and the electrical potentials which can be derived from these cells will be discussed in Chapter 14.) Another way to detect the equivalence point is to use an *indicator*, a compound which itself is an acid or base and which changes color with a change in pH. The color of each acid–base indicator will change over a specific range of pH values and this range depends upon the acid dissociation constant for the indicator. Let us represent the formula of the indicator by HIn. The dissociation of the indicator in aqueous solution is

$$HIn_{(aq)} \rightleftharpoons H^+_{(aq)} + In^-_{(aq)}$$

FIGURE 13–3   Apparatus for titration

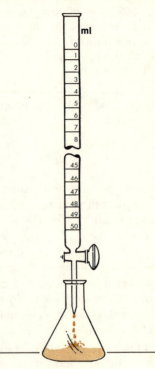

The HIn and/or In⁻ are colored and as the $H^+_{(aq)}$ concentration is decreased by the addition of OH^-, the color of the solution changes from the color of HIn to In⁻. There will be an equilibrium constant expression for the dissociation of the indicator just as for any acid

13–25
$$K_I = \frac{[H^+][In^-]}{[HIn]}$$

Taking the logarithm (base 10) of equation 13–25

$$\log K_I = \log [H^+] + \log \frac{[In^-]}{[HIn]}$$

and rearranging

$$-\log [H^+] = -\log K_I + \log \frac{[In^-]}{[HIn]}$$

Using the definitions of pK and pH we have

$$pH = pK_I + \log \frac{[In^-]}{[HIn]}$$

FIGURE 13-4 Titration of a strong base versus a strong acid (0.005 equivalents of HCl present initially at a concentration of 0.1 N). Note the rapid rise in pH in the region near the equivalence point.

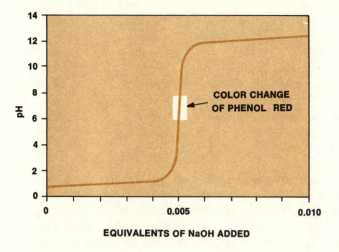

In order to detect visually a color change in the solution, the ratio $[In^-]/[HIn]$ must change from approximately 0.1 to 10, which corresponds to the change from $[HIn]$ being 10 times the $[In^-]$ to the case where the $[In^-]$ is 10 times the $[HIn]$. Since log 0.1 = −1 and log 10 = +1, the approximate pH range over which an indicator will change color is

13-26
$$pH = pK_I \pm 1$$

where the ±1 defines a range of approximately 2 pH units about the value of pK_I. For example, the indicator methyl red has $K_I = 5 \times 10^{-6}$ and $pK_I = 5.3$. The range over which methyl red changes from the red form (HIn) to the yellow form (In^-) is

$$pH = pK_I \pm 1 = 5.3 \pm 1$$

$$pH = 4.3 \text{ to } 6.3$$

The pH range for the indicator color change also depends upon the intensity of the colors of the HIn and In^- forms. Obviously if one form has a more intense color than the other its concentration need not be as large to be seen. A list of some indicators is given in Table 13-1 along with the acid and base colors of the indicator and the pH range over which the color change occurs. The color change for phenol red has been marked in Figure 13-4. The color change (from yellow at low pH values to red at high pH values) occurs approximately over the range pH 6.4-8.0.

TABLE 13–1   Acid-base indicators

INDICATOR	COLOR OF ACID FORM (HI)	COLOR OF BASE FORM (In⁻)	pH RANGE OF COLOR CHANGE
methyl orange	red	yellow	3.1–4.4
bromcresol green	yellow	blue	3.8–5.4
methyl red	red	yellow	4.2–6.3
bromthymol blue	yellow	blue	6.0–7.6
phenol red	yellow	red	6.4–8.0
phenolphthalein	colorless	red	8.0–9.6
thymolphthalein	colorless	blue	9.3–10.5

In Figure 13–4 we have shown the titration curve for a strong base against a strong acid. The equivalence point occurs at the neutral point, $pH = 7$. The pH of the solution at the equivalence point can be understood by considering the salt that is produced from the reaction and the hydrolysis of this salt in an aqueous solution. For example, in the reaction of $NaOH$ (strong base) with HCl (strong acid) the salt $NaCl$ is produced. The hydrolysis of $NaCl$ produces only solvated Na^+ and Cl^- with no change in pH in the aqueous solution.

$$NaCl_{(s)} \xrightarrow{aq} Na^+_{(aq)} + Cl^-_{(aq)}$$

Consequently the aqueous solution remains neutral at a $pH = 7$.

On the other hand if $NaOH$ is titrated against a weak acid, such as acetic acid (CH_3COOH), the salt CH_3COONa is produced at the equivalence point. According to our discussion in Section 13–G, the hydrolysis of CH_3COONa will give a basic solution. This is also evident from the titration curve of a strong base (such as $NaOH$) with a weak acid (such as CH_3COOH) as shown in Figure 13–5. Note that the equivalence point occurs at a $pH = 8.9$ which is a basic solution

$$[OH^-] = 7.5 \times 10^{-6} \text{ M} > [H^+] = 1.3 \times 10^{-9} \text{ M}$$

In order to detect this equivalence point an acid-base indicator must be chosen which undergoes a color change *in this pH region*. For the titration of CH_3COOH with $NaOH$ a logical selection from Table 13–1 is phenolphthalein which changes in the region pH 8.0 to 9.6.

PROBLEM 13–22   a. Sketch the titration curve that you would expect for the titration of a strong acid, such as HCl, against a weak base, such as NH_4OH. Recall your answer to Problem 13–21.   b. The pH of the equivalence point for the titration of 0.005 equivalents NH_4OH is 5.1. Select an acid-base indicator which is appropriate for this titration.

FIGURE 13–5   Titration of a strong base versus a weak acid (0.005 equivalents of CH_3COOH present initially at concentration of 0.1 N)

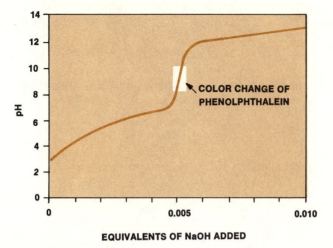

13–I. SOLUBILITY OF IONIC SOLIDS

SOLUBILITY PRODUCT CONSTANT

Most ionic solids undergo complete dissociation into their constituent ions upon dissolving in aqueous solution and there is an equilibrium between the solid and the solvated ions. For example, the equilibrium for $Mg(OH)_2$ would be

13–27
$$Mg(OH)_{2(s)} \xrightleftharpoons{\; aq \;} Mg^{2+}_{(aq)} + 2OH^-_{(aq)}$$

The solubility of $Mg(OH)_2$ depends upon the extent to which the solid dissociates into Mg^{2+} and OH^- ions. We can relate the concentrations of these ions to a constant that is similar to the equilibrium constant. As before, we consider the change in free energy for equation 13–27

$$\Delta G = (\Delta G_{f, Mg^{2+}_{(aq)}} + 2\Delta G_{f, OH^-_{(aq)}}) - \Delta G_{f, Mg(OH)_{2(s)}}$$

The free energy of the solid does not change during the reaction and if the pressure is constant at one temperature

13–28
$$\Delta G_{f, Mg(OH)_{2(s)}} = \Delta G^\circ_{f, Mg(OH)_{2(s)}}$$

The change in free energy is zero at equilibrium and we obtain

13–29
$$0 = (\Delta G_{f, Mg^{2+}_{(aq)}} + 2\Delta G_{f, OH^-_{(aq)}}) - \Delta G_{f, Mg(OH)_{2(s)}}$$

The equation for the free energy of formation of ions is similar to equation 12–26, provided that the ion concentrations are low:

13-30a
$$\Delta G_{f,Mg^{2+}_{(aq)}} = \Delta G^{\circ}_{f,Mg^{2+}_{(aq)}} + RT \ln [Mg^{2+}]$$

13-30b
$$\Delta G_{f,OH^{-}_{(aq)}} = \Delta G^{\circ}_{f,OH^{-}_{(aq)}} + RT \ln [OH^-]$$

Equation 13–28, 13–30a, and 13–30b can be substituted into equation 13–29 (see Appendix O) to obtain equation 13–31.

13-31
$$\Delta G^{\circ} = -RT \ln K_{sp}$$

where K_{sp} is called the *solubility product constant*

13-32
$$K_{sp} = [Mg^{2+}][OH^-]^2$$

and

$$\Delta G^{\circ} = \Delta G^{\circ}_{f,Mg^{2+}_{(aq)}} + 2\Delta G^{\circ}_{f,OH^{-}_{(aq)}} - \Delta G^{\circ}_{f,Mg(OH)_{2(s)}}$$

It should be noted that the concentration of $Mg(OH)_2$ does *not* appear in the denominator of equation 13–32. The reactant in this process is *solid* $Mg(OH)_2$ which does not exist as molecules of $Mg(OH)_2$ in solution; therefore no concentration units are applicable.

Since the solubility product constants (K_{sp}) are used frequently, tables of K_{sp} values have been constructed as in Appendix T, Table 15. Use of this table eliminates having to calculate K_{sp} from the calculated ΔG°.

SOLUBILITY CALCULATIONS WITH K_{sp}

The concentrations in equation 13–32 are of Mg^{2+} and OH^-, which are in equilibrium with solid $Mg(OH)_2$. Since the Mg^{2+} and OH^- do not react with the water, the concentrations of these ions will depend only on the solubility of $Mg(OH)_2$. Letting x represent the solubility of magnesium hydroxide, the ion concentrations are related to x by

$$[Mg^{2+}] = x$$

$$[OH^-] = 2x$$

The factor of 2 in the expression for $[OH^-]$ arises from the production of $2OH^-$ for each $Mg(OH)_2$ formula unit (equation 13–27). Substituting into equation 13–32

$$K_{sp} = (x)(2x)^2$$

$$K_{sp} = (x)(4x^2) = 4x^3$$

$$x^3 = \frac{K_{sp}}{4}$$

13-33
$$x = \left(\frac{K_{sp}}{4}\right)^{\frac{1}{3}}$$

For each ionic solid, the number of cations and anions produced may differ so that the mathematical relationship between solubility and K_{sp} may differ from equation 13-33.

Although we have shown how to calculate the solubility of solids from $\Delta G°$ and hence K_{sp} values, it should be pointed out that the $\Delta G_f°$ values for the solid and ions are determined by experiment. Frequently the solubility of a solid is experimentally measured and $\Delta G°$ then calculated. From $\Delta G°$ we can then calculate the $\Delta G_{f(aq)}°$ of the ions. The importance of determining the $\Delta G_{f(aq)}°$ for an ion has been emphasized before; it can be used to calculate $\Delta G°$ for *any* reaction involving that specific ion.

Finally, it should be remembered that the derivation of equations 13-31 and 13-32 assumed a low concentration of solute and these equations are applicable only for slightly soluble solutes. For more soluble solutes the activities and activity coefficients defined in equation 12-26 must be used.

EXAMPLE 13-7   *Calculate the solubility of copper(I) iodide at 25 °C using the $\Delta G_f°$ for $CuI_{(s)}$, Cu^+, and I^-.*

$$\Delta G_{f, CuI_{(s)}}° = -16.62 \ kcal/mole$$

$$\Delta G_{f, Cu_{(aq)}^+}° = 12.0 \ kcal/mole$$

$$\Delta G_{f, I_{(aq)}^-}° = -12.35 \ kcal/mole$$

The solution process involves the ionization

$$CuI_{(s)} \xrightleftharpoons{aq} Cu_{(aq)}^+ + I_{(aq)}^-$$

$$\Delta G° = \Delta G_{f, Cu_{(aq)}^+}° + \Delta G_{f, I_{(aq)}^-}° - \Delta G_{f, CuI_{(s)}}°$$

$$= (12.0 - 12.35) - (-16.62)$$

$$\Delta G° = 16.27 \ kcal$$

$$\Delta G° = -RT \ ln \ K_{sp}$$

$$16.27 \ kcal = -(1.987 \times 10^{-3} \ kcal/deg\text{-}mole)(298 \ deg)(2.303) \ log \ K_{sp}$$

$$log \ K_{sp} = -11.96 = -12 + 0.04$$

$$K_{sp} = 1.1 \times 10^{-12}$$

Let x = solubility of CuI. Therefore $[Cu^+] = x$ *and* $[I^-] = x$

$$K_{sp} = [Cu^+][I^-]$$

$$K_{sp} = (x)(x) = x^2 = 1.1 \times 10^{-12}$$

$$2 \log x = \log 1.1 \times 10^{-12} = 0.04 - 12 = -11.96$$

$$\log x = -5.98 = 0.02 - 6$$

$$x = 1.05 \times 10^{-6} \ mole/liter$$

If we wished to express the solubility in terms of grams/ml

$$solubility \ CuI = (1.05 \times 10^{-6} \ moles/liter)(MW)(10^{-3} \ liters/ml)$$

$$= (1.05 \times 10^{-6} \ moles/liter)(190.45 \ g/mole)(10^{-3} \ liters/ml)$$

$$= 1.99 \times 10^{-7} \ grams/ml$$

PROBLEM 13-23 The solubility of iron(II) hydroxide is 0.00015 g/100 ml at 18 °C.
a. Calculate K_{sp} and $\Delta G°$ at 18 °C for the equilibrium

$$Fe(OH)_{2(s)} \overset{aq}{\rightleftharpoons} Fe^{2+}_{(aq)} + 2OH^-_{(aq)}$$

b. Calculate the $\Delta G°_{f,Fe^{2+}}$ using $\Delta G°_{f,Fe(OH)_{2(s)}} = -115.6$ kcal/mole and $\Delta G°_{f,OH^-_{(aq)}} = -37.6$. Assume all values are at 18 °C.

PROBLEM 13-24 Calculate the $[Fe^{3+}]$ of an aqueous solution that is in contact with solid $Fe(OH)_3$ at 18 °C. K_{sp} for $Fe(OH)_{3(s)}$ at 18 °C is 1.1×10^{-36}.

The relationship between K_{sp} and ion concentrations, as shown in equation 13-32 for $Mg(OH)_2$, is more generally significant and useful than for the calculation of solubility. Equation 13-32, relating K_{sp} to the product of the ion concentrations $[Mg^{2+}]$ and $[OH^-]^2$, must be satisfied whenever there is solid $Mg(OH)_2$ in equilibrium with a solution containing Mg^{2+} and OH^- ions, irrespective of the source of the ions. For example, the OH^- ions may also come from other soluble or partially soluble salts that might have been added to the solution, such as $NaOH$. In the calculation of equation 13-33 the relationship $[OH^-] = 2[Mg^{2+}]$ was used. In general, this relationship between Mg^{2+} and OH^- does not exist unless the only source of ions is the solid $Mg(OH)_2$. Nevertheless, equation 13-32 remains valid regardless of the source of the Mg^{2+} and OH^-. It is this fact that makes the relationship between K_{sp} and ion concentrations so valuable.

EXAMPLE 13–8 *Calculate the solubility of $Mg(OH)_2$ if NaOH has been added to the solution to give a concentration of 0.05 M NaOH. Assume t = 18 °C. The K_{sp} for $Mg(OH)_{2(s)}$ at 18 °C is 1.2×10^{-11}.*

The solubility of $Mg(OH)_2$ in pure water (without any added NaOH) can be calculated from equation 13–32 and is found to be 1.44×10^{-4} moles/liter. The molar concentration of OH^- arising from the $Mg(OH)_{2(s)}$ is thus only 2.88×10^{-4}. Since this contribution to the concentration of OH^- is negligible compared to the 0.05 M from the added NaOH, we assume that the OH^- in solution arises only from the NaOH. Hence

$$[OH^-] = 0.05$$

The Mg^{2+} arises only from dissolving $Mg(OH)_2$. If we set

$$[Mg^{2+}] = x$$

then the K_{sp} expression becomes

$$K_{sp} = [Mg^{2+}][OH^-]^2$$

$$1.2 \times 10^{-11} = (x)(0.05)^2$$

$$x = 4.8 \times 10^{-9}$$

The solubility of $Mg(OH)_2$ has been greatly reduced by adding the common ion OH^- to the solution. This suppression of solubility by adding a common ion is similar to the common ion effect on acid dissociation that was illustrated in Example 13–5.

PROBLEM 13–25 Calculate the solubility of $Fe(OH)_2$ at 18 °C in a solution whose pH is buffered at 8.0. See the results of Problem 13–23 for the K_{sp}. Assume pK_W at 18 °C is 14.23. Compare the solubility at pH = 8 with the solubility of 0.00015 g/100 ml which was given in Problem 13–23.

SOLUBILITY OF METAL SALTS UNDERGOING HYDROLYSIS

Thus far we have considered the solubility of ionic solids where the dissociation produces essentially unreactive ions. In many cases this is not true and the ions react with water (hydrolysis) or with other ions in the solution. This is especially true for anions of weak acids which can react with either H_2O or $H^+_{(aq)}$ to give the weak acid. For example, $BaCO_{3(s)}$ dissociates to give CO_3^{2-} and Ba^{2+} ions, but the CO_3^{2-} reacts with water to give HCO_3^- and OH^-.

$$BaCO_{3(s)} \rightleftharpoons Ba^{2+} + CO_3^{2-}$$

$$CO_3^{2-} + H_2O \rightleftharpoons HCO_3^- + OH^-$$

The K_{sp} expression for $BaCO_{3(s)}$ is

$$K_{sp} = 8.1 \times 10^{-9} = [Ba^{2+}][CO_3^{2-}]$$

If the CO_3^{2-} did not react with water, the solubility of $BaCO_3$ would be calculated in the conventional manner

$$x = [Ba^{2+}] = [CO_3^{2-}]$$

$$8.1 \times 10^{-9} = x^2$$

$$x = 0.9 \times 10^{-4} \text{ mole/liter}$$

However, the CO_3^{2-} does react with water to give HCO_3^-; consequently the $BaCO_{3(s)}$ must continue to dissolve to satisfy the solubility relationship. When the solution is finally saturated, the concentration of Ba^{2+} will exceed the concentration of CO_3^{2-}. The solubility of $BaCO_3$ measured as $[Ba^{2+}]$ will exceed the solubility expected if no reaction with the water had occurred.

As a result of the reaction of CO_3^{2-} with water to give OH^-, the $[H^+]$ in the water must decrease since the equilibrium expression $K_w = [H^+][OH^-]$ must be satisfied. Consequently, a saturated solution of $BaCO_3$ will have a pH > 7. The pH of a saturated solution of $BaCO_3$ is 9.99 and the solubility is 1.52×10^{-4} moles/liter. The following concentrations of ions present in the saturated solution should give you an idea of the extent of hydrolysis.

$$[Ba^{2+}] = 1.52 \times 10^{-4} \text{ mole/liter}$$

$$[CO_3^{2-}] = 0.54 \times 10^{-4} \text{ mole/liter}$$

$$[HCO_3^-] = 0.98 \times 10^{-4} \text{ mole/liter}$$

As shown earlier, if no hydrolysis had occurred the $[Ba^{2+}] = [CO_3^{2-}]$ should have been 0.9×10^{-4} moles/liter, but the actual $[CO_3^{2-}]$ in the saturated solution was lowered due to the reaction with water to give HCO_3^-. The concentration of Ba^{2+} is necessarily increased in order to satisfy the K_{sp} relationship.

$$[Ba^{2+}][CO_3^{2-}] = (1.52 \times 10^{-4} \text{ moles/liter})(0.54 \times 10^{-4} \text{ moles/liter})$$

$$= 8.1 \times 10^{-9} = K_{sp}$$

PROBLEM 13–26 Copper(II) oxalate has a $K_{sp} = 2.87 \times 10^{-8}$.   a. Calculate the solubility of copper(II) oxalate assuming that the ions do not react with water.

b. The oxalate ion is an anion of a weak acid. Predict whether this will increase or decrease the solubility of the copper(II) oxalate compared to (a).

c. Predict whether the pH of the saturated solution in (b.) will increase or decrease compared to (a.).

The solubility of an ionic solid that contains the anion of a weak acid can also be affected by the pH of the solution. For example, if the pH of a $BaCO_3$ solution is low, the CO_3^{2-} will react extensively with the H^+ to form HCO_3^- and H_2CO_3. This reaction will keep occurring and more $BaCO_3$ will dissolve until the solubility product constant expression is satisfied. In fact the control of solubility by regulating the pH of a solution provides the basis for the selective precipitation of cations, as will be discussed in Section 13–J.

AMPHOTERISM

As defined in Chapter 3, an *amphoteric* compound is one which can act as both an acid and a base. For this reason dissociative equilibria both as an acid and a base must be considered in determining the solubility for these compounds. Most amphoteric compounds are metal hydroxides in the solid state which produce OH^- upon dissolving in solution. A K_{sp} is assigned to this equilibrium. The amphoteric metal hydroxide also produces H^+ by an acid dissociation from the solid. Some of the metal hydroxides also dissolve as molecular species, which in turn dissociate into OH^- and H^+. Since the concentration of the molecular species is small, we will neglect it in our discussion of equilibrium expressions and assume the solid metal hydroxide dissociates directly into ions in solution. Chapter 16, which discusses the amphoterism of a metal complex, assumes that the metal hydroxide is in solution as a molecular species.

$Pb(OH)_2$, which is amphoteric, can be used to illustrate the equilibria in which both $H^+_{(aq)}$ and $OH^-_{(aq)}$ are produced.

$$Pb^{+2}_{(aq)} + 2OH^-_{(aq)} \underset{aq}{\rightleftharpoons} Pb(OH)_{2(s)} \underset{aq}{\rightleftharpoons} H^+_{(aq)} + HPbO^-_{2(aq)}$$

The equilibrium expressions are

$$K_{sp} = [Pb^{+2}][OH^-]^2 = 4 \times 10^{-15}$$

$$K_a = [H^+][HPbO_2^-] = 2 \times 10^{-16}$$

In addition the relationship for the H^+ and OH^- equilibrium with water must be satisfied.

$$K_w = [H^+][OH^-]$$

The pH of the solution in equilibrium with the solid amphoteric hydroxide will depend upon the relative magnitudes of K_{sp} and K_a. For $Pb(OH)_2$, the $[OH^-]$ should be approximately equal to $(K_{sp})^{1/3} \approx 10^{-5}$ M and the $[H^+]$ should be approximately equal to $(K_a)^{1/2} \approx 10^{-8}$ M. Since $(K_{sp})^{1/3} > (K_a)^{1/2}$ for $Pb(OH)_2$, the OH^- concentration will exceed the H^+ concentration and the solution will be basic. These values of $[OH^-]$ and $[H^+]$ are not exact since the K_w expression was not considered. The exact calculation of pH is complicated since the three equilibrium expressions must be solved simultaneously.

* 13–J. METAL SULFIDE PRECIPITATION

In the previous section we stated that the solubility of an ionic solid is such that the solid will dissolve until the concentrations of the ions satisfies the expression for the solubility product constant. We can consider the reverse process of the precipitation of a solid to occur when the product of the ion concentrations in the K_{sp} expression *exceeds* the value of K_{sp}. For example, suppose CO_3^{2-} is present in a solution at a concentration of 10^{-5} M and we add Ba^{2+} ion to the solution. A precipitate will form when the Ba^{2+} concentration becomes such that

$$[Ba^{2+}][10^{-5} \text{ M}] > K_{sp} = 8.1 \times 10^{-9}$$

In solving for the Ba^{2+} concentration necessary to start precipitation we see that

$$[Ba^{2+}] > \frac{8.1 \times 10^{-9}}{10^{-5}} = 8.1 \times 10^{-4}$$

If a solution contains a mixture of two or more ions, such as CO_3^{2-} and $C_2O_4^{2-}$, addition of Ba^{2+} will cause precipitation of each ion as the product of concentrations exceeds the K_{sp} for that salt. For example, if the concentrations of CO_3^{2-} and $C_2O_4^{2-}$ are both 10^{-5} M, precipitation of $BaCO_3$ and $BaC_2O_4 \cdot 2H_2O$ will occur when

$$[Ba^{2+}][10^{-5}] > K_{sp, \text{ BaCO}_3} = 8.1 \times 10^{-9}$$

$$[Ba^{2+}][10^{-5}] > K_{sp, \text{ Ba C}_2\text{O}_4 \cdot 2 \text{ H}_2\text{O}} = 1.2 \times 10^{-7}$$

Solving for $[Ba^{2+}]$ we see that

when $\quad [Ba^{2+}] > \dfrac{8.1 \times 10^{-9}}{10^{-5}} = 8.1 \times 10^{-4}$, $BaCO_3$ will precipitate

when $\quad [Ba^{2+}] > \dfrac{1.2 \times 10^{-7}}{10^{-5}} = 1.2 \times 10^{-2}$, $BaC_2O_4 \cdot 2H_2O$ will precipitate

Since the Ba^{2+} concentration required to precipitate $BaCO_3$ is less than that required to precipitate $BaC_2O_4 \cdot 2H_2O$, $BaCO_3$ will precipitate first. As more Ba^{2+} is added, $BaCO_3$ will continue to precipitate and the CO_3^{2-} concentration will decrease. When the $[Ba^{2+}]$ reaches 1.2×10^{-2} M, $BaC_2O_4 \cdot 2H_2O$ will just begin to precipitate. At this point the $[CO_3^{2-}]$ will have been reduced to 0.067×10^{-5} M. Since the initial concentration of $[CO_3^{2-}]$ was 1×10^{-5} M, 93.3 % of the CO_3^{2-} has been precipitated as $BaCO_3$ by the time the $BaC_2O_4 \cdot 2H_2O$ begins to precipitate. If the $BaCO_3$ precipitate is removed by filtration we have carried out a separation of 93.3 % of the CO_3^{2-} from the mixture of CO_3^{2-} and $C_2O_4^{2-}$.

By this procedure of selective precipitation of ions we can effectively separate and identify metal cations. It is often convenient to carry out the precipitation as metal sulfides, since the S^{2-} concentration can be controlled by the pH of the solution. The $[S^{2-}]$ in a solution of H_2S can be varied from 10^{-22} to 0.07 M by going from a $[H^+]$ of 1.0 M to a highly basic solution. The $[H^+]$ can be controlled with a buffer solution or by addition of strong acid if the concentration of ions being precipitated is small. (If the concentration of ions is large, addition of strong acid will precipitate a large amount of S^{2-}, thereby changing $[HS^-]$, $[H_2S]$, and pH.)

The dissociation of H_2S (Section 13–D) occurs in two steps.

$$H_2S_{(aq)} \rightleftharpoons H^+_{(aq)} + HS^-_{(aq)} \qquad K_1 = \frac{[H^+][HS^-]}{[H_2S]} = 1.1 \times 10^{-7}$$

$$HS^-_{(aq)} \rightleftharpoons H^+_{(aq)} + S^{2-}_{(aq)} \qquad K_2 = \frac{[H^+][S^{2-}]}{[HS^-]} = 1 \times 10^{-14}$$

For a 0.1 M H_2S solution the $[S^{2-}]$ was shown $= K_2 = 1 \times 10^{-14}$. It was assumed for this calculation that the $[H^+]$ arose primarily from the first dissociation step. If H^+ is added to the solution, say as a strong acid, the equilibria in both steps would shift to the left and the $[HS^-]$ and $[S^{2-}]$ would decrease. In order to calculate the $[S^{2-}]$ in an acidic solution the $[HS^-]$ can be eliminated from the two equilibrium expressions by taking the product

$$K_1 K_2 = \frac{[H^+][HS^-]}{[H_2S]} \frac{[H^+][S^{2-}]}{[HS^-]} = \frac{[H^+]^2[S^{2-}]}{[H_2S]} = 1.1 \times 10^{-21}$$

This combined equilibrium expression corresponds to the equilibrium

$$H_2S \rightleftharpoons 2H^+_{(aq)} + S^{2-}_{(aq)}$$

which is the sum of the above two successive ionization steps. Solving for $[S^{2-}]$

$$[S^{2-}] = 1.1 \times 10^{-21} \frac{[H_2S]}{[H^+]^2}$$

TABLE 13-2 Sulfide concentrations for 0.1 M H_2S in acidic solutions of varying pH (given in molar concentrations)

[HCl] ADDED	[H$^+$]	pH	[S^{2-}]
1.0	1.0	0	1.1×10^{-22}
0.10	0.10	1	1.1×10^{-20}
0.01	0.01	2	1.1×10^{-18}
0.001	0.001	3	1.1×10^{-16}
0.0	0.0001	4	1.1×10^{-14}

The $[S^{2-}]$ can now be calculated for solutions of different $[H^+]$ and $[H_2S]$. These are shown in Table 13-2 for $[H_2S] = 0.1$ M.

The H_2S solutions can also be made basic, thereby increasing the $[S^{2-}]$. If a strong base such as NaOH is used, addition of sufficient base can convert the H_2S to Na_2S. For a 0.1 M Na_2S solution, the $[S^{2-}] = 0.07$ M. It can be seen then that the pH of the solution can be used to regulate the $[S^{2-}]$ from 10^{-20} M to 0.07 M, a range of about 10^{21}.

With the $[S^{2-}]$ specified by the pH of the solution, the calculation of the metal ion concentration required to cause precipitation is obtained directly from the K_{sp} relationship. As an example, consider the precipitation of Pb^{2+} as PbS. The K_{sp} expression is

$$K_{sp} = [Pb^{2+}][S^{2-}] = 4 \times 10^{-26}$$

In order for precipitation to occur, the product of concentrations must exceed the K_{sp} value:

$$[Pb^{2+}][S^{2-}] > 4 \times 10^{-26}$$

Hence the $[Pb^{2+}]$ required for precipitation from a solution of H_2S at a specified $[S^{2-}]$ is

$$[Pb^{2+}] > \frac{4 \times 10^{-26}}{[S^{2-}]}$$

If the solution contains 1.0 M HCl and 0.1 M H_2S (see Table 13-2) the $[Pb^{2+}]$ for precipitation must be

$$[Pb^{2+}] > \frac{4 \times 10^{-26}}{1 \times 10^{-22}} = 4 \times 10^{-4} \text{ M}$$

For Bi_2S_3 the K_{sp} equation is

$$K_{sp} = [Bi^{3+}]^2[S^{2-}]^3 = 1 \times 10^{-70}$$

In order for precipitation to occur, the $[Bi^{3+}]$ must be

$$[Bi^{3+}] > \left(\frac{1 \times 10^{-70}}{[S^{2-}]^3}\right)^{1/2}$$

If the solution contains 1.0 M HCl and 0.1 M H_2S then $[S^{2-}] = 1 \times 10^{-22}$ M (Table 13–2), and a precipitate will form if

$$[Bi^{2+}] > \left(\frac{1 \times 10^{-70}}{1 \times 10^{-66}}\right)^{1/2} = (1 \times 10^{-4})^{1/2} = 1 \times 10^{-2} \text{ M}$$

PROBLEM 13–27 The K_{sp} for NiS is 1×10^{-22}. What concentration of Ni^{2+} must be present in order to form a NiS precipitate in a solution containing 0.1 M HCl and 0.1 M H_2S?

PROBLEM 13–28 What is the solubility of $Tl_2S_{(s)}$ in a 0.1 M H_2S solution? $K_{sp} = 1 \times 10^{-22}$.

PROBLEM 13–29 How could you separate a mixture of Fe^{2+}, Tl^+, and Pb^{2+} ions (each at a concentration of 10^{-2} M) using the sulfide precipitates? The K_{sp} values are $FeS, \cdot 4 \times 10^{-17}$; $PbS, 4 \times 10^{-26}$; $Tl_2S, 1 \times 10^{-22}$.

13–K. GAS PHASE EQUILIBRIA

The equilibrium constant expression and the relationship to the change in standard free energy for a reaction of ideal gases was given in Section 13–A.

13–7
$$\Delta G° = - RT \ln K_P$$

Since the standard state for ideal gases is one atmosphere, the partial pressures of the reactants and products are necessarily expressed in atmospheres. For example, for the decomposition of N_2O_4:

$$N_2O_{4(g)} \rightleftharpoons 2NO_2$$

$$K_P = \frac{(P_{NO_2})^2}{P_{N_2O_4}}$$

Using the standard free energies given in Appendix T, Table 3

$$\Delta G° = 2\Delta G°_{f,NO_{2(g)}} - \Delta G°_{f,N_2O_{4(g)}}$$

$$\Delta G° = 2(12.39) - 23.49 = 1.29 \text{ kcal}$$

The equilibrium constant, K_P, is found from equation 13–7 to be 0.113 at 298 °K. In order to carry out the equilibrium calculations, we must specify whether the reaction will be carried out at constant volume or constant pressure. For reactions in solution this is not a problem, since the volume of the liquid solution remains approximately constant and the pressure is held constant as well. In the gas phase this is not true.

Generally gas reactions are carried out in some closed vessel, which means that the volume is necessarily held constant. At constant volume, the pressure of the equilibrium mixture will differ from the initial pressure if there is a *change in the number of moles* for the reaction. In the reaction N_2O_4 there are two moles of NO_2 produced for the decomposition of one mole of N_2O_4, an increase in the number of moles ($\Delta n = 1$). When the pressure changes during the course of the reaction, it is awkward to use the K_P expression in the equilibrium calculation. An equilibrium constant expressed in terms of molar concentrations (K) is more convenient for the reaction at constant volume. A simple relationship between K_P and K can be derived using the ideal gas law for each component at partial pressure, P_i.

13–34

$$P_i = \frac{n_i}{V} RT = [i]RT$$

where the concentration of species *i* is now in moles/liter. Substituting P_i for each partial pressure into K_P and then using equation 13–34 one obtains

13–35

$$K = K_P(RT)^{-\Delta n}$$

where Δn is the change in number of moles for the gas reaction. For the N_2O_4 reaction $\Delta n = +1$, and

$$K = 0.113(0.08205 \times 298)^{-1}$$

$$K = 0.00460 = \frac{[NO_2]^2}{[N_2O_4]}$$

With the equilibrium constant K expressed in molar concentrations, the procedure for calculating the equilibrium concentrations is identical to that for liquid solutions. Let *a* represent the initial concentration of N_2O_4 and *x* the concentration of NO_2 produced. Since two moles of NO_2 are produced for every mole of N_2O_4 decomposed, the amount of N_2O_4 at equilibrium is the initial concentration of *a* minus *x*/2. Summarizing the changes in concentration:

	INITIAL CONDITION	EQUILIBRIUM
$[N_2O_4]$	a	$\left(a - \frac{x}{2}\right)$
$[NO_2]$	0	x

The equilibrium expression is

$$0.00460 = \frac{x^2}{\left(a - \dfrac{x}{2}\right)}$$

This equation generally must be solved using the quadratic formula. The approximation of $(a - x) \approx a$ which was made for the dissociation of weak acids and weak bases can seldom be made in gas reactions unless K is exceptionally small. This complication is frequently encountered in calculating equilibrium concentrations for gas phase reactions. The problem is even more severe when the equilibrium expression is a cubic or quadratic equation in x. For the gas phase reactions that will be considered in this text, only quadratic equations need be solved.

PROBLEM 13–30 Calculate the equilibrium concentrations for the reaction

$$CO_{(g)} + Cl_{2(g)} \rightleftharpoons COCl_{2(g)}$$

Assume the volume is constant. The initial concentration of CO is 0.1 M and of Cl_2 0.2 M; $T = 298\ °K$. [Hint: Since the equilibrium constant is large, it is difficult to calculate $[COCl_2]$ directly. Assume alternatively that the reaction has gone to completion to give 0.1 M $COCl_2$ and then calculate the extent of the back reaction to equilibrium. Also, remember that $\Delta G° = - RT \ln K_P$.]

PROBLEM 13–31 Set up the equilibrium expression for the gas phase reaction

$$N_{2(g)} + 3H_{2(g)} \rightleftharpoons 2NH_{3(g)}$$

Assume the initial concentrations of N_2 and H_2 are 0.1 M. Let the equilibrium concentration of NH_3 be x. What is the order of the resulting equation with respect to the variable x?

13–L. TEMPERATURE DEPENDENCE OF CHEMICAL EQUILIBRIA

The temperature and pressure dependence of chemical equilibria can be understood from Le Chatelier's principle (Section 11–E). Recall that a system in equilibrium will be shifted to the right (toward products) by an increase in temperature if heat is absorbed in the process. Conversely the system in equilibrium will be shifted to the left (toward reactants) by an increase in temperature if heat is liberated in the process. For a chemical equilibrium the heat change is given by ΔH if the reaction is carried out at constant pressure. This is summarized for the general reaction:

$$aA + bB \rightleftharpoons cC + dD$$

$$\Delta H = c\Delta H_{f,C} + d\Delta H_{f,D} - a\Delta H_{f,A} - b\Delta H_{f,B}$$

$$\Delta H > 0; \text{ increase } T, \text{ shift to right}$$

$$\Delta H < 0; \text{ increase } T, \text{ shift to left}$$

This temperature-dependent shift in equilibrium arises from a change in the equilibrium constant, K. A shift in equilibrium to the right corresponds to an increase in K and a shift to the left results from a decrease in K. The equation which can be used to calculate the equilibrium constant, K_2, at some new temperature, T_2, is

13–36
$$\log \frac{K_2}{K_1} = \frac{\Delta H^\circ}{2.303 \, R} \left(\frac{1}{T_1} - \frac{1}{T_2} \right)$$

where K_1 is the equilibrium constant known for some temperature, T_1.

EXAMPLE 13–9   *Consider the acid dissociation of HCN which was discussed in Example 13–1.*

$$HCN_{(aq)} \rightleftharpoons H^+_{(aq)} + CN^-_{(aq)}$$

a. Will an increase in temperature from 25 °C to 45 °C shift the equilibrium to the right or left?
b. Calculate the equilibrium constant for this reaction at t = 45 °C.

a. In order to determine the change in equilibrium upon increasing the temperature we must calculate the standard change in enthalpy (25 °C) for the reaction

$$\Delta H^\circ = \Delta H^\circ_{f, H^+_{(aq)}} + \Delta H^\circ_{f, CN^-_{(aq)}} - \Delta H^\circ_{f, HCN_{(aq)}}$$

Using the enthalpies of formation from Appendix T, Tables 3 and 4

$$\Delta H^\circ = 0 + 36.1 - 25.2 = 10.9 \text{ kcal}$$

Since ΔH° is positive the equilibrium will be shifted to the right when the temperature is increased to 45 °C. In other words, HCN will be a stronger acid at 45 °C.

b. The equilibrium constant at 45 °C is calculated from equation 13–36

$$\log \frac{K_2}{K_1} = \frac{\Delta H^\circ}{2.303 \, R} \left(\frac{1}{T_1} - \frac{1}{T_2} \right)$$

$$T_1 = 298 \text{ °K} \qquad T_2 = 318 \text{ °K}$$

$$K_1 = 4.9 \times 10^{-10} \qquad K_2 = ?$$

$$\log \left(\frac{K_2}{4.9 \times 10^{-10}} \right) = \frac{10.9 \text{ kcal}}{(2.303)(1.987 \times 10^{-3} \text{ kcal/deg-mole})} \times \left(\frac{1}{298 \text{ deg}} - \frac{1}{318 \text{ deg}} \right)$$

$$log\ K_2 + 9.31 = (2.38 \times 10^3)(3.36 \times 10^{-3} - 3.15 \times 10^{-3})$$

$$log\ K_2 = -9.31 + .50 = -8.81$$

$$K_2 = 1.5 \times 10^{-9}$$

Note that K has increased from 4.9×10^{-10} to 1.5×10^{-9} in going from 25 °C to 45 °C.

PROBLEM 13–32   Consider the base dissociation of an aqueous ammonia solution (see Problem 13–1).

$$NH_{3(aq)} + H_2O \rightleftharpoons NH_{4(aq)}^+ + OH_{(aq)}^-$$

a. Predict from Le Chatelier's principle whether $NH_{3(aq)}$ will be a stronger base at 40 °C than at 25 °C.
b. Calculate the base dissociation constant for $NH_{3(aq)}$ at 40 °C.

PROBLEM 13–33   The solubility of CuI at 25 °C was calculated in Example 13–7. Calculate the solubility of CuI at 45 °C.

13–M. PRESSURE DEPENDENCE OF CHEMICAL EQUILIBRIA

The qualitative effect that a pressure change has on chemical equilibria can also be predicted directly from Le Chatelier's principle. Recall from Chapter 11 that for a system at equilibrium for which there is a decrease in volume [$\Delta V < 0$] in going from left to right (reactants to products), an increase in pressure will cause the equilibrium to shift to the right (toward products). Conversely, if there is an increase in volume, an increase in pressure will shift the equilibrium to the left (toward reactants). For chemical reactions which are carried out in solution, the ΔV for the reaction is small. Consequently, the effect of pressure on the chemical equilibrium is negligible unless extremely high pressures, on the order of thousands of atmospheres, are used. On the other hand, for chemical reactions involving gases, ΔV can be very large and the effect of pressure is significant. For this reason we will only consider the effect of pressure on reactions involving gases.

We will again use our general reaction

$$aA + bB \rightleftharpoons cC + dD$$

The change in volume for the reaction is given by

$$\Delta V = V_{products} - V_{reactants}$$

where the volumes are all measured at the same pressure. If we assume that the gaseous reactants and products obey the ideal gas equation, the volume of each reactant and product is proportional to the number of moles of each species

$$V_i = \frac{RT}{P} n_i$$

Consequently, ΔV for the gaseous reaction is proportional to the change in number of moles, Δn, for the reaction

$$\Delta n = (c + d) - (a + b)$$

$$\begin{matrix} \text{moles} & \text{moles} \\ \text{products} & \text{reactants} \end{matrix}$$

The effect which pressure has on the equilibrium of a gaseous reaction can be predicted simply by examining Δn for the reaction. A decrease in the number of moles, $\Delta n < 0$, means that there is a corresponding decrease in volume, $\Delta V < 0$ and the equilibrium will shift to the right (towards products) in order to relieve an increase in pressure. Of course, the converse is true if there is a decrease in pressure or if $\Delta n > 0$ for the reaction.

The shift in equilibrium caused by a change in pressure for a gas phase reaction does not arise from a change in either equilibrium constant, K_P or K. You will recall that the equilibrium constant expressed in terms of partial pressures is designated by K_P and the equilibrium constant expressed in terms of molar concentrations by K. These equilibrium constants are related by equation 13–35

$$K = K_P(RT)^{-\Delta n}$$

where Δn = change in number of moles of gaseous substances in the reaction. K_P is independent of the pressure applied to the equilibrium mixture, since at a given temperature K_P depends only on $\Delta G°$ (and $\Delta G°$ is defined for products and reactants at the standard pressure of 1 atm). Since K is related directly to K_P through equation 13–35, K is also independent of pressure. We might then ask how can the equilibrium mixture be affected by pressure if the equilibrium constant is not changed? The answer to this question will become apparent from the following discussion.

In order to calculate quantitatively the effect which pressure has on chemical equilibrium we must investigate the change in the equilibrium constant expression. Let us use for illustration

$$N_2O_4 \rightleftharpoons 2NO_2$$

The expression for the equilibrium constant for this reaction is

$$K = 0.00460 = \frac{[NO_2]^2}{[N_2O_4]}$$

$$0.00460 = \frac{x^2}{\left(a - \dfrac{x}{2}\right)}$$

where a = initial $[N_2O_4]$ and $x = [NO_2]$ at equilibrium. The unknown x must be solved by the quadratic formula. It is the value of a which will change with pressure, thus causing x and hence $[NO_2]$ and $[N_2O_4]$ to change. In other words, it is the effect of pressure on the initial concentration which changes what final equilibrium mixture will be obtained. For example, if the initial concentration of N_2O_4 were $a = 0.1$ mole/liter, the calculation of x gives

$$x = [NO_2] = 0.0193 \text{ mole/liter}$$

$$\left(a - \frac{x}{2}\right) = [N_2O_4] = 0.0807 \text{ mole/liter}$$

In other words, 19.3 % of the original N_2O_4 has dissociated. If the pressure is increased by decreasing the volume by a factor of 10, the initial concentration of N_2O_4, a, will be 1.0 mole/liter. Similar calculation of x gives at equilibrium

$$x = [NO_2] = 0.066 \text{ moles/liter}$$

$$\left(a - \frac{x}{2}\right) = [N_2O_4] = 0.943 \text{ moles/liter}$$

In other words, 6.6 % of the original N_2O_4 has now dissociated. Note that increasing the pressure by a factor of 10 (by decreasing the volume) decreases the percentage yield from 19.3 to 6.6.

PROBLEM 13–34   In Problem 13–30 the equilibrium concentrations were calculated for the reaction

$$CO_{(g)} + Cl_{2(g)} = COCl_{2(g)}$$

Predict on the basis of Le Chatelier's principle what qualitative effect an increase in pressure would have on this reaction.

EXERCISES

1. Write equilibrium expressions for the following reactions
   a. $2SO_{2(g)} + O_{2(g)} \rightleftharpoons 2SO_{3(g)}$
   b. $3H_{2(g)} + N_{2(g)} \rightleftharpoons 2NH_{3(g)}$
   c. $\frac{1}{2}H_{2(g)} + \frac{1}{2}I_{2(g)} \rightleftharpoons HI_{(g)}$
   d. $H_{2(g)} + C_2H_{4(g)} \rightleftharpoons C_2H_{6(g)}$
   e. $CaC_{2(s)} + H_2O_{(l)} \rightleftharpoons C_2H_{2(g)} + Ca(OH)_{2(aq)}$

f. $3NO_{2(g)} + H_2O_{(l)} \rightleftharpoons 2HNO_{3(aq)} + NO_{(g)}$

g. $Fe^{3+}_{(aq)} + V^{3+}_{(aq)} + H_2O \rightleftharpoons Fe^{2+}_{(aq)} + VO^{2+}_{(aq)} + 2H^+_{(aq)}$

h. $2Ce^{4+}_{(aq)} + HAsO_{2(aq)} + 2H_2O \rightleftharpoons 2Ce^{3+}_{(aq)} + H_3AsO_{4(aq)} + 2H^+_{(aq)}$

2. The hydrolysis of glucose-6-phosphate is shown in the following equation:

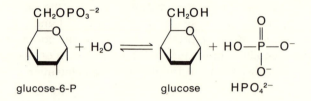

glucose-6-P   glucose HPO_4^{2-}

The hydrolysis is 99.95 % complete in a 0.1 M glucose-6-phosphate aqueous solution at 298 °K. Calculate K and $\Delta G°$ for this hydrolysis reaction. Assume the concentration of H_2O remains constant. The vertical lines in the formulas represent OH groups.

3. At 2000 °K the equilibrium constant (expressed in terms of molar concentrations) for the reaction

$$H_{2(g)} + CO_{2(g)} \rightleftharpoons H_2O_{(g)} + CO_{(g)}$$

is 4.40. If 0.15 moles of H_2 and CO_2 are placed initially in a 2 liter container at 2000 °K, what is the final concentration of each species?

4. A 10 liter container at 600 °K contains at equilibrium 2.5×10^{-3} moles of ammonia, 0.1 moles of N_2, and 0.1 moles of H_2. Using these equilibrium data, calculate the equilibrium constants (K and K_p) for the reaction

$$3H_{2(g)} + N_{2(g)} \rightleftharpoons 2NH_{3(g)}$$

5. An alcohol (ROH) and a carboxylic acid (R'COOH) react to give an ester (R'COOR) and water. At equilibrium the amounts of the various components in 20 ml of an acetone solution are

ROH	+	R'COOH	$\rightleftharpoons$	R'COOR	+ H_2O
5×10^{-3} moles		5×10^{-3} moles		3×10^{-2} moles	3×10^{-2} moles

a. Calculate the equilibrium constant (expressed in terms of molar concentrations).

b. If 3 ml of H_2O is added to the equilibrium mixture, calculate the concentration of each of the components when equilibrium is re-established. Assume the density of H_2O is 1 g/ml and the total volume is now 23 ml.

6. In Exercise 21 at the end of Chapter 10 the synthesis of sucrose from glucose and fructose was considered

a. Calculate and compare the equilibrium constants for the uncoupled and coupled reaction at 298 °K.

	$\Delta G°$
glucose + fructose $\longrightarrow$ sucrose + H_2O	5 kcal
glucose + fructose + ATP $\longrightarrow$ sucrose + ADP + HPO_4^{2-}	−7 kcal

b. If the initial concentrations of glucose and fructose are 10^{-2} M, what is the concentration of sucrose at equilibrium in the uncoupled reaction? Neglect the [H_2O] in the equilibrium expression.

c. If the initial concentrations of glucose and fructose are 10^{-2} M and of ATP 1 M, calculate what the equilibrium concentration of sucrose is in the coupled reaction. [Hint: Assume the reaction has gone to completion and calculate the amount of back reaction to glucose and fructose.]

7. Find the pH of the following solutions at 25 °C in water (X stands for halide).
   a. 0.1 M HNO_3 d. 0.01 M HX, $K_a = 1 \times 10^{-6}$
   b. 0.01 M KOH   e. 0.01 M HX + 0.1 M RbX
   c. 0.001 M HCl   f. 0.1 M XOH, 0.01 % ionized

8. 22 kcal are released in the exothermic reaction

$$3H_{2(g)} + N_{2(g)} \rightleftharpoons 2NH_{3(g)}$$

What affect will the following have on the equilibrium system:
   a. increase in temperature   c. addition of N_2
   b. increase in pressure   d. increase in time

9. At 90 °C pure water has a [H_3O^+] = 1×10^{-6} M. What is the value of K_W at 90 °C?

10. A volume of 0.51 ml CO_2 at 0 °C and 760 mm Hg pressure will dissolve in 1 ml of blood at 38 °C. Dissolved CO_2 is in equilibrium with ionic species according to the following two reactions

$$CO_{2(aq)} + H_2O \rightleftharpoons H^+_{(aq)} + HCO^-_{3(aq)} \qquad K_1 = \frac{[H^+][HCO_3^-]}{[CO_2]} = 4.4 \times 10^{-7}$$

$$HCO^-_{3(aq)} \rightleftharpoons H^+_{(aq)} + CO^{2-}_{3(aq)} \qquad K_2 = \frac{[H^+][CO_3^{2-}]}{[HCO_3^-]} = 6 \times 10^{-11}$$

a. The pH of blood is 7.2. Calculate the [HCO_3^-] using the first equilibrium step and the [CO_3^{2-}] using the second equilibrium step.

b. Using the first equilibrium step to calculate [HCO_3^-] neglects the loss of HCO_3^- in the second dissociation step. Use your calculations in (a) to justify doing this.

11. Glycine, NH_2-CH_2-COOH, is an amino acid whose hydrochloride salt $Cl^{-+}NH_3-CH_2-COOH$ can be considered a diprotic acid. The acid dissociation steps are

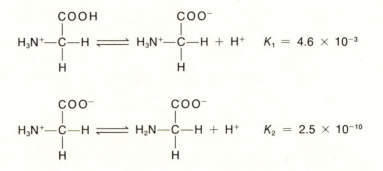

How could glycine be used as a buffer in the region of $pH = 2-3$ and again in another region, $pH = 9-10$? You may use HCl to form the hydrochloride salt or $NaOH$ to neutralize any portion of the glycine.

12. a. If 0.01 mole of $Mg(OH)_2$ is added to 1 liter of water, show that not all of the $Mg(OH)_2$ will dissolve. $K_{sp, Mg(OH)_2} = 8.9 \times 10^{-12}$.
   b. If the solution is buffered at $pH = 5$, will the 0.01 mole of $Mg(OH)_2$ then dissolve?

13. What is the concentration of all ionic species in an aqueous solution saturated with carbon dioxide at 0.1 atm pressure and 25 °C. The acid dissociation constants for H_2CO_3 are given in Exercise 10. See Table 12–4 (Section 12–I) for the solubility of CO_2 in water.

14. The solubility of $Ag_2CO_3(K_{sp} = 8.0 \times 10^{-12})$ would be greatest in one liter of
   a. 0.05 M Na_2CO_3 d. 0.05 M NH_3
   b. 0.05 M $AgNO_3$ e. 0.10 M K_2CO_3
   c. pure water
   (The silver ion, Ag^+, reacts with NH_3 to form the $Ag(NH_3)_2^+$ complex.)

$$Ag^+ + 2NH_3 \rightleftharpoons Ag(NH_3)_2^+ \quad K = 1.66 \times 10^7$$

15. a. Calculate the molar solubility of Ag_3PO_4 assuming that PO_4^{3-} does not hydrolyze.
   b. Give the reaction for the hydrolysis of PO_4^{3-}. What effect will these reactions have on the solubility of Ag_3PO_4 calculated in part a.?
   c. What effect should the addition of acid (other than H_3PO_4) have upon the solubility of Ag_3PO_4? (K_{sp} for Ag_3PO_4 is 1.8×10^{-18}.)

16. The equilibrium constant for the reaction

$$FeO_{(s)} + H_{2(g)} \rightleftharpoons Fe_{(s)} + H_2O_{(g)}$$

has been measured at 700 °C and found to be 0.422. When the temperature was increased to 900 °C the equilibrium constant increased to 0.594. Calculate $\Delta H°$ for the reaction from these data.

17. The equilibrium constant for the reaction

$$FeO_{(s)} + CO_{(g)} \rightleftharpoons Fe_{(s)} + CO_{2(g)}$$

has been found to be 0.678 at 700 °C. Using this information along with the data from the previous problem, calculate the equilibrium constant for the water–gas reaction at 700 °C.

$$H_{2(g)} + CO_{2(g)} \rightleftharpoons H_2O_{(g)} + CO_{(g)}$$

18. Acetic acid, CH_3COOH, can form a dimer, $(CH_3COOH)_2$, in the gas phase

$$2CH_3COOH \rightleftharpoons (CH_3COOH)_2$$

The dimer is held together through two hydrogen bonds (Chapter 8)

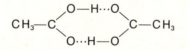

The strength of the bonds holding the dimer together is 15.9 kcal/mole. The equilibrium constant, K_P, for the reaction at 25 °C has been determined to be 1.3×10^3 where the partial pressures in the equilibrium expression are given in atmospheres. Assume acetic acid is initially at a pressure of 10 torr with no dimer present. Calculate the percentage of acetic acid which converts to the dimer in the equilibrium mixture at (a) 25 °C and (b) 50 °C.

19. Calculate the pH of a solution containing aniline, $C_6H_5NH_2$, at a concentration of 0.1 M and $HClO_4$ at 0.04 M. Aniline is a weak base and $HClO_4$ is a strong acid.

20. Suppose we have a solution of Zn^{2+}, Ca^{2+}, Cu^{2+}, Al^{3+} at a pH of 1.0. Describe a procedure for the separation of these ions using only $NaOH$, HCl, and NH_4OH. The K_{sp} for the hydroxides of these ions can be found in Appendix T, Table 15. The ions Zn^{2+} and Cu^{2+} also form complexes with ammonia. Assume all ions are present initially at a concentration 0.1 M. Recall that $Zn(OH)_2$ and $Al(OH)_3$ are amphoteric and will dissolve in strong base. Al^{3+} and Ca^{2+} do not form NH_3 complexes.

$$Zn^{2+} + 4NH_3 \rightleftharpoons Zn(NH_3)_4{}^{2+} \qquad K = 1.1 \times 10^9$$

$$Cu^{2+} + 4NH_3 \rightleftharpoons Cu(NH_3)_4{}^{2+} \qquad K = 1.5 \times 10^{12}$$

21. Describe how a buffer solution with a pH = 9.5 could be made.

22. Would an aqueous solution of NaF be acidic, basic, or neutral? Explain.

23. The following heats of neutralization have been measured experimentally at 25 °C.

$\Delta H°$

1. $HCl_{(aq)} + NaOH_{(aq)} \longrightarrow NaCl_{(aq)} + H_2O$ −13.36 kcal
2. $HI_{(aq)} + NaOH_{(aq)} \longrightarrow NaI_{(aq)} + H_2O$ −13.36 kcal
3. $HNO_{3(aq)} + NaOH_{(aq)} \longrightarrow NaNO_{3(aq)} + H_2O$ −13.36 kcal
4. $HCl_{(aq)} + KOH_{(aq)} \longrightarrow KCl_{(aq)} + H_2O$ −13.36 kcal
5. $HCl_{(aq)} + NH_{3(aq)} \longrightarrow NH_4Cl_{(aq)}$ −12.42 kcal
6. $HCN_{(aq)} + NaOH_{(aq)} \longrightarrow NaCN_{(aq)} + H_2O_{(aq)}$ − 2.36 kcal

a. Account for the fact that the first four heats of neutralization have the same value.

b. Explain why the neutralizations of NH_3 and HCN in reactions 5. and 6. are less exothermic than the first three neutralization reactions above.

24. The hydrochloride of glycine undergoes the following acid dissociations

Sketch out a titration curve that you would expect in titrating 25 ml of 0.1 M glycine hydrochloride with 0.1 M NaOH.

25. An essential reaction in the production of smog is

$$O_3 + NO \rightleftharpoons O_2 + NO_2$$

a. Calculate the K_P for this reaction at 25 °C using thermodynamic data from Appendix T.

b. If the initial concentrations are $[O_3] = 1 \times 10^{-8}$ moles/liter and $[NO] = 1 \times 10^{-5}$ moles/liter with $[O_2] = 8.2 \times 10^{-3}$ moles/liter (equivalent to 0.2 atm), what are the approximate concentrations of O_2 and NO_2 at equilibrium?

c. Will the concentrations of O_2 and NO_2 increase or decrease if the temperature is increased to 35 °C?

26. Hardness in water is caused by the presence of Ca^{2+} and Mg^{2+} ions which react with soaps (see Section 12–G), such as sodium stearate, $NaC_{18}H_{35}O_2$, to form an insoluble precipitate. The reaction of Ca^{2+} with the stearate ion of sodium stearate gives

$$Ca^{2+}_{(aq)} + 2C_{18}H_{35}O^-_{2(aq)} \longrightarrow Ca(C_{18}H_{35}O_2)_{2(s)}$$

In order to soften water, the ions Ca^{2+} and Mg^{2+} must be removed from the solution and this can be accomplished in different ways.

a. If HCO_3^- is also present in the water (temporary hardness), or added to the solution, the Ca^{2+} and Mg^{2+} ions can be removed by boiling the water and thereby causing the following reaction to occur:

$$Ca^{2+}_{(aq)} + 2HCO^-_{3(aq)} \rightleftharpoons CaCO_{3(s)} + H_2O + CO_{2(g)}$$

The precipitation of $CaCO_3$ removes the Ca^{2+} from solution. The reaction is reversible but it can be made to go essentially to completion by boiling water exposed to the atmosphere. Explain how this can occur.

b. Temporary hardness can also be removed by adding lime water, $Ca(OH)_2$. The lime water produces Ca^{2+} and OH^- ions in solution

$$Ca(OH)_2 \longrightarrow Ca^{2+} + 2OH^-$$

which obviously increases the Ca^{2+} concentration. Explain then how lime water can be effective in removing Ca^{2+} or Mg^{2+} from water containing HCO_3^-.

c. A third way to remove Ca^{2+} and Mg^{2+} from solution is by the addition of Na_2CO_3 to form the precipitates of $CaCO_3$ and $MgCO_3$ directly. If your water contained Ca^{2+} at a concentration of 0.01 M, what percentage of the Ca^{2+} would be removed by adding enough Na_2CO_3 to make the solution 0.05 M in CO_3^{2-} before reaction with Ca^{2+}? [Hint: Let x represent the concentration of Ca^{2+} which reacts to give $CaCO_{3(s)}$. Then the $[Ca^{2+}] = (0.01 - x)$ and $[CO_3^{2-}] = (0.05 - x)$. You may neglect the hydrolysis of CO_3^{2-} and assume that all excess CO_3^{2-} remains as the free carbonate.]

27. The dissolution of a non-ionic solid can be thought to occur in two more fundamental steps as shown in the following Born-Haber cycle

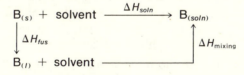

a. Show that the heat of solution for a solid, ΔH_{soln}, can be expressed as:

$$\Delta H_{soln} = \Delta H_{fus} + \Delta H_{mixing}$$

b. Considering the various types of intermolecular interactions, list various factors which might influence the ΔH_{fus} and ΔH_{mixing} and consequently would influence the value of ΔH_{soln}.

c. Considering the signs of ΔH_{fus} and ΔH_{mixing}, what possible signs can the ΔH_{soln} have?

d. The ΔH_{soln} is usually positive. How would you account for this? [Hint: If the substance formed an ideal solution with the solvent, what is the ΔH_{mixing}?]

e. Do you expect the solubility of most non-ionic solids to increase or decrease with increasing temperature? Why?

28. The dissociation of an ionic solid in water can be thought to occur in several fundamental steps as shown in the following Born-Haber cycle (*sub* means sublimation).

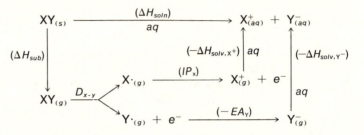

a. Show that the heat of solution for an ionic solid, ΔH_{soln}, can be expressed as

$$\Delta H_{soln} = \Delta H_{sub} + D_{X-Y} + IP_X - EA_Y - \Delta H_{solv,X^+} - \Delta H_{solv,Y^-}$$

b. Considering the various types of interactions which ions can have with other ions or molecules, what several factors can influence the ΔH_{soln} for an ionic solid?

c. Calculate the ΔH_{sub} for $LiCl_{(s)}$ and $KCl_{(s)}$ using

$$\Delta H^{\circ}_{f,LiCl_{(s)}} = -97.70 \text{ kcal/mole}$$
$$\Delta H^{\circ}_{f,LiCl_{(g)}} = -53.00 \text{ kcal/mole}$$
$$\Delta H^{\circ}_{f,KCl_{(s)}} = -104.18 \text{ kcal/mole}$$
$$\Delta H^{\circ}_{f,KCl_{(g)}} = -51.6 \text{ kcal/mole}$$

d. Calculate ΔH° of solution for $LiCl_{(s)}$ and $KCl_{(s)}$ using the ΔH_{sub} calculated in a. Refer to Chapter 5 for *IP* and *EA* and Chapter 8 for ΔH_{solv}. The bond dissociation energies are given in Appendix T, Table 6.

e. Based only on the results of part d., do you expect the solubility of $LiCl$ or KCl to be greatest? Which compound will show the greatest increase in solubility with temperature? Why?

29. In order to understand why some substances are weak acids whereas other substances are strong acids, we can express the ΔH for acid dissociation in terms of several fundamental steps as shown in the following Born-Haber cycle.

a. Show that

$$\Delta H = \Delta H_{solv,HA} + D_{HA} + IP_H - EA_A - \Delta H_{solv,H^+} - \Delta H_{solv,A^-}$$

b. Calculate ΔH for HF and HCl using the above expression. All of the information can be found from previous chapters except $\Delta H_{solv,HA}$ which is 11.5 kcal/mole for HF and 4.2 kcal/mole for HCl.

c. On the basis of your calculations in b. what is the dominant factor which allows proton dissociation of an acid to be an energetically favorable process. In other words, what is so unique about acids?

30. Citric acid undergoes the following acid dissociation steps

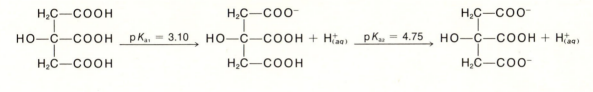

The titration of 10 ml of 0.1 M citric acid with sodium hydroxide gives the following pH curve:

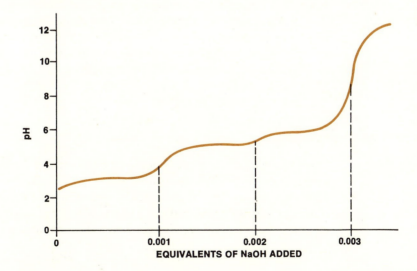

a. Account for the fact that there is very little change in pH in the region 0.0002 eq to 0.0008 eq.

b. Explain why there are three equivalence points in the titration curve.

CHAPTER FOURTEEN

ELECTROCHEMISTRY

In the previous four chapters we saw how the change in free energy (ΔG) enables us to predict the feasibility of both chemical reactions and phase transitions. In addition, the $\Delta G°$ has been related to the equilibrium constant. In this chapter we will use ΔG to predict the feasibility of an oxidation–reduction reaction and $\Delta G°$ to calculate the equilibrium concentrations of reactants and products. We will see, also, that free energy can be converted to *electrical work* or electrical energy by means of an *electrochemical cell*, and we will use the electrical potential of the electrochemical cell to develop a very convenient criterion for the feasibility of oxidation-reduction reactions. Finally, we will discuss the reverse process of converting electrical energy into chemical free energy. This process, which uses an electrical current to induce a chemical reaction, is called *electrolysis*.

14–A. ELECTROCHEMICAL CELLS

Examples of oxidation-reduction reactions, first discussed in Chapter 3, have been encountered throughout the preceding chapters. In each oxidation-reduction reaction, one chemical species gives up one or more electrons and another chemical species accepts these electrons. For example, copper(II) ions accept 2 electrons from metallic zinc according to the reaction:*

$$Zn_{(s)} + Cu^{2+}_{(aq)} \longrightarrow Zn^{2+}_{(aq)} + Cu_{(s)}$$

*Recall from Section 3–C that we will use a single arrow even for reversible reactions unless we especially want to emphasize the equilibrium condition.

FIGURE 14–1 $Zn_{(s)} + Cu^{2+}_{(aq)}$ reaction in solution

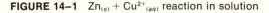
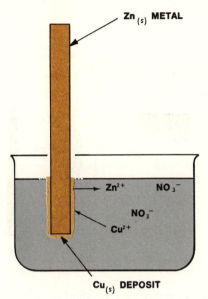

When a solid zinc rod is placed in an aqueous solution containing a copper (II) salt, such as $Cu(NO_3)_2$, the Cu^{2+} ions in solution come in contact with the zinc metal, accepting $2e^-$ from a zinc atom and forming the reduced $Cu_{(s)}$ (copper metal) and the Zn^{2+} ion (Figure 14–1). In this reaction there is a *decrease* in free energy, ($\Delta G < 0$). Since there is no means by which the free energy released in this process can be utilized, it is "lost."

Instead of allowing Cu^{2+} ions to come in direct contact with the $Zn_{(s)}$, an alternative experimental arrangement can be the transfer of electrons through a wire, as illustrated in Figure 14–2. The $Zn_{(s)} \longrightarrow Zn^{2+}_{(aq)} + 2e^-$ and $Cu^{2+}_{(aq)} + 2e^- \longrightarrow Cu_{(s)}$ reactions are physically separated here, with the electrons flowing through an external wire. These reactions are called *half-cell reactions* since each constitutes half of the total reaction of the *electrochemical cell*. The flow of electrons in the wire can be made to do electrical work and as we will see in Section 14–C, this electrical work arises from the conversion of the change in free energy for the reaction, ($\Delta G < 0$).

FIGURE 14–2   Transfer of e^- from $Zn_{(s)}$ to $Cu^{2+}_{(aq)}$

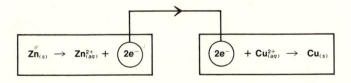

FIGURE 14–3 Electrochemical cell for Zn^{2+}, Zn and Cu^{2+}, Cu half cells with a KNO_3 salt bridge (V represents a voltmeter)

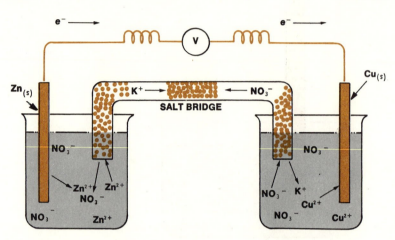

An electrochemical cell utilizing the Zn, Zn^{2+} and Cu, Cu^{2+} half cells is shown in Figure 14–3. In this electrochemical cell it is essential that the half cell reactions be carried out in separate containers. The $Zn_{(s)}$ rod constitutes the *electrode* in the Zn^{2+}, Zn half cell. The external wire connects the $Zn_{(s)}$ electrode to the $Cu_{(s)}$ electrode in the Cu^{2+}, Cu half cell. The electrode at which oxidation occurs is called the *anode* and the one at which reduction occurs is called the *cathode*.* The cell reactions are

$$Cu^{2+}_{(aq)} + 2e^- \longrightarrow Cu_{(s)} \quad \text{(reduction at cathode)}$$

$$Zn_{(s)} \longrightarrow Zn^{2+}_{(aq)} + 2e^- \quad \text{(oxidation at anode)}$$

If these reactions are allowed to continue, the anode compartment will generate an excess of $Zn^{2+}_{(aq)}$ (relative to the NO_3^- present) whereas the cathode compartment will deplete the $Cu^{2+}_{(aq)}$ (leaving an excess of $NO^-_{3(aq)}$). In essence we would thereby form a positively charged solution on the left of Figure 14–3 and a negatively charged solution on the right. In practice this does not occur, since charge separation requires a tremendous amount of energy. Instead, the solutions in both electrode compartments are kept neutral by the flow of ions between the compartments via a *salt bridge*. The salt bridge consists of a gelatinous substance which prevents the diffusion of liquids between the two compartments but does allow a flow of ions. The salt bridge contains a salt that is completely dissociated into ions that do not react with the ions existing in either of the compartments. In Figure 14–3 the salt bridge contains KNO_3. A Zn^{2+} ion generated in the anode compartment will enter the salt bridge displacing a K^+ into the cathode compartment

*A convenient mnemonic device for recalling this is the recognition that the process and electrode are both in alphabetical order: oxidation comes before reduction; anode comes before cathode.

at the other end. The excess of NO_3^- formed in the cathode compartment enters the salt bridge displacing a NO_3^- at the other end into the anode compartment. The flow of cations and anions is summarized as follows:

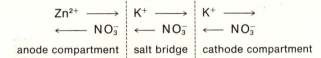

anode compartment ┊ salt bridge ┊ cathode compartment

Though the flow of cations and anions is not at the same rate, the *total* flow of cations from left to right and anions from right to left is such as to maintain neutrality in both the anode and cathode compartments.

In Figure 14–3 the external wire is connected to a voltmeter* (V), which measures the *electrical potential* or *electromotive force* ($\mathcal{E}$) generated between the anode and cathode. This electrical potential $\mathcal{E}$ is the driving force which causes the electrons to flow from the anode to the cathode. The size of the electrical potential will be different when different half cell reactions are used.

The half cells shown in Figure 14–3 consist of a metal electrode in contact with an aqueous solution containing the cation of that metal, but there are also other kinds of electrodes and half cell reactions. In general, electrodes can be classified into four basic types as in the following list. The vertical line in a half-cell designation indicates a different phase. For example, $Cu|Cu^{2+}$ indicates that the copper and copper(II) ion are in different phases.

TYPE OF ELECTRODE	EXAMPLES	REACTION
1. Metal-solution electrode metal|metal ion$_{(aq)}$	$Ag\|Ag^+_{(aq)}$	$Ag^+_{(aq)} + e^- \longrightarrow Ag_{(s)}$
2. Gas electrode inert electrode|gas|gas ion$_{(aq)}$	$Pt\|H_{2(g)}\|H^+_{(aq)}$ $Pt\|Cl_{2(g)}\|Cl^-_{(aq)}$	$H^+_{(aq)} + e^- \longrightarrow \frac{1}{2}H_{2(g)}$ $\frac{1}{2}Cl_{2(g)} + e^- \longrightarrow Cl^-_{(aq)}$
3. Redox electrode inert electrode|ion (reduced state)$_{(aq)}$, ion (oxidized state)$_{(aq)}$	$Pt\|Fe^{2+}_{(aq)}, Fe^{3+}_{(aq)}$	$Fe^{3+}_{(aq)} + e^- \longrightarrow Fe^{2+}_{(aq)}$
4. Composite electrode metal|metal salt (slightly soluble)|anion of salt$_{(aq)}$	$Ag\|AgCl_{(s)}\|Cl^-_{(aq)}$	$AgCl_{(s)} + e^- \longrightarrow Ag + Cl^-_{(aq)}$

An example of an electrochemical cell utilizing electrodes of types 2 and 3 is shown in Figure 14–4. The cell reactions are

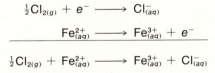

FIGURE 14–4 Electrochemical cell with gas and redox electrodes

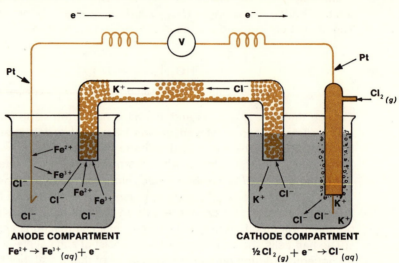

ANODE COMPARTMENT

$Fe^{2+} \rightarrow Fe^{3+}_{(aq)} + e^-$

CATHODE COMPARTMENT

$\frac{1}{2}Cl_{2(g)} + e^- \rightarrow Cl^-_{(aq)}$

In this system the solutions in the two half cells must be separated to prevent reaction of $Fe^{2+}_{(aq)}$ directly with $Cl_{2(g)}$. A salt bridge of KCl can be used, since there are $FeCl_2$ and $FeCl_3$ solutions in the anode compartment and an HCl solution in the cathode compartment.

In those cases where reactants are localized in the vicinity of each electrode and cannot come into direct contact with each other, a single solution can be used (eliminating the necessity of a salt bridge). Figure 14–5 shows a silver-silver chloride electrode of type 4 and a hydrogen gas electrode of type 2 combined in an electrochemical cell. The cell reactions are

$$\frac{1}{2}H_{2(g)} \longrightarrow H^+_{(aq)} + e^-$$

$$\underline{AgCl_{(s)} + e^- \longrightarrow Ag_{(s)} + Cl^-_{(aq)}}$$

$$\frac{1}{2}H_{2(g)} + AgCl_{(s)} \longrightarrow H^+_{(aq)} + Ag_{(s)} + Cl^-_{(aq)}$$

Note that the reactants $H_{2(g)}$ and $AgCl_{(s)}$ are restricted to the vicinity of the electrodes and do not come in contact with each other to react directly. A solution of HCl can supply the necessary $H^+_{(aq)}$ and $Cl^-_{(aq)}$ for both half-reactions.

PROBLEM 14–1 Consider the oxidation-reduction reaction

$$Cr^{2+}_{(aq)} + Fe^{3+}_{(aq)} \longrightarrow Cr^{3+}_{(aq)} + Fe^{2+}_{(aq)}$$

a. Write half-cell reactions showing the oxidation and reduction reactions.
b. Sketch an electrochemical cell that could be set up using these oxidation and reduction reactions as half cells.

FIGURE 14-5 Electrochemical cell with gas and composite electrodes

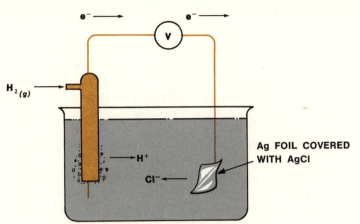

PROBLEM 14-2 Write the half-cell reactions for the oxidation-reduction reaction

$$Pb_{(s)} + Sn^{4+}_{(aq)} + SO^{2-}_{4(aq)} \longrightarrow PbSO_{4(s)} + Sn^{2+}_{(aq)}$$

14-B. WRITING AND BALANCING OXIDATION-REDUCTION EQUATIONS BY THE HALF-CELL METHOD

The equation representing an oxidation-reduction reaction can be balanced quite easily when all reactants and products of the reaction are known. The electron-transfer method for balancing oxidation-reduction equations was presented in Chapter 3. In this method the reactants which are oxidized or reduced are identified. The number of electrons lost and gained in the processes is then determined, and the oxidized and reduced reactants and products are balanced by making

$$e^- \text{ lost} = e^- \text{ gained}$$

The remaining elements involved in the reaction are then balanced. The following example illustrates the electron-transfer method and can be used for comparison with the half-cell method.

EXAMPLE 14-1 *Balance the following oxidation-reduction reaction using the electron-transfer method.*

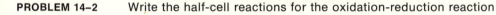

The $Cu_{(s)}$ is oxidized to $Cu^{2+}_{(aq)}$, losing $2e^-$ in the process. The NO_3^- is reduced to NO, gaining $3e^-$ in the process. The electrons lost can be made to equal those gained by

$$electrons\ lost = electrons\ gained$$
$$3(-2e^-) \qquad 2(3e^-)$$

Thus a coefficient of 3 before Cu and Cu^{2+} and a coefficient of 2 before NO_3^- and NO balances the oxidized and reduced reactants and products.

$$3Cu_{(s)} + 2NO_{3(aq)}^- + H_{(aq)}^+ \longrightarrow 3Cu^{2+}_{(aq)} + 2NO_{(g)} + H_2O \qquad (unbalanced)$$

Without changing the coefficient on the species involved in the oxidation-reduction, we must balance the remaining elements. In checking the oxygen atoms, we see that there are six in $2NO_3^-$ and two in $2NO$. Therefore, we must have $4H_2O$ to balance the equation. We note, however, that there are eight hydrogens in the $4H_2O$ but only one in $H_{(aq)}^+$. We must therefore have a coefficient of 8 for $H_{(aq)}^+$. The final balanced equation is

$$3Cu_{(s)} + 2NO_{3(aq)}^- + 8H_{(aq)}^+ \longrightarrow 3Cu^{2+}_{(aq)} + 2NO_{(g)} + 4H_2O$$

The balancing of the equation can be checked by the conservation of charge; that is, the sum of the charges on the left (reactants) must equal the sum of charges on the right (products).

$$3(0) + 2(-1) + 8(+1) \overset{?}{=} 3(+2) + 2(0) + 4(0)$$
$$-2 + 8 \overset{?}{=} +6$$
$$+6 = +6$$

PROBLEM 14-3 Balance the following oxidation-reduction reactions
a. $Fe^{2+}_{(aq)} + ClO_{3(aq)}^- + H_{(aq)}^+ \longrightarrow Fe^{3+}_{(aq)} + HClO_2 + H_2O$
b. $UO_{2(aq)}^+ + V^{3+}_{(aq)} + H_{(aq)}^+ \longrightarrow U^{4+}_{(aq)} + VO^{2+}_{(aq)} + H_2O$
c. $Zn_{(s)} + Ni(OH)_{2(s)} + OH_{(aq)}^- \longrightarrow ZnO_{2(aq)}^{2-} + Ni_{(s)} + H_2O$

In Example 14-1 and Problem 14-3 all the reactants and products were properly identified. Even $H_{(aq)}^+$ and H_2O were given in Example 14-1 as reactant and product, respectively. However, when a reaction is carried out in solution, frequently only the reactants and products actually involved in the oxidation and reduction processes are known, and the solution is specified as either acidic or basic or possibly neutral. With only this information, it is more difficult to write and balance the complete oxidation-reduction equation. Instead, it is simpler to write and balance the equation if it is first broken up into half-cell reactions and each of them is then properly balanced. This technique is called the *half-cell method* of balancing oxidation-reduction equations.

The half-cell method is outlined below as a series of specific steps. In order to clarify how this method is applied, the following oxidation-reduction reaction is carried through each step of the procedure.

$$Cu_{(s)} + NO_{3(aq)}^- \longrightarrow Cu_{(aq)}^{2+} + NO_{(g)} \qquad \text{(acid solution)}$$

1. Identify the individual processes of oxidation and reduction and write the half-cell reactions. Write the electrons lost and gained in each of the half cells based on the change in oxidation number.

$$Cu \longrightarrow Cu_{(aq)}^{2+} + 2e^-$$

$$\overset{(+5)}{NO_{3(aq)}^-} + 3e^- \longrightarrow \overset{(+2)}{NO_{(g)}}$$

2. If the reaction is being carried out in an acidic solution, add $H_{(aq)}^+$ to whichever side of the half-cell equation is necessary to conserve charge. If the reaction is being carried out in a basic solution, add $OH_{(aq)}^-$ to whichever side of the half-cell equation is necessary to conserve charge. Then add H_2O to either side of the equation if needed to balance the oxygen atoms. The balance of the hydrogen atoms can be used as a check to see if the half-cells are balanced. The $Cu|Cu^{2+}$ half cell has the charge balanced. For the $NO_3^-|NO$ half cell there are -4 charges on the left with no charge on the right. Therefore, we must add $4H^+$ to the left side.

$$Cu \longrightarrow Cu_{(aq)}^{2+} + 2e^-$$

$$NO_{3(aq)}^- + 3e^- + 4H_{(aq)}^+ \longrightarrow NO_{(g)} + 2H_2O$$

Then $2H_2O$ are added to the right to balance the oxygen atoms. The half-cell equation is properly balanced since the H atoms balance.

3. Now multiply the half cells by the proper factors to equalize the e^- lost and gained.

$$3 \times [Cu \longrightarrow Cu_{(aq)}^{2+} + 2e^-]$$

$$\underline{2 \times [NO_{3(aq)}^- + 3e^- + 4H_{(aq)}^+ \longrightarrow NO_{(g)} + 2H_2O]}$$

$$3Cu + 2NO_{3(aq)}^- + 8H_{(aq)}^+ \longrightarrow 3Cu_{(aq)}^{2+} + 2NO + 4H_2O$$

4. Add half-cell equations and eliminate any H^+, OH^-, or H_2O that may be common to both sides of the equation.
   There are no H^+, OH^-, or H_2O common to both sides of the equation, so the equation is balanced.

PROBLEM 14–4   Write and balance oxidation-reduction equations for:
a. $Fe_{(aq)}^{2+}$ and $ClO_{3(aq)}^-$, producing $Fe_{(aq)}^{3+}$ and $HClO_2$ in acidic solution.
b. $UO_{2(aq)}^+$ and $V_{(aq)}^{3+}$, producing $U_{(aq)}^{4+}$ and $VO_{(aq)}^{2+}$ in acidic solution.
c. $Zn_{(s)}$ and $Ni(OH)_{2(s)}$, producing $ZnO_{2(aq)}^{2-}$ and $Ni_{(s)}$ in basic solution.

14–C. FREE ENERGY CHANGE (ΔG) AND CELL POTENTIAL ($\mathcal{E}_{cell}$)

The mathematical relationship between the ΔG for an oxidation-reduction reaction and the $\mathcal{E}$ for the corresponding electrochemical cell is derived from the electrical work that can be performed by the cell. As you will recall from Chapter 10, *electrical work* (W_{el}) is the displacement of charge (q) against an electrical potential:

10–2
$$W_{el} = (\mathcal{E})(q)$$

The work that can be performed by an electrochemical cell is the displacement of charge (q) in the form of electrons against an *opposing* electrical potential ($\mathcal{E}$). Recall from Chapter 10 that the expansion of a gas is carried out reversibly when the pressure of the gas is equal to the external pressure, P_{ex}. Similarly, an electrochemical cell can be operated reversibly when the potential of the cell, $\mathcal{E}_{cell}$, is equal to the opposing potential, $\mathcal{E}$. The work done under reversible conditions is the maximum work which can be performed. Therefore, the maximum reversible electrical work, $W_{el,rev}$, which can be performed by an electrochemical cell is:

14–1
$$W_{el,rev} = (\mathcal{E}_{cell})(q)$$

The charge displaced in an oxidation-reduction reaction involves transfer of electrons from the substance oxidized (at the anode) to the substance reduced (at the cathode). For example, in the reaction

$$\overset{+2e^-}{Cu^{2+}_{(aq)} + 2Ag_{(s)} + 2Cl^-_{(aq)} \longrightarrow Cu_{(s)} + 2AgCl_{(s)}} \atop {-2e^-}$$

there are two electrons transferred from two Ag atoms to a Cu^{2+} ion. We designate the number of electrons transferred by the symbol n. However, in making thermodynamic calculations of ΔG, ΔH, or ΔS for a chemical reaction, we have assumed that the coefficients in the balanced equation designate mole or g-atom quantities. Since a mole contains 6.023×10^{23} molecules and a g-atom contains 6.023×10^{23} atoms, we will express the transfer of electrons in units of 6.023×10^{23} electrons. For this reason we define a faraday (F) as the charge on 6.023×10^{23} electrons. Since the charge on a single electron equals 1.602×10^{-19} coulombs, a faraday is

$$1 \text{ faraday } (F) = (6.023 \times 10^{23} \text{ electrons})\left(1.602 \times 10^{-19} \frac{\text{coulombs}}{\text{electron}}\right)$$

$$1 \text{ faraday } (F) = 96,500 \text{ coulombs}$$

The charge (q) transferred in a reaction is the product of n times F

14–2 $$q = nF$$

Substituting equation 14–2 into equation 14–1, we obtain

14–3 $$W_{el,rev} = nF\mathcal{E}_{cell}$$

Calculation of the change in free energy for the reaction (ΔG) is straightforward, since the electrochemical cell is operated reversibly and at constant temperature and pressure. At constant temperature we know that

10–22 $$\Delta G = \Delta H - T\Delta S$$

At constant pressure, $\Delta H = Q$ if only PV work is done (see equation 10–10). However, in this case we also have electrical work, hence

$$\Delta H = Q - W_{el}$$

Furthermore, if the process is reversible,

$$\Delta H = Q - W_{el,rev}$$

and substitution into equation 10–22 gives

14–4 $$\Delta G = Q_{rev} - W_{el,rev} - T\Delta S$$

However, since the process is reversible, the second law of thermodynamics gives

10–17 $$\Delta S = \frac{Q_{rev}}{T}$$

Substitution of equation 10–17 into equation 14–4 gives

$$\Delta G = Q_{rev} - W_{el,rev} - \cancel{T}\frac{Q_{rev}}{\cancel{T}}$$

and therefore,

14–5 $$\Delta G = -W_{el,rev}$$

The relationship between ΔG and $\mathcal{E}_{cell}$ is now obtained by setting equation 14–3 equal to equation 14–5.

14–6 $$\Delta G = -nF\mathcal{E}_{cell}$$

If standard conditions are used in the electrochemical cell

14–7 $$\Delta G^\circ = -nF\mathcal{E}^\circ_{cell}$$

The standard conditions are the same as those defined earlier for free energy: for a solid, liquid, or gas, pressure (P) = 1 atm, and for a solute, activity (a_i) = 1 (If γ_i = 1 then concentration is 1 M). As you can see, there is a very simple relationship between the cell potential and the change in standard free energy for the oxidation-reduction reaction. The significance of this relationship will become apparent in Sections 14–D, 14–E, and 14–F. The conversion of units to give ΔG° in kilocalories when the cell potential has been expressed in volts is illustrated in the following example.

EXAMPLE 14–2 *An electrochemical cell was set up utilizing the following half-cell reactions:*

$$Ni_{(s)} \longrightarrow Ni^{2+}_{(aq)} + 2e^- \qquad (anode)$$

$$2Ce^{4+}_{(aq)} + 2e^- \longrightarrow 2Ce^{3+}_{(aq)} \qquad (cathode)$$

$$\overline{Ni_{(s)} + 2Ce^{4+}_{(aq)} \longrightarrow Ni^{2+}_{(aq)} + 2Ce^{3+}_{(aq)}}$$

The cell had a potential of 1.86 volts when all substances were in their standard states. Calculate the ΔG° for this oxidation-reduction reaction.
Since 2 electrons are transferred from a Ni atom to $2Ce^{4+}$, then n = 2. Substituting the known information into equation 14–7 for ΔG°

$$\Delta G^\circ = -nF\mathcal{E}^\circ_{cell}$$

$$\Delta G^\circ = -(2)(96{,}500 \; coulombs)(1.86 \; volts)$$

$$\Delta G^\circ = -3.59 \times 10^{+5} \; volt\text{-}coulomb$$

Since 1 volt-coulomb = 1 joule, and 1 kcal = 4.1840×10^3 joules, then

$$\Delta G^\circ = -3.59 \times 10^5 \; joules \left(\frac{1 \; kcal}{4.1840 \times 10^3 \; joules} \right)$$

$$\Delta G^\circ = -0.86 \times 10^2 \; kcal = -86 \; kcal$$

PROBLEM 14–5 a. Calculate the standard cell potential ($\mathcal{E}$) expected for an electrochemical cell constructed from the oxidation-reduction reaction

$$2Fe^{2+}_{(aq)} + 2Hg^{2+}_{(aq)} \longrightarrow 2Fe^{3+}_{(aq)} + Hg^{2+}_{2(aq)}$$

The standard free energies of formation are

ION	ΔG_f° (kcal/mole)
$Fe_{(aq)}^{2+}$	-20.30
$Fe_{(aq)}^{3+}$	-2.52
$Hg_{2(aq)}^{2+}$	36.79
$Hg_{(aq)}^{2+}$	39.38

b. Would you obtain the same cell potential if the reaction were written

$$Fe_{(aq)}^{2+} + Hg_{(aq)}^{2+} \longrightarrow Fe_{(aq)}^{3+} + \tfrac{1}{2}Hg_{2(aq)}^{2+}$$

c. As a result of parts (a) and (b) would you conclude that $\Delta G°$ for a reaction is dependent on the number of electrons transferred whereas the $\mathcal{E}_{cell}°$ is independent of the number of electrons transferred? Explain.

14–D. STANDARD HALF-CELL POTENTIALS AND FREE ENERGY CHANGE

As we have seen, an oxidation-reduction reaction can be written as the sum of two half-cell reactions. Since G is a state function, the $\Delta G°$ can be taken as the sum of the change in free energy for each of the half-cell reactions. We can define a standard half-cell potential, $\mathcal{E}_{\frac{1}{2}}°$, for each of the half cells and, as we will see shortly, the sum of the half-cell potentials gives the cell potential.

The free energy change for a half cell is related to the half-cell potential just as in equation 14–7.

14–8
$$\Delta G_{\frac{1}{2}}° = -nF\mathcal{E}_{\frac{1}{2}}°$$

In Problem 14–5, we have seen that the standard cell potential, $\mathcal{E}_{cell}°$, is independent of the number of electrons transferred in the reaction. For the same reason $\mathcal{E}_{\frac{1}{2}}°$ is independent of the number of electrons lost or gained in a half-cell reaction. We can show the relationship between $\mathcal{E}_{cell}°$ and $\mathcal{E}_{\frac{1}{2}}°$ using a $Cu_{(aq)}^{2+}|Cu$ and $AgCl|Ag|Cl^-$ cell. The standard half-cell reduction potentials are

$$Cu_{(aq)}^{2+} + 2e^- \longrightarrow Cu_{(s)} \qquad \mathcal{E}_{Cu^{2+},Cu}° = 0.340 \text{ V}$$

$$AgCl_{(s)} + e^- \longrightarrow Ag_{(s)} + Cl_{(aq)}^- \qquad \mathcal{E}_{AgCl,Ag,Cl^-}° = 0.222 \text{ V}$$

The cell reaction can be formed from the two half-cell reactions by reversing the direction of the $AgCl|Ag|Cl^-$ half cell. When the direction of a half-cell reaction is reversed, the sign of $\Delta G_{\frac{1}{2}}°$ is reversed as is the sign of $\mathcal{E}_{\frac{1}{2}}°$.

$$\begin{array}{ll} & \Delta G^{\circ} \\ Cu^{2+}_{(aq)} + 2e^- \longrightarrow Cu_{(s)} & -2F(0.340 \text{ V}) \\ 2Ag_{(s)} + 2Cl^- \longrightarrow 2AgCl_{(s)} + 2e^- & -2F(-0.222 \text{ V}) \\ \hline Cu^{2+}_{(aq)} + 2Ag_{(s)} + 2Cl^-_{(aq)} \longrightarrow Cu_{(s)} + 2AgCl_{(s)} & -2F\,\mathcal{E}^{\circ}_{cell} \end{array}$$

The free energy change is

$$\Delta G^{\circ} = -2F\mathcal{E}^{\circ}_{cell} = -2F(0.340 \text{ V}) - 2F(-0.222\text{V})$$

Solving for $\mathcal{E}^{\circ}_{cell}$ we see that the cell potential is simply the algebraic sum of the half-cell potentials

$$\mathcal{E}^{\circ}_{cell} = 0.340 \text{ V} - 0.222 \text{ V}$$

Recall that the half-cell potentials here apply to the half cells as written. *If the half cell is an oxidation step, the half-cell potential is the negative of that half cell's reduction potential.* Also note that $\mathcal{E}^{\circ}_{cell}$ is the sum of the $\mathcal{E}^{\circ}_{\frac{1}{2}}$'s regardless of the number of electrons oxidized or reduced per molecule or ion. In summary $\mathcal{E}^{\circ}_{\frac{1}{2}}$ and $\mathcal{E}^{\circ}_{cell}$ are independent of the number of electrons transferred.

PROBLEM 14–6   The half-cell potential for the following reaction is $\mathcal{E}^{\circ}_{\frac{1}{2}} = 0.910$ volts

$$2Hg^{2+}_{(aq)} + 2e^- \longrightarrow Hg^{2+}_{2(aq)}$$

What is the half-cell potential for

$$Hg^{2+}_{(aq)} + e^- \longrightarrow \tfrac{1}{2}Hg^{2+}_{2(aq)}$$

PROBLEM 14–7   Calculate the cell potential for the electrochemical cell utilizing the reaction:

$$2Fe^{2+}_{(aq)} + 2Hg^{2+}_{(aq)} \longrightarrow 2Fe^{3+}_{(aq)} + Hg^{2+}_{2(aq)}$$

$$\mathcal{E}^{\circ}_{Fe^{3+},Fe^{2+}} = 0.7704 \text{ volts}$$

Although we have defined a standard half-cell potential in terms of the standard free energy change for the half cell, we cannot measure either $\mathcal{E}^{\circ}_{\frac{1}{2}}$ or $\Delta G^{\circ}_{\frac{1}{2}}$ directly. We can only make measurements on electrochemical cells which are necessarily composed of *two* half-cell reactions. For this reason we can only establish a set of *relative* half-cell potentials in terms of a potential assigned arbitrarily to some half cell. The half cell taken as the reference is the $H^+|H_{2(g)}$ half cell, which is assigned a standard half-cell potential of zero.

$$H^+_{(aq)} + e^- \longrightarrow \tfrac{1}{2}H_{2(g)} \qquad\qquad \mathcal{E}^{\circ}_{H^+,H_{2(g)}} = 0$$

$$\Delta G^{\circ}_{H^+,H_{2(g)}} = 0$$

A complete table of standard half-cell or electrode potentials is given in Appendix T, Table 16. All $\mathcal{E}_{\frac{1}{2}}^{\circ}$ values in Appendix T are relative to a $\mathcal{E}_{H^+,H_{2(g)}}^{\circ} = 0$.

Since the electron is common to all half-cell reactions, it is customary to assign it a value:

$$\Delta G_{f,e^-}^{\circ} = 0$$

You will recall from Chapter 10 that the standard free energies of the elements in their most stable form are assigned a value of zero; consequently $\Delta G_{f,H_{2(g)}}^{\circ} = 0$. Since

$$\Delta G_{H^+,H_{2(g)}}^{\circ} = \Delta G_{f,H_{2(g)}}^{\circ} - \Delta G_{f,e^-}^{\circ} - \Delta G_{f,H_{(aq)}^+}^{\circ}$$

it follows that the standard free energy of $H_{(aq)}^+$ must also be zero. The standard free energies of the ions given in Appendix T, Tables 3 and 4 are all relative to this reference zero.

PROBLEM 14–8   The standard cell potentials have been measured for the reactions

$$2AgCl_{(s)} + H_{2(g)} \longrightarrow 2Ag_{(s)} + 2Cl_{(aq)}^- + 2H_{(aq)}^+ \qquad \mathcal{E}^{\circ} = 0.2223 \text{ volts}$$

$$Ag_{(s)} + Cl_{(aq)}^- + \tfrac{1}{2}Br_{2(l)} \longrightarrow AgCl_{(s)} + Br_{(aq)}^- \qquad \mathcal{E}^{\circ} = 0.8432 \text{ volts}$$

Calculate the $\mathcal{E}_{Br_2,Br^-}^{\circ}$ from this data. Do *not* use the half-cell potentials in the Appendix T. [Hint: $\mathcal{E}_{H^+,H_2}^{\circ} = 0$.]

PROBLEM 14–9   Using your result from Problem 14–8, calculate the $\Delta G_{f,Br^-}^{\circ}$ and $\Delta G_{f,Cl^-}^{\circ}$. You will need to use $\Delta G_{f,AgCl}^{\circ} = -26.224$ kcal/mole.

Some elements have more than one oxidation state and half-cell reduction potentials can be obtained for the reaction between any two of the oxidation states. For example, iron has a +2 and a +3 oxidation state in addition to the pure metal. The standard half-cell potentials for the reduction to successive oxidation states are conveniently shown above the arrows in the following diagram

$$Fe^{3+} \xrightarrow{\;0.770\;} Fe^{2+} \xrightarrow{\;-0.440\;} Fe$$
$$\underset{-0.037}{\underline{\hspace{6cm}}}$$

Offhand you might expect the $Fe^{3+}|Fe$ potential to be the *sum* of the $Fe^{3+} \longrightarrow Fe^{2+}$ and the $Fe^{2+} \longrightarrow Fe$ potentials. However, this is *not* the case. The $\mathcal{E}^{\circ}$ for the overall step $Fe^{3+} \longrightarrow Fe$ can be calculated using the half-cell potentials $\mathcal{E}_{Fe^{3+},Fe^{2+}}^{\circ}$ and $\mathcal{E}_{Fe^{2+},Fe}^{\circ}$. Remembering that the ΔG° values for the half-cell reactions are added, the $\mathcal{E}^{\circ}$ is calculated accordingly.

$$Fe^{3+} + e^- \longrightarrow Fe^{2+} \qquad \Delta G_1^{\circ} = -F(+0.770 \text{ V})$$

$$\underline{Fe^{2+} + 2e^- \longrightarrow Fe \qquad \Delta G_2^{\circ} = -2F(-0.440 \text{ V})}$$

$$Fe^{3+} + 3e^- \longrightarrow Fe \qquad \Delta G_3^{\circ} = -3F\,\mathcal{E}_{Fe^{3+},Fe}^{\circ}$$

$$\Delta G_3^\circ = \Delta G_1^\circ + \Delta G_2^\circ$$

$$-3F\mathcal{E}^\circ_{Fe^{3+},Fe} = -F(0.770 \text{ V}) - 2F(-0.440 \text{ V})$$

$$-3F\mathcal{E}^\circ_{Fe^{3+},Fe} = F(-0.770 + 0.880) \text{ V}$$

$$\mathcal{E}^\circ_{Fe^{3+},Fe} = \frac{-0.110}{3} \text{ V} = -0.037 \text{ V}$$

The above sequence of oxidation states and reduction potentials for iron applies as well to acid solutions of the ions. In a basic solution hydroxides can form, so a new set of potentials must be used. For iron in basic solution we have

$$\text{Fe(OH)}_3 \xrightarrow{-0.56 \text{ V}} \text{Fe(OH)}_2 \xrightarrow{-0.89} \text{Fe}$$
$$\underset{-0.78 \text{ V}}{\underline{}} \uparrow$$

For elements such as manganese which can have several oxidation states, there are numerous $\mathcal{E}^\circ$ for the transition between these oxidation states. The half-cell potentials for manganese in acid solution are shown below.

$$\text{MnO}_4^- \xrightarrow{0.6} \text{MnO}_4^{2-} \xrightarrow{2.3} \text{MnO}_2 \xrightarrow{1.0} \text{Mn}^{3+} \xrightarrow{1.5} \text{Mn}^{2+} \xrightarrow{-1.2} \text{Mn}$$

with the spanning values 1.7 (MnO$_4^-$ to MnO$_2$), 1.25 (MnO$_2$ to Mn^{2+}), 1.5 (MnO$_4^{2-}$ to MnO$_2$)

These reactions, along with half-cell potentials, are very useful in understanding the stability of the various oxidation states as well as the reducing and oxidizing power of each state. They will be used in Chapter 16 when we discuss the transition elements.

PROBLEM 14–10   The half-cell potentials for copper are:

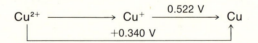

$$\text{Cu}^{2+} \xrightarrow{} \text{Cu}^+ \xrightarrow{0.522 \text{ V}} \text{Cu}$$
$$\underset{+0.340 \text{ V}}{\underline{}} \uparrow$$

Calculate the half-cell potential for Cu^{2+} $\longrightarrow$ Cu$^+$ using the values given.

14–E. PREDICTION OF OXIDATION-REDUCTION REACTIONS—STANDARD CELL POTENTIAL AND EQUILIBRIUM CONSTANT

The standard half-cell potentials can easily be used to predict the feasibility of an oxidation-reduction reaction. We know from thermodynamics that the criterion for the feasibility of a reaction at constant temperature and pressure is

$$10-23 \qquad\qquad\qquad \Delta G < 0$$

If we have standard conditions, then

$$\Delta G° < 0$$

According to equation 14–7, the criterion for spontaneity becomes

$$-nF\mathcal{E}°_{cell} < 0$$

Since n and F are positive constants, the criterion for spontaneity can be stated as

14–9 $-\mathcal{E}°_{cell} < 0$ or $\mathcal{E}°_{cell} > 0$ (at standard conditions)

In Section 14–D we saw that the $\mathcal{E}°_{cell}$ is simply the algebraic sum of the standard half-cell potentials. Therefore, the sign of $\mathcal{E}°_{cell}$ can be found by inspection of the standard half-cell potentials (Appendix T, Table 16). It is important to note that the prediction of feasibility based on $\mathcal{E}°_{cell}$ has the same limitations as the prediction using $\Delta G° < 0$. That is, the *rate* of the reaction has not been taken into account. However, if the oxidation-reduction reaction is an electron transfer between simple ions without any bond breaking or rearrangement involved, the rate of the reaction is probably not a limiting factor, and the reaction probably will proceed when $\mathcal{E}°_{cell} > 0$. In general, neglect of the kinetics of a reaction is not as serious for oxidation-reduction reactions as it is for other types of reactions.

EXAMPLE 14–3 *Predict whether the following reactions are feasible at 25 °C under standard conditions:*

a. $Zn_{(s)} + Fe^{3+}_{(aq)} \longrightarrow Zn^{2+}_{(aq)} + Fe^{2+}_{(aq)}$
b. $Cu + 2H^+_{(aq)} \longrightarrow Cu^{2+}_{(aq)} + H_{2(g)}$
c. $2Cu^+_{(aq)} \longrightarrow Cu_{(s)} + Cu^{2+}_{(aq)}$

Looking up the half-cell potentials in Appendix T, Table 16,
a. $Fe^{3+}_{(aq)} + e^- \longrightarrow Fe^{2+}_{(aq)}$ *0.770 V*
   $Zn^{2+}_{(aq)} + 2e^- \longrightarrow Zn_{(s)}$ *−0.7628 V*
   $\mathcal{E}°_{cell} = 0.770\ V + 0.7628\ V = 1.533\ V$
Since $\mathcal{E}°_{cell} > 0$, the reaction is feasible at standard conditions.
b. $2H^+_{(aq)} + 2e^- \longrightarrow H_{2(g)}$ *0.0 V*
   $Cu^{2+}_{(aq)} + 2e^- \longrightarrow Cu_{(s)}$ *0.340 V*
   $\mathcal{E}°_{cell} = 0.0\ V - 0.340\ V = -0.340\ V$
Since $\mathcal{E}°_{cell} < 0$, the reaction is not feasible at standard conditions.
c. $Cu^+_{(aq)} + e^- \longrightarrow Cu_{(s)}$ *0.522 V*
   $Cu^{2+}_{(aq)} + e^- \longrightarrow Cu^+_{(aq)}$ *0.158 V*
   $\mathcal{E}°_{cell} = 0.522\ V - 0.158\ V = 0.364\ V$
Since $\mathcal{E}°_{cell} > 0$, the reaction is feasible and Cu$^+$ undergoes simultaneous reduction and oxidation to Cu$_{(s)}$ and Cu$^{2+}_{(aq)}$. Consequently, the Cu$^+$ is not stable in aqueous solutions.

PROBLEM 14–11 Predict whether the following reactions are feasible at 25 °C under standard conditions. Use the $\mathcal{E}°$ from Appendix T, Table 16.
a. $Au_{(s)} + 3H^+_{(aq)} \longrightarrow Au^{3+} + \frac{3}{2}H_{2(g)}$
b. $Hg_2SO_{4(s)} + 2Cr^{2+}_{(aq)} \longrightarrow 2Hg_{(l)} + SO^{2-}_{4(aq)} + 2Cr^{3+}_{(aq)}$
c. $Hg_2{}^{2+} \longrightarrow Hg_{(l)} + Hg^{2+}$
d. $2Fe_{(s)} + O_{2(g)} + 4H^+_{(aq)} \longrightarrow 2Fe^{2+}_{(aq)} + 2H_2O$

PROBLEM 14–12 If you have a table of twenty half-cell potentials, how many oxidation-reduction reactions could you make predictions for, barring complications such as side reactions?

Since $\Delta G°$ is related to both $\mathcal{E}°_{cell}$ and the equilibrium constant, K, by equations 14–7 and 13–3

$$\Delta G° = -nF\mathcal{E}°_{cell}$$

$$\Delta G° = -RT \ln K$$

the equilibrium constant for an oxidation-reduction reaction can be calculated from the $\mathcal{E}°_{cell}$:

$$\ln K = \frac{nF\mathcal{E}°}{RT}$$

Converting to logarithms to the base 10:

14–10
$$\log K = \frac{nF\mathcal{E}°}{2.303\ RT}$$

At 25 °C the constant term $(F/2.303\ RT)$ can be evaluated and is found to be 16.8 volt^{-1}. The relationship between K and $\mathcal{E}°_{cell}$ is then specifically

14–11 $\log K = (16.9\ \text{volt}^{-1})n\mathcal{E}°_{cell}$ $(t = 25\ °C)$

It is worth noting that K is large when the $\mathcal{E}°_{cell}$ is large and positive. However, when the $\mathcal{E}°_{cell}$ is negative, K is less than unity, and very little product will be formed at equilibrium.

We have said that when $\mathcal{E}°_{cell}$ is negative, a reaction is not possible, yet in the above paragraph we state that when $\mathcal{E}°_{cell}$ is negative, K is less than unity, meaning that some small amount of product will be formed at equilibrium. These statements may appear contradictory, but they are not. (See the discussion of the same apparent contradiction between $\Delta G° < 0$ and product formation in Section 13–B.) The criterion for feasibility of a reaction under standard conditions assumes that *both reactants and products are present in the mixture at standard conditions.* When $\mathcal{E}°_{cell} > 0$, the criterion predicts that the reaction will proceed from left to right to produce *more products.* However, when $\mathcal{E}°_{cell} < 0$, the reverse reaction occurs, and products react to give *more reactants.*

When we have a reaction where $\mathcal{E}^\circ_{cell} < 0$, this does not mean that we cannot produce some products under other conditions different from the standard conditions. For example, if we were to start the reaction with *only reactants* at standard conditions, we would produce some products and the reaction would proceed until the equilibrium constant expression is satisfied. In this case, however, the reaction is not considered to occur at standard conditions since the products are not in their standard states (which for gases, liquids, and solids is $P = 1$ atm and for solutes in solution is $a_i = 1$). The following example will illustrate a reaction starting with (a) reactants and products at standard conditions and (b) with only reactants at standard conditions. It can be seen from the results of Example 14–4 that the amount of product at equilibrium is very small when $\mathcal{E}^\circ_{cell}$ is negative by as little as only -0.2 volts. If the cell potential is more negative, even less product is present at equilibrium.

EXAMPLE 14–4 *For the following reaction*

$$Cu_{(s)} + Sn^{4+} \longrightarrow Cu^{2+} + Sn^{2+}$$

Calculate the concentrations for all species at equilibrium at 25 °C when the reaction is initiated with
a. both reactants and products at standard conditions and
b. only reactants at standard condition.
For this reaction $\mathcal{E}^\circ = -0.20$ V and $\Delta G^\circ = +9.25$ kcal. Since $\Delta G^\circ > 0$ and $\mathcal{E}^\circ_{cell} < 0$, this reaction would not be feasible or spontaneous at standard conditions.
a. In order to calculate the equilibrium concentrations we need to evaluate the equilibrium constant, which is calculated from equation 14–11

$$\log K = (16.9 \text{ volt}^{-1})n\mathcal{E}^\circ = (16.9 \text{ volt}^{-1})(2)(-0.20 \text{ V}) = -6.76$$

$$K = 1.7 \times 10^{-7}$$

Initially all species are in their standard state ($a_i = 1$). For the present discussion, let us assume the activity coefficients are one, so that the initial concentrations of all ions are 1 M. The reaction will proceed from right to left, since $\mathcal{E}^\circ_{cell} < 0$ and $\Delta G^\circ > 0$. Let x equal the amount of Sn^{4+} produced to establish equilibrium. The Cu^{2+} and Sn^{2+} concentrations will be reduced by x, and at equilibrium we have

$$Cu_{(s)} + Sn^{4+}_{(aq)} \rightleftharpoons Cu^{2+}_{(aq)} + Sn^{2+}_{(aq)}$$
pure $(1 + x)$ $(1 - x)$ $(1 - x)$
solid

The equilibrium expression is

$$K = \frac{[Cu^{2+}][Sn^{2+}]}{[Sn^{4+}]}$$

$$1.7 \times 10^{-7} = \frac{(1 - x)(1 - x)}{(1 + x)} = \frac{(1 - x)^2}{1 + x}$$

In order for the right hand side of this equation to be small, the $(1 - x)$ must be very small or $x \approx 1$. Let us assume $x = 1$ in the denominator alone and solve for x from the numerator

$$1.7 \times 10^{-7} = \frac{(1 - x)^2}{1 + 1} = \frac{(1 - x)^2}{2}$$

$$(1 - x)^2 = 3.4 \times 10^{-7} = 34 \times 10^{-8}$$

$$(1 - x) = 5.8 \times 10^{-4}$$

$$x = 1 - 5.8 \times 10^{-4} = 0.99942$$

Therefore, the concentrations at equilibrium are

$$[Sn^{4+}] = 1 + x = 1.99942 \ M$$

$$[Cu^{2+}] = 1 - x = 0.00058 \ M$$

$$[Sn^{2+}] = 1 - x = 0.00058 \ M$$

Obviously, when $\mathcal{E}°_{cell} < 0$ or $\Delta G° > 0$ the reaction will proceed spontaneously from right to left and the reaction from left to right will not be feasible. There will always be some products and reactants at equilibrium; however, when $\mathcal{E}°_{cell}$ is negative, with a magnitude greater than a few tenths of a volt, the amount of product at equilibrium will be small. In this example only 0.00058 M Cu^{2+} and Sn^{2+} products are present at equilibrium.

b. *If we start out with only reactants at standard conditions, we will have a pure Cu metal in a solution of 1 M Sn^{4+} (assuming $\gamma_{Sn^{4+}} = 1$). No Cu^{2+} or Sn^{2+} are present initially. The calculation of the equilibrium concentrations proceeds in the usual manner. We can represent the unknown concentrations of Cu^{2+} and Sn^{2+} with x, and the concentration of Sn^{4+} at equilibrium with $(1 - x)$.*

$$Cu_{(s)} + Sn^{4+} \rightleftharpoons Cu^{2+}_{(aq)} + Sn^{2+}$$

pure $(1 - x)$ x   x
solid

Substituting into the equilibrium expression

$$K = \frac{[Cu^{2+}][Sn^{2+}]}{[Sn^{4+}]}$$

$$1.7 \times 10^{-7} = \frac{(x)\ (x)}{(1 - x)}$$

Since K is small, x will be small and we can assume $(1 - x) \approx 1$. The unknown x can then be calculated

$$1.7 \times 10^{-7} = \frac{x^2}{1}$$

$$x = 4.1 \times 10^{-4}$$

The final equilibrium mixture will contain

$$[Cu^{2+}] = 0.00041 \ M$$

$$[Sn^{2+}] = 0.00041 \ M$$

$$[Sn^{4+}] = 0.99959 \ M$$

pure Cu metal

Obviously some product of Cu^{2+} and Sn^{2+} is produced starting with only reactants Cu metal and Sn^{4+}; however, the amount of product at equilibrium is small.

PROBLEM 14–13 a. Calculate K for reactions b. and c. in Problem 14–11.
b. If you started the reactions in b. and c. with only reactants at standard conditions in 1 liter of solution, which reaction would produce the most $Hg_{(l)}$?

14–F. CELL POTENTIAL AND CONCENTRATION—NERNST EQUATION

So far our discussion of electrochemical cells has been restricted to standard conditions. However, the cell potential and the half-cell potentials depend upon the *concentrations* of the species in solution, as well as upon the *pressure* of any gases involved in the reaction. The relationship between the cell potential and nonstandard concentrations and/or pressure can be derived as was the $\Delta G°$ $= -RT \ln K$ relationship (see Appendix P). For the simple reaction

$$bB_{(aq)} \rightleftharpoons cC_{(aq)}$$

the cell potential is given by

14–12 $\mathcal{E}_{cell} = \mathcal{E}°_{cell} - \dfrac{RT}{nF} \ln \dfrac{[C]^c}{[B]^b}$ (low solute concentration region)

where [C] and [B] are the actual molar concentrations in the cell. It is convenient to convert the natural logarithm to base ten and to calculate the resulting constant

2.303 RT/F to be 0.0592 V at 25 °C. Equation 14–12 can be simplified to

14–13
$$\mathcal{E}_{cell} = \mathcal{E}^{\circ}_{cell} - \frac{0.0592}{n} \log \frac{[C]^c}{[B]^b}$$

In oxidation-reduction reactions there is generally more than one reactant and product species. However, the extension of equations 14–12 and 14–13 to more than one product or reactant is obvious from the stoichiometric equation. If a gas is involved in the reaction, a pressure term will also appear in the logarithmic term.

An equation analogous to equation 14–12 can be written for a half-cell reaction:

14–14
$$\mathcal{E}_{\frac{1}{2}} = \mathcal{E}^{\circ}_{\frac{1}{2}} - \frac{RT}{nF} \ln \frac{[\text{products}]}{[\text{reactants}]}$$

where the product and reactant concentrations are raised to the appropriate powers according to the coefficients in the equation for the half cell. The concentration of electrons is neglected in this term.

Equations 14–12 through 14–14 can be used to calculate the $\mathcal{E}$ for a cell in which the reactants and products are not at standard conditions. If the concentrations differ by several orders of magnitude, $\mathcal{E}$ can differ significantly from $\mathcal{E}^{\circ}$. However, if the concentration differs by only a factor of 10, the difference is only on the order of 0.1 V, which is rather small. The calculation of $\mathcal{E}_{cell}$ can also be used as a criterion for the feasibility of the reaction when the species are not at standard conditions.

EXAMPLE 14–5

Calculate the half-cell potential for a $H_{2(g)}$ electrode at a pressure of 740 mm in contact with pure H_2O at 25 °C.

The half-cell expression is given by

$$2H^+_{(aq)} + 2e^- \longrightarrow H_{2(g)}$$

$$\mathcal{E} = \mathcal{E}^{\circ} - \frac{0.0592}{2} \log \frac{P_{H_2}}{[H^+]^2}$$

The $P_{H_2} = 740/760$ atm = 0.973 atm. The $[H^+]$ in pure H_2O is 10^{-7} M. Then

$$\mathcal{E} = 0.0 - \frac{0.0592}{2} \log \frac{0.973}{[10^{-7}]^2} = -0.41 \text{ V}$$

The change from 0 V to −0.41 is quite significant, since it corresponds to a large H^+ concentration change from 1 M to 10^{-7} M.

PROBLEM 14–14

Calculate the half-cell potential for a $H_{2(g)}$ electrode at a pressure of 760 mm in contact with a 0.1 M HCl solution.

PROBLEM 14–15   Predict whether the following reaction is feasible:

$$Fe_{(s)} + 2H^+_{(aq)} \longrightarrow Fe^{2+}_{(aq)} + H_{2(g)}$$

when   a. all species are at standard conditions,
b. $[H^+] = 0.17$ M, $[Fe^{2+}] = 0.01$ M, $P_{H_2} = 0.01$ atm, or
c. H^+ comes from pure H_2O, $[Fe^{2+}] = 0.01$ M, $P_{H_2} = 0.01$ atm.

* 14–G. BATTERIES

In the previous sections we have seen how electrochemical cells and measurements of cell potentials can eventually be used to predict (on a thermodynamic basis) oxidation-reduction reactions. Electrochemical cells can also be used practically to produce electrical energy in the form of batteries. Since the potential of a single electrochemical cell seldom exceeds 2–3 V, electrochemical cells are connected in series in a battery to produce the desired electrical potential. For example, in the common lead storage battery, frequently used in automobiles, each lead cell has a potential ≈ 2 V and three of these must be connected in series to give a 6 V battery or six cells to give a 12 V battery. The potentials of batteries are made as high as 200 V.

The construction of a battery is quite different from that of the electrochemical cells discussed previously. So far, we have been concerned with the determination of precise cell potentials where the cell reactions are well-defined and little or no current is consumed in making the measurements. However, the purpose of a battery is to produce a desired electrical current and it is so constructed that the desired current can be maintained for some desired number of hours. The electrical currents from batteries can range from a few milliamps to run a transistor radio to 400 amps to start a jet airplane engine. Batteries have a wide range of uses from miniature electronic equipment, where the electrical energy requirements are low, all the way to ignition of diesel engines. Batteries are also used in motorized vehicles, such as fork lifts, where it is important to minimize exhaust fumes, noise, and accidents. However, battery-operated vehicles, in contrast to those powered by an internal combustion engine, are more expensive to operate and have limited working periods before the batteries need recharging. There are numerous types of batteries designed for various applications but only a few of the more common ones will be described in this section.

LEAD STORAGE BATTERY

Most automobiles use this type of storage battery at 6 or 12 V with currents of 100 amps. The individual cells in the battery contain lead and lead dioxide plates as electrodes that are immersed in a solution of sulfuric acid. The first acid dissociation in H_2SO_4 is essentially complete whereas the second is only partial.

$$H_2SO_4 \longrightarrow H^+_{(aq)} + HSO^-_{4(aq)}$$

$$HSO^-_{4(aq)} \rightleftharpoons H^+_{(aq)} + SO^{2-}_{4(aq)}$$

At the concentration of H_2SO_4 in a fully charged battery, HSO^-_4 is the dominant negative ion species. The cell reactions are

		$\mathcal{E}^\circ$
anode (oxidation)	$Pb_{(s)} + HSO^-_{4(aq)} \longrightarrow PbSO_{4(s)} + H^+_{(aq)} + 2e^-$	0.295
cathode (reduction)	$PbO_{2(s)} + HSO^-_{4(aq)} + 3H^+_{(aq)} + 2e^- \longrightarrow PbSO_{4(s)} + 2H_2O$	1.625
discharge reaction	$Pb_{(s)} + PbO_{2(s)} + 2H^+_{(aq)} + 2HSO^-_{4(aq)} \longrightarrow 2PbSO_{4(s)} + 2H_2O$	1.920

Note in the overall *discharge* reaction that H_2SO_4 is used up and $PbSO_{4(s)}$ is formed. The extent of discharge can thus be measured by the loss of H_2SO_4 in solution. Since the density of H_2SO_4 is greater than that of H_2O, the amount of H_2SO_4 in solution can be determined from the measured density of the solution. The cell reaction is reversible and, by applying a potential to the battery in excess of its discharge potential, the battery can be charged—the reaction reversed to restore the electrodes and the H_2SO_4.

NICKEL–CADMIUM STORAGE BATTERY

The nickel-cadmium battery is also used in automobiles. It has a longer expected lifetime; however, it is also more expensive. The reaction is carried out in a basic or alkaline medium and the battery is referred to as an alkaline storage battery. The cell reactions, especially the nickel half cell, are complicated and research is still being carried out to determine the mechanism. The oxidation states of nickel in the oxide and hydroxides are probably mixed, but the most common reaction is thought to go from Ni(III) in $NiO(OH)$ to Ni(II) in $Ni(OH)_2$. The cell reaction can best be represented by

anode (oxidation)	$Cd_{(s)} + 2OH^- \longrightarrow Cd(OH)_{2(s)} + 2\ e^-$
cathode (reduction)	$2NiO(OH)_{(s)} + 2H_2O + 2e^- \longrightarrow 2Ni(OH)_{2(s)} + 2OH^-$
	$Cd_{(s)} + 2NiO(OH)_{(s)} + 2H_2O \longrightarrow Cd(OH)_{2(s)} + 2Ni(OH)_{2(s)}$

Note that the OH^- ion does not appear in the overall cell reaction and for this reason the voltage of the cell does not change with use as was the case for the lead storage cell. The cell reaction is a function of the manner in which the electrodes are prepared, and the above cell reaction is only the best representation.[*]

[*]For a more complete discussion of this type of battery and other types of alkaline storage batteries, a recent text reviewing the subject should be consulted: S. U. Falk and A. J. Salkind, *Alkaline Storage Batteries*, John Wiley and Sons, Inc., New York, 1969.

DRY CELL OR LECLANCHÉ CELL

There are various types of dry cells but one of the most common is the common flashlight battery or Leclanché dry cell. Dry cells generally contain an electrolyte in a paste rather than a liquid solution of the electrolyte—hence the name "dry" cell. In the Leclanché cell the metal container is made of zinc and acts as the anode (oxidation). A graphite rod in the middle of the cylindrical container acts as the cathode where MnO_2, in a paste of $ZnCl_2$ and NH_4Cl, is reduced. The cell reaction for manganese is still not certain and may involve a mixture of products. The product of the MnO_2 reduction is thought to be the Mn(III) oxidation state. One possible cell reaction is

anode (oxidation)

$$Zn_{(s)} \longrightarrow Zn^{2+} + 2e^-$$

cathode (reduction)

$$2\,MnO_{2(s)} + 2NH_4^+ + 2e^- \longrightarrow Mn_2O_{3(s)} + H_2O + 2NH_3$$

$$Zn_{(s)} + 2MnO_{2(s)} + 2NH_4^+ \longrightarrow Zn^{2+} + Mn_2O_{3(s)} + H_2O + 2NH_3$$

The $Mn_2O_{3(s)}$ can also be present as a mixed oxide such as $ZnOMn_2O_{3(s)}$. The potential produced by this cell is 1.5 V in an open circuit. The voltage is lowered upon discharging.

MANGANESE–ZINC ALKALINE DRY CELL

The zinc–MnO_2 electrode materials can also be used in an alkaline medium with superior performance. Apparently in an alkaline medium the dry cell can retain a higher voltage for a longer period of time. The cell reactions are complicated but the best representation is

anode (oxidation)

$$2Zn_{(s)} + 4OH^- \longrightarrow 2Zn(OH)_2 + 4e^-$$

cathode (reduction)

$$3\,MnO_{2(s)} + 2H_2O + 4e^- \longrightarrow Mn_3O_{4(s)} + 4OH^-$$

$$2Zn_{(s)} + 3MnO_{2(s)} + 2H_2O \longrightarrow 2Zn(OH)_2 + Mn_3O_{4(s)}$$

Mn_3O_4 is actually a mixture $+2$ and $+3$ oxidation states of manganese: $Mn(II)Mn(III)_2O_4$.

Other alkaline dry cells employing $Zn + HgO$ or $Cd + HgO$ are also manufactured. The Zn–HgO cell is exceptionally good at maintaining a constant potential over long periods of time where very low currents are drawn. Due to this stability they are excellent reference batteries. The Cd–HgO dry cell has an exceptionally long shelf life (10 years) since the electrodes do not react except during discharge of the battery.

PROBLEM 14–16   a Write cell reactions for the Zn-HgO battery giving solid $Zn(OH)_2$ and liquid Hg as products in an alkaline medium.

b. Estimate the voltage from this dry cell. Use Appendix T, Table 16 for additional half-cell potentials if they are required.

c. Is OH^- consumed in this reaction?

d. Why should the potential of the cell remain constant even after some use?

PROBLEM 14–17   Write cell reactions for the Cd–HgO dry cell in an alkaline medium and estimate the potential of the cell.

* 14–H FUEL CELLS

Batteries convert chemical energy to electrical energy with high efficiency. However, the substances used as electrode material or reactants are not generally found as such in nature and these cells do not represent a true generation of power from natural raw materials. Electrical power today is generated in three basic steps:

1. Production of heat by burning a fuel such as natural gas (CH_4) or coal.
2. Conversion of heat into mechanical energy as in a steam turbine.
3. Conversion of mechanical energy into electrical energy with a generator.

The sum of these three processes results in the conversion of chemical energy in the fuel (CH_4) and in O_2 into electrical energy with an efficiency of only 35–40 %. In addition there is considerable air pollution from the burning of the fuel and thermal pollution from the steam. Also, the noise generated is often excessive.

A much more efficient conversion of chemical energy to electrical energy can be accomplished in an electrochemical cell where the efficiency can be 70% or higher. Since the burning of fuel is an oxidation-reduction reaction, in theory at least, the oxidation of a fuel can be used in an electrochemical cell to generate electrical power. Electrochemical cells using fuels are called *fuel cells* and they differ from electrochemical cells only in that the reaction is carried out irreversibly with the products of the reaction discharged into the atmosphere. The oxidation of fuels in fuel cells is essentially complete and air pollution from hydrocarbons and partially oxidized hydrocarbons (Section 19–E) is essentially eliminated. Furthermore, no excessive heat is produced and no moving parts are involved as in a turbine. For these reasons fuel cells appear to be very promising for generating electrical power in the future. Only the hydrogen–oxygen fuel cell has been sufficiently well-developed to work efficiently. Since H_2 is not a natural fuel, this fuel cell is not useful for large scale power generation. The H_2–O_2 fuel cell has been useful in attaining maximum electrical power for a minimum weight as required in the space program. Much research is being carried out on fuel cells employing natural fuels such as methane (CH_4) and propane (C_3H_8). Other non-hydrocarbon fuels are also being considered.

Any of the following reactions involving fuel oxidation can be employed, at least potentially, in a fuel cell.

$$H_2 + \tfrac{1}{2}O_2 \longrightarrow H_2O$$

$$CH_4 + 2O_2 \longrightarrow CO_2 + 2H_2O$$

$$C_3H_8 + 5O_2 \longrightarrow 3CO_2 + 4H_2O$$

$$CH_3OH + \tfrac{3}{2}O_2 \longrightarrow CO_2 + 2H_2O$$

The hydrogen–oxygen fuel cells consist of the following electrode and cell reactions

anode (oxidation) $H_{2(g)} \longrightarrow 2H_{(aq)}^{+} + 2e^{-}$

cathode (reduction) $\underline{2H_{(aq)}^{+} + \tfrac{1}{2}O_{2(g)} + 2e^{-} \longrightarrow H_2O_{(g)}}$

$$H_{2(g)} + \tfrac{1}{2}O_{2(g)} \longrightarrow H_2O_{(g)}$$

The reaction can also be carried out in an alkaline medium which causes less deterioration of the electrodes. The cell reactions in alkaline medium are

anode (oxidation) $H_{2(g)} + 2OH_{(aq)}^{-} \longrightarrow 2H_2O + 2e^{-}$

cathode (reduction) $\underline{\tfrac{1}{2}O_{2(g)} + H_2O + 2e^{-} \longrightarrow 2OH_{(aq)}^{-}}$

$$H_{2(g)} + \tfrac{1}{2}O_{2(g)} \longrightarrow H_2O_{(g)}$$

A schematic diagram of the hydrogen-oxygen fuel cell in alkaline medium is shown in Figure 14–6. Note that the overall process shown in Figure 14–6 is the intake of H_2 and O_2 and the output of H_2O. The electron flow could run a direct current motor.

The fuel cell utilizing methane in acid medium would operate according to the reactions

anode (oxidation) $CH_4 + 2H_2O \longrightarrow CO_2 + 8H_{(aq)}^{+} + 8e^{-}$

cathode (reduction) $\underline{2O_2 + 8H^{+} + 8e^{-} \longrightarrow 4H_2O}$

$$CH_4 + 2O_2 \longrightarrow CO_2 + 2H_2O$$

PROBLEM 14–18   Draw a schematic diagram showing the operation of a CH_4 fuel cell.

PROBLEM 14–19   Write the cell reactions and draw a schematic diagram for the cell construction of a propane fuel cell in an acid medium.

FIGURE 14-6 Hydrogen–oxygen fuel cell in an alkaline electrolyte (M represents an electric motor)

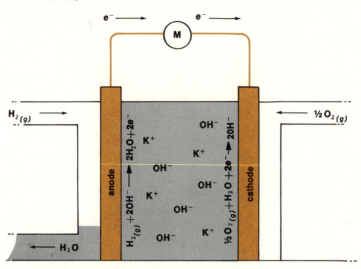

PROBLEM 14-20 Calculate the maximum cell potential that can be produced from a. H_2-O_2 fuel cell, b. CH_4-O_2 fuel cell. [Hint: Calculate $\Delta G°$ for the reaction; $\Delta G° = -nF\mathcal{E}°$.]

The difficulty in the technology of fuel cells employing hydrocarbons has been the speed or rate of the electrode reactions. For example, the reaction of CH_4 with $2H_2O$ to give CO_2 and $8H^+$ is very complex and the reaction is quite slow. Suitable catalysts can increase the rate of a reaction and much of the fuel cell research today is a search for such a catalyst. The tremendous possibilities that fuel cells offer for efficient power generation with little or no pollution has kept research on fuel cells very active. In the future, we may see automobiles with electric motors powered by fuel cells.

* 14- I. ELECTROLYSIS

Electrolysis is the process whereby an electrical potential applied between two electrodes causes a chemical change of the oxidation-reduction type to occur. Since electrical energy is being converted to chemical free energy, the processes occurring in electrolysis is opposite to those occurring in any electrochemical cell.

In order for electrolysis to occur the applied electrical potential must exceed a specific value which is a function of the materials being electrolyzed and the electrode materials. The electrolysis of molten alkali metal salts, used for the preparation of the alkali metals (Section 9–G), can be used as an example. A

FIGURE 14–7   Electrolysis of molten NaCl

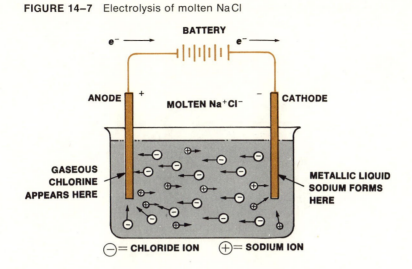

molten alkali salt, such as NaCl, contains Na^+ and Cl^- ions, and these ions will migrate to the electrodes under an applied potential as shown in Figure 14–7.

$$Na_{(s)} \longrightarrow Na^+_{(molten\,NaCl)} + e^- \qquad\qquad \mathcal{E}_1$$

$$\tfrac{1}{2}Cl_{2(g)} + e^- \longrightarrow Cl^-_{(molten\,NaCl)} \qquad\qquad \mathcal{E}_2$$

$$Na_{(s)} + \tfrac{1}{2}Cl_{2(g)} \longrightarrow Na^+_{(molten\,NaCl)} + Cl^-_{(molten\,NaCl)} \qquad (\mathcal{E}_1 + \mathcal{E}_2)$$

Since the reaction is reversible, the application of a potential greater than the potential of the electrochemical cell $(\mathcal{E}_1 + \mathcal{E}_2)$ will cause electrolysis to occur

$$Na^+_{(molten\,NaCl)} + Cl^-_{(molten\,NaCl)} \xrightarrow{\;\mathcal{E} > (\mathcal{E}_1 + \mathcal{E}_2)\;} Na_{(s)} + \tfrac{1}{2}Cl_{2(g)}$$

In some cases the electrolysis reaction will proceed very slowly, or not at all, when the applied electrical potential $\mathcal{E}$ just exceeds $(\mathcal{E}_1 + \mathcal{E}_2)$. This is especially true when a gas, such as H_2, is produced at a solid electrode. In these cases the $\mathcal{E}$ must be increased in order to accelerate the electrode processes in the electrolysis. This extra voltage that must be applied is called the *overvoltage* $(\mathcal{E}_{ov})$. Overvoltages range anywhere from approximately zero to on the order of 1.5 V.

In the electrolysis of molten NaCl there is little doubt concerning the nature of the reaction that takes place since there are only Na^+ and Cl^- ions in the liquid phase. However, in the electrolysis of a solution other ions and the solvent molecules are present which might also participate in the electrode reactions. When alternative reactions can occur at an electrode, the process which occurs is the

FIGURE 14–8 Electrolysis of a 1.0 M aqueous NaCl solution

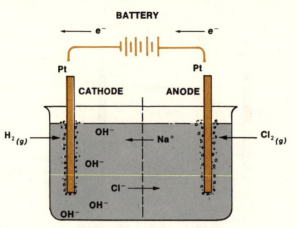

one which requires the lowest potential. The required potential consists of the half-cell potential plus the overvoltage if such exists at that electrode.

$$\mathcal{E} = \mathcal{E}_{\frac{1}{2}} + \mathcal{E}_{ov}$$

Since both $\mathcal{E}_{\frac{1}{2}}$ and $\mathcal{E}_{ov}$ must be considered, it is not always easy to predict what electrolysis reaction will occur in a solution.

For example, let us consider the electrolysis of a 1.0 M aqueous solution of NaCl instead of molten NaCl. Experimentally we observe that $H_{2(g)}$ is produced at the cathode (reduction) and Cl_2 is produced at the anode (oxidation). Also the solution in the vicinity of the cathode becomes basic during electrolysis. A diagram showing the electrolysis reaction is shown in Figure 14–8 where the dashed line represents an imaginary boundary between the cathode compartment and the anode compartment. When the potential is applied to the electrodes, an electric field gradient is set up and the charged ions will migrate under the influence of the field gradient: $Na^+_{(aq)}$ to the cathode and $Cl^-_{(aq)}$ to the anode.

With this information on the electrolysis of aqueous NaCl we might ask what reactions occur at the electrodes and why do these reactions occur in preference to others? At the cathode there are two reactions to be considered and these are given below ($[Na^+] = 1$ M, $[Cl^-] = 1$ M, $[H^+] = [OH^-] = 10^{-7}$ M).

HALF-CELL REACTION	$\mathcal{E}^\circ_{\frac{1}{2}}$	$\mathcal{E}_{\frac{1}{2}}$	$\mathcal{E}_{ov}$	$\mathcal{E}_{\frac{1}{2}} + \mathcal{E}_{ov}$
$Na^+_{(aq)} + e^- \longrightarrow Na_{(s)}$	−2.71	−2.71	?	−2.71
$H_2O + e^- \longrightarrow \frac{1}{2}H_{2(g)} + OH^-_{(aq)}$	−0.82	−0.41	−0.1 to −0.3	−0.5 to −0.7

An alternative to the latter reaction is the reduction of $H^+_{(aq)}$ to $H_{2(g)}$ which will also produce an excess of OH^- since $H^+_{(aq)}$ ions are used in the reaction. However, since

H_2O is far in excess of $H^+_{(aq)}$, the reduction of H_2O given above is preferred. The $H_2O|H_{2(g)}$ half-cell potential is actually -0.41 rather than -0.82 V since the concentration of the product OH^- is lowered from standard conditions of $[OH^-] = 1$ M to $[OH^-] = 10^{-7}$ M. The overvoltage for H_2 on platinum depends upon the amount of current applied to the electrolysis cell and for this reason the overvoltage can range from -0.1 to -0.3 volts. The overvoltage for sodium cannot be measured in this system but it is probably negligible. Taking the sum $(\mathcal{E}_{\frac{1}{2}} + \mathcal{E}_{ov})$ we see that the second cathode reaction would require the lowest potential for reduction to occur. Note that $OH^-_{(aq)}$ is also produced so the solution in the cathode compartment should be alkaline. The Na^+ ions migrate into the cathode compartment whereas the Cl^- ions migrate out leaving an excess of $Na^+ + OH^-$. Possible reactions at the anode are

HALF-CELL REACTION	$\mathcal{E}^\circ_{\frac{1}{2}}$	$\mathcal{E}_{\frac{1}{2}}$	$\mathcal{E}_{ov}$	$\mathcal{E}_{\frac{1}{2}} + \mathcal{E}_{ov}$
$H_2O \longrightarrow \frac{1}{2}O_{2(g)} + 2H^+_{(aq)} + 2e^-$	-1.23	-0.82	-0.8	-1.62
$Cl^-_{(aq)} \longrightarrow \frac{1}{2}Cl_2 + e^-$	-1.36	-1.36	$-$	-1.36

The $\mathcal{E}_{\frac{1}{2}}$ for the formation of O_2 in the first reaction is more favorable than the formation of Cl_2 from Cl^- at 1 M concentration. However, the evolution of $O_{2(g)}$ on platinum has a significant overvoltage $(\sim -0.8$ V) and when this is taken into account the formation of $Cl_{2(g)}$ is favored. Therefore the overall cell reaction which would require the smallest potential is

$$\begin{array}{rl} & \mathcal{E}_{\frac{1}{2}} + \mathcal{E}_{ov} \\ H_2O + e^- \longrightarrow \frac{1}{2}H_{2(g)} + OH^-_{(aq)} & \overline{-0.5 \text{ to } -0.7} \\ Cl^-_{(aq)} \longrightarrow \frac{1}{2}Cl_{2(g)} + e^- & \underline{-1.36} \\ H_2O + Cl^-_{(aq)} \longrightarrow \frac{1}{2}H_{2(g)} + \frac{1}{2}Cl_{2(g)} + OH^-_{(aq)} & -1.9 \text{ to } -2.1 \text{ V} \end{array}$$

A negative sign on the potential means that the potential must be supplied rather than produced by the cell. The cell potential required is on the order of 1.9 to 2.1 volts.

PROBLEM 14–21 a. Predict the electrolysis products and the reactions for a 0.1 M K_2SO_4 aqueous solution using platinum electrodes. Use the data given in the discussion of the electrolysis of aqueous $NaCl$ and the following half-cell potentials for SO_4^{2-} in acid solution

$$\begin{array}{cc} & \mathcal{E}^\circ \text{ (volts)} \\ 2SO_4^{2-} + 4H^+ + 2e^- \longrightarrow S_2O_6^{2-} + 2H_2O & -0.22 \\ \\ S_2O_8^{2-} + 2e^- \longrightarrow 2SO_4^{2-} & +2.01 \end{array}$$

Assume the SO_4^{2-} reactions do not require an overvoltage and the solution of K_2SO_4 is neutral (pH $= 7.0$). Further assume that the concentration of $S_2O_6^{2-}$

and $S_2O_8{}^{2-}$ must be 10^{-4} M in order that electrolysis to these products be significant.

b. Estimate the potential required for this electrolysis.

PROBLEM 14–22   a. Predict the electrolysis products and the reactions for a 1 M $CuSO_4$ aqueous solution. The half-cell potentials for $Cu^{2+}_{(aq)}$ are

$$\mathcal{E}° \text{ (volts)}$$

$$Cu^{2+} + 2e^- \longrightarrow Cu \qquad\qquad 0.34$$

$$Cu^{2+} + e^- \longrightarrow Cu^+ \qquad\qquad 0.158$$

[Hint: Is Cu^+ stable? See Problem 14–10.]

b. Estimate the potential required for the electrolysis.

The calculation of the amount of product formed during an electrolysis reaction is given by Faraday's laws which state: The weight of a substance produced at an electrode is dependent on two factors: (1) amount of electricity passed through the cell and (2) the equivalent weight of the material. The two factors are related quantitatively through the *faraday* which is the amount of electricity that will produce 1 equivalent of a substance. One equivalent of material reacts with 96,500 coulombs. The amount of electricity, *q* (in coulombs), passed through an electrolytic cell is given by the product of the current, I (in amps), and the time, *t* (in seconds). An ampere is defined as the flow of one coulomb of charge past a given point in one second.

$$q(\text{coulombs}) = I\ (\text{amps})\ t\ (\text{sec})$$

The calculation of the weight or volume of material from equivalents uses the conversions summarized in Figure 3–7.

$$\text{equivalents} \longrightarrow \text{moles} \dashrightarrow \text{volume}$$
$$\downarrow$$
$$\text{mass}$$

EXAMPLE 14–6   *Calculate a. equivalents, b. weight, and c. volume at standard temperature and pressure of H_2 produced in the electrolysis of aqueous $NaCl$ with a current of 0.05 amps for 1 hour.*

a. The amount of electricity, q, is calculated

$$q\ (\text{coulombs}) = I\ (\text{amps})\ t\ (\text{sec})$$

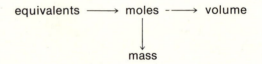

$$q = (0.05\ \text{amps})(1\ \text{hour})\left(\frac{60\ \text{min}}{1\ \text{hr}}\right)\left(\frac{60\ \text{sec}}{1\ \text{min}}\right)$$

$$q = 180\ \text{coulombs}$$

The number of equivalents of H_2 produced is then calculated from

$$(180 \text{ coulombs})\left(\frac{1 \text{ eq } H_2}{96,500 \text{ coulombs}}\right) = 0.001865 \text{ eq } H_2$$

b. Since two electrons are involved in the electrolysis to produce one molecule of H_2

$$2H_2O + 2e^- \longrightarrow H_{2(g)} + 2OH^-_{(aq)}$$

$$(0.001865 \text{ eq } H_2)\left(\frac{1 \text{ mole } H_2}{2 \text{ eq } H_2}\right) = 0.000933 \text{ moles } H_2$$

The molecular weight of $H_2 = 2.016$ g

$$(0.000933 \text{ moles})\left(\frac{2.016 \text{ g } H_2}{mole}\right) = 0.00188 \text{ g } H_2$$

c. The calculation of the volume of H_2 produced at STP requires the moles of H_2 produced, which was calculated above to be 0.000933.

$$(0.000933 \text{ moles})\left(\frac{22.4 \text{ liters } H_2 \text{ at STP}}{mole}\right) = 0.0209 \text{ liters } H_2 \text{ at STP}$$

PROBLEM 14-23   Calculate the equivalents, moles, grams, and volume at STP of Cl_2 produced in the electrolysis of aqueous $NaCl$ when a current of 2 amps have been passed through the cell for 2 minutes.

PROBLEM 14-24   Calculate the mass in grams of Cu deposited in Problem 14-22 if the electrolysis were carried out for 5 minutes at 0.75 amps.

* 14-J. CORROSION

Corrosion is the destructive attack of a metal by chemical or electrochemical reactions with its environment. The destructive chemical reactions involve oxidation-reduction reactions and are therefore related to electrochemistry. The importance of understanding and preventing of corrosion is best emphasized by the fact that approximately 1/3 of all iron ever produced has been lost by corrosion.* This amounts to a loss of billions of dollars per year because of corrosion. Since iron is the principal metal used in construction, most of our discussion will be concerned with the corrosion of iron.

Before discussing corrosive reactions, we will first examine a phase diagram (Figure 14-9) depicting the forms of iron which are stable at various combina-

*G. Kortum, Treatise in Electrochemistry, Elsevier Publishing Co., New York, 1965.

tions of reduction potential, $\mathcal{E}$, and solution pH. When the combination of $\mathcal{E}$ and pH favor the production of Fe^{2+} and Fe^{3+}, corrosion will occur. When the combination of $\mathcal{E}$ and pH give various iron oxides and hydroxides (Fe_2O_3, Fe_3O_4, $Fe(OH)^{2+}$ or FeO_4^{2-}), the term *passivity* is used to designate the fact that the rate of corrosion becomes infinitesimally small. When the reduction potential is sufficiently low, the iron is unreactive regardless of pH (up to approximately pH 10). This region is designated as *immunity* in Figure 14–9. The boundaries between corrosion and immunity and between corrosion and passivity are indistinct (represented by the shaded areas in Figure 14–9), since $\mathcal{E}$ is a function of the concentrations of the ions present. For example, the line between corrosion and immunity can move from −0.62 to −0.44 V when the concentration of Fe^{2+} changes from 10^{-6} M to 1 M.

The corrosion of iron can be classified into three different types based upon the nature of the electrochemical reaction taking place. Each of these will be discussed separately.

ANODIC REACTION INDUCED BY METAL IONS

The standard reduction potential for the $Fe^{2+}|Fe_{(s)}$ half cell is −0.44 V but the reduction potential is, of course, dependent upon the concentration of Fe^{2+}. The following table shows the concentration-dependent potentials for the *oxidation* reactions (note opposite sign for potentials).

$$Fe_{(s)} \longrightarrow Fe^{2+} + 2e^-$$

$\mathcal{E}$ (volts)	$[Fe^{2+}]$
0.44	1 M
0.50	10^{-2} M
0.56	10^{-4} M
0.62	10^{-6} M

Note in Figure 14–9 that $Fe_{(s)}$ is stable in the presence of an oxidizing agent whose reduction potential is less than −0.44 to −0.62 V and that corrosion can occur at potentials greater than −0.44 to −0.62 V. This potential is independent of pH up to pH of about 6 to 9. Corrosion by oxidation to Fe^{2+} can potentially occur when $Fe_{(s)}$ is in contact with a metal ion (M^{n+}) whose reaction potential for reaction to the metal or other oxidation state is greater than approximately −0.6 to −0.4 V.

$$M^{n+} + ne^- \longrightarrow M_{(s)} \qquad \mathcal{E} > -0.4 \text{ V}$$

or

$$M^{n+} + (n - m)e^- \longrightarrow M^{m+} \qquad \mathcal{E} > -0.4 \text{ V}$$

For example, $Cu^{2+}|Cu_{(s)}$ has a standard reduction potential of 0.34 V; therefore, Cu^{2+} will corrode $Fe_{(s)}$

FIGURE 14–9 Range of $\mathcal{E}$ and pH to give immunity, corrosion, and passivity for iron

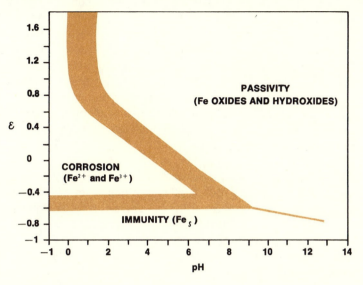

		volts
cathode (reduction)	$Cu^{2+} + 2e^- \longrightarrow Cu_{(s)}$	0.1 to 0.3
anode (oxidation)	$Fe_{(s)} \longrightarrow Fe^{2+} + 2e-$	0.4 to 0.6
	$Fe_{(s)} + Cu^{2+} \longrightarrow Fe^{2+} + Cu_{(s)}$	0.5 to 0.9

The range of potentials given correspond to changes in $[Fe^{2+}]$ and $[Cu^{2+}]$ from 1 M to 10^{-6} M. This prediction, based on the cell potential, is subject to the restriction of any thermodynamic prediction; that is, the reaction is feasible, but, it may be very slow.

PROBLEM 14–25   Which of the following ions could cause iron to corrode by anodic reaction: Zn^{2+}, Pb^{2+}, Hg_2^{2+}, Al^{3+}, Sn^{2+}? Assume the concentration of each ion is 10^{-2} M.

CORROSION WITH EVOLUTION OF H_2

The $H^+|H_{2(g)}$ half cell has a potential ranging from -0.41 V when $[H^+] = 10^{-7}$ M (pure water) to 0.0 V when $[H^+] = 1$ M.

$$H^+_{(aq)} + e^- \longrightarrow \tfrac{1}{2}H_{2(g)}$$

$[H^+]$	$\mathcal{E}$
1 M	0
10^{-2} M	-0.12
10^{-4} M	-0.24
10^{-6} M	-0.36
10^{-7} M	-0.41

Therefore, according to thermodynamic principles, H_2O with $[H^+] = 10^{-7}$ M can oxidize (corrode) any metal whose reduction potential is less than -0.4 volts. This would include iron for which the reduction potential ranges from -0.44 to -0.62. However, from the discussion in Section 14–I, we know that H_2 has a considerable overvoltage on most metallic electrodes and the potential required for H_2 evolution is considerably greater than the half-cell potential. On iron electrodes this overvoltage increases from 0.1 V to 0.8 V as the current flow increases. Because of the overvoltage of H_2 on iron, iron will not corrode in *pure* H_2O. If the water is made acidic, Fe can be oxidized to Fe^{2+}; however, highly acidic conditions are not normally present in most environmental conditions and this is not a common way for the corrosion of iron to occur.

PROBLEM 14–26   The overvoltage of H_2 on zinc ranges from 0.7 to 1.1 V. Predict whether water can cause zinc to corrode. Will acidic solutions react with zinc? Assume that $[Zn^{2+}]$ must become 10^{-2} M in order for the corrosion to be significant.

PROBLEM 14–27   The overvoltage of H_2 on copper ranges from 0.2 to 0.82 V. Predict the corrosive or noncorrosive properties of copper in water and in acidic conditions ($[H^+] = 1$ M). Assume that $[Cu^{2+}]$ must become 10^{-2} M in order for the corrosion to be significant.

CORROSION WITH THE CONSUMPTION OF O_2

This is by far the most common mechanism by which the corrosion of iron occurs. Oxygen in the atmosphere dissolves to a certain extent in water (see Section 12–I) and the following electrode reactions can occur.

$$\tfrac{1}{2}O_{2(g)} + H_2O + 2e^- \longrightarrow 2OH^-$$

$[OH^-]$	$\mathcal{E}$
1 M	0.40
10^{-2} M	0.52
10^{-4} M	0.64
10^{-6} M	0.76
10^{-7} M	0.81

$$\tfrac{1}{2}O_{2(g)} + 2H^+ + 2e^- \longrightarrow H_2O$$

$[H^+]$	$\mathcal{E}$
10^{-7} M	0.81
10^{-6} M	0.87
10^{-4} M	0.99
10^{-2} M	1.11
1 M	1.23

The reduction potential for O_2 is very sensitive to pH and varies from 0.40 V in 1 M base to 0.81 V in neutral solution to 1.23 V in 1 M acid. These potentials assume $P_{O_2} = 1$ atm. However, since atmospheric oxygen has $P_{O_2} \approx 0.2$ atm, the potential is reduced somewhat. In any event the potentials are sufficient to oxidize iron to Fe^{2+}. The sequence of reaction steps involved in corrosion with consumption of O_2 in a basic or neutral solution are:

1. oxidation of Fe to Fe^{2+}

cathode (reduction) $\frac{1}{2}O_{2(g)} + H_2O + 2e^- \longrightarrow 2OH^-$ 0.4 to 0.8

anode (oxidation) $Fe_{(s)} \longrightarrow Fe^{2+} + 2e^-$ 0.4 to 0.6

$Fe_{(s)} + \frac{1}{2}O_{2(g)} + H_2O \longrightarrow Fe^{2+} + 2OH^-$ 0.8 to 1.4

2. formation of a precipitate when the $[Fe^{2+}][OH^-]^2$ product exceeds the K_{sp} of $Fe(OH)_2$ which is 1.6×10^{-14}.

$$Fe^{2+} + 2OH^- \rightleftharpoons Fe(OH)_2$$

3. oxidation of the $Fe(OH)_2$ by oxygen to form rust ($Fe_2O_3 \cdot H_2O$).

$$2Fe(OH)_2 + \frac{1}{2}O_{2(g)} \longrightarrow Fe_2O_3 \cdot H_2O \text{ (reddish brown)} + H_2O$$

4. formation of magnetite, Fe_3O_4, as an intermediate step, if insufficient O_2 is dissolved.

$$3Fe(OH)_2 + \frac{1}{2}O_{2(g)} \longrightarrow Fe_3O_4 \cdot H_2O \text{ (green)}$$

$$Fe_3O_4 \cdot H_2O \longrightarrow H_2O + Fe_3O_4 \text{ (black)}$$

Steps 1 through 4 are the reactions most commonly encountered in the corrosion of iron under normal atmospheric conditions. The need for both O_2 and H_2O in order to cause corrosion of iron is obvious.

Evidence for these various steps in the oxidation process can be found by examination of iron which has undergone extensive corrosion. Underneath the exterior reddish-brown rust, there often are layers of different oxidized products. Since the oxygen must diffuse through the water and through the oxide layers, the concentration of O_2 will be diminished nearest the iron surface. Consequently the lower oxidation states of iron are found at the greatest penetration of the corrosion. The layers of successive oxidation products are shown in Figure 14-10.

Salts such as NaCl generally accelerate the corrosion process by supplying Na^+ and Cl^- ions. Ion migration increases the conductance of the water in the same manner as the salt bridge of an electrochemical cell. This allows the cathode and anode reactions to function when separated over greater distances. Since the Fe^{2+} produced at the anode may be separated by substantial distances from the OH^- produced at the cathode, the ions must diffuse into the solution to cause the

FIGURE 14–10   The surface of corroded iron

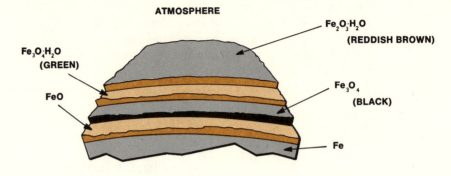

precipitation of Fe(OH)$_2$. Since the Fe(OH)$_2$ is formed away from the anode and cathode, it will not protect the metal from further corrosion (passivity).

Because of the immense economic importance of corrosion, a great deal of research has been carried out to develop methods for its prevention. The preventive measures which are most widely used are:

1. metallic coatings
2. nonmetallic coatings
3. addition inhibitors
4. cathodic protection

Metallic coatings, such as zinc, gold, silver, tin, or chromium, can be applied to a metal surface by electrolysis, spraying, deposition from the vapor phase, or dipping into the hot molten metal. The outer metallic coating is chosen to be more resistant to corrosion and so protects the metal being coated by excluding the atmosphere and moisture. In the case of iron, corrosion can be prevented by electroplating a thin layer of zinc on the surface. This is called *galvanized* iron. Both pipe and sheet metal are commonly galvanized. Iron can also be protected from corrosion with a thin coating of tin. This is usually accomplished by dipping the object in hot, molten tin. The so-called tin can is actually iron with a thin coating of tin applied in this manner. Extreme care should be exercised that the protective tin coating is not scratched or punctured since this will lead to rapid and extensive corrosion. In the damaged area where the protective metal has been broken, an electrochemical cell is set up between the two metals and the more easily oxidized iron metal becomes the anode. Since only a small area of the iron (anode) is exposed compared to the large cathode surface (metallic coating), the anode reacts rapidly. This is shown in Figure 14–11 where the cathodic reaction is assumed to be the reduction of H$_2$O to give H$_2$. Though the protection from corrosion fails rapidly if there is the slightest abrasion or crack in the tin coating, this problem does not arise from zinc coatings; the Zn^{2+}|Zn$_{(s)}$ half cell has a lower reduction potential than the Fe^{2+}|Fe$_{(s)}$ half cell and the zinc will act as the anode.

FIGURE 14–11   Corrosion at a break in tin-coated iron

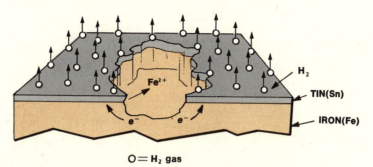

$O = H_2$ gas

Nonmetallic coatings such as paint or grease can be applied, but these are even more subject to abrasion than metallic coatings. A rather effective nonmetallic coating for some metals is the oxide of the material. As discussed in Chapter 9, aluminum forms a very stable oxide Al_2O_3 which adheres to the aluminum metal and prevents further corrosion. This aluminum oxide coating can also be put on the aluminum surface by electrolysis in a dilute H_2SO_4 solution where $Al_{(s)}$ is the anode. The process is called *anodizing*. The oxide so formed must be hydrated with steam or hot water to improve its protective properties. The oxide may be dyed various colors. This process is frequently done with aluminum and alloys of aluminum and it is called *anodized aluminum*. The resulting coating is a firm, thick Al_2O_3 which is a nonconductor of electricity. An electrolyte of phosphate ions and phosphoric acid can form strongly adhering phosphates on iron, steel, zinc, and cadmium. These phosphates will also protect the metal from corrosion.

Corrosion can also be prevented or slowed down by adding appropriate *inhibitors* to the metal. The type of inhibitor depends upon the nature of the metal and whether you want to inhibit an anode or cathode reaction. Cathodic inhibitors prevent the formation of H_2 by increasing the overvoltage. Some organic substances can also act as inhibitors when adsorbed on the surface of the metal, by preventing the surface from being exposed to the atmosphere. Red lead (Pb_3O_4) and zinc chromate ($ZnCrO_4$) are added to paints as inhibitors. It is thought that the ions PbO_4^{4-} and CrO_4^{2-} create passivity on the iron surface, thus preventing corrosion.

One of the most effective means of preventing corrosion of metals in underground metal structures, such as pipelines or piers, is called *cathodic protection*. The protection is furnished by a metal which is more easily oxidized than the substance being protected. When the substances are placed in contact underground an electrochemical cell is set up where the more readily oxidizable metal becomes the anode forcing the substance being protected to become the cathode. For example, zinc or magnesium are used to protect iron. Recall from Figure 14–8 that the $Fe^{2+}|Fe_{(s)}$ half cell has a reduction potential in the range of -0.44 to -0.62 V. In order for iron to act as the cathode, the other metal must have a reduction potential less than -0.62 V (equivalent to an oxidation potential greater than $+0.62$ V).

FIGURE 14–12 Cathodic protection of an underground pipe

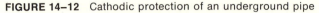
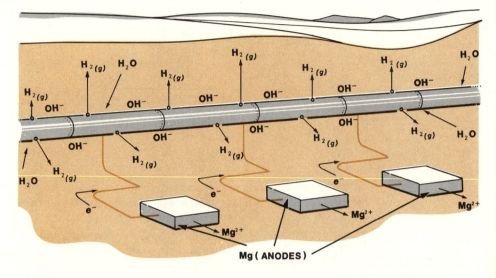

Mg (ANODES)

		$\mathcal{E}$ (volts)
anode (oxidation)	$M_{(s)} \longrightarrow M^{2+}_{(aq)} + 2e^-$	$> +0.62$
cathode (reduction)	$Fe^{2+}_{(aq)} + 2e^- \longrightarrow Fe_{(s)}$	-0.44 to -0.62
	$M_{(s)} + Fe^{2+}_{(aq)} \longrightarrow M^{2+}_{(aq)} + Fe_{(s)}$	> 0

Since Fe^{2+} is being reduced, this prevents iron metal from being oxidized to Fe^{2+}. Magnesium is commonly used as the anodic material for iron. In order to protect an iron pipe magnesium rods are placed at intervals along the pipe (Figure 14–12). The distance between rods depends upon the conductance of the soil or water. Since the anodic material is oxidized, it must be replaced periodically.

An alternative way in which to provide cathodic protection is to apply an external direct current such that the material being protected acts as the cathode. The anode can be scrap iron or graphite rods.

PROBLEM 14–28 In order to supply cathodic protection to a lead pipe, what oxidization potential should the anodic metal have? Select a substance that would provide protection to lead.

PROBLEM 14–29 Iron can be protected from corrosion with a tin coating.
a. If a scratch is made in the tin coating, thereby exposing the iron to the atmosphere, what electrochemical cell would be created? Estimate its potential assuming $[H^+] = [OH^-] = 10^{-7}$ M and $O_{2(g)}$ is present from the atmosphere ($P_{O_2} = 0.2$ atm). Further assume that the corrosion of Fe is significant when $[Fe^{2+}] = 10^{-2}$ M.
b. What happens to the iron in this reaction?

c. If an anodized aluminum can is scratched, what reactions will take place?

d. Will the corrosion of the scratched anodized aluminum be extensive?

e. Why does the disposal of aluminum cans present more of a problem than "tin" cans?

EXERCISES

1. Balance the following equations. Add additional products or reactants such as H_2O, H^+, or OH^- as required.

   a. $HNO_3 + HCl \longrightarrow NO + Cl_{2(g)}$ (acid solution)

   b. $Sn_{(s)} + HNO_3 \longrightarrow SnO_{2(s)} + NO_2$ (acid solution)

   c. $FeCl_3 + H_2S \longrightarrow FeCl_2 + HCl + S_{(s)}$ (acid solution)

   d. $I_{2(s)} + HNO_3 \longrightarrow HIO_3 + NO_{2(g)} + H_2O$ (acid solution)

   e. $Bi(OH)_3 + Na_2SnO_2 \longrightarrow Bi + Na_2SnO_3$ (basic solution)

   f. $MnO_{4(aq)}^- + NO_{2(aq)}^- \longrightarrow MnO_{2(s)} + NO_{3(aq)}^-$ (basic solution)

2. Balance the following reactions in which a single substance is both oxidized and reduced (disproportionation).

   $$Cu_{(aq)}^+ \longrightarrow Cu_{(s)} + Cu_{(aq)}^{2+} \qquad \text{(acid solution)}$$
   $$Cl_{2(g)} \longrightarrow OCl_{(aq)}^- + Cl_{(aq)}^- \qquad \text{(basic solution)}$$
   $$ClO_{2(aq)} \longrightarrow ClO_{2(aq)}^- + ClO_{3(aq)}^- \qquad \text{(acid solution)}$$

   Break up each reaction into two half-cell reactions where the same substance is oxidized in one half-cell reaction and reduced in the other half-cell reaction.

3. Using standard half-cell potentials from Appendix T, Table 16, predict whether the reactions in Exercises 1 (a, c, d, f) and 2 are feasible at standard conditions.

4. Consider the following electrochemical cell.

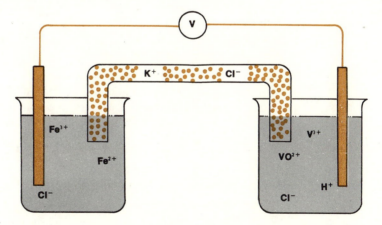

a. Write the half-cell reactions that occur in this electrochemical cell.

b. What potential will this cell generate if all reactants and products are at standard conditions? In which direction would the electrons flow?

c. Show the direction of flow of ions in and out of the salt bridge.

5. The half-cell reaction for the reduction of pyruvic acid to lactic acid is given by

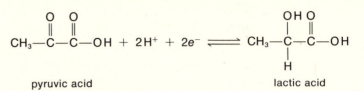

pyruvic acid   lactic acid

The standard potential for this half cell at 35 °C is 0.289 V. The half cell for the reduction of $O_{2(g)}$ can be written as

$$O_{2(g)} + 4H^+_{(aq)} + 4e^- \rightleftharpoons 2H_2O_{(l)}$$

The standard potential for this half cell is 1.23 V at 35 °C.

a. Write the cell reaction for the oxidation of lactic acid to pyruvic acid using O_2 gas. Is the potential for this cell dependent upon the pH of the solution?

b. Calculate the cell potential for this reaction and determine whether the oxidation of lactic acid by O_2 gas is feasible at 35 °C under standard conditions.

6. For Problem 10–18 in Chapter 10, we calculated $\Delta G°$ for the formation of glucose from CO_2 and H_2O. The oxidation of glucose is the reverse of that reaction

$$C_6H_{12}O_{6(aq)} + 6O_{2(g)} \longrightarrow 6CO_{2(g)} + 6H_2O_{(l)}$$

This reaction can be thought of as the combination of two half-cell reactions

    1. $C_6H_{12}O_{6(aq)} + 6H_2O_{(l)} \rightleftharpoons 6CO_{2(g)} + 24H^+_{(aq)} + 24e^-$
    2. $O_{2(g)} + 4H^+_{(aq)} + 4e^- \rightleftharpoons 2H_2O_{(l)}$

a. Calculate the standard potential for the cell composed of these two half cells using the $\Delta G°$ from Problem 10–19.

b. Calculate $\mathcal{E}^\circ_{\frac{1}{2}}$ for the oxidation of glucose according to half-cell reaction 1. Use the standard reduction potential for O_2 found in Appendix T, Table 16.

7. a. List the following metals in the order of increasing resistivity towards corrosion in an acidic solution: Fe, Cu, Cs, Be, Au, Ce, Al, Pb, Pd, Mg, Tl, V, Sr, Pt.

b. Which elements in part a. would be stable in the presence of $O_{2(g)}$ ($P = 1$ atm) in water? The half-cell potential for O_2 in water is

$$O_{2(g)} + 4H^+ (10^{-7}\ M) + 4e^- \rightleftharpoons 2H_2O_{(l)} \qquad \mathcal{E} = 0.82\ V$$

Assume an O_2 overvoltage of 0.8 V.

c. For the metals in part a., note their position in the periodic table. In what groups in the periodic table do you find the metals which are resistant to corrosion?

8.
$$Fe^{3+} + e^- \longrightarrow Fe^{2+} \qquad\qquad 0.770\ V$$
$$Cu^{2+} + e^- \longrightarrow Cu^+ \qquad\qquad 0.158\ V$$
$$VO^{2+} + 2H^+ + e^- \longrightarrow V^{3+} + H_2O \qquad\qquad -0.49\ V$$

a. Using these half-cell reduction potentials, show that in water Cu^+ can reduce Fe^{3+} to Fe^{2+} and the Cu^{2+} which is produced can oxidize V^{3+} to VO^{2+}. The reduction potential for VO^{2+} to V^{3+} is for pure water where $[H^+] = 10^{-7}$ M.

b. Write the chemical equation for the overall reaction.

c. Is Cu^+ reduced at the completion of the reaction? As we will see in Chapter 15, the Cu^+ in this reaction is called a catalyst. A catalyst is a substance which speeds up a reaction, in this case the reaction of Fe^{3+} with VO^{2+}, but does not change itself in the process.

9. Hemoglobin contains a large porphyrin ring similar to chlorophyll (see Section 19–D) which is bonded to a Fe^{2+} ion. The Fe^{2+} ion is also bonded to the amino acid histidine and a water molecule. The Fe^{2+} bonded to the porphyrin and histidine is called a complex ion and the complex ion, $Fe^{2+}_{(porphyrin)}$, can undergo oxidation to $Fe^{3+}_{(porphyrin)}$. The standard reduction potential is 0.14 V for

$$Fe^{3+}_{(porphyrin)} + e^- \rightleftharpoons Fe^{2+}_{(porphyrin)}$$

The reduction potential for the uncomplexed Fe^{3+} is

$$Fe^{3+}_{(aq)} + e^- \longrightarrow Fe^{2+}_{(aq)} \qquad \mathcal{E}° = 0.77 \text{ V}$$

The oxidation of a substance by dissolved $O_{2(g)}$ in water can be thought to use either of the half-cell reactions

$$\tfrac{1}{2}O_{2(g)} + H_2O_{(l)} + 2e^- \rightleftharpoons 2OH^-_{(aq)} \qquad \mathcal{E}° = 0.40 \text{ V}$$

$$O_{2(g)} + 4H^+ + 4e^- \rightleftharpoons 2H_2O_{(l)} \qquad \mathcal{E}° = 1.23 \text{ V}$$

a. Show that the half-cell reduction potential for O_2 in pure water ($[H^+] = [OH^-] = 10^{-7}$ M) is the same for either of the above O_2 half-cell reactions.

b. Calculate the half-cell potential for O_2 in water where the O_2 comes from air containing 21 mole percent O_2. Assume the air is at a *total pressure* of one atmosphere.

c. Using the half-cell potential calculated in part b., determine whether Fe^{2+} and $Fe^{2+}_{(porphyrin)}$ are stable to air oxidation in pure water. [Hint: Calculate the cell potential.]

10. Using the potential diagram (in volts) in acid solution

$$ClO_4^- \xrightarrow{1.19} ClO_3^- \xrightarrow{1.21} HClO_2 \xrightarrow{1.64} HClO \xrightarrow{1.63} Cl_2 \xrightarrow{1.36} Cl^-$$

show that a. $HClO$, b. $HClO_2$, and c. ClO_3^- should disproportionate in an acid solution.

11. Quite often the amount of charge passed through a cell is determined by measuring the mass of silver deposited by electrolysis of Ag^+ solution. If the

mass at the cathode increases by 0.430 g of $Ag_{(s)}$, how many coulombs have been transferred?

12. Aluminum is produced commercially by electrolysis of bauxite (Al_2O_3) dissolved in molten cryolite (Na_3AlF_6) at approximately 1000 °C. This process is called the Hall process and the electrolysis cell is shown in the following figure.

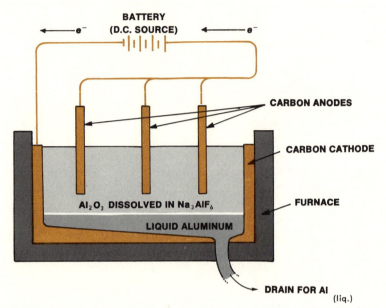

The electrode reactions are complicated but the overall reaction is

$$2Al_2O_3 \longrightarrow 4Al_{(l)} + 3O_{2(g)}$$

Calculate the amount of electricity in coulombs required to make 1 kg of aluminum.

13. For the nonmetals of groups VIA and VIIA, correlate the relative oxidizing power in neutral solution with position in the periodic table and note any trends that exist.

14. The reduction potentials for oxalacetate to malate and acetate to acetaldehyde in neutral aqueous solution are

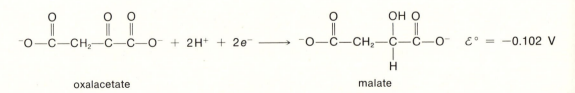

$$\underset{\text{acetate}}{CH_3-\overset{\overset{\displaystyle O}{\|}}{C}-O^-} + 3H^+ + 2e^- \longrightarrow \underset{\text{acetaldehyde}}{CH_3-\overset{\overset{\displaystyle O}{\|}}{C}-H} + H_2O \qquad \mathcal{E}° = -0.600 \text{ V}$$

The standard half-cell potentials given above have been adjusted for the fact that $[H^+] = 10^{-7}$ M in neutral water. a. If a solution contains oxalacetate, malate, acetate, and acetaldehyde all at 1 M concentration, predict what reaction will take place. b. Calculate $\Delta G°$ and K for the reaction. c. Would the addition of acid change your answers to parts a. and b.?

15. The direction in which an amino acid moves in an aqueous solution due to an external electric field depends upon the net charge on the amino acid. The formula for an amino acid is

$$NH_2-R-COOH$$

The amino group is basic and capable of accepting a proton whereas the carboxyl group is acidic. If a strong base were added to the solution, would the amino acid move towards the anode (positive) or cathode (negative). The difference in rate of migration of ions (such as from amino acids) is used as a method of separation; the technique is called electrophoresis.

CHAPTER FIFTEEN

CHEMICAL KINETICS

15–A. OBJECTIVES AND DEFINITIONS

Before discussing chemical kinetics, we should first outline the objectives of chemical kinetics and see how kinetics relates to thermodynamics and to molecular structure. In Chapter 10, we used the motion of a boulder (Figures 10–1 and 10–11), reproduced here as Figure 15–1, to illustrate the significance of ΔG in predicting the feasibility or nonfeasibility of a chemical reaction. The same analogy can be used to illustrate the relationship between chemical *kinetics* and chemical *thermodynamics*.

As we emphasized in Chapter 10, thermodynamics considers only the initial and final states of a process and is able to predict, on the basis of the change in free energy for a process, whether the process is feasible. In Figure 15–1 the initial state A has been assigned to the reactants $Cl_{2(g)}$ and $CO_{(g)}$. Since the reaction to produce $COCl_2$ has a negative ΔG, the reaction is feasible. The final state B in Figure 15–1 has therefore been associated with the product $COCl_{2(g)}$. However, if the ΔG were positive (ΔG_{AC} in Figure 15–1), the reaction would not be feasible.

The thermodynamic concept of feasibility ignores the pathway and obstacles which may occur between the initial state and the final state B. In the analogy of the boulder, an obstacle such as the mountain shown by the dashed line in Figure 15–1 would prevent the boulder from rolling by itself from A to B, and energy would have to be supplied to roll the boulder to state D, at which point the boulder would pass of its own accord to state B. A similar situation occurs in a chemical

FIGURE 15–1   Free energy change for a reaction path

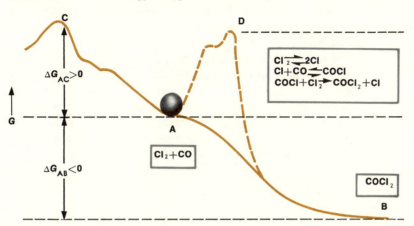

reaction such as the one shown between Cl_2 and CO. Molecules of Cl_2 and CO do not just come together and react to give molecules of $COCl_2$. The reaction pathway is more complicated, as the series of three reactions in the box at state D illustrate. This series of reactions, describing the *intermediate* species, is called the *reaction mechanism* and this will be taken up in greater detail later. The reaction steps involved in the reaction mechanism are frequently called *elementary reactions* because they explicitly describe the molecules involved in each specific reaction step. The term reaction mechanism is also used to describe the specific structural and electronic aspects of the collision between two reactants or intermediate species.

Note that in the mechanism for the reaction between Cl_2 and CO, the reaction is initiated by the dissociation of Cl_2 into atoms

15–1 $$Cl_2 \longrightarrow 2Cl$$

The chlorine atoms can then react with CO

15–2 $$Cl + CO \longrightarrow COCl$$

and the COCl that is formed as an intermediate can react with a second molecule of Cl_2:

15–3 $$COCl + Cl_2 \longrightarrow COCl_2 + Cl$$

The chlorine atoms that are produced in equation 15–3 can then go back to react with another CO in equation 15–2, etc. Because the elementary reaction steps 15–2 and 15–3 make up a repeating cycle or chain, this type of mechanism is called a *chain reaction*. Not all reactions proceed by a chain mechanism.

In the complete mechanism shown in Figure 15–1 the reverse reactions of equations 15–1 to 15–3 also occur, starting with $COCl_2$ to produce CO and Cl_2. The reverse reaction must pass through all the intermediate steps shown in equations 15–1 to 3, but in the reverse direction.

The rate of a reaction depends upon the mechanism through which the reaction proceeds. Moreover, the specific *way* in which the rate of the reaction depends upon the concentrations of the reactants also is a function of the mechanism involved. A kinetic investigation therefore generally involves a study of the relationship between the *rate of the reaction and the concentrations of the reactants* (Section 15–D), and this information is then used to evaluate a proposed mechanism for the reaction (Section 15–C).

15–B. ELEMENTARY REACTIONS

Most chemical reactions proceed by a *complex mechanism* involving several elementary reaction steps accompanied by the breakage and formation of several chemical bonds. As we will see later, some reactions that appear to be very simple, such as the reaction of H_2 and O_2 to give H_2O, are actually very complicated and consist of several elementary reaction steps. However, in order to understand and analyze reactions involving complex mechanisms, we must first be able to write rate expressions for elementary reactions.

An *elementary reaction* is one in which the reaction occurs in a *single collision* of the reactant molecules. This does not mean to imply that all collisions of reactants lead to reaction. On the contrary, usually only a small fraction of such collisions leads to products, but when the reaction does occur it results from one of these collisions.

Generally only two molecules are involved in a collision since the probability of the simultaneous collision of three or more molecules is very low. The *molecularity* of an elementary reaction is the number of molecules which react as a result of the collision. If only one of the molecules reacts, the elementary reaction is called *unimolecular*. The reaction can be represented by

15–4
$$A \longrightarrow P$$

where P may be one or more products of the reaction. If both molecules involved in the collision enter into the reaction, the elementary reaction is called *bimolecular*. In this case, the reaction can involve different molecules, A and B, or two molecules of the same type, say A and A.

15–5
$$A + B \longrightarrow P$$

15–6
$$2A \longrightarrow P$$

The rate of a reaction expresses either the rate at which the reactants disappear or the rate at which products appear. The rate of the reaction is commonly expressed as the change in concentration (moles/liter) which occurs in one second, thus the units of the rate are moles/liter-sec. In general the rate expression for a reaction must be determined using experimental results (Section 15–D). However, the rate expression for an elementary reaction is a simple product of concentrations of reactants and can be written by inspection of the stoichiometric equation. For example, for equation 15–5 the rate expression is (see Appendix Q for derivation)

15–7
$$\text{rate} = k[A][B]$$

where k is a constant called the rate constant. Note that the concentrations of A and B in the rate expression for a bimolecular reaction are to the first power because one A molecule and one B molecule are involved. For the elementary reaction involving two molecules of A (equation 15–6), the rate expression is

15–8
$$\text{rate} = k[A]^2$$

In other words, the rate expression for an elementary reaction is the product of concentrations of reactants each raised to a power corresponding to the number of reacting molecules of that type.

The units for the rate constant for a reaction can be evaluated from a dimensional analysis of the rate expression. For the bimolecular reactions, the units of k are evaluated in the following way:

$$\text{rate} = k(\text{concentration})(\text{concentration})$$

$$\overline{\text{moles/liter}}\text{-sec} = k(\overline{\text{moles/liter}})(\text{moles/liter})$$

$$k = (\text{liter/mole-sec})$$

The following list summarizes the rate expressions for essentially all types of elementary reactions and gives the units for the rate constant.

ELEMENTARY REACTION	RATE EXPRESSION	UNITS OF k	MOLECULARITY
A $\longrightarrow$ P	rate = $k[A]$	sec^{-1}	unimolecular
A + B $\longrightarrow$ P	rate = $k[A][B]$	liter-mole^{-1}sec^{-1}	bimolecular
2A $\longrightarrow$ P	rate = $k[A]^2$	liter-mole^{-1}sec^{-1}	bimolecular
2A + B $\longrightarrow$ P	rate = $k[A]^2[B]$	liter^2-moles^{-2}sec^{-1}	termolecular

It should be emphasized that writing rate expressions by inspection can be done only for *elementary reactions. For most reactions this cannot be done, unless the overall reaction is expected to occur in a single step and therefore is the equivalent of an elementary reaction step.* For example the reaction

$$NO + O_3 \longrightarrow NO_2 + O_2$$

is thought to occur in a single reaction step. Therefore,

$$rate = k[NO][O_3]$$

As we will see in Chapter 18, this reaction is important in the kinetics of air pollution. This reaction is very simple, involving a simple transfer of an oxygen atom from O_3 to NO. Because the bond dissociation energy D_{O_2-O} is low, the bond can be readily broken. On the other hand, the large D_{ON-O} bond energy of the product further facilitates the reaction.

Even though it may appear that a reaction should occur through a single elementary reaction step, the rate expression should be tested on the basis of experimental data. For example, it must be confirmed experimentally for the above reaction that the rate depends upon the first powers of [NO] and [O_3]. The details of this experimental analysis are discussed in Section 15–D.

PROBLEM 15–1   Write rate expressions for the following elementary reactions:
a. $O_3 \longrightarrow O_2 + O$
b. $Cl_2 \longrightarrow 2Cl$
c. $O_3 + O \longrightarrow 2O_2$
d. $Cl + CO \longrightarrow COCl$
e. $Cl_2 + COCl \longrightarrow COCl_2 + Cl$

15–C. COMPLEX REACTIONS—MECHANISMS

As mentioned in Section 15–B, a chemical reaction may occur in a single, elementary reaction step or it may involve a more complex mechanism consisting of several elementary steps. The reaction between Cl_2 and CO, which involves two intermediate reaction steps before the product $COCl_2$ is produced (equations 15–1 to 15–3), is considered a complex mechanism. In Section 15–D, we will see how the dependence of rate on the concentration of each of the species involved in the reaction can be determined experimentally. At this time we wish to derive the rate expression for a mechanism that is proposed for a chemical reaction. Once this is obtained, we can then compare the derived rate expression for the proposed mechanism with the experimental rate expression to see if they agree. If they disagree, then we know the proposed mechanism is in error. However, if they agree, then the proposed mechanism *may* be correct. As we will see, different mechanisms can have the same rate expressions; for this reason, *agreement with the experimental rate expression is not unequivocal proof that the mechanism is correct.*

As we have seen in Section 15–B, if the mechanism is thought to consist of a single elementary reaction step, then the predicted rate expression can be written down by inspection of the chemical equation. For example, if we proposed that the reaction between CO and Cl_2 occurred by a single reaction step

15–9
$$CO + Cl_2 \longrightarrow COCl_2$$

then the *predicted* rate expression would be

15–10
$$\text{rate} = k[CO][Cl_2]$$

However, we know *experimentally* that the rate expression is

15–11
$$\text{rate} = k_f[CO][Cl_2]^{3/2}$$

where k_f is a general rate constant for the forward reaction. This disagreement in rate expressions disproves the proposition that the reaction occurs by the single, elementary reaction step of equation 15–9.

If a proposed mechanism is complex and consists of several elementary reaction steps, the derivation of the expected rate expression is more complicated. Fortunately, there are certain approximations which we can frequently make which simplify the problem and help us avoid complex mathematical procedures. The major approximation is the assumption that one reaction step in the complex mechanism is slow relative to the other steps. If one reaction step is much slower than the other reaction steps, it should limit the rate at which the reaction proceeds and for this reason is called the *rate-determining step*.

We will illustrate the procedure for the analysis of a complex mechanism using the reaction between CO and Cl_2 as illustrated in Figure 15–1. The proposed mechanism for the reaction is

15–1
$$Cl_2 \underset{}{\overset{K_1}{\rightleftharpoons}} 2Cl \qquad \text{(fast)}$$

15–2
$$Cl + CO \underset{}{\overset{K_2}{\rightleftharpoons}} COCl \qquad \text{(fast)}$$

15–3
$$COCl + Cl_2 \overset{k_3}{\longrightarrow} COCl_2 + Cl \qquad \text{(slow)}$$

The steps in arriving at the rate expression are outlined as follows:

1. *Identify the rate-determining step and write the rate expression*. In this mechanism the third step, equation 15–3, is the rate determining step. The rate for this step is

15–12
$$\text{rate} = k_3[COCl][Cl_2]$$

The actual identification of the slow, rate-determining step is difficult even for an experienced chemist. In this text, the reaction step expected or assumed to be slow will always be identified for you.

2. *Set up expressions for concentrations of intermediates.* The rate expression for the slow, rate-determining step will generally contain concentrations of intermediates such as [COCl] in equation 15–12. However, we wish to obtain the rate expression in terms of concentrations of reactants of the reaction, CO and Cl_2. In order to express the concentrations of the intermediates in terms of the reactants we assume that any reversible reactions preceding the rate-determining step are at equilibrium and we solve for the concentrations of intermediates from the equilibrium expressions. In the proposed mechanism for $CO + Cl_2$, the first two steps, equations 15–1 and 15–2, are assumed to be at equilibrium. The equilibrium expressions are

15–13
$$K_1 = \frac{[Cl]^2}{[Cl_2]}$$

15–14
$$K_2 = \frac{[COCl]}{[Cl][CO]}$$

We can solve for [COCl] from equation 15–14

15–15
$$[COCl] = K_2[Cl][CO]$$

The concentration of atomic chlorine can be obtained from equation 15–13

15–16
$$[Cl] = (K_1[Cl_2])^{1/2} = K_1^{1/2}[Cl_2]^{1/2}$$

Substituting equation 15–16 into equation 15–15

15–17
$$[COCl] = K_2 K_1^{1/2}[Cl_2]^{1/2}[CO]$$

3. *Eliminate concentrations of intermediates in rate expression.* The concentration of COCl can now be eliminated by substituting equation 15–17 into equation 15–12

$$\text{rate} = k_3[COCl][Cl_2] = k_3 K_2 K_1^{1/2}[Cl_2]^{1/2}[CO][Cl_2]$$

15–18
$$\text{rate} = k_3 K_2 K_1^{1/2}[CO][Cl_2]^{3/2}$$

Equation 15–18 expresses the rate in terms of the reactants CO and Cl_2.

As stated earlier, the experimentally determined rate expression for this reaction is

15–11
$$\text{rate} = k_f[CO][Cl_2]^{3/2}$$

Comparing equations 15–11 and 15–18, we see they are essentially equivalent. The proposed mechanism predicts (equation 15–18) that the rate of the reaction will depend upon the first power of $[CO]$ and the 3/2 power of $[Cl_2]$. This is in agreement with the experimentally determined rate expression in equation 15–11 which thus supports the proposed mechanism. Comparing equations 15–11 and 15–18, we see that the general rate constant k_f is composed of

15–19 $$k_f = k_3 K_2 K_1^{1/2}$$

If thermodynamic data are available for equations 15–1 and 15–2, K_1 and K_2 can be calculated and k_3 can be evaluated from k_f and equation 15–19.

Again, agreement between the derived rate expression for the proposed mechanism is *not* unequivocal proof that the mechanism is correct. For example, the following proposed mechanism also gives a rate equation for the $CO + Cl_2$ reaction which is in agreement with the experimental rate equation 15–11.

$$Cl_2 \underset{}{\overset{K_1}{\rightleftharpoons}} 2Cl \quad \text{(fast)} \qquad K_1 = \frac{[Cl]^2}{[Cl_2]}$$

$$Cl + Cl_2 \underset{}{\overset{K_2}{\rightleftharpoons}} Cl_3 \quad \text{(fast)} \qquad K_2 = \frac{[Cl_3]}{[Cl][Cl_2]}$$

$$Cl_3 + CO \xrightarrow{k_3} COCl_2 + Cl \quad \text{(slow)} \qquad \text{rate} = k_3[Cl_3][CO]$$

The rate expression is derived as before. Since the first two reversible reactions are fast, we can assume equilibrium is established. We eliminate the concentrations of intermediates such as $[Cl_3]$ from the rate equation:

$$K_2 = \frac{[Cl_3]}{[Cl][Cl_2]}$$

$$[Cl_3] = K_2[Cl][Cl_2]$$

and substitute into the rate equation for the slow step:

$$\text{rate} = k_3[Cl_3][CO] = k_3 K_2[Cl][Cl_2][CO]$$

We also solve for $[Cl]$ from the first equilibrium expression:

$$K_1 = \frac{[Cl]^2}{[Cl_2]}$$

$$[Cl] = K_1^{1/2}[Cl_2]^{1/2}$$

and substitute into the rate expression

$$\text{rate} = k_3K_2[Cl][Cl_2][CO] = k_3K_2K_1^{1/2}[Cl_2]^{1/2}[Cl_2][CO]$$

15-20 $$\text{rate} = k_3K_2K_1^{1/2}[CO][Cl_2]^{3/2}$$

Equation 15-20 should be compared with the experimentally determined rate expression, equation 15-11. Again we see agreement between these expressions, and on the basis of the rate expression and concentration dependence alone, we are unable to decide which proposed mechanism is correct. In general, we will have to obtain additional experimental information to differentiate between any two proposed mechanisms if both result in rate expressions that are in agreement with the experimental one. In the case of the reaction between CO and Cl_2 there is additional experimental evidence that COCl is produced as an intermediate rather than Cl_3; this suggests that the first mechanism, equations 15-1 through 15-3, is correct.

The reverse reaction

$$COCl_2 \longrightarrow CO + Cl_2$$

has a mechanism which involves the same elementary reaction steps as the forward reaction, except that the rate-determining step is the reverse of the third step:

15-21 $$Cl + COCl_2 \xrightarrow{k_{-3}} COCl + Cl_2$$

The chlorine radicals produced in equation 15-3 will be in equilibrium with the Cl_2 produced in equation 15-21. When the reverse reaction is initially started with pure $COCl_2$, there is no Cl_2 present and the reaction cannot follow equation 15-21. However, $COCl_2$ can dissociate to give $COCl + Cl$ to initiate the reaction. Very shortly thereafter there will be sufficient Cl_2 produced so that the reaction can follow the mechanism

15-1 $$Cl_2 \underset{}{\overset{K_1}{\rightleftharpoons}} 2Cl \qquad\qquad K_1 = \frac{[Cl]^2}{[Cl_2]}$$

15-21 $$Cl + COCl_2 \xrightarrow{k_{-3}} COCl + Cl_2 \quad \text{rate} = k_3[Cl][COCl_2]$$

15-22 $$COCl \rightleftharpoons CO + Cl$$

The COCl will be in equilibrium with CO and Cl in reaction 15-22, but this reaction will not enter into the rate expression since it *follows* the rate-determining step, equation 15-21. We can obtain the rate expression for this proposed mechanism for the reverse reaction by using the equilibrium expression to eliminate [Cl] in the rate equation for reaction 15-21

$$K_1 = \frac{[Cl]^2}{[Cl_2]}$$

$$[Cl] = K_1^{1/2}[Cl_2]^{1/2}$$

15-21 $$\text{rate} = k_{-3}[Cl][COCl_2]$$

15-23 $$\text{rate} = k_{-3}K_1^{1/2}[Cl_2]^{1/2}[COCl_2]$$

Comparing this derived rate equation with the experimentally established equation

15-24 $$\text{rate}_{reverse} = k_r[COCl_2][Cl_2]^{1/2}$$

we note that both expressions have the same dependence on $[COCl_2]$ and $[Cl_2]$. That is, the predicted and experimental orders of the reaction with respect to $COCl_2$ and Cl_2 are in agreement, and the proposed mechanism is supported. Note that the rate of the reverse reaction (equation 15-23) is dependent on the concentration of Cl_2, which is a *product* for this reaction. As we will see in Section 15-D, the possibility of rate dependence on products must be considered in experimental determinations of rate expressions.

PROBLEM 15-2 a. Derive the rate expression for the reverse reaction

$$COCl_2 \longrightarrow CO + Cl_2$$

based on the following mechanism

$$Cl_2 \xrightarrow{K_1} 2Cl \qquad \text{(fast)}$$

$$Cl + COCl_2 \xrightarrow{k_2} Cl_3 + CO \qquad \text{(slow)}$$

$$Cl_3 \rightleftharpoons Cl_2 + Cl \qquad \text{(fast)}$$

b. Does the derived rate equation agree with the experimental rate expression?

$$\text{rate}_{reverse} = k_r[COCl_2][Cl_2]^{1/2}$$

c. Can you prove or disprove the proposed mechanism with this information? Explain.

PROBLEM 15-3 Ozone decomposes to oxygen according to the chemical reaction

$$2O_3 \longrightarrow 3O_2$$

The reaction is thought to occur by the following mechanism:

$$O_3 \underset{}{\overset{K_1}{\rightleftharpoons}} O_2 + O \qquad \text{(fast)}$$

$$O + O_3 \xrightarrow{k_2} 2O_2 \text{(slow)}$$

a. Derive the rate expression, assuming the first elementary reaction step is at equilibrium.

b. Can the above mechanism be differentiated (on the basis of the experimental rate equation) from the mechanism which assumes a single elementary reaction step?

$$2O_3 \longrightarrow 3O_2$$

PROBLEM 15–4   a. Derive the rate expression for the reaction

$$H_2 + I_2 \longrightarrow 2HI$$

which is proposed to go by the following mechanism

$$I_2 \underset{}{\overset{K_1}{\rightleftharpoons}} 2I \qquad \text{(fast)}$$

$$I + H_2 \underset{}{\overset{K_2}{\rightleftharpoons}} H_2I \qquad \text{(fast)}$$

$$H_2I + I \xrightarrow{k_3} 2HI \qquad \text{(slow)}$$

Assume the first two reaction steps are at equilibrium.

b. Can the above mechanism be differentiated (on the basis of the experimental rate expression) from the mechanism which assumes a single elementary reaction step?

$$H_2 + I_2 \longrightarrow 2HI$$

In the previously proposed mechanisms for the $CO + Cl_2 \longrightarrow COCl_2$ reaction, the $2O_3 \longrightarrow 3O_2$ reaction in Problem 15–8, and the $H_2 + I_2 \longrightarrow 2HI$ in Problem 15–9, you should note the extreme simplicity of the elementary reaction steps. In each case a *single* bond is formed or broken. Furthermore, the geometrical requirements are generally not very stringent. For example, the recombination of two chlorine atoms is not restricted to any specified approach by the atoms. For the reaction

$$CO + Cl \longrightarrow COCl$$

the chlorine radical must approach CO at the carbon end, since the structure for COCl is

but this approach is not restricted to a single direction. On the other hand, if we expect Cl_2 to react directly with CO, the molecules have to approach one another in a unique way to give a highly specific geometrical arrangement

$$\begin{array}{c} Cl \\ | \quad :C\equiv O \\ Cl \end{array} \longrightarrow \begin{array}{c} Cl \\ \diagdown C=O \\ Cl \diagup \end{array}$$

Apparently it is this rather stringent geometrical requirement which prevents the reaction of $Cl_2 + CO$ from proceeding in a single reaction step. The more complex mechanism consists of a series of simple elementary steps; however, each step is still quite fast. Any reaction will always proceed along the path that allows the least resistance in terms of energy barriers or geometrical constraints, regardless of the number of elementary steps this path will require.

Though some reactions appear quite simple from their overall chemical equations the actual mechanisms are frequently quite complicated and in some cases not completely understood. A chemical reaction that might appear to be quite simple on the basis of its chemical equation is the burning of hydrogen to give water

$$2H_2 + O_2 \longrightarrow 2H_2O$$

The detailed mechanism for this reaction actually involves numerous elementary steps. As you know, H_2 and O_2 will not react unless the reaction is started by a flame, spark, or high temperature. Once the reaction has been started the reaction sustains itself, as evidenced by the burning of a jet of H_2 in a flame. If a mixture of H_2 and O_2 are enclosed in a container, an explosion can occur if the temperature reaches 600 °C. The reaction of H_2 and O_2 is highly exothermic, and unless the heat is removed the reaction mixture will rapidly attain a temperature high enough for an explosion to occur. Extreme care must be exercised in handling mixtures of H_2 and O_2. The reaction of H_2 and O_2 is thought to occur by the following chain mechanism, but much research work must still be done to establish this mechanism with greater certainty.

15–25a $\qquad\qquad\qquad\qquad H_2 \longrightarrow 2H\cdot$

15–25b $\qquad\qquad\qquad H^{\cdot} + O_2 \longrightarrow {}^{\cdot}OH + O^{\cdot}$

15–25c $\qquad\qquad\qquad O^{\cdot} + H_2 \longrightarrow {}^{\cdot}OH + H^{\cdot}$

15–25d $\qquad\qquad H^{\cdot} + O_2 + M \longrightarrow HO_2^{\cdot} + M$ (continued)

15–25e $\qquad\qquad\qquad\qquad HO_2^{\cdot} + H_2 \longrightarrow H_2O + {}^{\cdot}OH$

15–25f $\qquad\qquad\qquad\qquad {}^{\cdot}OH + H_2 \longrightarrow H_2O + H^{\cdot}$

If the reaction is carried out in a vessel, the radicals $H^{\cdot}$, $\overset{\cdot}{O}H$, and $H\overset{\cdot}{O}_2$ can be removed from the chain by reaction at the surface of a container. The M in equation 15–25d is any molecular species that is involved in the collision but does not react itself. In this mechanism, M can be H_2, O_2 or H_2O. The function of M is to remove the energy which is released when a bond is formed, as in $H^{\cdot} + O_2 \longrightarrow H\overset{\cdot}{O}_2$. Note that the above mechanism is initiated by the dissociation of H_2 in equation 15–25a. Since the H—H bond dissociation energy is 104.18 kcal/mole, whereas the O=O bond dissociation energy is 118.06 kcal/mole, it requires less energy to initiate the reaction by dissociating H_2 molecules. Reaction steps 15–25b and 15–25c produce $\overset{\cdot}{O}H$ radicals through a chain mechanism. The $\overset{\cdot}{O}H$ can then react with H_2 to produce H_2O (equation 15–25f). H_2O can also be produced by reaction 15–25d followed by reaction 15–25e. Again the simplicity of each reaction step in the above mechanism should be noted; with the exception of equation 15–25e, each reaction step involves at most the breakage of one bond and the formation of another bond. Only the initial step (equation 15–25a) requires a great deal of energy, so that after initiation, the reaction proceeds rapidly through the remaining elementary reaction steps. Moreover, if one takes into consideration the structure of H_2O, it is almost *impossible* to visualize how H_2O could be produced by a single reaction step. If H_2O were produced by a single reaction step, it would involve a very complicated approach of $2H_2$ molecules to an O_2 molecule.

$$
\begin{array}{c}
\text{H}\cdot\text{.}\qquad\quad\text{.}\cdot\text{H} \\
|\quad\overset{\cdot\cdot}{\text{O}}=\overset{\cdot}{\text{O}}\text{.}\quad| \\
\text{H}\cdot\text{.}\qquad\quad\text{.}\cdot\text{H}
\end{array}
$$

The likelihood of such a collision occurring is extremely remote and therefore the reaction of $2H_2 + O_2$ does *not* occur in a one-step mechanism.

15–D. GENERAL RATE EXPRESSION – CONCENTRATION DEPENDENCE

For most chemical reactions the dependence of the rate of the reaction on the concentration of reactants and possibly products can be described explicitly by a general rate expression. Consider the general chemical reaction involving reactants A and B to give products C and D:

$$aA + bB \longrightarrow cC + dD$$

The rate of the reaction is given by the general rate expression

15–26 $\qquad\qquad\qquad rate = k[A]^{\alpha}\ [B]^{\beta}\ [C]^{\gamma}\ [D]^{\delta}$

where k is the rate constant and α, β, γ, and δ are constants. The constants α, β, γ, and δ are small whole numbers or fractions and they are referred to as the *order of the reaction* for each species. For illustration let us assume a rate expression was found to be

$$\text{rate} = k[A]\ [B]^2$$

We would say that the reaction was first order with respect to A and second order with respect to B.

In the reaction of Cl_2 and CO, which we considered earlier, it has been found that

$$Cl_2 + CO \longrightarrow COCl_2$$

15–11
$$\text{rate} = k_f[CO][Cl_2]^{3/2}$$

This reaction is first order with respect to CO and three-halves order with respect to Cl_2. The rate expression for the reverse reaction, the dissociation of $COCl_2$, has also been established experimentally and found to be

$$COCl_2 \longrightarrow CO + Cl_2$$

15–24
$$\text{rate} = k_r[COCl_2][Cl_2]^{1/2}$$

For this reaction we would say that the rate is first order with respect to $COCl_2$ and one-half order with respect to Cl_2. Note that in this case the rate depends upon the concentration of Cl_2 which is a *product* of the reaction.

The rate of the reaction is expressed as the change in concentration of either product or reactant per unit of time; therefore, the units of the rate are some concentration unit divided by time. As we discussed in Section 15–B, we will express the concentration change in moles/liter and the time in seconds; the units of R are thereby moles/liter-sec. The units of k must be determined from a dimensional analysis of the rate expression as described in Section 15–B for elementary reactions. For example, substituting the appropriate units into the rate expression

$$\text{rate} = k[A]\ [B]^2$$

$$\text{moles/liter-sec} = k(\text{moles/liter})(\text{moles/liter})^2$$

we find that k has the units

$$k = \text{liter}^2/\text{moles}^2\text{-sec}$$

For the reaction of Cl_2 and CO the units of k_f can be found in a similar manner

15–11
$$\text{rate} = k_f[CO][Cl_2]^{3/2}$$

$$\text{moles/liter-sec} = k_f \, (\text{moles/liter})(\text{moles/liter})^{3/2}$$

$$k_f = \text{liter}^{3/2}/\text{moles}^{3/2}\text{-sec}$$

PROBLEM 15–5   What are the units of k_r in equation 15–24?

The order of a reaction (α, β, etc.) can be evaluated by measuring the rate of a reaction at different concentrations of reactants and products. The evaluation of α and β is simplified if the concentration of only one species is varied, while the concentrations of the other species are held constant. If the rates R_1 and R_2 are measured at two different concentrations of A, $[A]_1$ and $[A]_2$, holding the concentrations of B, C and D constant, the rate expressions are

$$R_1 = k[A]_1{}^\alpha \; [B]^\beta \; [C]^\gamma \; [D]^\delta$$

$$R_2 = k[A]_2{}^\alpha \; [B]^\beta \; [C]^\gamma \; [D]^\delta$$

Since $[B]$, $[C]$ and $[D]$ are constant, the terms involving these concentrations will cancel in the ratio

15–27
$$\frac{R_1}{R_2} = \frac{[A]_1{}^\alpha}{[A]_2{}^\alpha} = \left(\frac{[A]_1}{[A]_2}\right)^\alpha$$

Generally α can be evaluated by comparing the R_1/R_2 ratio measured to the $[A]_1/[A]_2$ ratio. If $[A]_1/[A]_2$ is 1/2, the α will depend on R_1/R_2 according to the following table

α	R_1/R_2
1	1/2
2	1/4
3	1/8

If α is a fraction, the calculation is not as simple and α is best obtained from the logarithmic expression:

15–28
$$\ln(R_1/R_2) = \alpha \ln([A]_1/[A]_2)$$

$$\alpha = \frac{\ln(R_1/R_2)}{\ln([A]_1/[A]_2)}$$

Either natural or common logarithms can be used. β can be evaluated in a similar manner if $[B]$ is varied while the concentrations of the other species are held constant, and the same procedure can be used to determine the orders γ and δ. After α, β, etc., have been evaluated, the rate constant k can be calculated from any one known rate and the concentrations $[A]$ and $[B]$.

EXAMPLE 15–1   *Evaluate the rate expression (α, β, . . . and k) for the reaction*

$$2A + B \longrightarrow 2C$$

using the following data.

EXPERIMENT NO.	[A]	[B]	[C]	RATE(moles/liter-sec)
1	1×10^{-3}	1×10^{-3}	1×10^{-4}	5.6×10^{-2}
2	2×10^{-3}	1×10^{-3}	1×10^{-4}	11.2×10^{-2}
3	1×10^{-3}	2×10^{-3}	1×10^{-4}	22.4×10^{-2}
4	1×10^{-3}	1×10^{-3}	2×10^{-4}	2.8×10^{-2}

Our general rate expression for this reaction would be

$$R = k[A]^{\alpha}[B]^{\beta}[C]^{\gamma}$$

where R = rate at which product C is produced in units of moles/liter-sec. Note that in experiments no. 1 and no. 2 only the [A] has changed, thus the ratio of rate expressions for these experiments gives

$$\frac{R_1}{R_2} = \frac{5.6 \times 10^{-2}}{11.2 \times 10^{-2}} = \frac{k(1 \times 10^{-3})^{\alpha}(1 \times 10^{-3})^{\beta}(1 \times 10^{-4})^{\gamma}}{k(2 \times 10^{-3})^{\alpha}(1 \times 10^{-3})^{\beta}(1 \times 10^{-4})^{\gamma}} = \frac{(1 \times 10^{-3})^{\alpha}}{(2 \times 10^{-3})^{\alpha}}$$

$$\frac{1}{2} = \left(\frac{1 \times 10^{-3}}{2 \times 10^{-3}}\right)^{\alpha} = \left(\frac{1}{2}\right)^{\alpha}$$

$$\alpha = 1$$

We may use data from experiments no. 1 and no. 3, where only [B] has changed, to solve for β.

$$\frac{R_1}{R_3} = \frac{5.6 \times 10^{-2}}{22.4 \times 10^{-2}} = \frac{k(1 \times 10^{-3})(1 \times 10^{-3})^{\beta}(1 \times 10^{-4})^{\gamma}}{k(1 \times 10^{-3})(2 \times 10^{-3})^{\beta}(1 \times 10^{-4})^{\gamma}} = \frac{(1 \times 10^{-3})^{\beta}}{(2 \times 10^{-3})^{\beta}}$$

$$\frac{1}{4} = \left(\frac{1 \times 10^{-3}}{2 \times 10^{-3}}\right)^{\beta} = \left(\frac{1}{2}\right)^{\beta}$$

$$\beta = 2$$

We may use the data in experiments no. 1 and no. 4 to solve for γ since only [C] has been changed.

$$\frac{R_1}{R_4} = \frac{5.6 \times 10^{-2}}{2.8 \times 10^{-2}} = \frac{k(1 \times 10^{-3})(1 \times 10^{-3})^2(1 \times 10^{-4})^{\gamma}}{k(1 \times 10^{-3})(1 \times 10^{-3})^2(2 \times 10^{-4})^{\gamma}} = \frac{(1 \times 10^{-4})^{\gamma}}{(2 \times 10^{-4})^{\gamma}}$$

$$\frac{2}{1} = \left(\frac{1 \times 10^{-4}}{2 \times 10^{-4}}\right)^{\gamma} = \left(\frac{1}{2}\right)^{\gamma}$$

$$\gamma = -1$$

Our rate expression is therefore

$$R = k[A][B]^2[C]^{-1} = k\,\frac{[A][B]^2}{[C]}$$

The rate constant can now be evaluated from the data in any of the experiments, such as experiment no. 1

$$5.6 \times 10^{-2}\ moles/liter\text{-}sec = \frac{k(1 \times 10^{-3}\ moles/liter)(1 \times 10^{-3}\ moles/liter)^2}{(1 \times 10^{-4}\ moles/liter)}$$

$$5.6 \times 10^{-2}\ moles/liter\text{-}sec = k(1 \times 10^{-5}\ moles^2/liter^2)$$

$$k = 5.6 \times 10^3\ liter/mole\text{-}sec$$

PROBLEM 15–6   Determine the rate expression and calculate the rate constant for the reaction

$$2NO_2 + F_2 \longrightarrow 2NO_2F$$

using the following data

EXPERIMENT NO.	$[NO_2]$	$[F_2]$	$[NO_2F]$	RATE(moles/liter-sec)
1	1×10^{-3}	5×10^{-3}	1×10^{-3}	2×10^{-4}
2	2×10^{-3}	5×10^{-3}	1×10^{-3}	4×10^{-4}
3	6×10^{-3}	2×10^{-3}	1×10^{-3}	4.8×10^{-4}
4	6×10^{-3}	4×10^{-3}	1×10^{-3}	9.6×10^{-4}
5	1×10^{-3}	1×10^{-3}	1×10^{-3}	4×10^{-5}
6	1×10^{-3}	1×10^{-3}	2×10^{-3}	4×10^{-5}

PROBLEM 15–7   If the initial $[Cl_2]$ is doubled for the reaction between Cl_2 and CO, by what factor will the rate of reaction vary?

When the order of the reaction has been determined for a given reaction, the rate expression can be integrated (using the principles of integral calculus) to derive an equation that relates concentration and time. This equation can then be used to test whether the experimental data agree with this reaction order throughout the reaction. Numerous such equations, many quite complex, can be derived from various rate equations. In this text we will consider only the simple first order equation.

$$A \longrightarrow P$$

$$rate = k\,[A]$$

FIGURE 15–2   Graph of concentration versus time for a first order reaction

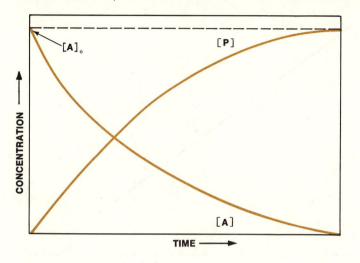

The rate in this case can either be for the loss of A or for production of P. The equation relating [A] and time can be derived from the rate expression (see Appendix R):

15–29
$$\ln \frac{[A]}{[A]_0} = -kt$$

or

15–30
$$[A] = [A]_0\, e^{-kt}$$

where $[A]_0$ is the concentration of A at the beginning of the reaction ($t = 0$). As time increases, e^{-kt} will decrease, since the exponential is negative. A plot of [A] versus time will be an exponentially decreasing curve as shown in Figure 15–2. The concentration of products [P] is given by

$$[P] = [A]_0 - [A] = [A]_0 - [A]_0\, e^{-kt}$$

$$[P] = [A]_0\,(1 - e^{-kt})$$

Note in Figure 15–2 that the concentration of P increases at the same rate at which the concentration of A decreases.

The logarithmic form of the equation relating [A] with time is useful in graphing experimental rate data of [A] and time, t, for a reaction suspected of being first order. The logarithmic equation 15–29 can be rewritten in a linear form

15–31
$$\ln [A] = \ln [A]_0 - kt$$

$$y = b + ax$$

FIGURE 15–3 Graph of ln [A] versus time for a first order reaction

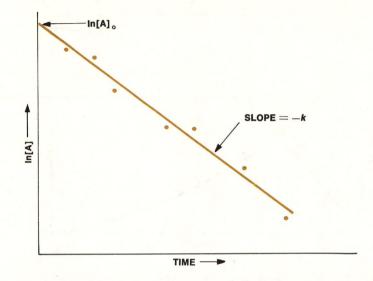

where $y = \ln [A]$, $b = \ln [A]_0$, $a = -k$, and $x = t$. Plotting ln [A] versus t should give a linear graph if the reaction is first order, as Figure 15–3 shows.

15–E. TEMPERATURE DEPENDENCE

The rates of most reactions increase with an increase in temperature with only a few exceptions. The increase in the rate of reaction is due to an increase in the rate constant, k, with increasing temperature. The temperature dependence of k generally follows the *Arrhenius equation*

15–32
$$k = Ae^{-\frac{E^*}{RT}}$$

where A and E^* are constants characteristic of the reaction. E^* is called the *activation energy* and A is called the pre-exponential or frequency factor.

For an elementary reaction, the activation energy E^* can be understood in terms of an increase in internal energy required to form an *activated molecular complex*. The activated molecular complex is a transient species formed from the colliding molecules which then proceeds to give the products of that elementary reaction step. The activation energy, E^*, for a multi-step reaction is complicated and there is no simple interpretation of its meaning. A graph of k as a function of temperature (T) according to the Arrhenius equation (15–32) is shown in Figure 15–4 for different values of E^* and a given value of A. Note that k is significantly larger at lower temperatures *when the activation energy is low*. For example, at $T = 300 \,°K$,

FIGURE 15-4   Rate constant as function of temperature (Arrhenius equation). $A = 10^{11}$ liter/mole-sec

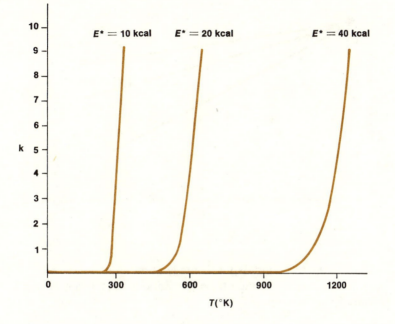

the rate constant decreases by a factor of 10^{-7} when E^* is doubled from 10 kcal to 20 kcal.

The Arrhenius equation can be put into a linear form by taking the logarithm of equation 15-32

15-33
$$\ln k = -\frac{E^*}{R}\frac{1}{T} + \ln A$$

Converting to logarithm of base 10

15-34
$$\log k = -\frac{E^*}{2.303\,R}\frac{1}{T} + \log A$$

A graph of $\log k$ versus $1/T$ is shown in Figure 15-5 for the same A and E^* values as were used in Figure 15-4.

According to the form of equation 15-34

$$\text{slope} = -\frac{E^*}{2.303\,R}$$

The slope is negative since E^* is usually positive. The magnitude of the slope depends upon the value of E^*, as shown in Figure 15-5. When the rate constant k

FIGURE 15–5 Log *k* versus 1/*T* (Arrhenius equation). $A = 10^{11}$ liter/mole-sec, $E^* = 10$, 20, and 40 kcal

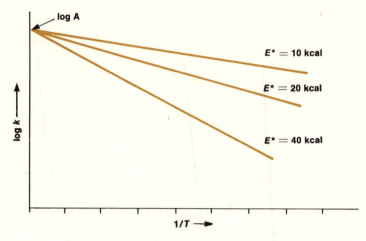

has been determined experimentally at different values of *T*, and a graph of log *k* versus 1/*T* is constructed from this data, the activation energy can then be calculated from the slope.

EXAMPLE 15–2   *Calculate the activation energy from the following experimental rate constants for the reaction*

$$2N_2O \longrightarrow 2N_2 + O_2$$

T (°K)	k(liter/mole-sec)
1,125	11.59
1,053	1.67
1,001	0.380
838	0.0011

The logarithm of k and the reciprocal of T must be calculated in order to plot a linear graph (equation 15–34)

$1/T \times 10^4$ deg^{-1}	log k
8.890	1.064
9.497	0.223
9.991	−0.420
11.930	−2.960

The graph of log k versus T is as follows

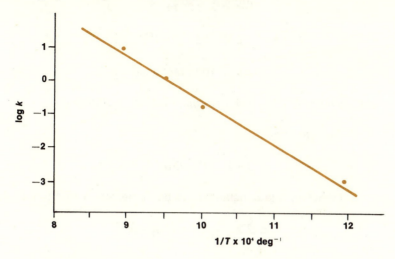

The slope is evaluated from the graph using data points at 1/T = 9.0 × 10⁻⁴ deg⁻¹,
log k = 0.8, and 1/T = 11.0 × 10⁻⁴ deg⁻¹, log k = −1.7.

$$slope = \frac{0.8 - (-1.7)}{9.0 \times 10^{-4} \ deg^{-1} - 11.0 \times 10^{-4} \ deg^{-1}} = \frac{-2.5}{2.0 \times 10^{-4}} \ deg$$

$$= -1.25 \times 10^4 \ deg$$

The activation energy is evaluated from

$$slope = -\frac{E^*}{2.303 \ R}$$

$$-1.25 \times 10^4 \ deg = -\frac{E^*}{(2.303)(1.987 \times 10^{-3} \ kcal/deg\text{-}mole)}$$

$$E^* = 57 \ kcal/mole$$

The pre-exponential factor A can be obtained from the y intercept of a log k versus
1/T graph. However, our graph for this example does not extend to 1/T = 0, and
the log A cannot be read off the graph directly. Alternatively, the pre-exponential
factor A can be calculated from equation 15–34 with the data point 1/T = 9.0 ×
10⁻⁴ deg⁻¹, log k = 0.8.

$$log \ k = -\frac{E^*}{2.303 \ R} \frac{1}{T} + log \ A$$

$$log\ A = log\ k + \frac{E^*}{2.303\ R}\ \frac{1}{T}$$

$$log\ A = 0.8 + (1.25 \times 10^4\ deg)(9.0 \times 10^{-4}\ deg^{-1}) = 12.0$$

$$A = 1.0 \times 10^{12}\ liter/mole\text{-}sec$$

Note that the units of k determine the units of A.

PROBLEM 15–8   For the reaction

$$CH_3I + C_2H_5ONa \longrightarrow CH_3OC_2H_5 + NaI$$

the following rate constants have been measured as a function of temperature

$t(°C)$	k(liter/mole-sec)
0	5.60×10^{-5}
6	1.18×10^{-4}
12	2.45×10^{-4}
18	4.88×10^{-4}
24	1.00×10^{-3}
30	2.08×10^{-3}

Calculate the activation energy and the pre-exponential factor, *A*, in the Arrhenius equation for this reaction.

15–F. RELATIONSHIP BETWEEN KINETICS AND THERMODYNAMICS

The equilibrium expression in Chapter 13 was derived on the basis of thermodynamic principles alone. The concept of equilibrium can also be approached from a kinetic viewpoint. In reversible chemical reactions, the reaction proceeds both in the forward direction and in the reverse direction. With a mixture of reactants and products both reactions are occurring simultaneously. *At equilibrium there is no net change in the concentrations of the reactants and products; therefore, the forward and reverse rates must be equal* (Section 3–B).

$$rate_f = rate_r$$

Referring once again to our reaction of $Cl_2 + CO$:

15–11 $$rate_f = k_f\,[CO][Cl_2]^{3/2}$$

15–24 $$rate_r = k_r\,[COCl_2][Cl_2]^{1/2}$$

Therefore,

$$k_f \, [CO][Cl_2]^{3/2} = k_r \, [COCl_2][Cl_2]^{1/2}$$

$$\frac{k_f}{k_r} = \frac{[COCl_2][Cl_2]^{1/2}}{[CO][Cl_2]^{3/2}} = \frac{[COCl_2]}{[CO][Cl_2]}$$

The right-hand side of the above equation is exactly the same as the equilibrium constant expression, K. The ratio of the rate constants, k_f/k_r, must therefore be equal to K:

15–35
$$K = \frac{k_f}{k_r} = \frac{[COCl_2]}{[CO][Cl_2]}$$

The derivation of equation 15–35 is based on the assumption that both the forward and reverse reactions go through the same intermediates (Figure 15–1). If the forward and reverse reactions take different paths, the relationship between K and k_f/k_r is not valid.

If the forward and reverse reactions go through the same intermediates, it follows that the change in internal energy for the reaction, ΔE, is the activation energy for the forward step minus the activation energy for the reverse step.

15–36
$$\Delta E = E_f^* - E_r^*$$

This is shown in Figure 15–6 where the internal energy, E, is plotted as a function of the reaction path. Equation 15–36 relates a thermodynamic quantity, ΔE, with

FIGURE 15–6   Internal energy as a function of reaction path

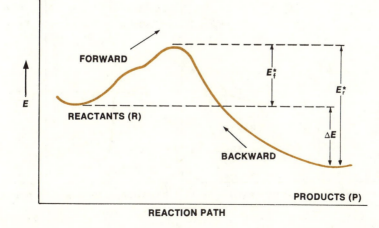

kinetic quantities, E_f^* and E_r^*. Obviously the magnitude of ΔE does not depend upon the individual activation energies but only on their difference. On this basis, we say that predictions based on thermodynamics are independent of the kinetics or rate of the reaction. On the other hand, it should be noted that the activation energy for a reaction must always be greater than or equal to ΔE for the reaction.

$$E^* \geq \Delta E$$

This is shown in Figure 15–7 for different combinations of E^* and ΔE where the thermodynamic quantity ΔE does place a lower limit on E^*.

PROBLEM 15–9   The following gas phase reaction has an activation energy of 31.6 kcal and a pre-exponential factor of $A = 1.2 \times 10^{10}$ liter/mole-sec.

$$NO_2 + CO \longrightarrow NO + CO_2$$

a. Calculate the rate constant at 25 °C for this reaction from the Arrhenius equation.

b. Using thermodynamic data from the appendices along with the activation energy and rate constant from part a., calculate the rate constant and activation energy at 25 °C for the reverse reaction

$$NO + CO_2 \longrightarrow NO_2 + CO$$

[Hint: For this reaction the change in moles is zero, $\Delta E = \Delta H°$, and $K = K_P$. Also recall that $\Delta G° = -RT \ln K_P$.]

FIGURE 15–7   Possible relationship between E^* and ΔE. R = reactants; P = products

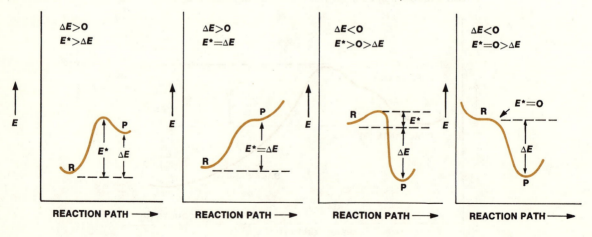

* 15–G. TRANSITION STATE THEORY

One theory of chemical kinetics assumes a unique transition state through which the reactants must pass in order to give the products of the reaction. For the elementary reaction between A and B, we designate the intermediate transition state by $AB^{\ddagger}$, $AB^{\ddagger}$ is a unique species in that it can always react to give the products. In other words, $AB^{\ddagger}$ has A and B oriented in the proper geometry and with the necessary activation energy to spontaneously give the products. For this reason $AB^{\ddagger}$ is commonly referred to as the *activated molecular complex.*

Transition state theory assumes that *the reactants are in equilibrium with the activated molecular complex*, $AB^{\ddagger}$. The kinetic scheme is thus

15–37
$$A + B \underset{}{\overset{K^{\ddagger}}{\rightleftharpoons}} AB^{\ddagger} \overset{k^{\ddagger}}{\longrightarrow} \text{Products}$$

The rate of the reaction is the rate at which $AB^{\ddagger}$ goes to products

15–38
$$\text{rate} = k^{\ddagger}[AB]$$

Since A and B are in equilibrium with $AB^{\ddagger}$, we can solve for $[AB^{\ddagger}]$ by the equilibrium expression

$$K^{\ddagger} = \frac{[AB^{\ddagger}]}{[A][B]}$$

$$[AB^{\ddagger}] = K^{\ddagger}[A][B]$$

Substitution into our rate expression, equation 15–38 gives

15–39
$$\text{rate} = k^{\ddagger} K^{\ddagger} [A][B]$$

Since $K^{\ddagger}$ is an equilibrium constant, it is related to the thermodynamic quantities $\Delta G^{\ddagger}$, $\Delta H^{\ddagger}$, $\Delta S^{\ddagger}$, $\Delta E^{\ddagger}$ through the equations

15–40
$$\Delta G^{\ddagger} = - RT \ln K^{\ddagger}$$

15–41
$$\Delta G^{\ddagger} = \Delta H^{\ddagger} - T\Delta S^{\ddagger}$$

15–42
$$\Delta H^{\ddagger} = \Delta E^{\ddagger} + \Delta(PV)$$

Equations 15–40 to 15–42 can be solved for $K^{\ddagger}$ (see Appendix S) and substituted into equation 15–39, to give

$$\text{rate} = k^{\ddagger} (e^{\Delta S^{\ddagger}/R})(e^{-\Delta E^{\ddagger}/RT}) [A][B]$$

The rate expression for a bimolecular reaction is

$$\text{rate} = k\,[A][B]$$

Comparing these rate expressions, the rate constant according to the transition state theory is

15–43
$$k = (k^{\ddagger}e^{\Delta S^{\ddagger}/R})e^{-\Delta E^{\ddagger}/RT}$$

Comparing equation 15–43 with the experimental Arrhenius equation $k = A\,e^{-E^*/RT}$ (equation 15–32)

$$k = \underbrace{(k^{\ddagger}e^{\Delta S^{\ddagger}/R})}_{A}\underbrace{(e^{-\Delta E^{\ddagger}/RT})}_{e^{-E^{\ddagger}/RT}}$$

we see that the transition state theory does predict the correct form for the rate constant as a function of T.[§] The constant term, *A*, according to the transition state theory is

15–44
$$A = k^{\ddagger}\,e^{\Delta S^{\ddagger}/R}$$

The activation energy, E^*, is the internal energy of activation $\Delta E^{\ddagger}$, to give the activated complex, $AB^{\ddagger}$.

15–45
$$E^* = \Delta E^{\ddagger}$$

The internal energy of the activated molecular complex corresponds to the highest point in Figure 15–6 between the reactants and products.

Note that equation 15–44 for *A* includes a term with $\Delta S^{\ddagger}$, the change in standard entropy in going from A + B to $AB^{\ddagger}$. Recall from Section 10–G that entropy is a measure of the order of a system. For a system with a high degree of order the entropy is low. Therefore, when the activated complex $AB^{\ddagger}$ has a unique and specific configuration, there will be a high degree of order in the complex and $\Delta S^{\ddagger}$ will be low, resulting in a small rate constant. Likewise when the structure of the activated complex is less rigid with alternate geometrical arrangements allowable, there will be more disorder in $AB^{\ddagger}$ and $\Delta S^{\ddagger}$ will be larger, resulting in a large rate constant. The dependence of the $\Delta S^{\ddagger}$ term upon the geometrical configuration of $AB^{\ddagger}$ can therefore account for the effect that any specific geometrical requirement of the reaction step may have on the reaction rate.

The usefulness of transition state theory in predicting the rates of chemical reactions is restricted because of the difficulty in calculating $\Delta S^{\ddagger}$ and $\Delta E^{\ddagger}$. For reactions which involve only simple molecules, such as diatomic molecules,

[§]The $k^{\ddagger}$ varies slightly with temperature so *A* is not expected to be absolutely constant. However, the dominant temperature-dependent term is $e^{-\Delta E^{\ddagger}/RT}$.

theoretical calculation of $\Delta S^{\ddagger}$ and $\Delta E^{\ddagger}$ can be carried out, but even then the calculations are quite difficult. The transition state theory finds its greatest value in correlating kinetic data obtained experimentally for different, but related, reactions. The A and E^{*} from the Arrhenius equation can be derived from experimental data, as described in Section 15–E, and then $\Delta S^{\ddagger}$ and $\Delta E^{\ddagger}$ can be calculated using equations 15–44 and 15–45. The $k^{\ddagger}$ in equation 15–44 can also be evaluated theoretically.

To illustrate the nature of an activated molecular complex, we will consider the reaction between an alkyl halide such as CH_3Cl and sodium in the gas phase

$$CH_3\text{---}Cl + Na \longrightarrow \cdot CH_3 + NaCl$$

This reaction is thought to occur in a single elementary reaction step where a sodium atom approaches the chlorine of CH_3Cl. In the activated complex the $CH_3\text{---}Cl$ bond is elongated, and the $\text{---}Cl\cdots Na$ separation is small enough so that the chlorine can be transferred to the sodium.

$$CH_3\text{---}Cl + Na \rightleftharpoons CH_3\cdots\cdots Cl\cdots Na \longrightarrow \cdot CH_3 + NaCl$$

The internal energy can be plotted as a function of the $CH_3\text{---}Cl$ distance (Figure 15–8). Note that $\Delta E^{\ddagger}$, the internal energy for the activated molecular complex, is at the maximum in the curve.

FIGURE 15–8 Change in internal energy for the reaction of CH_3Cl with Na

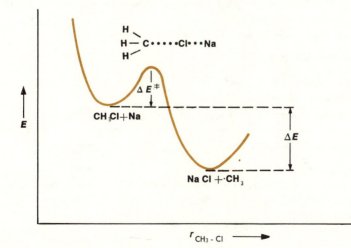

*15–H CATALYSIS

A *catalyst* is a substance which increases the rate at which a chemical reaction approaches equilibrium without being permanently changed by the reaction, as

$$\text{reactants} + \text{catalyst} \longrightarrow \text{products} + \text{catalyst}$$

Before proceeding to a more comprehensive discussion, it is important to point out the significance of catalysis in chemistry. By 1975 catalysis will be involved in 25 % of all U.S. manufactured goods, 4.7 billion barrels of petroleum products per year, and 80 % of the chemical industry's products. Furthermore, $750 million per year will be invested in research and development of catalysts. In addition to these synthetic applications, catalytic agents in the form of naturally occurring *enzymes* are extremely important in biological systems.

A catalytic reaction can occur in the gas, liquid, or solid phases or at the interface of any two phases. When a catalytic reaction occurs within a single phase, it is known as *homogeneous catalysis*. When a catalytic reaction occurs at an interface, it is known as *heterogeneous catalysis*. Table 15–1 gives a classification of some reactions into these categories and shows the phases involved.

PROPERTIES OF CATALYST SYSTEMS

A catalyst increases the rate at which a reaction reaches equilibrium. Such a reaction, however, must be thermodynamically possible without a catalyst (ΔG negative) and able to proceed without a catalyst, even if only very slowly. A catalyst cannot make a reaction occur which has been predicted by thermodynamics to be impossible.

Recall from Chapter 10 that the change in any of the thermodynamic functions is dependent only on the difference of the final state (products) and the initial state (reactants) and is independent of the path taken. A catalyst alters the path taken but does not change the initial and final states. *Therefore, a catalyst does*

TABLE 15–1 Classification of some catalyzed reactions

HOMOGENEOUS CATALYSIS	PHASE
decomposition of $KClO_3$ by MnO_2	solid
oxidation of SO_2 to SO_3 by NO	gas
hydrolysis of esters by acids and bases	liquid

HETEROGENEOUS CATALYSIS	PHASE
oxidation of CO to CO_2 with ZnO	solid–gas
hydrogenation[a] of benzene to cyclohexane by Ni	solid–liquid–gas
polymerization of alkenes by H_3PO_4	liquid–gas

[a]involves addition of hydrogen to a molecule

not alter the ΔG for a reaction. Similarly, a catalyst cannot alter the change in the standard free energy, $\Delta G°$, the change in standard enthalpy, $\Delta H°$, or the change in standard entropy, $\Delta S°$. Furthermore, since

13–3
$$\Delta G° = - RT \ln K$$

and $\Delta G°$ is not altered by a catalyst, *the equilibrium constant also is not altered by a catalyst.*

Because none of the thermodynamic functions are altered by a catalyst, it is possible to determine heats of reaction for very slow reactions at low temperatures by using a catalyst. Ordinarily, the heats of reaction of very slow reactions must be determined at high temperatures where the rate is faster; however, this is often difficult to do. The ΔH for the hydrogenation of ethylene has been determined by measurement of the heat of the reaction at low temperature (82 °C) with a catalyst and also by calculation from equilibrium data taken at ~400 °C (see equation 13–36 in Section 13–L). The results are

calorimetry with catalyst (82 °C) $\Delta H° = -32.57$ kcal/mole

calculation from equilibrium constant (~400 °C) $\Delta H° = -32.6$ kcal/mole

These data provide an experimental confirmation of our statement that ΔH of reaction does not change in the presence of a catalyst.

Since the equilibrium constant K is unaffected by a catalyst, *an increase in the forward rate, k_f, induced by a catalyst is accompanied by an equal increase in the reverse rate, k_r.*

15–35
$$K = \frac{k_f}{k_r}$$

A catalyst, although not *permanently* changed, must be involved temporarily in the reaction. The most likely way in which a catalyst participates is via some intermediate molecular complex which is able to be transformed into products more easily than the one formed in the absence of the catalyst. Thus, a catalyst converts reactants into products more rapidly by opening up a new reaction path which is energetically lower than the path for the uncatalyzed reaction. For a given reaction, there may be several possible paths but a catalyst will cause the one with the lowest activation energy to dominate. For example, two different catalysts can alter the reaction path for the decomposition of HCOOH to give different products:

$$HCOOH \xrightarrow{\text{Al}_2\text{O}_3 \text{ catalyst}} H_2O + CO$$

$$HCOOH \xrightarrow{\text{metal catalysts}} H_2 + CO_2$$

Earlier we mentioned enzyme catalysts. Such catalysts contain a protein molecule. Although we shall not go into the details concerning this here, it may be well to point out the efficiency of such catalysts. The following rates have been determined:

hydrogenation of benzene by Ni-Al$_2$O$_3$ catalyst	4×10^{17} molecules/sec/g catalyst
decomposition of H$_2$O$_2$ by catalase enzyme	6×10^{26} molecules/sec/g catalyst

Note that the rate for enzyme decomposition is more than 10 million times faster than the rate for catalytic hydrogenation of benzene. It is apparent that without biological catalysts life would not exist, since critical reactions would simply be too slow.

ACTIVATION ENERGY AND CATALYSIS

Essentially all chemical reactions obey the Arrhenius equation

15–32
$$k = A\,e^{-E^*/RT}$$

In order for a reaction to proceed more rapidly at a given temperature, the activation energy must be decreased and/or A increased. A catalyst decreases the activation energy and/or increases A for a reaction, thereby increasing k according to equation 15–32. It has been shown experimentally that the activation energy is generally lowered by a catalyst. The energy changes for an exothermic reaction with and without a catalyst are shown in Figure 15–9. Some activation energies for reactions with and without a catalyst are given in Table 15–2. Note that the values of E^* with a catalyst do depend upon the nature of the catalyst.

TABLE 15–2 Activation energies of some reactions with and without a catalyst

REACTION	E^* WITHOUT CATALYST	E^* WITH CATALYST
pyrolysis (breakdown by heat) of C$_2$H$_5$—O—C$_2$H$_5$	53.5 kcal/mole	34.3 kcal/mole
2HI $\longrightarrow$ H$_2$ + I$_2$	44 kcal/mole	25 kcal/mole (Au) 14 kcal/mole (Pt)
2N$_2$O $\longrightarrow$ 2N$_2$ + O$_2$	58.5 kcal/mole	29 kcal/mole (Au) 32.5 kcal/mole (Pt)

FIGURE 15–9   Changes of internal energy for a reaction with and without a catalyst

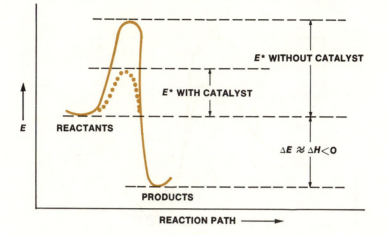

HOMOGENEOUS CATALYSIS

As noted at the beginning of our discussion, there are two types of catalysis: homogeneous and heterogeneous. In homogeneous catalysis, a new path is opened up by the formation of a complex between a reactant and the catalyst. As we shall see shortly, in heterogeneous catalysis, the new path arises as a result of a reactant being activated by adsorption (adherence) and complex formation on a solid surface. For homogeneous catalysis the mechanism is

$$\text{reactant} + \text{catalyst} \underset{}{\overset{K_1}{\rightleftharpoons}} \text{complex} \xrightarrow{k_2} \text{product} + \text{catalyst}$$

The rate expression for this mechanism can be derived by assuming that the first reaction is in equilibrium

15–46 $$\text{rate} = k_2\,[\text{complex}]$$

15–47 $$K_1 = \frac{[\text{complex}]}{[\text{reactants}][\text{catalyst}]}$$

Solving for [complex]

15–48 $$[\text{complex}] = K_1\,[\text{reactants}][\text{catalyst}]$$

equation 15–46 then becomes

15–49 $$\text{rate} = k_2 K_1\,[\text{reactants}]\,[\text{catalyst}]$$

FIGURE 15–10 Free energy changes during catalyzed and uncatalyzed reactions. — uncatalyzed reaction, ••• catalyzed reaction.

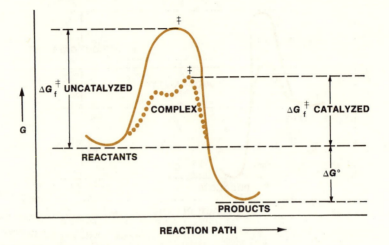

Note that the rate of the catalyzed reaction depends upon the concentration of catalyst and furthermore that the rate constant is the product of k_2K_1. The rate constant for the catalyzed reaction is larger than that for the uncatalyzed reaction; the increase of the rate constant corresponds to a lowering of the standard free energy of activation, as shown in Figure 15–10. Note that while the presence of the catalyst lowers the standard free energy of activation for both the forward and the reverse reactions, $\Delta G°$ remains unchanged.

Table 15–3 presents some general classes of homogeneous catalytic reactions and we shall consider a few specific examples of these.

Two examples involving acids and bases as catalysts are the hydrolysis of esters and the halogenation of acetone. The mechanism for the acidic halogenation of acetone is proposed to be

15–50

$$CH_3-\overset{O}{\overset{\|}{C}}-CH_3 + HA \rightleftharpoons CH_3-\overset{OH^+}{\overset{\|}{C}}-CH_3 + A^-$$

$$CH_3-\overset{OH}{\underset{X}{\overset{|}{\underset{|}{C}}}}-CH_2X \xleftarrow{+X_2} CH_3-\overset{OH}{\overset{|}{C}}=CH_2 + H^+$$

$$CH_3-\overset{}{\underset{\|}{\overset{|}{C}}}-CH_2X + HX$$
$$\overset{\|}{O}$$

TABLE 15-3   Homogeneous catalytic reactions

REACTION CLASS	CATALYST	EXAMPLES OF CATALYST
hydrolysis, oxidation, reduction, isomerization	acids and bases	H^+, NH_3, OH^-
hydrogenation of unsaturated compounds	metal salts and complexes	$Cu(C_2H_3O_2)_2$
polymerization of alkenes	metal salts and complexes	$TiCl_3$—$Al(C_2H_3)_3$
oxidation of ethylene to acetaldehyde (CH_3CHO)	metal salts and complexes	$PdCl_2$
oxidation SO_2	gases	NO

and for the basic halogenation

15–51

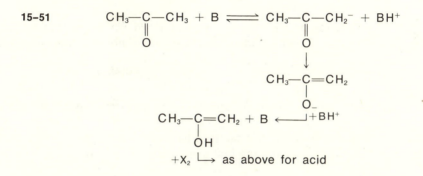

For the acidic hydrolysis of an ester, the proposed mechanism is

15–52

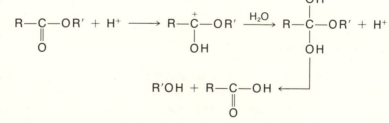

and for the basic hydrolysis

15–53

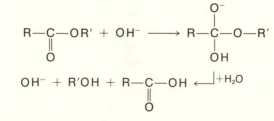

In equations 15–50 through 15–53, H^+, or HA, and OH^-, or B (the catalysts) are all regenerated (reformed).

An example of the use of a transition metal complex as a catalyst is the oxidation of ethylene

15–54

$$PdCl_4{}^{2-} \text{ (from } PdCl_2 + \text{aq } HCl) + H_2C{=}CH_2 \longrightarrow [PdCl_3(C_2H_4)]^- + Cl^-$$

$$[PdCl_2(OH)(C_2H_4)]^- + H^+ + Cl^- \longleftarrow {}^{+H_2O}$$

$$PdCl(CH_2CH_2OH) + Cl^-$$

$$\longrightarrow Pd^\circ + CH_3{-}\overset{\overset{\displaystyle H}{|}}{C}{=}O + Cl^- + H^+$$

During the process Pd(II) is reduced to Pd, which can then be reoxidized by air or Cu^{2+} ion. If acetic acid and acetate ion are present, then vinyl acetate

$$CH_3COOCH{=}CH_2$$

is produced. This is a commercial process for the production of vinyl acetate.

PROBLEM 15–10 The molecule NO can be used as a catalyst for the eventual oxidation of SO_2 to SO_3. Assuming that the first step in the reaction is

$$NO + \tfrac{1}{2}O_2 \longrightarrow NO_2$$

complete the reaction sequence.

HETEROGENEOUS CATALYSIS

In heterogeneous catalysis, the catalyst is generally a solid. The specificity of the catalyst surface is a function of specific chemical or physical characteristics. These include, for example, the ability to dissociate H_2, the ability to gain or lose O_2, and affinity for H_2O. Table 15–4 presents some reactions involving heterogeneous catalysis. Note that metal oxides appear in two places in Table 15–4. However, the metal oxides used in dehydrations and isomerizations are (electrical) non-conductors, whereas those used in oxidations, etc. are semiconductors (Chapter 5). In the case of nonconductor metal oxides, O_2 cannot be lost or gained by heating, whereas O_2 can be lost or gained in the case of the semiconductors. These factors affect the specific chemical function the catalysts can perform.

An important step in heterogeneous catalysis is the adsorption of molecules on the surface of the catalyst, producing a complex of some sort followed by the

TABLE 15–4 Heterogeneous catalysis reactions

REACTION	CATALYST	EXAMPLES
hydrogenation, dehydrogenation[a]	metals	Fe, Ni, Pt, Pd
oxidation, reduction, dehydrogenation	metal oxides and sulfides	NiO, WS_2, CuO, Cr_2O_3 (semiconductors)
dehydration, isomerization	metal oxides	Al_2O_3, SiO_2, MgO (nonconductor)
polymerization, cracking[b], isomerization	acids	H_3PO_4, SiO_2–Al_2O_3, H_2SO_4

[a]involves removal of hydrogen from a molecule
[b]involves breakdown of a molecule into smaller components

formation of the products. There are two types of adsorptions: physical and chemical (chemisorption). In physical adsorption the attractive forces are relatively weak and of the type discussed in Chapter 8. In chemisorption, new chemical bonds are formed which represent strong forces. For example, the chemisorption of H_2 and Ni

$$H_2 + 2Ni \longrightarrow 2\ H\!-\!Ni$$

has a $\Delta H \sim 30$ kcal/mole and the hydrogens are only about 1.6 Å from the surface.

The catalytic surface can activate molecules in two principal ways: by dissociation or by association. Some examples of dissociation are:

15–55
$$H_2 + 2M \longrightarrow \begin{matrix} H & H \\ | & | \\ -M\!-\!M- \end{matrix}$$

15–56
$$CH_4 + 2M \longrightarrow \begin{matrix} CH_3 & H \\ | & | \\ -M\!-\!M- \end{matrix}$$

15–57
$$O_2 + 2M \longrightarrow \begin{matrix} O & O \\ \| & \| \\ -M & M- \end{matrix} \quad \text{or} \quad \begin{matrix} O^- & O^- \\ | & | \\ -M^+ & M^+- \end{matrix}$$

where M is the metal catalyst.
Some examples of association are:

15–58
$$O_2 + 2M \longrightarrow \begin{matrix} O\!-\!O \\ | \quad | \\ -M\!-\!M- \end{matrix}$$

15–59
$$H_2C\!=\!CH_2 \longrightarrow \begin{matrix} H_2C\!-\!CH_2 \\ | \quad\quad | \\ -M\!-\!M- \end{matrix}$$

FIGURE 15–11 Potential energy curves for the dissociation of H_2 on a metal surface. ΔH_{CA} and ΔH_{PA} are enthalpies of chemisorption and physical absorption respectively and $E\ddagger$ is the activation energy for $H_2 \longrightarrow 2H$ on the surface. $\ddagger$ is the transition state for the reaction of complex $\longrightarrow$ products.

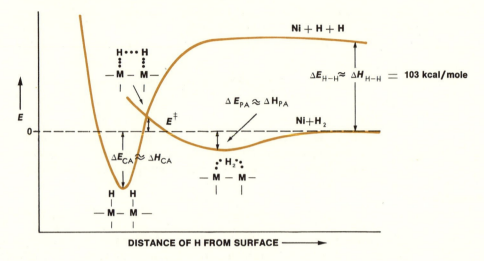

DISTANCE OF H FROM SURFACE ⟶

Typically, the dissociation of H_2 involves both physical and chemical adsorption. Figure 15–11 will provide us with a basis for discussion. Note from the diagram that because the potential energy curve for $Ni + H_2$ crosses that of $Ni + H + H$, only a relatively small activation energy is required for the reaction

$$H_2 \xrightarrow{\text{surface}} 2H_{\text{surface}}$$

This low activation energy is in marked contrast to that required if the curves did not cross ($\Delta H_{H-H} = 103$ kcal/mole).

We shall also consider the mechanism for the class of reaction involving the hydrogenation of an alkene. We already have the format for considering this reaction in the form of equations 15–55 and 15–59 as well as Figure 15–11.

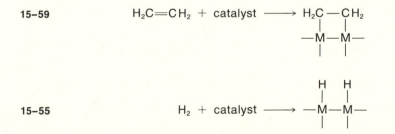

If the adsorbed ethylene and hydrogen occupy adjacent sites on the catalyst, the following reactions can occur.

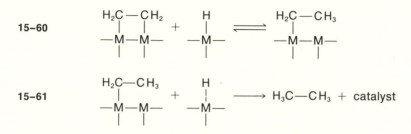

15–60

15–61

Note that reaction 15–60 involves the formation of the chemisorbed alkyl radical $H_2\dot{C}$—CH_3. This mechanism applies when Ni, Fe, Co, Ru, Rh, and Pd are the catalysts.

EXAMPLE 15–3   *The molecule C_2H_5OH can undergo dehydration to $H_2C{=}CH_2$ plus H_2O. Develop a mechanism for this reaction on a surface considering CH_3—$CH_2\cdot$, $\cdot OH$ and $\cdot CH_2$—$CH_2 \cdot$ attached to the surface as intermediates.*

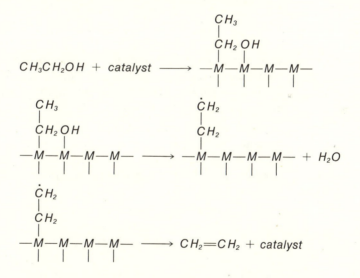

PROBLEM 15–11   The molecule 1-butene reacts with hydrogen over nickel to produce both butane and 2-butene. Write out a mechanism for both of these reactions.

PROBLEM 15–12   The molecule C_2H_5OH can undergo dehydrogenation to form

$$CH_3 - \underset{\underset{H}{|}}{C}{=}O + H_2$$

Develop a mechanism for this reaction on a surface, considering CH_3—CH_2—$O\cdot$, $H\cdot$ and CH_3—$\overset{\displaystyle |}{\underset{\displaystyle H}{C}}$—$O\cdot$ attached to the surface as intermediates.

EXERCISES

1. The kinetic data listed in the table were obtained for the following reaction

$$A_{(g)} + B_{(g)} \longrightarrow C_{(g)}$$

EXPERIMENT	[A]	[B]	INITIAL RATE OF FORMATION OF [C] (moles/liter-min) or (moles liter^{-1} min^{-1})
1	1.0	32.0	3.0
2	2.0	32.0	24.0
3	3.0	32.0	81.0
4	32.0	6.0	5.3
5	32.0	18.0	15.9
6	32.0	24.0	21.2

a. The rate was found to be independent of the concentration of C. Using the above data determine the rate law expression for this reaction.
b. What would you predict the order of the overall reaction to be?
c. Calculate the rate constant.

2. Suppose you are employed by a chemical firm that produces some chemical product, P. In order to make production more economical your boss has asked you to speed up the reaction. What factors or conditions would you suggest that might speed up the reaction? Briefly discuss the effect of each factor.

3. Hypophosphite decomposes according to the reaction

$$H_2PO_2^- + OH^- \longrightarrow HPO_3^{2-} + H_2$$

The following initial concentrations of $H_2PO_2^-$ and OH^- produce the corresponding volumes of $H_2(V_{H_2})$ in time, t.

[$H_2PO_2^-$]	[OH^-]	V_{H_2} (cc at STP)	time(min)
0.50	1.28	19.5	20
0.25	1.28	25.0	50
0.25	3.94	135.0	30

a. The moles of H_2 produced can be calculated from the volume of H_2 produced at STP. Calculate the initial rates of the reaction from the moles of H_2 produced at time, t, from

$$\text{Rate} = \frac{n_{H_2}}{t}$$

b. Using the rates calculated in part a., determine the order of the reaction with respect to $H_2PO_2^-$ and OH^-.

4. Near 225 °C diethylperoxide decomposes according to the equation

$$C_2H_5-O-O-C_2H_5 \xrightarrow{\;k\;} 2C_2H_5O\cdot$$

where the rate constant* is given by

$$k = 2 \times 10^{13} \exp(-31.7 \times 10^3 \text{ cal}/RT) \quad (\text{sec}^{-1})$$

If diethylperoxide were used to initiate a polymerization reaction, by what factor would the rate of the initiation step increase in raising the temperature from 225° to 300 °C? ($R = 1.987$ cal/mole-deg)

5. Consider the reaction $2NO_2 \longrightarrow N_2O_4$

a. Calculate $\Delta H°$ using $\Delta H_f°$ in the appendices.
b. Calculate $\Delta E°$ from the equation at 298 °K

$$\Delta H° = \Delta E° + \Delta(PV) = \Delta E° + \Delta n(RT)$$

where Δn = change in the number of moles during the reaction according to the stoichiometric equation (products minus reactants).
c. The activation energy for the reaction is 35.28 kcal. Sketch the curve of E as the reaction proceeds similar to Figures 15–6 and 15–7. What is the activation energy for the reaction

$$N_2O_4 \longrightarrow 2NO_2$$

Show this on the curve.

6. A reaction which is thought to be involved in the mechanism of smog formation (see Chapter 18) is

$$O_{3(g)} + NO_{(g)} \xrightarrow{\;k_f\;} O_{2(g)} + NO_{2(g)}$$

The rate expression for this reaction is

$$\text{rate}_f = k_f[O_3][NO]$$

*R. E. Rebbert and K. J. Laidler, J. Chem. Phys., *20*, 574 (1952).

a. With the above information can you state whether the reaction of O_3 and NO might consist of a single elementary reaction step or must proceed by a complex mechanism? Explain briefly.

b. Rationalize, on the basis of molecular structure, that the reaction of O_3 and NO can proceed by a single reaction step.

c. The reverse reaction

$$O_{2(g)} + NO_{2(g)} \longrightarrow O_{3(g)} + NO_{(g)}$$

has the rate expression

$$\text{rate}_r = k_r[O_2][NO_2]$$

Assuming that the reverse reaction can also proceed by a single elementary reaction step, write the equation relating the equilibrium constant to the forward and reverse rate constants and the concentrations of reactants and products.

d. If k_f were known experimentally, how could you determine k_r using data from the appendices?

7. A reaction that might be involved in the production of smog is

$$2NO + O_2 \longrightarrow 2NO_2$$

Experimentally this reaction has been studied and the rate expression is found to be

$$\text{rate} = k_f[NO]^2[O_2]$$

a. On the basis of this information can you eliminate the possibility that the reaction might proceed in a single reaction step?

b. Which of the following mechanisms could explain the experimental rate expression

Mechanism I: $2NO \xrightleftharpoons{K_1} N_2O_2$ (fast)

$N_2O_2 + O_2 \xrightarrow{k_2} 2NO_2$ (slow)

Mechanism II: $NO + O_2 \xrightleftharpoons{K_1} NO_3$ (fast)

$NO_3 + NO \xrightarrow{k_2} 2NO_2$ (slow)

Mechanism III: $NO + O_2 \xrightarrow{k_1} NO_3$ (slow)

$NO_3 + NO \xrightarrow{k_2} 2NO_2$ (fast)

c. According to mechanism II show that $k_f = K_1 k_2$

d. Calculate ΔE_{298}° for the reaction using ΔH_f° from the appendices and the relationship

$$\Delta H^\circ = \Delta E^\circ + \Delta n \, (RT)$$

where Δn = change in moles for the reaction. In this reaction $\Delta n = -1$.

e. Sketch E versus reaction path identifying ΔE_1°, E_2^*, E^*, and ΔE° and intermediate species according to mechanism II.

8. Which of the following mechanisms are in agreement with the rate expression for the reaction

$$2H_2 + 2NO \longrightarrow N_2 + 2H_2O$$

$$\text{rate} = k \, [NO]^2 [H_2]$$

Mechanism I: $2H_2 + 2NO \longrightarrow N_2 + 2H_2O$

Mechanism II:

$H_2 + NO \longrightarrow H_2O + N$ (slow)

$N + NO \longrightarrow N_2 + O$ (fast)

$O + H_2 \longrightarrow H_2O$ (fast)

Mechanism III:

$H_2 + 2NO \longrightarrow N_2O + H_2O$ (slow)

$N_2O + H_2 \longrightarrow N_2 + H_2O$ (fast)

Mechanism IV:

$2NO \rightleftharpoons N_2O_2$ (fast)

$N_2O_2 + H_2 \longrightarrow N_2O + H_2O$ (slow)

$N_2O + H_2 \longrightarrow N_2 + H_2O$ (fast)

9. Consider the reaction involving the decomposition of hydrogen peroxide in water solution,

$$2H_2O_{2(aq)} \longrightarrow 2H_2O_{(l)} + O_{2(g)}$$

a. According to thermodynamic principles, is H_2O_2 stable with respect to decomposition to H_2O and O_2? Use thermodynamic data from the appendices.

b. Why is it possible to store H_2O_2 in a 3 % aqueous solution (the state in which it is sold commercially)?

c. In the presence of Fe^{2+} the decomposition of H_2O_2 occurs quite rapidly. At the end of the reaction Fe^{2+} is unchanged. What is Fe^{2+} called?

d. Hydrogen peroxide can be involved in the following half cell reactions

	$\mathcal{E}^\circ$
$H_2O_2 + 2H^+ + 2e^- \longrightarrow 2H_2O$	1.77
$O_2 + 2H^+ + 2e^- \longrightarrow H_2O_2$	0.68

Show that the decomposition of H_2O_2 can be thought of in terms of a combination of these half-cell reactions and show that the process is thermodynamically feasible (*i.e.*, $\mathcal{E} > 0$).

e. Why might you expect the decomposition of H_2O_2 to be slow, considering the structure is

f. The half-cell potential for $Fe^{3+}|Fe^{2+}$ is intermediate to the two half-cell potentials for H_2O_2

$$Fe^{3+} + e^- \longrightarrow Fe^{2+} \qquad\qquad \mathcal{E}° = 0.77$$

Propose a mechanism whereby Fe^{2+} and Fe^{3+} act as intermediates in reacting with H_2O_2 according to the previous half-cell reactions. Make each reaction step thermodynamically favorable. Will Fe^{2+} be changed at the end of the reaction?

10. Dinitrogen pentoxide decomposes in the gas phase according to the chemical equation

$$2N_2O_5 \longrightarrow 4NO_2 + O_2$$

The reaction rate expression is $\quad$ rate $= k[N_2O_5]$

a. Explain why the reaction cannot consist of a single elementary reaction step.

b. A mechanism which has been proposed for this reaction is

$$N_2O_5 \longrightarrow NO_2 + NO_3 \qquad \text{(slow)}$$
$$NO_3 \longrightarrow NO + O_2 \qquad \text{(fast)}$$
$$NO + N_2O_5 \longrightarrow NO_2 + N_2O_4 \qquad \text{(fast)}$$
$$N_2O_4 \longrightarrow 2NO_2 \qquad \text{(fast)}$$

Explain why this mechanism would be consistent with the experimental rate expression given above where the order with respect to N_2O_5 is first order.

c. The structure for N_2O_5 is thought to be

On the basis of this structure discuss why you would not expect this reaction to proceed in a single reaction step.

d. Show that the following mechanism can also account for the observed rate expression

$$N_2O_5 \longrightarrow NO_2 + NO_3 \quad \text{(slow)}$$
$$NO_3 + N_2O_5 \longrightarrow 3NO_2 + O_2 \quad \text{(fast)}$$

11. Hydrogen and bromine react to give hydrogen bromide

$$H_2 + Br_2 \longrightarrow 2HBr$$

The mechanism which has been proposed for this reaction is

$$Br_2 \longrightarrow 2Br\cdot$$
$$Br\cdot + H_2 \longrightarrow HBr + H\cdot$$
$$H\cdot + Br_2 \longrightarrow HBr + Br\cdot$$
$$H\cdot + HBr \longrightarrow H_2 + Br\cdot$$
$$Br\cdot + Br\cdot \longrightarrow Br_2$$

Identify the elementary reaction step(s) which   a. initiate the reaction, b. form a chain mechanism,   c. exhibit or slow down the reaction, and d. terminate the chain reaction.

12. Chloroform, $CHCl_3$, can be chlorinated to carbon tetrachloride by the photochemical reaction

$$CHCl_3 + Cl_2 \xrightarrow{h\nu} CCl_4 + HCl$$

The light quanta, $h\nu$, initiate the reaction by dissociating Cl_2

$$Cl_2 \xrightarrow{h\nu} 2Cl\cdot$$

Suggest a mechanism for this reaction using the photochemical step to produce a chlorine radical as the initiating step. [Hint: A chain mechanism can be set up.]

13. Biochemical reactions are generally catalyzed by enzymes. The reaction is shown in general as the reaction of a substrate (S) with the enzyme (E) to form an intermediate substrate-enzyme complex (ES). The complex can then react to give product (P), liberating the enzyme in its original state.

$$S + E \underset{}{\overset{K_1}{\rightleftharpoons}} ES \qquad \frac{1}{K_m} = K_1 = \frac{[ES]}{[S][E]}$$

$$ES \xrightarrow{k_2} P + E \qquad R = \text{rate} = k_2[ES]$$

Generally equilibrium is assumed in the first reaction step. The rate-determining step is the second step.

a. The initial concentration of enzyme $[E]_0$ must equal the sum of the concentration of free enzyme, $[E]$, and complexed enzyme, $[ES]$.

$$[E]_0 = [E] + [ES]$$

Using this expression for $[E]_0$ and the equilibrium expression, show that

$$[ES] = \frac{[E]_0[S]}{[S] + K_m}$$

where K_m is known as the Michaelis constant.*

b. Using this expression for $[ES]$ and the rate expression, show that

$$\frac{[S]}{R} = \left(\frac{1}{k_2[E]_0}\right)[S] + K_m$$

c. Recognizing that the equation in part (b) is of the linear form

$$y = ax + b$$

identify the variables y and x and the constants a and b.

d. The following rates have been determined experimentally for different initial concentrations of substrate, S. The initial concentration of enzyme, $[E]_0$, is the same in all reactions.

[S] (moles/liter)	R(moles/liter-min) or (moles liter^{-1} min^{-1}) × 10^6
1×10^{-2}	75.0
1×10^{-4}	60.0
7.5×10^{-5}	56.3
6.25×10^{-6}	15.0

Graph the linear function $[S]/R$ versus $[S]$ and evaluate the constants ($k_2[E]_0$) and K_m from the slope and intercept. Use only the last three data points in your graph.

14. Benzene diazonium chloride, C_6H_5—N≡N—Cl, dissolved in isoamyl alcohol decomposes according to the reaction

$$C_6H_5-N{\equiv}N-Cl_{(sol)} \longrightarrow C_6H_5Cl_{(sol)} + N_{2(g)}$$

*The Michaelis constant is actually defined as $(k_{-1} + k_2)/k_1$. However, frequently $k_{-1} \gg k_2$ so $K_m \approx k_{-1}/k_1$.

This reaction was followed by measuring the volume of N_2 produced.* The following volumes (V) of N_2 at STP (0 °C and 1 atm) were measured at various times.

t(min)	V(ml)
0	0.00
50	8.21
100	15.76
140	21.08
160	23.57
200	28.17
∞	69.84

a. At infinite time the reaction has gone to completion to produce 69.84 ml N_2. If there was no N_2 present at the start of the reaction ($t = 0$), calculate the number of moles of C_6H_5—N≡N—Cl (A) that must have been present initially ($n_A°$).

b. The number of moles of A that have reacted at any time, t, is equal to the moles of N_2 produced. The remaining moles of A at time t is given by

$$n_A = (n_A° - n_{N_2})$$

Using the equation for a first order rate

$$\ln [A] = \ln [A]° - kt$$

show that the decomposition of C_6H_5—N≡N—Cl is first order with respect to C_6H_5—N≡N—Cl.

[Hint: $[A] = n_A/V$ and $\ln [A] = \ln n_A - \ln V$. Substitute $\ln [A]$ and similar expression for $\ln [A]°$ into the rate equation. What happens to $\ln V$?]

15. Frequently two electron transfer oxidation-reduction reactions are slow such as

$$\text{As (III)} + 2\,\text{Ce (IV)} \longrightarrow \text{As (V)} + 2\,\text{Ce (III)}$$

The addition of I_2 speeds this reaction up and the rate is

$$\text{rate} = k\,[\text{As (III)}]\,[I_2]$$

Propose a mechanism in which I_2 acts as a catalyst through the half cell reaction I_2, I^-. The reduction potential for As(V), As(III) in neutral solution is 0.11 volts.

*C. E. Waring and J. R. Abrams, Journal of the American Chemical Society, *63*, 2757 (1941).

16. Some oxidation-reduction reactions in solution may appear to be very simple according to their stoichiometric equations. For example,

$$H_2O + V^{3+} + Fe^{3+} \longrightarrow VO^{2+} + Fe^{2+} + 2H^+$$

appears to involve a single electron transfer. However, the reaction is catalyzed by Cu^{2+} and the rate expression is

$$rate = k[V^{3+}][Cu^{2+}]$$

Propose a mechanism whereby Cu^{2+} acts as a catalyst utilizing the half cell reaction $Cu^2|Cu^+$. [Hint: See Exercise 8 at the end of Chapter 14.] The reduction potential for VO^{2+}, V^{3+} is 0.07 volts in neutral solution.

17. The molecule 1-butene can react with H_2 over nickel to produce 2-butene. Develop a mechanism for this reaction considering $CH_3-CH_2-CH-CH_3$

    with a bond mark below the CH

    attached to the catalyst as one of the intermediates. [Hint: Consider the mode of attachment to the surface as shown by ethylene in equations 15–59 to 15–61. Remember that hydrogen may be lost as well as gained.]

18. The molecule $CH_3\overset{OH}{\underset{|}{C}}HCH_3$ can undergo dehydration to form $CH_3CH{=}CH_2$. Develop a mechanism for this reaction on a surface considering $CH_3\dot{C}HCH_3$, $\cdot OH$, $\dot{C}H_2\dot{C}HCH_3$ to be attached to the catalyst as intermediates.

CHAPTER SIXTEEN

TRANSITION ELEMENTS

16–A. GENERAL FEATURES OF TRANSITION ELEMENTS

All of the transition elements have partially filled $3d$, $4d$, or $5d$ subshells. For example, the configuration of cobalt is $...3d^7\ 4s^2$. It is particularly this characteristic of the transition metals and their ions which is responsible for many of their unique properties—a wide variety of oxidation states, colored salts and complexes in the solid state as well as in solution, and paramagnetism (Section 6–E) of the ions and complexes. We shall only briefly comment on the origin of these three properties in this first section. More details on each of these will be developed later in this chapter.

The variable oxidation states arise from the varying numbers of electrons that can be lost from both the ns and $(n-1)d$ subshells of any metallic element. The ionic charge of the individual ions of transition metals is commonly +2 or +3. Electrons are lost from the ns subshell to give the +2 ions and an additional electron is lost from the $(n-1)d$ subshell to give the +3 ions. However, in covalently bonded molecules or complex ions, the oxidation state can be as high as +8. For example,

$(MnO_4)^-$	Mn (VII)
$(CrO_4)^{2-}$	Cr (VI)
$Ni(CO)_4$	Ni (IV)
OsO_4	Os (VIII)
$CuCl_2(ionic)$	Cu (II)

Most of the transition metal ions, as well as their salts and complexes, are colored (with the exception of Zn^{2+}, Cd^{2+}, and Hg^{2+} ions, in which the *d* subshells are filled). This unusual property is restricted to the transition elements, including the lanthanides. Other ions such as Na^+, Ca^{2+}, and Cl^- are not colored. In the case where a transition metal ion is free from all interactions with other ions or molecules, as in the gas phase, the five *d* orbitals are degenerate in energy. However, when ions or molecules closely surround a transition metal ion, the *d* orbitals are no longer degenerate; the *d* orbitals spread apart or split in energy according to the nature and geometrical arrangement of the ions or molecules surrounding the transition metal ion. The electrons originally present in the *d* orbitals occupy these split *d* levels according to the Pauli principle and with the consideration that the total electronic energy of the system be at a minimum. In general this situation results in electrons being excited between split *d* energy levels which are relatively close in energy as compared to levels in nontransition metal ions. Absorption of visible light is sufficient for the excitation and since a portion of the visible spectrum is absorbed, this causes the salts, ions, and complexes to be colored. We shall discuss shortly in more quantitative terms the dependence of energy splitting upon the nature and geometry of the surrounding ions and molecules.

It is worth a reminder that if an element is paramagnetic, it means that a sample is attracted into a magnetic field. This magnetic state most commonly arises from the presence of unpaired electrons. An overall magnetic moment arises from the spin of unpaired electrons. When all electrons are paired, a sample is diamagnetic and is repelled out of a magnetic field. It is common for the *d* orbitals of transition metal ions to be partially filled. Further, since the *d* orbitals of transition metal ions are split in energy, individual unpaired electrons can exist in any or all of the *d* orbitals, resulting in paramagnetic behavior. The magnitude of the magnetic moment depends upon the number of unpaired electrons. It can be seen that both the existence of color and the magnitude of the magnetic moments depend upon the nature of splitting of the *d* orbitals. We shall shortly discuss the crystal field theory which has been quite successfully applied in the interpretation and prediction of color and magnetic moments for compounds of transition metal ions.

16–B. COORDINATION COMPOUNDS

One of the outstanding properties of the transition metal ions is their ability to form a wide variety of complex ions and compounds. In fact it is essentially impossible to prevent the formation of such ions and compounds. For example, when a transition metal salt dissolves in water it results in the formation of a complex ion between the transition metal and water. A wide variety of molecules and ions are capable of combining with transition metal ions. The combining molecules and ions are called the *coordinating species* or the *ligands*. A common feature of all ligands is that at least one atom in the ligand is capable of donating

TABLE 16–1   Ligands and their coordination ability

MONODENTATE[a]	BIDENTATE[b]	
:Cl⁻ is written as :Cl⁻		

TABLE 16–1   Ligands and their coordination ability

MONODENTATE[a]

:Cl⁻

:CN⁻

H₂O:

:NH₃

BIDENTATE[b]

oxalate anion ($C_2O_4^{2-}$)

$CH_2—CH_2$
H_2N NH_2

ethylenediamine (en)

$H_3C—C—C—CH_3$
$HO—N$ $N=O$

dimethylglioxime anion (−1 charge)

[a]The electron pair is located on the atom acting as the coordinating site.

[b]Arrows show the atoms which coordinate to the central atom or ion.

an electron pair to the transition metal ion when they combine. A wide variety of molecules and ions are ligands: CN^-, CO, NH_3, H_2O, Cl^-, F^- and ethylene diamine ($H_2N—CH_2—CH_2—NH_2$). Examples of complex ions are $[Cu(NH_3)_4]^{2+}$, $[Zn(CN)_4]^{2-}$ and $[FeCl_4]^-$, where all coordinated groups or ligands and the transition metal ion are within the square brackets.

Ligands are commonly divided into two types based upon the number of donor atoms or coordination sites which the ligand contains. Ligands are known as *mono-* or *unidentate* if they contain one coordination site or atom and *bidentate* if they contain two coordination sites or atoms. Table 16–1 gives a few examples of each.

PROBLEM 16–1   Classify the following as uni- or bidentate

a. Br⁻
   bromide

b. OH⁻
   hydroxide

c.
1,10 phenanthroline
or *o*-phen group

d. NH₂⁻
   amide group

e.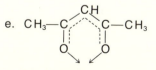

acetylacetonato   group   or
AcAc⁻ (the dashed line indicates charge delocalization over the whole molecule)

f. NO
   nitrosyl

g.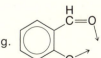

salicylaldehyde anion

TABLE 16–2 Coordination number of some ions

METAL	PREDICTED COORDINATION NUMBER	EXPERIMENTAL EXAMPLES
Ag^+	2	$[Ag(NH_3)_2]^+$
Cu^{2+}	4	$[Cu(NH_3)_4]^{2+}$
Co^{3+}	6	$[Co(NH_3)_6]^{3+}, [CoF_6]^{3-}$
W(IV)	8	$[W(CN)_8]^{4-}$
Cd^{2+}	4	$[Cd(CN)_4]^{2-}$

There are many organic ligands. It is possible to design specific ligands that are useful in the selective extraction of certain metal ions from sea water, even when these metals are present only at very low concentrations. For example, it is possible to extract gold selectively from the ocean, where the concentration is only 1.4×10^{-6} g/liter. With this method, essentially all of the metal can be removed, which has important implications when we remember that the total amount of gold in the seas is estimated at 70,000,000 tons. It is also possible to design ligands selective for uranium and copper which are able to extract essentially 100 % of these metals from sea water where the ion concentrations are $4.5\text{-}5 \times 10^{-6}$ g/liter and 5×10^{-7} g/liter respectively. It is estimated that the total amount of copper and uranium in the seas is greater than 10^9 tons and 10^8 tons, respectively. Finally, it is important to note that in biological systems, there are naturally occurring ligands with the same ability to selectively concentrate metals (ions). For example, the copper in the cuttlefish *Octopus vulgaris* (a multi-armed mollusk resembling a squid but with a calcified inner shell) is concentrated by a factor of 100,000 over that in the surrounding sea water. Another example is the tunicate *Phallusia mamillata* (a primitive marine animal) that concentrates vanadium by a factor of 1,000,000 over that found in the surrounding sea water. Obviously, it would be useful to be able to duplicate or mimic these biological ligands synthetically.

The coordination number of the central metal atom or ion is equal to the number of binding positions available to ligands. An approximate rule is that the coordination number of a transition metal ion is twice its oxidation number. Table 16–2 gives some examples. Although other coordination numbers are possible for a metal in a given oxidation state, this rule serves as a useful guide.

PROBLEM 16–2 Predict the coordination number for the following ions:
a. Cr^{3+} b. V^{3+} c. Zn^{2+} d. Au^+ e. Au^{3+} f. Ag^+.

NOMENCLATURE

The naming of coordination compounds is more complicated than that of other compounds. The following rules can be used to name most coordination compounds.

1. Ligands can be of two types, anionic and neutral.
   a. The name of anionic ligands is derived from the stem portion of the name of the atom or group and end in -o, as Cl^- (chloro), CN^- (cyano), F^- (fluoro), and $C_2O_4^{2-}$ (oxalato).
   b. Neutral ligands have no consistent names. The more common ones are H_2O (aquo), NH_3 (ammine) and CO (carbonyl).
2. For complexes with a single central atom:
   a. When the compound is ionic, the cation is named first.
   b. The number of ligands is denoted by the prefixes *di*-(2), *tri*-(3), *tetra*-(4), *penta*-(5) and *hexa*-(6).
   c. If the complex ion is cationic or neutral, the ligands are given first in the word name, beginning with the anionic ones, followed by the name of the central metal and a roman numeral in parenthesis equal to the oxidation number. Examples of these are:

$$[Co(NH_3)_6]Cl_3$$
hexammine cobalt (III) chloride

$$[Cr(H_2O)_4Cl_2]Cl$$
dichlorotetraaquochromium (III) chloride

   d. If the complex ion is anionic, the suffix -*ate* is added to the name of the central atom and is followed by the oxidation number (with roman numeral in parenthesis). Examples are:

$$K[PtCl_3(NH_3)]$$
potassium trichloroammineplatinate (II)

$$K_3[Co(CN)_6]$$
potassium hexacyanocobaltate (III)

   e. If the ligand is of a more complicated nature, such as ethylenediamine (en), the prefixes giving the number of such ligands are modified from b as *bis*- (two), *tris*- (three) and *tetrakis*- (four). An example is

$$[Co(NH_2—(CH_2)_2—NH_2)_3]\ Cl_3 \quad \text{or} \quad [Co\ en_3]\ Cl_3$$
tris(ethylenediamine) cobalt (III) chloride

PROBLEM 16–3   Name the following compounds:
a. $[CoBr_2(NH_3)_4]Cl$ c. $[Cr(H_2O)_6]^{3+}$
b. K_4FeF_6

PROBLEM 16–4   Write formulas for the following compounds:
a. dichlorotetramminecobalt (III) bromide
b. tris(ethylenediamine) chromium (III) ion
c. lithium hexacyanotitanate (IV)

GEOMETRIC AND IONIZATION ISOMERS

We mentioned earlier that different geometrical arrangements of the ligands are possible around the same central atom. The form of this geometric arrangement in part is dependent upon the number of ligands. However, for some particular numbers, such as four, alternative geometries exist. Table 16–3 summarizes some frequently encountered cases.

TABLE 16–3 Geometry and coordination number of central atom

COORDINATION NUMBER	GEOMETRY	SPATIAL REPRESENTATION[a]	EXAMPLE
2	linear	L—M—L	$[H_3N—Ag—NH_3]^{2+}$
4	tetrahedral		
	square-planar		
6	octahedral		

[a]The symbol M always represents the central metal atom or ion and L represents the ligand.

It is also possible to have *cis* and *trans* geometric isomers of complexes as well as of pure organic compounds (see the discussion of 2-butene in Section 7–H). We shall consider two specific examples. The first is $Pt(NH_3)_2Cl_2$, where the geometric isomers are shown in Figure 16–1. Note that the chloro and ammine groups are diagonally opposed in the *trans* isomer but adjacent in the *cis*. We can use $[Co(NH_3)_4Cl_2]^+$ ion as an example of isomers of an octahedral complex (Figure 16–2).

Coordination compounds with the same chemical constitution relative to a particular coordinating ion may release different numbers of that ion into solution. This is called ionization isomerization. The ions that can dissociate from the complex in solution depend upon the type of coordination between the ion and the transition metal. If the ion is directly coordinated to the transition metal, it is considered to be nonionizable. The distinction is variously made between inner sphere (coordinated directly to metal ion) and outer sphere (ionic) binding or between primary (coordinated directly to metal ion) and secondary binding. For example, a series of compounds of Pt, NH_3, and Cl^- exist, each of which has a different number of ionizable (outer sphere) Cl^- ions. The number of the latter can be determined by the moles of insoluble $AgCl$ obtained per mole of platinum compound when treated with $AgNO_3$. Table 16–4 gives examples of several possibilities.

FIGURE 16–1 Geometric isomers of $Pt(NH_3)_2Cl_2$

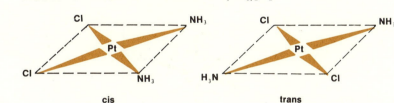

cis   trans

FIGURE 16–2 Geometric isomers of $[Co(NH_3)_4Cl_2]^+$

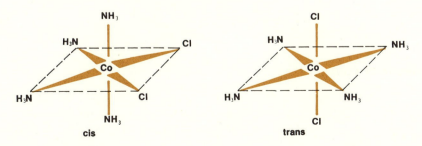

cis   trans

TABLE 16–4 Structure of a series of Pt complexes depicting nature of coordination of Cl^-

COMPOUND	STRUCTURE	MOLES AgCl MOLE Pt COMPOUND
$[Pt(NH_3)_6]Cl_4$		4
$[Pt(NH_3)_4Cl_2]Cl_2$		2
$Pt(NH_3)_2Cl_4$		0

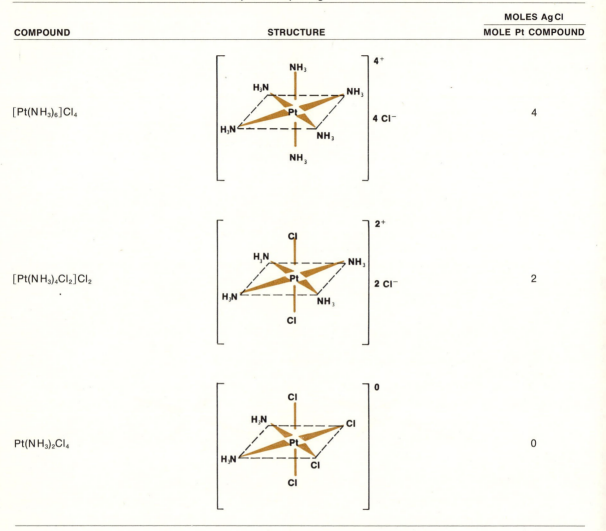

16–C. STABILITY OF COMPLEX IONS

In this discussion we will be concerned only with the thermodynamic stability of complex ions. We shall first consider equilibria involving the association-dissociation of the metal ion and coordinating species. The simplest case involves a metal ion M of arbitrary charge and a unidentate ligand L.

16–1 $$M + L \rightleftharpoons ML \qquad K_1 = \frac{[ML]}{[M][L]}$$

16–2 $$ML + L \rightleftharpoons ML_2 \qquad K_2 = \frac{[ML_2]}{[ML][L]}$$

$$\vdots \qquad\qquad \vdots$$

16–3 $$ML_{n-1} + L \rightleftharpoons ML_n \qquad K_n = \frac{[ML_n]}{[ML_{n-1}][L]}$$

For example, $n = 6$ for the octahedral complex $[Co(CN)_6]^{3-}$ and $n = 4$ for the tetrahedral complex $[Ni(CN)_4]^{2-}$. In equations 16–1 to 16–3, the K's represent the association constants for the individual complex species ML, ML_2, ML_3, etc. which lead stepwise to the final complex. This is an inverse but parallel situation to the stepwise ionization of a polyprotic acid such as H_3PO_4:

$$H_3PO_4 \rightleftharpoons H^+ + H_2PO_4^- \qquad K_1 = \frac{[H^+][H_2PO_4^-]}{[H_3PO_4]}$$

$$H_2PO_4^- \rightleftharpoons H^+ + HPO_4^{2-} \qquad K_2 = \frac{[H^+][HPO_4^{2-}]}{[H_2PO_4^-]}$$

$$HPO_4^{2-} \rightleftharpoons H^+ + PO_4^{3-} \qquad K_3 = \frac{[H^+][PO_4^{3-}]}{[HPO_4^{2-}]}$$

It is also possible to write an association-dissociation equilibrium for any final complex species with the metal ion and ligand themselves as reactants:

16–4 $$M + nL \rightleftharpoons ML_n \qquad K(M,L) = \frac{[ML_n]}{[M][L]^n}$$

For equation 16–4 $K(M,L)$ is the constant for the association beginning with the metal ion and ligand as reactants. The association constant for the equilibrium beginning with metal ion and ligand as reactants is the product of the stepwise association constants.

16–5 $$K(M,L) = K_1 K_2 K_3 \cdots K_n$$

If we consider as an example the complex $[Cd(NH_3)_4]^{2+}$, the appropriate stepwise reactions are:

16–6 $$Cd^{2+} + NH_3 \rightleftharpoons [Cd(NH_3)]^{2+} \qquad K_1 = 10^{2.65}$$

16–7 $$[Cd(NH_3)]^{2+} + NH_3 \rightleftharpoons [Cd(NH_3)_2]^{2+} \qquad K_2 = 10^{2.10}$$

16–8 $$[Cd(NH_3)_2]^{2+} + NH_3 \rightleftharpoons [Cd(NH_3)_3]^{2+} \qquad K_3 = 10^{1.44}$$

16–9 $$[Cd(NH_3)_3]^{2+} + NH_3 \rightleftharpoons [Cd(NH_3)_4]^{2+} \qquad K_4 = 10^{0.93}$$

The overall reaction is

16–10 $$Cd^{2+} + 4NH_3 \rightleftharpoons [Cd(NH_3)_4]^{2+}$$

and the association constant is

$$K(M,L) = K(Cd^{2+}, NH_3) = K_1 K_2 K_3 K_4 = 10^{7.12}$$

We can see that it is possible to have several complex ion species present at the same time, whether the complex $[Cd(NH_3)_4]^{2+}$ were being prepared from the reaction of Cd^{2+} and NH_3 or a solution were prepared by dissolving $[Cd(NH_3)_4]Cl_2$ in water. It is obvious that as the concentration of ligand NH_3 is increased, the concentration of the final complex $[Cd(NH_3)_4]^{2+}$ will increase to become the dominant species at high ligand concentrations. The magnitudes of the individual K_n values are important in determining the stability of the complex ions in general. In the case of $[Cd(NH_3)_4]^{2+}$, a large K_4 value would mean that the extent of dissociation of $[Cd(NH_3)_4]^{2+}$ is small and this complex ion is the dominant species. Since K_4 is actually small, the $[Cd(NH_3)_4]^{2+}$ ion dissociates appreciably to $[Cd(NH_3)_3]^{2+}$ and NH_3 (equation 16–9).

It is worth noting that individual K_n and $K(M,L)$ values vary according to the type of ligand for a given metal ion. For example:

16–11 $$[Cd(CN)]^+ + CN^- \rightleftharpoons [Cd(CN)_2] \qquad K_2 = 10^{5.12}$$

16–12 $$[Cd(CN)_3]^- + CN^- \rightleftharpoons [Cd(CN)_4]^{2-} \qquad K_4 = 10^{3.55}$$

16–13 $$Cd^{2+} + 4CN^- \rightleftharpoons [Cd(CN)_4]^{2-} \qquad K(M,L) = 10^{18.8}$$

where equations 16–11 to 16–13 should be compared with equations 16–7, 16–9, and 16–10 respectively. This dependence of the association constants on the type of ligand also applies to bidentate ligands such as ethylenediame (en) in the complex ion $[Cd(en)_2]^{2+}$.

The special case of the hydrolysis of metal ions is best understood in terms of aquo complexes where the hydrolysis reaction occurs with the ionization of a coordinated H_2O molecule:

16–14 $$[Fe(H_2O)_6]^{3+} + H_2O \rightleftharpoons [Fe(H_2O)_5OH]^{2+} + H_3O^+$$

This is also true for the hydrolysis of ions of nontransition metals, such as Al^{3+}:

16–15 $$[Al(H_2O)_6]^{3+} + H_2O \rightleftharpoons [Al(H_2O)_5OH]^{2+} + H_3O^+$$

Equations 16–14 and 16–16 are only the first steps in hydrolysis; the products of these reactions can be hydrolyzed further. The values of K_h measured for equations 16–14 and 16–15 depend upon the metal ion. For example, $K_h = 10^{-5}$ for Al^{3+} and 6×10^{-3} for Fe^{3+}. Thus, water solutions of both Al^{3+} and Fe^{3+} are acidic, with the Fe^{3+} solution having the higher H_3O^+ concentration (lower pH).

A related concept is amphoterism. Consider hydrous $Al(OH)_3$ or $Al(OH)_3(H_2O)_3$ which can produce either H_3O^+ or OH^- upon hydrolysis, thus acting either as an acid or a base.

16–16

$$[Al(OH)_4(H_2O)_2]^- + H_3O^+ \rightleftharpoons Al(OH)_3(H_2O)_3 + H_2O \rightleftharpoons [Al(OH)_2(H_2O)_4]^+ + OH^-$$

Another useful way to view the ability of $Al(OH)_3(H_2O)_3$ to exhibit amphoterism is the manner in which it can react with an acid or base.

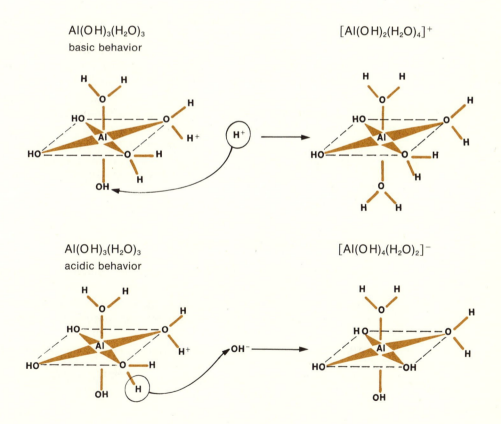

Certain transition metal hydroxides and oxides show a similar behavior.

16–D. PARAMAGNETISM

It is possible to predict the magnetic moment of an ion based upon the number of unpaired electrons. This is known as the spin-only moment and for a single electron is

16–17
$$\mu_s = 2 \sqrt{m_s (m_s + 1)}$$

where μ_s is the spin-only magnetic moment and m_s is the absolute value of the spin quantum number (1/2). The unit of measure of magnetic moment is the *Bohr magneton* (BM) (9.273×10^{-21} erg gauss^{-1}). If there is 1 unpaired electron then $\mu_s = 1.73$ BM. If there is more than 1 electron, then $\mu_s = 2 \sqrt{M_s(M_s + 1)}$ where the total spin quantum number is the sum of the individual spin quantum numbers, $M_s = m_{s_1} + m_{s_2} + m_{s_3} + \cdots$. (The symbol S is often used instead of M_s.) Inversely, we can predict the number of unpaired electrons from the experimentally determined magnetic moment. Table 16–5 gives some examples.

PROBLEM 16–5   Calculate the μ_s expected for Mn^{2+} ion which contains 5 unpaired electrons.

The range of values in the last column of Table 16–5 comes about because the measured experimental value includes not only the spin moment but also the moment arising from the orbital motion of the electron. The magnitude of the latter varies depending upon the nature of the atoms, ions or molecules surrounding the central transition metal ion. Despite this complication, Table 16–5 does show that the observed moment is very often the same or very close to the spin-only moment. Thus, from the observed moment, the number of unpaired electrons can be predicted which can directly aid in determining the electronic configuration of the ion. We shall discuss this in more detail shortly when we consider crystal field theory.

16–E. CRYSTAL FIELD THEORY (LIGAND FIELD THEORY)

Earlier we mentioned that the splitting of the *d* electronic levels depends upon the geometry of the ligands around the transition metal ion and upon the specific ligand involved. In fact, we should actually say that the *pattern of splitting depends upon the geometry* and the *magnitude of splitting within a given pattern depends*

TABLE 16–5   Predicted and observed moments of some transition metal ions

ION	NUMBER UNPAIRED e's	M_s	PREDICTED μ_s (BM)	OBSERVED MOMENT (BM)
Cu^{2+}	1	1/2	1.73	1.7–2.2
Ni^{2+}	2	1	2.83	2.8–4.0
Cr^{3+}	3	1 1/2	3.87	~3.8

FIGURE 16–3   The octahedral distribution of an anion L⁻ around a central ion

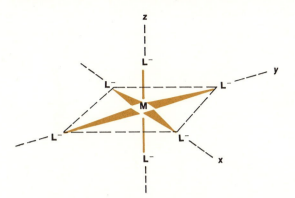

upon which specific ligand is complexed. We shall immediately proceed to a quantitative discussion of these points.

First of all, recall that in a *free* ion (as in the gas phase) the five d orbitals are energetically equivalent or degenerate. Thus, for a transition metal ion with a single electron, the electron has an equal probability of being in any one of the five d orbitals. Will the situation be the same, however, if the same ion is octahedrally surrounded by six anions, such as Cl^-? To answer this question we must recall the geometric disposition of six ligands in an octahedron and also establish the shape of each of the five d orbitals. The geometry of the placement of the ligands is shown schematically in Figure 16–3. The shape of the individual d orbitals is shown in Figure 16–4. Note that there are five d orbitals, each with a subscript related to the orientation in a rectangular coordinate system. The d_{xy}, d_{yz}, and d_{xz} orbitals are identical in shape but are oriented in different planes. It is important to remember that the lobes which characterize the electron charge density distribution for these three orbitals are oriented at an angle of 45° to the coordinate axes. In other words, the highest concentration of negative charge density is *in between* the coordinate axes. The $d_{x^2-y^2}$ orbital has the same shape as the previous three d orbitals, but the lobes are oriented precisely *along* the x and y axes (rather than between the axes, as are the d_{xy}, d_{xz}, d_{yz} orbitals). Finally, the lobes of the d_{z^2} orbital are oriented precisely along the z axis. (The doughnut shaped portion is of no importance in our present discussion.)

Utilizing Figure 16–4, we shall now analyze the consequences of the placement of the anions in the octahedral arrangement in Figure 16–3 upon the energy of the d orbitals. In order to do this we assume that the charge of the anion is concentrated in a tiny ball known as a point charge. This assumption constitutes the basis of *crystal field theory*. First, the d_{z^2} and $d_{x^2-y^2}$ orbitals have their lobes directed precisely toward the negative ion ligands. On the other hand, all three of the d_{xy}, d_{xz}, and d_{yz} orbitals have their lobes directed at an angle of 45° to the negative ion ligands. Thus two of the d orbitals, the d_{z^2} and $d_{x^2-y^2}$, have equivalent environments in that the lobes are directed towards the negative ion. The three

FIGURE 16–4 The shapes of the five *d* orbitals (with anions L⁻ shown arranged with an octahedral geometry)

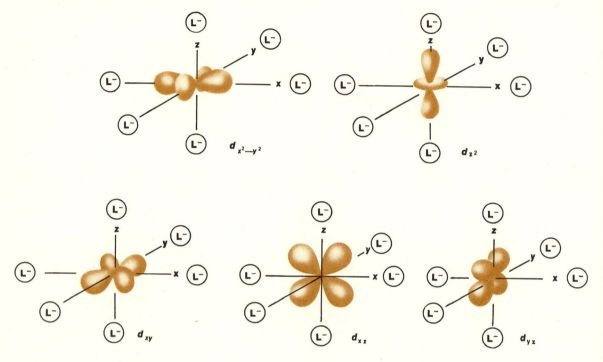

others, d_{xy}, d_{xz}, and d_{yz}, have equivalent environments different from the other two in that the lobes are directed halfway between the negative ion. Because like charges repel each other, electrostatic repulsion raises (relatively) the energy of both the d_{z^2} and $d_{x^2-y^2}$ orbitals but lowers the energy of the three other orbitals (Figure 16–5). Recall that the term *degeneracy* is used when two or more levels have the same energy. Thus, we can say that the original fivefold degeneracy of the *d* orbitals is split into two groups as in Figure 16–5, one of threefold (three *d* orbitals) and one of twofold (two *d* orbitals) degeneracy.

The term *octahedral field* refers to the electrostatic potential associated with negative point charges placed at the corners of an octahedron. The symbol $\Delta_\circ$ is the energy difference between the two degenerate *d* orbital sets in an octahedral field; its value depends upon the nature of the anion. To avoid the awkwardness of referring to all of the *d* orbitals, the upper set is referred to as the *e* orbitals (two) and the lower set as the *t* orbitals (three). The relative magnitude of $\Delta_\circ$ for some common transition metal ions in an octahedral field is

$$CN^- > C_2O_4^{2-} > OH^- > F^- > Cl^- > Br^- > I^-$$

FIGURE 16–5 Energy level diagram showing the splitting of the d orbitals about the free ion reference energy by electrostatic interaction in an octahedral field.

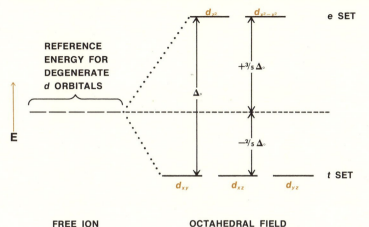

We have been dealing with anions at the corners of an octahedron, a negative point charge approximation. However, in many neutral molecules with a large dipole, an atom with a relatively high negative charge exists in the molecule:

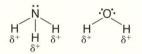

It is possible to *approximate* the action of such whole neutral molecules as *point charges*. The same crystal field model is also appropriate for complexes involving such molecules where the atom with the relatively high negative charge acts as the coordination site to the transition metal ion. In these cases, the term *ligand field theory* is often used. The relative magnitude of the $\Delta_\circ$ values for some ions and molecules is:

$$CN^- > NH_3 > H_2O \approx C_2O_4^{2-} > OH^- > F^- > Cl^- > Br^- > I^-$$

In the CN^- and H_2O octahedral complexes of Fe^{2+} the $\Delta_\circ$ for CN^- is about 94 kcal/mole (4.2 e.v. or 33,000 cm^{-1})[†]; whereas for H_2O, $\Delta_\circ$ is about 37 kcal/mole (1.6 e.v. or 10,400 cm^{-1}). All three units — kcal, cm^{-1}, and e.v. — are energy units (8066 cm^{-1} = 1 e.v. = 23 kcal/mole).

So far we have seen that the transition metal ions form complexes very easily with a variety of ligands. In addition, we know that the presence of such ligands alters the energy of the d orbitals and results in a specific pattern of d orbitals

[†]When e.v. or cm^{-1} are used, reference is to the individual molecule, ion, or atom.

when arranged in an octahedral geometry around any transition metal ion. This creates the problem of how to assign the electrons of the *d* orbitals to the *e* set and *t* set orbitals and correlate this arrangement with the experimental paramagnetic nature (the magnitude of the magnetic moment) and color characteristics of transition metal ions.

ELECTRONIC CONFIGURATION OF TRANSITION METAL IN COMPLEX ION

We are now in a position to consider the electron configuration of octahedral complexes in which the transition metal ion can have from 1 to 9 *d* electrons. Again, the specific configuration will obviously have a direct bearing on the magnetic and spectral characteristics of the complex. The configuration is developed in the same general manner as for atoms (Sections 4–L and 4–M), whereby the lowest energy orbital is first filled with one electron. The Pauli principle (Section 4–L) applies as for atoms.

One problem not present in atoms arises when we apply Hund's rule (Section 4–L) to transition metal complexes. Recall that Hund's rule states that when electrons are to occupy degenerate orbitals, they will occupy them with parallel spins until all orbitals are singly occupied. In the free atom or ion, all five of the *d* orbitals are degenerate, but this is not true when the ion is coordinated. By Hund's rule, the configuration of four *d* electrons in the free atom or ion would be:

$$(\uparrow) \quad (\uparrow) \quad (\uparrow) \quad (\uparrow) \quad (\,)$$
$$d \qquad d \qquad d \qquad d \qquad d$$

In the case of the same metal ion in an octahedral field, however, the solution is not obvious. As can be seen from Figure 16–6, two alternatives exist and the actual configuration will depend upon the relative magnitudes of the spin pairing energy (E_{SP}) and the energy splitting of the *e* and *t* orbital sets ($\Delta_\circ$). The E_{SP} is the energy required per electron pair to pair spins ($\uparrow\downarrow$) in the same orbital. Thus, if

$$E_{SP} > \Delta_\circ$$

more energy is required to pair two electrons than to move one of the two electrons from the *t* orbital to an empty *e* orbital. In other words, the total energy of configuration B is lower than that of configuration A. Consequently, configuration B will be the stable configuration. On the other hand if

$$\Delta_\circ > E_{SP}$$

then the total energy of configuration A will be lower than that of configuration B, and configuration A will be the stable one. Since the total spin, M_s, is greater in configuration B–4/2–than in configuration A–2/2–configuration B is called the high-spin configuration or state and configuration A is called the low-spin con-

FIGURE 16–6 Alternative configurations for d^4 in an octahedral field   A. low spin B. high spin

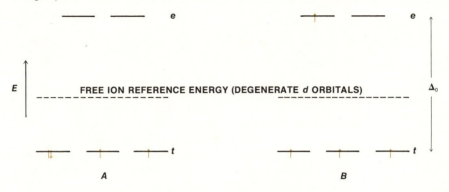

figuration or state. Furthermore, since the magnitude of the magnetic moment depends upon the number of unpaired electrons, experimental evaluation of the moment can determine whether the complex has the high- or low-spin configuration.

The problem of the electronic arrangement just discussed does not exist for all d configurations. For example, if the d configurations are d^1, d^2, d^3, d^8, d^9, or d^{10}, no alternatives exist. Figure 16–7 shows the electronic arrangement for these cases in an octahedral field.

In the case of the d^1, d^2, and d^3 configurations, 1, 2, and 3 electrons will go into each of the three t degenerate orbitals. This occurs by virtue of the t orbitals being lower in energy than are the e orbitals and by Hund's rule.

In the case of the d^8, d^9, and d^{10} configurations, both the high-spin and low-spin configurations are identical. To show this, we will consider the d^8 configuration. If the splitting between the t and e levels were such that the high-spin complex formed, we would place the first 3 electrons in the three t orbitals as in the d^3

FIGURE 16–7 Unique electronic arrangements for certain d configurations

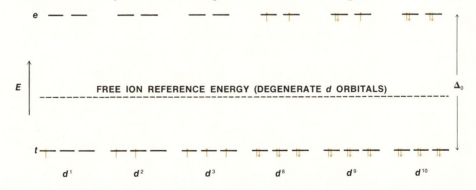

configuration and the next 2 electrons in the two *e* orbitals. The remaining 3 electrons would then be placed in the three *t* orbitals with spins opposite to the 3 electrons already there. The result is the d^8 configuration shown in Figure 16–7 with 6 electrons in the *t* orbitals and 2 in the *e* orbitals. If a low-spin complex is formed, we would first fill all the *t* orbitals with a total of 6 electrons and then place the remaining 2 electrons in the two *e* orbitals with parallel spins. This configuration is identical to that obtained from the high-spin complex just discussed. A similar situation occurs with the d^9 and d^{10} configurations. Thus, the d^1, d^2, d^3, d^8, d^9, and d^{10} configurations are uniquely determined, and no high- and low-spin alternatives can exist.

PROBLEM 16–6   Write out the electronic configuration for the transition metal ion in the following complexes and draw a schematic energy level diagram showing the electron arrangement of the *d* electrons.
a. $[TiF_6]^{3-}$
b. $[Zn(NH_3)_6]^{2+}$
c. $[Cu(H_2O)_6]^{2+}$

PROBLEM 16–7   Predict μ_s for each of the complex ions in Problem 16–6.

PROBLEM 16–8   Construct a schematic energy level diagram and give the electronic configurations for the high- and low-spin alternatives for d^5.

CRYSTAL FIELD STABILIZATION ENERGY

Recall that for an octahedral arrangement, the crystal field results in the *t* and *e* levels being respectively below and above the free ion reference energy of degenerate *d* orbitals (Figures 16–5, 16–6 and 16–7). For a complexed metal ion with a d^1 configuration for example, the one electron goes into a *t* orbital which is $\frac{2}{5}\Delta_\circ$ below the free ion reference energy. The energy lowering for the configuration is known as the *crystal field stabilization energy*. Table 16–6 gives some other examples of stabilization resulting from complexing of ions for which there is a unique configuration (Figure 16–7).

TABLE 16–6   Energy of stabilization of some complexed metal ions in an octahedral field

ION	CONFIGURATION	NUMBER OF *t* ELECTRONS	NUMBER OF *e* ELECTRONS	CRYSTAL FIELD STABILIZATION E
Ti^{3+}	d^1	1	0	$1(-2/5\Delta_\circ)=-2/5\Delta_\circ$
V^{3+}	d^2	2	0	$2(-2/5\Delta_\circ)=-4/5\Delta_\circ$
Cr^{3+}	d^3	3	0	$3(-2/5\Delta_\circ)=-6/5\Delta_\circ$
Ni^{2+}	d^8	6	2	$6(-2/5\Delta_\circ)+2(+3/5\Delta_\circ)=-6/5\Delta_\circ$
Cu^{2+}	d^9	6	3	$6(-2/5\Delta_\circ)+3(+3/5\Delta_\circ)=-3/5\Delta_\circ$

Note that ions with either no d or 10 d electrons (such as Zn^{2+}) do not have any crystal field stabilization energy.

High-spin versus low-spin alternatives do exist for the d^4, d^5, d^6, and d^7 cases. However, it is clear (Figures 16–5 and 16–6) that the smaller the magnitude of $\Delta_\circ$, the more likely it is that the electrons will be in the upper e orbitals, that is, in the high-spin state. Furthermore, a decision in an individual case involving d^4–d^7 configurations depends upon the total energy of the electronic configuration, which is a combination of the crystal field energy and the spin pairing energy. For example, reconsider the d^4 configuration in Figure 16–6. Since the t levels are below the free ion reference level of degenerate d levels, the energy is lowered (minus) 2/5 $\Delta_\circ$ per electron. The energy is relatively increased for electrons in the e level by (plus) 3/5$\Delta_\circ$ per electron. Also, the sign of E_{SP} is always positive since energy is required for pairing. Thus, the energy of the configuration t^4 (Figure 16–6A) relative to that of the free ion is

16–18
$$E(t^4) = 4(-2/5\Delta_\circ) + 1\ E_{SP} = -8/5\Delta_\circ + E_{SP}$$

while that for the $t^3 e^1$ configuration (Figure 16–6B) is

16–19
$$E(t^3e^1) = 3(-2/5\Delta_\circ) + 1\ (+3/5\Delta_\circ) = -3/5\Delta_\circ$$

Using equations 16–18 and 16–19, we shall consider a specific case of a d^4 configuration, $[Cr(H_2O)_6]^{2+}$, for which the required data is given in Table 16–6.

From Table 16–7, it can be seen that for $Cr(H_2O)_6^{2+}$

$$E(t^4) = -8/5(13900) + 23500 = +1460\ cm^{-1}$$

$$E(t^3e^1) = -3/5(13900) = -8340\ cm^{-1}$$

Thus, the total energy of the configuration t^3e^1 for this particular complex is considerably lower (about 9,800 cm^{-1} or 25 kcal) than that for the t^4 configuration, and the high-spin state can be predicted to be the ground state. This is in agreement with experimental observation based on magnetic moment measurements.

TABLE 16–7 Tables of $\Delta_\circ$, E_{SP}, and spin state predictions for d^4 configurations

COMPLEX	ION	CONFIGURATION	$\Delta_\circ{}^a(cm^{-1})$	$E_{SP}^b(cm^{-1})$	SPIN STATE PREDICTED	OBSERVED
$Cr(H_2O)_6^{2+}$	Cr^{2+}	d^4	13900(H_2O)	23500	high	high
$Mn(H_2O)_6^{3+}$	Mn^{3+}	d^4	21000(H_2O)	28000	high	high

[a]The values for $\Delta_\circ$ depend upon both the ligand and the transition metal ion (including its oxidation state); therefore, $\Delta_\circ$ does not have a constant value even for the same ligand.
[b]These values are theoretically calculated and their magnitudes depend upon the individual ion.

Another example of a d^4 configuration is given in Table 16–7. Predictions for each of d^4, d^5, d^6, and d^7 may be done in the same way.

It is possible to perform an experiment measuring the magnetic moment, to use this to establish the number of unpaired electrons, and from this, to determine the configuration of a particular ion. However, if the theoretical crystal field model had not previously been determined, the interpretation of such experimental data might not be possible.

PROBLEM 16–9   Quantitatively show that for $[Zn(NH_3)_6]^{2+}$, no crystal field stabilization energy is present.

PROBLEM 16–10   Predict using the given data whether the ground state of the octahedral complex will be a low- or high-spin state.

a. $[Fe(H_2O)_6]^{2+}$ $E_{SP} = 17{,}600$ cm^{-1}

   $\Delta_\circ(H_2O) = 10{,}400$ cm^{-1}

b. $[CoF_6]^{3-}$ $E_{SP} = 21{,}000$ cm^{-1}

   $\Delta_\circ(F^-) = 13{,}000$ cm^{-1}

COLOR OF TRANSITION METAL IONS

The color of transition metal ions, salts, and complexes depends upon the pattern of splitting of the d orbitals (as in an octahedral field), the number of electrons, and the $\Delta_\circ$. Let us consider one of the simplest cases, a d^1 type of ion complex such as $[Ti(H_2O)_6]^{3+}$. The excitation process is

$$t^1 \xrightarrow{\;h\nu\;} e^1$$

where the energy required is $\Delta_\circ$. The value of $\Delta_\circ$ is 20,000 cm^{-1}. Thus the wavelength of light absorbed is

$$\lambda = \frac{1}{\bar{\nu}} \times 10^8 = \frac{1 \times 10^8}{2 \times 10^4} = 5000 \text{ Å}$$

where the wave number $\bar{\nu}$ has units of cm^{-1} and the wavelength λ is in angstroms (Å). The absorption is not a line absorption as for free atoms and ions but has a breadth, resulting from the many motions of the other atoms of the complex accompanying the electronic excitation (Section 19–A). The resulting absorption is as shown in Figure 16–8. Note that the absorption band maximum is centered at 5,000 Å and relatively more green and yellow light are absorbed. Therefore,

FIGURE 16–8 Absorption of light by the $Ti(H_2O)_6^{3+}$ complex

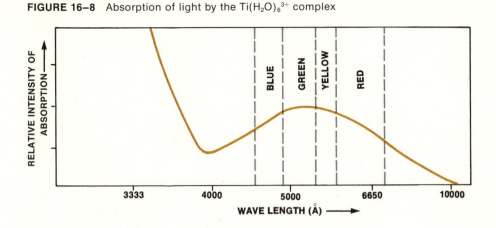

what remains to be transmitted is primarily a mixture of blue and red light, and the solution containing $Ti(H_2O)_6^{+3}$ is a violet color. Actually the reverse process is more common—using the location of the absorption band to evaluate $\Delta\circ$.

CRYSTAL FIELD THEORY AND THERMODYNAMICS

If we were to measure the heat of hydration, ΔH_{hyd}, of some ions, we would expect those containing partially filled *d* orbitals and having a crystal field stabilization energy to have abnormally high ΔH_{hyd} compared to ions that do not have partially filled *d* orbitals (no crystal field stabilization energy). Figure 16–9 gives experimental data showing that this is indeed true. The straight line is an approximation of what would be expected based on the ionic size *alone* (that is, the ionic size decreases from Ca^{+2} to Zn^{+2} so the hydration energies should increase). We can say, therefore, that the crystal field effect has a direct effect on the thermodynamic properties as represented by the effect on ΔH_{hyd}. Subtraction of the crystal field stabilization energies from the actual hydration energies would result in the points falling on the line in Figure 16–9. The same effect is found in other *d* configurations for both high- and low-spin cases.

Such results also have a direct bearing on the equilibrium constants (Chapter 13). If we assume that the $\Delta S°$ for complex formation is constant for a series of ions, then according to

$$\Delta G° = \Delta H° - T\Delta S°$$

the values of $\Delta G°$ will depend only on the values of $\Delta H°$, and since

$$\Delta G° = -RT \ln K$$

then the stability of the complexes, as measured by $K(M,L)$ of equation 16–5, depends only on the value of $\Delta H°$.

FIGURE 16–9 Heats of hydration for some +2 ions. The number of *d* electrons is given in parenthesis beside the symbol.

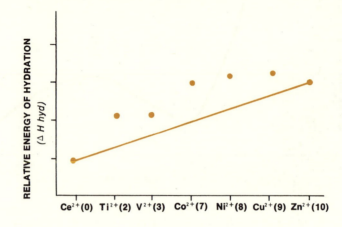

$Ce^{2+}(0)$ $Ti^{2+}(2)$ $V^{2+}(3)$ $Co^{2+}(7)$ $Ni^{2+}(8)$ $Cu^{2+}(9)$ $Zn^{2+}(10)$

PROBLEM 16–11 Assume that $[Co(H_2O)_6]^{2+}$ has a high-spin state as its ground state. Calculate the crystal field stabilization energy for the cobalt ion in terms of $\Delta_\circ$.

PROBLEM 16–12 Assume that the hydration energy for a divalent ion having no crystal field is 650 kcal/mole. Also, assume $\Delta_\circ$ for H_2O is 9300 cm^{-1}. Calculate the heat of hydration for the $[Co(H_2O)_6]^{2+}$ ion in the previous problem.

CRYSTAL FIELD EFFECT IN TETRAHEDRAL AND PLANAR GEOMETRIES

We have limited our discussion to the particular case of an octahedral field. However, all of the aspects considered are applicable to other fields, the principal differences being that the pattern of splittings and the magnitude of $\Delta_\circ$ changes. Figure 16–10 compares an octahedral and tetrahedral field. The splitting Δ_t in the tetrahedral field is approximately 1/2 of that in an octahedral field. Also, note that the relative location of the *t* and *e* orbitals is opposite to that present in an octahedral field. The pattern of *d* orbitals is still different if the ligands are at the corners of a square (Figure 16–10).

PROBLEM 16–13 There is a unique electronic configuration for a tetrahedral complex with a d^7 configuration.
a. Draw a schematic energy diagram showing this and predict whether the complex will be paramagnetic.
b. Compare the diagram with those of a d^7 configuration for an octahedral complex.

FIGURE 16–10 Splitting of d orbitals in three different fields

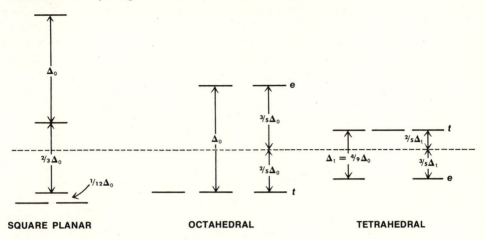

SQUARE PLANAR OCTAHEDRAL TETRAHEDRAL

16–F. PROPERTIES OF THE TRANSITION ELEMENTS

It would be impossible to discuss all of the physical and chemical properties of the transition elements and their compounds. Therefore, in this and the following section, we shall attempt to summarize some of their more salient characteristics.

There are three series of transition elements, each containing d electrons.

$$3d \qquad {}_{21}Sc, {}_{22}Ti...{}_{30}Zn$$

$$4d \qquad {}_{39}Y, {}_{40}Zr...{}_{48}Cd$$

$$5d \qquad {}_{57}La...{}_{72}Hf...{}_{80}Hg$$

Two points are worth making. The last member of each series, zinc, cadmium, and mercury, all have $ns^2(n-1)d^{10}$ configurations, and therefore are not formally transition elements. Also, note in the third series that there is the interjection of the lanthanides or rare earth elements with atomic numbers 57 (counting lanthanum) through 71. This interjection is accompanied by a phenomenon known as the lanthanide contraction. In the lanthanides the $4f$ orbitals are being filled, and the valence shell configuration is $6s^2$, remaining so during the addition of the fourteen $4f$ electrons. The common charge of the ions of the lanthanides is +3 resulting from the loss of the two $6s$ and one $4f$ electrons. As the atomic number increases, since the electrons are all going into the same subshell ($4f$), there is a corresponding increase in Z_{eff}, and thereby a progressive decrease in ionic size. The same is generally true for atomic size. Because of the presence of the lanthanide contraction in the third transition metal series, the sizes of the ions of the same charge in the second and third series are very similar. For example, La^{3+} is approximately 0.2 Å larger than Y^{+3}; and therefore, Hf(IV) would be expected also to be about 0.2 Å larger than Zr(IV). However, the lanthanide contraction almost perfectly counterbalances the expected 0.2 Å increase, and the size of Zr(IV) (0.79 Å) and Hf(IV) (0.78 Å) are nearly identical. Because of the foregoing considerations concerning atomic and ionic size, the chemical properties of the members of the second and third transition metal series are parallel.

Figures 16–11 and 16–12 give data about the atomic radii and ionic radii of the transition elements. Figure 16–12 is limited to the first transition series because 2+ ions are uncommon for the second and third series. The general decrease in size occurs because of an increase in Z_{eff}. Note in Figure 16–11, the near identity of the atomic (single bond) radii of the elements of the second and third series.

The transition elements are all metals with generally high melting points, ranging from about 1700 °C to 3400 °C but the boiling points can be as high as 6000 °C. The first ionization potentials of the transition elements are relatively high compared with most of the nontransition metals. A plot of the first ionization potential for the three series is shown in Figure 16–13. Note a trend toward increasing energy required for ionization going from left to right in a series. This occurs because of a general increase in Z_{eff} in the same direction.

FIGURE 16–11 Atomic radii of the three transition metal series

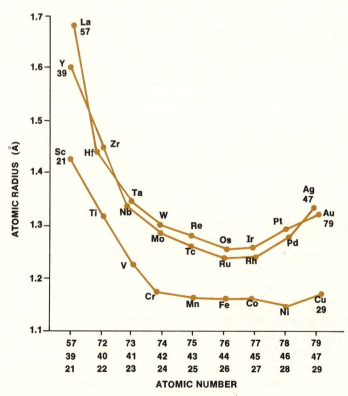

FIGURE 16–12   Ionic radii (for the +2 ions) of the first transition metal series

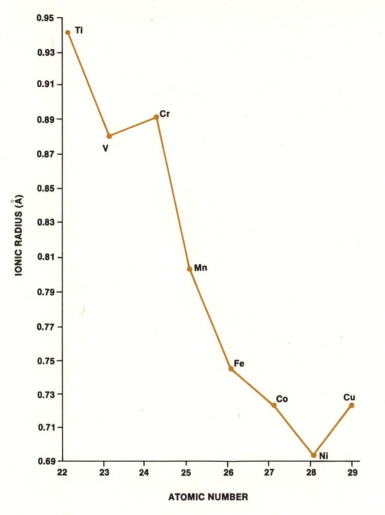

16–G. CHEMISTRY OF THE TRANSITION ELEMENTS

In this discussion, we shall be concerned primarily with the first series of transition elements and refer to the two other series from time to time when the comparison seems worthwhile; we will be limiting our discussion principally because the chemistry of the elements of the first series deviates so much from that of the elements of the second and third series.

As we discussed in Section 16–B, the ions of all the transition metals form complex ions with a wide variety of ligands. Since there is relatively little difference between the energy of the ns and the $(n-1)$ d electrons, both types of electrons

FIGURE 16–13   First ionization potentials of the three transition metal series

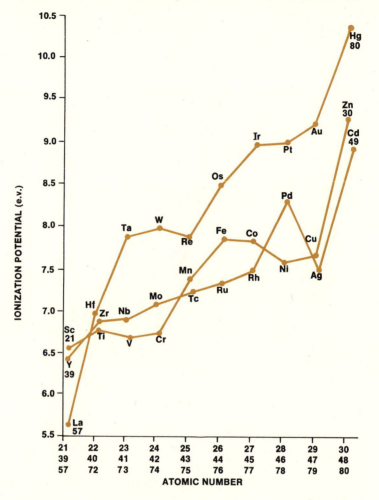

can be involved in chemical reactions so that, as a result, a large number of oxidation states can exist for almost all of the elements (in all series except for the scandium and zinc groups). The maximum oxidation state corresponds to the sum of the ns and the $(n-1)$ d electrons for the scandium through iron groups, but not for the others. Nickel, for example, has eight d- and two s-electrons and yet its maximum oxidation state is +4. For the zinc group (Zn, Cd, Hg), only the number of s electrons is important. The maximum oxidation state of +8 is shown by ruthenium and osmium in RuO_4 and OsO_4. It is a common feature that members of the second and third series exhibit higher oxidation states than those of the first series. However, the highest oxidation state of a given transition metal is not necessarily its most common or most stable oxidation state. The higher oxidation states are more likely to be found in complexes; the lower oxidation states are uncommon in the third series. Since the elements of the second and third series resemble each other in size, both as atoms and ions, their chemistry and compounds are similar; this is particularly noticeable for the sets of elements ruthenium, rhodium, and palladium as well as osmium, iridium, and platinum.

In the first series, manganese in MnO_4^- has the highest oxidation state of $+7$. Following manganese, the higher oxidation states are difficult to obtain. Except for a relatively unstable state of $+6$ for iron, the oxidation state common to iron, cobalt, and nickel is $+2$. (Iron and cobalt also can be $+3$.)

The higher the oxidation state of a transition metal in an oxide, the more acidic is the oxide, as for example:

$$CrO \qquad \text{basic}$$

$$Cr_2O_3 \qquad \text{amphoteric}$$

$$CrO_3 \qquad \text{acidic}$$

Compounds which contain a transition element in its highest oxidation state are generally covalent in nature like $TiCl_4$, a relatively low-boiling liquid (137 °C).

Table 16–8 summarizes some of the properties of the first series transition elements. Some properties of the elements or their compounds which deserve special mention will be considered now.

TITANIUM — $3d^2 4s^2$

One of the most important compounds of titanium is TiO_2, which exists in three crystal modifications. TiO_2 is heavily used as a white pigment in paint. Titanates exist and generally are mixed metal oxides. One of these, strontium titanate, is known as counterfeit diamond and sells for about $40 a carat.

EMF values for reactions involving the simple ions are

$$Ti^{3+} \xrightarrow{-2.0\text{ V}} Ti^{2+} \xrightarrow{-1.63\text{ V}} Ti$$

Ti^{2+} is a powerful reducing agent and can react with water to produce H_2 (>0.42 V required).

Ti (IV) forms complexes with many ligands, including H_2O, Cl^-, and F^-, giving for example $[TiF_6]^-$. Although generally insoluble in non-oxidizing acids, titanium does react with HF (containing fluoride ions) because of the great stability of $[TiF_6]^{2-}$.

TABLE 16–8 Some properties of the first series transition elements

	Ti $3d^2 4s^2$	V $3d^3 4s^2$	Cr $3d^5 4s^1$	Mn $3d^5 4s^2$	Fe $3d^6 4s^2$	Co $3d^7 4s^2$	Ni $3d^8 4s^2$	Cu $3d^{10} 4s^1$
Common oxidation state	+4	+3,+4, +5	+3	+2	+2,+3	+2,+3	+2	+1,+2
Common oxide	TiO_2	V_2O_5	Cr_2O_3	MnO_2	Fe_2O_3	CoO	NiO	CuO
Acid-base character of oxide	amphoteric	acidic	amphoteric	acidic	basic	—	basic	basic
Reaction with:								
halogens	at high T	at high T	at high T	at high T	at mod^e T	at high T	at high T	at mod T
oxygen	at high T	at high T	at high T	at high T	at mod T	at high T	at high T	at high T
nitrogen	at high T	at high T	at high T	at high T		no		
sulfur	at high T	at high T	at high T	at high T	at mod T	at high T	at high T	at high T
hydrogen	at high T			no		no	c	
Reaction with:								
non-oxidizing acids^a	no	no	yes	yes	yes	yes	yes	no
oxidizing acids^b	hot	yes	no		d		d	yes
alkali	no	no			hot			

^{a}Normally excluding HF but including cold dilute sulfuric.
^{b}Commonly including nitric, hot sulfuric and aqua regia (1 HNO_3 : 3 HCl) acids.
^{c}Finely divided metal absorbs large quantities.
^{d}Reacts with warm dilute nitric acid but the use of concentrated acid causes the metal to be unreactive.
^{e}mod signifies moderate.

VANADIUM — $3d^3 4s^2$

Vanadium forms complexes with many ligands, such as H_2O, F^-, Cl^-, and NH_3, giving for example $[VF_6]^-$ and $[V(NH_3)_6]^{2+}$. The metal is also attacked by HF for the same reasons as is titanium. A common oxy ion is VO^{2+}, which readily complexes. Some EMF values for reactions of vanadium in acid solution are

$$VO_2^+ \xrightarrow{\text{0.36 V}} V^{3+} \xrightarrow{-0.25 \text{ V}} V^{2+} \xrightarrow{-1.2 \text{ V}} V$$

CHROMIUM — $3d^5 4s^1$

An important material containing Cr^{3+} ions is ruby. Ruby is a matrix of Al_2O_3, with a scattering of Cr^{3+} ions replacing Al^{3+} ions, thereby giving ruby a red color. Very sharp electronic transitions (lines) occur from a particular excited state of the Cr^{3+} ion. Emission from this state is the basis for the ruby laser. We shall discuss the properties and uses of lasers in Chapter 19.

Many complexes of chromium exist such as $[Cr(NH_3)_6]^{3+}$, CrO_4^{2-}, $[CrX_6]^{3-}$, where X is a halogen, and $[Cr(H_2O)_6]^{3+}$.

Some EMF values for reactions of chromium in acid solution are:

$$Cr_2O_7^{2-} \xrightarrow{\;+1.33\ V\;} Cr^{3+} \xrightarrow{\;-0.41\ V\;} Cr^{2+} \xrightarrow{\;-0.91\ V\;} Cr$$

with $Cr^{3+} \xrightarrow{\;-0.74\ V\;} Cr$

Acidic solutions of $Cr_2O_7^{2-}$ are powerful oxidizing agents, as can be inferred from the potential of the first reaction given above. In detail the reaction in which $Cr_2O_7^{2-}$ acts as an oxidizing agent is

$$Cr_2O_7^{2-} + 14H^+ + 6e^- \longrightarrow 2Cr^{3+} + 7H_2O \qquad \mathcal{E} = +1.33\ ev$$

The Cr^{2+} ion is a good reducing agent and is oxidized by O_2. The reaction:

$$2Cr + 6HCl \longrightarrow 2CrCl_3 + 3H_2$$

is typical of the reactions noted in Table 16–8 to occur with non-oxidizing acids.

MANGANESE — $3d^5 4s^2$

Manganese is similar to iron in regard to its chemical and physical properties. Manganese complexes with a wide variety of ligands, including CO, Cl^-, H_2O, and $C_2O_4^{2-}$, giving for example $[MnCl_4]^-$ and $[Mn(H_2O)_6]^{2+}$.

Some half-cell potential values for reactions of manganese in acid solutions are

$$MnO_4^- \xrightarrow{\;+0.56\;} MnO_4^{2-} \xrightarrow{\;+2.26\ V\;} MnO_2 \xrightarrow{\;+0.95\;} Mn^{3+} \xrightarrow{\;+1.51\ V\;} Mn^{2+} \xrightarrow{\;-1.18\ V\;} Mn$$

with $MnO_4^- \xrightarrow{\;+1.51\ V\;} Mn^{2+}$

From the potentials it can be seen that Mn^{2+} is quite stable to oxidation, whereas Mn^{3+} is quite easily reduced to Mn^{2+}. MnO_2 is an oxidizing agent which can, for example, react with HCl to give Cl_2. The MnO_4^{2-} ion is only stable in strong bases and disproportionates in acid or neutral solution to give MnO_2 and MnO_4^-. A *disproportionation* reaction is one in which a single chemical species simultaneously undergoes both oxidation and reduction. In the previous example, Mn(VI), as MnO_4^{2-}, goes to Mn(IV) as MnO_2, and Mn(VII) as MnO_4^-. The MnO_4^- is also a strong oxidizing agent (giving Mn^{2+}).

IRON — $3d^6 4s^2$

Iron is the second most abundant metal. Its practical usefulness suffers from the fact that in humid air it oxidizes to an Fe_2O_3 hydrate (rust), which forms a loosely packed cover which enables further oxidation to take place (Chapter 14).

Iron can complex with a wide variety of ligands, including H_2O, CO, Cl^-, CN^- and $C_2O_4{}^{2-}$, giving for example $Fe(CO)_5$, $[FeCl_4]^-$, and $[Fe(CN)_6]^{4-}$. One ligand of significant biological importance is porphyrin. When complexed, the molecule is known as an iron porphyrin; it is a planar molecule:

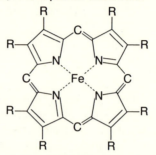

where R can be a wide variety of substituents. An iron(II) porphyrin (heme) is complexed with a protein globin in the important biological molecule hemoglobin, the O_2-carrying vehicle in the blood stream (Chapter 18). The O_2 likely complexes to the iron site, but CO can readily displace O_2, resulting in carbon monoxide poisoning. Our diet must provide sufficient iron for hemoglobin to be made; a deficiency results in anemia.

The most important acid solution reactions of simple ions of iron are

$$Fe^{3+} \xrightarrow{\ +0.77\ V\ } Fe^{2+} \xrightarrow{\ -0.41\ V\ } Fe$$

The potential of the first reaction indicates that in acid solution, O_2 can oxidize Fe^{2+} to Fe^{3+} as:

$$2\,Fe^{2+} + \tfrac{1}{2}O_2 + 2H^+ \longrightarrow 2\,Fe^{3+} + H_2O$$

The Fe^{3+} is relatively unstable unless complexed, and even then, in water solution it undergoes hydrolysis. The higher oxidation states of iron exist in ferrates such as $[FeO_4]^{2-}$.

COBALT — $3d^7 4s^2$

Cobalt can complex with a wide variety of ligands, including NH_3, F^-, CN^-, and Cl^-, to give for example $[CoF_6]^-$ and $[Co(NH_3)_6]^{2+}$. The most important solution reactions of simple ions of cobalt are:

$$Co^{3+} \xrightarrow{\ +1.84\ V\ } Co^{2+} \xrightarrow{\ -0.23\ V\ } Co$$

The potential of Co^{2+} to Co^{3+} indicates that it is very difficult to oxidize Co^{2+}. Co^{3+} reacts with water by oxidizing it to O_2. These two facts account for the rarity of simple salts containing Co^{3+} ion.

One of the important biological compounds of cobalt is Vitamin B_{12}:

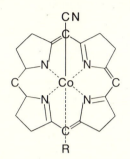

Only the core structure is shown in this diagram. Note that Vitamin B has a por-phyrin-like structure, except that the degree of bond unsaturation is considerably less than in heme. Note also that the cobalt is hexa-coordinated and that, in-terestingly enough, CN^- ion is one of the ligands. The CN^- ion can be replaced and biological activity is still maintained.

NICKEL — $3d^8 4s^2$

Nickel and iron are the prime constituents in meteorites. Finely divided nickel can absorb a large amount of H_2 and is used as a hydrogenation catalyst.

Nickel can complex with a wide variety of ligands including CO, CN^-, and NH_3 to produce, for example, $Ni(CO)_4$, $[Ni(NH_3)_6]^{2+}$ and $[Ni(CN)_4]^{4-}$. Ni (II) salts are very common but no Ni (III) salts are known. The most important solution reaction of simple Ni ion is

$$Ni^{2+} \xrightarrow{\;-0.25\text{ V}\;} Ni$$

With certain ligands, nickel forms complexes with a particular geometry and a particular color at a given temperature. If the temperature is changed, the color changes because of the fact that the geometry changes, causing in turn a change in the splitting of the d orbitals (Figure 16–10) and thereby also changing the energy of the electronic transition. For example,

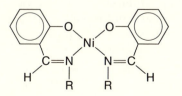

can exist in a planar bright green form near room temperature but when heated to higher temperatures, the geometry changes to tetrahedral with an accompanying color change (brownish). This phenomenon is known as thermochromism, that is, color changes by thermal means. Such changes are commonly reversible by decreasing the temperature (see also Section 19–I).

COPPER – $3d^{10}4s^1$

Compared with the other elements in the series, copper is unique in that it is soft and ductile. It has a very high thermal and electrical conductivity. Copper can complex with a number of ligands including NH_3, F^-, and CN^-, as for example in $[CuF_4]^{2-}$ and $[Cu(NH_3)_4]^{2+}$.

Copper can exist as both Cu^+ and Cu^{2+} ions. The half-cell potentials of such reactions are:

$$Cu^{2+} \xrightarrow{\text{+0.15 V}} Cu^+ \xrightarrow{\text{+0.42 V}} Cu$$
$$\uparrow \underset{\text{+0.34 V}}{\longleftarrow}$$

The Cu^+ is unstable to disproportionation in water, giving Cu^{+2} ions and free Cu. In the presence of complexing ions or molecules, the Cu(I) can be more stable than Cu(II).

EXERCISES

1. Predict the μ_s value for ions that would contain 4, 6, and 7 unpaired electrons.

2. Is it possible for a transition metal ion of the first transition metal series to have 6 or 7 unpaired electrons? Briefly justify your answer.

3. Make a brief general statement concerning the number of oxidation states proceeding from left to right in the first transition metal series.

4. What is the lanthanide contraction and where in the periodic table does it have an effect?

5. Name the following:
   a. $Cs_2[TiCl_5(H_2O)]$
   b. $[Pt\ en_2]Cl_2$
   c. $[Pd(NH_3)_6]^{4+}$

6. Write formulas for the following:
   a. potassium trisoxalatoferrate (III).
   b. tetraiodomercury (II) ion
   c. tetracyanozinc (II) ion

7. Which of the following are Lewis acids and bases?
   a. Cu^{2+} b. NH_3 c. CN^- d. Co^{3+} e. Rh^{2+} f. Au^+ g. H_2O h. F^-

8. Assuming the coordination number of the metal ions is twice the oxidation number, write the formulas for the Cl^- complex of the following: [Hint: Recall that the ns electrons are always lost]:
   a. $Ti - d^1$ d. $V - d^2$
   b. $Fe - d^6$ e. $Co - d^7$
   c. $Cu^+ - d^{10}$ f. $Co - d^6$

9. Which of the following could exist as *cis* and *trans* isomers?
   a. $[AuCl_2]^-$
   b. $[Cr(NH_3)_2(H_2O)_4]^{2+}$
   c. $[Pt\ en_2]Cl_2$
   d. $[FeF_6]^{4-}$
   e. $[CoBr_2(NH_3)_4]^+$

10. Sketch structural diagrams for the *cis* and *trans* isomers of those ions of Exercise 9 capable of existing as geometric isomers.

11. Construct a schematic energy level diagram and give the electronic configuration in an octahedral field for the high- and low-spin configurations for d^6 and d^7.

12. What is the d electronic configuration for the transition metal ion in the following complexes? Draw a schematic energy level diagram showing the electronic arrangement of the d-electrons.
    a. $[VCl_4]^-$, tetrahedral
    b. $[Cr(AcAc)_3]$

13. Predict μ_s for each of the complex ions in Exercise 8 assuming octahedral complexation and both high and low spin configurations where applicable.

14. Given that a hexacoordinated compound of iron in a $+2$ state has a magnetic moment of 4.8 BM. What can be said about the relative magnitude of the electrostatic field potential associated with the ligands (all ligands are identical)?

15. Given that the experimental μ_s values for the $[Fe(H_2O)_6]Cl_3$ is 5.9 BM
    a. Write the electronic configuration for the Fe ion.
    b. What would be the predicted value for the magnetic moment if the Fe ion had a t^5 configuration? Is this the high- or low-spin state?

16. Consider the compound K_2VCl_6. Assume the value of Δ_o for Cl^- is 12,000 cm^{-1}.
    a. What will be the nature of the field in which the vanadium ion resides?
    b. What will be the configuration and schematic energy level diagram for the vanadium ion?
    c. Predict μ_s for the compound.
    d. At what wavelength would absorption occur presuming a pure electronic transition with no vibrational broadening.

17. Consider the complex $[Mn(H_2O)_6]^{2+}$.
    a. Give the configuration for the low- and high-spin states.
    b. Assuming the high-spin state is the ground state, determine the crystal field stabilization energy for the manganese ion. (continued)

c. Would you expect the hydration energy of the manganese ion to fall on or above the line of Figure 16–9?

18. Assume the magnetic moment for $[CoBr_4]^{2-}$ is 3.9 BM.

    a. What would be the nature of the field surrounding the cobalt ion (non-planar).

    b. Write the configuration for the cobalt ion.

    c. Calculate the crystal field stabilization energy for the cobalt ion in this complex.

19. Construct a schematic energy level diagram and give all of the possible electronic configurations in a tetrahedral field for the d^6 and d^8 cases.

20. Assume you had a complex of a transition metal of d^6 configuration but you did not originally know whether it was octahedrally or tetrahedrally coordinated (geometry). A magnetic measurement shows that the complex is diamagnetic. Based on the foregoing data, can you decide whether or not it is an octahedral or tetrahedral complex and how?

CHAPTER SEVENTEEN

NUCLEAR CHEMISTRY

In Chapters 2 and 4 we briefly discussed some aspects of the nucleus. We noted that (1) essentially all the mass of an atom is concentrated in the nucleus, (2) the nucleus is very small compared with the extranuclear region, and (3) the principal components of the nucleus are protons and neutrons. In this chapter we discuss the nucleus in terms of its stability and radioactivity, and the applications of radioactivity to chemistry and other areas.

17–A. SIZE AND SHAPE OF THE NUCLEUS

In Chapter 2, we commented that the nucleus was generally spherical in shape and very small. Indeed, the nucleus is small, ranging between 10^{-13} and 10^{-12} cm in diameter. A range exists because nuclei of different mass have different sizes. The situation is complicated by the fact that a nucleus does not have a "hard" boundary with a set shape. In fact, the nucleus has a fuzzy boundary and is not spherical at all times.

We can determine the size and shape of the nucleus by scattering experiments. High energy electrons bombard a target and are deflected by the nuclei at various angles. The ratio of the number of those electrons scattered at various angles to the number of incident electrons permits the generation of a charge distribution picture of the nucleus. In this approach the distribution is revealed by the inter-action of the positively charged protons in the nucleus with the negatively charged

FIGURE 17–1   Nuclear shapes

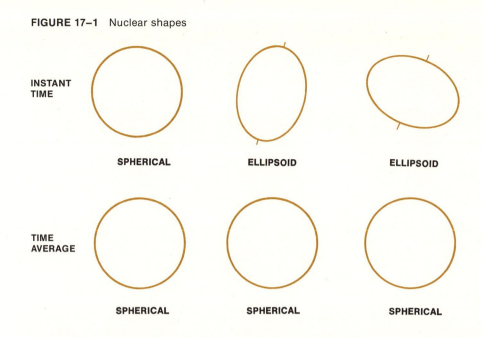

bombarding electrons. The picture thereby produced of the charge distribution of the nucleus is of charge density versus distance from the center. From such studies, it is now known that at any given time the shape of nuclei may be spherical or ellipsoidal (pear shaped can also exist), as in Figure 17–1. It is important to realize that the time averaged charge distribution of all nuclei is spherical; that is, the equivalent of a time exposure would show all nuclei to have the same spherical shape (see Figure 17–1). Some nuclei having particular numbers of protons and neutrons are always spherical, but the majority of nuclei at any instant of time are not spherical.

As we have mentioned, the boundary of a nucleus is not hard. Scattering experiments indicate that all nuclei but the lightest have a constant surface shell thickness of about 2.5×10^{-13} cm. This shell is defined as the distance over which the charge density drops from 90 % to 10 % of its value at the center. On the other hand, the overall size of the nucleus depends on the nuclear mass. The nuclear radius is approximately

17–1 $R = \sqrt[3]{A}\ 1.1 \times 10^{-13}$ cm

where A is the mass number of the nucleus, Å (the angstrom unit) $= 10^{-8}$ cm. The nuclear radius is defined as the distance in centimeters over which the charge density drops to 50 % of its value at the center.

PROBLEM 17–1   Calculate the nuclear radius of
a. $^4_2$He   b. $^{108}_{47}$Ag   and c. $^{238}_{92}$U.

17–B. FUNDAMENTAL PARTICLES

We have already noted in Chapter 2 that the principal components of the nucleus are protons and neutrons. However, over thirty nuclear particles have been defined. These are either emitted from a nucleus or created in other nuclear reactions. Some of these particles are quite stable but many others have very short lifetimes, on the order of 10^{-11} sec. These particles are divided into classes according to their mass: photons, leptons, mesons, and baryons (Table 17–1).

The particles in Table 17–1 are called *fundamental particles*, which means that these particles cannot be divided further into component particles. In fact, there is evidence that this is not true for all the particles given. Therefore, the term fundamental, as rigorously defined, should not be taken literally for all of the particles in Table 17–1.

Note that Table 17–1 contains *antiparticles*. These are the unstable counterparts of the stable particles—a positron is the antiparticle of an electron. Antiparticles have the same mass as their stable particle counterparts but inverse charge. When particles and antiparticles come into contact, annihilation of each occurs along with the creation of large amounts of energy (all mass converted into energy).

PROBLEM 17–2 Calculate the mass of a muon and a pion in atomic mass units based on the data in Table 17–1.

TABLE 17–1 Some fundamental particles

CLASS	PARTICLE	SYMBOL	CHARGE	NUCLEAR REST MASS[a]		DECAY LIFETIME (sec)
				amu	MeV	
baryon	proton	p	+	1.007277	938.256	∞
	neutron	n	0	1.008665	939.550	7.2×10^2
	antiproton	p^-	−	1.008665	938.256	
lepton	electron	β^-, e^-	−	0.00005486	0.511	∞
	muon	μ^-	−		105.66	2.2×10^{-6}
	positron (antielectron)	β^+, e^+	+	0.00005486	0.511	
meson	pion	π^+	+		139.6	2.55×10^{-8}
	pion	π^-	−		139.6	2.55×10^{-8}
photon	photon	γ^b	0		0	∞

[a]One a.m.u. is equal to 931.4 MeV, where 1 MeV is 1 million electron volts (see Section 17–C). 1 g is equal to 6.02×10^{23} amu.
[b]A γ-ray is not a particle but a high energy photon.

17–C. NUCLEAR BINDING, STABILITY, AND STRUCTURE

BINDING ENERGY

The energies involved in chemical reactions are at most a few hundred kcal/mole. On the other hand, the energies involved in nuclear binding and nuclear reactions involve millions of kcal/mole. Therefore, we commonly use the unit MeV or million electron volts for nuclear studies. Mass and energy are interchangeable and equivalent. Einstein developed an equation relating these two variables

4–10
$$E = mc^2$$

where E is in ergs, m is in grams and c is the velocity of light equal to 3×10^{10} cm/sec. One amu is equivalent to 1.5×10^{-3} ergs (1 erg $= 2.39 \times 10^{-8}$ cal) or 931.4 MeV (1 ev $= 1.6 \times 10^{-12}$ ergs).

PROBLEM 17–3   Calculate the energy in calories equivalent to a mass of 1 g.

We shall first consider what is known as the mass defect. The *mass defect* is the difference between the actual mass of an atomic nucleus and the sum of its constituent parts:

mass defect = (masses protons + neutrons) − nuclear mass

The term *nuclide* is used as a general term to refer to any particular nucleus or nuclear species. A nuclide is defined by the symbol of an element, the atomic number (Z), and the mass number (A) in the same manner as for an isotope (Section 2–A). The nuclear mass (atomic mass less the mass of the electrons) of the nuclide $^{16}_{8}O$ is 15.994911 amu while the sum of the masses of 8 protons and 8 neutrons is 16.131925 amu. The difference of 0.137010 amu, the mass defect, is equivalent of 127.6 MeV *per atom*, which is the binding energy holding the nucleons (the protons and neutrons) together in the nucleus. On a per mole or per g-at-wt. basis, this is about 15 million times the energy evolved in the chemical reaction

$$2O \longrightarrow O_2$$

The binding energy per nucleon in the $^{16}_{8}O$ is approximately 8 MeV. Interestingly enough, this value is not a constant for all nuclei of different masses. The binding energy in $^{2}_{1}H$ is approximately 1 MeV, with a rapid increase across the first period to about 8 MeV for $^{16}_{8}O$, but then a variation of only about ±0.7 MeV for all remaining nuclei through $^{239}_{92}U$ (Figure 17–2).

PROBLEM 17–4   What is the energy in calories equivalent to the mass defect in $^{16}_{8}O$ per g-at-wt?

PROBLEM 17–5   Calculate the total binding energy and the binding energy per nucleon for the nuclides $^{4}_{2}He$, $^{12}_{6}C$ and $^{3}_{1}H$. The nuclear masses in amu involved are $^{1}_{1}H$ (p) $= 1.007277$, $^{1}_{0}n = 1.008665$, $^{4}_{2}He = 4.002604$, $^{12}_{6}C = 12.000000$, and $^{3}_{1}H = 3.0160$.

FIGURE 17–2 Binding energy of nucleons as a function of mass number

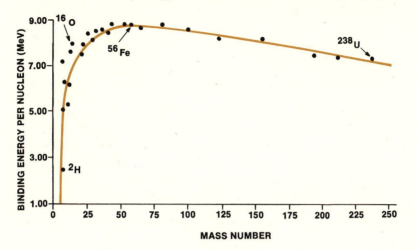

NUCLEAR STABILITY

At first thought, the preceding values for the binding energy per nucleon seem extraordinarily large. However, remember that protons with 1 unit each of positive charge are confined to nuclear dimensions, approximately 10^{-13} cm. Thus, the force of repulsion between them

17–2
$$F = \frac{q_1 q_2}{r^2}$$

becomes exceedingly great. A current question of importance is the nature of the interaction binding the protons and neutrons together. At least one good possibility is that pi mesons (see Table 17–1) are very rapidly exchanged as follows:

and that this pi meson exchange provides the source of the binding energy.

Nuclear stability also depends upon (1) the number of neutrons present for high Z atoms, (2) whether even or odd numbers of protons and neutrons are present and (3) the number of each that are present. By "stable" we mean lack of radioactive decay involving ejection of a particle from the nucleus or the capture of an electron from the $n = 1$ or K shell (see Section 17–D for more details on radioactivity). Figure 17–3 is a plot of the number of neutrons and protons for the stable (nonradioactive) nuclides. Note that as the Z value increases the neutron/proton ratio deviates significantly from 1. For nuclides with a Z value up to about 20, the ratio is about 1 but for higher Z values more neutrons than protons must be present in order that a nuclide be stable. Table 17–2 presents data showing the

FIGURE 17–3 Neutron to proton ratio for stable nuclides

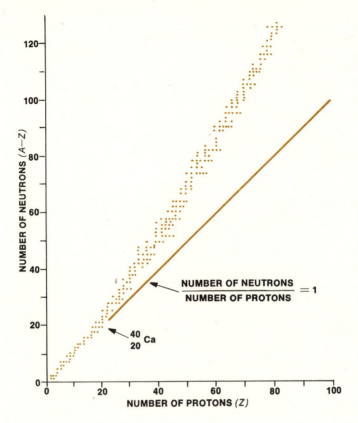

stability of nuclei or nuclides to be a function of the odd and even numbers of protons and neutrons present. Note that the nuclides containing an even number of both protons and neutrons are particularly stable while those with an odd number of each are particularly unstable. Examples of the former would be $^4_2$He and $^{208}_{82}$Pb. The five nuclides of an odd-odd type which are stable are: $^1_1$H, $^6_3$Li, $^{10}_5$B, $^{14}_7$N, and $^{180}_{73}$Ta.

The third factor affecting nuclear stability is the actual number of protons and neutrons present. Thus, it has been found that if 2, 8, 20, 28, 50, 82, 126, and 184 protons or neutrons are present, the nuclide is actually or is predicted to be particularly stable. Stable is used broadly to indicate no radioactive decay or that the decay process occurs over a longer time compared to nuclides without the above numbers of nucleons. These numbers have been called "magic" numbers. A nuclide having both a magic number of protons and neutrons is often called "doubly magic" and an even higher degree of stability is exemplified by these, as for $^{40}_{20}$Ca.

TABLE 17–2 Nuclear stability and odd-even number of nucleons

MASS NUMBER (A)	NEUTRON NUMBER	PROTON NUMBER	NUMBER OF STABLE NUCLIDES
even	even	even	157
even	odd	odd	5
odd	even	odd	50
odd	odd	even	52

NUCLEAR STRUCTURE

One structural model that has been proposed for the nucleus is the shell model. In this model the nucleons move in individual orbits essentially independent of one another. This model is based on quantum mechanics in which a particle restricted to a finite volume is limited to occupying a discrete set of energy orbitals. We applied this same principle to electrons (Chapter 4) and can apply it equally well to nucleons within the nucleus (which here is the finite volume). The energy orbitals are grouped together in shells that are separated by rather large energy gaps (similar to electron shells). If either the number of protons or neutrons is magic then the nucleus has a closed shell (all orbitals filled). If the nucleus has doubly closed shells, both one for the protons and one for the neutrons, it is doubly magic. Magic numbers of nucleons are thus associated with nuclear stability as certain numbers of electrons are associated with atomic stability, for example, 2 (He) and 10 (Ne). Both helium and neon, for example, have high ionization potentials and no chemical reactivity. Thus magic numbers of electrons are likewise associated with stability in a physical and chemical context.

What evidence exists for the idea of energy orbitals and shells for nucleons? One excellent piece of evidence comes from the experiments described in Section 17–A. Electron scattering experiments, as in Figure 17–4, show that the nuclear charge density does not fall smoothly with distance from the center. The wiggly region between 0 and approximately 3×10^{-13} cm is positive evidence for shell structure within the nucleus.

PROBLEM 17–6 What are the other magic numbers of electrons besides 2 and 10?

PROBLEM 17–7 Give explanations for the difference in lifetimes of the following pairs of nuclides:
a. $^{16}_{8}O$, ∞; $^{15}_{8}O$, 71 seconds.
b. $^{40}_{20}Ca$, ∞; $^{45}_{20}Ca$, 165 days.
c. $^{288}_{82}Pb$, ∞; $^{211}_{82}Pb$, 36.1 minutes.

17–D. RADIOACTIVITY

DECAY PROCESSES

Radioactive decay involves the ejection of a particle of high energy or the capture of an electron. Two kinds of radioactivity exists: natural and induced. In the case

FIGURE 17–4 Charge density as a function of distance

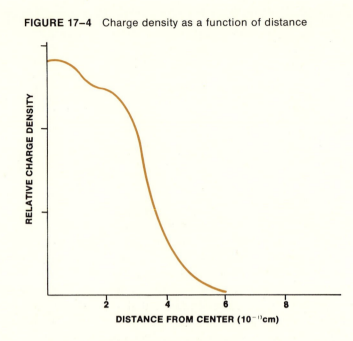

DISTANCE FROM CENTER (10^{-13}cm)

of induced radioactivity, a normally stable nucleus is induced into a radioactive state by bombardment with particles such as protons or neutrons. The principal types of radioactivity are

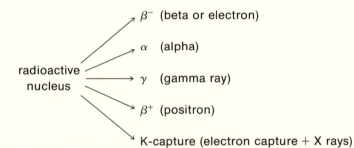

where the β^- and β^+ particles and γ rays are as defined in Table 17–1, an α particle is a helium nucleus ($^4_2$He^{2+}), and K-capture involves capture of an electron by a nucleus from the $n = 1$ or K electron shell. In all types of radioactivity, the nucleus changes charge or energy and a new element or energy state is produced. Since an emission of some type is associated with this nuclear change, the term *radioactive decay* is used. Examples of each of the foregoing radioactive processes are

β^- decay: $\qquad\qquad\qquad\qquad\qquad ^{28}_{13}\text{Al} \longrightarrow {}^{28}_{14}\text{Si} + \beta^-$

α decay: $\qquad\qquad\qquad\qquad\qquad ^{238}_{92}\text{U} \longrightarrow {}^{234}_{90}\text{Th} + \alpha\ (^4_2\text{He}^{2+})$

γ decay:

$$(_{27}^{60}\mathrm{Co})^* \longrightarrow _{27}^{60}\mathrm{Co} + \gamma$$

β^+ decay:

$$_{6}^{11}\mathrm{C} \longrightarrow _{5}^{11}\mathrm{B} + \beta^+$$

K-capture: $_{4}^{7}\mathrm{Be} + e^-$ (from $n = 1$ or K shell) $\longrightarrow _{3}^{7}\mathrm{Li} + $ X rays

where ( )* indicates an excited energy state. Some examples of induced radio-activity are

$$_{11}^{23}\mathrm{Na} + _{0}^{1}n \longrightarrow _{11}^{24}\mathrm{Na} + \gamma$$

$$_{3}^{6}\mathrm{Li} + _{0}^{1}n \longrightarrow _{1}^{3}\mathrm{H} + _{2}^{4}\alpha$$

$$_{17}^{35}\mathrm{Cl} + _{0}^{1}n \longrightarrow _{16}^{35}\mathrm{S} + _{1}^{1}\mathrm{H}$$

$$_{27}^{59}\mathrm{Co} + _{0}^{1}n \longrightarrow _{27}^{60}\mathrm{Co} + \gamma$$

$$_{92}^{238}\mathrm{U} + _{6}^{12}\mathrm{C} \longrightarrow _{98}^{246}\mathrm{Cf} + 4\ _{0}^{1}n$$

Note that there is always an equality of mass, based on the mass numbers, and charge in the balanced equation.

Each of the β^-, α and γ particles (or *rays* as they are sometimes called) have different *penetrating power*. By penetrating power we mean the relative ability to pass through a substance. The α rays have poor penetrating power but create many pairs of ions (ion pairs) as they pass through a substance, air for example. An ion pair is commonly composed of a positive ion and electron. The formation of these can be viewed as

$$\alpha + \mathrm{M} \longrightarrow \mathrm{M}^+ + e^- + \alpha \text{(lower energy)}$$

where M represents a molecule. β rays have about 100 times the penetration power of α rays but create considerably fewer ion pairs. γ rays are highly penetrating but produce very few ion pairs. Alpha particles are stopped by a very thin piece of aluminum foil; β particles require a thicker piece to be stopped; and γ rays require many inches of shielding to be stopped. These factors are obviously important in producing radiation damage of particular concern to humans. In Chapter 19 we shall consider the effects of both ultraviolet and higher energy radiation.

PROBLEM 17–8   Properly balance the following nuclear reactions:

a. $_{56}^{140}\mathrm{Ba} \longrightarrow \beta^- + ?$

b. $_{8}^{13}\mathrm{O} \longrightarrow \beta^+ + ?$

c. $_{4}^{9}\mathrm{Be} + _{2}^{4}\mathrm{He} \longrightarrow _{6}^{12}\mathrm{C} + ?$

d. $_{3}^{6}\mathrm{Li} + _{0}^{1}n \longrightarrow _{1}^{3}\mathrm{H} + ?$

e. $_{12}^{24}\mathrm{Mg} + ? \longrightarrow _{14}^{27}\mathrm{Si} + _{0}^{1}n$

f. $_{20}^{43}\mathrm{Ca} + ? \longrightarrow _{21}^{46}\mathrm{Sc} + _{1}^{1}\mathrm{H}$

g. $_{52}^{130}\mathrm{Te} + _{1}^{2}\mathrm{H} \longrightarrow 2_{0}^{1}n + ?$

FISSION AND FUSION PROCESSES

Two important nuclear reactions are the processes of *fission* and *fusion*. In a *fission* process, a high mass nuclide divides into two principal parts after having absorbed or reacted with a neutron. The nuclides $^{235}_{92}U$ and $^{239}_{94}Pu$ can undergo fission, for example,

$$^{235}_{92}U + ^1_0n \longrightarrow ^{236}_{92}U$$

$$\longrightarrow ^{90}_{38}Sr + ^{143}_{54}Xe + 3\ ^1_0n + energy$$

$$\longrightarrow ^{93}_{36}Kr + ^{140}_{56}Ba + 3\ ^1_0n + energy$$

$$\longrightarrow ^{97}_{40}Zr + ^{137}_{52}Te + 2\ ^1_0n + energy$$

Note that there are several alternative paths by which a fission process can proceed. A schematic diagram of the last process is shown in Figure 17-5. In the fission process, more than 2×10^{10} cal/g of uranium are released. This is about equal to the energy released by burning 2 million g of coal.

A *fusion* process is one in which two nuclides fuse to make a new nuclide with the release of enormous amounts of energy, for example

$$^1_1H + ^2_1H \longrightarrow ^3_2He + \gamma + 3.2\ MeV/atom$$

$$^2_1H + ^2_1H \longrightarrow ^3_1H + ^1_1H + 4.0\ MeV/atom$$

$$^3_1H + ^2_1H \longrightarrow ^4_2He + ^1_0n + 17.6\ MeV/atom$$

$$^1_1H + ^7_3Li \longrightarrow 2\ ^4_2He + 17.2\ MeV/atom$$

Remember that 1 MeV is about 4×10^{-14} cal. Thus, fusion of 1 mole each of ^3_1H (3 g) and ^2_1H (2 g) would release more than 4×10^{11} cal. Unfortunately the temperature required for fusion of nuclei is very high, about 1 million degrees. The fission process does release enough energy to reach this temperature, but it is obviously a potentially dangerous approach to use fission to initiate fusion. It is possible to produce plasmas which have very high temperatures over a very short time period. A *plasma* is a gas containing species, all or nearly all of which are ionized.

FIGURE 17–5 The fission process

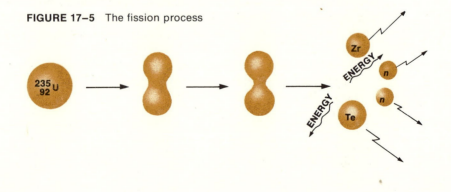

This method of producing the temperatures required for fusion is potentially a good procedure for *controlled* fusion reactions.

If the neutrons released in the fission process are allowed to react with other $^{235}_{92}U$ nuclei, a chain reaction occurs when 2 or 3 neutrons produce 2 or 3 more, and these produce 2 or 3 more, and so forth. The *critical mass* is that mass sufficient for the neutrons released in each fission reaction to be absorbed by other $^{235}_{92}U$ nuclides and not escape. In an atomic bomb, two chunks of $^{235}_{92}U$ are jammed together quickly such that the critical mass is attained or surpassed. In a very short time, approximately 1/1000 of a second, a violent explosion occurs. Because the energy released in an atomic bomb is so great, the energy is measured in terms of that released in the chemical reaction (explosion) of 1 ton of TNT (trinitrotoluene). The first fission-type A-bomb was the equivalent of approximately 20 thousand tons of TNT and higher energy yields have been attained in more recent times. In an explosion involving a fusion process the equivalent of 20–100 million tons of TNT is released.

It is obviously in man's best interests to be able to control both fission and fusion processes. Incidentally, lest we leave the wrong impression, a bomb does not necessarily connote destruction. For example, underground explosions of A-bombs have been used to markedly improve oil recovery from oil-bearing geologic formations and to blast out harbor sites.

It is possible to control or moderate the chain reaction in a fission process. Such materials as D_2O (heavy water, where D is deuterium, ^2_1H) and graphite modify the energy of the neutrons released, and cadmium can be used to absorb neutrons. Thus with the energy and number of neutrons controlled, a controlled fission process can proceed with no explosion (actually fast neutrons are less effective in causing fission). Of course, a substantial amount of energy, largely in the form of heat, is released even in controlled fission. This heat can be used for all sorts of purposes, including running engines and turbines. The turbines can be used to generate electricity; thus controlled fission reactions are the basis of the operation of nuclear power plants.

In view of the importance of $^{235}_{92}U$, we shall describe a method to obtain it. In naturally occurring uranium, only 0.7% is the isotope $^{235}_{92}U$. In order to concentrate $^{235}_{92}U$, natural uranium is first converted to UF_6 (a liquid) which will contain a mixture of $^{235}_{92}UF_6$ and $^{238}_{92}UF_6$. The law of diffusion discussed in Section 11–A can then be utilized. The mixture is heated to put it in the gas phase and passed through a porous disk. Since the $^{235}_{92}UF_6$ is of lower molecular weight it will diffuse at a rate 1.004 times faster than the $^{238}_{92}UF_6$. This will result in a slightly higher concentration of the $^{235}_{92}UF_6$ in the emerging gas. If this process is repeated many times eventually the $^{235}_{92}UF_6$ can be separated and converted back to $^{235}_{92}U$ metal.

PROBLEM 17–9   In the nuclear fission reaction

$$^{235}_{92}U + {}^1_0n \longrightarrow {}^{90}_{38}Sr + {}^{143}_{54}Xe + 3\,{}^1_0n$$

energy is released. Calculate the energy released per gram $^{235}_{92}U$. The nuclear masses in amu are: $^1_0n = 1.0087$, $^{235}_{92}U = 235.0439$, $^{90}_{38}Sr = 89.8864$. Assume the mass of $^{143}_{54}Xe = 142.8800$.

RATE OF RADIOACTIVE DECAY

All radioactive decay processes occur over some finite time period:

17-3
$$\text{rate of disappearance} = -\frac{\Delta N}{\Delta t} = kN$$

where N is the number of atoms of a nuclide, t is time, k is the rate constant, and the $(-)$ sign indicates a decrease or disappearance. Note that N is to the first power, so that the decay is a first order process (identical in form to that of a first order process for a chemical reaction, Chapter 15).

The quantity of radioactivity coming from some source is known as its activity $(\Delta N/\Delta t)$ and is determined by equation 17-3. This is commonly expressed in terms of counts or disintegrations (ΔN) per some time unit (Δt) (e.g., 150 counts per sec of β^- particles). Activity is commonly expressed in terms of curies: 1 curie (C) is 3.70×10^{15} disintegrations per sec. In the limit as Δt approaches zero, the term $\Delta N/\Delta t$ becomes dN/dt and integration of this expression (see Appendix R) gives:

17-4
$$\log \left(\frac{N}{N_0}\right) = -\frac{kt}{2.30}$$

where N_0 is the amount originally present at $t = 0$, and N is the amount present at time t.

The most commonly used time period is called the half-life. The half-life $(t_{1/2})$ is the time it takes for the activity of a given number of atoms to decay to one-half of that originally present. Thus, $N = 1/2\ N_0$ and substitution into equation 17-4 gives

17-5
$$t_{1/2} = \frac{0.693}{k}$$

Radioactive decay is a process analogous to that of a frog successively jumping one-half the distance of each previous jump. Although the frog continues to move forward, it does not take long before it is not going forward very fast. Similarly in radioactive decay, the amount of material or radioactivity remaining after several half-lives of decay levels off. This is characteristic of an exponentially dependent process. A plot of the amount of original nuclide A remaining and of the nuclide B produced for the reaction

$$A \longrightarrow B + \text{radioactive particle}$$

is shown as a function of time in Figure 17-6. Some values of $t_{1/2}$ for decay processes are given in Table 17-3. These are but a few examples to point out the vast range of $t_{1/2}$ values.

EXAMPLE 17-1   *The nuclide $^{87}_{37}Rb$ has a $t_{1/2}$ of 6×10^{10} years. How much $^{87}_{37}Rb$ will be present after 1×10^9 years?*

TABLE 17–3 Half-lives

NUCLIDE	TYPE OF RADIOACTIVITY	$t_{1/2}$
^3_1H	β^- decay	12.5 years
^7_4Be	K-capture	54.5 days
^8_4Be	α decay	$\sim 10^{-16}$ seconds
$^{14}_6C$	β^- decay	5700 years
$^{13}_7N$	β^+ decay	10.1 minutes
$^{73}_{34}Se$	β^+ decay	7.08 hours
$^{87}_{37}Rb$	β^- decay	6×10^{10} years

The rate constant can be determined from equation 17–5.

$$k = \frac{0.693}{t_{1/2}} = \frac{0.693}{6 \times 10^{10}} = 1.15 \times 10^{-11}/year$$

The fraction remaining after 1×10^9 years is then determined from equation 17–4.

$$\log \left(\frac{N}{N_0}\right) = -\frac{(1.15 \times 10^{-11})(1 \times 10^9)}{2.30}$$

$$= -0.0050 = -1 + 0.995$$

$$\frac{N}{N_0} = 0.988$$

or 98.8 % remains undecomposed.

FIGURE 17–6 Relationships of the amounts of a nuclide present, A, and of a nuclide produced, B, versus the number of half-lives of decay

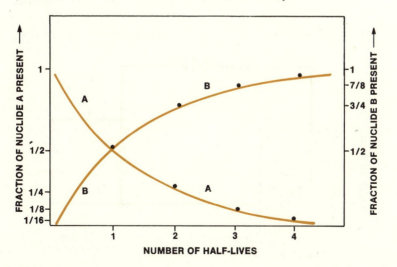

PROBLEM 17–10   Assume that a sample of $^{24}_{11}$Na at time $= 0$ had an activity of 500 counts of β^- particles per minute. After 30 hours, the same sample showed 125 counts per minute.
Determine the $t_{1/2}$ for $^{24}_{11}$Na. [Hint: Consider the meaning or definition of $t_{1/2}$.]

17–E. APPLICATIONS OF RADIOACTIVITY

DETECTION

There have been some very exciting uses of radioactivity in the fields of chemistry,
physics, biology, geology, engineering, medicine, and metallury. However, before
we proceed to these, it is worthwhile to discuss how radioactivity is detected and
evaluated. One of the most common methods makes use of the fact that α, β, and
γ rays produce ion pairs when passing through a gas. In most instances the ion
pair consists of a positive ion and an electron. An ion-collector or Geiger-Müller
tube can be used for detection. A schematic of such a detection system is shown
in Figure 17–7. The electrons produced are collected on the anode wire. A current
then flows and is detected externally. Because γ rays produce so few ion pairs
another type of detector is preferred that consists of a scintillating material that
gives off photons when penetrated by a γ ray; the photons are then counted. Such
a device may use scintillators such as NaI which have a high absorption power
for γ rays as compared with air. This type of detector can also be used for α and
β rays, but the scintillator is commonly then a solution of an organic compound.

DATING

All living things have some carbon in them and one isotope or nuclide of carbon
$^{14}_{6}$C, is radioactive with a $t_{1/2}$ of 5,700 years. While an organism is alive the amount

FIGURE 17–7   Ion-collector system for the detection of radioactivity

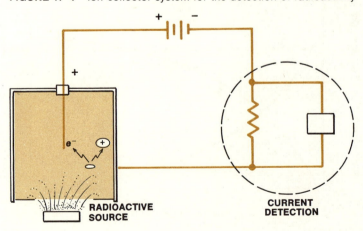

RADIOACTIVE
SOURCE

CURRENT
DETECTION

of $^{14}_{6}C$ reaches some equilibrium concentration. However, after the organism is dead, the equilibrium condition no longer exists and the $^{14}_{6}C$ decays away. (It is assumed that the amount of $^{14}_{6}C$ in the CO_2 of the air does not change with time and that atmospheric CO_2 is the ultimate source of $^{14}_{6}C$ in plants and animals. There is experimental evidence for this assumption.) If an old sample is found to contain, for example, one-half of the $^{14}_{6}C$ radioactivity of a new sample, then the old sample is 5,700 years old. It has been possible, using this method, to date the Dead Sea Scrolls as approximately 1,920 years old, cave drawings in France as about 16,000 years old, and Sequoia trees as approximately 2,900 years old. Also, it has been established that the retreat of the last ice sheets in North America occurred 10,000 years ago, based on carbon dating of wood, peat, and charred bones among other items.

Water is also radioactive because of the presence of T_2O molecules (where T is tritium, $^{3}_{1}H$). Thus such things as wines, brandies, glaciers, snowfields, and icebergs can be dated.

Several procedures can be used for dating of very old objects, such as rocks, on the earth and moon. Based on such procedures, it is possible to date the origin of the earth or of the moon or of wherever else we may go. Several decay schemes that are appropriate for dating purposes are:

$$^{40}_{19}K \xrightarrow{\text{K-capture}} {}^{40}_{18}A \qquad t_{1/2} = 1.3 \times 10^9 \text{ years}$$

$$^{87}_{37}Rb \longrightarrow {}^{87}_{38}Sr + \beta^- \qquad t_{1/2} = 6 \times 10^{10} \text{ years}$$

$$^{238}_{92}U \longrightarrow {}^{234}_{90}Th \xrightarrow{\text{14 fast steps}} {}^{206}_{82}Pb \qquad t_{1/2} = 4.5 \times 10^9 \text{ years}$$

$$^{232}_{90}Th \longrightarrow {}^{228}_{90}Ra \xrightarrow{\text{fast steps}} {}^{208}_{82}Pb \qquad t_{1/2} = 1.39 \times 10^{10} \text{ years}$$

In the $^{238}_{92}U$ and $^{232}_{90}Th$ sequences, the ratios of the $^{238}_{92}U$ to $^{206}_{82}Pb$ and of the $^{232}_{92}Th$ to $^{208}_{82}Pb$ can be used to evaluate when the uranium and thorium were laid down in the source. For example, if the number of atoms of $^{208}_{82}Pb$ were equal to those of $^{232}_{90}Th$, then one half-life has passed (see Figure 17–6) and the thorium was deposited 1.39×10^{10} years ago. A parallel procedure is followed for the other radioactive decays where the ratios of argon to potassium and of strontium to rubidium (in terms of number of atoms) are the ones of concern. Of course, inherent in all of these dating methods is the assumption that the only source of the final nuclide which is being measured is that formed from the beginning nuclide; that is, no $^{87}_{38}Sr$ should have been simultaneously deposited with the $^{87}_{37}Rb$ parent nuclide.

PROBLEM 17–11 Assume that in a rock sample the ratio of the number of gram-atomic-weights of $^{40}_{18}Ar$ to $^{40}_{19}K$ was 7. What is the age of the rock? [Hint: Consider Figure 17–6 or equation 17–4.]

Based on one or more of the foregoing dating methods the earth has been estimated to be about 4.5 billion years old. Samples, taken on the Apollo 11 landing,[*] have been analyzed from the Sea of Tranquility on the moon. Three types of samples were involved: basaltic igneous rocks (formed from molten rock), breccias (a mechanical mixture of soil and small rock fragments compacted into a coherent rock) with fragments of anorthosite (granular igneous rock), and lunar soil. The rocks, except for a small number of crystalline fragments, are considered to have come from underlying bedrock. The age of the basaltic rocks was 3.7×10^9 years but a single rock fragment had an age of 4.4×10^9 years. The soil and breccia were 4.6×10^9 years old. These data in conjunction with other information suggest:

1. The igneous rocks were melted and crystallized about 10^9 years following formation of the moon.

2. The moon has not been dead but has undergone differentiation at least near the surface.

3. The soil and breccia are good representations of the lunar crust. Further, the original crust of substances 4.6×10^9 years old underwent differentiation at later times.

4. The time span *prior* to the occurrence 3.7×10^9 years ago is very likely recorded in the Highland area of the moon and represents a time span blotted out in the earth's history.

It is certain that the materials on the lunar surface are the result of much geochemical and petrological[†] evolution. Sampling in the Sea of Tranquility area alone gives a previously unattainable picture of early evolutionary processes of a terrestrial body. It is also worth noting that analysis of the rare gases (relative abundance and isotopic distribution) of lunar material clearly shows that a considerable quantity of these gases has been entrapped from the solar wind.

Later Apollo landings including Apollo 14 and 15 have added further information about the moon.[**] In the Fra Mauro region, rock samples indicate that the formation of the basin occurred about 3.9×10^9 years ago. A lunar rock from near the rim of Hadley Rille indicates the final lava flow that spilled into Hadley Rille occurred about 3.3×10^9 years ago. One anorthosite rock from the lunar highlands has been found to be about 4.1×10^9 years old. This plus other data indicates that the primitive lunar crust was molten (at least a portion of the crust formed 4.1×10^9 years ago) and that the lunar crust formed within the first $3-4 \times 10^8$ years of the moon's existence.

[*]Lunar Sample Analysis Planning Team. Summary of Apollo 11 Lunar Science Conference. *Science, 167,* 449–451 (1970). See also other articles in this issue.
[†]Petrology deals with the history, occurrence, structure, chemical composition and classification of rocks.
[**]Data from several articles in *Science, 175,* pp. 417–419, 419–421, and 428–430 (1972).

TRACERS

Another large area of application of radioactivity is as tracers. Recall from Chapter 2 that an isotope of an element is an atom having the same atomic number but a different mass number. Tracers are atoms that are single isotopes of an element. For example, deuterium (D or ^2_1H), is an isotope of hydrogen, and we can make DCl instead of HCl where the deuterium atom is the tracer element. Commonly but not always tracers are radioactive and are used to determine the course of events in which the tracers can participate. In the cases where the tracer is not radioactive, it can be located and identified by its unique mass compared with all other isotopes of the same element. We shall examine several areas showing the purpose and function of a tracer.

Tracers are important tools in investigating the mechanism of a chemical reaction. For example, one method of preparing esters (Section 7–H) is by the direct reaction of an alcohol and an acid:

17–6
$$R-C\underset{OH}{\overset{O}{\diagup}} + HO-R' \rightleftharpoons R-C\underset{OR'}{\overset{O}{\diagup}} + H_2O$$

An important question is whether the hydroxyl group in the molecule of H_2O formed comes from the acid or the alcohol. The —OH group of the alcohol may be labeled by using the $^{18}_8O$ isotope, $R'-^{18}O-H$. If the ester produced in reaction 17–6 is enriched in ^{18}O, then the —OR' portion would come from the alcohol, having displaced the —OH group of the acid to give

$$R-C\underset{^{18}O-R'}{\overset{O}{\diagup}}$$

Indeed this is the case experimentally and it is the H (from the alcohol) and OH (from the acid) which form H_2O. This mechanism can be verified by examining the reverse reaction, hydrolysis of the ^{18}O-labeled ester:

$$CH_3CH_2-C\underset{^{18}OC_2H_5}{\overset{O}{\diagup}} + H_2O \longrightarrow CH_3CH_2\ C\underset{OH}{\overset{O}{\diagup}} + C_2H_5{}^{18}OH$$

It has been found that the ethyl alcohol is enriched in ^{18}O while the acid contains only the normal percentage of ^{18}O present in natural oxygen, showing that bond breakage occurs at the dashed line in equation 17–7. This is the same place as for the acid in reaction equation 17–6.

In the photosynthesis process there is a sequence of reactions in which CO_2 is ultimately fixed in carbohydrates; that is, the carbon atom of CO_2 is incorporated into the carbohydrates (sugars). This important sequence of reactions is called the *dark cycle* of photosynthesis, in contrast to the *light cycle* which will be discussed in Section 19–D. Although more detail will be given on the dark cycle in Chapter 18, we shall make a few points here. Radioactive CO_2 can be made by using the isotope $^{14}_{6}C$. Algae or leaves are then exposed to $^{14}CO_2$ for various time periods and then killed. The cellular compounds are then extracted and separated and the compounds containing the ^{14}C can be identified. By varying the time allowed for CO_2-fixation, the first compound and succeeding compounds formed in the path to carbohydrates can be elucidated. The $^{14}CO_2$ is incorporated principally into the compound 2-phosphoglyceric acid within the first few seconds:

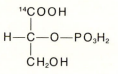

Within an hour or so, most of the ^{14}C is found in carbohydrates. It has also been possible to show that the O_2 evolved in the photosynthetic process comes from H_2O not CO_2. This was proved by using H_2O labeled with ^{18}O as a tracer and showing that the O_2 evolved was enriched in ^{18}O.

Another extremely important use of tracers is to follow the movement of radioactive nuclides in food chains. For example $^{137}_{55}Cs$ is one of the nuclides occurring from the fallout of atomic explosions. Moreover, it is of concern because it has an appreciable half-life of 30 years. Also, recall that cesium is a Group IA element and therefore its compounds have properties similar to those of sodium which are important in biological systems. In Alaska ^{137}Cs is concentrated in lichens which are eaten by caribou which in turn are eaten by man (or wolf). It can be found that ^{137}Cs is concentrated by a factor of about 2 for the lichen-caribou system. Interestingly enough, ^{22}Na ($t_{1/2} = 2.6$ yrs) is concentrated by a factor of 10 for the lichen-caribou system but only by a factor of 2 for the caribou-wolf system. This happens because of localization of the radioactive nuclide in bone of the caribou where it is effectively removed from the continuing food cycle.

Similar food chain studies have been done in Finland, in Italy, on the Olympic Peninsula in Washington, and on lakes and streams in various locations. These have involved many types of radioactive nuclides other than cesium. For example, the liver deoxyribonucleic acid (DNA) of kangaroo rats living near a Nevada atomic explosion test site shows a concentration of tritium ($^{3}_{1}H$) of about 2 times that present in their body water. The radioactive nuclides of principal biological concern from nuclear accidents are ^{131}I, ^{90}Sr, and ^{137}Cs, because of their substantial half-lives and their chemical similarity to elements of biological importance. Extensive studies indicate that milk constitutes the major pathway for ^{131}I into humans. Techniques such as eliminating milk and other foods from the human diet could substantially reduce the intake of such nuclides, if this ever became necessary.

Tracers have been heavily used to diagnose various medical problems. For example, a very recent development* shows that the ^{67}Ga ion (from a salt) localizes in various types of malignant tissue. The ^{67}Ga localizes particularly in the thymus and lymph nodes of mice with two different types of leukemia. It is known that it is the free gallium ion that is involved in the localization, and that the binding agent for the gallium ion is highly specific. This binding agent is located within the malignant cell and it is likely that it is a protein.

EXERCISES

1. Which of the following statements are correct regarding an isotope ^A_ZE of an element E?
   a. A is the number of protons
   b. Z is the mass number
   c. Another isotope of E will have A the same and Z different
   d. Another isotope of E will have Z the same and A different
   e. Another isotope of E will have both Z and A different

2. What particles would result in annihilation upon contact?

3. Calculate the total binding energy and binding energy per nucleon for ^{3}He, ^{7}Li, ^{31}P, ^{127}I, and ^{243}Cm. See Problem 17–1 in the text for the nuclear masses of $^1_1$H and $^1_0 n$. The other masses in amu involved are $^3_2$He $= 3.01603$, $^7_3$Li $= 7.01600$, $^{31}_{15}$P $= 30.99376$, $^{127}_{53}$I $= 126.9004$, and $^{243}_{96}$Cm $= 243.0614$.

4. Compare the total binding energy and binding energy per nucleon for $^3_2$He (Exercise 3 above) and $^4_2$He (Problem 17–5 in text).

5. Compare the binding energy per nucleon for the nuclides of Exercise 3 above. Do you notice any trend?

6. Write balanced equations for the following reactions:
   a. $^3_1$H $\longrightarrow \beta^- + ?$
   b. $^{115}_{49}$In $\longrightarrow \gamma + ?$
   c. $^{211}_{85}$At $+$ K-capture $\longrightarrow ?$
   d. $^{59}_{27}$Co $+ ? \longrightarrow ^{56}_{25}$Mn $+ ^4_2$He
   e. $^{55}_{25}$Mn $+ ^1_0 n \longrightarrow ^{56}_{25}$Mn $+ ?$
   f. $^{27}_{14}$Si $\longrightarrow ^{27}_{13}$Al $+ ?$
   g. $^{245}_{97}$Bk $\longrightarrow ^4_2$H $+ ?$
   h. $^{44}_{20}$Ca $+ ? \longrightarrow ^{45}_{21}$Se $+ ^1_0 n$

7. A possible fusion reaction is

$$^1_1\text{H} + ^2_1\text{H} \longrightarrow ^3_2\text{He}$$

and energy is released. Calculate the amount of energy released by reaction of one g-at-wt of each of $^1_1$H and $^2_1$H. The nuclear masses in amu are $^1_1$H $= 1.00728$, $^2_1$H $= 2.01355$, and $^3_2$He $= 3.01493$.

*U.S. Atomic Energy Commission. Fundamental Nuclear Research, A Supplemental Report to the Annual Report to Congress for 1970 (January 1971).

8. Suppose we had a sample of $^{157}_{65}Tb$ which weighed 1.57×10^{-6} g. What would be the rate of radioactive decay (activity) per sec? The $t_{1/2} = 150$ years.

9. The nuclide $^{70}_{34}Se$ has a $t_{1/2} = 44$ minutes. How many grams of this nuclide will have disappeared after 1 day if there was 10 g present initially?

10. A sample of radioactive material had a disintegration rate of 5,400 counts per minute. After 12 hours, the activity has decreased to 540 counts per minute. Calculate the specific rate constant k and the $t_{1/2}$ for this material. [Hint: Consider equation 17–3, which shows number of atoms is directly proportional to disintegration rate.]

11. Suppose you placed a dish of water in an enclosed glass case. As time proceeds you note the water level drops and then finally it becomes constant. The question is how would you prove that there is a dynamic and not a static equilibrium involved after the water level becomes constant? (Assume you have a method of entry into the enclosed glass case through a rubber diaphragm.)

12. Suppose we wanted to send an unmanned spacecraft to Pluto. In order to determine the time it took to get there, we placed some $^{155}_{63}Eu$ on board. The disintegration rate at the time it left the orbit of earth was 5.0×10^{10} counts/minute and just before it crashed on Pluto it was 15.6×10^{8} counts/minute. The $t_{1/2}$ for $^{155}_{63}Eu$ is 1.8 years. How long did it take the spacecraft to reach Pluto? [Hint: Consider equation 17–3.]

13. A sample of cloth from an archeological find was found to decay at the rate of 210 counts/sec/g of carbon. The carbon from modern cloth disintegrates at the rate of 920 counts/sec/g of carbon. Estimate the age of the cloth in the archeological find. The $t_{1/2}$ of $^{12}_{6}C$ is 5,770 years.

14. A geologist finds a rock in which the amount of $^{87}_{38}Sr$ is 80×10^{-6} g/g of rock and the amount of $^{87}_{37}Rb$ is 2×10^{-6} g/g of rock. What is the age of the rock? For the nuclear reaction $^{87}_{37}Rb \longrightarrow {}^{87}_{38}Sr + \beta^-$, $t_{1/2} = 6 \times 10^{10}$ years.

15. What fraction of the $^{40}_{19}K$ originally present when the earth was formed is now present? The $t_{1/2}$ of $^{40}_{19}K$ is 1.3×10^{9} years (see Section 17–E for the age of earth).

16. Prehistoric relics can be dated by measuring the quantity of $^{14}_{6}C$ ($t_{1/2} = 5,700$ years) present. Obviously, hydrogen must also be present in many cases yet dating using $^{3}_{1}H$ ($t_{1/2} = 12.5$ years) is not done. Give an explanation for this.

17. The half-life of decay of $^{13}_{7}N$ is 10.1 minutes. What is the weight of a sample that has an activity of 0.2 curies? Assume the atomic weight of $^{13}_{7}N$ is 13.

18. Silver is a precious metal and is also widely used in industry as in photography. Obviously, it would be prudent to remove as much of any silver from solution as possible before dumping it down the drain. Suppose there were three methods available to remove the silver (as Ag^+ in solution). How could you quantitatively determine which way was best without going through all kinds of elaborate procedures of precipitation, drying, weighing, etc.?

CHAPTER EIGHTEEN

BIOCHEMISTRY

Biochemistry is that area of chemistry which deals with the reactions and compounds involved in the processes of life. Biochemistry seeks the answers to some of the basic questions about the chemistry of life; interest in the origin of life (see Chapter 19) and the maintenance of life is primary. A great deal has happened in the field of biochemistry in the last two decades. Perhaps most significant has been the exchange of ideas, techniques, and theories that has taken place between chemists, physicists, and biologists.

In Chapter 7 we discussed many of the important organic molecules. In biological systems, simple molecules are the basis for the complex, large molecular weight molecules involved in biochemical reactions. The principles of thermodynamics discussed in Chapter 10 and the principles of kinetics discussed in Chapter 15 form a basis for beginning to understand the complicated chemistry of life. In this chapter, we shall discuss some aspects of chemistry and structure that are responsible for the specificity of reactions in biochemistry. Stress will be placed on the molecular aspects of biology.

18–A. BIOLOGICAL MACROMOLECULES

The study of biochemistry is complicated by the large molecules found in living systems. These large molecules are chain-like and contain large numbers of atoms, with the smallest containing about 800 atoms. Such molecules are called *macromolecules* and have molecular weights of approximately 5000 and greater. The two major classes of macromolecules that determine the functioning of biological organisms are nucleic acids and proteins (including enzymes). The nucleic acids include DNA (deoxyribonucleic acid), which ordinarily carries the hereditary information appropriate to each organism, and RNA (ribonucleic acid), which is involved in the translation of the information into the production of thousands of various proteins. The proteins act both as structural materials and as chemically active agents in metabolic reactions—for example, oxygen transport, antibody action, and catalysis.

Both nucleic acids and proteins are composed of a number of molecular subunits that have been polymerized. The number and kinds of molecular subunits, their order, and the three-dimensional geometrical arrangement of the final macromolecule dictate the highly specific functions of each macromolecule.

TABLE 18–1 Amino acids in proteins

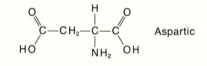 Aspartic

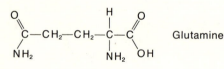 Glutamic

Asparagine

Glutamine

Proline

Lysine

Arginine

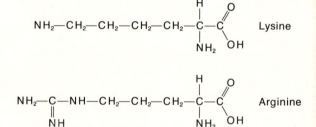 Histidine

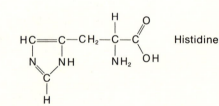 Cysteine

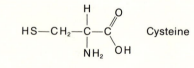 Methionine

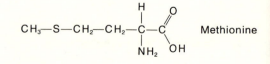

18–B. PROTEINS

STRUCTURE OF AMINO ACIDS

All proteins contain a number of amino acid units. The general structural formula for amino acids was described in Chapter 7 as:

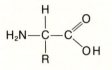

About 20 amino acids are found commonly in proteins. The formulas of these amino acids and their names are given in Table 18–1.

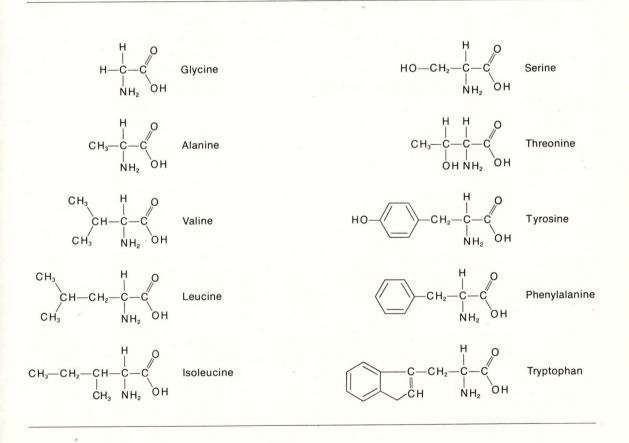

Because of the large number of amino acid sequences possible, more than 5 million different proteins theoretically can be made from 20 amino acids. The precise structure of only a very few of these proteins is known. Fewer proteins yet have been synthesized outside of living organisms.

PROPERTIES OF AMINO ACIDS

In aqueous solution, amino acids have a large dipole moment, and the following equilibrium exists with the highly polar *zwitterion*:

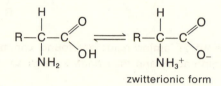

zwitterionic form

The zwitterionic form is involved in the acid-base behavior of amino acids. If we titrate a solution of a protonated amino acid with a base, we will get a characteristic curve. We discussed in Chapter 3 the neutralization of acids having more than one ionizable hydrogen. The titration curves of most amino acids have more than one plateau, each indicating the point where equimolar amounts of the proton donor and proton acceptor exist. Each equivalence point corresponds to a definite pK value of each proton donated during neutralization.
At pK_1

$$\underset{\underset{+}{NH_3}}{H_3C-CH-\overset{\overset{O}{\|}}{C}-OH} \rightleftharpoons \underset{\underset{+}{NH_3}}{H_3C-CH-\overset{\overset{O}{\|}}{C}-O^-} + H^+$$

At pK_2

$$\underset{\underset{+}{NH_3}}{H_3C-CH-\overset{\overset{O}{\|}}{C}-O^-} \rightleftharpoons \underset{NH_2}{H_3C-CH-\overset{\overset{O}{\|}}{C}-O^-} + H^+$$

A curve such as Figure 18–1 can be derived for all of the amino acids. The shape of the curve and the pK values vary depending on the R group attached. The point where there is no *net* electrical charge on the molecules is the first inflection point of the titration curve; it is called the *isoelectric point*. The zwitterion exists at this point.

FIGURE 18-1 Titration curve for alanine. The existence of a zwitterion and of two ionizable hydrogens is indicated.

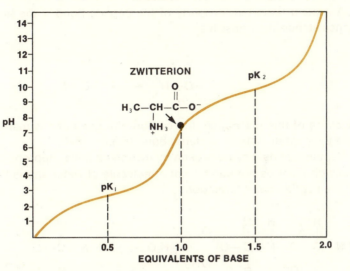

REACTIONS OF AMINO ACIDS TO FORM PEPTIDE BONDS

The reaction of the carboxylic acid end group of an amino acid with the amine end group of another amino acid is of particular interest in protein chemistry. This reaction forms an amide

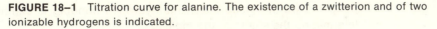

or peptide linkage

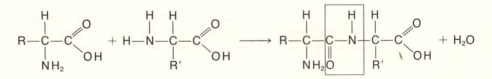

The carbon-nitrogen bond of the peptide linkage is called the *peptide bond*. Amino acids form long-chain polymers by reacting head-to-tail with a number of other amino acids. Such polymers are called polypeptides because they have multiple peptide bonds. One of the smallest proteins, insulin, is a polypeptide containing 51 amino acids (with 777 atoms). Some proteins are believed to be hundreds or more times larger.

The importance of the peptide bond in biochemical structures is based on two properties of this bond. First, in polypeptides and proteins, there is the possibility of a hydrogen bond (Section 8-G) between the amino group and the carbonyl group. As we shall see later, this hydrogen bonding is extremely important in determining the shape of the molecule.

A second important property of the peptide bond is the high degree of resonance stabilization possible

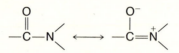

Because of this resonance, the carbon-nitrogen bond is restricted in its rotation and the peptide bond is a planar bond (Figure 18–2).

Peptide bonds may be broken by *hydrolysis* under appropriate conditions. This reaction involves the addition of a molecule of water and is the reverse reaction of the peptide bond formation

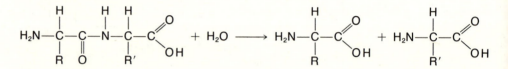

The technique for identifying the amino acid sequence of proteins is based on this reaction. By analyzing the fragments of a series of controlled hydrolysis reactions, it has been possible to determine the structure of a number of proteins containing hundreds of amino acids.

SIDE-CHAIN REACTIONS OF AMINO ACIDS

In considering reactions of amino acids, we must also discuss possible side-chain reactions between various R groups. We shall consider one reaction in particular which involves the R group of cysteine (Table 18–1)

FIGURE 18–2 The peptide bond

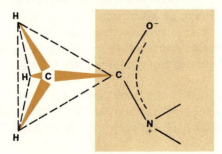

$$HS-CH_2-$$

which contains a thiol group, —SH. This group is capable of being oxidized by removal of the hydrogen. If another oxidized cysteine molecule is adjacent, the two sulfurs can link together to form a covalent disulfide bond. This reaction is involved in linking sections of a protein together. The amino acid cystine contains a disulfide bond that may be considered as having originated from two cysteine molecules

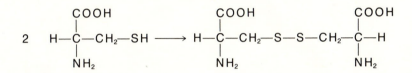

The disulfide linkage is important in determining the overall structure of a protein, thereby affecting its biological function. The disulfide linkage can be broken by hydrolysis.

STRUCTURE OF PROTEINS

When many amino acids are linked together in a particular sequence with a particular geometry, a specific protein molecule is formed. A particular three-dimensional structure must be preserved if the molecule is to be biologically important. The complex structure of a protein may be described in terms of a primary, secondary, tertiary and quaternary structure.

The *primary structure* of proteins is determined by the sequence of amino acid units in the molecule. As suggested earlier, one way of determining the primary structure is by hydrolysis of the peptide bonds in a protein polypeptide chain. The primary structure of proteins having more than 300 amino acid units has been analyzed.

The *secondary structure* of proteins is based on the number and position of hydrogen bonds in the molecule. Proteins may have a helical structure. The helix in proteins appears to be exclusively right-handed (Figure 18–3). The planar peptide bonds are parallel to the axis of the polypeptide chain. The helix is held rigid by hydrogen bonds between the oxygen of the carbonyl group and the hydrogen attached to a nitrogen three amide groups down the chain (Figure 18–3). Another kind of structure is possible if a polypeptide hydrogen bonds with a second chain. In this case the hydrogen bonds no longer contribute to the internal structure of the chain as in the helix. These interchain hydrogen bonds give rise to the so-called pleated sheet (Figure 18–4). In this structure, the amino groups and carboxyl groups alternately point up and down.

Globular proteins, such as myoglobin, the oxygen carrier in muscle, have a *tertiary* structure where a chain is coiled and folded into a three-dimensional

FIGURE 18–3 Helical structure of proteins.
A. The form of a right-handed helix and B. hydrogen bonding within helix.

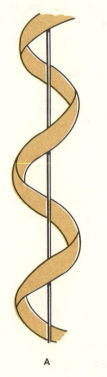

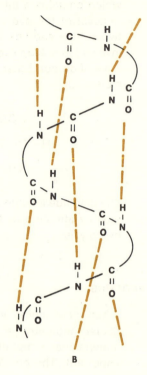

A   B

pattern. This pattern is governed by the bonds formed between the R groups of the polypeptide chain. Bonds which must be considered include hydrogen, ionic, covalent, and apolar bonds. Examples of such bonds are shown in Figure 18–5.

If a protein has two or more separate polypeptide chains, it has a *quaternary* structure. That is, the chains are oriented in a three-dimensional pattern. For example, globin, the protein portion of the oxygen carrier hemoglobin in the blood, has four component polypeptide chains, a molecular weight of about 64,500, and about 10,000 atoms.

18–C. NUCLEIC ACIDS

NUCLEOTIDES

Nucleic acids are made up of molecular subunits known as *nucleotides*. A nucleotide is composed of a sugar, a phosphate group, and a cyclic organic molecule. The latter are broadly known as purine and pyrimidine bases. These names are

FIGURE 18–4 Pleated sheet structure. Notice that the R groups point either out of or into the plane.

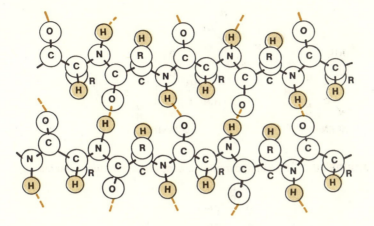

derived from the principal core structure of these molecules—purine for the purine bases and pyrimidine for the pyrimidine bases—see Figures 18–6 and 18–7. There are principally two purine and three pyrimidine bases found in nucleotides. Note that these compounds are cyclic and contain nitrogen atoms in place of carbon at some locations in the rings. The purines and pyrimidines are called bases because the nitrogen in the ring is an electron pair donor (and a proton acceptor), and is therefore basic.

When a base is attached to a five-carbon sugar (pentose) in its cyclic form, a nucleoside is formed, as in Figure 18–8. Sugars will be discussed more in Section 18–F. The addition of a phosphate group to the sugar-base system produces a nucleotide (Figure 18–8). The phosphate group can potentially bond to any of the carbons on the sugar bearing a hydroxyl group but in biological molecules it is often found in the 5' position. The phosphate group gives a strongly acidic character to the nucleotide.

FIGURE 18–5 Bonds which determine the tertiary structure of proteins

FIGURE 18–6 Principal purine bases found in nucleic acids

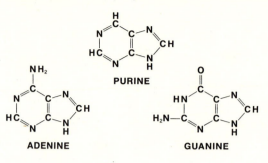

FIGURE 18–7 Principal pyrimidine bases found in nucleic acids

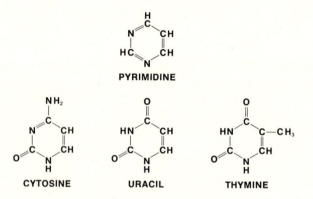

FIGURE 18–8 Components of nucleic acids   A. schematic and   B. the nucleotide
adenine-5′-phosphate or adenylic acid

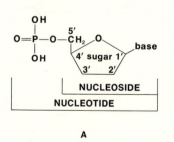

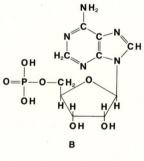

STRUCTURE OF NUCLEIC ACIDS

When nucleotides are condensed or polymerized, a polynucleotide is produced as shown in Figure 18–9. Note that each nucleotide is linked to the other via the phosphate group through the 3′ and 5′ positions of the sugars. A portion of the structures of DNA and RNA is shown in Figure 18–10. Recall that DNA and RNA are

FIGURE 18–9 Schematic structure of a polynucleotide

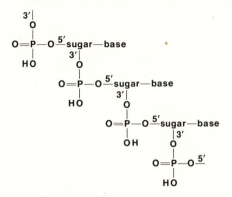

of particular interest because they are responsible respectively for carrying hereditary information and for participation in the translation of the information into proteins.

Even though thousands of nucleotide subunits may polymerize to form the nucleic acid, only four different nucleotides are found in a nucleic acid molecule. In DNA, the two pyrimidine bases cytosine and thymine occur, while in RNA, uracil replaces thymine (Figure 18–10). In both DNA and RNA, the two purine bases that occur are adenine and guanine. The major difference between DNA and RNA is that DNA contains the five-carbon sugar deoxyribose while RNA contains the five-carbon sugar ribose (Figure 18–10). A complete DNA molecule can be several million angstroms in length but it is only about 20 Å in diameter (1 Å $= 10^{-8}$ cm). DNA is localized in the chromosomes* during cell division.

Both DNA and some types of RNA contain double strands of the appropriate polynucleotide in the form of a double helix wound around a vertical axis as shown in Figures 18–11 and 18–12. Note that within each strand of polynucleotide there is a sequence of bases (in the nucleotide subunits). This sequence determines the sequence of bases in the other strand because of the specific hydrogen bonding that can occur only between certain complementary base pairs (each base in a pair is in a different strand). This is shown in more detail in Figure 18–13. The particular pair combinations allowed in DNA are

adenine (A) — thymine (T)

guanine (G) — cytosine (C)

The hydrogen bonding involved is responsible for the stability of the double helix and for the base sequence in one strand relative to the other, because of the necessity of base pairing as already noted.

---

*Chromosomes constitute a group of dense bodies containing genes and are principally responsible for the differentiation and the kind of activities of a cell.

FIGURE 18–10 Polynucleotides   A. DNA and   B. RNA

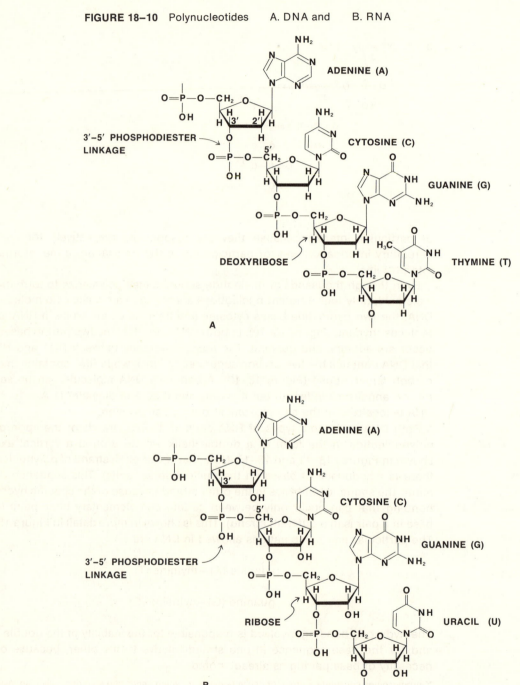

FIGURE 18–11   The double helical structure of DNA showing hydrogen bonding between complementary base pairs. G is guanine, C is cytosine, A is adenine, and T is thymine.

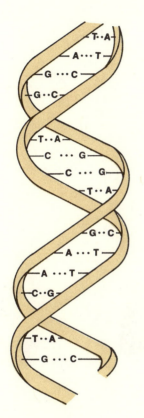

Replication (self-duplication) of DNA apparently involves the partial untwisting of the double strand. While this is occurring, a new complementary strand is polymerized so that it is hydrogen bonded to an old strand; at the end of the process, the two DNA molecules each contain one old and one new strand. In the synthesis it appears that base pairing dictates the proper sequence of bases on the newly synthesized strand.

DNA itself is not involved in the synthesis of proteins. The information from the DNA template (pattern) is transcribed into RNA and carried by RNA to the site of the protein synthesis—this type of RNA is known as *messenger* RNA. The messenger RNA in tobacco mosaic virus contains 6,400 nucleotides. While all the details of the mechanism of the transcription of DNA to give messenger RNA is

FIGURE 18–12 DNA helix constructed with atomic space-filling models (Courtesy of the Ealing Corporation, Cambridge, Massachusetts)

FIGURE 18–13 Hydrogen bonding of base pairs in DNA

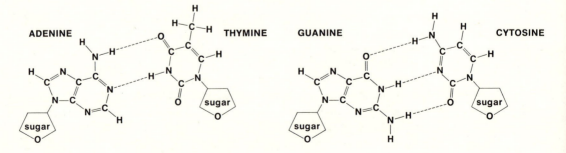

unknown, it is known that in going from DNA to RNA the complementary base pairs are formed on an untwisted DNA strand.

DNA strand – RNA strand

adenine – uracil

thymine – adenine

cytosine – guanine

guanine – cytosine

The newly formed RNA strand is complementary to one of the two strands of the template DNA. Given a certain sequence of bases (nucleotides) in DNA, the order of bases (nucleotides) in RNA is determined as illustrated in Figure 18–14. Another type of polynucleotide that is important is a modified RNA known as *transfer* RNA. There are several kinds of transfer RNA, each capable of recognizing a specific amino acid. A transfer RNA picks up its specific amino acid and brings it to the cellular site of protein synthesis where the amino acids are assembled into a protein. Figure 18–15 shows the interrelation among the nucleotides in protein synthesis.

One important question that finally remains is how the sequence of amino acids is determined by the messenger RNA. Recall that there is a sequence of bases in a single strand of RNA (Figure 18–14). There is a code in which a sequence of three bases – a *codon* – determines which amino acid is appropriate. The codon of a messenger RNA is in turn related to the sequence of a triplet in the transfer RNA such that ultimately a triplet of the transfer RNA recognizes a triplet of the messenger RNA. The complementary base pairing between transfer RNA triplets and messenger RNA triplets can then order the amino acid sequence in the developing protein as in Figure 18–16.

FIGURE 18–14 Formation of a messenger RNA strand from a DNA template with a given base sequence. A is adenine, G is guanine, C is cytosine, T is thymine and U is uracil; —S—P— is the sugar-phosphate chain. The bases shown in circles represent the complementary base pairs of DNA and RNA.

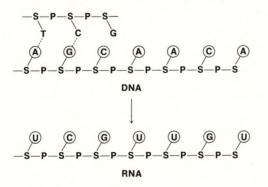

FIGURE 18–15 Interrelationship among the nucleotides in protein synthesis

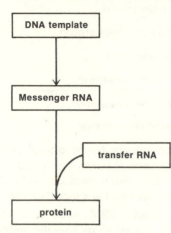

18–D. ENZYMES

Biochemical reactions utilize more than a thousand different catalysts called *enzymes*. Enzymes are specialized protein molecules which exhibit highly specific abilities to catalyze particular types of reaction. The high specificity of the enzyme, as well as its effectiveness, far exceeds that obtained in chemical systems outside of living organisms. An example of the relative effectiveness of an enzyme compared to a catalyst for a typical chemical reaction was given in Section 15–H. The

FIGURE 18–16 Triplet code base pairing of transfer RNA and messenger RNA determining order of amino acids in protein being synthesized

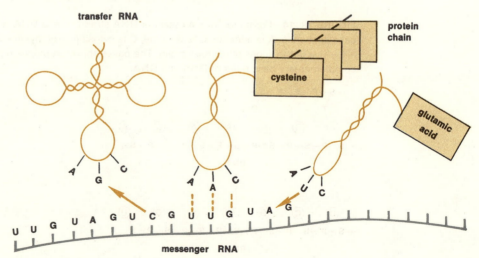

general mechanism for most enzyme catalyzed reactions was given in Section 15–H. The complex in the biochemical mechanism is the enzyme complexed to the reactant(s). The structure of the enzyme is extremely specific since it must combine in some way with a reactant or substrate to form a transition state whose energy is below that existing in the absence of the enzyme catalyst (see Figures 15–9 and 15–10). We shall discuss more on this point shortly.

The true significance of the high specificity of some enzymes can be shown by the following examples. There exists a group of enzymes known as *proteolytic* or protein-digesting. These enzymes catalyze the hydrolysis of peptide bonds of proteins (Section 18–A). One of these hydrolyzes only a peptide bond whose carbonyl ($>C=O$) group originated from an amino acid with a positively charged side chain. Another acts preferentially on peptide bonds whose carbonyl group originated from an amino acid containing a benzene ring (tyrosine or phenylalanine, Table 18–1). Finally, another acts only on the last peptide bond of a protein chain. There is also an enzyme which specifically breaks down RNA by attacking a phosphorous atom (of phosphate) attached to a cytosine or uracil but not to an adenine or guanine.

The specific binding of substrate to enzyme and the way the activation energy of a reaction is altered are under intensive current investigation. Some insight can be gained from the following consideration. Sugars can be linked together to form what is known as polysaccharides. An enzyme called lysozyme (containing 129 amino acid units) can cleave a bond between two of these sugars* (ultimately causing the rupture of a bacterial cell wall containing the substituted polysaccharides). The suggested nature of the interactions and reactions occurring are shown in Figure 18–17. It is supposed that the interaction of the substrate with the lysozyme is localized, distorting the normally unplanar ring *y* and causing four of the six atoms (including oxygen) to become planar. This results in a strain of the C—O bond (to be cleaved) such that addition of H^+ to the colored oxygen atom results in C—O bond cleavage. The catalytic effect is likely the result of strain which causes a lowering of the activation energy for bond cleavage.

A number of ions or functional groups may act as *activators* of enzymes. These groups when combined in various ways with the enzyme cause the reaction rate to increase more than would ordinarily be expected. *Inhibitors* have been useful in studying metabolic pathways and enzyme mechanisms—they usually are drugs or poisons.

Enzyme inhibition may be considered in terms of reversible and irreversible reactions. In irreversible inhibition, the enzyme is destroyed or modified so that it cannot catalyze further reactions. Reversible inhibition may, however, be of two types—competitive and noncompetitive.

Competitive inhibition takes place when an enzyme inhibitor prevents access to the site where activation generally occurs. If the substrate concentration is increased so that there is more substrate than inhibitor, the reaction will again proceed. One example of a competitive inhibition is the inhibition by sulfanilamide

*D. C. Phillips, Bio-Organic Chemistry. Readings from Scientific American, pp. 62–74 (1968).

FIGURE 18–17 Schematic of polysaccharide being held by lysozyme showing C—O bond breakage. The OH⁻ and H⁺ come from water.

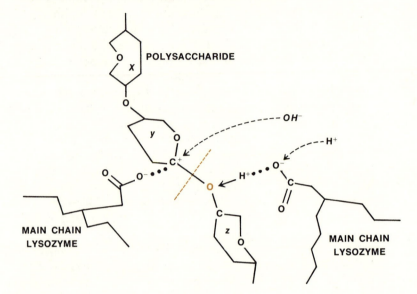

of a coenzyme† synthesis. Sulfanilamide occupies the site generally occupied by the normal reactant *p*-aminobenzoic acid:

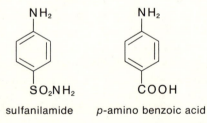

sulfanilamide   *p*-amino benzoic acid

Notice the similarity in the structures of these two molecules.

Noncompetitive inhibition, on the other hand, cannot be reversed by simply adding more substrate. The inhibitor may bond to the enzyme itself and change the shape of the enzyme or block a specific binding site. The latter kind of inhibition involves the formation of covalent bonds between inhibitor and enzyme. In this way the nerve gases inhibit an enzyme involved in the conduction of nerve impulses, and thereby effectively induce paralysis. It is now also possible to displace the attached nerve gas and reactivate the enzyme.

†A coenzyme is much smaller than an enzyme and is not a protein.

18–E. THE ROLE OF STRUCTURE IN BIOCHEMICAL ACTIVITY

In this section we shall point out some special, less familiar roles certain molecules play in biological reactions. In most instances we will be interested in the existence of common core molecular structures and we will emphasize the important role structure and shape play in dictating biological responses.

ODOR

One of the more outstanding examples of the importance of molecular size and shape in determining biological response is in the detection of odors. There are seven odors that appear to be what could be called primary—camphor-like, musky, floral, pepperminty, ethereal (ether-like), pungent, and putrid. Other odors can be considered as derived from these—in the case of the odor fruit, from a combination of floral, pepperminty, and ethereal.

One current theory explaining the ability to smell differences in odors suggests the existence of seven different kinds of olfactory receptor sites. At a molecular level, the receptor sites are thought to be depressions of particular size and shape each of which is capable of accepting a molecule of a certain size and shape, thereby resulting in an odor. In some cases size and shape are not important and the charge (δ^+ or δ^-) on the molecule is the dominant factor, as for pungent and putrid odors. Some molecules may be able to fit more than one site, giving combinations of primary odors.

It is worth pointing out that many molecules whose chemical structures are unrelated still have a common odor. For example

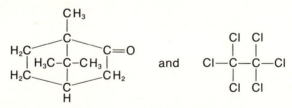

both have a camphor-like odor. The common feature among such molecules is the closeness of their size and shape. Thus, if the same type of receptor site can accept two molecules of a certain size and shape, these molecules will have similar odors. Examples of some receptor sites and molecules that can fit them is given in Figure 18–18. Recall that because of the tetrahedral structure of carbon, menthol (Figure 18–18B) is not planar. Furthermore, the existence of a group capable of hydrogen bonding to the binding site is important in at least one case—pepperminty. It is also possible to change the odor of a molecule by relatively minor alteration. For example

$$H_3COOC-(CH_2)_2-\underset{\underset{COOCH_3}{|}}{\overset{\overset{H}{|}}{C}}-(CH_2)_2COOCH_3$$

FIGURE 18–18 Shape of receptor sites for different odors and examples of molecules giving an odor: A. camphor-like—hexachloroethane,   B. pepperminty—menthol, and   C. ethereal—diethyl ether [Redrawn from John E. Amoore, James W. Johnston, Jr. and Martin Rubin, "The Stereochemical Theory of Odor;" Copyright © 1964 by *Scientific American Inc.* All rights reserved.

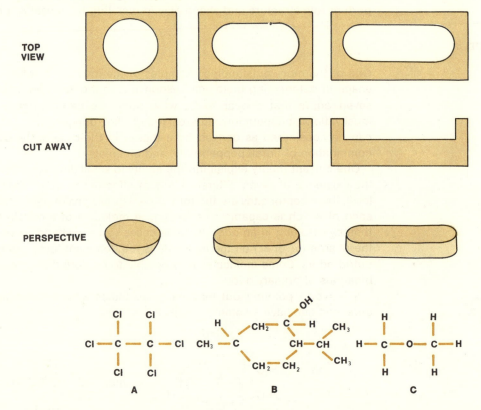

smells fruity (fits three sites) but substitution by a methyl group of the lone hydrogen attached to the central carbon atom changes the odor to ethereal. This presumably occurs because the altered molecule will now fit only one site.

PHEROMONES

Pheromones are chemicals used in communication between species. The pheromones of insects are most studied. The most powerful physiological pheromones are the sex attractants of insects where only several hundred molecules can be detected by a male. The method of detection is by odor and therefore this discussion is related to the previous discussion. There is great ecological importance in understanding pheromones. If we were able to use chemicals to attract unwanted insects, traps could be made which contain other chemicals or devices as well to exterminate the insects. Obviously this procedure would avoid the massive dissemination of insecticides that now appears to be necessary to destroy harmful insects.

Size, shape and chemical structure are again important in pheromones. That is, these factors determine which species responds to the chemical stimulus. For example, the natural attractant for the gypsy moth is

$$H_3C—(CH_2)_5—\overset{\displaystyle H}{\underset{\displaystyle O—COOCH_3}{C}}—CH_2—CH{=}CH—(CH_2)_4—CH_2—CH_2—OH$$

If the two carbons adjacent to the hydroxy or alcohol group (—OH) are removed, the resulting compound is inactive. On the other hand if two additional carbons are inserted in this location, an active synthetic attractant results. An even more dramatic example of the sensitivity of chemical stimuli to structure is exemplified by attractants for the Mediterranean fruit fly. The following compounds are two of eight possible spatial arrangements of the groups attached to the cyclohexane ring

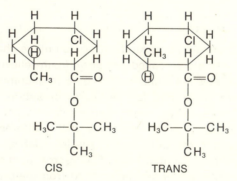

CIS   TRANS

The *trans* compound is considerably more active than the *cis*. This is indeed a delicate differentiation.

Generally insect sex attractants range between 10 to 17 carbon atoms with molecular weights in the range of 180 to 300. The matter of size is important in terms of being able to have a compound sufficiently specific that only a single species will respond. On the other hand, the compound cannot be too complicated to be biologically synthesized and finally, it must be volatile enough to be carried over a distance. It has been possible to eradicate insects from entire islands by a combination of use of attractants and insecticides put together on absorbent material. The development of synthetic compounds that will act as pheromones offers a unique and rewarding challenge to chemistry students and professionals alike.

FIGURE 18–19 Chemical hallucinogens with a molecular core structure containing indole. Note that mescaline does not have an indole core structure but does resemble it. The neurotransmitter serotonin (which is not a drug) is presented for comparison.

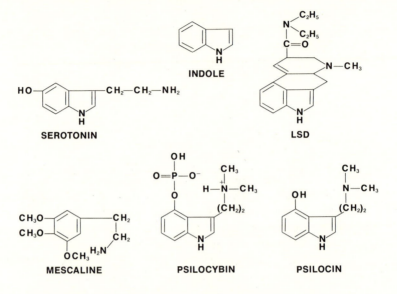

DRUGS

There are many kinds of drugs but we shall limit our discussion to the *hallucinogens* and so called *analgesic* (pain-relieving) drugs. Here again for each of these classes the emphasis will be on the common structural aspects which apparently are responsible for their common action.

Many of the hallucinogens have a core molecular structure of indole (Figure 18–19). The minimal dose of the different hallucinogens required to produce a certain response in humans is variable, for example, 500 mg mescaline, 20 mg psilocybin and 0.1 mg LSD (lysergic acid diethylamide). All of the hallucinogens in Figure 18–19 can be derived from natural sources, such as cactus, mushroom, and fungus. Some can now be synthetically prepared. The precise chemical role these drugs play is under very intensive current investigation. LSD is known to be an inhibitor of serotonin which is chemically important in the central nervous system. LSD and similar molecules containing the indole ring (Figure 18–19) can interact with serotonin receptor sites at nerve endings in the central nervous system.

The analgesic drugs include, for example, heroin, morphine, and codeine. These are quite similar in structure as shown in Figure 18–20. Methadone, now used as a substitute for heroin in clinical treatment, bears a resemblance to heroin but has a number of structural differences. Here again, the mechanism of action is of intense current interest. It is known that morphine and related drugs inhibit the enzymatic modification of a compound important in the central nervous system.

FIGURE 18–20 Structures of some analgesic drugs. Codeine differs from morphine only in the substitution of one —OCH₃ group for one —OH group

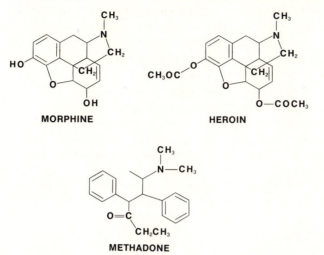

MORPHINE   HEROIN

METHADONE

18–F. BIOENERGETICS

The basic principles of chemical thermodynamics, which were developed in Chapter 10, apply equally well to biochemical reactions. Also, since most biochemical reactions are reversible, the considerations of chemical equilibrium in Chapter 13 can be applied to biochemical processes. For example, an important reaction in the metabolism of glucose in the body is the conversion of glucose-1-phosphate (G-1-P) to glucose-6-phosphate (G-6-P).

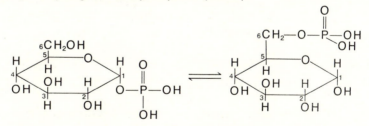

Under standard conditions the reaction proceeds with a decrease in free energy of −1.8 kcal/mole.

$$\Delta G^\circ = \Delta G^\circ_{f,\text{G-6-P}} - \Delta G^\circ_{f,\text{G-1-P}} = -1.8 \text{ kcal}$$

The equilibrium constant for the reaction is calculated using equation 13–3, $\Delta G^\circ = -RT \ln K$, and is found to be $K = 19$. If the initial concentration of glucose-1-P is 0.01 M, at equilibrium the concentration of glucose-6-P will be 0.0095 M and the concentration of glucose-1-P 0.0005 M. In other words, 95 % of the glucose-1-P would be converted to glucose-6-P. Evidently the 6 position for the phosphate group is more stable than the 1 position and this slight difference in stability ($\Delta G° = -1.8$ kcal/mole) makes the reaction forming glucose-6-P favored. Even though this reaction is thermodynamically favorable on the basis of a decrease in free energy, the reaction proceeds exceedingly slowly without catalysis by the enzyme phosphoglucomutase. The mechanism for enzyme catalysis was discussed in the Sections 15–H and 18–D. Most biochemical reactions require an enzyme catalyst which is specific for that reaction.

Many biochemical reactions have a positive ΔG and are therefore not favorable from a thermodynamic standpoint. In other words there should be very little yield of product at equilibrium. For example, one of the reactions involved in the metabolic breakdown of sugars has a change in free energy of +5.73 kcal/mole.

$$\text{fructose-1,6-diP} \rightleftharpoons \text{glyceraldehyde-3-P} + \text{dihydroxyacetone-P}$$

where fructose is a 6-carbon sugar and glyceraldehyde and dihydroxyacetone are 3-carbon sugars. In order to make a reaction such as this one more favorable (increase yield) a reaction can be thought of as being *coupled* with another reaction which undergoes a decrease in free energy. When the overall ΔG for the process is negative, the coupled reaction is said to drive the other reaction. A reaction commonly utilized for coupling is the hydrolysis of the nucleotide adenosine triphosphate (ATP), whose structure is shown in Figure 18–21. The hydrolysis of one phosphate group causes a decrease in free energy of 7 kcal

$$\text{ATP} + H_2O \longrightarrow \text{ADP} + H_3PO_4 \qquad \Delta G° = -7 \text{ kcal}$$

The P—O bond cleavage which gives ADP is shown in Figure 18–21. The net coupled reaction is simply the sum of the previously unfavorable reaction and the hydrolysis of ATP

	$\Delta G°$
fructose-1,6-diP $\rightleftharpoons$ glyceraldehyde-3-P + dihydroxyacetone-P	5.7
ATP + H_2O $\rightleftharpoons$ ADP + H_3PO_4	−7.0
fructose-1,6-diP + ATP $\rightleftharpoons$ glyceraldehyde-3-P + dihydroxyacetone-P + ADP + H_3PO_4	−3.3

From the net reaction it is clear that the hydrolysis of ATP to ADP and phosphoric acid supplies the necessary decrease in free energy to drive the reaction from left to right. Since the hydrolysis of ATP to ADP involves the breaking of a phosphate bond with a substantial decrease in free energy, the phosphate bonds in ATP are frequently referred to as *high energy bonds*. The ATP is considered to be a storage place for free energy which is released by the hydrolysis of these high energy bonds. The free energy stored in ATP arose from the oxidation of glucose, which gives off 686 kcal. If the oxidation of glucose is coupled to the

FIGURE 18–21 Structure of adenosine triphosphate (ATP) and adenosine diphosphate (ADP)

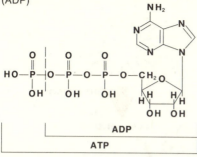

phosphorylation of ADP, 38 equivalents of ADP can be converted to ATP per mole of glucose.

$$C_6H_{12}O_6 + 6O_2 + 38\,ADP + 38\,H_3PO_4 \longrightarrow 6CO_2 + 44\,H_2O + 38\,ATP$$

18–G. THE DARK CYCLE IN PHOTOSYNTHESIS

Photosynthesis involves the conversion of energy from a light source—the sun—to chemical energy within a living system. Most plants and many microorganisms participate in photosynthetic reactions. Higher organisms, such as man, are dependent on the energy-fixing powers of these photosynthetic organisms. Figure 18–22 schematically illustrates the energy cycle. The overall reaction in photosynthesis is:

$$2n\,H_2O + n\,CO_2 \xrightarrow{\text{light}} \text{sugars } (CH_2O)_n + n\,H_2O + n\,O_2$$

In actual fact, water need not be a reactant; some other hydrogen donor, such as hydrogen sulfide or an organic compound, may be the starting material. All photosynthetic reactions, however, have several features in common. They all must have a hydrogen donor and a hydrogen or electron acceptor. In this sense, photosynthetic reactions can be thought of as oxidation-reduction reactions (see Chapter 14).

One of the most important aspects of photosynthesis is that the electron flow from donor to acceptor moves *toward the less easily reduced species*. This is the reverse of the reactions we studied previously in Chapter 14. There must be a strong driving force in this reaction and that driving force is light.

The photosynthetic pathway occurs in two parts. The first part is called the *light cycle* because it requires the absorption of light energy from the sun in

FIGURE 18–22 The exchange of energy from the sun to plant to animals

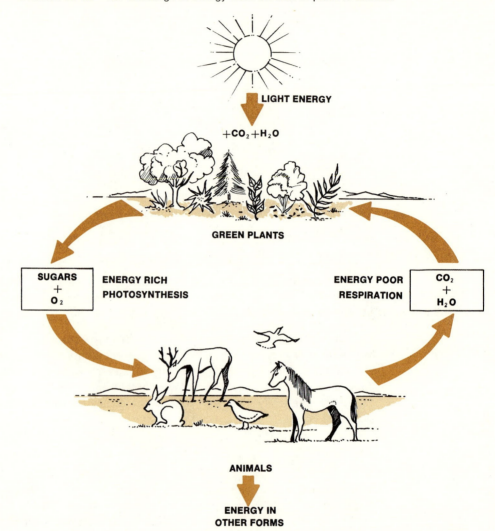

order to take place. This reaction is a photochemical reaction and will be discussed in detail in Section 19–D. In this chapter, we will be more concerned with the second part of the photosynthetic pathway, *the dark cycle*. The products of the light reaction are utilized in the dark cycle to reduce carbon dioxide to sugars.

The major product of the dark reaction is a hexose—a 6-carbon sugar which may be represented in the open chain form* as

*The parentheses indicate that although each carbon has a hydrogen and a hydroxyl group, any particular type of sugar has these groups positioned in a specific way.

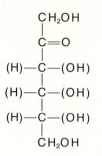

or in the ring form* as

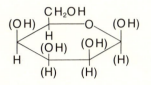

Radioactive carbon has been used to study the reactions involved in the reduction of carbon dioxide to sugar during the dark reaction (Chapter 17). When labeled carbon, in the tracer $^{14}CO_2$, is used as the starting material and the reaction is stopped and analyzed at various stages, the first stable intermediate in which labeled carbon can be found is phosphoglyceric acid (in the carboxyl group):

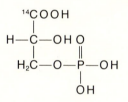

This, however, is only one step in the dark reaction mechanism. A cell starting with only CO_2 and a hydrogen donor will form a 6-carbon molecule. If the reaction is allowed to proceed with labeled $^{14}CO_2$, labeled carbon will be found eventually in all six carbons of the hexose product. The most recent research indicates that no fewer than eleven different enzymes and fifteen reactions may be involved in the dark reaction. Figure 18–23 gives some of the important steps in the dark cycle. Each reaction must obtain enough energy chemically to pass an activation energy barrier. Each reaction can be considered to be both energy requiring and energy producing in that energy is passed along chemically from one molecule to another. Recall that in biochemical reactions compounds which can carry energy from one reaction to another are known as high-energy compounds (Section 18–F). Two such compounds are adenosine triphosphate (ATP) and reduced nicotinamide adenine dinucleotide phosphate (NADPH). The structure of ATP is given in Figure 18–21 and NADPH is shown in Figure 18–24. NADPH acts as a reductant in the dark reaction and is essential chemically for the conversion of CO_2 to carbohydrates. Each energy-carrying molecule must be accounted for in writing the equation for the overall dark reaction:

$$12H^+ + 6CO_2 + 18ATP + 12NADPH \longrightarrow Hexose + 18ADP + 12NADP + 18H_3PO_4$$

Note that the diphosphate ADP is formed, indicating that one phosphate group has been cleaved from ATP, and NADPH is oxidized in this reaction. The hydrolysis of the 18 ATP's produces the energy necessary for the reaction to proceed.

FIGURE 18–23 Principal steps in carbon dark cycle. H is hydrogen and P is phosphate. Ribulose (sugar) phosphate represents the starting point. NADP and NADPH as well as ATP and ADP are described in the text and in Figures 18–21 and 18–24.

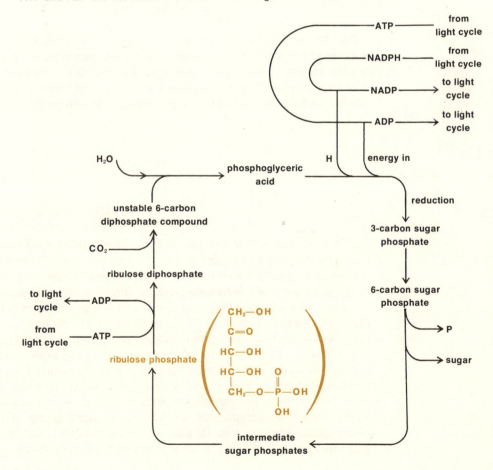

FIGURE 18–24 Structures of NADP (A, left) and NADPH (B, right). The only difference between A and B is in the reduction of the nicotinamide portion of NADP with hydrogen to form NADPH.

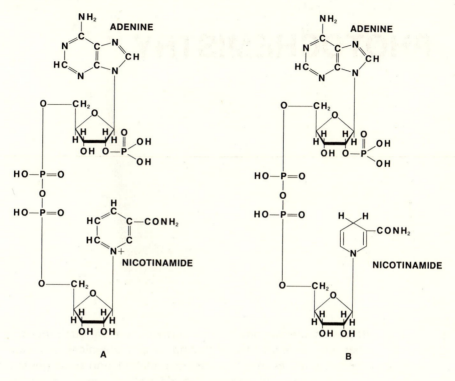

CHAPTER NINETEEN

PHOTOCHEMISTRY

In this chapter we shall be concerned about various processes in which photo-chemistry is important including vision, chemical evolution, photosynthesis, polymer (natural and synthetic) degradation, and smog generation. In essentially all of these examples molecules, rather than atoms, are involved. The processes of absorption and emission of light by molecules are more complex than for atoms. The only excitation process that can take place in an atom is excitation of electrons from one energy level or state to another. However, in a molecule with multi-bonded atoms other excitation processes can occur. The only photo-reaction involving a single atom is ionization of an electron; however, for a molecule many photochemical reactions are possible.

Since the prerequisite for photochemical reactions is electronic excitation, we shall first consider the fundamentals of spectroscopy of molecular systems.

19–A. MOLECULAR SPECTROSCOPY

Excitation of an electron from one atomic orbital to another results in an excited state for the atom; for example, in beryllium

19–1

$$1s^2\,2s^2 \xrightarrow[\text{excitation}]{\text{electronic}} 1s^2\,2s^1\,2p^1$$

ground state   excited state

The atom in its normal lowest energy configuration is referred to as the *ground state*. Excitation to the first, second, and successively higher energy orbitals results in formation of the first, second, and successively higher excited states. Of concern to us is the relative orientation of the spins of the electrons in the ground and excited states. In the ground state all electrons are paired and therefore each $+\frac{1}{2}$ spin is canceled by a $-\frac{1}{2}$ spin to give a resultant of zero total spin. For beryllium (equation 19–1)

$$\text{Total spin } (M_s) = 1s(+\tfrac{1}{2}) + 1s(-\tfrac{1}{2}) + 2s(+\tfrac{1}{2}) + 2s(-\tfrac{1}{2}) = 0$$

It can be seen that there is a total spin of zero when all electrons are in filled orbitals of any atom since the spins must be opposed because of the Pauli principle (Section 4–L). In the first or lowest excited state of beryllium, the contribution of the $1s^2$ electrons to the total spin also will be zero but the electrons in the $2s$ and $2p$ orbitals can be antiparallel or parallel:

$$M_s = 2s(+\tfrac{1}{2}) + 2p(+\tfrac{1}{2}) = 1 \qquad \text{(parallel in } 2s \text{ and } 2p \text{ orbitals)}$$
$$\text{or}$$
$$M_s = 2s(+\tfrac{1}{2}) + 2p(-\tfrac{1}{2}) = 0 \qquad \text{(antiparallel in } 2s \text{ and } 2p \text{ orbitals)}$$

Thus, two types of first excited states are possible depending upon the relative spin orientation of the electrons. In order to differentiate these the term multiplicity (M) is used which is defined as

$$2M_s + 1$$

In the case $M_s = 0$ $\qquad M = 2M_s + 1 = 1 \qquad$ singlet (S)

and in the case $M_s = 1$ $\quad M = 2M_s + 1 = 3 \qquad$ triplet (T)

The lowest excited singlet and triplet states differ in important ways:

1. Triplet states are lower in absolute energy than are singlet states.
2. The lifetime prior to light emission is considerably longer (approximately 1 million times) for the triplet state compared to the singlet state.

In di- and polyatomic molecules, a comparable situation exists. It is most common that each molecular orbital is filled with two electrons so $M_s = 0$ and the ground state will be a singlet state, as for Li_2:

$$(1s\sigma)^2(1s\sigma^*)^2(2s\sigma)^2$$

Excitation of an electron to an empty molecular orbital results in the formation of an excited state. The electron excited can maintain the same spin or change it, giving the possibility of $M_s = 0$ again or $M_s = 1$. If $M_s = 0$, then the excited state is again a singlet but if $M_s = 1$, then the excited state is a triplet. The energy of the excited triplet state is always less than that of the excited singlet state. In excitation of an electron in Li_2:

$$(1s\sigma)^2(1s\sigma^*)^2(2s\sigma)^2 \xrightarrow{\text{light}} (1s\sigma)^2(1s\sigma^*)^2(2s\sigma)^1(2s\sigma^*)^1$$
$$\text{ground state } (S_0) \qquad\qquad \text{first excited state } (S_1 \text{ or } T_1)$$

$$\xrightarrow{\text{light}} (1s\sigma)^2(1s\sigma^*)^2(2s\sigma)^1(2p\sigma)^1$$
$$\text{second excited state } (S_2 \text{ or } T_2)$$

If the electron in the $2s\sigma^*$ orbital is still antiparallel to the one remaining in the $2s\sigma$ orbital, an excited singlet (S_1) state will result; however, if it is parallel then an excited triplet (T_1) state will result. A general energy diagram for any polyatomic molecule would appear as in Figure 19–1.

Note in Figure 19–1 that it is possible for an excited molecule to go from any S_n state to the lowest excited singlet state (S_1) and from S_1 to the lowest excited triplet state (T_1) *without emission of light*. Also, following these non-light-emitting transitions, two kinds of light-emitting transitions or emissions are possible and these are distinguished by the nature of the excited state from which they originate. *Fluorescence* occurs from S_1 to S_0; it is at shorter wavelengths and generally has a substantially shorter half-life than *phosphorescence which occurs from T_1 to S_0*. The half-life of fluorescence is commonly about 10^{-8} sec while for phosphorescence it is 10^{-1} sec or longer. Thus, anytime there is a change in multiplicity of a state from singlet to triplet *or* from triplet to singlet, the half-life is much longer than in the situation where both initial and final states in the transition have the same multiplicity.

FIGURE 19–1 Energy level diagram for a polyatomic molecule. S and T indicate singlet and triplet states respectively, and the subscripts indicate the energy ordering, with S_0 the ground state. A wobbly arrow indicates that no emission of light occurs.

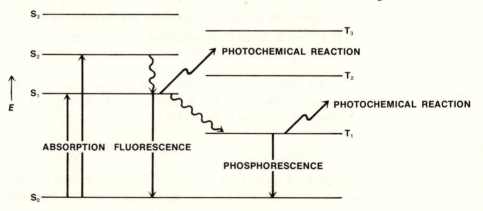

The substantial difference in lifetime is important in terms of the state from which photochemical reactions such as bond breakage or dissociation can most readily occur. That is, these reactions must compete with emission processes and the longer the lifetime of an excited state, the better the chance for a photochemical reaction to occur from that state. Since the triplet state, T_1, has a longer lifetime than the singlet state, S_1, many photochemical reactions occur from the T_1 state. The product(s) produced directly after the absorption of light are known as the primary products and are the result of the primary photochemical process.

Unfortunately the situation in molecules is not actually as simple as we have made it thus far. As mentioned in the introductory comments to this chapter, other excitations, in addition to the purely electronic excitation, accompany absorption or emission processes in molecules. These other excitations involve vibrational and rotational motions of the molecule.

A molecule in any electronic state can rotate about its center of mass. The *center of mass* of an object is that position in space at which one may think of the mass as being concentrated. For a diatomic molecule the center of mass lies along the axis joining the two nuclei (internuclear axis). In Figure 19–2 the internuclear axis is taken in the *y* direction and rotation of the molecule can take place about the *x* and *z* axes. The origin of the rectangular coordinate system is at the center of mass. For polyatomic molecules which are nonlinear there are three axes of rotation, *x*, *y*, and *z*. Generally, each of the rotations occurs with quantized energies so that a series of rotational energy levels exist.

In addition to rotational motion, molecules can also vibrate; that is, the atoms can undergo displacements as shown in Figure 19–3. The orientation of the arrows shows the relative direction of displacement and the length of arrows is proportional to the magnitude of the atomic displacement. These vibrational displacements are oscillatory or periodic in nature. After the molecule has stretched to its extreme position in the direction of the arrows in Figure 19–3, all the atoms will simultaneously reverse direction until the minimum internuclear separation is reached. This can be shown as

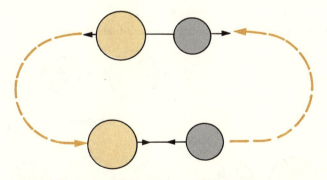

In general, the greater the displacement of the atoms in any vibration, the greater the energy required to cause that displacement. As in the case of rotational mo-

FIGURE 19–2 Rotational motions for a linear diatomic molecule

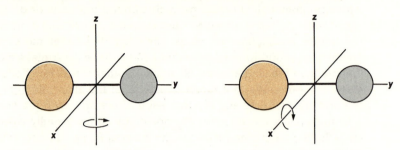

tion, the energy levels for vibrational motion are quantized and a series of vibrational energy levels exist for each vibrational mode. It should be noted that in both the rotational and vibrational processes, the center of mass of the molecule does not translate in space; that is, the molecule does not move as a whole.

The wavelength regions for observing the various kinds of absorption spectra are

pure rotational   1,000,000 Å and longer (microwave region)
pure vibrational   20,000 Å – 1,000,000 Å (infrared region)
pure electronic   1,000 Å – 12,000 Å (ultraviolet-visible-near infrared regions)

Since shorter wavelengths have greater energy, it is clear that electronic excitations involve greater energy than do vibrational and rotational excitations.

PROBLEM 19–1   Assume a pure rotational transition occurs at a wavelength of 1,000,000 Å, a pure vibrational transition occurs at 50,000 Å and a pure electronic transition occurs at 5000 Å. Calculate the energy difference in wave number (in cm⁻¹) and in kcal/mole

FIGURE 19–3 Vibrational motions for linear di- and triatomic molecules

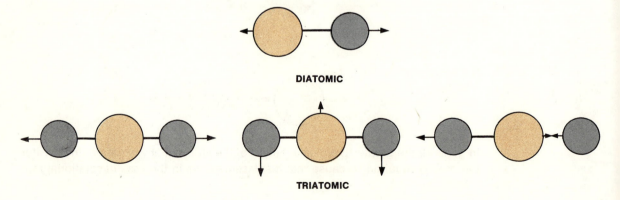

DIATOMIC

TRIATOMIC

between rotational, vibrational and electronic energy levels. [Hint: See Chapter 4 and Appendix A for the necessary formulas and conversion factors.]

It will be worthwhile to explore briefly a schematic diagram showing electronic, vibrational, and rotational states (Figure 19–4). The spectral band representing a pure electronic transition such as $E_0 \longrightarrow E_1$ of Figure 19–4 is broadened because the rotational levels in E_1 are so close in energy that they become simultaneously excited. When the energy of the exciting light is increased somewhat, then the electronic transition occurs to the $E_1 + v_1$ levels, giving a second band. Again because the rotational levels are so close in energy, they are simultaneously excited giving breadth to the second band. Further bands arise in a similar fashion. The remainder of the electronic transitions and their shape arise from excitations to electronic states E_2 and E_3 and contain components as described above. Remember that pure electronic transitions occur between levels of relatively high energy and therefore occur at relatively short wavelength (1,000 Å – 15,000 Å).

Even if the molecule is kept in its ground electronic state, E_0, it is possible to excite it to various vibrational levels (v_1, v_2, etc.) giving a vibrational spectrum. Again each vibrational band is broadened by the rotational components. Recall that vibrational energy levels are relatively close in energy and therefore, the pure vibrational spectra occur at lower energy or longer wavelength (15,000–1,000,000 Å) than do electronic transitions. Again, even if the molecule is in the ground electronic state, E_0, and ground vibrational state, v_0, it is still possible to excite it to various rotational levels (r_1, r_2, r_3, etc.) to give a rotational spectrum. The pure rotational spectra occur at very long wavelengths (1,000,000 Å and longer) as noted earlier, because the energy spacing between each level is much smaller than for vibrational or electronic levels.

The light used to cause electronic transitions commonly contains many wavelengths and therefore is composed of a spectrum of energies of excitation. This causes many rotational and some vibrational levels to be simultaneously excited, although it is ideally possible to excite individual vibrational components and rotational components, particularly the former.

To summarize, in polyatomic molecules the electronic states are commonly identified only in terms of their relative energy and relative multiplicity. Any molecule with an even number of electrons has filled molecular orbitals, resulting in $M_s = 0$, $M = 1$ and a ground state singlet. Excitation can occur to any one of a series of excited singlet states. The molecule rapidly ends up in the lowest excited singlet state (S_1) from which fluorescence can occur, photochemical reactions can occur, or the triplet state (T_1) can be formed (Figure 19–1). From the triplet state (T_1), phosphorescence and also photochemical reactions can occur.

EXPERIMENTAL DETERMINATION OF SPECTRA

The same apparatus can be used to perform emission spectral work as well as to measure absorption spectra (Figure 19–5). Light sources are commonly lamps

FIGURE 19–4   Energy level diagram showing electronic, vibrational, and rotational energy levels. E stands for an electronic state, v for a vibrational state, and r for a rotational state. The subscripts indicate the energy level ordering where 0 is the ground state and energy increases with increasing magnitude of the number. An absorption spectrum resulting from electron excitation processes is shown at the left.

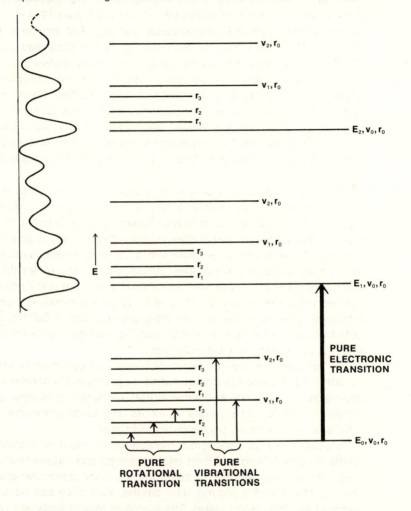

filled with gases or gaseous substances such as mercury or xenon through which an electrical discharge is continuously passed, causing excitation of the mercury or xenon atoms. The light emitted from the lamp then arises from the de-excitation of the excited atoms. A monochromator is a device that can select a narrow band of wavelengths out of a continuum of wavelengths originating from the light source. The exciting light then falls on the sample which can be in solution, a gas, or even a solid. The light then emitted from the sample falls on another monochromator which scans over all the emitted light, and the detector responds to those wavelengths present. The signal from the detector is then recorded.

LASER

The term *laser* is an acronymn for *l*ight *a*mplification by *s*timulated *e*mission of *r*adiation. Both gas and crystal *lasers* exist. The unique properties of light from a *laser* compared with a conventional source such as an arc discharge or heat filament lamp are (a) coherence, or coordinated waves, (b) monochromaticity, or emission of a single wavelength, and (c) directionality and intensity. The latter two properties are the important ones for our consideration here. Light from a laser is of one wavelength, nondivergent, and intense. For example, it is possible to obtain intensities of greater than 100 million watts/cm² in a pulse approximately 20×10^{-9} sec long. Recently, it has been possible to obtain picosecond (10^{-12} sec) pulses in which the intensity is about a trillion watts. (Interestingly enough, in the latter case laser light interacting with a deuterium nucleus in LiD has resulted in neutron ejection.)

To generate the intensities noted above, a laser uses crystals such as ruby as a source. Gas lasers also exist and can give a continuous output. For example, it is possible to obtain 8800 watts of continuous power from a CO_2 laser. This is sufficient to cut through thin steel in seconds and can be used in welding. It is obvious that both the pulsed and continuous laser offer exciting new light sources for spectroscopy and photochemistry. With lasers, it is potentially possible to excite specific vibrational and rotational levels as well as electronic levels.

FIGURE 19–5 Schematic experimental set-up for obtaining emission spectra

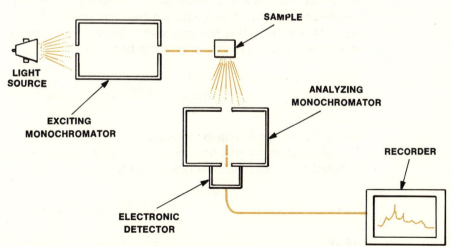

PHOTOCHEMISTRY OF SIMPLE MOLECULES

Before proceeding on to special topics in photochemistry, we shall consider the photochemistry of a few simple molecules. Note that although more than one reaction is sometimes possible, one reaction predominantly occurs. Sometimes the nature of the dominant reaction depends upon the wavelength of the exciting light. In the reactions that follow, the wavelengths given are those used to excite the reactant to an excited electronic state, followed by a photochemical reaction to give the products shown:

19–2 $$CH_4 \xrightarrow{1440\,\text{Å}} :CH_2 + H_2$$

19–3 $$NH_3 \xrightarrow{2130\,\text{Å}} \cdot NH_2 + H\cdot \qquad NH_3 \xrightarrow{1550\,\text{Å}} :NH + H_2$$

19–4 $$H_2O \xrightarrow{2400\,\text{Å}} \cdot OH + \cdot H \qquad H_2O \xrightarrow{1240\,\text{Å}} H_2 + O$$

19–5 $$H_2CO \xrightarrow{3200\,\text{Å}} HCO\cdot + H\cdot \quad \text{and} \quad H_2 + CO$$

19–6a $$(CH_3)_2C{=}O \xrightarrow{3100\,\text{Å}} (CH_3)_2C{=}O \text{ (singlet)}$$

19–6b $$(CH_3)_2C{=}O \text{ (singlet)} \rightsquigarrow (CH_3)_2C{=}O \text{ (triplet)}$$

19–6c
$$(CH_3)_2C{=}O \text{ (triplet)} \longrightarrow H_3C{-}\overset{\cdot}{C}{=}O + \cdot CH_3$$
$$\longrightarrow CO + \cdot CH_3 \longrightarrow C_2H_6$$

Note that in all the above reactions free radicals such as $:CH_2$, $HCO\cdot$ and $\cdot OH$ are produced. These are highly reactive chemical species. It can be seen from equations 19–2 to 19–6 that the radicals are formed by removing one or more atoms with one of the two electrons from the bond, as

$$H{:}\overset{\cdot\cdot}{C}{:}H \xrightarrow{1440\text{Å}} \cdot \overset{\cdot\cdot}{C}{:}H + 2H\cdot$$

where the $2H\cdot$ immediately forms H_2.

The production of HCl from H_2 and Cl_2 proceeds by a radical chain reaction initiated by a photochemical reaction as follows

19–8a $$Cl_2 \xrightarrow{3300\,\text{Å}} Cl\cdot + Cl\cdot$$

19–8b $$Cl\cdot + H_2 \longrightarrow HCl + H\cdot$$

19–8c
$$H\cdot + Cl_2 \longrightarrow HCl + Cl\cdot$$

A chain mechanism is set up by the reactions in equation 19–8 since the $Cl\cdot$ produced in the third step can be used in the second step to react with H_2, etc. In the sections that follow, it is often not known whether the photochemical reaction occurs in the excited S_1 or T_1 state or both.

19–B. VISION PROCESS

The visual pigments are located in the rod and cone cells in the eye. There are more rods than cones in the human retina by a factor of about $20(125 \times 10^6$ compared to $6.5 \times 10^6)$. The rods of humans commonly contain only a single vision pigment known as *rhodopsin* which has an absorption maximum near 5000 Å at room temperature (the green portion of the visible spectrum). Most research on the chemistry of vision has centered on rhodopsin or models for it.

Rhodopsin contains a protein (Section 18–B) and an unsaturated nonprotein part chemically bonded to the protein. The structure of the unsaturated nonprotein pigment group is known to be a long-chain polyene (Section 7–G). Researchers believe that a long-chain polyene aldehyde called retinal (equation 19–9a) condenses with the protein opsin to make rhodopsin (equation 19–9b). The detailed structure of the protein is not known. At low temperatures prelumirhodopsin, a somewhat different rhodopsin from metarhodopsin, is produced (differing in the shape of the protein). The structure of 11-*cis* retinal and its photochemical reaction and a schematic structure for rhodopsin and its photochemical reaction are given in equations 19–9.

19–9a

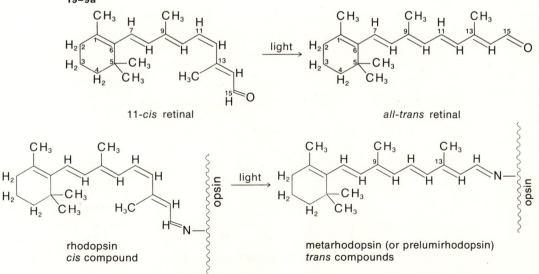

11-*cis* retinal   all-*trans* retinal

19–9b

rhodopsin
cis compound   metarhodopsin (or prelumirhodopsin)
trans compounds

The primary process in this reaction involves the internal rotation by 180° about a double bond, equation 19–9b (see Section 7–G). Since light is involved, this process is known as *photoisomerization*.

Following this primary process, a series of thermal reactions producing other types of rhodopsin molecules occurs which involve, at least in part, conformational changes of the protein. The sequence presumably terminates with a bond scission to give opsin and *all-trans* retinal. It is not known at which step or how the nerve trigger occurs.

Current data from studies of a model of prelumirhodopsin indicate that some degree of protonation of the nitrogen atom is necessary to cause the long wavelength band to be at 5000 Å in the naturally occurring *cis* compound (rhodopsin). In fact low temperature studies of the model molecule with HCl gas gives an absorption maximum essentially identical to that of the photochemically produced all-*trans* isomer (prelumirhodopsin).

Recall from Section 7–G (Figure 7–4) that twisting around the double bond in the ground state requires considerable energy. Theoretical calculations clearly tell us that in the excited states S_1 and T_1, the barrier to twisting is extremely small (Figure 19–6). The photoisomerization likely occurs in both the singlet and the triplet states. The importance of the light absorption is to create an excited state

FIGURE 19–6 Potential energy curves for twisting around a double bond in the ground and excited states (S_1 and T_1). Note the high energy required to overcome the barrier near 90° in the ground state and the existence of only very low barriers in the excited states (near 60°).

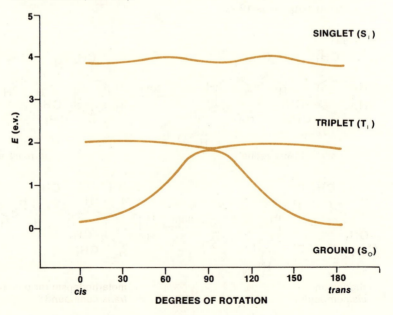

wherein the twisting about the double bond is much easier. Thus, the vision process can be said to require light in order that the initiating reaction, photoisomerization be possible.

19–C. CHEMICAL EVOLUTION

The prime goal of space chemistry and biology is the search for compounds important to life as well as the search for life itself. In studying prebiotic chemistry and the beginning of life on earth, models are made of events which probably have occurred and will occur in the universe many times over. Theories of chemical evolution attempt to explain how compounds of biological significance could have arisen. One of the principal goals is to unravel the nature of the compounds required for the production of proteins and nucleic acids (Sections 18–B and 18–C). A chemical beginning point would be with the compounds present in the earth's primitive or primordial atmosphere. It is generally agreed that the primitive atmosphere was a reducing one; that is, oxidation reactions were not possible. Two of the possible mixtures could have been NH_3, H_2, CH_4, and N_2, or CH_4, NH_3, and H_2O, the first of which could presumably have existed for up to 1 billion years. Free oxygen may have arisen from the photodissociation of water, equation 19–4, in the upper atmosphere and from the photosynthetic process (Section 19–D). The latter process probably could only begin when the ozone layer that filters out killing ultraviolet light was formed in the upper atmosphere (Section 6–E). There is also reason to believe that formaldehyde $H_2C{=}O$ was formed relatively early in the earth's history.

The principal energy sources for synthesis of organic molecules at the beginning were ultraviolet light, ionizing radiation (Chapter 17), electric discharge and heat. The sources of these include the sun, radioactive sources on the earth, lightning, volcanoes, and even meteorites. The energy available from ultraviolet light is some 100, or more, times greater than that of the source next highest in amount. Also, the flux (amount of light per unit time) of ultraviolet light of various wavelengths has been estimated to have been high in the primordial atmosphere. The molecules that have been found in interstellar space are given below.

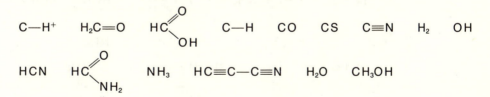

It is very interesting to note how complex some of these are and that many contain carbon—the basis of our life. Most of these molecules have been found in dust clouds in the Milky Way. The temperature in these clouds is estimated to be in the region of $-243\,°C$ to $-269\,°C$. Furthermore, it is estimated that there are only 1 to 10 molecules/cc in interstellar space whereas on earth the average density in the atmosphere is about 10^{16} molecules per cc—a near perfect vacuum in space! In any event, there are molecules in space that could provide a basis for chemical evolution anywhere. On or near planets, more types of molecules exist and at higher concentrations.

In the following discussion we shall primarily be concerned with photochemical reactions but shall allude to other types of reactions from time to time for comparison. Mixtures of CH_4, NH_3, and H_2O as well as CH_2O, NH_3, and CO_2 have been irradiated with ultraviolet light. Some amino acids—the building blocks for proteins—have been produced including glycine, alanine, and valine (Section 18–B). Electric discharge, ionizing radiation, and heating of similar mixtures have resulted in the production of amino acids as well. Little is known about the mechanism of formation of the amino acids under these conditions. It has been suggested that aldehydes and HCN are made first, followed by reaction of these with ammonia in the aqueous solution:

$$CH_4 + NH_3 + H_2O \xrightarrow[\text{or spark}]{\text{ultraviolet light}} R\overset{\displaystyle H}{\underset{}{C}}{=}O + HCN$$

$$R\overset{\displaystyle H}{\underset{}{C}}{=}O + HCN + NH_3 \rightleftharpoons RCH(NH_2)CN + H_2O$$

$$RCH(NH_2)CN + 2H_2O \longrightarrow RCH(NH_2)COOH + NH_3$$
$$\text{an amino acid}$$

Purine and pyrimidine bases (Section 18–C) have been synthesized from CH_4, NH_3, and H_2O mixtures and HCN solutions by ultraviolet irradiation and ionizing irradiation. These bases are the building blocks for nucleosides and nucleotides, which in turn are the basic units in nucleic acids. Again, little is known about the mechanisms of formation under these conditions. However, it appears that HCN and products of HCN polymerization are important intermediates:

19-10a $\quad CH_4 + NH_3 + H_2O \xrightarrow[\text{or spark}]{\text{ultraviolet}} HCN + \text{ other products}$

19-10b $\quad 3HCN \longrightarrow NC\overset{\displaystyle NH_2}{\underset{\displaystyle H}{-}C-}CN$

19-10c $\quad 2HCN + 2NH_3 \longrightarrow 2 \overset{HN}{\underset{H_2N}{>}}CH$

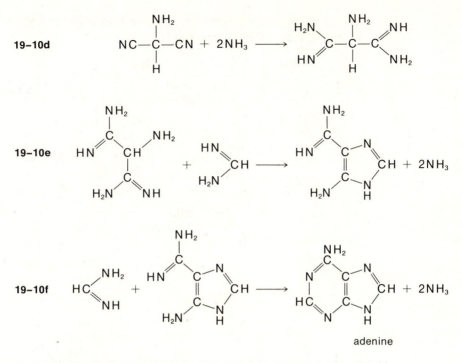

19–10d

19–10e

19–10f

adenine

The overall reaction could be simply written as in equation 19–11.

19–11

$$5HCN \xrightarrow[\text{of NH}_3]{\text{presence}} \text{adenine}$$

A tetramer of HCN, either *cis* or *trans*

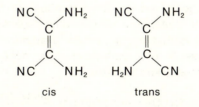

cis   trans

is also an important potential precursor. It now seems likely that photons cause both interconversion between the *cis* and *trans* forms as well as formation of an intermediate imidazole.

The imidazole can be further reacted to give some purine bases. The reaction beginning with the HCN tetramer is likely to be

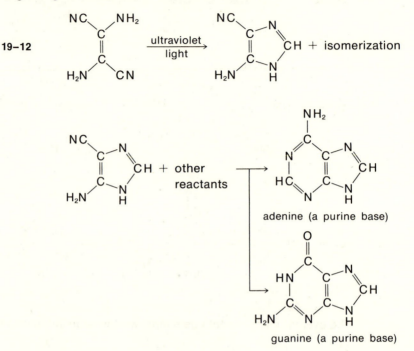

adenine (a purine base)

guanine (a purine base)

Notice particularly in equations 19–10 to 19–12 the apparent importance of HCN. Recall that this molecule has been found in interstellar space and apparently is present in the tails of comets. Also, cyanide is highly stable to thermal dissociation because of the carbon-nitrogen triple bond.

The knowledge of the synthesis of monosaccharides (sugars) is not as advanced as for the amino acids and nucleosides discussed above. Very recent information shows that the *ultraviolet irradiation* of formaldehyde in water solution in the presence of a catalyst such as MgO produces monosaccharides consisting of pentoses and hexoses

$$\text{CH}_2\text{O (water solution)} \xrightarrow{\text{ultraviolet light and catalyst}} \text{pentoses + hexoses}$$

Heating of solutions of formaldehyde in the presence of alkaline earth hydroxides also has produced a mixture of sugars. The mechanism is not known but a possible pathway is

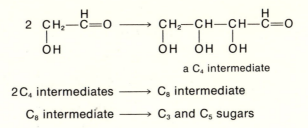

a C$_4$ intermediate

$$2C_4 \text{ intermediates} \longrightarrow C_8 \text{ intermediate}$$

$$C_8 \text{ intermediate} \longrightarrow C_3 \text{ and } C_5 \text{ sugars}$$

However, very recent evidence shows that *heating* formaldehyde gas in the presence of a material such as MgO or ZnO does *not* produce sugars. Instead, other interesting organic molecules are produced which in themselves could be important in chemical evolution

There has been only very limited success in synthesizing nucleosides and mononucleotides (Section 18–C). A nucleoside has been made by the ultraviolet irradiation of a sugar and a purine base in an aqueous phosphate solution.

19–13

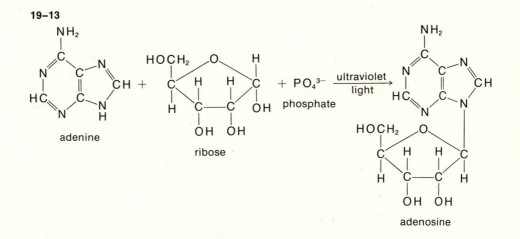

Ribose has been shown in the cyclic form in equation 19–13 although in water solution it is in equilibrium with a small amount of the open chain form:

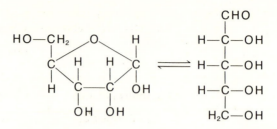

In most cases, the conditions used to produce the nucleosides and nucleotides appear not to be compatible with the environment present on the primitive earth.

In the case of polypeptides, many of the efforts have involved heat as the energy source. There has been some success in producing polymers in the absence of water by heating amino acids. Some small polypeptides have been produced by electrical discharge and ionizing irradiation of some amino acids. Di- and tri-peptides have been made by ultraviolet irradiation of amino acids. Ionizing and ultraviolet irradiation as well as electrical discharge of methane–ammonia–water mixtures have produced polypeptides. The mechanisms in most cases are not known.

19–D. PHOTOSYNTHESIS

Photosynthesis occurs in two parts, the *light cycle* and the *dark cycle*. We shall not consider the dark cycle here, since it was included in Section 18–G; it includes the reactions fixing CO_2 and ultimately leading to sugars. The dark cycle reactions have been worked out in detail. However, the light cycle, involving some photochemical reactions, has not been fully elucidated. The essential reaction involves the absorption of light by chlorophyll. The light energy is eventually changed into chemical energy. Furthermore, there are apparently two photochemical systems: in one, the oxygen-evolving process occurs, and in the other, a photoreduction process occurs ultimately leading to fixation of CO_2. The ideas known, thus far, regarding these two processes will be discussed shortly. The overall important reaction in photosynthesis is then

19–14 $$6\,CO_2 + 12\,H_2{}^{18}O \xrightarrow[\text{chlorophyll}]{\text{light and}} C_6H_{12}O_6 + 6\,{}^{18}O_2 + 6\,H_2O$$
$$\text{glucose (and other sugars)}$$

Equation 19–14 has been written with the ^{18}O isotope of oxygen to indicate that the source of the O_2 is H_2O and not CO_2. An interesting point regarding the photosynthetic process is that the ratio of O_2 evolved to CO_2 fixed is 1. Photosynthesis is the essential source of the O_2 which humans breathe. The heat of combustion (Chapter 10) of glucose (a sugar) is about 670 kcal per mole. The light energy has ultimately been used and stored as chemical energy in the sugars. Only about

0.5 to 3.5 % of the light that hits a leaf is utilized in photosynthesis. It should be noted that the bulk of the photosynthesis that occurs on earth is done by algae, not the higher plants.

The pigment chlorophyll is a metal complex—a large molecule or ligand (a porphyrin ring) coordinated with a magnesium ion—as shown below.

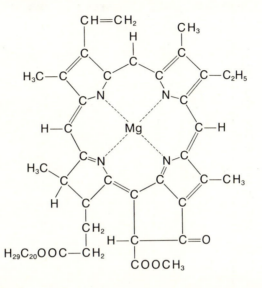

There are actually two major kinds of chlorophyll differing primarily only in substitution of —CH_2O for a —CH_3 group. Chlorophylls are green, and Figure 19–7 shows the absorption spectra of chlorophyll-*a* and chlorophyll-*b*.

Recall that we mentioned earlier that there were two photochemical systems functioning in the overall photosynthetic process. These are connected by electron carriers. A schematic of the principal occurrences is presented in Figure 19–8.

Our knowledge of the reactions and carriers in system II is particularly poor although several of the carriers connecting system II to I are known. Also, the true electron acceptor (carrier) out of system I is not known. NADPH and ATP are required in the dark cycle to fix CO_2 (Section 18–G).

19–E. AIR POLLUTION

In our discussion of air pollution we shall be primarily concerned with the photochemical reactions involved. However, where appropriate, relevant thermal reactions will be discussed.

It has been estimated that some 10^{12} tons of substances of all kinds enter the earth's atmosphere each year. Of this amount, only about 0.05 % arise from activities of man. In other words, some 99.95 % arise from processes involving

FIGURE 19-7 Absorption spectrum of chlorophyll-*a* as a solid line, chlorophyll-*b* as a dotted line. Note that the principal absorption is in the blue-violet (4000–4800 Å) and yellow and red (5700–7000 Å) regions, leaving green as the dominant color for the chlorophylls.

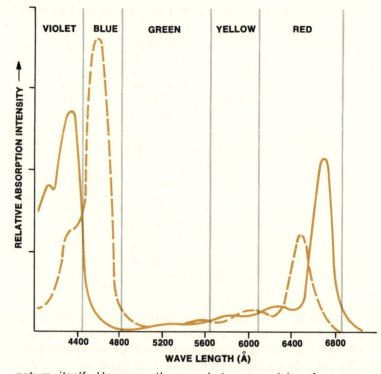

nature itself. However, those substances arising from man-made origins are generally concentrated in relatively small geographical areas. Generally, the chemical compounds emitted into the atmosphere by man are relatively simple, but the reactions involving these compounds may be very complex. Five classes of pollutants account for 90 % or more of those in the atmosphere. Table 19–1 presents data concerning the five: CO; hydrocarbons; NO_x, principally NO and NO_2; SO_x, mainly SO_2; and solid particulate matter.

It is generally agreed that smog commonly found over large cities is formed through initial reactions which are photochemical in nature. Although we do not understand all of the steps leading to the production of photochemical smog, certain facts are known. The three principal requirements are (1) the presence of nitrogen oxides, principally NO and NO_2, (2) the presence of hydrocarbons, particularly olefins but also including aldehydes, ketones and substituted aromatic hydrocarbons (Chapter 7), and (3) the presence of light of wavelengths between 3000 Å and 4200 Å. Obviously the light comes from the sun. There is very recent evidence that CO is important in the acceleration of the photochemical reactions of smog. We shall discuss this aspect later.

TABLE 19–1 Classes of pollutants and their relative abundance

TYPE POLLUTANT	% OF TOTAL[1]
CO	43.6
hydrocarbons	13.5
NO_x	7.2
SO_x	13.4
solid particulates	12.3

[1]The sum of the five shown here equals 90 % of total pollutants produced by man.

FIGURE 19–8 The light cycle of photosynthesis showing the two photosystems. The area within the dashed line represents system I and the area within the dash-dotted line represents system II. The nature of ADP, ATP, NADP, and NADPH are discussed in Sections 18–F and 18–G.

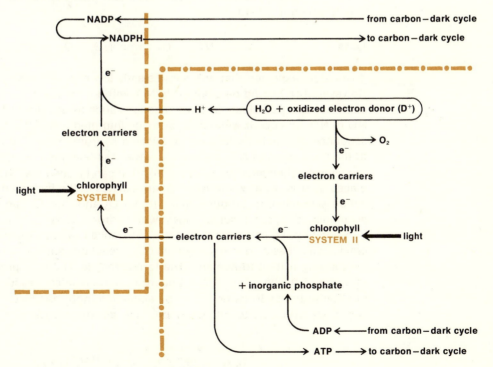

The best evidence to date clearly indicates that the first reaction in the reaction chain leading to smog involves a photochemical reaction of NO_2

19–15
$$NO_2 \xrightarrow{\text{light}} NO + O$$

which is followed by thermal reactions (equations 19–16 and 19–17)

19–16
$$O + O_2 \xrightarrow{\Delta} O_3$$

19–17
$$O_3 \text{ and } O + \text{hydrocarbon} \xrightarrow{\Delta} \text{oxidized hydrocarbons}$$

Recall that ozone is a strong oxidizing agent. It appears that a portion of it reacts with the NO produced in equation 19–15 to give more NO_2

19–18
$$O_3 + NO \longrightarrow NO_2 + O_2$$

The O_3 concentration in significantly polluted air can be several tenths of a part per million. Figure 19–9 gives the concentration of ozone in the air over the Los Angeles area as a function of the time of day. Note that the concentration increases with time and then decreases, consistent with the ozone being of photochemical origin. Based on the combination of equations 19–15, 19–16, and 19–18 to give equation 19–19

19–19
$$NO_2 + O_2 \underset{\Delta}{\overset{\text{uv}}{\rightleftharpoons}} NO + O_3$$

it would be expected that only some steady low concentration of O_3 would exist. However, if NO could be oxidized to NO_2 without the use of O_3, then O_3 could accumulate and participate in other reactions such as equation 19–17. Oxidation of NO to NO_2 can occur without O_3, by using intermediates from reaction 19–17, the sequence of reactions leading to oxidized hydrocarbons. The reactions leading to the production of NO_2 without O_3 as well as reactions 19–15, 19–16, and 19–18 are shown schematically in Figure 19–10. It can be seen that NO can react at two places, that is, with two different intermediates in the hydrocarbon oxidation reaction sequence, to produce NO_2 without O_3 and thus O_3 can accumulate. The accumulated O_3 can react as shown with hydrocarbons.

In addition, O_3 can react with other substances such as rubber and cause it to crack, can act as a general oxidant in other reactions such as with paints, and can be a strong irritant to humans. The formaldehyde, $H_2C{=}O$, produced is also an irritant and furthermore, it can undergo photochemical reactions. Some of the radicals in the hydrocarbon reaction sequence (right hand side of Figure 19–10) can react with SO_2 to ultimately produce an aerosol (Section 12–J) of H_2SO_4

$$SO_2 \xrightarrow[\text{radicals}]{\text{hydrocarbon}} SO_3 \xrightarrow{H_2O} H_2SO_4$$

FIGURE 19–9 Computer-produced maps of air pollution in the Los Angeles area. The grey represents an ozone count below 0.08 parts per million of air; the light brown represents 0.08 to 0.27 ppm; the dark brown represents an ozone concentration greater than 0.27 ppm. (Adapted from the Sunday Press-Enterprise of August 15, 1971, Riverside, California.)

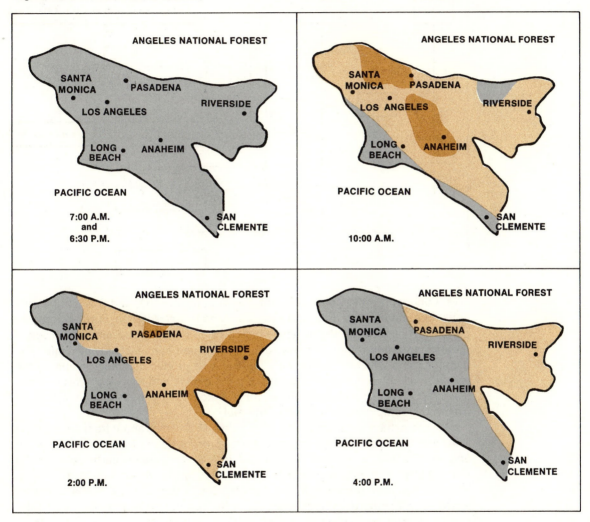

Of course, H_2SO_4 is a strong acid and can damage substances and be an irritant to humans.

Finally, one of the classes of complex organic compounds formed in smog is a peroxyacetyl nitrate (compound B in equation 19–20).

19–20 $CH_3{-}\overset{\displaystyle}{\underset{\displaystyle\underset{O}{\|}}{C}}{-}O{-}O{\cdot}\ +\ NO_2\ \longrightarrow\ CH_3{-}\overset{\displaystyle}{\underset{\displaystyle\underset{O}{\|}}{C}}{-}O{-}O{-}NO_2$

A   B

FIGURE 19–10 Photochemical reactions involved in the production of smog (All possible reactions in the hydrocarbon sequence are not shown.)

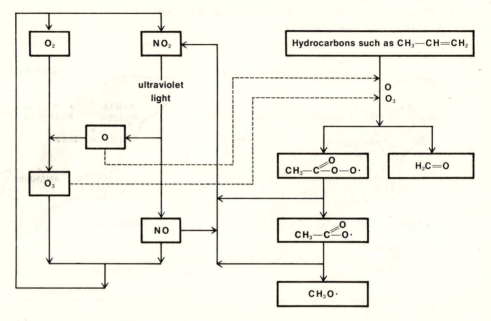

This along with other peroxyacyl nitrates, O_3, and the NO_x and SO_x compounds constitute the principal cause of eye, nose, and throat irritation. Furthermore, the peroxyacyl nitrates are highly active biologically; they can cause crop damage in the concentration region of parts per 10 million and can inactivate enzymes. The hydrocarbons which show the most reactivity in producing smog, including the ability to form the peroxyacylnitrates, are the olefins and aromatic hydrocarbons (Chapter 7). These hydrocarbons occur in automobile exhausts (as one source) from the incomplete combustion of the fuel.

NO is an exhaust product from automobiles. This comes about from the reaction of N_2 and O_2 under the higher temperatures present during fuel combustion in an engine. The NO is very rapidly oxidized to NO_2 which then participates in the primary step in smog production. Of course, NO_2 is then regenerated via equation 19–18. It has been proposed that an excited singlet state of O_2, which is obviously energy-rich compared with ordinary O_2 in the ground state and also has a relatively long lifetime, initially reacts with NO to produce NO_2. Recall that O_2 in its ground state is in a triplet state (Section 6–E). Thus, deactivation from the excited singlet state to the triplet ground state is very slow, since there is a change in multiplicity of states involved (see Section 19–A).

As mentioned earlier, very recent evidence shows that carbon monoxide can also accelerate the production of smog. This occurs via a reaction which causes the production of NO_2. The net reaction is

19–21 $$CO + O_2 + NO \longrightarrow CO_2 + NO_2$$

Both the reaction in equation 19–21 and the one involving excited singlet state O_2 produce more NO_2, which in turn initiate smog production via equation 19–15.

There are already records of deaths occurring when abnormally high levels of pollutants are present in the atmosphere. There is growing evidence that pollutants lead or contribute to chronic respiratory ailments, heart disease, and lung cancer. It obviously is important that some aspiring scientists put their minds and talents to the task of further elucidating the nature of the reactions which create smog and air pollution in general and proposing solutions for the prevention of such reactions.

19–F. RADIATION EFFECTS ON SOME BIOMOLECULES

In this section we shall be concerned with the effects of ultraviolet and higher-energy radiation, such as x rays and γ rays, on molecules important in biological processes.

The ultraviolet irradiation of DNA causes several photochemical alterations of the DNA which are known to have a biological effect. These include the formation of hydration products of constituent bases, dimerization of pyrimidine bases (Section 18–C), local denaturation (change in configuration), chain breaks, and intermolecular cross-linking such as to protein molecules. Several of the alterations in proteins and DNA are shown schematically in Figure 19–11. The most pronounced effects of higher energy x-ray and γ-ray radiation are hydrogen bond breakage, chain breaks, and degradation of the constituent bases. The principal consequences of such damages to DNA are the impairment of the replication of DNA or continued replication but with a high frequency of error in the arrangements of the bases. Thus, it is possible to produce mutants, impair mobility (as in Paramecia), delay or modify cell division, affect cell metabolism, and kill cells. Interestingly enough, it is possible for the damage to be repaired, in vivo (in the living system). Sometimes external influences such as irradiation with a wavelength different from that of the damaging one can cause repair. RNA also undergoes damage upon irradiation, including hydration, dimerization, and protein cross-linking. These result in inactivation of the RNA (or of a whole virus, for example).

Before discussing the photochemistry of proteins, we shall briefly consider the photochemistry of some amino acids which are the components of proteins. Cystine which contains a disulfide linkage, shows photochemical cleavage of the S—S bond:

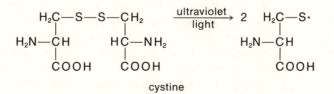

cystine

FIGURE 19-11 A. Photochemical damage to DNA, where S = sugar, P = phosphate group, A = adenine, G = guanine, T = thymine, and C = cytosine. The hydrogen bonds are indicated by dotted lines.

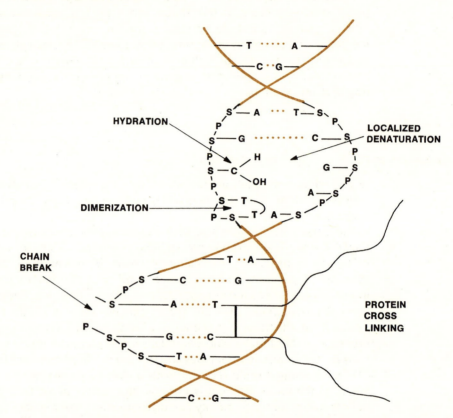

Phenylalanine undergoes substantial photochemical alteration but liberation of CO_2 is dominant (by splitting of the —COOH group on the side chain).

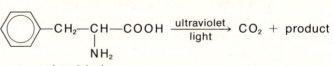

phenylalanine

Other amino acids such as tryptophan, tyrosine, and histidine undergo photo-chemical reactions (when irradiated at 2537 Å), although the degree is particularly small for histidine.

We can categorize the reactions resulting from the irradiation of proteins. These include (1) depolymerization (from chain breaks) as well as polymerization, (2) enzyme deactivation, and (3) denaturation. The peptide bond itself can be broken by absorption of light. Furthermore, if a protein contains cystine, photochemical cleavage, involving mostly the S—S bond, accounts for most of the damage to the protein by irradiation at wavelengths ≥ 2537 Å. The degree of inactivation of

FIGURE 19-11   B. Photochemical damage to proteins.

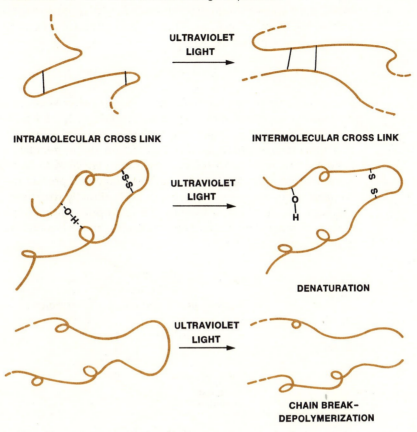

an enzyme containing cystine is approximately proportional to the amount of cystine present. It appears that modification of proteins may be more important in biological inactivation than is modification of DNA.

The main effect of high energy irradiation by x rays and γ rays, and α and β particles is ionization, rather than production of excited states as by irradiation with light. In the ionization process, positive ions and electrons are produced. The ejected electrons can cause further ionization and can be captured. The positive and negative molecular ions produced are highly reactive, especially the positive ones. Often the damage by the irradiation is not done directly upon the solute molecule, but rather there is first decomposition of a solvent whose fragments attack the solute. For example, H_2O in the presence of ionizing radiation gives the following reactions

19-22 $$H_2O + \text{ionizing irradiation} \longrightarrow H\cdot + \cdot OH$$

19-23 $$H_2O + O_2 + \text{ionizing irradiation} \longrightarrow HO_2\cdot + \cdot OH$$

Note that there is a difference between the reaction in the presence (equation 19–23) and absence (equation 19–22) of O_2. Also note that the radicals produced are highly reactive species. The radicals produced from the solvent can then attack a solute molecule, RH

$$3RH + HO_2\cdot \longrightarrow 3R\cdot + 2H_2O$$

and in turn, the R· can further react. The sensitivity of cells to radiation damage increases in the presence of oxygen because the $HO_2\cdot$ radical is so highly oxidative.

An important aspect relative to radiation damage, is radiation protection. Recall that the radicals produced from water are highly reactive and can attack biologically functioning molecules. However, if we were to have present other molecules that could react with the radicals, we could protect the biomolecules from damage. Also, even if a biomolecule were directly damaged, it is possible that a protector molecule could transfer the appropriate atom to the damaged biomolecule and reactivate it. General representations of such reactions are given in equations 19–24 to 19–27:

19–24

$$RH \xrightarrow[\text{radiation}]{\text{high energy}} R\cdot + H\cdot$$

biomolecule (α,β,x- and γ rays)   free radicals   (reaction of biomolecule)

19–25

$$R\cdot \left\langle \begin{array}{l} \xrightarrow{+R\cdot} R\text{---}R \\ \xrightarrow{+O_2} RO_2\cdot \end{array} \right.$$

(cross linking)

(oxidation)

19–26

$$R\cdot + P\text{---}H \longrightarrow RH + P\cdot$$

protector molecule   (restoration of biomolecule)

In the case of equation 19–26 the protector molecule protects from damage by donating a hydrogen atom to restore the directly attacked biomolecule (Equation 19–24) which otherwise would undergo destructive reactions (equation 19–25). Other possibilities exist, including the protector molecule reacting with free radicals such as the ·OH produced from water:

19–27

$$\cdot OH + P\text{---}H \longrightarrow \cdot P\text{---}H(OH)$$

stable unreactive free radical

The reaction of equation 19–27 protects biomolecules from damage originating via the indirect effect of irradiation (not direct attack on biomolecule). Several types of molecules act as protectors, for example

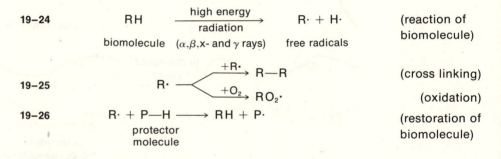

$$HS\text{---}CH_2\text{---}CH_2\text{---}NH_2$$

cysteamine

2-mercaptoethylguanidine

19–G. RADIATION EFFECTS ON SYNTHETIC POLYMERS

In view of the discussion of biopolymers, it will be worthwhile to consider briefly the action of light and ionization radiation on synthetic polymers. In Section 7–N, we discussed some synthetic polymers and gave the structure of some polymers in Table 7–6.

The polymers in Table 7–6 are affected to varying degrees by light ranging from 3000–3800 Å, depending upon the individual structure. This is important since sunlight contains these wavelengths, but at first sight it is unexpected; polymers like polyethylene, polypropylene, and polyvinylchloride are alkane-like compounds and should not absorb light in the region of 3000–3800 Å, but at much shorter wavelengths. We will discuss this shortly.

Ultraviolet and ionization radiation can cause chain breaks, elimination of small molecules, formation of unsaturation in the polymer chain, depolymerization, increased polymerization, and oxidation of the polymer in the presence of O_2. Many of the same reactions also occur for natural polymers such as DNA and proteins (Section 19–F). In most cases where the light should *not* have been absorbed by the polymer, it is quite certain that impurity molecules are present. Often the impurities are only slightly modified polymers, such as

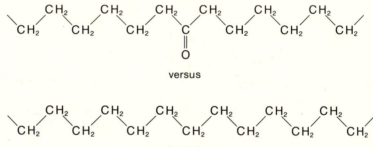

versus

polyethylene

The impurities *are* excited by the light in the 3000 Å–3800 Å region and proceed to attack polymer chains.

Just as in the case of natural polymers, it is possible to protect synthetic polymers from light (and thermal) degradation. In many cases it is desirable to keep the polymers clear and colorless, so the protective compounds which are added (dissolved) absorb only ultraviolet light (and therefore are not colored). Such a compound is

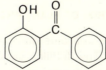

These protector molecules, also called stabilizers, generally have a dual function. They absorb ultraviolet light and thus act as a screening agent to protect the polymer and lower the amount of light absorbed by the harmful impurities. Stabilizers

also can accept energy from a polymer molecule or polymer impurity that may have been excited; the energy is transferred from the polymer molecule or polymer impurity to the stabilizer and degradation of the polymer is prevented. In both cases, the protecting molecule must be able to dissipate the exciting energy so that it will not be destroyed itself. If some degradation of a polymer to radicals does occur (equation 19–28), either by light or heat, other types of protector molecules must be present to react with such radicals to protect the other polymer molecules from attack (equation 19–29).

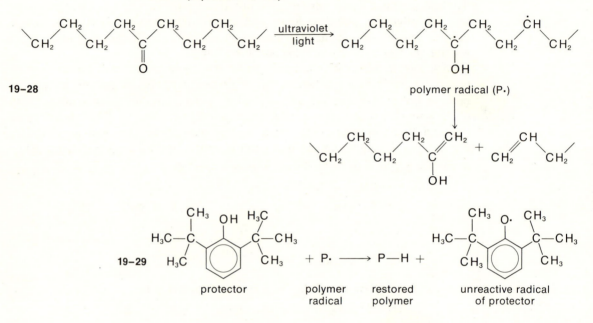

19–28

polymer radical (P·)

19–29

protector polymer radical restored polymer unreactive radical of protector

This is similar in concept to the protection mechanism necessary for biomolecular polymers.

Since so many items, such as outdoor-indoor rugs, are made from synthetic polymers, it is of great importance to find methods to prevent their degradation by sunlight.

19–H. PHYTOCHROME AND CONTROL OF PLANT PROCESSES

Phytochrome is a blue photopigment consisting of a chromophore(s) (dye part) and protein. The detailed structure of phytochrome is not presently known. Two forms of this pigment seem to predominate—one absorbs principally at 6600 Å and the other at 7300 Å. The 6600 Å form can be converted to the 7300 Å form by red light. Also, the 7300 Å form can be changed to the 6600 Å form with far-red light or with heat in the absence of light.

$$19\text{-}30 \qquad P_{6600} \underset{\text{light (7000–8000 Å) or heat}}{\overset{\text{light (5800–7000 Å)}}{\rightleftarrows}} P_{7300}$$

P_{6600} and P_{7300} in equation 19–30 represent the phytochromes with principal absorptions at 6600 Å and 7300 Å. The exciting aspect of phytochromes is that the relative quantities of these forms apparently determine or trigger seed germination, budding, flowering, and other phenomena of botanical importance. When the days are long, the 7300 Å form predominates. However, when the days are short and the nights are long, the increased dark time allows the 6600 Å form to predominate. Of course, these situations in their extreme correspond to summer and winter, and a plant can thereby recognize the season and the changes of seasons by the relative amounts of P_{6600} and P_{7300}, and then respond accordingly by budding and flowering in the summer.

Germination was mentioned above as one factor controlled by phytochromes. For example, lettuce seed kept in the dark will germinate only very poorly ($< 5\,\%$). However, if they are irradiated with 5800–7000 Å light, nearly 100 % of the seeds germinate. On the other hand if seeds are irradiated with 7000–8000 Å light before being allowed to germinate, then only very few germinate. Also, if the seeds originally irradiated with 5800–7000 Å light are re-irradiated with 7000–8000 Å light, then again only very few germinate.

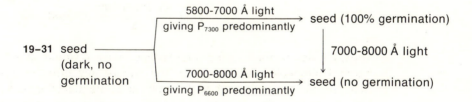

It can be seen from these experiments, as summarized in equation 19–31, that the 7300 Å form is the form responsible for triggering the germination process.

19-I. PHOTOCHROMISM AND THERMOCHROMISM

Photochromism usually involves a reversible photochemical conversion of an uncolored molecule to a colored one.

$$19\text{-}32 \qquad \text{uncolored molecule} \underset{\substack{\text{light}_2 \\ \text{or heat}}}{\overset{\text{light}_1}{\rightleftarrows}} \text{colored molecule}$$

Note that the reverse reaction in equation 19–32 requires light at a different wavelength (as indicated by the subscript) but can be caused by thermal means as well. The process of *thermochromism* is similar, except that heat is used for the forward reaction instead of light.

19–33   uncolored molecule $\underset{-\Delta}{\overset{+\Delta}{\rightleftharpoons}}$ colored molecule

In either equation 19–32 or 19–33 it is possible for *both* the reactant and product molecules to be colored, but with different colors (as the two forms of phytochrome, equation 19–30).

Considerable progress has been made in very recent years with respect to the kinds of molecules that show either photo- or thermochromism, the mechanism of the processes, and the nature of the colored products. The process has many potential applications including photography, micro-imaging, automatic light attenuation (such as automatic sunglasses), and information storage (as in a computer). Recall that the photo- and thermochromic processes occur on a molecular level, whereas conventional photography involves grains of a silver halide of finite size. Thus, the resolution or sharpness of a photochromic film can be increased potentially to the order of molecular dimensions (10 Å).

The intensity of color in the photochromic process depends upon the number of molecules converted and thereby upon the intensity of light that strikes the photochromic material. It is therefore possible to have glasses that would change color to varying degrees depending upon the intensity of light—automatic sunglasses that would be essentially clear indoors but change to a dark color when exposed to outdoor sunlight. Ultraviolet light is usually required for the photochromic process

uncolored $\underset{\Delta}{\overset{\text{ultraviolet light}}{\rightleftharpoons}}$ colored

Thus, glasses containing a photochromic material would specifically screen out harmful ultraviolet radiation from the sun or any other source. Ultraviolet irradiation can cause damage to the retina and potential blindness. Of course, since the glasses become colored, the intensity of visible light reaching the eye also is reduced. Therefore, the photochromic material ultimately performs or serves a dual function.

The use of photochromism or thermochromism for information storage in a computer would be based in part upon the rapid and reversible color change that can be attained with photochromic or thermochromic materials. The change in color supplies "bit" information in the same manner as magnetized iron in a magnetic tape. Also, since the thermo- or photochromic process is a molecular phenomenon, potentially vast amounts of information could be packed into a very small spatial area. Figure 19–12 shows how an entire 1245-page Bible can be recorded on a film only two inches square. For example, *if* a water molecule were photochromic, 18 g or about 1 teaspoon of water would be potentially capable of storing 6×10^{23} "bits" of information.

Table 19–2 lists some molecules exhibiting photochromism (and their substituted derivatives), the nature of the photocolored product, and the color involved. Some derivatives of

chromene

occur in nature, including in leaves. It has been suggested that such compounds may play roles in several different biological processes.

FIGURE 19–12   Photochromic micro-image process illustrated by reproduction of an entire Bible on a film chip which is only two inches square. The pile of film chips represents 1900 books. (Courtesy of the Electronics Division of the National Cash Register Company.)

TABLE 19-2 Some photochromic molecules

PHOTOCHROMIC MOLECULE	PHOTOCOLORED PRODUCT	COLORS

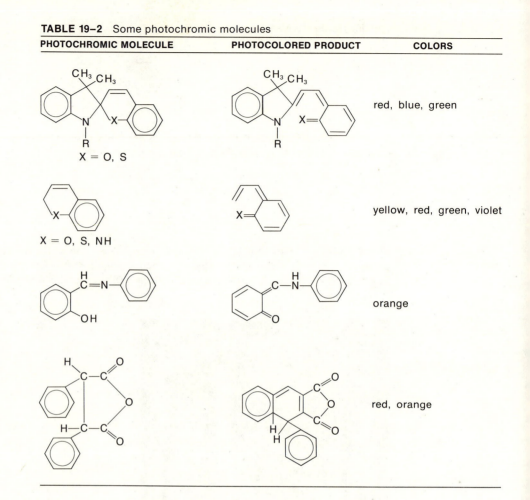

red, blue, green

yellow, red, green, violet

orange

red, orange

APPENDICES

A. THE METRIC SYSTEM

The metric system of mass, length, and volume is much simpler than the English system of measurement since it is a decimal system; that is, larger units are multiples of 10 times some basic unit. An example of a decimal system is our American currency: 10 cents = 1 dime; 10^2 cents = 10 dimes = 1 dollar; 10^3 cents = 10^2 dimes = 10 dollars, etc.

The basic units of mass, length, and volume in the metric system are the *gram* (g), *meter* (m), and *liter*. Conversion factors between these metric units and English units will be given shortly. Multiples or fractions of these basic units are designated by prefixes. Thus, we have the following metric units of weight, length, and volume.

DECIMAL EQUIVALENT	PREFIX (ABBREVIATION)	VOLUME	DISTANCE	MASS
$10^{-6} = 0.000001$	micro- (μ)	microliter (μl)	—	microgram (μg)
$10^{-3} = 0.001$	milli- (m)	milliliter (ml)	millimeter (mm)	milligram (mg)
$10^{-2} = 0.01$	centi- (c)	—	centimeter (cm)	centigram (cg)
$10^{-1} = 0.1$	deci- (d)	—	decimeter (dm)	—
1	*unit*	liter (l)	meter (m)	gram (g)
$10^3 = 1000$	kilo- (k)	—	kilometer (km)	kilogram (kg)

A dash indicates that there is no common term.

It is often necessary to convert metric units to English units or vice versa. Some conversion factors are:

METRIC TO ENGLISH	ENGLISH TO METRIC
1 kilogram = 2.2046 pounds*	1 pound* = 0.4536 kilograms
1 meter = 39.37 inches	1 inch = 2.540 centimeters
1 liter = 1.05671 quarts	1 quart = 0.94633 liters

*The pound is a unit of *weight* in the English system, and these conversion factors properly take into account the conversion to *mass* in kilograms.

Another unit of volume commonly used in the metric system is the cubic centimeter (cm^3 or cc) which is the volume contained in a cube with an edge of 1 cm. The cubic centimeter is equal to 1 milliliter.

$$1 \text{ ml} = 1 \text{ cc} = 1 \text{ cm}^3$$

These units are commonly used interchangeably. Similarly,

$$1 \text{ liter} = 1,000 \text{ cc}$$

There are many different types of units with which it is convenient to express energy. Frequently, it is necessary to convert from one energy unit to another and the following table of conversion factors has been constructed for that purpose. The use of the table can be understood from the following example. To find the conversion factor from liter-atm/mole to kcal/mole, we look in the left-hand column for 1 liter-atm/mole. The conversion factor, 0.024218, lies on that line under the column heading kcal/mole.

$$1 \text{ liter-atm/mole} = 0.024218 \text{ kcal/mole}$$

	liter-atm/mole	kcal/mole	erg/mole	eV/molecule
1 liter-atm/mole	1	0.024218	1.013×10^9	1.050×10^{-3}
1 kcal/mole	41.29	1	4.184×10^{10}	4.338×10^{-2}
1 erg/mole	9.869×10^{-10}	2.390×10^{-11}	1	1.036×10^{-12}
1 e.v./molecule	952.2	23.05	9.648×10^{11}	1

There are also several conversion factors within the English system of measurement that are frequently useful.

$$1 \text{ pound} = 16 \text{ ounces}$$

$$1 \text{ ton (short)} = 2,000 \text{ pounds}$$

$$1 \text{ mile} = 5,280 \text{ feet}$$

B. DIMENSIONAL ANALYSIS AND CONVERSION OF UNITS

Probably the most significant difference between working problems in mathematics and in the physical sciences such as chemistry is the units or dimensions associated with most numbers. In solving an algebraic equation in mathematics, you work with pure numbers and your answer is a dimensionless number. However, in the physical sciences, most numbers have come from experimental measurements and each measurement generally is in units. For example, the measurement may be velocity and the units can be miles/hour, feet/second, meters/second or centimeters/second.

When you are solving an algebraic equation in a chemical problem, it is essential that you write down the units for each quantity in the equation. You may cancel or combine units in a manner similar to the normal mathematical treatment of numbers. For example, the equation for the kinetic energy of an object of mass m moving at a velocity v is

$$KE = \tfrac{1}{2}mv^2$$

Suppose an object weighs 5.0 g and travels at a speed of 300 cm/sec. The kinetic energy of this object is

$$KE = \tfrac{1}{2}(5 \text{ g})(300 \text{ cm/sec})^2$$

When we square the 300, we must also square the units of velocity to give cm^2/sec^2.

$$KE = \frac{5 \times 300^2}{2} \frac{g \text{ } cm^2}{sec^2}$$

$$KE = 225{,}000 \frac{g \text{ } cm^2}{sec^2} \text{ or } 225{,}000 \text{ g } cm^2 sec^{-2}$$

Note that sec^2 in the denominator can be written in the numerator with a power of -2 since

$$\frac{1}{sec^2} = sec^{-2}$$

The combination of units g cm^2/sec^2 is defined as an erg

$$1 \text{ erg} = \frac{g \text{ } cm^2}{sec^2}$$

so the kinetic energy of the object can be expressed as

$$KE = 225{,}000 \text{ ergs}$$

If, on the other hand, we are given the energy and mass of a moving object, we can calculate the velocity, v. The units of v will depend on the units of m and of *KE*. Suppose the kinetic energy is 500 ergs and the mass 10 g. The calculation of *v* and the dimensional analysis will be as follows

$$KE = \tfrac{1}{2}mv^2$$

$$500 \text{ ergs} = (\tfrac{1}{2})(10 \text{ g}) \, v^2$$

$$v^2 = \frac{2(500)}{10} \frac{\text{ergs}}{\text{g}} = \frac{2(500)}{10} \frac{\cancel{g} \, \text{cm}^2/\text{sec}^2}{\cancel{g}}$$

$$v^2 = 100 \text{ cm}^2/\text{sec}^2$$

$$v = 10 \text{ cm/sec}$$

In order to convert from one unit or set of units to another, we use what is called the *factor method*. In Appendix A, we gave several conversion factors which stated the equivalence between two different units or dimensions. For example,

$$1 \text{ ft} = 12 \text{ in}$$

$$1 \text{ mile} = 5,280 \text{ ft}$$

$$1 \text{ hr} = 60 \text{ min}$$

$$1 \text{ min} = 60 \text{ sec}$$

$$1 \text{ in} = 2.54 \text{ cm}$$

In the factor method, a change in units is accomplished by multiplying by a ratio derived from the appropriate conversion factor. For example, if we wished to convert 54 inches to feet, we multiply the 54 inches by the ratio 1 ft/12 in and cancel the inches

$$\left(54 \text{ in}\right)\left(\frac{1 \text{ ft}}{12 \text{ in}}\right) = \frac{54}{12} \text{ ft} = 4.5 \text{ ft}$$

Note that the ratio 1 ft/12 in was formed rather than 12 in/1 ft. The proper choice of ratio is confirmed by the correct cancellation of units. If the inverse ratio is used, the units will not cancel, and immediately you know that the other ratio must be used. For example, in the above conversion of 54 inches to ft., if we had multiplied by the ratio 12 in/1 ft the result would be

$$\left(54 \text{ in}\right)\left(12 \frac{\text{in}}{1 \text{ ft}}\right) = (54)(12) \frac{\text{in}^2}{\text{ft}}$$

Obviously we did not want the units inches²/foot. Writing down the units for any number is therefore very important.

Another example will show how more complex units, such as for velocity, can be converted using this same procedure. Suppose we wished to convert a velocity of 60.0 miles/hr to cm/sec.

$$\left(\frac{60.0 \text{ miles}}{\text{hr}}\right)\left(\frac{5,280 \text{ ft}}{1 \text{ mile}}\right)\left(\frac{12 \text{ in}}{1 \text{ ft}}\right)\left(\frac{2.54 \text{ cm}}{1 \text{ in}}\right)\left(\frac{1 \text{ hr}}{60 \text{ min}}\right)\left(\frac{1 \text{ min}}{60 \text{ sec}}\right) = 2,680 \frac{\text{cm}}{\text{sec}}$$

The reason for expressing the numerical answer as 2,680 rather than as the more precise 2,682.24 will be discussed in Appendix E.

C. EXPONENTS

Exponents may be used to express very large or small numbers. For example, it is very awkward to write the number 100,000,000,000,000., and to keep account of the 14 zeros between the decimal point and the one. A more concise expression of this number is the exponential form

$$\underbrace{100{,}000{,}000{,}000{,}000.}_{14 \text{ places}} = 1 \times 10^{14}$$

The exponent of 14 was obtained by counting the number of places the decimal point was moved to the left until it was next to the one. Most commonly numbers with an exponential component are put in the *normal form*, that is, with the decimal point to the right of the first non-zero digit (X.XXX). For example, the following number is written in *normal form*

$$56{,}800{,}000. = 5.68 \times 10^7$$

where the decimal point has been moved to the left 7 places.

For writing fractional numbers with an exponent, the same procedure is used except that the decimal point is moved to the right and the exponent is a negative. For example,

$$0.000000367 = 3.67 \times 10^{-7}$$

where the decimal point has been moved 7 places to the right.

You will also have occasion to solve algebraic equations utilizing numbers with exponential components. In order to do this, you will need to know certain basic rules of multiplication and division, summarized in the following general equations.

$$\frac{1}{10^a} = 10^{-a}$$

$$10^a \times 10^b = 10^{a+b}$$

$$\frac{10^a}{10^b} = 10^{a-b}$$

A few examples utilizing these general equations are as follows

$$\frac{1}{2.5 \times 10^3} = \left(\frac{1}{2.5}\right) \times 10^{-3} = 0.4 \times 10^{-3} = 4.0 \times 10^{-4}$$

$$(5 \times 10^5) \times (8.4 \times 10^2) = 42 \times 10^7 = 4.2 \times 10^8$$

$$\frac{12.2 \times 10^3}{2.0 \times 10^{14}} = \frac{12.2}{2.0} \times 10^{(3-14)} = 6.1 \times 10^{-11}$$

D. LOGARITHMS

The logarithm is useful in chemistry to simplify certain equations. In order to do this, it is necessary that you know the properties of logarithms and be able to find logarithms of numbers. There are two types of logarithms: *common logarithms* (log or $\log_{10}$) using a base 10 and *natural logarithms* (ln) using a base $e = 2.718\cdots$. *A logarithm of a number x is simply the exponent to which the designated base (10 or e) must be raised in order to equal that number.*

$$x = 10^{\log x} \qquad\qquad x = e^{\ln x}$$

common logarithm natural logarithm

For example, the logarithms of some numbers are

$\log 1 = 0; 1 = 10^0$ $\ln 1 = 0; 1 = e^0$
$\log 5 = 0.699; 5 = 10^{0.699}$ $\ln 2.718\cdots = 1; 2.718\cdots = e^1$
$\log 10 = 1.0; 10 = 10^1$ $\ln 5 = 1.609; 5 = e^{1.609}$
$\log 50 = 1.699; 50 = 10^{1.699}$ $\ln 50 = 3.913; 50 = e^{3.93}$
$\log 100 = 2.0; 100 = 10^2$
$\log 1,000 = 3.0; 1,000 = 10^3$

The natural logarithm is related to the common logarithm by

$$\ln x = 2.303 \log x$$

Tables of natural logarithms are not as available as those of common logarithms so it is a common procedure to find natural logarithms by looking up common logarithms in a table and converting by the previous equation.

Because logarithms are exponents, it follows that the logarithm of the product of two numbers is the sum of the logarithms of the numbers.

D–1 $$\log xy = \log x + \log y$$

D–2 $$\ln xy = \ln x + \ln y$$

According to the list above log 50 = 1.699. By equation D–1, we have

$$\log 50 = \log (5 \times 10) = \log 5 + \log 10$$
$$= 0.699 + 1.00 = 1.699$$

The procedure for taking the common logarithm of a number is first to put the number into its normal form (Appendix C) — a number X.XXX times an exponential term — and then to take the logarithm of this product. A table of common logarithms is given for the range 0 to 10 in Appendix U. For example, suppose we wanted the logarithm of 56,800,000.

$$\log (56{,}800{,}000) = \log (5.68 \times 10^7) = \log 5.68 + \log 10^7$$

The logarithm of 5.68 is found from the table to be 0.7543 and

$$\log (56{,}800{,}000) = 0.7543 + 7.00 = 7.7543$$

If a number is less than one, its logarithm is negative; nevertheless, the procedure for taking the logarithm is the same. For example, we will take the logarithm of 0.000000367.

$$\log (0.000000367) = \log (3.67 \times 10^{-7}) = \log 3.67 + \log 10^{-7}$$
$$= (0.5647) - 7.00$$
$$= -6.4353$$

The reverse procedure of finding a number when the logarithm of the number is known is called taking the antilogarithm. The procedure in this case is to break the logarithm up into two parts, one of which must consist of a *positive* number less than one in the form of decimal point followed by digits and the second a whole number, either positive or negative. The antilogarithm is then found by taking the antilogarithm of the two parts, the product of which gives the desired answer. We will illustrate this by first using the logarithms previously obtained. In the first example,

$$\log x = 7.7543$$

This can be broken into the sum of the whole number 7.00 and the decimal 0.7543.

$$\log x = 0.7543 + 7.00$$

Taking the antilogarithm of both numbers and multiplying them, we have

$$x = \text{antilog } (0.7543) \times \text{antilog } (7.000)$$
$$x = 3.67 \times 10^7$$

Similarly for the negative logarithm −6.4353

$$\log x = -6.4353$$

We break this up into a sum of the next highest negative whole number, −7.0, and a positive number 0.5647.

$$\log x = 0.5647 - 7.0$$

$$x = \text{antilog } (0.5647) \times \text{antilog } (-7.0)$$

$$x = 5.67 \times 10^{-7}$$

Other examples of taking antilogarithms are

$$\log x = -2.370 = +0.630 - 3.00$$
$$x = 4.27 \times 10^{-3}$$

$$\log x = -0.757 = +0.243 - 1.00$$
$$x = 1.75 \times 10^{-1} = 0.175$$

$$\log x = 0.917 \text{ (already in proper form)}$$
$$x = 8.26$$

$$\log x = 5.917 = 0.917 + 5.0$$
$$x = 8.26 \times 10^{5}$$

E. ERRORS AND SIGNIFICANT FIGURES

All experimental measurements have errors associated with them as a consequence of the errors in the equipment being used to make the measurement. The error is usually expressed by a ±(number) following the actual value of the measurement. For example, we might measure the length of a glass tube and find it to be 31.28 cm with an error of ±0.2 cm. We would write the number as 31.28 ± 0.2 cm. Since there is error in the tenths digit, the 8 in the hundredths position has no real significance and we round off the number to 31.3. The number written in this manner is meant to suggest that there is error in the last digit. Since there are three digits in the number 31.3 we say that this number contains three *significant figures*.

The number 0.0042 has error in the fourth position beyond the decimal point and we say that this number has two significant figures. When the last digit in the number is a zero but it occurs to the right of the decimal point, this zero is considered to be a significant figure. For example, the number 0.004200 has four significant figures. The last two zeros in this number are not necessary to specify the magnitude of the number but rather to specify that the error is in the sixth position beyond the decimal point. If the last digit in a number is a zero to the left of the decimal point, you cannot be certain if this zero constitutes a significant figure. For example the number 550 may have error in the third position or the second position and we do not know whether there are two or three significant figures. Alternatively, if we write the number in *normal form* we can designate the proper number of significant figures:

two significant figures 5.5×10^2

three significant figures 5.50×10^2

When carrying out a multiplication or division it is important that the answer be written with the proper number of significant figures. The rule to follow is: *Round off the result of either multiplication or division to contain the number of significant figures equal to the term containing the least number of significant figures.* For example, the product of 5.5×10^2 times 0.004200 should contain only two significant figures since 5.5×10^2 has only two significant figures.

$$(5.5 \times 10^2)(0.004200) = 2.3$$

Similarly the quotient of these two numbers has only two significant figures.

$$\frac{5.5 \times 10^2}{0.004200} = 1.3 \times 10^5$$

The rule to follow in addition or subtraction is: *The result of either addition or subtraction should contain the same number of decimal places as the least reliable number.* For example, suppose we add the numbers

36.8
4.420
+ 7.89
──────
49.110

The least reliable number is 36.8 with error in the tenths position, so the answer should be rounded off to the tenths position, 49.1. Similarly, in subtraction

54.873
− 3.40
──────
51.473

The least reliable number is 3.40 and the answer should be rounded off to the hundredths position, 51.47.

F. TECHNIQUES OF GRAPHING—EVALUATION OF LINEAR FUNCTIONS

Frequently in chemistry we wish to graph experimental data to see if the data satisfy an equation or to evaluate certain constants in the equation. The equation may have been developed on the basis of some theory and we wish to test the validity of the equation. For example, the data might be experimental vapor pressure measurements, p, measured at various absolute temperatures, T. An equation can be derived which relates these variables

F–1
$$p = Ae^{-\Delta H/RT}$$

where A, ΔH, and $R = 1.987 \times 10^{-3}$ kcal/deg-mole are constants. If the experimental values of p and T fit this equation, then the next step is to evaluate one or more of the constants, such as A and ΔH.

In order to both test the data to see if they satisfy the equation and to evaluate some of the constants in the equation, we try to rearrange the equation into a *linear* form (straight line). A general equation for a linear function is

F–2
$$y = ax + b$$

where $a = slope$, $b = y\text{-}intercept$ ($x = 0$). For example, the equation F–1 relating p and T can be put into a linear form by taking the natural logarithm of both sides of the equation

$$\ln p = \ln A + \ln e^{-\Delta H/RT}$$

$$\ln p = \frac{-\Delta H}{RT} + \ln A$$

Since it is more convenient to work with logarithms to the base 10, we use the conversion $\ln x = 2.303 \log x$, and the equation becomes

$$2.303 \log p = - \frac{\Delta H}{RT} + 2.303 \log A$$

Rearranging we obtain the linear function where the variables are $y = \log p$ and $x = 1/T$

F–3
$$\underbrace{\log p}_{y} = - \underbrace{\frac{\Delta H}{2.303\,R}}_{a} \underbrace{\frac{1}{T}}_{x} + \underbrace{\log A}_{b}$$

The slope and intercept are thus identified as

F–4
$$a = - \frac{\Delta H}{2.303\,R}$$

F–5
$$b = \log A$$

The constants ΔH and A can then be evaluated from the slope, a, and intercept, b.

A graph is constructed by first calculating from the experimental data the designated y and x to be graphed. For the experimental data of vapor pressure, p, and absolute temperature, T, we calculate log p and $1/T$ as shown in the following table:

$T(°K)$	$x = 1/T$ (deg^{-1})	p(torr)	$y = \log p$
269.3	3.73×10^{-3}	1.00	0.00
306.1	3.266×10^{-3}	10.0	1.00
333.9	2.994×10^{-3}	40.0	1.60
356.3	2.806×10^{-3}	100.0	2.00
397.1	2.518×10^{-3}	400.0	2.60
419.0	2.386×10^{-3}	760.0	2.88

We must now examine the data to see the range of the x and y values. Note that y goes from zero to 2.88 (a range of 2.88) and x from 2.386×10^{-3} to 3.713×10^{-3} (corresponding to a range of 1.327×10^{-3}). If we have graph paper with 18 major divisions along one axis and 25 divisions along the other we can conveniently use the 18 divisions for the x-axis with each division equal to 0.1×10^{-3} units. The axis with the 25 divisions can be used to represent the y-axis with each division equal to 0.2 units. The graph of log p versus $1/T$ will be as follows:

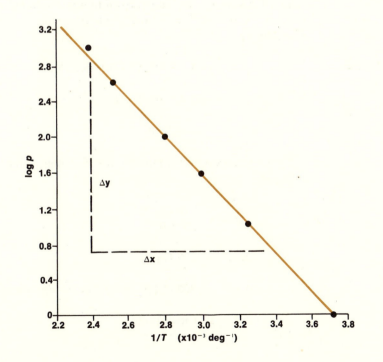

The *slope* of a straight line is equal to the ratio of the change in $y(\Delta y)$ for a corresponding change in $x(\Delta x)$. For example, on our graph Δx corresponds to the change from 2.4×10^{-3} deg^{-1} to 3.4×10^{-3} deg^{-1} or

$$\Delta x = (3.4 - 2.4) \times 10^{-3} \text{ deg}^{-1} = 1.0 \times 10^{-3} \text{ deg}^{-1}$$

The corresponding y values are 2.89 and 0.69 and the change in y is

$$\Delta y = 0.69 - 2.89 = -2.20$$

The slope is thus

$$\text{slope} = \frac{\Delta y}{\Delta x} = \frac{-2.20}{1.0 \times 10^{-3} \text{ deg}^{-1}} = -2.20 \times 10^3 \text{ deg}$$

If the x-axis goes to zero on the graph, the intercept can be found directly from the y value on the graph where $x = 0$. However, in the previous graph of log p versus $1/T$ the $1/T$ axis does not go to zero. In this case the intercept, b, must be calculated from the linear equation

$$y = ax + b$$

We have found the slope a to be -2.20×10^3 deg. If one data point of x and y is read off the graph, we can substitute into the linear equation and solve for b. At $x = 1/T = 3.0 \times 10^{-3}$ deg^{-1} the value of $y = $ log $p = 1.58$ and b can be calculated accordingly.

$$1.58 = (-2.20 \times 10^3 \text{ deg})(3.0 \times 10^{-3} \text{ deg}^{-1}) + b$$

$$1.58 = -6.60 + b$$

$$b = 8.18$$

The constants A and ΔH in equation F–1 can then be determined using equation F–4 and F–5.

$$a = - \frac{\Delta H}{2.303 \, R}$$

$$\Delta H = -2.303 \, R \, (a)$$

$$\Delta H = -2.303 \, (1.987 \times 10^{-3} \text{ kcal/deg-mole})(-2.20 \times 10^3 \text{ deg})$$

$$\Delta H = 10.1 \text{ kcal/mole}$$

and

$$b = \log A$$

$$A = \text{antilog } b$$

$$A = \text{antilog } 8.18 = (\text{antilog } 0.18) \times (\text{antilog } 8.0)$$

$$A = 1.51 \times 10^8$$

G. CENTIGRADE AND FAHRENHEIT TEMPERATURE

In science the centigrade (Celsius) temperature scale is used universally and will be the one employed throughout this text. However, in many countries, such as the United States, the temperature in weather reporting and in household appliances is expressed on the Fahrenheit scale. For this reason we must be able to convert temperature from one temperature scale to the other.

The conversion between temperature scales can be understood quite readily by noting two defined temperatures on the two scales. These two temperatures are: (1) the freezing point of water which is assigned a value of 0 degrees on the centigrade scale (0 °C) and 32 degrees on the Fahrenheit scale (32 °F); (2) the normal boiling point of water which is assigned a value of 100 degrees on the centigrade scale (100 °C) and 212 degrees on the Fahrenheit scale (212 °F). These points are shown on the two thermometers below:

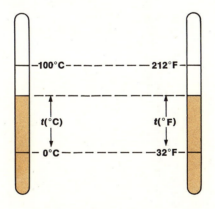

The conversion between scales is derived by setting up the proportion

$$\frac{t(°C) - 0}{100 - 0} = \frac{t(°F) - 32}{212 - 32}$$

The equation used to convert °F to °C is obtained by solving for $t(°C)$.

$$t(°C) = \left(\frac{100-0}{212-32}\right)[t(°F) - 32] + 0$$

G-1
$$t(°C) = \frac{5}{9}[t(°F) - 32]$$

Similarly the equation used to convert °C to °F is obtained by solving for $t(°F)$

$$\left(\frac{212-32}{100-0}\right)[t(°C) - 0] = t(°F) - 32$$

G-2
$$t(°F) = \frac{9}{5}t(°C) + 32$$

H. ELECTRONIC CONFIGURATIONS OF THE ELEMENTS

In the following table a question mark indicates that the electronic configuration which is given could be in error.

Z	ELEMENT	1	2		3			4				5				6				7
		s	s	p	s	p	d	s	p	d	f	s	p	d	f	s	p	d	f	s
1	H	1																		
2	He	2																		
3	Li	2	1																	
4	Be	2	2																	
5	B	2	2	1																
6	C	2	2	2																
7	N	2	2	3																
8	O	2	2	4																
9	F	2	2	5																
10	Ne	2	2	6																
11	Na	2	2	6	1															
12	Mg	2	2	6	2															
13	Al	2	2	6	2	1														
14	Si	2	2	6	2	2														
15	P	2	2	6	2	3														
16	S	2	2	6	2	4														
17	Cl	2	2	6	2	5														
18	Ar	2	2	6	2	6														
19	K	2	2	6	2	6		1												
20	Ca	2	2	6	2	6		2												
21	Sc	2	2	6	2	6	1	2												

Z	ELEMENT	1	2		3			4				5				6				7
		s	s	p	s	p	d	s	p	d	f	s	p	d	f	s	p	d	f	s
22	Ti	2	2	6	2	6	2	2												
23	V	2	2	6	2	6	3	2												
24	Cr	2	2	6	2	6	5	1												
25	Mn	2	2	6	2	6	5	2												
26	Fe	2	2	6	2	6	6	2												
27	Co	2	2	6	2	6	7	2												
28	Ni	2	2	6	2	6	8	2												
29	Cu	2	2	6	2	6	10	1												
30	Zn	2	2	6	2	6	10	2												
31	Ga	2	2	6	2	6	10	2	1											
32	Ge	2	2	6	2	6	10	2	2											
33	As	2	2	6	2	6	10	2	3											
34	Se	2	2	6	2	6	10	2	4											
35	Br	2	2	6	2	6	10	2	5											
36	Kr	2	2	6	2	6	10	2	6											
37	Rb	2	2	6	2	6	10	2	6			1								
38	Sr	2	2	6	2	6	10	2	6			2								
39	Y	2	2	6	2	6	10	2	6	1		2								
40	Zr	2	2	6	2	6	10	2	6	2		2								
41	Nb	2	2	6	2	6	10	2	6	4		1								
42	Mo	2	2	6	2	6	10	2	6	5		1								
43	Tc	2	2	6	2	6	10	2	6	6		1?								
44	Ru	2	2	6	2	6	10	2	6	7		1								
45	Rh	2	2	6	2	6	10	2	6	8		1								
46	Pd	2	2	6	2	6	10	2	6	10										
47	Ag	2	2	6	2	6	10	2	6	10		1								
48	Cd	2	2	6	2	6	10	2	6	10		2								
49	In	2	2	6	2	6	10	2	6	10		2	1							
50	Sn	2	2	6	2	6	10	2	6	10		2	2							
51	Sb	2	2	6	2	6	10	2	6	10		2	3							
52	Te	2	2	6	2	6	10	2	6	10		2	4							
53	I	2	2	6	2	6	10	2	6	10		2	5							
54	Xe	2	2	6	2	6	10	2	6	10		2	6							
55	Cs	2	2	6	2	6	10	2	6	10		2	6			1				
56	Ba	2	2	6	2	6	10	2	6	10		2	6			2				
57	La	2	2	6	2	6	10	2	6	10		2	6	1		2				
58	Ce	2	2	6	2	6	10	2	6	10	2	2	6			2?				
59	Pr	2	2	6	2	6	10	2	6	10	3	2	6			2?				
60	Nd	2	2	6	2	6	10	2	6	10	4	2	6			2				
61	Pm	2	2	6	2	6	10	2	6	10	5	2	6			2?				
62	Sm	2	2	6	2	6	10	2	6	10	6	2	6			2				

Z	ELEMENT	1	2		3			4				5				6				7
		s	s	p	s	p	d	s	p	d	f	s	p	d	f	s	p	d	f	s
63	Eu	2	2	6	2	6	10	2	6	10	7	2	6			2				
64	Gd	2	2	6	2	6	10	2	6	10	7	2	6	1		2				
65	Tb	2	2	6	2	6	10	2	6	10	9	2	6			2?				
66	Dy	2	2	6	2	6	10	2	6	10	10	2	6			2?				
67	Ho	2	2	6	2	6	10	2	6	10	11	2	6			2?				
68	Er	2	2	6	2	6	10	2	6	10	12	2	6			2?				
69	Tm	2	2	6	2	6	10	2	6	10	13	2	6			2				
70	Yb	2	2	6	2	6	10	2	6	10	14	2	6			2				
71	Lu	2	2	6	2	6	10	2	6	10	14	2	6	1		2				
72	Hf	2	2	6	2	6	10	2	6	10	14	2	6	2		2				
73	Ta	2	2	6	2	6	10	2	6	10	14	2	6	3		2				
74	W	2	2	6	2	6	10	2	6	10	14	2	6	4		2				
75	Re	2	2	6	2	6	10	2	6	10	14	2	6	5		2				
76	Os	2	2	6	2	6	10	2	6	10	14	2	6	6		2				
77	Ir	2	2	6	2	6	10	2	6	10	14	2	6	7		2				
78	Pt	2	2	6	2	6	10	2	6	10	14	2	6	9		1				
79	Au	2	2	6	2	6	10	2	6	10	14	2	6	10		1				
80	Hg	2	2	6	2	6	10	2	6	10	14	2	6	10		2				
81	Tl	2	2	6	2	6	10	2	6	10	14	2	6	10		2	1			
82	Pb	2	2	6	2	6	10	2	6	10	14	2	6	10		2	2			
83	Bi	2	2	6	2	6	10	2	6	10	14	2	6	10		2	3			
84	Po	2	2	6	2	6	10	2	6	10	14	2	6	10		2	4?			
85	At	2	2	6	2	6	10	2	6	10	14	2	6	10		2	5?			
86	Rn	2	2	6	2	6	10	2	6	10	14	2	6	10		2	6			
87	Fr	2	2	6	2	6	10	2	6	10	14	2	6	10		2	6			1
88	Ra	2	2	6	2	6	10	2	6	10	14	2	6	10		2	6			2
89	Ac	2	2	6	2	6	10	2	6	10	14	2	6	10		2	6	1		2?
90	Th	2	2	6	2	6	10	2	6	10	14	2	6	10		2	6	2		2
91	Pa	2	2	6	2	6	10	2	6	10	14	2	6	10	2	2	6	1		2?
92	U	2	2	6	2	6	10	2	6	10	14	2	6	10	3	2	6	1		2
93	Np	2	2	6	2	6	10	2	6	10	14	2	6	10	4	2	6	1		2?
94	Pu	2	2	6	2	6	10	2	6	10	14	2	6	10	6	2	6			2?
95	Am	2	2	6	2	6	10	2	6	10	14	2	6	10	7	2	6			2?
96	Cm	2	2	6	2	6	10	2	6	10	14	2	6	10	7	2	6	1		2?
97	Bk	2	2	6	2	6	10	2	6	10	14	2	6	10	8	2	6	1		2?
98	Cf	2	2	6	2	6	10	2	6	10	14	2	6	10	10	2	6			2?
99	Es	2	2	6	2	6	10	2	6	10	14	2	6	10	11	2	6			2?
100	Fm	2	2	6	2	6	10	2	6	10	14	2	6	10	12	2	6			2?
101	Md	2	2	6	2	6	10	2	6	10	14	2	6	10	13	2	6			2?
102	No	2	2	6	2	6	10	2	6	10	14	2	6	10	14	2	6			2?
103	Lr	2	2	6	2	6	10	2	6	10	14	2	6	10	14	2	6	1		2?

I. COULOMB'S LAW

The coulomb potential energy of attraction between monocharged ions in an ion pair molecule is

I–1 $$\text{coulomb energy} = -\frac{(e)(e)}{r}$$

where e is equal to the electronic charge of 4.80×10^{-10} esu and r is the distance in centimeters. Since

I–2 $$\text{esu} = g^{1/2}\,cm^{3/2}\,sec^{-1}$$

the units for coulomb energy are

$$\frac{\text{coulomb energy}}{\text{per ion pair}} = \frac{(g^{1/2}\,cm^{3/2}\,sec^{-1})^2}{cm} = \frac{g\,cm^3\,sec^{-2}}{cm}$$

I–3 $$\frac{\text{coulomb energy}}{\text{per ion pair}} = \frac{g\,cm^2}{sec^2} = erg$$

To obtain the coulomb energy for 1 mole of ion pair molecules we must multiply the right hand side of equation I–1 by Avogadro's number N where $N = 6.02 \times 10^{23}$ molecules per mole. Furthermore for ions of arbitrary charge n_+ and n_-, we must multiply the electronic charge (e) by this charge, finally giving

6–2 $$\text{coulomb energy (ergs/mole)} = -N\frac{(n_+e)(n_-e)}{r}$$

J. WORK OF EXPANSION OR COMPRESSION

We can visualize the work of expansion or compression by considering a gas contained by a piston in a cylinder.

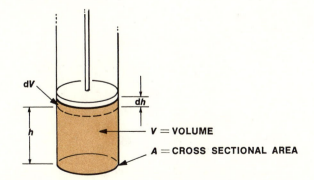

The cross sectional area of the cylinder is represented by A and the volume of the gas is then the product of the height of the piston, h, times A

$$V = Ah$$

If the gas is expanded by raising the piston a small distance, dh, the change in volume, dV, is given by

$$dV = A\,dh$$

or

$$dh = dV/A$$

By definition, pressure is the force per unit area. Therefore the pressure created by the piston (the external pressure) is the force, f, divided by the cross-sectional area, A

$$P_{ex} = f/A$$

or

$$f = P_{ex}\,A$$

Since the change in mechanical work is the product of the force times a displacement in the direction of the force,

$$dW = f\,dh$$

Substituting our previous expressions for f and dh into the equation for dW, we obtain

$$dW = (P_{ex}\,A)(dV/A)$$

$$dW = P_{ex}\,dV$$

If the external pressure is held constant during the expansion, the differential expression for work can be integrated from the initial volume V_1 to the final volume V_2

$$\int dW = \int_{V_1}^{V_2} P_{ex}\,dV = P_{ex} \int_{V_1}^{V_2} dV$$

$$W = P_{ex}\,(V_2 - V_1)$$

10–3 $\qquad\qquad W = P_{ex}\,\Delta V \qquad\qquad\qquad (P_{ex} = \text{constant})$

where the $\Delta V = (V_2 - V_1) = $ change in volume.

K. ISOTHERMAL, REVERSIBLE WORK FOR AN IDEAL GAS

The equation for work given in equation 10–3 is restricted to a process at constant pressure. In the present derivation we do not want this restriction and must start with the general definition of work in the differential form

$$dW = P_{ex}\, dV$$

For a reversible process the external pressure must equal the pressure of the gas at all times

$$P_{ex} = P_g = P$$

and

$$dW = P\, dV$$

Since the gas is assumed to be an ideal gas, then by the ideal gas equation

$$PV = nRT$$

$$P = \frac{nRT}{V}$$

and

$$dW = \frac{nRT}{V}\, dV$$

In order to find the work, W, we must integrate this expression from V_1 to V_2

$$dW = \int_{V_1}^{V_2} \frac{nRT}{V}\, dV$$

When the temperature is held constant (isothermal), T can be taken out of the integral sign along with the constants n and R. Thus

$$dW = nRT \int_{V_1}^{V_2} \frac{dV}{V}$$

10–4
$$W = nRT \ln \frac{V_2}{V_1}$$

L. RELATIONSHIP BETWEEN FREE ENERGY AND PRESSURE

From the discussion of thermodynamics in Chapter 10, we can derive (as below) a very important relationship between the free energy and pressure of an ideal gas.

From Chapter 11 we know that the free energy of a liquid is equal to the free energy of a gas in equilibrium with the liquid. This then allows us to obtain a relationship between the free energy of the liquid and the vapor pressure.

Let us consider the reversible expansion or compression of an ideal gas at constant temperature. Since the change is carried out at constant temperature, the change in free energy is

10–23
$$\Delta G = \Delta H - T\Delta S$$

For an ideal gas, the enthalpy (*H*) depends only on temperature and not on pressure. (An illustration of this was given in Example 10–4.) Both ΔH_{ADB} from Example 10–4 and ΔH_{AEB} from Problem 10–7 are zero, and in both cases the initial temperature equals the final temperature. Since the process is carried out isothermally, ΔH is zero, and

$$\Delta G = -T\Delta S$$

An expression for ΔS for a reversible isothermal expansion of an ideal gas was derived in Section 10–G. We will replace the final volume ($V_1 + V_2$) with V_f and the initial volume V_1 with V_i.

$$\Delta S = \frac{Q_{rev}}{T} = nR \ln \frac{V_f}{V_i}$$

The change in free energy for this process is then

L–1
$$\Delta G = -T\Delta S = -n\,RT \ln \frac{V_f}{V_i}$$

Usually we want this equation to be expressed in terms of pressure. Because the process is isothermal (T = constant), the ideal gas law gives

$$P_i V_i = n\,RT = P_f V_f$$

or

$$\frac{V_f}{V_i} = \frac{P_i}{P_f}$$

Substitution into equation L–1 gives

L–2
$$\Delta G = -n\,RT \ln \frac{P_i}{P_f}$$

Since

$$-\ln x = \ln \frac{1}{x}$$

then equation L–2 can be written

L–3
$$\Delta G = G_f - G_i = n\,RT \ln \frac{P_f}{P_i}$$

where G_f and G_i are the free energies of the final and initial states respectively.

Equation L–3 gives the change in free energy for an ideal gas from pressure P_i to P_f. If we want an expression for the free energy of formation of one mole of the ideal gas at pressure P_f, then we must use as a reference the free energy of the ideal gas in the standard state. You will recall from Chapter 10 that the standard state for an ideal gas is one atmosphere pressure. Let us represent the pressure of the final state by simply P with no subscript. The change from the initial state (i) to the final state (f) is

$$P_i = 1 \text{ atm} \qquad P_f = P$$
$$G_i = \Delta G_f^\circ \qquad G_f = \Delta G_f$$

and for one mole of an ideal gas, equation L–3 becomes

$$\Delta G_f - \Delta G_f^\circ = RT \ln \frac{P}{1 \text{ atm}}$$

The free energy of formation of an ideal gas at pressure P is then

$$\Delta G_f = \Delta G_f^\circ + RT \ln P$$

where P must be expressed in atmospheres in order to cancel the 1 atm in the denominator (above). The logarithm term is therefore unitless.

If we have a mixture of ideal gases, each at a partial pressure P_i, the molar free energy of each gas is

12–27
$$\Delta G_{f,i} = \Delta G_{f,i}^\circ + RT \ln P_i$$

Note in equation 12–27 that the term $RT \ln P_i$ can be considered as a correction term that changes the standard free energy, $\Delta G_{f,i}^\circ$, at $P_i = 1$ atm to the free energy, $\Delta G_{f,i}$, at a pressure P_i.

M. SOLUTE FREE ENERGY AND CONCENTRATION—ACTIVITY

For our consideration of the solubility of solids in liquids and chemical equilibria in solution we need an expression relating the free energy of a solute and its concentration. We know that for a vapor in equilibrium with its liquid

$$\Delta G_{f,l} = \Delta G_{f,vap}$$

Similarly for each component in a solution which is in equilibrium with its vapor

12–7 $$\Delta G_{f,i_{(soln)}} = \Delta G_{f,i_{(vap)}}$$

Using equation 12–27, which we derived in Appendix L, we can express the free energy for each component in the gas phase by

M–1 $$\Delta G_{f,i_{(vap)}} = \Delta G^{\circ}_{f,i_{(vap)}} + RT \ln p_i$$

where p_i = vapor pressure for component i. Substituting equation M–1 into equation 12–7 we can relate the free energy of each component in solution to its vapor pressure

M–2 $$\Delta G_{f,i_{(soln)}} = \Delta G^{\circ}_{f,i_{(vap)}} + RT \ln p_i$$

Now in order to relate the free energy in solution to the concentration we must express the vapor pressure, p_i, in terms of concentration.

DILUTE SOLUTIONS

At low concentrations of solute B we know that the vapor pressure of the solute obeys Henry's law as given in equation 12–9

12–9 $$p_B = K_B^H X_B$$

At low concentrations of solute we can equate the mole fraction of solute with its molarity and can assume that the volume of the solution is the volume of solvent. The moles of solvent, n_A, is given by

M–3 $$n_A = \frac{(1000 \text{ ml})(\rho_A)}{MW_A}$$

where ρ_A = density of solvent, MW_A = molecular weight of solvent. If we assume one liter of solution, the moles of solute is the molarity [B] of the solute and the mole fraction of solute becomes

M–4 $$X_B = \frac{n_B}{n_A + n_B} = \frac{[B]}{\dfrac{(1000 \text{ ml})(\rho_A)}{MW_A} + [B]}$$

At low concentrations the number of moles of solvent greatly exceed the number of moles of solute

$$n_A = \frac{(1000 \text{ ml})\rho_A}{MW_A} >> [B]$$

and equation M–4 reduces to

$$X_B = \frac{MW_A}{(1000 \text{ ml})\rho_A} [B]$$

Substitution into Henry's law (equation 12–9) gives

M–5
$$p_B = \frac{K_B^H \, MW_A}{(1000 \text{ ml})\rho_A} [B]$$

The relationship between free energy and solute concentration is then obtained by substituting equation M–5 into equation M–2

$$\Delta G_{f,B(soln)} = \Delta G^\circ_{f,B(vap)} + RT \ln \left(\frac{K_B^H \, MW_A}{(1000 \text{ ml})\rho_A} \right) [B]$$

$$\Delta G_{f,B(soln)} = \underbrace{\Delta G_{f,B(vap)} + RT \ln \left(\frac{K_B^H \, MW_A}{(1000 \text{ ml})\rho_A} \right)}_{\Delta G^\circ_{f,B(soln)}} + RT \ln [B]$$

The first two terms on the right consist of constants (at constant temperature) and for simplicity we define the sum of these terms as the *standard free energy for the solute in solution*.

For simplicity we will drop the (*soln*) notation and it will be understood that the equation applies only to a solute in a dilute solution. Our equation relating the free energy of solute with its molar concentration in dilute solutions becomes

$$\Delta G_{f,B} = \Delta G^\circ_{f,B} + RT \ln [B|$$

For any solute i

M–6 $\Delta G_{f,i} = \Delta G^\circ_{f,i} + RT \ln [i]$ (solute in dilute solution)

ACTIVITY

You should note the similarity in form between the equation relating the free energy of an ideal gas to its pressure, equation 12–27 in Appendix L, and the equation relating the free energy of a solute to its concentration, equation M–6 above. Both expressions have a term for standard free energy, $\Delta G^\circ_{f,i}$ and an ($RT \ln$) term. As you will see in Appendix O, these equations lead to a very simple expression which can be used to calculate the concentrations when a chemical reaction has come to equilibrium.

Since the form of equations 12–27 and M–6 is so convenient we would also like to have a similar expression for the free energy of a solute in any solution, not restricted to low concentrations. To accomplish this we define a term *activity*, a_i; the activity of a solute is related to its free energy by

12–26
$$\Delta G_{fi} = \Delta G_{fi}^\circ + RT \ln a_i$$

where the activity is related to the molar concentration by

$$a_i = \gamma_i[i]$$

γ_i is called the *activity coefficient*. Note that equations 12–26 and M–6 differ only in that molar concentration is replaced by activity. At low solute concentrations, equation 12–26 becomes identical to M–6 and the activity of the solute becomes equal to the molar concentration in dilute solutions (Henry's law region). It necessarily follows that the activity coefficient, γ_i, equals one in dilute solutions.

From the above discussion we can see that the activity of a solute becomes the molar concentration if the solute had the same properties or effectiveness as in the dilute solution. However, solutes at higher concentrations do not usually have the same properties or effectiveness as in the dilute solution, and to account for this discrepancy we introduced the factor γ_i, the activity coefficient, to correct the actual molar concentration to give its effective concentration, activity. For this reason the activity of a solute is frequently referred to as the *effective concentration* of the solute.

N. DERIVATION OF THE EQUILIBRIUM EXPRESSION

If we substitute equation 13–2 into equation 13–1 we obtain

$$c(\Delta G_{f,C_{(aq)}}^\circ + RT \ln a_C) = b(\Delta G_{f,B_{(aq)}}^\circ + RT \ln a_B)$$

N–1
$$c\Delta G_{f,C_{(aq)}}^\circ - b\Delta G_{f,B_{(aq)}}^\circ = -(c\,RT \ln a_C - b\,RT \ln a_B)$$

The change in standard free energy (ΔG°) for a reaction is the free energy of the products minus the free energy of the reactants.

N–2
$$\Delta G^\circ = c\Delta G_{f,C_{(aq)}}^\circ - b\Delta G_{f,B_{(aq)}}^\circ$$

Since
$$a \ln x = \ln x^a$$

and
$$\ln x - \ln y = \ln \frac{x}{y}$$

then

$$c \, RT \ln a_C - bRT \ln a_B = RT \ln a_C{}^c - RT \ln a_B{}^b$$

$$= RT \, (\ln a_C{}^c - \ln a_B{}^b)$$

N–3
$$cRT \ln a_C - bRT \ln a_B = RT \ln \frac{(a_C)^c}{(a_B)^b}$$

Substituting equations N–2 and N–3 into equation N–1

N–4
$$\Delta G^\circ = - \, RT \ln \frac{(a_C)^c}{(a_B)^b}$$

For a given reaction at constant temperature T, ΔG° has some specific value. Therefore, the right-hand side of equation N–4 must equal a constant. Since R and T are constant, the $\ln (a_C)^c/(a_B)^b$ must be constant and

$$\frac{(a_C)^c}{(a_B)^b} = \text{constant} = K$$

where K is the equilibrium constant. Therefore, equation N–4 can be rewritten as

13–3
$$\Delta G^\circ = -RT \ln K$$

where

13–4
$$K = \frac{(a_C)^c}{(a_B)^b}$$

O. DERIVATION OF THE SOLUBILITY PRODUCT EXPRESSION

The solubility product expression for solid $Mg(OH)_2$ can be obtained by substituting equations 13–30 into equation 13–29.

$$0 = \Delta G^\circ_{f, Mg^{2+}} + RT \ln [Mg^{2+}] + 2\Delta G^\circ_{f, OH^-} + 2RT \ln [OH^-] - \Delta G^\circ_{f, Mg(OH)_{2(s)}}$$

$$\Delta G^\circ_{f, Mg^{2+}} + 2\Delta G^\circ_{f, OH^-} - \Delta G^\circ_{f, Mg(OH)_{2(s)}} = RT \ln [Mg^{2+}] + RT \ln [OH^-]^2$$

Combine the logarithmic terms on the right and note that the left-hand side of the equation is the change in standard free energy, ΔG°, for the process,

O–1
$$\Delta G^\circ = - \, RT \ln [Mg^{2+}][OH^-]^2$$

Since ΔG° is a constant for the process (constant temperature), the right-hand side of equation O–1 must also be constant. Since R and T are constant, the product of the concentrations must also be constant which we designate as K_{sp}. Therefore equation O–1 becomes

13–31
$$\Delta G^\circ = -RT \ln K_{sp}$$

where

13–32
$$K_{sp} = [\text{Mg}^{2+}][\text{OH}^-]^2$$

P. DERIVATION OF THE RELATIONSHIP BETWEEN $\mathcal{E}$ AND CONCENTRATION— NERNST EQUATION

In order to derive the equation relating the cell potential to the concentrations of the species in solution, we first must obtain a relationship between ΔG and the concentrations. Then using the relationship $\Delta G = -nF\mathcal{E}$ we can obtain the Nernst equation relating $\mathcal{E}$ with concentrations.

Let us assume the cell reaction involves the reaction

$$bB \rightleftharpoons cC$$

The change in free energy for this process, not necessarily at equilibrium, is

$$\Delta G = c\Delta G_{f,C} - b\Delta G_{f,B}$$

If the reaction happens to be at equilibrium, ΔG equals zero; however, prior to reaching equilibrium ΔG is not zero and this non-equilibrium situation is the one of concern to us. Substituting in the expressions for $\Delta G_{f,B}$ and $\Delta G_{f,C}$ as given by equation 12–26, and assuming $\gamma_B = \gamma_C = 1$,

$$\Delta G = c\Delta G^\circ_{f,C} + cRT\ln[\text{C}] - b\Delta G^\circ_{f,B} - bRT\ln[\text{B}]$$

P–1
$$\Delta G = c\Delta G^\circ_{f,C} - b\Delta G^\circ_{f,B} + RT\ln[\text{C}]^c - RT\ln[\text{B}]^b$$

The change in standard free energy for the process is

$$\Delta G^\circ = c\Delta G^\circ_{f,C} - b\Delta G^\circ_{f,B}$$

Substituting this expression for ΔG° into equation P–1 and combining the logarithmic terms, we have

P–2
$$\Delta G = \Delta G^\circ + RT\ln \frac{[\text{C}]^c}{[\text{B}]^b}$$

Equation P–2 expresses the change in free energy for a reaction of b moles at the concentration [B] to give c moles of C at the concentration [C].

Using equations 14–6 and 14–7 relating ΔG and $\Delta G°$ to cell potentials

$$\Delta G = -nF\mathcal{E}$$

$$\Delta G° = -nF\mathcal{E}°$$

we can substitute into equation P–2

$$-nF\mathcal{E} = -nF\mathcal{E}° + RT \ln \frac{[C]^c}{[B]^b}$$

and solve for $\mathcal{E}$

14–12
$$\mathcal{E} = \mathcal{E}° - \frac{RT}{nF} \ln \frac{[C]^c}{[B]^b}$$

Equation 14–12 is called the Nernst equation and it expresses the cell potential that occurs when C is at concentration [C] and B is at concentration [B].

Q. DERIVATION OF THE BIMOLECULAR RATE EXPRESSION

Consider an elementary reaction between two reactants, A and B, which involves a bimolecular collision between a molecule of A and a molecule of B

$$A + B \longrightarrow \text{products}$$

Consider the collisions of a single molecule of A with molecules of B in a given volume *V*. Regardless of whether the mixture is in the liquid or the vapor phase, the molecules present will be moving about randomly and, hence, will be homogeneously distributed within the system. If we follow the single molecule of A, it will undergo a certain number of collisions in a given period of time. Dividing the total number of collisions by the time would give an average collision rate, collisions per second, for the single molecule of A. If the total number of molecules of B is doubled or tripled without changing the total volume of the system, the average collision rate for the single molecule of A will likewise double or triple, respectively. However, if the volume of the system is also doubled or tripled as the number of molecules of B is doubled or tripled, the average collision rate for the single molecule of A will remain the same. This argument suggests that the average collision rate of a single molecule of A with molecules of B is proportional to the *concentration* of molecules of B.

Q–1   collision rate (single molecule of A) α [B]

Now suppose that instead of a single molecule of A, two or three molecules of A are admitted to the system, and the bimolecular collisions with molecules of B are considered collectively. If the average rate of collisions between molecules of A and molecules of B is again determined, counting the number of collisions of all molecules of A present, it will either be doubled, in the case of two molecules of A, or tripled, in the case of three molecules of A. For N molecules of A, the average rate of collisions between molecules of A and molecules of B will be N times as great as the average rate of collisions for a single molecule of A. Thus it can be concluded that the total rate of collisions between molecules of A and molecules of B is also proportional to the total number of molecules of A in the system, N_A.

Q–2 $$\text{collision rate } (N \text{ molecules of A}) \; \alpha \; N_A$$

Combining equations Q–1 and Q–2 results in the following expression of proportionality.

Q–3 $$\text{collision rate } (N \text{ molecules of A}) \; \alpha \; N_A[B]$$

However, the rate we want is the average number of collisions per second per unit volume. Dividing both sides of equation Q–3 by the total volume of the system, and taking note of the fact that molecules per unit volume is proportional to moles per liter, we obtain

$$\text{collision rate } \left(\frac{N \text{ molecules of A}}{V}\right) \; \alpha \; \frac{N_A}{V}[B] \; \alpha \; [A][B]$$

$$\text{rate } \alpha \; [A][B]$$

or

Q–4 $$\text{rate} = k[A][B]$$

A similar argument can be employed in the case of other elementary reaction processes, such as a homogeneous bimolecular collision process represented by the equation

$$A + A \longrightarrow \text{products}$$

In this case, the rate of the reaction will be found to be proportional to the square of the concentration of the reactant.

$$\text{rate } \alpha \; [A]^2$$

or

Q–5 $$\text{rate} = k[A]^2$$

R. DERIVATION OF THE FIRST ORDER INTEGRATED RATE EQUATION

For the reaction $\qquad A \longrightarrow P$

the first order rate equation for the rate of loss of A is

R–1 $$\text{rate} = -\frac{d[A]}{dt} = k[A]$$

Equation R–1 can be substituted into the differential expression

$$d[A] = \frac{d[A]}{dt}\, dt$$

$$d[A] = -k[A]dt$$

Separating the variables $[A]$ and t by dividing by $[A]$ we have

R–2 $$\frac{d[A]}{[A]} = -kdt$$

The integration of equation R–2 is carried out from the initial time $t = 0$ where $[A] = [A]_0$

$$\int_{[A]_0}^{[A]} \frac{d[A]}{[A]} = \int_0^t -kdt$$

R–3 $$\ln[A] - \ln[A]_0 = -kt$$

or

15–29 $$\ln \frac{[A]}{[A]_0} = -kt$$

Equation 15–29 can also be rearranged to give

$$\ln \frac{[A]_0}{[A]} = kt$$

The natural logarithm can be expressed as the common logarithm

$$2.303 \log \frac{[A]_0}{[A]} = kt$$

or

R-4
$$\log \frac{[A]_0}{[A]} = \frac{k}{2.303} t$$

The rate expression given in equation R–1 in terms of concentration can also be expressed in terms of the number of moles or number of molecules of A present. Letting N represent the molecules of A, equation R–5 gives the rate of loss of A in terms of molecules per unit of time.

R-5
$$\text{rate} = -\frac{dN}{dt} = kN$$

Integration of equation R–5 proceeds in the same manner as for equation R–1. The integrated expression corresponding to equation R–4 is

17-4
$$\log \frac{N_0}{N} = \frac{k}{2.303} t$$

where N_0 is the number of A molecules present at $t = 0$.

S. DERIVATION OF $K^{\ddagger}$ IN TERMS OF $\Delta H^{\ddagger}$, $\Delta S^{\ddagger}$, AND $\Delta G^{\ddagger}$

According to transition state theory the rate expression for a bimolecular reaction is

15-39
$$\text{rate} = k^{\ddagger} K^{\ddagger} [A][B]$$

Since $K^{\ddagger}$ is an equilibrium constant, the thermodynamic expression, equation 13–3 is applicable (in modified form)

15-40
$$\Delta G^{\ddagger} = -RT \ln K^{\ddagger}$$

where $\Delta G^{\ddagger}$ is the change in standard free energy in going from A + B to AB‡. The standard state in this case corresponds to a concentration of one mole per liter. The change in free energy at constant temperature is related to ΔH and ΔS by

10-23
$$\Delta G = \Delta H - T\Delta S$$

For the process A plus B to AB‡ we have

15-41
$$\Delta G^{\ddagger} = \Delta H^{\ddagger} - T\Delta S^{\ddagger}$$

Setting equations 15–39 and 15–40 equal

S–1
$$-RT\ln K^{\ddagger} = \Delta H^{\ddagger} - T\Delta S^{\ddagger}$$

Since
$$H = E + PV$$

then

15–42
$$\Delta H^{\ddagger} = \Delta E^{\ddagger} + \Delta(PV)$$

However, we know that $\Delta(PV)$ is small for most reactions and we can assume for this discussion that

$$\Delta H^{\ddagger} = \Delta E^{\ddagger}$$

and

$$-RT\ln K^{\ddagger} = \Delta E^{\ddagger} - T\Delta S^{\ddagger}$$

Solving for $K^{\ddagger}$

$$\ln K^{\ddagger} = \frac{\Delta S^{\ddagger}}{R} - \frac{\Delta E^{\ddagger}}{RT}$$

$$K^{\ddagger} = e^{\left(\frac{\Delta S^{\ddagger}}{R} - \frac{\Delta E^{\ddagger}}{RT}\right)}$$

Since

$$e^a e^b = e^{a+b}$$

we can break the exponential term into two parts

S–2
$$K^{\ddagger} = (e^{\frac{\Delta S^{\ddagger}}{R}})(e^{\frac{-\Delta E^{\ddagger}}{RT}})$$

Substitution of equation S–2 into equation 15–38 gives

$$\text{rate} = K^{\ddagger}(e^{\frac{\Delta S^{\ddagger}}{R}})(e^{\frac{-\Delta E^{\ddagger}}{RT}})[A][B]$$

T. TABLES OF CONSTANTS AND DATA

TABLE T–1   Physical constants*

SYMBOL	NAME	VALUE
c	velocity of light	2.998×10^{10} cm sec^{-1}
h	Planck's constant	6.626×10^{-27} erg sec
N	Avogadro's number	6.023×10^{23} molecules/mole, atoms/g-atom, formula units/mole
F	Faraday's constant	2.8926×10^{14} esu mole^{-1}[a]
		9.648×10^{4} C mole^{-1}[a]
		2.306×10^{4} cal V^{-1} mole^{-1}[a]
V_0	molar volume of gas at STP	22.4136 liters mole^{-1}
P_0	standard pressure	1.000 atm
		760 mm Hg (Torr)
		14.696 lb in^{-2}
k	Boltzmann's constant	1.3806×10^{-16} erg °K^{-1} molecule^{-1}
e	charge on the electron	4.803×10^{-10} esu
		1.602×10^{-19} C
m_e	rest mass of the electron	9.109×10^{-28} g
		5.4859×10^{-4} amu
amu	atomic mass unit	1.660×10^{-24} g
m_P	rest mass of the proton	1.6726×10^{-24} g
		1.00727 amu
m_n	rest mass of the neutron	1.6749×10^{-24} g
		1.00866 amu
R	gas constant	0.082056 liter atm °K^{-1} mole^{-1}
		8.3143 J°K^{-1} mole^{-1}[a]
		1.9872 cal °K^{-1} mole^{-1}
eV	electron volt	1.602×10^{-12} erg molecule^{-1}
		2.3061×10^{4} cal mole^{-1}
P_{air}	density of air at STP	1.2929 g liter^{-1}
P_{H_2O}	density of water at STP	0.99987 g cm^{-3}

[a] esu = electrostatic unit; C = coulomb; V = volt; J = joule.

Handbook of Chemistry and Physics, Fiftieth edition. The Chemical Rubber Co., Cleveland, Ohio, 1969.

TABLE T–2 Dipole moments

COMPOUND	DIPOLE MOMENT $\times 10^{-18}$ esu
$CsCl$	10.5
HF	1.82
HCl	1.09
HBr	0.79
HI	0.38
H_2O	1.87
H_2S	1.10
NH_3	1.3
N_2H_4	1.84
NO	0.16
NO_2	0.29
CO	0.10
FCl	0.83
FBr	1.29
$ClBr$	0.57
ClI	0.65

TABLE T-3   Values of chemical thermodynamic properties of inorganic compounds at 25 °C*

ΔH_f° = the standard heat of formation of a given substance from its elements at 25 °C in kcal/mole

ΔG_f° = the standard free energy of formation of a given substance from its elements at 25 °C in kcal/mole

S° = the entropy of the given substance at the reference temperature at 25 °C in cal/deg-mole

c = crystalline; in certain cases where a substance exists in more than one crystalline form there is an indication as to which form is concerned

g = gaseous

am = amorphous

aq = aqueous; unless otherwise indicated the aqueous solution is taken as the hypothetical ideal state of unit molality

g/s = glass

l = liquid

SUBSTANCE	STATE	ΔH_f°	ΔG_f°	S°
ammonium				
NH_3	g	−11.04	−3.976	46.01
	aq	−19.32	−6.36	26.3
	$aq\ \infty$	−19.32	−	−
NH_4^+	aq	−31.74	−19.00	26.97
NH_4OH	aq	−87.64	−	−
NH_4Cl	c	−75.38	−48.73	22.6
NH_4NO_2	c	−63.1	−	−
NH_4NO_3	c	−87.27	−	−
bromine				
Br_2	g	7.34	0.75	58.64
	l	0.00	0.00	36.4
	aq	−1.1	−	−
$BrCl$	g	3.51	−0.21	57.34
HBr	g	−8.66	−12.72	47.44
calcium				
Ca	g	46.04	37.98	36.99
	c	0.00	0.00	9.95
$Ca(OH)_2$	c	−235.80	−214.33	18.2
	aq	−239.68	−207.37	−18.2

*The data in this table has been taken from the *Handbook of Chemistry and Physics*, Fiftieth edition, The Chemical Rubber Co., Cleveland, Ohio, 1969. The student should consult this reference for a more complete listing if he should need additional data.

TABLE T–3 (Cont.)

SUBSTANCE	STATE	ΔH_f°	ΔG_f°	S°
carbon				
C	*g*	171.70	160.85	37.76
C(diamond)	*c*	0.45	0.69	0.58
C(graphite)	*c*	0.00	0.00	1.36
CO_2	*g*	−94.05	−94.26	51.06
	aq	−98.69	−92.31	29.0
CO	*g*	−26.42	−32.81	47.30
chlorine				
Cl_2	*g*	0.00	0.00	53.29
Cl^-	*aq*	−40.023	−31.350	13.17
ClF	*g*	−13.3	−13.6	52.05
HCl	*g*	−22.06	−22.77	44.62
copper				
Cu	*g*	81.52	72.04	39.74
	c	0.00	0.00	7.96
Cu^+	*aq*	12.4	12.0	−6.3
Cu^{2+}	*aq*	15.39	15.53	−23.6
CuI	*g*	62.	50.	61.06
	c	−16.2	−16.62	23.1
hydrogen				
H_2	*g*	0.00	0.00	31.21
H^+	*aq*	0.00	0.00	0.00
H	*g*	52.09	48.58	27.39
H_2O	*g*	−57.80	−54.64	45.11
	l	−68.32	−56.69	16.72
H_2O_2	*g*	−31.83	—	—
	l	−44.84	−28.2	—
	aq	−45.68	−31.47	
iodine				
I_2	*g*	14.88	4.63	62.28
	c	0.00	0.00	27.9
I^-	*aq*	−13.37	−12.35	26.14
IBr	*g*	9.75	0.91	61.80
ICl	*g*	4.20	−1.32	59.12
HI	*g*	6.20	0.31	49.31
	aq	−13.37	−12.35	26.14
iron				
Fe	*g*	96.68	85.76	43.11
	c	0.00	0.00	6.49
Fe^{2+}	*aq*	−21.0	−20.30	−27.1
Fe^{3+}	*aq*	−11.4	−2.52	−70.1

TABLE T–3 (Cont.)

SUBSTANCE	STATE	ΔH_f°	ΔG_f°	S°
lead				
Pb	g	46.34	38.47	41.89
	c	0.00	0.00	15.51
$Pb(C_2H_5)_4$	l	52	–	–
lithium				
Li	g	37.07	29.19	33.14
	c	0.00	0.00	6.70
Li^+	aq	−66.54	−70.22	3.4
LiBr	g	−41.	−50.	53.78
	c	−83.72	–	–
	aq	−95.45	−94.69	22.7
LiCl	g	−53.	−58.	51.01
	c	−97.70	–	–
	aq	−106.58	−101.57	16.6
mercury				
Hg	g	14.54	7.59	41.8
	l	0.00	0.00	18.5
Hg^{2+}	aq	–	39.38	–
Hg_2^{2+}	aq	–	36.79	–
nitrogen (NH_4OH; NH_2OH; N_2H_4; and N_3H; see under ammonia)				
N_2	g	0.00	0.00	45.77
NO	g	21.60	20.72	50.34
NO_2	g	8.09	12.39	57.47
N_2O_4	g	2.31	23.49	72.73
N_2O	g	19.49	24.76	52.58
HNO_2	g	−28.9	–	–
HNO_3	l	−41.40	−19.10	37.19
	aq	−49.37	−26.41	35.0
oxygen				
O_2	g	0.00	0.00	49.003
O_3	g	34.0	39.06	56.8
OH^-	aq	−54.957	−37.595	−2.519
potassium				
K	g	21.51	14.62	38.30
	c	0.00	0.00	15.2
K^+	aq	−60.04	−67.466	24.5
KCl	g	−51.6	−56.2	57.24
	c	−104.18	−97.592	19.76
	aq	−100.06	−98.82	37.7

TABLE T–3 (Cont.)

SUBSTANCE	STATE	ΔH_f°	ΔG_f°	S°
silver				
Ag	*g*	69.12	59.84	41.32
	c	0.00	0.00	10.21
Ag^+	*aq*	25.31	18.43	17.67
Ag Cl	*g*	23.23	16.79	58.5
	c	−30.36	−26.22	22.97
sodium				
Na	*g*	25.98	18.67	36.72
	c	0.00	0.00	12.2
Na^+	*aq*	−57.28	−62.59	14.4
Na Br	*c*	−86.03	—	—
	aq	−86.18	−87.16	33.7
Na Cl	*c*	−98.23	−91.79	17.30
	aq	−97.302	−93.94	27.6
Na OH	*c*	−101.99	—	—
	aq	−112.24	−100.184	11.9
sulfur				
S	*g*	53.25	43.57	40.085
(rhb)	*c*	0.00	0.00	7.62
(mon)	*c*	0.071	0.023	7.78
SO_2	*g*	−70.96	−71.79	59.40
SO_3	*g*	−94.45	−88.52	61.24

TABLE T–4 Values of chemical thermodynamic properties of carbon containing compounds at 25 °C*

FORMULA	NAME	STATE	ΔH_f°	ΔG_f°	S°
CO_3^{2-}	carbonate ion	aq	−161.63	−126.22	−12.7
CH_3	methyl	g	32.0	−	−
CH_4	methane	g	−17.889	−12.140	44.50
$HCOO^-$	formate ion	aq	−98.0	−80.0	21.9
HCO_3^-	bicarbonate ion	aq	−165.18	−140.31	22.7
CH_2O_2	formic acid, monomer	g	−86.67	−80.24	60.0
	formic acid	l	−97.8	−82.7	30.82
H_2CO_3	carbonic acid	aq	−167.0	−149.00	45.7
$COCl_2$	carbonyl chloride	g	−52.30	−50.31	69.13
CN^-	cyanide ion	aq	36.1	39.6	28.2
HCN	hydrogen cyanide	g	31.2	28.7	48.23
		l	25.2	29.0	26.97
	(hydrocyanic acid)	aq	25.2	26.8	30.8
CH_5N	methylamine	g	−6.7	6.6	57.73
CH_3O_2N	nitromethane	l	−21.28	2.26	41.1
$C_2O_4^{2-}$	oxalate ion	aq	−197.0	−161.3	12.2
$HC_2O_4^-$	bioxalate ion	aq	−195.5	−167.1	36.7
C_2H_2O	ketene	g	−14.6	−	−
C_2H_5Cl	chloroethane	g	−25.1	−12.7	65.90
$C_2H_2O_4$	oxalic acid	c	−197.6	−166.8	28.7
$C_2H_3O_2^-$	acetate ion	aq	−116.843	−87.32	20.8
C_2H_4O	acetaldehyde	g	−39.76	−31.96	63.5
$C_2H_4O_2$	acetic acid	aq	−116.4	−93.8	38.2
C_2H_6O	ethanol	g	−56.24	−40.30	67.4
C_2H_7N	ethylamine	g	−11.6	−	−
C_3H_6O	acetone	g	−51.72	−	−
C_6H_6	benzene	l	11.63	29.4	−
C_6H_5Cl	chlorobenzene	l	−	27.8	−
$C_6H_5NO_2$	nitrobenzene	g	15.8	−	−
C_6H_5COOH	benzoic acid	s	−92.1	−	−
$C_6H_{12}O_6$	glucose	aq	−301.	−219.6	69.1

*Handbook of Chemistry and Physics, Fiftieth edition. The Chemical Rubber Co., Cleveland, Ohio, 1969.

TABLE T–5.   Values of chemical thermodynamic properties of hydrocarbons (gas phase)*

FORMULA	NAME	ΔH_f°	ΔG_f°	$\log_{10} K_f$	S°
CH_4	methane	−17.889	−12.140	8.8985	44.50
C_2H_2	ethyne (acetylene)	54.194	50.000	−36.6490	47.997
C_2H_4	ethene (ethylene)	12.496	16.282	−11.9345	52.54
C_2H_6	ethane	−20.236	−7.860	5.7613	54.85
C_3H_6	propene	4.879	14.990	−10.9875	63.80
C_3H_8	propane	−24.820	−5.614	4.1150	64.51
C_4H_8	1-butene	0.280	17.217	−12.6199	73.48
C_4H_8	cis-2-butene	−1.362	16.046	−11.7618	71.90
C_4H_8	trans-2-butene	−2.405	15.315	−11.2255	70.86
C_4H_{10}	n-butane	−29.812	−3.754	2.7516	74.10
C_5H_{10}	cis-2-pentene	−6.710	17.173	−12.5874	82.76
C_5H_{10}	trans-2-pentene	−7.590	16.575	−12.1495	81.81
C_5H_{12}	n-pentane	−35.00	−1.96	1.4366	83.27
C_6H_6	benzene	19.820	30.989	−22.7143	64.34
C_6H_{12}	cyclohexane	−29.43	7.59	−5.5605	71.28
C_7H_8	methylbenzene	11.950	29.229	−21.4236	76.42
C_7H_{14}	methylcyclohexane	−36.99	6.52	−4.7819	82.06

TABLE T–6   Bond strength in diatomic molecules at 25 °C*

MOLECULE	kcal mole⁻¹	MOLECULE	kcal mole⁻¹	MOLECULE	kcal mole⁻¹
H—H	104.18	C—N	179	F—F	37
H—C	80.9	C—O	256.7	F—Si	116
H—N	75	C—F	128	F—Cl	59.9
H—O	102.4	C—Cl	93	Cl—Cl	58
H—F	135.0	C—Br	67	Cl—Br	52.3
H—Cl	103.1	C—I	50	Cl—I	50.5
H—Br	87.4	N—N	226.8	Br—Br	46.34
H—I	71.4	N—O	149.7	Br—I	42.8
C—C	144	O—O	118.86	I—I	36.503
Li—Cl	111.9			K—Cl	101.30

Handbook of Chemistry and Physics, Fiftieth edition. The Chemical Rubber Co., Cleveland, Ohio, 1969.

TABLE T–7 Bond strengths in polyatomic molecules (bond dissociation energies)*

BOND	kcal/mole^{-1}	BOND	kcal/mole^{-1}
$H-CH$	107	$H-N(CH_3)C_6H_5$	74
$H-CH_2$	105	$H-OCH_3$	102
$H-CH_3$	104.0	H_3C-CH_3	88
$H-C_2H_5$	98.0	CH_3-CF_3	100
$H-nC_3H_7$	98	CF_3-CF_3	97
$H-iC_3H_7$	94.5	$O=CO$	128
$H-NH_2$	103	CH_3-F	108
$H-NHCH_3$	92	CH_3-I	56.3
$H-N(CH_3)_2$	86	O_2N-NO_2	13.6
$H-NHC_6H_5$	80		

TABLE T–8 Heats of formation of gaseous atoms from the elements in their standard states at 25 °C*

For elements which exist as diatomic gases in their standard states, the heats of formation are readily obtained from the bond strength. For elements which are crystalline in their standard states the heats of formation are derived from vapor pressure data. All values are given in kcal mole^{-1} at 25 °C.

ELEMENTS	GASEOUS ATOM	kcal mole^{-1}
$H_{2(g)}$	H	52.10
$C_{(c)}$	C	170.9
$N_{2(g)}$	N	113.0
$O_{2(g)}$	O	59.56
$F_{2(g)}$	F	18.86
$S_{(c)}$	S	65.65
$Cl_{2(g)}$	Cl	28.92
$K_{(c)}$	K	21.3
$Ca_{(c)}$	Ca	42.81
$Sc_{(c)}$	Sc	88.
$Ti_{(c)}$	Ti	112.5
$Fe_{(c)}$	Fe	48.
$I_{2(c)}$	I	25.54
$Br_{2(l)}$	Br	26.74

Handbook of Chemistry and Physics, Fiftieth edition. The Chemical Rubber Co., Cleveland, Ohio, 1969.

TABLE T–9   Heats of formation of free radicals at 25 °C*

RADICAL	kcal mole^{-1}	RADICAL	kcal mole^{-1}
CH_2	91	CH_3O	2
CH_3	34.0	CH_3CO	−5.4
C_2H_3	64	CH_3CO_2	−45
C_2H_5	25.7	$C_2H_5CO_2$	−54
C_3H_5	38	$(CH_3)_3CO$	−24
nC_3H_7	21	C_6H_5CO	16
iC_3H_7	17.6	$C_6H_5CO_2$	−64
tC_4H_9	6.7	CF_2	−39
$neoC_5H_{11}$	7.5	CF_3	−111
C_6H_5	80.0	CCl_3	19
$C_6H_5CH_2$	44.6	NH_2	40
CN	109	NF_2	9
CH_3NH	34	OH	9.3
$(CH_3)_2N$	29	HO_2	5
C_6H_5NH	48	HS	33?
CHO	7	CH_3S	30?

Handbook of Chemistry and Physics, Fiftieth edition. The Chemical Rubber Co., Cleveland, Ohio, 1969.

TABLE T–10 Average bond energies*

BOND	ϵ	BOND	ϵ
H—H	104	F—F	37
H—C	99	F—Si	136
H—N	84	Si—Si	50
H—O	111	Si—S	60
H—F	135	Si—Cl	90
H—P	76	Si—Br	73
H—S	81	Si—I	53
H—Cl	103	P—P	51
H—Se	66	P—Cl	78
H—Br	88	P—Br	64
H—Te	57	P—I	49
H—I	71	S—S	49
C—C	80	S—Cl	61
C—N	64	S—Br	51
C—O	81	Cl—Cl	58
C—F	102	Br—Br	46
C—Si	75	I—I	36
C—S	61	C=C	146
C—Cl	79	C≡C	200
C—Br	66	C=N	147
C—I	52	C≡N	213
N—N	32	C=O	177
N—F	56	C≡O	256
N—Cl	37	C=S	115
O—O	33	N=N	100
O—F	45	N≡N	226
O—Si	106		
O—Cl	50		

*Energies of single bonds from J. Kleinberg, W. J. Argersinger, Jr., and E. Griswold. *Inorganic Chemistry*. D. C. Heath and Co., Boston, 1960. Energies of double and triple bonds from F. A. Cotton and G. Wilkinson. *Advanced Inorganic Chemistry*. Interscience Publications, New York, 1962.

TABLE T–11 Heat capacity. The C_P for the substances in the gas phase are at 298.1 °K.

SUBSTANCE (FORMULAS)	STATE	C_P (cal/deg/mole)
N_2	g	6.897
H_2	g	6.930
NH_3	g	8.895
O_2	g	6.991
CO_2	g	8.999
H_2O	g	7.966
	s	8.7
	l	18.0
C_2H_2	g	13.25
Br_2	g	8.682
Cl_2	g	8.212
$BrCl$	g	8.50[a]
Al	s	4.94
Fe	s	6.1

[a]estimated

TABLE T–12   Acid dissociation constants (K_a)*

ACIDS	EQUILIBRIUM EQUATION	TEMPERATURE (°C)	K_a	pK_a
acetic	$HC_2H_3O_2 + H_2O \rightleftharpoons H_3O^+ + C_2H_3O_2^-$	25	1.76×10^{-5}	4.75
ammonium ion	$NH_4^+ + H_2O \rightleftharpoons H_3O^+ + NH_3$		$5.6 \times 10^{-10}(K_h)$	
arsenic	$H_3AsO_4 + H_2O \rightleftharpoons H_3O^+ + H_2AsO_4^-$	18	5.62×10^{-3}	2.25
	$H_2AsO_4^- + H_2O \rightleftharpoons H_3O^+ + HAsO_4^{2-}$	18	1.70×10^{-7}	6.77
	$HAsO_4^- + H_2O \rightleftharpoons H_3O^+ + AsO_4^{3-}$	18	3.95×10^{-12}	11.60
benzoic	$HC_7H_5O_2 + H_2O \rightleftharpoons H_3O^+ + C_7H_5O_2^-$		6.46×10^{-5}	4.19
carbonic	$H_2CO_3 + H_2O \rightleftharpoons H_3O^+ + HCO_3^-$	25	4.3×10^{-7}	6.37
	$HCO_3^- + H_2O \rightleftharpoons H_3O^+ + CO_3^{2-}$	25	5.6×10^{-11}	10.25
hydrocyanic	$HCN + H_2O \rightleftharpoons H_3O^+ + CN^-$		4×10^{-10}	
hydrofluoric	$HF + H_2O \rightleftharpoons H_3O^+ + F^-$	25	3.53×10^{-4}	3.45
hydrogen sulfide	$H_2S + H_2O \rightleftharpoons H_3O^+ + HS^-$	25	1.1×10^{-7}	6.94
	$HS^- + H_2O \rightleftharpoons H_3O^+ + S^{2-}$	25	1.0×10^{-14}	14.0
hypochlorous	$HClO + H_2O \rightleftharpoons H_3O^+ + ClO^-$		3.2×10^{-8}	
nitrous	$HNO_2 + H_2O \rightleftharpoons H_3O^+ + NO_2^-$		4.5×10^{-4}	
oxalic	$H_2C_2O_4 + H_2O \rightleftharpoons H_3O^+ + HC_2O_4^-$		3.8×10^{-2}	
	$HC_2O_4^- + H_2O \rightleftharpoons H_3O^+ + C_2O_4^{2-}$		5.0×10^{-5}	
phenol	$C_6H_5OH + H_2O \rightleftharpoons H_3O^+ + C_6H_5O^-$		1.28×10^{-10}	9.8
p-nitrophenol	$NO_2C_6H_4OH + H_2O \rightleftharpoons H_3O^+ + NO_2C_6H_4O^-$		7×10^{-8}	7.15
2,4-dinitrophenol	$(NO_2)_2C_6H_3OH + H_2O \rightleftharpoons H_3O^+ + (NO_2)_2C_6H_3O^-$		1.1×10^{-4}	3.96
2,4,6-trinitrophenol	$(NO_2)_3C_6H_2OH + H_2O \rightleftharpoons H_3O^+ + (NO_2)_3C_6H_2O^-$		4.2×10^{-1}	0.38
phosphoric (ortho)	$H_3PO_4 + H_2O \rightleftharpoons H_3O^+ + H_2PO_4^-$		7.5×10^{-3}	2.12
	$H_2PO_4^- + H_2O \rightleftharpoons H_3O^+ + HPO_4^{2-}$		6.2×10^{-8}	7.21
	$HPO_4^{2-} + H_2O \rightleftharpoons H_3O^+ + PO_4^{3-}$		2.2×10^{-13}	12.67
plumbous hydroxide	$Pb(OH)_2 + H_2O \rightleftharpoons H_3O^+ + Pb(OH)O^-$		2×10^{-16}	
sulfurous	$H_2SO_3 + H_2O \rightleftharpoons H_3O^+ + HSO_3^-$		1.25×10^{-2}	
	$HSO_3^- + H_2O \rightleftharpoons H_3O^+ + SO_3^{2-}$		5.6×10^{-8}	

Handbook of Chemistry and Physics, Fiftieth edition. The Chemical Rubber Co., Cleveland, Ohio, 1969.

TABLE T–13 Base dissociation constants (K_b)*

BASE	EQUILIBRIUM EQUATIONS	K_b
acetate ion	$C_2H_3O_2^- + H_2O \rightleftharpoons HC_2H_3O_2 + OH^-$	5.6×10^{-10}
ammonia	$NH_3 + H_2O \rightleftharpoons NH_4^+ + OH^-$	1.80×10^{-5}
calcium hydroxide	$Ca(OH)_2(aq) \rightleftharpoons Ca(OH)^+_{aq} + OH^-_{(aq)}$	3.7×10^{-3}
	$Ca(OH)^+_{(aq)} \rightleftharpoons Ca^{2+}_{aq} + OH^-_{(aq)}$	4.0×10^{-2}
carbonate ion	$HCO_3^- + H_2O \rightleftharpoons H_2CO_3 + OH^-$	2.4×10^{-8}
cyanide ion	$CN^- + H_2O \rightleftharpoons HCN + OH^-$	2.5×10^{-5}
fluoride ion	$F^- + H_2O \rightleftharpoons HF + OH^-$	1.5×10^{-11}
hydrazine	$N_2H_{4(aq)} + H_2O \rightleftharpoons N_2H^+_{5(aq)} + OH^-_{(aq)}$	1.7×10^{-6}
hydroxylamine	$HONH_{2(aq)} + H_2O \rightleftharpoons HONH^+_{3(aq)} + OH^-_{(aq)}$	1.07×10^{-8}
nitrite ion	$NO_2^- + H_2O \rightleftharpoons HNO_2 + OH^-$	2.2×10^{-11}
oxalate ion	$C_2O_4^{2-} + H_2O \rightleftharpoons HC_2O_4^- + OH^-$	2.0×10^{-10}
phosphate ion (ortho)	$PO_4^{3-} + H_2O \rightleftharpoons HPO_4^{2-} + OH^-$	10^{-2}
	$HPO_4^{2-} + H_2O \rightleftharpoons H_2PO_4^- + OH^-$	1.6×10^{-7}
sulfite ion	$SO_3^{2-} + H_2O \rightleftharpoons HSO_3^- + OH^-$	1.8×10^{-7}
	$HSO_3^- + H_2O \rightleftharpoons H_2SO_3 + OH^-$	8.0×10^{-13}
sulfide ion	$S^{2-} + H_2O \rightleftharpoons HS^- + OH^-$	7.7×10^{-2}
	$HS + H_2O \rightleftharpoons H_2S + OH^-$	1.0×10^{-7}

TABLE T–14 Dissociation constants for organic bases in aqueous solutions*

$$BH^+ \xrightleftharpoons{K_a} B + H^+$$

COMPOUND (B)	TEMPERATURE (°C)	STEP	K_a	pK_a
aniline	25		2.34×10^{-5}	4.63
aniline, 4-nitro	25		1.00×10^{-1}	1.0
diethylamine	40		3.24×10^{-11}	10.489
ethylamine	20		1.56×10^{-11}	10.807
glycine	25	1	4.46×10^{-3}	2.3503
methylamine	25		2.70×10^{-11}	10.657
piperidine	25		7.53×10^{-12}	11.123
purine	20	1	5.01×10^{-3}	2.30
pyridine	25		5.62×10^{-6}	5.25
pyridine, 3-chloro	25		1.45×10^{-3}	2.84
o-toluidine	25		3.63×10^{-5}	4.44
m-toluidine	25		1.86×10^{-5}	4.73
p-toluidine	25		8.32×10^{-6}	5.08
triethylamine	18		9.77×10^{-12}	11.01

Handbook of Chemistry and Physics, Fiftieth edition. The Chemical Rubber Co., Cleveland, Ohio, 1969.

TABLE T–15   Solubility product constants (K_{sp})*

ANION	COMPOUND	TEMPERATURE (°C)	K_{sp}
bromate	$AgBrO_3$	25	6×10^{-5}
	$Ba(BrO_3)_2$	25	5.5×10^{-6}
bromide	$CuBr$	18–20	4.15×10^{-8}
	$PbBr_2$	—	4.6×10^{-6}
	Hg_2Br_2	25	1.3×10^{-21}
	$HgBr_2$	—	1.1×10^{-19}
	$AgBr$	25	7.7×10^{-13}
carbonate	$BaCO_3$	25	8.1×10^{-9}
	$CdCO_3$	—	5.2×10^{-12}
	$CaCO_3$	25	8.7×10^{-9}
	$PbCO_3$	18	3.3×10^{-14}
	$MgCO_3$	12	2.6×10^{-5}
	Ag_2CO_3	25	6.5×10^{-12}
	$SrCO_3$	25	1.6×10^{-9}
chloride	$CuCl$	18–20	1.02×10^{-6}
	$PbCl_2$	—	1.6×10^{-5}
	Hg_2Cl_2	25	2×10^{-18}
	$HgCl_2$	—	6.1×10^{-15}
	$AgCl$	25	1.6×10^{-10}
chromate	$BaCrO_4$	18	1.6×10^{-10}
	$PbCrO_4$	18	1.77×10^{-14}
fluoride	BaF_2	18	1.7×10^{-6}
	CaF_2	18	3.4×10^{-11}
hydroxide	$Al(OH)_3$	25	3.7×10^{-15}
	$Ca(OH)_2$	—	5.5×10^{-6}
	$Cu(OH)_2$	25	2.2×10^{-22}
	$Fe(OH)_2$	18	1.6×10^{-14}
	$Mg(OH)_2$	18	1.2×10^{-12}
	$Zn(OH)_2$	18	1.8×10^{-14}
iodate	$Ba(IO_3)_2 \cdot 2H_2O$	25	6.3×10^{-10}
	$Ca(IO_3)_2 \cdot 6H_2O$	18	6.4×10^{-7}
	$Pb(IO_3)_2$	25.8	2.6×10^{-13}
	$AgIO_3$	9.4	0.92×10^{-8}
iodide	CuI	18–20	3.06×10^{-12}
	PbI_2	15	7.47×10^{-9}
	AgI	25	1.5×10^{-16}

Handbook of Chemistry and Physics, Fiftieth edition. The Chemical Rubber Co., Cleveland, Ohio, 1969.

TABLE T–15 (Cont.)

ANION	COMPOUND	TEMPERATURE (°C)	K_{sp}
oxalate	$BaC_2O_4 \cdot 2H_2O$	18	1.2×10^{-7}
	$CaC_2O_4 \cdot H_2O$	25	2.57×10^{-9}
	PbC_2O_4	18	2.74×10^{-11}
sulfate	$BaSO_4$	25	1.08×10^{-10}
	$CaSO_4$	10	1.95×10^{-4}
	$PbSO_4$	18	1.06×10^{-8}
	Ag_2SO_4	—	1.7×10^{-5}
sulfide	CdS	18	3.6×10^{-29}
	CoS	18	3×10^{-26}
	CuS	18	8.5×10^{-45}
	Cu_2S	16–18	2×10^{-47}
	FeS	18	3.7×10^{-19}
	PbS	18	3.4×10^{-28}
	MnS	18	1.4×10^{-15}
	HgS	18	$\sim 3 \times 10^{-52}$
	NiS	18	1.4×10^{-24}
	Ag_2S	18	1.6×10^{-49}
	ZnS	18	1.2×10^{-23}

TABLE T–16 Standard reduction potentials*

REACTION	POTENTIAL (volts)
$\frac{1}{2}F_2 + e^- \longrightarrow F^-$	2.85
$MnO_4^{2-} + 4H^+ + 2e^- \longrightarrow MnO_2 + 2H_2O$	2.3
$S_2O_8^{2-} + 2e^- \longrightarrow 2SO_4^{2-}$	2.0
$PbO_2 + SO_4^{2-} + 4H^+ + 2e^- \longrightarrow PbSO_4 + 2H_2O$	1.685
$Au^+ + e^- \longrightarrow Au$	1.68
$Mn^{3+} + e^- \longrightarrow Mn^{2+}$	1.51
$MnO_4^- + 8H^+ + 5e^- \longrightarrow Mn^{2+} + 4H_2O$	1.491
$Ce^{4+} + e^- \longrightarrow Ce^{3+}$	1.4430
$Au^{3+} + 3e^- \longrightarrow Au$	1.42
$Cl_{2(g)} + 2e^- \longrightarrow 2Cl^-$	1.3583
$Cr_2O_7^{2-} + 14H^+ + 6e^- \longrightarrow 2Cr^{3+} + 7H_2O$	1.33
$O_2 + 4H^+ + 4e^- \longrightarrow 2H_2O$	1.229
$ClO_3^- + 3H^+ + 2e^- \longrightarrow HClO_2 + H_2O$	1.21
$MnO_2 + 4H^+ + 2e^- \longrightarrow Mn^{2+} + 2H_2O$	1.208
$Pt^{2+} + 2e^- \longrightarrow Pt$	~1.2
$HCrO_4^- + 7H^+ + 3e^- \longrightarrow Cr^{3+} + 4H_2O$	1.195
$2IO_3^- + 12H^+ + 10e^- \longrightarrow I_2 + 6H_2O$	1.19
$ClO_3^- + 2H^+ + e^- \longrightarrow ClO_2 + H_2O$	1.15
$Br_{2(l)} + 2e^- \longrightarrow 2Br^-$	1.065
$NO_3^- + 4H^+ + 3e^- \longrightarrow NO + 2H_2O$	0.96
$ClO_{2(aq)} + e^- \longrightarrow ClO_2^-$	0.954
$2Hg^{2+} + 2e^- \longrightarrow Hg_2^{2+}$	0.910
$ClO^- + H_2O + 2e^- \longrightarrow Cl^- + 2OH^-$	0.90
$Pd^{2+} + 2e^- \longrightarrow Pd$	0.83
$O_2 + 4H^+(10^{-7}\,M) + 4e^- \longrightarrow 2H_2O$	0.82
$2NO_3^- + 4H^+ + 2e^- \longrightarrow N_2O_4 + 2H_2O$	0.81
$Ag^+ + e^- \longrightarrow Ag$	0.7996
$\frac{1}{2}Hg_2^{2+} + e^- \longrightarrow Hg$	0.7986
$Fe^{3+} + e^- \longrightarrow Fe^{2+}$	0.770
$UO_2^+ + 4H^+ + e^- \longrightarrow U^{4+} + 2H_2O$	0.62
$Hg_2SO_4 + 2e^- \longrightarrow 2Hg + SO_4^{2-}$	0.6158
$MnO_4^- + 2H_2O + 3e^- \longrightarrow MnO_2 + 4OH^-$	0.588
$H_3AsO_4 + 2H^+ + 2e^- \longrightarrow HAsO_2 + 2H_2O\,(1\,M\;HCl)$	0.58
$MnO_4^- + e^- \longrightarrow MnO_4^{2-}$	0.564
$I_2 + 2e^- \longrightarrow 2I^-$	0.535
$Cu^+ + e^- \longrightarrow Cu$	0.522
$O_2 + 2H_2O + 4e^- \longrightarrow 4OH^-$	0.401
$Cu^{2+} + 2e^- \longrightarrow Cu$	0.3402
$VO^{2+} + 2H^+ + e^- \longrightarrow V^{3+} + H_2O$	0.337
$AgCl + e^- \longrightarrow Ag + Cl^-$	0.2223
$Cu^{2+} + e^- \longrightarrow Cu^+$	0.158
$Sn^{4+} + 2e^- \longrightarrow Sn^{2+}$	0.15
$S + 2e^- \longrightarrow S^{2-}$	0.141

Handbook of Chemistry and Physics, Fiftieth edition. The Chemical Rubber Co., Cleveland, Ohio, 1969.

REACTION	POTENTIAL (volts)
$HgO + H_2O + 2e^- \longrightarrow Hg + 2OH^-$	0.0984
$NO_3^- + H_2O + 2e^- \longrightarrow NO_2^- + 2OH^-$	0.01
$2H^+ + 2e^- \longrightarrow H_2$	0.0000
$Fe^{3+} + 3e^- \longrightarrow Fe$	−0.036
$P + 3H^+ + 3e^- \longrightarrow PH_{3(g)}$	−0.04
$CrO_4^{2-} + 4H_2O + 3e^- \longrightarrow Cr(OH)_3 + 5OH^-$	−0.12
$Pb^{2+} + 2e^- \longrightarrow Pb$	−0.1263
$Sn^{2+} + 2e^- \longrightarrow Sn$	−0.1364
$Ni^{2+} + 2e^- \longrightarrow Ni$	−0.23
$Tl^+ + e^- \longrightarrow Tl$	−0.3363
$PbSO_4 + 2e^- \longrightarrow Pb + SO_4^{2-}$	−0.356
$Cd^{2+} + 2e^- \longrightarrow Cd$	−0.4026
$Fe^{2+} + 2e^- \longrightarrow Fe$	−0.409
	(−0.440)**
$Cr^{3+} + e^- \longrightarrow Cr^{2+}$	−0.41
$S + 2e^- \longrightarrow S^{2-}$	−0.508
$As + 3H^+ + 3e^- \longrightarrow AsH_3$	−0.54
$Cr^{2+} + 2e \longrightarrow Cr$	−0.557
$Fe(OH)_3 + e^- \longrightarrow Fe(OH)_2 + OH^-$	−0.56
$U^{4+} + e^- \longrightarrow U^{3+}$	−0.61
$Ni(OH)_2 + 2e^- \longrightarrow Ni + 2OH^-$	−0.66
$AsO_4^{3-} + 2H_2O + 2e^- \longrightarrow AsO_2^- + 4OH^-$	−0.71
$Cr^{3+} + 3e^- \longrightarrow Cr$	−0.74
$Zn^{2+} + 2e^- \longrightarrow Zn$	−0.7628
$Se + 2e^- \longrightarrow Se^{2-}$	−0.78
$HSnO_2^- + H_2O + 2e^- \longrightarrow Sn + 3OH^-$	−0.79
$Cd(OH)_2 + 2e^- \longrightarrow Cd + 2OH^-$	−0.809
$2H_2O + 2e^- \longrightarrow H_2 + 2OH^-$	−0.8277
$Te + 2e^- \longrightarrow Te^{2-}$	−0.92
$Mn^{2+} + 2e^- \longrightarrow Mn$	−1.029
$V^{2+} + 2e^- \longrightarrow V$	−1.2
$ZnO_2^- + 2H_2O + 2e^- \longrightarrow Zn + 4OH^-$	−1.216
$Zn(OH)_2 + 2e^- \longrightarrow Zn + 2OH^-$	−1.245
$Be^{2+} + 2e^- \longrightarrow Be$	−1.70
$Al^{3+} + 3e^- \longrightarrow Al(0.1\ M\ NaOH)$	−1.706
$Ce^{3+} + 3e^- \longrightarrow Ce$	−2.335
$H_2AlO_3^- + H_2O + 3e^- \longrightarrow Al + 4OH^-$	−2.35
$Mg^{2+} + 2e^- \longrightarrow Mg$	−2.375
$Na^+ + e^- \longrightarrow Na$	−2.7109
$Ca^{2+} + 2e^- \longrightarrow Ca$	−2.76
$Sr^{2+} + 2e^- \longrightarrow Sr$	−2.89
$K^+ + e^- \longrightarrow K$	−2.92

**From W. M. Latimer, *The Oxidation States of the Elements and Their Potentials in Aqueous Solution*, Second edition. Prentice-Hall, Englewood Cliffs, N.J., 1952.

TABLE T–17 Electronegativities, ionization potentials and electron affinities of the elements

NAME	SYMBOL	ATOMIC NO.	ELEC. NEG.*	IONIZ. POTENTIAL (e.v.)*		ELECTRON AFFINITY (e.v.)
				1st	2nd	
hydrogen	H	1	2.1	13.60		0.75
helium	He	2		24.58	54.5	
lithium	Li	3	1.0	5.39	75.6	0.77
beryllium	Be	4	1.5	9.32	18.21	0.38
boron	B	5	2.0	8.30	25.15	0.18
carbon	C	6	2.5	11.26	24.38	1.29
nitrogen	N	7	3.0	14.54	29.61	−0.21
oxygen	O	8	3.5	13.61	35.15	1.46
fluorine	F	9	4.0	17.42	34.98	3.50
neon	Ne	10		21.60	41.07	−0.22
sodium	Na	11	0.9	5.14	47.29	0.54
magnesium	Mg	12	1.2	7.64	15.03	−0.22
aluminum	Al	13	1.5	5.98	18.82	0.20
silicon	Si	14	1.8	8.15	16.34	1.36
phosphorus	P	15	2.1	11.02	19.65	0.71
sulfur	S	16	2.5	10.36	23.4	2.04
chlorine	Cl	17	3.0	13.01	23.8	3.62
argon	Ar	18		15.76	27.6	−0.37
potassium	K	19	0.8	4.34	31.8	0.47
calcium	Ca	20	1.0	6.11	11.9	−1.93
scandium	Sc	21	1.3	6.56	12.89	
titanium	Ti	22	1.5	6.83	13.63	
vanadium	V	23	1.6	6.74	14.2	
chromium	Cr	24	1.6	6.76	16.6	
manganese	Mn	25	1.5	7.43	15.7	
iron	Fe	26	1.8	7.90	16.16	
cobalt	Co	27	1.8	7.86	17.3	
nickel	Ni	28	1.8	7.63	18.2	
copper	Cu	29	1.9	7.72	20.34	
zinc	Zn	30	1.6	9.39	17.89	~0.9
gallium	Ga	31	1.6	6.00	20.43	0.37
germanium	Ge	32	1.8	7.88	15.86	1.44
arsenic	As	33	2.0	9.81	20.1	1.07
selenium	Se	34	2.4	9.75	21.3	2.12
bromine	Br	35	2.8	11.84	19.1	3.36
krypton	Kr	36		14.00	26.4	−0.42
rubidium	Rb	37	0.8	4.18	27.36	0.42
strontium	Sr	38	1.0	5.69	10.98	−1.51
yttrium	Y	39	1.2	6.5	12.3	

*D. B. Summers. *Chemistry Handbook*. Willard Grant Press, Inc., Boston, 1970.

TABLE T–17 (Cont.)

NAME	SYMBOL	ATOMIC NO.	ELEC. NEG.*	IONIZ. POTENTIAL (e.v.)*		ELECTRON AFFINITY (e.v.)
				1st	2nd	
zirconium	Zr	40	1.4	6.95	13.97	
niobium	Nb	41	1.6	6.77	14.4	
molybdenum	Mo	42	1.8	7.10	16.2	
technetium	Tc	43	1.9	7.28	15.26	
ruthenium	Ru	44	2.2	7.36	16.8	
rhodium	Rh	45	2.2	7.46	18.1	
palladium	Pd	46	2.2	8.33	19.8	
silver	Ag	47	1.9	7.57	21.4	
cadmium	Cd	48	1.7	8.99	16.84	~.6
indium	In	49	1.7	5.79	18.79	0.20
tin	Sn	50	1.8	7.34	14.5	1.03
antimony	Sb	51	1.9	8.64	18	0.94
tellurium	Te	52	2.1	9.01		1.96
iodine	I	53	2.5	10.45	19.4	3.06
xenon	Xe	54		12.14	21.1	−0.45
cesium	Cs	55	0.7	3.89	23.4	0.39
barium	Ba	56	0.9	5.21	9.95	−0.48
Elements No.		57–71	1.1 1.2	(5–6.9)	(11.4–14.8)	
hafnium	Hf	72	1.3	5.5	14.8	
tantalum	Ta	73	1.5	7.88	16.1	
tungsten	W	74	1.7	7.98	17.7	
rhenium	Re	75	1.9	7.87	16.6	
osmium	Os	76	2.2	8.7	17.0	
iridium	Ir	77	2.2	9		
platinum	Pt	78	2.2	9.0	18.6	
gold	Au	79	2.4	9.22	19.95	
mercury	Hg	80	1.9	10.43	18.65	
thallium	Tl	81	1.8	6.11	20.32	0.32
lead	Pb	82	1.8	7.42	14.96	1.03
bismuth	Bi	83	1.9	7.29	16.6	0.95
polonium	Po	84	2.0	8.43		1.32
astatine	At	85	2.2	—		2.80
radon	Rn	86	—	10.75		
francium	Fr	87	0.7	—		
radium	Ra	88	0.9	5.28	10.10	
actinium	Ac	89	1.1	6.6	12.2	
thorium	Th	90	1.3	6.95		
proactinium	Pa	91	1.5	—		
uranium	U	92	1.7	6.1		
Elements No.		93–103	1.3	(6–?)		

TABLE T–18 Relative strengths of acids and bases*

CONJUGATE ACID	CONJUGATE BASE
$HClO_4$	ClO_4^-
HNO_3	NO_3^-
HCl	Cl^-
H_2SO_4	HSO_4^-
H_3O^+	H_2O
$H_2C_2O_4$	$HC_2O_4^-$
HSO_4^-	SO_4^{2-}
H_2SO_3	HSO_3^-
H_3PO_4	$H_2PO_4^-$
HF	F^-
HNO_2	NO_2^-
$HCOOH$	$HCOO^-$
$HC_2O_4^-$	$C_2O_4^{2-}$
$HC_2H_3O_2$	$C_2H_3O_2^-$
$Al(H_2O)_6^{3+}$	$Al(H_2O)_5OH^{2+}$
H_2CO_3	HCO_3^-
H_2S	HS^-
$H_2PO_4^-$	HPO_4^{2-}
HSO_3^-	SO_3^{2-}
HCN	CN^-
NH_4^+	NH_3
HCO_3^-	CO_3^{2-}
HPO_4^{2-}	PO_4^{3-}
H_2O	OH^-
HS^-	S^{2-}
NH_3	NH_2^-

INCREASING ACID STRENGTH →

INCREASING BASE STRENGTH →

TABLE T–19 Molal boiling and freezing point constants*

SUBSTANCE	$t_b(°C)$	$K_b(°C)$	$t_f(°C)$	$K_f(°C)$
water	100	0.52	0.00	1.86
acetic acid	118	2.93	16.7	3.90
benzene	80.1	2.64	5.53	4.90
carbon tetrachloride	76.8	5.03	−22.96	31.8
chloroform	61.2	3.88	−63.5	7.3
camphor	208	5.95	178.4	37.7
dioxane	101	—	10.5	4.9
ethanol	78.4	1.23	−115	1.99
cyclohexane	80.7	2.79	6.5	20.0
naphthalene	218.0	5.65	80.1	6.8

*D. B. Summers. *Chemistry Handbook.* Willard Grant Press, Inc., Boston, 1970.

x	0	1	2	3	4	5	6	7	8	9
1.0	.0000	.0043	.0086	.0128	.0170	.0212	.0253	.0294	.0334	.0374
1.1	.0414	.0453	.0492	.0531	.0569	.0607	.0645	.0682	.0719	.0755
1.2	.0792	.0828	.0864	.0899	.0934	.0969	.1004	.1038	.1072	.1106
1.3	.1139	.1173	.1206	.1239	.1271	.1303	.1335	.1367	.1399	.1430
1.4	.1461	.1492	.1523	.1553	.1584	.1614	.1644	.1673	.1703	.1732
1.5	.1761	.1790	.1818	.1847	.1875	.1903	.1931	.1959	.1987	.2014
1.6	.2041	.2068	.2095	.2122	.2148	.2175	.2201	.2227	.2253	.2279
1.7	.2304	.2330	.2355	.2380	.2405	.2430	.2455	.2480	.2504	.2529
1.8	.2553	.2577	.2601	.2625	.2648	.2672	.2695	.2718	.2742	.2765
1.9	.2788	.2810	.2833	.2856	.2878	.2900	.2923	.2945	.2967	.2989
2.0	.3010	.3032	.3054	.3075	.3096	.3118	.3139	.3160	.3181	.3201
2.1	.3222	.3243	.3263	.3284	.3304	.3324	.3345	.3365	.3385	.3404
2.2	.3424	.3444	.3464	.3483	.3502	.3522	.3541	.3560	.3579	.3598
2.3	.3617	.3636	.3655	.3674	.3692	.3711	.3729	.3747	.3766	.3784
2.4	.3802	.3820	.3838	.3856	.3874	.3892	.3909	.3927	.3945	.3962
2.5	.3979	.3997	.4014	.4031	.4048	.4065	.4082	.4099	.4116	.4133
2.6	.4150	.4166	.4183	.4200	.4216	.4232	.4249	.4265	.4281	.4298
2.7	.4314	.4330	.4346	.4362	.4378	.4393	.4409	.4425	.4440	.4456
2.8	.4472	.4487	.4502	.4518	.4533	.4548	.4564	.4579	.4594	.4609
2.9	.4624	.4639	.4654	.4669	.4683	.4698	.4713	.4728	.4742	.4757
3.0	.4771	.4786	.4800	.4814	.4829	.4843	.4857	.4871	.4886	.4900
3.1	.4914	.4928	.4942	.4955	.4969	.4983	.4997	.5011	.5024	.5038
3.2	.5051	.5065	.5079	.5092	.5105	.5119	.5132	.5145	.5159	.5172
3.3	.5185	.5198	.5211	.5224	.5237	.5250	.5263	.5276	.5289	.5302
3.4	.5315	.5328	.5340	.5353	.5366	.5378	.5391	.5403	.5416	.5428
3.5	.5441	.5453	.5465	.5478	.5490	.5502	.5514	.5527	.5539	.5551
3.6	.5563	.5575	.5587	.5599	.5611	.5623	.5635	.5647	.5658	.5670
3.7	.5682	.5694	.5705	.5717	.5729	.5740	.5752	.5763	.5775	.5786
3.8	.5798	.5809	.5821	.5832	.5843	.5855	.5866	.5877	.5888	.5899
3.9	.5911	.5922	.5933	.5944	.5955	.5966	.5977	.5988	.5999	.6010
4.0	.6021	.6031	.6042	.6053	.6064	.6075	.6085	.6096	.6107	.6117
4.1	.6128	.6138	.6149	.6160	.6170	.6180	.6191	.6201	.6212	.6222
4.2	.6232	.6243	.6253	.6263	.6274	.6284	.6294	.6304	.6314	.6325
4.3	.6335	.6345	.6355	.6365	.6375	.6385	.6395	.6405	.6415	.6425
4.4	.6435	.6444	.6454	.6464	.6474	.6484	.6493	.6503	.6513	.6522
4.5	.6532	.6542	.6551	.6561	.6571	.6580	.6590	.6599	.6609	.6618
4.6	.6628	.6637	.6646	.6656	.6665	.6675	.6684	.6693	.6702	.6712
4.7	.6721	.6730	.6739	.6749	.6758	.6767	.6776	.6785	.6794	.6803
4.8	.6812	.6821	.6830	.6839	.6848	.6857	.6866	.6875	.6884	.6893
4.9	.6902	.6911	.6920	.6928	.6937	.6946	.6955	.6964	.6972	.6981
5.0	.6990	.6998	.7007	.7016	.7024	.7033	.7042	.7050	.7059	.7067
5.1	.7076	.7084	.7093	.7101	.7110	.7118	.7126	.7135	.7143	.7152
5.2	.7160	.7168	.7177	.7185	.7193	.7202	.7210	.7218	.7226	.7235
5.3	.7243	.7251	.7259	.7267	.7275	.7284	.7292	.7300	.7308	.7316
5.4	.7324	.7332	.7340	.7348	.7356	.7364	.7372	.7380	.7388	.7396
x	0	1	2	3	4	5	6	7	8	9

x	0	1	2	3	4	5	6	7	8	9
5.5	.7404	.7412	.7419	.7427	.7435	.7443	.7451	.7459	.7466	.7474
5.6	.7482	.7490	.7497	.7505	.7513	.7520	.7528	.7536	.7543	.7551
5.7	.7559	.7566	.7574	.7582	.7589	.7597	.7604	.7612	.7619	.7627
5.8	.7634	.7642	.7649	.7657	.7664	.7672	.7679	.7686	.7694	.7701
5.9	.7709	.7716	.7723	.7731	.7738	.7745	.7752	.7760	.7767	.7774
6.0	.7782	.7789	.7796	.7803	.7810	.7818	.7825	.7832	.7839	.7846
6.1	.7853	.7860	.7868	.7875	.7882	.7889	.7896	.7903	.7910	.7917
6.2	.7924	.7931	.7938	.7945	.7952	.7959	.7966	.7973	.7980	.7987
6.3	.7993	.8000	.8007	.8014	.8021	.8028	.8035	.8041	.8048	.8055
6.4	.8062	.8069	.8075	.8082	.8089	.8096	.8102	.8109	.8116	.8122
6.5	.8129	.8136	.8142	.8149	.8156	.8162	.8169	.8176	.8182	.8189
6.6	.8195	.8202	.8209	.8215	.8222	.8228	.8235	.8241	.8248	.8254
6.7	.8261	.8267	.8274	.8280	.8287	.8293	.8299	.8306	.8312	.8319
6.8	.8325	.8331	.8338	.8344	.8351	.8357	.8363	.8370	.8376	.8382
6.9	.8388	.8395	.8401	.8407	.8414	.8420	.8426	.8432	.8439	.8445
7.0	.8451	.8457	.8463	.8470	.8476	.8482	.8488	.8494	.8500	.8506
7.1	.8513	.8519	.8525	.8531	.8537	.8543	.8549	.8555	.8561	.8567
7.2	.8573	.8579	.8585	.8591	.8597	.8603	.8609	.8615	.8621	.8627
7.3	.8633	.8639	.8645	.8651	.8657	.8663	.8869	.8675	.8681	.8686
7.4	.8692	.8698	.8704	.8710	.8716	.8722	.8727	.8733	.8739	.8745
7.5	.8751	.8756	.8762	.8768	.8774	.8779	.8785	.8791	.8797	.8802
7.6	.8808	.8814	.8820	.8825	.8831	.8837	.8842	.8848	.8854	.8859
7.7	.8865	.8871	.8876	.8882	.8887	.8893	.8899	.8904	.8910	.8915
7.8	.8921	.8927	.8932	.8938	.8943	.8949	.8954	.8960	.8965	.8971
7.9	.8976	.8982	.8987	.8993	.8998	.9004	.9009	.9015	.9020	.9025
8.0	.9031	.9036	.9042	.9047	.9053	.9058	.9063	.9069	.9074	.9079
8.1	.9085	.9090	.9096	.9101	.9106	.9112	.9117	.9122	.9128	.9133
8.2	.9138	.9143	.9149	.9154	.9159	.9165	.9170	.9175	.9180	.9186
8.3	.9191	.9196	.9201	.9206	.9212	.9217	.9222	.9227	.9232	.9238
8.4	.9243	.9248	.9253	.9258	.9263	.9269	.9274	.9279	.9284	.9289
8.5	.9294	.9299	.9304	.9309	.9315	.9320	.9325	.9330	.9335	.9340
8.6	.9345	.9350	.9355	.9360	.9365	.9370	.9375	.9380	.9385	.9390
8.7	.9395	.9400	.9405	.9410	.9415	.9420	.9425	.9430	.9435	.9440
8.8	.9445	.9450	.9455	.9460	.9465	.9469	.9474	.9479	.9484	.9489
8.9	.9494	.9499	.9504	.9509	.9513	.9518	.9523	.9528	.9533	.9538
9.0	.9542	.9547	.9552	.9557	.9562	.9566	.9571	.9576	.9581	.9586
9.1	.9590	.9595	.9600	.9605	.9609	.9614	.9619	.9624	.9628	.9633
9.2	.9638	.9643	.9647	.9652	.9657	.9661	.9666	.9671	.9675	.9680
9.3	.9685	.9689	.9694	.9699	.9703	.9708	.9713	.9717	.9722	.9727
9.4	.9731	.9736	.9741	.9745	.9750	.9754	.9759	.9763	.9768	.9773
9.5	.9777	.9782	.9786	.9791	.9795	.9800	.9805	.9809	.9814	.9818
9.6	.9823	.9827	.9832	.9836	.9841	.9845	.9850	.9854	.9859	.9863
9.7	.9868	.9872	.9877	.9881	.9886	.9890	.9894	.9899	.9903	.9908
9.8	.9912	.9917	.9921	.9926	.9930	.9934	.9939	.9943	.9948	.9952
9.9	.9956	.9961	.9965	.9969	.9974	.9978	.9983	.9987	.9991	.9996
x	0	1	2	3	4	5	6	7	8	9

ANSWERS TO PROBLEMS AND EXERCISES

CHAPTER TWO PROBLEMS

1.

atom	protons	neutrons	electrons
$^{18}_{8}O$	8	10	8
$^{19}_{9}F$	9	10	9
$^{7}_{3}Li$	3	4	3

2. 19.0 amu **3.** 35.45 **4.** 26 protons; 23 electrons
5. the number of protons or atomic number
6. $C_6H_{12}O_6$ **7.** $C_4H_6F_3Cl$; 3 CH_2 groups; 3 F groups; $CF_3(CH_2)_3Cl$ **8.** chromite ion; iodite ion; hypobromite ion **9. (a)** $AlCl_3$ **(b)** $(NH_4)_2SO_4$ **(c)** Na_2SiO_3
10. No; it is a mixture since it does not contain atoms in any definite or distinct proportion; Yes, it can be separated by evaporating the water; physical.
11. 180.15 **12.** 110.98 **13.** .574 g S per 1.000 g Fe
14. 1 g H per 8 g O **15.** %Mg = 27.7, %P = 23.6, %O = 48.7; %C = 40.0, %H = 6.7, %O = 53.3
16. CH_2O; no; the molecular weight **17.** rubidium bromide, beryllium fluoride, aluminum nitride, chromium (III) oxide, ammonium carbonate, hydrogen bromide, potassium dichromate, chromium (II) bisulfate or chromium (II) hydrogen sulfate, iron (III) cyanide **18.** $RbBrO$, Ca_3P_2, $FePO_4$, Na_2HPO_4, CaC_2, $Fe_2(C_2O_4)_3$, H_2Se **19.** tellurium tetrafluoride, sulfur dioxide, phosphorous trichloride, sulfur hexafluoride, chlorine dioxide, dichlorine heptaoxide, oxygen difluoride **20.** CCl_4, PCl_5, BrF_5, P_2O_5, BF_3, SiF_4 **21.** 0.312 g-atoms, 1.88×10^{23} atoms
22. 0.156 moles, 9.4×10^2 molecules **23.** 1.81×10^{24} ions Ca^{2+}; 3.61×10^{24} ions Cl^- **24.** mole fraction ethanol = 0.881; mole fraction H_2O = 0.119 **25.** 4.68 $\times 10^{-2}M$ **26.** 2.02×10^{-3} m Mg^{2+}; 4.04×10^{-3} m NO_3^-
27. 6.15 atm. **28.** 0.377 g **29.** P_{N_2} = 577 torr; P_{O_2} = 156 torr

CHAPTER TWO EXERCISES

1.

e	at. no.	p	e	n	mass no.
Mg	12	12	12	12	24
Cl	17	17	17	18	35
Sn	50	50	50	69	119
Ba^{2+}	56	56	54	81	137
S^{2-}	16	16	18	19	35

2. 10.81 **3. (a)** NO_2; **(b)** N_2O_4 **4. (a)** For a substance to evaporate, it must absorb its heat of vaporization from its surroundings. On your skin this absorption of heat causes a sensation of cooling. **(b)** Water would not be as effective. Its boiling point is much higher than body temperature and therefore would vaporize slowly compared to isopropyl alcohol. Cooking oil would be even less effective for the same reason. **5. (a)** calcium hydride **(b)** zinc acetate **(c)** strontium oxalate **(d)** tin (IV) oxide **(e)** silver thiosulfate **(f)** diiodine pentoxide **(g)** boron nitride **(h)** cesium hypoiodite **(i)** chromium (II) bromide **(j)** strontium dichromate
6. (a) CrI_2 **(b)** Na_2O_2 **(c)** MgH_2 **(d)** AuF_3
(e) $Sb(MnO_4)_3$ **(f)** Na_2SiO_3 **(g)** $Fe(MnO_4)_2$
(h) $CaSiO_3$ **(i)** Na_2Se **(j)** $Ca(H_2PO_4)_2$ **7.** NH_3; NH_3
8. $C_{14}H_9Cl_5$; 50.1%; $CH(C_6H_4Cl)_2CCl_3$ **9. (a)** 8.2×10^{-2} g; **(b)** $C_8H_{16}O$; **(c)** 75% C, 12.5% H, 12.5% O
10. (a) 19 protons, 19 electrons, 22 neutrons
(b) The atomic weight is a weighted average of the atomic weights of all isotopes present. The isotope in part **(a)** is a heavy isotope, heavier by approximately 2 neutrons than the average. **11.** $NaBO_2$
12. 15.9 g/mole **13. (a)** BH_3; **(b)** 27.6 g/mole;
(c) B_2H_6 **14. (a)** 3.08×10^{21} molecules; **(b)** 2.25 ml.
15. (a) 25.2 g; **(b)** 0.141 M **16.** P_{O_2} = 0.8 atm, P_{N_2O} = 0.2 atm **17. (a)** 9.75×10^{-2} gram-atoms/liter
(b) 0.224%; **(c)** 5.71 g **18.** 38.0% **19.** Fe_2S
20. 2.2×10^{-2} g **21. (a)** 3.95×10^9 g; **(b)** 6.17×10^{-6} M; **(c)** $C_5H_{11}PN_2O_2$; **(d)** 17.3%

22.

(a) mole fraction	(b) mole fraction	(b) wt. %
0.209 CO_2	0.128 C	29.4 C
0.114 SO_2	0.070 S	42.7 S
0.041 N_2	0.050 N	13.5 N
0.616 H_2O	0.752 H	14.4 H

23. (a) 2.31×10^{-3} mole; **(b)** 8.66×10^2 l. **24.** $^{40}_{19}K$
25. (a) 31.6 g; **(b)** 0.160 g-atoms; **(c)** 9.66×10^{22} electrons **26.** 0.54 mi. **27.** 80 g/mole; may be too large, could be any factor of 80 such as 2, 5, or 40. **28.** Dilute 27.2 ml. of the 60% solution to a volume of 500 ml. **29.** 78 g/mole **30.** 1.6×10^4 g/mole

CHAPTER THREE PROBLEMS

1. 2.4×10^{24} **2.** $4Al + 3O_2 \longrightarrow 2Al_2O_3$; $PCl_3 + 3H_2O \longrightarrow H_3PO_3 + 3HCl$; $Ca_3P_2 + 6H_2O \longrightarrow 3Ca(OH)_2 + 2PH_3$; $3Ca(OH)_2 + 2H_3PO_4 \longrightarrow Ca_3(PO_4)_2 + 6H_2O$ **3.** Direct combination; Displacement; Double displacement; Displacement; Displacement; Decomposition **4.** $\overset{+1+5-2}{HNO_3}$; $\overset{+\ -}{HCl}$; $\overset{+5-2}{(PO_4)^{3-}}$; $\overset{+2\ +5-2}{Ca_3(PO_4)_2}$; $\overset{+3}{Fe^{3+}}$

5.

Oxidized	Reduced	No. e
Au	F_2	6
H_2S	$Cr_2O_7^{2-}$	6
Cu	Cl_2	2
Fe^{2+}	Br_2	2

6. Between Zn^{2+} and Cu^{2+} **7. (a)** $2AuBr_4^- + 6Hg$ $\rightleftharpoons 2Au + 3Hg_2Br_2 + 2Br^-$ **(b)** $7Fe(CN)_6^{3-} + Re + 8OH^- \rightleftharpoons 7Fe(CN)_6^{4-} + ReO_4^- + 4H_2O$
(c) $3Zn + 2MnO_4^- + 4OH^- \rightleftharpoons 3ZnO_2^{2-} + 2MnO_2 + 2H_2O$ **8. (a)** 19.0 g/eq; **(b)** 2.63 eq, 1.32 mole; **(c)** 2.63 eq; **(d)** 0.877 g-atoms;
(e) 173 g **(f)** 2.63 equiv.; **(g)** Same; equivalent weight is the weight which will produce 1 mole of electrons in an oxidation-reduction reaction. The number of electrons gained must equal the number lost. Therefore, the number of equivalents of reactants is equal since they both produce or accept an equal number of electrons. **9.** potassium chlorate; chloric acid; hydrocyanic acid; hydrofluoric acid, sodium hydride; potassium periodate; periodic acid; calcium chlorite; nitrous acid; formic acid
10. $NaOH + HC_2H_3O_2 \longrightarrow NaC_2H_3O_2 + H_2O$; $Ca(OH)_2 + 2HCOOH \longrightarrow Ca(HCOO)_2 + 2H_2O$; $HCl + NH_4OH \longrightarrow NH_4Cl + H_2O$; $3H_2SO_4 + 2Al(OH)_3 \longrightarrow Al_2(SO_4)_3 + 6H_2O$; $H_2S + NaOH \longrightarrow NaHS + H_2O$; $3NaOH + H_3PO_4 \longrightarrow Na_3PO_4 + 3H_2O$ **11. (a)** eq. wt. $H_2SO_4 = 49$, eq. wt. $NaOH = 40$; **(b)** 0.51 equiv.; **(c)** 0.26 equiv. **12. (a)** base; base; acid; salt; acid; base; salt; base **(b)** $HClO_4 + NH_4Ac \longrightarrow HAc + NH_4ClO_4$ **13.** acid; base; salt; salt; base; acid **14.** $HNO_3 + NH_3 \rightleftharpoons NH_4^+ + NO_3^-$; $HCN + Cl^- \rightleftharpoons$ N.R.; $NH_4^+ + OH^- \rightleftharpoons NH_3 + H_2O$; $C_2H_5OH + C_6H_5O^- \rightleftharpoons$ N.R.; $HClO_4 + SO_4^{2-} \rightleftharpoons ClO_4^- + HSO_4^-$ **15.** 4.0 l. **16.** 4.82×10^{23} molecules **17.** 3.72 ml. **18.** 10 g SO_2
19. 6.02×10^{24} molecules **20.** 6.7 l. **21.** 3.9 g.
22. C_2H_6; 4.2 g C_2H_6 remaining **23.** 0.127 N
24. 19.4 ml. **25.** 0.061 g. **26.** 55.6 ml.

CHAPTER THREE EXERCISES

1. (a) $2H_3PO_4 + 3Mg(OH)_2 \longrightarrow Mg_3(PO_4)_2 + 6H_2O$
(b) $5SO_3^{2-} + 2MnO_4^- + 6H^+ \longrightarrow 5SO_4^{2-} + 2Mn^{2+} + 3H_2O$; SO_3^{2-}—reducing agent; MnO_4^-—oxidizing agent **(c)** $N_2O_5 + H_2O \longrightarrow 2HNO_3$ **(d)** $P_4O_{10} + 6H_2O \longrightarrow 4H_3PO_4$ **(e)** $PbCl_2 + 2CH_3COOH \longrightarrow Pb(CH_3COO)_2 + 2HCl$ **(f)** $Zn + H_2SO_4 \longrightarrow ZnSO_4 + H_2$; Zn—reducing agent; H_2SO_4—oxidizing agent
(g) $2N_2O \longrightarrow 2N_2 + O_2$; N_2O—oxidizing and reducing agent **(h)** $PBr_3 + 3H_2O \longrightarrow H_3PO_3 + 3HBr$
(i) $PbCO_3 + 2HCl \longrightarrow PbCl_2 + CO_2 + H_2O$ **(j)** Fe

$+ 4HNO_3 \longrightarrow Fe(NO_3)_3 + NO + 2H_2O$; Fe—reducing agent; HNO_3—oxidizing agent **(k)** $2HNO_3 + 6HCl \longrightarrow 2NO + 3Cl_2 + 4H_2O$; HCl—reducing agent; HNO_3—oxidizing agent **(l)** $2Al + 2NaOH + 2H_2O \longrightarrow 2NaAlO_2 + 3H_2$; Al—reducing agent + H_2O—oxidizing agent **2.** 21.0 l. **3. (a)** 2.38 moles; **(b)** 38.8 l. **4. (a)** 708 moles; **(b)** 1,440 hr.
5. (a) 5.0 M; **(b)** 0.6 mole; **(c)** 100 ml. **6.** 5.69 g.
7. 1.21 M
8.

(a) $HSO_3^- + H_2O \longrightarrow H_3O^+ + SO_3^{2-}$
    acid   base   acid   base

(b) $NH_4^+ + H_2O \longrightarrow NH_3 + H_3^+O$
    acid   base   base   acid

(c) $H_2O + H_2O \longrightarrow H_3O^+ + OH^-$
    acid   base   acid   base

(d) $H_2O + CO_3^{2-} \longrightarrow HCO_3^- + OH^-$
    acid   base   acid   base

9. $BH^+ - B$; $H_3^+O - H_2O$; $AH^+ - A$; $CH^+ - C$
10. Through dissociation and recombination of ions: $D_2O \rightleftharpoons D^+ + OD^-$; $H_2O \rightleftharpoons H^+ + OH^-$; $D^+ + OH^- \rightleftharpoons DHO$; $D^+ + OD^- \rightleftharpoons D_2O$;
11. (a) $2NaHCO_3 \longrightarrow Na_2CO_3 + H_2O + CO_2$
(b) 11.5 l. **(c)** A handful of baking soda when exposed to the heat of the fire will decompose, giving off about 12 liters of CO_2 which is heavier than air. This smothers the fire since CO_2 does not sustain oxidation. **12.** 327 mole NH_3; 150 torr N_2, 450 torr H_2 **13. (a)** $AgNO_3 + KNH_2 \longrightarrow KNO_{3(s)} + AgNH_2$ **(b)** $Pb(NO_3)_2 + 2KNH_2 \longrightarrow PbNH + NH_3 + 2KNO_{3(s)}$; $AgNO_3 + KOH \longrightarrow KNO_3 + AgOH$; $Pb(NO_3)_2 + 2KOH \longrightarrow PbO + H_2O + 2KNO_3$ **14. (a)** 3.07 g-atom Fe corroded **(b)** 8.87×10^{-4} cm. **15. (a)** NH_3 **(b)** $HCN + CH_3COO^- \longrightarrow$ N.R.; $HCN + NH_2^- \longrightarrow NH_3 + CN^-$; $HCN + OH^- \longrightarrow H_2O + CN^-$ **(c)** CH_3COOH **16. (a)** $CH_2=CH-CH_3 + NH_3 + 3/2O_2 \longrightarrow CH_2=CH-CN + 3H_2O$ **(b)** 2.6×10^3 kg. **17.** 7.63×10^3 ml. **18.** 1.34 hr. **19. (a)** 124 kg. ZnO and 18.4 kg. C **(b)** in reactor, above 1180 °K (to achieve vaporization); in condenser, less than 1180 °K but greater than 693 °K. (At less than 673 °K, it would solidify and clog the condenser.) **(c)** 1.50×10^5 l. **(d)** 3.74×10^4 l. **(e)** Breathing CO can endanger life. **20. (a)** 3.12×10^3 mole; **(b)** 1.66×10^5 ml.; **(c)** 31.2 m **21.** 103 l.

CHAPTER FOUR PROBLEMS

1. Yes ($7.95 \times 10^{-12} > 6.73 \times 10^{-12}$) **2.** No ($5.68 \times 10^{-12} < 6.73 \times 10^{-12}$) **3.** 3.156×10^{15} sec^{-1}; 949.8 Å; Lyman series **4.** $H_\infty = 3,647.1$ Å; As n_U becomes larger, $1/n_U^2$ becomes small and approaches zero and $\bar{\nu}$ and λ approach a constant value. **5.** 2.178×10^{-11} erg **6.** Incorrect because hydrogen atoms (or any atoms) can only absorb discrete wavelengths of light which correspond to energy transitions within the atom. **7.** 3.87 Å **8.** 1.91×10^{-28} Å; no effect since wavelength is very small.

9.

subshell	*l*	*m*
4s	0	0
4p	1	+1, 0, −1
4d	2	+2, +1, 0, −1, −2
4f	3	+3, +2, +1, 0, −1, −2, −3

There are 16 orbitals.

10. No **11.** 3

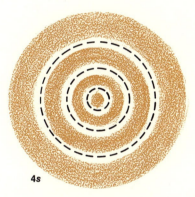

4s

12. A 3*d* has 2 nodal planes and 0 spherical nodes. A 4*p* has 1 nodal plane and 2 spherical nodes.

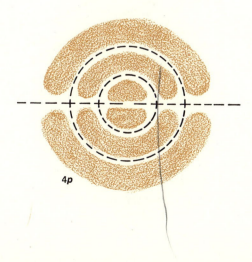

4p

13. The number of degenerate orbitals is n^2 and each orbital contains 2 electrons.

$$\therefore 2(2)^2 = 8$$

	n	*l*	*m*	m_s
1.	2	0	0	1/2
2.	2	0	0	−1/2
3.	2	1	−1	1/2
4.	2	1	−1	−1/2
5.	2	1	+1	1/2
6.	2	1	+1	−1/2
7.	2	1	0	1/2
8.	2	1	0	−1/2

14. 14 **15.** Cl: $1s^2 2s^2 2p^6 3s^2 3p^5$; Fe: $1s^2 2s^2 2p^6 3s^2 3p^6 3d^6 4s^2$
16. Mn: $1s^2 2s^2 2p^6 3s^2 3p^6 3d^5 4s^2$; Mn^{2+}: $1s^2 2s^2 2p^6 3s^2 3p^6 3d^5$; Mn^{3+}. $1s^2 2s^2 2p^6 3s^2 3p^6 3d^4$

CHAPTER FOUR EXERCISES

1. (a) $1s^2 2s^2 2p^6 3s^2 3p^6 3d^{10} 4s^2 4p^6 5s^1$

(b)

	n	*l*	*m*	m_s
$1s^2$	1	0	0	±1/2
$2s^2$	2	0	0	±1/2
$2p^6$	2	1	0, ±1	±1/2
$3s^2$	3	0	0	±1/2
$3p^6$	3	1	0, ±1	±1/2
$3d^{10}$	3	2	0, ±1, ±2	±1/2
$4s^2$	4	0	0	±1/2
$4p^6$	4	1	0, ±1	±1/2
$5s^1$	5	0	0	±1/2

2. Classical description of an electron treats it as if it obeys Newton's laws of motion while the quantum mechanical description assumes it has wave motion. Classical mechanics describes the exact position and velocity of the electron whereas wave mechanics describes the probability of finding the electron at a given position. **3. (a)** 19; **(b)** 38; **(c)** 7; **(d)** 12; **(e)** none; **(f)** 19; **(g)** cannot say **4. (a)** As the orbitals are filled, the 4*d* orbital must be lower in energy than the 5*s* and therefore they are filled first (Half-filled orbitals and filled orbitals are more stable than other partially filled orbitals). **(b)** Ag$^+$: $1s^2 2s^2 2p^6 3s^2 3p^6 3d^{10} 4s^2 4s^6 4d^{10}$ **5.** Less energy is required to remove the 3*d* electron from Fe^{2+} than from Co^{3+}. Co has 27 protons, Fe has 26, and therefore more positive charge is holding the 24 Co electrons than the 24 Fe electrons. **6. (a)** planar nodes: one each; spherical nodes: two (4*p*) and three (5*p*)

(b)

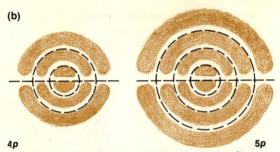

4p   **5p**

5*p* is larger since size increases with n.   **(c)** Charge densities get smaller in the outer region as orbitals get larger. That is, size increases without an increase in the number of electrons. Electrons are attracted toward the inner regions of the orbitals by the positively charged nucleus.

7.

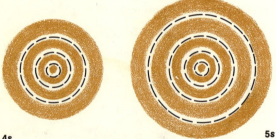

4s   **5s**

As *n* becomes larger the *p* orbitals have more of a spherical shape and the *s* and *p* orbitals are more similar. The effect of the nodal plane on the shape of the *p* orbital is less significant as *n* becomes large.   **8.** Yes, consistently increase.   **9.** 3.637×10^{-12} erg/photon; 52.35 kcal/mole   **10. (b)** 2,000 °K; **(c)** at 3,000 °K, 1.51×10^3; at 4,000 °K, 2.42×10^2 **13.**

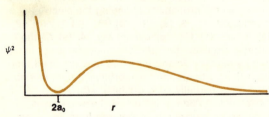

When $r/a_0 = 2$, $\psi_{2s} = 0$ (this corresponds to a radial node and happens only at one point).

CHAPTER FIVE PROBLEMS

1. H $1s^1$; Li $2s^1$; Na $3s^1$; K $4s^1$; Rb $5s^1$; Cs $6s^1$; Fr $7s^1$

2. Outermost orbital configuration is s^1.   **3.** Ca; Te; Al; Tl   **4.** Cl; Al; S; Tl; P   **5.** Z_{eff} significantly increased   **6.** Cs < Ca < S < F   **7.** Cl—because it has the highest Z_{eff} and addition of 1e$^-$ produces a highly stable configuration of 8 electrons.   **8.** C $1s^2 2s^2 2p^2$; N $1s^2 2s^2 2p^3$; N has 1e$^-$ in each of the three *p* orbitals giving a stable half filled orbital configuration and low electron affinity. C will have a half filled orbital configuration *after* 1 e$^-$ addition and therefore its electron affinity is relatively high.   **9.** Cl; O; Se; F   **10.** a < c < d < b $1s^2$ is in first period and remainder in second period. For the second period elements, the size would decrease with increasing atomic number or Z_{eff}.   **11.** Going from the top to the bottom the n value of the valence shell increases and therefore so does the size.   **12.** A negative ion has excess electrons which *decreases* Z_{eff} and increases the size while for a positive ion, the Z_{eff} is *increased* and the size decreases.

CHAPTER FIVE EXERCISES

1. Ca: Metal, Good, Low, Low, +; O: Nonmetal, Poor, High, High, −; Rb: Metal, Good, Low, Low, +; Br: Nonmetal, Poor, High, High, −;   **2.** decrease in Z_{eff}

3.

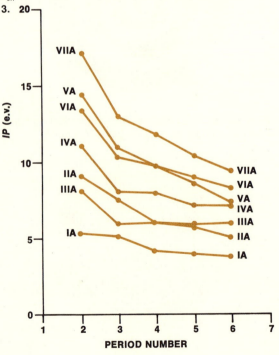

Within a period, the *IP* increases from left to right and within a group, the *IP* decreases from the top to the bottom.   **4.**

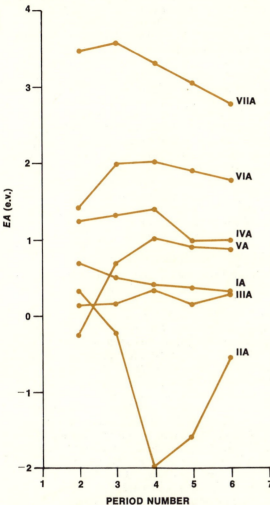

Within a period, the *EA* increases from left to right and the trend within different groups varies.
5. High *IP* and low *EA* because of highly stable completed valence shell electronic configuration.
6. *IP* of N high because of stable half filled *p* orbital configuration ($2p_x^1 2p_y^1 2p_z^1$), but 0 does not have this stable half filled configuration ($2p^2 2p^1 2p^1$) so electron more easily removable.   **7.** Second *IP* if Na greater than Mg because second e^- from Na$^+$ will have to come from a stable $s^2 p^6$ configuration.
8. (a) Nonmetals **(b)** Because nonmetals have a higher Z_{eff} than do metals.   **9.** Because the added

e^- goes into a shell of higher *n* value.   **10. (a)** HI > HBr > HCl > HF   **(b)** K$_2$ > Na$_2$ > Li$_2$ **(c)** I$_2$ > Br$_2$ > Cl$_2$ > F$_2$ **11.** 3.85 Å, 1.96 Å, 3.43 Å, 2.39 Å, 2.49 Å   **12.** 0^{2-} larger than Ne because Z_{eff} of 0^{2-} < Ne   **13.** Na$^+$ smaller than Ne because Z_{eff} of Na$^+$ > Ne   **14.** Mg^{+2} smallest, 0^{2-} largest. The ions are isoelectronic and Mg has highest + charge (greatest Z_{eff}). 0^{2-} largest since has largest negative charge (smallest Z_{eff}).   **15.** In the cases of Ca−Ga as well as Sr−In comparison, *d* orbitals are being filled thus reducing expected increase in Z_{eff} which is not the case for the Mg−Al comparison.
16. (a) As atomic size decreases, *IP* increases, both of which are related to Z_{eff} (relatively high value).
(b) In general, as atomic size decreases electron affinity increases, because of relatively high value of Z_{eff}.   **17.** VIIA because of highest positive *EA*.
18. (a) High *IP* and high positive *EA*.   **(b)** IA because of low *IP* and *EA*.   **(c)** 10.45 ev, 8.31 ev, 7.58 ev, 6.78 ev   **19.** From the top to the bottom in a group both *IP* and *EA* decrease so χ would decrease.   **20. (a)** (1) metal; (2) nonmetal **(b)** (1) Low *IP*, low *EA*, low χ; (2) High *IP*, high *EA*, high χ   **(c)** (2)   **(d)** (1)   **(e)** (1) = +2; (2) = −1
21. The Z_{eff} progressively increases from S^{2-} to Ca^{2+} because of the ratio of the number of electrons to the number of protons.   **22.** Cs < Na < F < Ne; Group IA elements have lowest *IP* which increases left to right in period and decreases top to bottom in group. Also rare gases highest *IP* because of stable ns^{2}np^6 electronic configuration.   **23.** Li$^+$ < Na$^+$ < Ar < Cl$^-$ < S^{2-}; Positive ions smaller than neutral atoms of same period which are smaller than negative ions of same period. Also Li$^+$ from second period and Na$^+$ from third (larger *n* value). The S^{2-} and Cl$^-$ from same period but Z_{eff} less for S^{2-}.   **24.** Cl$^-$; has smallest Z_{eff} **25.** IA

CHAPTER SIX PROBLEMS

1. 93.0 kcal which compares with the experimental value of 99 kcal (4.3 ev)   **2.** The term

$$\frac{N(n_+ e)(n_- e)}{r}$$

becomes larger since *r* for LiF is smaller than for NaF.
3. :N:::N: H:Br: :I:Cl: :C:::O: S::C::S Li$^+$:F:$^-$
4. N$_2$−triple; HBr−single; ICl−single; CO−triple; CS$_2$−double; LiF−ionic   **5.** Na$_2$ − σ; HBr − σ; ICl − σ; N$_2$ − 1σ + 2π; I$_2$ − σ;   **6.** σ bonds are

stronger than π bonds. **7.** Li_2: $(1s\sigma)^2$ $(1s\sigma^*)^2$ $(2s\sigma)^2$; He_2^+: $(1s\sigma)^2$ $(1s\sigma^*)^1$; B_2: $(1s\sigma)^2$ $(1s\sigma^*)^2$ $(2s\sigma)^2$ $(2s\sigma^*)^2$ $(2p\pi_y)^1$ $(2p\pi_z)^1$ **8.** He_2^+: $(1s\sigma)^2$ $(1s\sigma^*)^1$, stable by $\frac{1}{2}$ bond; Li_2: $(1s\sigma)^2$ $(1s\sigma^*)^2$ $(2s\sigma)^2$, stable by 1 bond; Be_2: $(1s\sigma)^2$ $(1s\sigma^*)^2$ $(2s\sigma)^2$ $(2s\sigma^*)^2$, unstable, zero bond; Be_2^+: $(1s\sigma)^2$ $(1s\sigma^*)^2$ $(2s\sigma)^2$ $(2s\sigma^*)^1$, stable by $\frac{1}{2}$ bond **9.** He_2: $(1s\sigma)^2$ $(1s\sigma^*)^2$ $\longrightarrow$ He_2^+: $(1s\sigma)^2$ $(1s\sigma^*)^1$; B_2: $(1s\sigma)^2$ $(1s\sigma^*)^2$ $(2s\sigma)^2$ $(2s\sigma^*)^2$ $(2p\pi)^2$ $\longrightarrow$ B_2^+: $(1s\sigma)^2$ $(1s\sigma^*)^2$ $(2s\sigma)^2$ $(2s\sigma^*)^2$ $(2p\pi)^1$; H_2: $(1s\sigma)^2$ $\longrightarrow$ H_2^-: $(1s\sigma)^2$ $(1s\sigma^*)^1$; Be_2: $(1s\sigma)^2$ $(1s\sigma^*)^2$ $(2s\sigma)^2$ $(2s\sigma^*)^2$ $\longrightarrow$ Be_2^-: $(1s\sigma)^2$ $(1s\sigma^*)^2$ $(2s\sigma)^2$ $(2s\sigma^*)^2$ $(2p\pi)^1$
10. (a) N_2—No dipole moment **(b)** dipole moment towards Cl **(c)** F_2—No dipole moment **(d)** dipole moment towards F **11.** CaO: ionic, Ca(+2) O(−2); ClF: covalent, Cl(+1), F(−1); NO: covalent, N(+2), O(−2); CO: covalent, C(+2), O(−2); KCl: ionic, K(+1), Cl(−1); HI: covalent, H(+1), I(−1); SrO: ionic, Sr(+2), O(−2); NaH: ionic, Na(+1), H(−1);
12. (a) ICl expected to have higher % ionic character. **(b)** HBr higher % ionic character than HI and less than HCl. $| \chi_H - \chi_{Br} | = | 2.1 - 2.8 | = 0.7$ and approximately 10% ionic character. **(c)** NaCl has greater % ionic character than LiCl, as $\chi_{Li} > \chi_{Na}$ and Cl is common to both compounds.

CHAPTER SIX EXERCISES

1. In valence bond theory, whole atoms with the electrons are brought together to form molecules where a bond is an electron pair. In molecular orbital theory, the nuclei are put at the distance they will finally be in the molecule and then electrons are added to polycentric molecular orbitals. **2.** The molecule has unsymmetrical charge distribution such that one end is relatively positive and the other negative. **3.** −37 kcal/mole **4.** −220 kcal/mole **5. (a)** −124 kcal/mole **(b)** 108 kcal/mole **6.** 84 kcal/mole **7.** Alkali halides have +1 and −1 ions. E of formation from the ions is approximately four times greater since Ca^{2+} and O^{2-};

$$E = \frac{6.023 \times 10^{23} \times (2 \times 4.8 \times 10^{10})^2 \times 2.39 \times 10^{-8}}{1.8 \times 10^{-8} \times 10^3}$$

$$= 737 \text{ kcal/mole}$$

8.

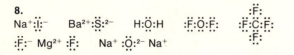

$Na^+ \, \ddot{:}\ddot{I}\ddot{:}^-$ $Ba^{2+} \, \ddot{:}\ddot{S}\ddot{:}^{2-}$ $H:\ddot{O}:H$ $\ddot{:}\ddot{F}\ddot{:}\ddot{O}\ddot{:}\ddot{F}\ddot{:}$ $\ddot{:}\ddot{F}\ddot{:}\ddot{C}\ddot{:}\ddot{F}\ddot{:}$

$\ddot{:}\ddot{F}\ddot{:}^- \, Mg^{2+} \, \ddot{:}\ddot{F}\ddot{:}$ $Na^+ \, \ddot{:}\ddot{O}\ddot{:}^{2-} \, Na^+$

9. H(+1) I(−1); Li(+1) F(−1); both O are zero; C(+4) S(−2); S(+6) O(−2); K(+1) Br(−1); Ca(+2) Cl(−1); Al(+3) O(−2); Ni(+2) O(−2); Zn(+2) Cl(−1); Ag (+1) O(−2); H(+1) P(+5) O(−2); C(+4) O(−2); S(+6) O(−2); Cl(+3) O(−2); N(−3) H(+1)
10. Cl_2—No; HI—Yes, I, large; CsBr—Yes, Br, large; LiF—Yes, F, large; IBr—Yes, Br, small; O_2—No; MgO—Yes, O, large; CuF—Yes, F, large **11.** Cl_2—covalent; HI—covalent; CsBr—ionic; LiF—ionic; IBr—covalent; O_2—covalent; MgO—ionic; CuF—ionic
12. N_2^*—$(1s\sigma)^2$ $(1s\sigma^*)^2$ $(2s\sigma)^2$ $(2s\sigma^*)^2$ $(2p\pi_y)^2$ $(2p\pi_z)^2$ $(2p\sigma)^1$ $(2p\pi^*)^1$; Li_2^*—$(1s\sigma)^2$ $(1s\sigma^*)^2$ $(2s\sigma)^1$ $(2s\sigma^*)^1$; C_2^*—$(1s\sigma)^2$ $(1s\sigma^*)^2$ $(2s\sigma)^2$ $(2s\sigma^*)^2$ $(2p\pi_y)^2$ $(2p\pi_z)^1$ $(2p\sigma)^1$
13. He_2^+: $(1s\sigma)^2$ $(1s\sigma^*)^1$—stable; C_2: $(1s\sigma)^2$ $(1s\sigma^*)^2$ $(2s\sigma)^2$ $(2s\sigma^*)^2$ $(2p\pi)^2$ $(2s\pi)^2$—stable; Ne_2: $(1s\sigma)^2$ $(1s\sigma^*)^2$ $(2s\sigma)^2$ $(2s\sigma^*)^2$ $(2p\pi)^2$ $(2p\pi)^2$ $(2p\sigma)^2$ $(2p\pi^*)^2$ $(2p\pi^*)^2$ $(2p\sigma^*)^2$—unstable; F_2: $(1s\sigma)^2$ $(1s\sigma^*)^2$ $(2s\sigma)^2$ $(2s\sigma^*)^2$ $(2p\pi)^2$ $(2p\pi)^2$ $(2p\sigma)^2$ $(2p\pi^*)^2$ $(2p\pi^*)^2$—stable; F_2^+: $(1s\sigma)^2$ $(1s\sigma^*)^2$ $(2s\sigma)^2$ $(2s\sigma^*)^2$ $(2p\pi)^2$ $(2p\pi)^2$ $(2p\sigma)^2$ $(2p\pi^*)^2$ $(2p\pi^*)^1$—stable **14.** H—I $\longleftrightarrow$ H^+I^- $\longleftrightarrow$ H^-I^+; H—I dominant since χ_I and χ_H are similar. Na—F $\longleftrightarrow$ Na^+F^- $\longleftrightarrow$ Na^-F^+; Na^+F^- dominant since $\chi_{Na} \ll \chi_F$. I—Br $\longleftrightarrow$ I^-Br^+ $\longleftrightarrow$ I^+Br^-; I—Br dominant since χ_I and χ_{Br} similar. Rb—Cl $\longleftrightarrow$ Rb^+Cl^- $\longleftrightarrow$ Rb^-Cl^+; Rb^+Cl^- since $\chi_{Rb} \ll \chi_{Cl}$ Br—Cl $\longleftrightarrow$ Br^+Cl^- $\longleftrightarrow$ Br^-Cl^+; BrCl since χ_{Br} and χ_{Cl} similar. **15.** BrF since $| \chi_{Br} - \chi_F |$ is largest, and deviation from mean of A—A and B—B is maximum for largest difference of χ_A and χ_B.
16. (a) B_2^+: $(1s\sigma)^2$ $(1s\sigma^*)^2$ $(2s\sigma)^2$ $(2s\sigma^*)^2$ $(2p\pi)^1$; B_2^-: $(2s\sigma)^2$ $(1s\sigma^*)^2$ $(2s\sigma)^2$ $(2s\sigma^*)^2$ $(2p\pi)^2$ $(2p\pi)^1$ **(b)** C_2^+: $(1s\sigma)^2$ $(1s\sigma^*)^2$ $(2s\sigma)^2$ $(2s\sigma^*)^2$ $(2p\pi)^2$ $(2p\pi)^1$; C_2^-: $(1s\sigma)^2$ $(1s\sigma^*)^2$ $(2s\sigma)^2$ $(2s\sigma^*)^2$ $(2p\pi)^2$ $(2p\pi)^2$ $(2p\sigma)^1$ **(c)** O_2^+: $(1s\sigma)^2$ $(1s\sigma^*)^2$ $(2s\sigma)^2$ $(2s\sigma^*)^2$ $(2p\pi)^2$ $(2p\pi)^2$ $(2p\sigma)^2$ $(2p\pi^*)^1$; O_2^-: $(1s\sigma)^2$ $(1s\sigma^*)^2$ $(2s\sigma)^2$ $(2s\sigma^*)^2$ $(2p\pi)^2$ $(2p\pi)^2$ $(2p\sigma)^2$ $(2p\pi^*)^2$ $(2p\pi^*)^1$; Note that in the O_2 case, experimental data indicates the $2p\sigma$ orbital is at lower energy than the $2p\pi$ orbitals.
17. (a) B_2: Bond order of 1; B_2^+: 1/2; B_2: 1 1/2. C_2: Bond order of 2; C_2^+: 1 1/2; C_2^-: 2 1/2. O_2: Bond order of 2; O_2^+: 2 1/2; O_2^-: 1 1/2.
(b) D of $O_2^+ > O_2$ because of greater bond order.
(c) If e^- add to bonding orbital or if electron removed from anti-bonding orbital. **18.** $(He—He)^*$ $(1s\sigma)^2$ $(1s\sigma^*)^1$ $(2s\sigma)^1$—stable **19.** Charge on ions; electron affinity; ionization potential; bond distance **20.** Multiple bond character or bond order; electronegativity of atoms; resonance **21.** −140 kcal/mole; −133 kcal/mole; −122 kcal/mole. The decrease in EA and increase in r_- of halogen.

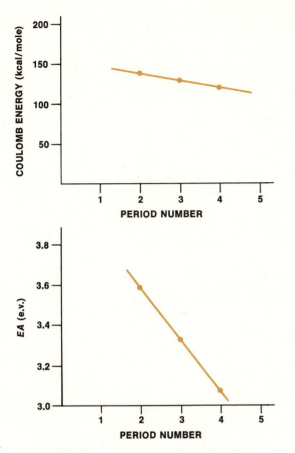

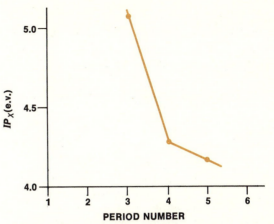

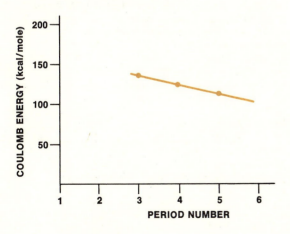

22. −140 kcal/mole; −124 kcal/mole; −119 kcal/mole. The decrease in *IP* and increase of r_+ of alkali metals approximately compensate.

23. Difference of electronegativity is related to % ionic character. $|\chi_{Li} - \chi_I| < |\chi_{Li} - \chi_{Br}|$; $|\chi_{Li} - \chi_I| < |\chi_{Na} - \chi_{Br}|$; $|\chi_{Li} - \chi_I| < |\chi_{Na} - \chi_{Cl}|$. Therefore L: I has less ionic character than any of the other 3 salts. **24.** HCl because product of charge separation (determined by % ionic character) times distance is greater.

CHAPTER SEVEN PROBLEMS

1. (a) 90° **(b)** 90° **(c)** 90° **(d)** 90° **2.** Cd: ···· $4d^{10}$ $5s^2$ where hybridization of $5s$ and $5p$ orbitals occurs to give two *sp* hybrid orbitals. **3. (a)** Covalent bonds because it is a liquid at low temperature and the electronegativity difference of Si and Cl is relatively small. Also, Si in same group as C and CCl_4 covalent. **(b)** Si is sp^3 hybridized. **4.** F_2: no dipole moment; HF: dipole moment towards F along bond; H_2S: dipole moment towards S between H's; NH_3: dipole moment towards N from middle of plane of H's; H_2CCl_2: dipole moment towards Cl's and between them
5. (a) Triangular (plane) **(b)** Tetrahedral **(c)** Trigonal bipyramidal **(d)** Linear **(e)** Octahedral **(f)** Linear **(g)** V-shaped **6. (a)** PI_3: greater χ of Br causes contraction and lowering of bonding orbital repulsion **(b)** SbI_3: greater χ of Cl causes contraction and lowering of bonding orbital repulsion
7.

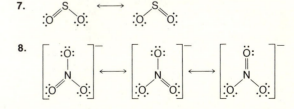

In NO_3^-, the N supplies five electrons for the bonding while in CO_3^{2-} the C supplies only 4 electrons.
9. (a) Yes, because they all are σ bonds. **(b)** σ bonds
10. (a) $CH_3—CH_2—CH_2—CH_2—CH_3$;

$$CH_3—\overset{\overset{\displaystyle CH_3}{|}}{\underset{\underset{\displaystyle CH_3}{|}}{C}}—CH_3; \quad CH_3—\overset{\overset{\displaystyle CH_3}{|}}{CH}—CH_2CH_3$$

11.

$$\overset{H}{\underset{H}{}}C=C\overset{CH_2CH_3}{\underset{H}{}}$$

No, since rotation about the $C=C$ by 180° does not generate a molecule distinguishable from the original. **12.** Proceeding in either direction causes increase in overlap of the p orbitals and ultimately a π bond at 0° or 180°
13. (a) $C_2H_5—O—C_2H_5$ **(b)** $CH_3—O—C_3H_7$
(c) $CH_3—CH_2—CH_2—OH$ **(d)** $CH_3—CH_2—CH_2—CH_2—OH$ **(e)** $CH_3—\overset{}{\underset{\underset{\displaystyle OH}{|}}{CH}}—CH_3$ **14.** $C_2H_5—\overset{}{\underset{\underset{\displaystyle O}{\|}}{C}}—C_2H_5$
15. $CH_3—CH_2—CH_2—COOH$; all C—C and C—H bond angles of the alkyl chain are $\sim$109° and the

bond angle $\overset{O}{\underset{OH}{C}}$ is 120°. **16.** $CH_3—\overset{\overset{\displaystyle H}{|}}{N}—C_2H_5$

17. (a) Bromoethane or ethyl bromide
(b) Iodomethane or methyl iodide

CHAPTER SEVEN EXERCISES

1. An np^1np^1 configuration for both O and S
2. Yes, CH_3^- replaces H but the same orbitals of O and S will participate in bond formation. **3.** Yes, as long as I value of main or central atom is the same. **4. (a)** 90° **(b)** 90° **(c)** 90° **(d)** 90°
5. Single atom **6.** Highest degree of overlap results in strongest bonds and determines bond angles (assuming a given set of atomic orbitals).
7. It is a covalent compound; expect all bonds to be σ bonds and angles of $\sim$109° because Pb will be sp^3 hybridized. **8.** There is a symmetrical distribution of Cl atoms around CCl_4 while in CH_3Cl, there is an electronegativity difference between the Cl and H atoms. **9.** Covalent and tetrahedral **10.** In p-dibromobenzene, there are equal and opposite dipoles or vectors to the Br atoms. In o-dibromobenzene, the dipoles or vectors to the Br atoms do not cancel and a resultant dipole moment exists. **11. (a)** $H—C\equiv N$

(b) $S=C=S$ **(c)** $C\equiv O$ **(d)** $F—O—F$ **(e)**

$$\overset{H}{\underset{H}{}}C=O$$

12. (a) Tetrahedral **(b)** Tetrahedral **(c)** Tetrahedral (pyramidal) **(d)** V-shaped **(e)** Linear **(f)** Trigonal bipyramid **(g)** Octahedral **(h)** Linear
(i) Octahedral **13. (a)** AsI_3 **(b)** PBr_3 **(c)** CF_4
(d) NF_3

14. $H—\overset{\overset{\displaystyle O}{\|}}{C}\overset{}{\underset{\displaystyle O^-}{}} \longleftrightarrow H—\overset{\overset{\displaystyle O^-}{|}}{C}\overset{}{\underset{\displaystyle O}{\|}}$ **15.**

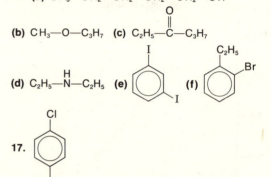

not planar because carbon atoms are sp^3 hybridized.
16. (a) $CH_3—CH_2—CH_2—CH_2—CH_2—OH$

(b) $CH_3—O—C_3H_7$ **(c)** $C_2H_5—\overset{\overset{\displaystyle O}{\|}}{C}—C_3H_7$

(d) $C_2H_5—\overset{\overset{\displaystyle H}{|}}{N}—C_2H_5$ **(e)** **(f)**

17.

18. $\sim$105° **19.** The carbon-carbon double bonds
20. (a) No, since all substituents are the same.
(b) No, since it is symmetrical all C—Cl dipoles or vectors cancel. **21.** sp^2 **22. (a)** order of overlap is $sp > sp^2 > sp^3 > p > s$, the extension into space of the individual hybrid orbitals is $sp > sp^2 > sp^3 > p > s$ resulting in maximum overlap for sp and minimum for s. **(b)** The bond strengths should be in the same order since their magnitudes depend largely upon the degree of overlap of the orbitals involved.
23.

$$\overset{Br}{\underset{NH_3}{}}\overset{}{\underset{}{Ni}}\overset{NH_3}{\underset{Br}{}}$$

planar, trans

24. (a)

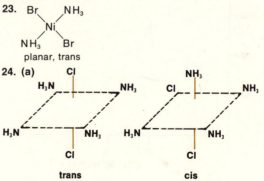

trans   cis

(b) the cis isomer

CHAPTER EIGHT PROBLEMS

1. The *IP*'s of K and Rb are not significantly different; *IP*'s do differ for K and Na. Also, the ionic radii of K and Rb are considerably closer in magnitude than for Na and K.   **2.** Less than S^{2-} since S^{2-} is smaller and therefore charge density is greater than for Se^{2-}.   **3.** The very small size results in a very high charge density.   **4.** Heat of hydration of K^+ is 75 kcal/mole and of Ba^{2+} is 308 kcal/mole. Since Ba^{2+} has twice the charge of K^+, the hydration energy should be (2^2) times larger or 300 kcal/mole.
5.

$$E = \frac{-2}{3kT}\frac{\mu_1^2\,\mu_2^2}{r^6}$$

since molecular rotation occurs preventing ideal orientation.   **6.** Yes, increase in *T* would decrease interaction.   **7.** Dipole-dipole interaction between H_2O and the hydroxyl groups of the glycol.   **8.** H_2 less polarizable. The boiling point of H_2 is lower than that of N_2.   **9. (a)** atomic electrons in Kr and molecular electrons in CH_4 **(b)** They are about equally polarizable based on the boiling points.
10. As the carbon chain length increases, the number of induced dipole-induced dipole interactions between CH_2 groups of adjacent chains increases as does the boiling point.   **11.** The difference in χ of C and H is too small.

CHAPTER EIGHT EXERCISES

1. $RbI < RbF < LiF < CaF_2$; The strongest ion-ion interaction is in CaF_2 because of +2 charge on Ca. The remaining order should approximately follow heats of formation which is largest in LiF and smallest in RbI.   **2.** LiCl higher since it is ionic, whereas HCl is covalent.   **3.** $He < Br_2 < NaCl$; He lowest because only induced dipole-induced dipole interaction which is weaker than between Br_2 molecules. NaCl highest because of strong ion-ion interaction.   **4.** The ions at the beginning and end of a series each have a single charge (either + or −).
5. Lesser since the size of Tl^{3+} is larger than Al^{3+} and therefore charge density is greater for Al^{3+}.
6. Gasoline is a mixture of alkane hydrocarbons which have no dipole moment. Alcohol and glycol have dipole-dipole interaction and can hydrogen bond.   **7.** Yes, because of ionization resulting from the high heat of hydration of H^+.   **8.** $CoSO_4$ and $CuSO_4$ hydrates are formed by ion-dipole (H_2O)

interaction while for butane only weak dipole (H_2O)-induced dipole (butane) interaction exists.   **9.** The electrons of H_2 are not as polarizable as Cl_2 and therefore the dipole-induced dipole interaction is greater for Cl_2.   **10.** The ethyl ester has no H-bonding between molecules such as is possible for acetic acid.   **11.** The ortho derivative forms hydrogen bonds within the molecule (intramolecular) while the paraderivative cannot; this limitation permits significant H bonding between molecules of the latter.
12. (a)

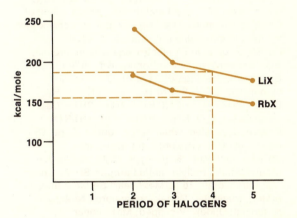

(b) LiBr: ~185 kcal/mole; RbBr: ~153 kcal/mole
13. Hydrogen bonding can occur between HCl molecules but not between ClF molecules.
14. (a) Octane—practically none because octane is not polar.   **(b)** Ethyl alcohol—less than water but more than for octane since ion-dipole forces could exist.   **15.** High degree of ionization resulting from large heat of hydration of H^+ (more than that of Na^+).   **16.** CH_4 has lowest melting point since it is the least polarizable member of the series. H_2O has abnormally high melting point because of extensive hydrogen bonding between molecules which is not present for other members of the series.   **17.** The alcohols are relatively non-volatile and form a layer on the surface preventing escape of water. The —OH part of the alcohol hydrogen bonds with water and the hydrocarbon part sticks above the surface. The hydrocarbon chains are held close together by induced dipole-induced dipole interaction.   **18.** For ionic compounds *E* is proportional to $1/r$ and for induced dipole-induced dipole *E* is proportional to $1/r^6$. For latter case, *E* decreases very fast as *r* increases and interaction is large only at small *r* values (short range).   **19.** $Cs^+ < Na^+ < Mg^{2+} < Al^{3+}$

Al^{3+} has the highest charge density and Mg^{2+} is next. Even though Na^+ and Cs^+ have same charge, Na^+ is smaller so its charge density is greater.

CHAPTER NINE PROBLEMS

1. $BaBr_2$: ionic, large electronegativity differences; PBr_3: covalent, small electronegativity difference
2. K_2O: ionic, large electronegativity difference; NO: covalent, small electronegativity difference
3. $MgO + SO_3 \longrightarrow MgSO_4$; $Na_2O + CO_2 \longrightarrow Na_2CO_3$; $Ca(OH)_2 + 2HClO_3 \longrightarrow Ca(ClO_3)_2 + 2H_2O$; $Li(OH) + Mg(OH)_2 \longrightarrow$ no reaction **4. (a)** Rb_2O: basic, oxide of Group I metal; SO_3: acidic, oxide of Group VI nonmetal. Dissolve in water and check the solution to determine if it is acidic or basic.
(b) Rb_2O: ionic, large electronegativity difference; SO_3: covalent, small electronegativity difference. Boiling and melting points of Rb_2O are much higher than SO_3. **5.** $2NaH_{(l)} \longrightarrow 2Na_{(l)} + H_2$; $CaH_2 + 2H_2O \longrightarrow Ca(OH)_2 + 2H_2$; $2HI + CaO \longrightarrow CaI_2 + H_2O$; $NH_3 + HNO_3 \longrightarrow NH_4NO_3$
6. $AlBr_3$ is covalent while AlF_3 is ionic. **7. (a)** CsI, ionic; **(b)** CF_4, covalent; **(c)** MgF_2, ionic; **(d)** PBr_3, covalent **8.** $(CH_3)_4Si$, $(C_2H_5)_4Si$; Alkyl groups tetrahedrally arranged around Si **9.** Group IVA; tetrahedral **10.** LiMe: linear; $Be(Me)_2$: linear; $B(Me)_3$: triangular; $C(Me)_4$: tetrahedral; $N(Me)_3$: pyramidal; $O(Me)_2$: V-shaped; MeF: linear
11. (a) $Li_2 = 2.46$ Å; $Cs_2 = 4.70$ Å **(b)** Li_2; shorter bond length indicates greater overlap and greater bond strength. **12.** NaBr; Na^+ has higher charge density than Rb^+. **13. (a)** $Li_2O + H_2O \longrightarrow 2LiOH$ **(b)** $2CsOH + H_2SO_4 \longrightarrow Cs_2SO_4 + 2H_2O$ **(c)** $KH + H_2O \longrightarrow KOH + H_2$ **(d)** $2Na + S \longrightarrow Na_2S$ **(e)** $2Na + 2NH_3 \longrightarrow 2NaNH_2 + H_2$ **(f)** $2Cs + H_2 \longrightarrow 2CsH$ **(g)** $RbBr \xrightarrow{aq.} Rb^+ + Br^-$ **(h)** $2NaBr_{(l)} \longrightarrow 2Na_{(l)} + Br_2$ **(i)** $3Cs + P \longrightarrow Cs_3P$ **14. (a)** Li^+, O^{2-} **(b)** Na^+, SO_4^{2-} **(c)** Rb^+, I^- **(d)** K^+, MnO_4^- **(e)** Na^+, $Cr_2O_7^{2-}$ **(f)** Cs^+, H^- **(g)** K^+, OH^- **15. (a)** Higher—stable s^2 configuration and greater Z_{eff}. **(b)** Ba less than Li: Z_{eff} less even though Ba in group IIA, because ionized electron of Ba has high n value ($6s^2$). About equal to Na: Z_{eff} about equal even though Ba in group IIA since ionized electron of Ba has higher n value.
16. (a) 541 kcal/mole; 177 kcal/mole **(b)** Ca^{2+} has higher charge density because of +2 charge.
17. K: 1 mole of electrons; Ba: 2 mole of electrons
18. (a) $BaO + 2HCl \longrightarrow BaCl_2 + H_2O$ **(b)** $Na_2O + CaO \longrightarrow$ no reaction **(c)** $Sr(OH)_2 + 2HBr \longrightarrow SrBr_2 + 2H_2O$ **(d)** $2Mg + O_2 \longrightarrow 2MgO$

(e) $Ba(OH)_2 + Rb(OH) \longrightarrow$ no reaction **(f)** $Ba + 2NH_3 \longrightarrow Ba(NH_2)_2 + H_2$ **(g)** $Be + 2H_2O \longrightarrow Be(OH)_2 + H_2$ **(h)** $Ca + Br_2 \longrightarrow CaBr_2$ **(i)** $BaCl_2 \xrightarrow{aq} Ba^{2+} + 2Cl^-$ **(j)** $3Ca + N_2 \longrightarrow Ca_3N_2$
19. Group IIA elements and Br have comparatively larger electronegativity differences than do the Group IIIA elements and Br. **20.** Higher Z_{eff}
21. (a) $4Ga + 3O_2 \longrightarrow 2Ga_2O_3$ **(b)** $Al(OH)_3 + 3HCl \longrightarrow AlCl_3 + 3H_2O$ **(c)** $In(OH)_3 + 3HCl \longrightarrow InCl_3 + 3H_2O$ **(d)** $In(OH)_3 + NaOH \longrightarrow$ No reaction **(e)** $Tl + H_2 \longrightarrow$ No reaction **(f)** $BF_3 + (CH_3)_3N \longrightarrow (CH_3)_3N—BF_3$ **(g)** $Al_2O_3 + 6HCl \longrightarrow 2AlCl_3 + 3H_2O$ **(h)** $2In + 3I_2 \longrightarrow 2InI_3$ **(i)** $3Ga + 3H_2SO_4 \longrightarrow Ga_2(SO_4)_3 + 3H_2$ **22.** The weakly bonded layers in graphite can slide past one another, whereas in diamond all carbon atoms are strongly bonded. **23.** Sn(II) has higher Z_{eff} because of filling of $4d$ orbitals **24. (a)** $CO(OH)_2 + HNO_3 \longrightarrow$ No reaction **(b)** $Na_2CO_3 + 2HI \longrightarrow 2NaI + H_2O + CO_2$ **(c)** $Pb + 2NaOH \longrightarrow Na_2PbO_2 + H_2$ **(d)** $Si + H_2O \longrightarrow$ No reaction **(e)** $SnO + 2HCl \longrightarrow SnCl_2 + H_2O$ **(f)** $SnO + 2LiOH \longrightarrow Li_2SnO_2 + H_2O$ **(g)** $SiO_2 + 2OH^- \longrightarrow SiO_3^{2-} + H_2O$ **(h)** $C + HNO_3 \longrightarrow$ No reaction **(i)** $C + H_2 \longrightarrow$ No reaction
25. $\left[\begin{array}{c} :\overset{..}{O}: \\ :\overset{..}{O}::N:\overset{..}{O}: \end{array} \right]^-$ and two resonance forms.

26. Since it is non-metallic, there could be molecules with consequent small interaction (actually P_4 molecules do exist in the solid).

27. (a) $3NaOH + PO(OH)_3 \xrightarrow{(complete)} Na_3PO_4 + 3H_2O$ **(b)** $PO(OH)_3 + SO_2(OH)_2 \longrightarrow$ No reaction **(c)** $NH_3 + HNO_3 \longrightarrow NH_4NO_3$ **(d)** $Sb_4O_6 + 4HBr \xrightarrow{aq} 4SbO^+ + 4Br^- + 2H_2O$ **(e)** $Sb_4O_6 + 4KOH \longrightarrow 4K^+ + 4SbO_2^- + 2H_2O$ **(f)** $2K + 2NH_{3(l)} \longrightarrow 2KNH_2 + H_2$ **(g)** $2P + 3Cl_2 \longrightarrow 2PCl_3$ **(h)** $As + HCl \longrightarrow$ No reaction **(i)** $P_4O_6 + 6H_2O \longrightarrow 4H_3PO_3$ **(j)** $4As + 3O_2 \longrightarrow As_4O_6$ **28.** Much smaller Z_{eff} because of excess electrons
29. $Pb(NO_3)_2 + H_2S \longrightarrow 2HNO_3 + PbS$
30. (a) $SO_3 + K_2O \longrightarrow K_2SO_4$ **(b)** $H_2Se + H_2O \rightleftharpoons HSe^- + H^+ \rightleftharpoons 2H^+ + Se^{2-}$ **(c)** $SeO_2(OH)_2 + HCl \longrightarrow$ No reaction **(d)** $SeO_2(OH)_2 + 2LiOH \longrightarrow Li_2SeO_4 + 2H_2O$ **(e)** $SO_2 + H_2O \longrightarrow H_2SO_3$ **(f)** $S + O_2 \longrightarrow SO_2$ **(g)** $Se + 3F_2 \longrightarrow SeF_6$ **(h)** $SO_2 + HCl \longrightarrow$ No reaction **(i)** $Na_2S + 2HCl \longrightarrow 2NaCl + H_2S$ **(j)** $3S + 4HNO_3 \longrightarrow 3SO_2 + 2H_2O + 4NO$ **31.** Increased size of the charge cloud increases polarizability and thereby magnitude of induced dipole-induced dipole interaction. **32.** Both

have same magnitude of charge (but opposite signs). Also, even though charge density of $+$ ion is greater, the $-$ ion is larger so more molecules can surround latter ion. **33. (a)** $HI \xrightarrow{aq} H^+ + I^-$
(b) $2HF + Cs_2O \longrightarrow 2CsF + H_2O$ **(c)** $2(ClO_3)OH + Ca(OH)_2 \longrightarrow Ca(ClO_4)_2 + 2H_2O$ **(d)** $I_2 + H_2 \longrightarrow 2HI$ **(e)** $4HF + SiO_2 \longrightarrow SiF_4 + 2H_2O$

(f) $3Br_2 + 6OH^- \xrightarrow{(hot)} BrO_3^- + 5Br^- + 3H_2O$

(g) $Cl_2 + H_2O \longrightarrow HCl + HOCl$ **(h)** $2P + 3Cl_2 \longrightarrow 2PCl_3$ **(i)** $Ca + Br_2 \longrightarrow CaBr_2$

(j) $CsBr \xrightarrow{aq} Cs^+ + Br^-$

CHAPTER NINE EXERCISES

1. Ion-ion interaction much stronger than molecule-molecule interaction. **2.** Yes, K dissolves to form K^+ and solvated e^- **3.** A compound where there are insufficient valence electrons to bond all atoms together. **4.** Halogens occur as diatomic molecules which are more polarizable than atoms of rare gases. **5.** Same element existing in different forms but in the same physical state; diamond and graphite. **6.** Decrease of number of hydrogen bonds causes a loss of open structure present in ice; therefore in liquid H_2O the molecules are closer together on average causing an increase in density. **7.** RbCl is ionic and PCl_5 covalent; melting point, boiling point, and conductivity as liquids. **8.** Wash with acid like HCl, since C does not react with HCl. **9.** H_2Te is covalent while LiH is ionic. **10.** No; heats of hydration of ions are exothermic, but breaking the crystal lattice is endothermic and so solution may cool.
11. Electronegativity of Na is less than H.
12. NF_3 is approximately tetrahedral with one position being occupied by the lone pair. Low dipole moment because negative lone pair on N balances dipole towards the F atoms. **13.** 52 g
14. 0.5 g 5.6 l **15.** 3.01×10^{23} electrons
16. CO_2, P_2O_5, SeO **17.** 0.12 g **18.** HF—largest electronegativity difference of atoms **19.** $P(OH)_3$

20. Br; Br; $\overset{\cdot\cdot}{\underset{\cdot\cdot}{:Br}}:\overset{\cdot\cdot}{\underset{\cdot\cdot}{Ga}}:\overset{\cdot\cdot}{\underset{\cdot\cdot}{Br}}:$; basic

21. CsI; SrI_2; InI_3; Cs_2Se; $SrSe$; In_2Se_3; Cs_2O; SrO; In_2O_3 **22.** Na **23.** $BeCl_2$ **24. (a)** $Se + H_2O \longrightarrow$ No reaction **(b)** $GeO + 2HCl \longrightarrow GeCl_2 + H_2O$ **(c)** $LiNO_3 \xrightarrow{aq} Li^+ + NO_3^-$ **(d)** $2In + 6HBr \longrightarrow 2InBr_3 + 3H_2$ **(e)** $2As + 3I_2 \longrightarrow 2AsI_3$ **(f)** $2NaI + H_3PO_4 \longrightarrow Na_2HPO_4 + 2HI$ **(g)** $4P + 3O_2 \longrightarrow P_4O_6$ **(h)** $Rb_2O + 2HCl \longrightarrow 2RbCl + H_2O$ **(i)** $NaH + H_2O \longrightarrow NaOH + H_2$ **(j)** $3Ca + N_2 \longrightarrow Ca_3N_2$ **(k)** $2Al + Fe_2O_3 \longrightarrow 2Fe + Al_2O_3$ **(l)** $2Al + 2KOH + 2H_2O \longrightarrow 2KAlO_2 + 3H_2$ **(m)** $3Sn + 4HNO_3 \longrightarrow 3SnO_2 + 4NO + 2H_2O$ **(n)** $As + 5HNO_3 \longrightarrow H_3AsO_4 + 5NO_2 + H_2O$ **(o)** $SbCl_3 + H_2O \longrightarrow SbOCl + 2HCl$ **(p)** $3Se + 4HNO_3 \longrightarrow 3SeO_2 + 2H_2O + 4NO$

CHAPTER TEN PROBLEMS

1. System: rock formation; Boundary: interface between rock and water; Surrounding: water; Properties of the system: temperature, pressure, rock density, rock composition, etc.; State of the system: a specific condition of a system that is completely described through its properties. For example, the rock at a specified temperature, pressure and volume. **2.** $W_{AEB} = 0.271$ kcal/mole; $W_{AEB} = W_{ACB} \neq W_{ADB}$ **3.** Not a state function, because work depends upon the path. **4. (a)** $Q_{ADB} = 0.542$ kcal/mole; $Q_{AEB} = 0.271$ kcal/mole **(b)** $Q_{ADB} \neq Q_{AEB}$; Q is not a state function. **5. (a)** $\Delta E_{ADB} = 0$; $\Delta E_{AEB} = 0$ **(b)** $\Delta E_{ADB} = \Delta E_{AEB} = 0$. No, ΔE is a state function. If slide rule was used to calculate the answer, the ΔE may differ by as much as 0.5 to 1 calorie. **6.** $W = 0.739$ kcal/mole; $\Delta E = 8.97$ kcal/mole; **7.** $\Delta H_{AEB} = 0$ **8.** $\Delta H_v = 9.71$ kcal/mole **9. (a)** $\Delta H_{AD} = 1.35$ kcal/mole; $\Delta H_{EB} = 0.678$ kcal/mole **(b)** $\Delta H_v = Q = 9.71$ kcal/mole **10.** $\Delta H° = 673$ kcal/mole **11.** $\Delta H° = 10.52$ kcal/mole **12.** $\Delta H_A = \Delta H_B = -2.48$ kcal **13.** $D_{CH_3-H} = 104.0$ kcal/mole; $D_{C_2H_5-H} = 98.0$ kcal/mole; $D_{CH_3CHCH_3} = 94.5$ kcal/mole. Yes, we would expect

$$\overset{|}{\underset{H}{}}$$

them to be similar because the C—H bond in all cases is formed by an overlap of H $1s$ and C sp^3 orbitals to form a bond. The difference among the different compounds is apparently due to steric and other effects from the neighboring CH_3 and C_2H_5 groups bonded to the C atom. **14.** $\Delta H_A = \Delta H_B = D_{I-I} + D_{H-H} - 2D_{H-I}$ **15.** $D_{R-Br} = 66$ kcal/mole. The largest deviation of the D_{R-Br} value from the mean is 14 kcal/mole. **16.** $\Delta H° = 29$ kcal/mole **17. (a)** Breaking: $H_2C = CH_2 \longrightarrow \cdot H_2C-CH_2\cdot$; Forming: $\cdot H_2C-CH_2\cdot + H\cdot + H\cdot \longrightarrow CH_3-CH_3$ **(b)** $\Delta H = -28$ kcal/mole **(c)** $\Delta H = -32.732$ kcal/mole The difference of 4.7 kcal is small, considering the fact that average bond energies were used in (b). **18.** $\Delta S° = -43.6$ cal/mole °K **19.** $\Delta G° = 686$ kcal/mole **20.** $\Delta G° = 686$ kcal/mole

CHAPTER TEN EXERCISES

1. (a) no **(b)** yes **(c)** yes **(d)** no **(e)** no **(f)** no

(g) no **(h)** yes **(i)** no **(j)** no **2. (a)** Yes.
$\Delta G = -682$ cal/mole < 0. The reaction is feasible.
(b) Because the answer in (a) only tells the feasibility
of this reaction but does not show the rate of the
reaction. The rate in this reaction is actually very
slow, so women do not need to worry about it.
3. $\Delta G° = -17.5$ kcal/mole. The reaction is feasible.
4. (1) negative **(2)** zero **(3)** positive **5.** The
investigation is doomed to fail since ΔE for the
reaction cannot be changed regardless of the
reaction sequence. **6.** $\Delta H_{CH_4} = -11.98$ kcal/g;
$\Delta H_{C_4H_{10}} = -10.95$ kcal/g; CH_4 is more economical.
7. cis 2-butene, $\Delta H° = -28.45$ kcal/mole; trans
2-butene, $\Delta H° = -27.407$ kcal/mole. From the $\Delta H°$
it is obvious that trans 2-butene is more stable since
less heat was released upon hydrogenation. This can
be reasoned from the molecular structures by the
fact that the double bond requires that the CH_3
groups be in the same plane, and the repulsive
force would be less in the trans than in the cis form.
8. (a) cis 2-pentene, $\Delta H = -28.29$ kcal/mole. The
two heats of hydrogenation are very close and
apparently the replacement of CH_3 by C_2H_5 has little
effect. **(b)** $\Delta H° = -85.2$ kcal/mole **(c)** $\Delta H° =$
-49.25 kcal/mole **(d)** Resonance of benzene
stabilizes the molecule, so the heat of formation
of the benzene is lower than the three alternate
double and single bond structure. **9.** The trend
of this sequence corresponds to the increase in the
number of CH_3 groups attached to the carbon to
which Br is bonded. The size of CH_3 makes the
repulsive force between them appreciable so as to
cause the decrease in bond dissociation energies.
10. 16.4 moles = 262 g **11.** oxy-acetylene flame:
8,400 °C; hydrogen-oxygen flame: 5,800 °C. The
temperature of a flame using air instead of pure
oxygen will be much lower because the nitrogen
in air absorbs a large amount of heat. **12.** $\Delta H =$
-2.6 kcal/mole **13.** $\Delta H°_{f\,Mg_2SiO_4} = -508.1$
kcal/mole **14.** $\Delta H = 10,970$ cal/mole **15.** $\Delta H°_c =$
-310.6 kcal/mole **16.** $\Delta H° = -42.5$ kcal/mole
17. (a) $\Delta H° = -44.12$ kcal/mole; **(b)** Heat required
= 525 cal/mole; **(c)** $\Delta H° = -43.07$ kcal/mole;
(d) 70.50 kcal/min.; **(e)** $T = 48.5$ °C **18.** $\Delta S_A =$
-0.236 kcal/°K mole; $\Delta S_B = +0.366$ kcal/°K mole;
$\Delta S = 0.13$ kcal/°K mole > 0. Since $Q_{act} = 0$,
$Q_{act} - T\Delta S < 0$ and $\Delta S > Q_{act}/T$. Therefore, since
ΔS is greater than Q_{act}/T, according to the second
law of thermodynamics the process may occur.
19. (a) $\Delta H° = -57.8$ kcal/mole; $\Delta S° = -10.6$
kcal/mole °K **(b)** $D_{H-OH} = 119.2$ kcal/mole **(c)** ΔH
$= -2D_{H-O} + D_{H-H} + \frac{1}{2}D_{O=O} = -74.8$ kcal/mole
(d) $2H_2O$ is more orderly than is $(2H_2 + O_2)$. The
reaction changes from 3 molecules to 2 molecules.

(e) $\Delta H° = -57.8$ kcal; $-T\Delta S° = 3.16$ kcal. The $\Delta H°$
is dominant, making $\Delta G° = -54.6$ kcal. **20. (a)** ΔH
$= \Delta H_{vap}$ (absorb energy) $+ \frac{1}{2}D_{Cl-Cl}$ (absorb
energy) $+ IP_{Na}$ (absorb energy) $- EA_{Cl}$ (release
energy) $- U_{NaCl}$ (release energy); **(b)** F_2 has low
D_{F-F}; fluorides have high lattice energy. **21.** ΔG
$= -7$ kcal/mole

CHAPTER ELEVEN PROBLEMS

1. $R_{235}/R_{238} = 1.0042$ **2.** 1.24 **3.** They will combine
because the surface of the combined drop is smaller
than the sum of the surface areas of the two
separate drops. Since the area is smaller, the ΔG
surface must be smaller; processes at constant T
and P can occur when $\Delta G < 0$. **4. (a)** -50 °C to
0 °C, solid; $t = 0$ °C, solid + liq. equil.; 0 °C to 121 °C,
liq.; $t = 121$ °C, liq. + vapor equil.; 121 °C to 200 °C,
vapor **(b)** -50 °C to 200 °C, vapor (no phase
changes) **5.** Yes, because the isothermal process
at 0 °C goes into the liquid region at high pressure.
6. -1.6 °C **7.** $t = 96$ °C. The lower temperature at
650 mm causes the chemical reactions in the
cooking process to occur at a slower rate to 100 °C
at 1 atmosphere pressure. **8.** At $P = 1$ atm
(a) -85 °C to -80 °C, solid; $t = -80$ °C, solid + vapor;
-80 °C to 100 °C, vapor **(b)** Because dry ice and
CO_2 vapor equilibrate at -80 °C whereas ice and
water equilibrate at 0 °C.

(c)

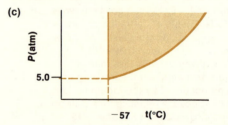

Conditions where T, P are in the shaded area.
$P > P_{triple\,point} = 5$ atm. **9.** 200 °C to 120 °C, liquid;
$t = 120$ °C, monoclinic + liq.; 120 °C to 95 °C,
monoclinic; $t = 95$ °C, monoclinic + rhombic; $t <$
95 °C, rhombic **10. (a)** Graphite

(b)

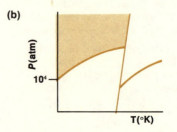

Conditions where T, P are in the shaded area. $P >$ 10^4 atm, $T < \approx 4,000\ °K$. **(c)** At 3800 °C graphite and carbon vapor coexist. At $T > 3800$ °C carbon vapor exists. **(d)** No. Because thermodynamic data can only be used to predict the feasibility of the reaction but not the rate of the reaction.

11.

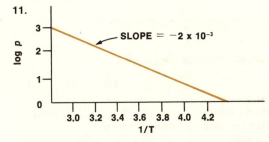

$\Delta H_{vap} = 9.18$ kcal/mole   **12.** $\Delta H_{vap} = 2.04$ kcal/mole. It is obvious that ΔH_{vap} of $CH_3OH \gg \Delta H_{vap}$ of CH_4 which can be explained by the fact that CH_3OH is a polar compound and can form a hydrogen bond, whereas CH_4 is non-polar.   **13. (a)** Since $V_{vapor} >$ V_{solid}, decreasing the pressure will cause the solid to sublime.   **(b)** Upon reducing the pressure, the isothermal process along $t = 0$ °C changes from the solid phase to the vapor region.   **14. (a)** Since $V_{solid} > V_{liquid}$, increasing the pressure causes the solid to melt.   **(b)** Along the isothermal process at $t = 0$ °C, increasing pressure changes from the solid phase to the liquid region.   **15. (a)** Pressure exerted by the blade causes the phase change from ice to water.   **(b)** No, the pressure exerted by the blade should not melt the solid benzene.   **16. (a)** $\Delta H_{melt} =$ $+$; increase in T causes ice to melt.   **(b)** At $P = 1$ atm the stable region changes from solid to liquid at 0 °C.   **17.** $\Delta H_{sub} = +$, decrease in T will produce solid.   **18. (c)**, because it is highly unsymmetrical with a long axis through the benzene ring and $C{=}C$ groups, and because it contains various functional groups which can have different types of intermolecular interactions.

CHAPTER ELEVEN EXERCISES

1. $HF > Cl_2 > CH_2Cl_2 > Xe > XeF_4 > Pb(CH_3)_4$
2. (a) Xe; greater induced dipole-induced dipole interactions due to greater number of electrons.
(b) CH_3Cl; polar   **(c)** $NaCl$; ion-ion interactions
(d) CH_3OH; polar; hydrogen bonding
(e) CH_3CH_2OH; polar; hydrogen bonding   **(f)** I_2; greater induced dipole-induced dipole interactions due to greater number of electrons.   **3. (a)** $V_{H_2}/V_{HD} =$ 1.222;   **(b)** $V_{235UF_6}/V_{238UF_6} = 1.00505$;

(c) $V_{CH_4}/V_{CD_4} = 1.126$; Therefore a $>$ c $>$ b.
4. CH_3OH; polar $+$ hydrogen bonding
5. (a) Van der Waal's crystal: Atoms or molecules are held together by Van der Waal's attractive force. Covalent crystal: Atoms are held together by covalent bonding. Ionic crystal: For example, the $NaCl$ crystal is composed of Na^+ and Cl^-. Metallic crystal: Conducting electrons are shared among the elements of the crystal and offer the bonding forces.
(b) diamond—covalent; iron—metallic; methane—Van der Waal's; benzene—Van der Waal's; cesium floride—ionic; CS_2—Van der Waal's; Cu—metallic; Xe—Van der Waal's; B—covalent.   **6.** Increasing the pressure will increase the boiling point and, therefore, shorten the time of cooking. Decreasing the time of cooking reduces the consumption of fuel.
7. The critical temperature of CH_4 is -82.1 °C, and CH_4 will remain in the gas phase at room temperature regardless of the pressure. On the other hand, the critical pressure for propane is 96.8 °C and at sufficiently high pressures it will condense to the liquid phase.   **8.** Ethyl ether is much more volatile then ethyl alcohol, and therefore the higher concentration of ethyl ether causes it to be more flammable. The lower vapor pressure of ethyl alcohol is due to the hydrogen bonding that occurs in the liquid phase.   **9. (a)** Diamond   **(b)** No
10. (a) 120 °C   **(b)** 1.956 atm; 120 °C   **(c)** 129.2 °C; 2.6 atm   **11. (a)** The intersection corresponds to the triple point since all three phases are at equilibrium with each other. Triple point is: $t \approx -185$ °C; $P \approx 76$ mm Hg.

(b)

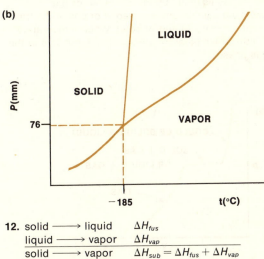

12. solid $\longrightarrow$ liquid ΔH_{fus}
liquid $\longrightarrow$ vapor ΔH_{vap}
solid $\longrightarrow$ vapor $\Delta H_{sub} = \Delta H_{fus} + \Delta H_{vap}$
13. $\Delta H_{sub} = 2.30$ kcal/mole; $\Delta H_{vap} = 2.08$ kcal/mole; $\Delta H_{fus} = 0.22$ kcal/mole   **14. (a)** 500 °C to 573, quartz

α; $t = 573$, quartz α + quartz β; 573 °C to 770 °C, quartz β; $t = 770$ °C, quartz α + tridymite; 770 °C to 1450 °C, tridymite; $t = 1450$ °C, tridymite + crystobalite; 1450 °C to 1810 °C, crystobalite; $t = 1810$ °C, crystobalite + liq.; 1810 °C to 2100 °C, liq. **(b)** From the diagram, the equilibrium line of quartz β and quartz α has a positive slope. This indicates that V of quartz β > V of quartz α, so that increasing pressure tends to produce more quartz α. Increasing temperature also shifts equilibrium to give quartz α. **15.** Point A: coesite + quartz α + quartz β; Point B: quartz β + tridymite + crystobalite; Point C: quartz β + crystobalite + liq. **16. (a)** ΔH = 2.388×10^{-4} kcal/mole **(b)** Kr: Van der Waal's attractive force only; non-polar. HBr: dipole-dipole interaction plus Van der Waal's attraction. NaCl: ion-ion attraction. Ion-ion > dipole-dipole > induced dipole-induced dipole (Van der Waal's). Parameter a decreases in the same order as above. NaCl > HBr > Kr **17. (a)** C_3H_7OH, C_2H_5COOH: H-bonding; CH_3CHO, CH_3OCH_3: dipole-dipole; C_3H_8, CH_4: induced dipole-induced dipole; C_3H_8 is more polarizable due to more electrons **(b)** T_{BP} = $0.5223\ a/b - 2.881$ = boiling point (°K); High value of a corresponds to high b.p.; High value of b corresponds to low b.p.; a term is dominant; Yes, high ΔH_{vap} corresponds to high b.p. **18.** CH_4, C_2H_6, C_3H_8 non-polar, Van der Waal's interaction; CH_3OCH_3, CH_3—CHO polar, dipole-dipole interaction; CH_3OH, CH_3COOH H-bonding **19. (a)** When critical temperature > actual temperature, increasing pressure results in: gas $\longrightarrow$ gas-liq. equilibrium $\longrightarrow$ liq. or gas $\longrightarrow$ gas-solid equilibrium $\longrightarrow$ solid or gas $\longrightarrow$ triple point $\longrightarrow$ liquid and/or solid. When temperature > critical temperature the substance remains in the gas phase.

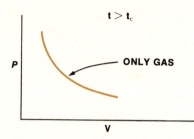

CHAPTER TWELVE PROBLEMS

		ΔH	ΔS	ΔG
1. (a)	ideal soln.	≈ 0	+	−
(b)	no soln.	+	+	+
(c)	soln. (approx. ideal)	small	+	−
(d)	soln.	−	+	−
(e)	no soln.	+	+	+
(f)	ideal soln.	small	+	−
(g)	no soln.	+	+	+
(h)	ideal soln.	≈ 0	+	−

2. $p_{hep} = 11.375$ Torr; $p_{hex} = 114.75$ Torr. The solution is an ideal solution since both substances are similar (non-polar). Intermolecular interaction is due to Van der Waal's attractive force.

3.

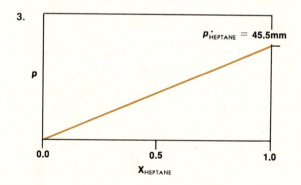

4. (a) $p_A^{\cdot} \cdot X_A = 345 \times 0.5 = 172.5 \neq 160$ mm; $p_C^{\cdot} \cdot X_C = 293 \times 0.5 = 146.5 \neq 100$ mm. No, they do not obey Raoult's law since the actual vapor pressures are lower than that prediced by Raoult's law. **(b)** Acetone and chloroform can form a H-bond

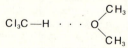

which causes the vapor pressure to be lower than expected in an ideal solution. **5.** 3.0 mm Hg **6.** No

(b)

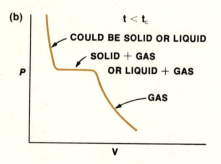

7. (a)

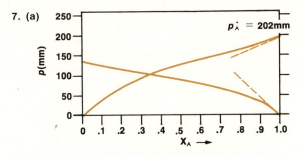

$p_A^* = 202mm$

(b) Region $X_B = 0$ to approximately $X_B = .04$ (more data points at lower concentrations would be required to determine this accurately).   (c) 391 Torr. (d) $p_{toluene} = X_{toluene(202\ mm)}$ 8. 6.84 torr.   9. $\Delta H_{fus} =$ 1,630 cal/mole; high MW_A 10. (a) 0.372 °C; (b) 7.54 °C   11. 0.0021 °C   12. 4.88×10^{-3} atm = 3.7 mm   13. 10,000 g/mole   14. 0.13 °C 15. increase   16. negative   17. (a) O_2, 2.58×10^{-4}; N_2, 5.2×10^{-4} (b) O_2, 5.77 ml; N_2, 11.6 ml   18. N_2, 13.0×10^{-3} mole/l; He, 8.2×10^{-3} mole/l. Helium has a lower solubility in the blood and body fluids. Furthermore, helium will diffuse out of the fluids more rapidly.   19. (a) Yes;   (b) No. $\Delta G°$ and K are independent of pressure and can only be changed by changing temperature.

CHAPTER TWELVE EXERCISES

1.

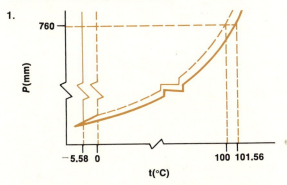

t(°C)

2. (a) 17,941;   (b) For $\pi = 0.06805$, MW = 17,928 and error in MW is 13;   (c) 0.001195 mm; No, the error would be too large.   3. $C_6H_{12}O_6$ 4. 252 cm ≈ 100 inches   5. increase; endothermic process, $\Delta H_{soln.}$ = positive   6. NaCl dissolved in H_2O lowers the freezing point. When NaCl is added to ice, some of the ice will melt, thereby absorbing heat and lowering the temperature to the point where ice is in equilibrium with the NaCl solution.   7. 0.011 °C

8. (a) 0.202 °C;   (b) 6.45%   9. $CaCl_{2(s)}$ lowers the freezing point of H_2O by dissolving in the water. When $CaCl_2$ is added to ice, the ice must melt (absorb heat) in order to lower the temperature to the freezing point with the $CaCl_2$ present. The ice will completely melt if the temperature is not below the freezing point in the presence of $CaCl_2$. 10. (a) 0.310 molal;   (b) 7.57 atm.   11. (a) $K = 0.94 \times 10^{-3}$ mole/l-atm at 48.5 °C   (b) 1.97×10^{-4} mole/l   (c) Yes, since at 48.5 °C $[O_2] = 1.97 \times 10^{-4}$ $> 1.34 \times 10^{-4}$ (d) 77.5 °C   12. −2.23 °C   13. Add a small piece of $Pb(ClO_4)_2$ crystal to the solution: If the solution is supersaturated, the solution will precipitate immediately. If the solution is saturated, nothing will happen. If the solution is unsaturated, the crystal will dissolve.   14. 5.5%   15. (a) Vapor pressure of solute, concentration, ionization (b) Na_2SO_4, $C_6H_{12}O_6$, CsBr, CH_3COOH, CH_3OH 16. $\bar{M} = 1,920$; $\bar{n} = 11.8 \approx 12$, $(C_6H_{10}O_5)_{12}$ 17. At the beginning when $BaSO_4$ crystals are formed, they exist as very fine particles and are suspended homogeneously in the water. The solution is classified as a colloidal solution. In time the colloidal particles combine to form a precipitate of $BaSO_4$. 18. The hydrocarbon portion of casein will dissolve in the fat particles and the COO^- and NH_3^+ groups of the casein will point outward into the water portion. The emulsified fat is soluble because of the ionic nature of COO^- and NH_3^+ and their solubility in water.   19. $Na_2CrO_4 \cdot 10H_2O$

CHAPTER THIRTEEN PROBLEMS

1. $K = 1.8 \times 10^{-5}$ 2. $K = 2.4 \times 10^{-5}$ 3. (a) $\Delta G° = 16.68$ kcal/mole; $K_P = 5.7 \times 10^{-13}$ (b) $\Delta G° = 33.36$ kcal; $K_P = 3.2 \times 10^{-25}$; $\Delta G°$ in (b) is twice as large as that in (a). Yes, the K_P in (a) is the square root of the K_P in (b)   (c) It would be more practical to make a table of $\Delta G°_{f,i}$ values.   4. $\Delta G° = \Delta G°_{f, CO_2(g)} + \Delta G°_{f, CaO(g)} - \Delta G°_{f, CaCO_3(g)}$; $K_P = P_{CO_2}$ 5. $K = 0.86$ 6. (a) $[OH^-] = 2 \times 10^{-12}$ M   (b) No. The product of $[H^+]$ and $[OH^-]$ must remain constant at $K_w = 10^{-14}$. The answer will be the same.   7. pOH = 2.48; pH = 11.52   8. pH = 1.70; $[OH^-] = 5 \times 10^{-13}$ 9. (a) $[H^+] = 3.58 \times 10^{-6}$ M   (b) $[H^+] \approx 3.58 \times 10^{-6}$ M; For very small K_a, the difference of the two calculations is negligible, so method (a) is satisfactory.   10. (a) $[H^+] = 0.21$ M   (b) $[H^+] = 0.083$ M; For large K_a, we should use the rigorous calculation in (b).   11. (a) $[H^+] = 1.13 \times 10^{-7}$ M (b) $[H^+] = 1.13 \times 10^{-7}$ M; For weak acids even at

low concentrations the approximation is satisfactory. **12.** $[OH^-] = 3.5 \times 10^{-6}$ M; $pH = 8.54$ **13. (a)** $[OH^-] = 4.2 \times 10^{-4}$ M; **(b)** $[OH^-] = 1.8 \times 10^{-6}$ M **14.** $[H^+] = [HCO_3^-] = 6.56 \times 10^{-5}$; $[CO_3^{2-}] = 5.61 \times 10^{-11}$ **15. (a)** $[H^+] = [H_2PO_4^-] = 2.39 \times 10^{-2}$ M; $[HPO_4^{2-}] = 6.23 \times 10^{-8}$ M; $[PO_4^{3-}] = 5.8 \times 10^{-19}$ M **(b)** $[H^+] = [H_2PO_4^-] = 5.69 \times 10^{-3}$ M; $[HPO_4^{2-}] = 6.23 \times 10^{-8}$ M; $[PO_4^{3-}] = 2.4 \times 10^{-18}$ M **16.** $[HS^-] = 1.1 \times 10^{-6}$ M; $[S^{2-}] = 1.1 \times 10^{-18}$ M **17.** $[H^+] = 4.3 \times 10^{-5}$ M **18. (a)** $[H^+] = 5.6 \times 10^{-10}$; **(b)** $[H^+] = 8.4 \times 10^{-10}$ **19. (a)** $[H^+] = 6.23 \times 10^{-8}$ M before adding; $[H^+] = 7.6 \times 10^{-8}$ M after adding; % increase = 22% **(b)** $[H^+] = 5.07 \times 10^{-8}$; % decrease = 18% **20.** If $[H_3PO_4] = [H_2PO_4^-]$, the solution will act as a buffer. $pH = 2.12$.

21.
$$NH_4ClO_4 + \xrightarrow{aq} \boxed{NH_4^+} + ClO_4^-$$
$$H_2O \rightleftharpoons \boxed{OH^-} + H^+$$
$$\Updownarrow$$
$$NH_{3(aq)} + H_2O$$

Since $HClO_4$ is a strong acid and NH_4OH is a weak base, the hydrolyzed solution is acidic.

22. (a)

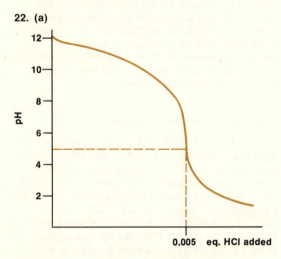

(b) $pK = 4.1 \sim 6.1$; Methyl red is the best.
23. (a) $K_{sp} = 1.9 \times 10^{-16}$; $\Delta G° = 20.94$ kcal/mole
(b) $\Delta G°_{f, Fe^{2+}} = -19.5$ kcal/mole **24.** $[Fe^{3+}] = 4.5 \times 10^{-10}$ **25.** $[Fe^{2+}] = 1.36 \times 10^{-4}$ compared to $[Fe^{2+}] = 1.67 \times 10^{-5}$ in Problem 13–23; the solubility increases by a factor of 8.
26. (a) $[Cu^{2+}] = 1.69 \times 10^{-4}$ M **(b)** $C_2O_4^{2-} + H_2O \rightleftharpoons HC_2O_4^- + OH^-$; This will increase the solubility of CuC_2O_4 due to the fact that $[C_2O_4^{2-}]$ decreases; thus the reaction shifts to the right in order to keep the product equal to the constant,

K_{sp}. **(c)** Since OH^- is produced, the pH will increase. **27.** $[Ni^{2+}] = 9 \times 10^{-3}$ M **28.** Solubility = 5×10^{-5} moles Tl_2S/liter **29.** Control the $[S^{2-}]$ by varying the amount of HCl added:

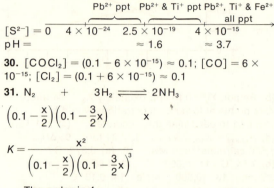

30. $[COCl_2] = (0.1 - 6 \times 10^{-15}) \approx 0.1$; $[CO] = 6 \times 10^{-15}$; $[Cl_2] = (0.1 + 6 \times 10^{-15}) \approx 0.1$
31.
$$N_2 \quad + \quad 3H_2 \rightleftharpoons 2NH_3$$
$$\left(0.1 - \frac{x}{2}\right)\left(0.1 - \frac{3}{2}x\right) \qquad x$$

$$K = \frac{x^2}{\left(0.1 - \frac{x}{2}\right)\left(0.1 - \frac{3}{2}x\right)^3}$$

The order is 4
32. (a) $\Delta H° = +0.94$ kcal. Since $\Delta H°$ is positive, the reaction will shift to the right as the temperature is increased to 40 °C and NH_3 is then a stronger base. However, since $\Delta H°$ is a small quantity the effect should be small. **(b)** $K_{b_2} = 1.9 \times 10^{-5}$ **33.** 2.36×10^{-6} M **34.** The reaction will shift to the right because there is a decrease in the number of moles and hence a decrease in volume. The effect is small, however, since the reaction is so close to 100% yield (see answer to problem 30).

CHAPTER THIRTEEN EXERCISES

1. (a) $K = \dfrac{[SO_3]^2}{[SO_2]^2[O_2]}$ or $K_P = \dfrac{P_{SO_3}^2}{P_{SO_2}^2 P_{O_2}}$

(b) $K = \dfrac{[NH_3]^2}{[N_2][H_2]^3}$ or $K_P = \dfrac{P_{NH_3}^2}{P_{H_2}^3 P_{N_2}}$

(c) $K = \dfrac{[HI]}{[H_2]^{\frac{1}{2}}[I_2]^{\frac{1}{2}}}$ or $K_P = \dfrac{P_{HI}}{P_{H_2}^{\frac{1}{2}} P_{I_2}^{\frac{1}{2}}}$

(d) $K = \dfrac{[C_2H_6]}{[H_2][C_2H_4]}$ or $K_P = \dfrac{P_{C_2H_6}}{P_{H_2} P_{C_2H_4}}$

(e) $K = P_{C_2H_2}[Ca(OH)_2]$

(f) $K = \dfrac{[HNO_3]^2 P_{NO}}{P_{NO_2}^3}$

(g) $K = \dfrac{[Fe^{2+}][VO^{2+}][H^+]^2}{[Fe^{3+}][V^{3+}]}$

(h) $K = \dfrac{[Ce^{3+}]^2[H_3AsO_4][H^+]^2}{[Ce^{4+}]^2[HAsO_2]}$

2. $K = 200$; $\Delta G° = -3.1$ kcal/mole **3.** $[H_2O_{(g)} =$

0.0508 M; $[CO] = 0.0508$ M; $[H_2] = 0.0242$ M; $[CO_2] = 0.0242$ M   **4.** $K = 6.25$ **5. (a)** $K = 36$
(b) $[R'COOR] = 1.03$ M; $[H_2O] = 8.28$ M; $[ROH] = 0.49$ M; $[R'COOH] = 0.49$ M
6. (a) uncoupled $K = 2.15 \times 10^{-4}$; coupled $K = 1.36 \times 10^5$ **(b)** [sucrose] $= 2.15 \times 10^{-8}$ M   **(c)** [sucrose] $= 0.01 - 2.72 \times 10^{-6} \approx 0.01$ M   **7. (a)** pH $= 1$
(b) pH $= 12$ **(c)** pH $= 3$ **(d)** pH $= 4$ **(e)** pH $= 7$
(f) pH $= 9$ **8. (a)** T increases: reaction shifts to left
(b) P increases: reaction shifts to right   **(c)** Addition
of N_2: reaction shifts to right   **(d)** No effect on
equilibrium—reaction approaches equilibrium as
time increases.   **9.** $K_w = 10^{-12}$ **10. (a)** $[HCO_3^-] = 0.020$; $[CO_3^{2-}] = 1.9 \times 10^{-5}$ **(b)** Change would be
in the fifth place and would not change $[HCO_3^-]$
within the 3 significant figures.   **11.** $pK_1 = 2.34$; If
25 ml 0.1 M HCl is added to 50 ml of 0.1 M glycine,
a buffer solution with pH $= pK_1 = 2.34$ will be
formed. Similarly if 25 ml .1 M NaOH is added to
50 ml of 0.1 M glycine, a buffer with pH $= pK_2 = 9.64$ will be formed.   **12. (a)** $[Mg^{2+}] = 1.306 \times 10^{-4}$ M $=$ the solubility; therefore, 0.01 moles cannot
dissolve in 1 liter.   **(b)** Yes   **13.** $[H^+] = [HCO_3^-] = 3.8 \times 10^{-5}$ M; $[CO_3^{2-}] = 6 \times 10^{-11}$ M   **14.** (d)
15. (a) 1.607×10^{-5} **(b)** $PO_4^{3-} + HOH \longrightarrow HPO_4^{2-} + OH^-$; $HPO_4^{2-} + HOH \longrightarrow H_2PO_4^- + OH^-$;
$H_2PO_4^- + HOH \longrightarrow H_3PO_4 + OH^-$; The effect
of hydrolysis is to lower the concentration of PO_4^{3-}
in the solution, so more Ag_3PO_4 is dissolved.
(c) Adding acid will neutralize OH^- which is
produced during the hydrolysis, so it increases the
hydrolysis and the solubility of Ag_3PO_4.   **16.** $\Delta H° = 3.88$ kcal/mole   **17.** $K = 0.622$ **18. (a)** 84%   **(b)** 62%
19. pH $= 4.81$
20.

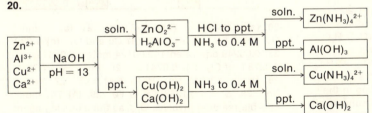

21. Phenol and sodium phenolate at concentrations
so that $[AH]/[A^-] = 2.00$; e.g. [phenol] $= 0.200$ and
[phenolate] $= 0.100$
22.

$$NaF \rightleftharpoons Na^+ + F^-$$
$$H_2O \rightleftharpoons OH^- + H^+$$
$$\downarrow$$
$$HF$$

NaOH is a strong base and HF is a weak acid, so the
solution is basic.   **23. (a)** 1–4 are strong acid +

strong base; reaction is $H_{(aq)}^+ + OH_{(aq)}^- \longrightarrow H_2O_{(aq)}$
(b) 5 is strong acid + weak base; reaction is
$H_{(aq)}^+ + NH_{3(aq)} \longrightarrow NH_{4(aq)}^+$; 6 is weak acid + strong
base; reaction is $HCN_{(aq)} + OH_{(aq)}^- \longrightarrow H_2O + CN_{(aq)}^-$

24.

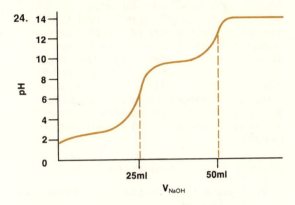

25. (a) $K_P = 5.5 \times 10^{34}$ **(b)** $[O_2] \approx$ initial
concentration $= 8.2 \times 10^{-3}$ moles/liter; $[NO_2] \approx 10^{-8}$ moles/liter   **(c)** $\Delta H = -46.5$ kcal/mole < 0, so
when temperature increases, $[O_2]$ and $[NO_2]$ will
decrease. However, the effect will be negligible
since the K_P at 35 °C will be large.   **26. (a)** Boiling
will drive off the CO_2 and shift the equilibrium to the
right and the Ca^{2+} will be removed as $CaCO_{3(s)}$.
(b) $OH^- + HCO_3^- \longrightarrow H_2O + CO_3^{2-}$; $CO_3^{2-} + Ca^{2+} \longrightarrow CaCO_{3(s)}$; Each mole of $Ca(OH)_2$ produces
2 moles of $[OH^-]$ and each mole of $[OH^-]$ causes
precipitation of 1 mole $CaCO_{3(s)}$. Therefore, one
mole $Ca(OH)_2$ will precipitate one mole (g-atom) of
Ca^{2+} ions.   **(c)** 99.78% Ca^{2+} will be removed as
$CaCO_{3(s)}$ **27. (b)** ΔH_{fus} depends on the type of
crystal: ionic, covalent, molecular. ΔH_{mixing} depends
on the nature of the solute: ionic, polar, or non-polar
and on the nature of the solvent: polar or non-polar.
(c) ΔH_{fus} (+), ΔH_{mix} (−); If $|\Delta H_{fus}| > |\Delta H_{mix}|$ then
ΔH_{soln} is (+) and vice versa   **(d)** The ΔH_{mixing} is
usually small compared to ΔH_{fus} which is always
positive. For example, if the solute and solvent are
similar they may form an ideal solution where
$\Delta H_{mixing} \approx 0$.   **(e)** Increase, because ΔH_{soln} is

usually positive. **28. (b)** X⁺, Y⁻: ion-ion attraction in the crystal; X⁺, H_2O: ion-dipole interaction (solvation energy); Y⁻, H_2O; ion-dipole interaction (solvation energy) **(c)** $\Delta H_{sub, LiCl_{(s)}} = 4.5$ kcal/mole; $\Delta H_{sub, KCl_{(s)}} = 52.6$ kcal/mole **(d)** $\Delta H^\circ_{soln, LiCl} = -14$ kcal/mole; $\Delta H^\circ_{soln, KCl} = 5$ kcal/mole **(e)** LiCl should have the greatest solubility. KCl solubility increases as temperature increases since ΔH is positive. **29. (a)** Since ΔH is a state function, $\Delta H = \Delta H_{solv, HA} + D_{HA} + IP_H - EA_A - \Delta H_{solv, H^+} - \Delta H_{solv, A^-}$ **(b)** HF: 0.305 kcal/mole; HCl: -11.0 kcal/mole **(c)** Solvation energy of H⁺ is extremely large. **30. (a)** Buffer solution composed of citric acid and the anion with one proton removed. **(b)** Triprotic acid (3 ionizable H⁺)

CHAPTER FOURTEEN PROBLEMS

1. (a) Oxidation: $Cr^{2+}_{(aq)} \rightleftharpoons Cr^{3+}_{(aq)} + e^-$; Reduction: $Fe^{3+}_{(aq)} + e^- \rightleftharpoons Fe^{2+}_{(aq)}$

(b)

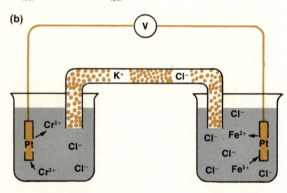

2. Oxidation: $Pb_{(s)} + SO_4^{2-} \longrightarrow PbSO_{4(s)} + 2e^-$; Reduction: $Sn^{4+}_{(aq)} + 2e^- \rightleftharpoons Sn^{2+}_{(aq)}$ **3. and 4.** **(a)** $2Fe^{2+}_{(aq)} + ClO^-_{3(aq)} + 3H^+_{(aq)} \, 2Fe^{3+}_{(aq)} + HClO_2 + H_2O$ **(b)** $UO^+_{2(aq)} + V^{+3}_{(aq)} + 2H^+_{(aq)} \, U^{4+}_{(aq)} + VO^{2+}_{(aq)} + H_2O$ **(c)** $Zn_{(s)} + Ni(OH)_{2(s)} + 2OH^-_{(aq)} \, ZnO_2^{2-}_{(aq)} + Ni_{(s)} + 2H_2O$ **5. (a)** 0.139 volts **(b)** Yes **(c)** Yes; ΔG° is dependent on the number of moles involved in the reaction and therefore depends upon the number of electrons transferred. $\mathcal{E}^\circ$ is independent of the number of electrons transferred and the n in equation 14–8 takes this into account. **6.** $\mathcal{E}^\circ_{\frac{1}{2}} = 0.910$ volts **7.** $\mathcal{E}^\circ_{\frac{1}{2}} = 0.140$ volts **8.** $\mathcal{E}^\circ_{Br_2, Br^-} = 1.0655$ volts **9.** $\Delta G^\circ_{f, Cl^-} = -31.35$ kcal/mole; $\Delta G^\circ_{f, Br^-} = -24.6$ kcal/mole **10.** 0.158 volts **11. (a)** No; **(b)** Yes; **(c)** No; **(d)** Yes **12.** 190 **13. (a)** for b., $K = 7.95 \times 10^{35}$; for c., $K = 1.82 \times 10^{-4}$ **(b)** Reaction (b) will produce the most $Hg_{(l)}$ **14.** -0.0592 V. **15. (a)** Yes; **(b)** Yes; **(c)** Yes **16. (a)** $Zn_{(s)} + HgO_{(s)} + H_2O$

$\longrightarrow Zn(OH)_{2(s)} + Hg_{(l)}$ **(b)** $\mathcal{E}^\circ = 1.34$ V **(c)** No **(d)** The overall cell reaction involves only the pure electrode materials and these will not change with use of the cell until one of the electrode materials is depleted. **17.** $Cd_{(s)} + HgO + H_2O \longrightarrow Cd(OH)_2 + Hg$; $\mathcal{E}^\circ = 0.907$ V.

18.

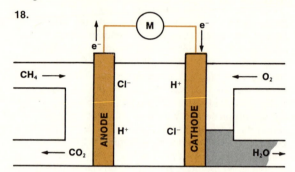

19. $C_3H_8 + 6H_2O \longrightarrow 3CO_2 + (20)H^+ + 20\ e^-$

$$\frac{5O_2 + (20)H^+ + 20\ e^- \longrightarrow 10H_2O}{C_3H_8 + 5O_2 \longrightarrow 3CO_2 + 4H_2O}$$

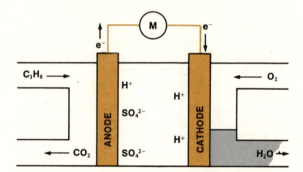

20. (a) 1.182 volts **(b)** 1.037 volts **21. (a)** H_2 and O_2 **(b)** 2.1 V ~ 2.3 V **22. (a)** Cu and O_2 **(b)** 1.28 V **23.** 0.00249 equivalents; 0.00124 moles; 0.0881 g.; 0.0278 l. of Cl_2 **24.** 0.0741 g. **25.** Pb^{2+}, Hg_2^{2+}, Sn^{2+} **26.** $[H^+] = 10^{-7}$ M, No; $[H^+] = 1$ M, Yes. **27.** No; No. **28.** $\mathcal{E}^\circ < -0.1263$; Cd or Fe **29. (a)** The most feasible reaction involves $O_{2(g)}$ as the oxidizing agent.

$$
\begin{array}{lr}
 & \mathcal{E}_{\frac{1}{2}} \\
\frac{1}{2}O_{2(g)} + H_2O + 2e^- \longrightarrow 2OH^- & 0.79 \\
Fe_{(s)} \longrightarrow Fe^{2+} + 2e^- & \approx +0.5 \\
\hline
Fe_{(s)} + \frac{1}{2}O_{2(g)} + H_2O \longrightarrow Fe^{2+} + 2OH^- & \mathcal{E} \approx 1.3 \text{ V}
\end{array}
$$

(b) The iron becomes ionized and eventually would react further with OH⁻ and O_2. $Fe^{2+} + 2OH^- \longrightarrow Fe(OH)_2$; $2Fe(OH)_2 + \frac{1}{2}O_2 \longrightarrow Fe_2O_3 \cdot H_2O + H_2O$ **(c)** Since Al_2O_3 is a non-conducting metal oxide, the contact area where electron transfer can occur is restricted to the scratched area. Therefore, the corrosion is greatly reduced. **(d)** No. **(e)** Because the corrosive effect on aluminum is much slower than that of "tin" cans.

CHAPTER FOURTEEN EXERCISES

1. (a) $2HNO_3 + 6HCl \longrightarrow 2NO + 3Cl_2 + 4H_2O$
(b) $Sn + 4HNO_3 \longrightarrow SnO_2 + 4NO_2 + 2H_2O$
(c) $2FeCl_3 + H_2S \longrightarrow 2FeCl_2 + 2HCl + S$ **(d)** $I_2 + 10HNO_3 \longrightarrow 2HIO_3 + 10NO_{2(g)} + 4H_2O$
(e) $2Bi(OH)_3 + 3Na_2SnO_2 \longrightarrow 2Bi + 3Na_2SnO_3 + 3H_2O$ **(f)** $2MnO_{4(aq)}^- + 3NO_{2(aq)}^- + H_2O \longrightarrow 2MnO_{2(s)} + 3NO_{3(aq)}^- + 2OH_{(aq)}^-$ **2. (a)** $2Cu^+ \longrightarrow Cu + Cu^{2+}$ **(b)** $Cl_2 + 2OH^- \longrightarrow OCl^- + Cl^- + H_2O$
(c) $2ClO_2 + H_2O \longrightarrow ClO_2^- + ClO_3^- + 2H^+$
3. Exercise 1: **(a)** -0.3983 V, No.; **(c)** $+0.629$ V, Yes.; **(d)** -0.39 V, No.; **(f)** $+0.587$ V, Yes. Exercise 2: $+0.364$ V, Yes.; $+0.4583$ V, Yes.; -0.196 V, No. **4. (a)** $Fe^{3+} + e^- \longrightarrow Fe^{2+}$; $V^{3+} + H_2O \longrightarrow VO^{2+} + 2H^+ + e^-$ **(b)** $\mathcal{E}° = 0.433$ V; electrons would flow from right to left

(c)

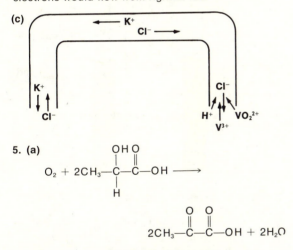

5. (a)

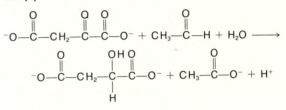

$\mathcal{E}°$ is independent of the pH of the solution.
(b) 0.941 V.; Yes. **6. (a)** 1.239 V **(b)** 0.010 V (standard reduction potential would be -0.010 V)
7. (a) Cs, Sr, Mg, Ce, Al, Be, V, Fe, Tl, Pb, Cu, Pd, Pt, Au **(b)** Au, Pt, Pd, Cu **(c)** VIII and IB
8. (a) $Fe^{3+} + Cu^+ \longrightarrow Fe^{2+} + Cu^{2+}$, $\mathcal{E} = 0.612$ V, feasible; $Cu^{2+} + V^{3+} + H_2O \longrightarrow Cu^+ + VO^2 + 2H^+$, $\mathcal{E} = 0.65$ V, feasible; **(b)** $Fe^{3+} + V^{3+} + H_2O \longrightarrow Fe^{2+} + VO^{2+} + 2H^+$, $\mathcal{E} = 1.26$ V **(c)** No.
9. (a) 0.81 V; **(b)** 0.79 V; **(c)** Neither Fe^{2+} nor $Fe^{2+}_{(porphyrin)}$ is stable to air oxidation; however, the potential for Fe^{2+} is more positive and the extent of

Fe^{2+} oxidation would far exceed the oxidation of $Fe^{2+}_{(porphyrin)}$.
10. (a) $3HClO \longrightarrow HClO_2 + Cl_2 + H_2O$, $\mathcal{E}° = -0.01$ V ≈ 0; **(b)** $2HClO_2 \longrightarrow ClO_3^- + HClO + H^+$, $\mathcal{E}° = 0.57$ V; **(c)** $2ClO_3^- + H^+ \longrightarrow ClO_4^- + HClO_2$, $\mathcal{E}° = 0.02$ V; Dissociation to other species is also possible. For example $7HClO \longrightarrow ClO_4^- + 3Cl_2 + H^+ + 3H_2O$, $\mathcal{E}° = 0.028$ V **11.** 384 coulombs
12. 1.07×10^7 coulombs **13.** The reduction potential is a measure of the oxidizing power of the non-metal. Note in these groups that $\mathcal{E}°_{\frac{1}{2}}$ increases from the bottom to the top of the group: VIA: Te(-0.92); Se(-0.78); S(-0.508); VIIA: I_2(0.535); Br_2 (1.065); Cl_2(1.36); F_2(2.87). Note also that $\mathcal{E}°_{\frac{1}{2}}$ increases as we go from group VIA to group VIIA.
14. (a)

$$^-O-\overset{O}{\underset{\|}{C}}-CH_2-\overset{O}{\underset{\|}{C}}-\overset{O}{\underset{\|}{C}}-O^- + CH_3-\overset{O}{\underset{\|}{C}}-H + H_2O \longrightarrow$$

$$^-O-\overset{O}{\underset{\|}{C}}-CH_2-\overset{OH}{\underset{\underset{H}{|}}{\overset{|}{C}}}-\overset{O}{\underset{\|}{C}}-O^- + CH_3-\overset{O}{\underset{\|}{C}}-O^- + H^+$$

(b) $\Delta G° = -22.971$ kcal/mole.; $K = 8 \times 10^{16}$
(c) Yes, but $[H^+]$ would have to be large to change the direction of the reaction in (a).
15. Anode

CHAPTER FIFTEEN PROBLEMS

1. (a) rate $= k[O_3]$ **(b)** rate $= k[Cl_2]$ **(c)** rate $= k[O_3][O]$ **(d)** rate $= k[Cl][CO]$ **(e)** rate $= k[COCl][Cl_2]$ **2. (a)** $Cl_2 \overset{K_1}{\rightleftharpoons} 2Cl$ (Fast); $Cl + COCl_2 \overset{k_{-3}}{\longrightarrow} Cl_3 + CO$; Rate reverse $= k_{-3}[Cl][COCl_2]$. $K_1 = [Cl]^2/[Cl_2]$ or $[Cl] = K_1^{\frac{1}{2}}[Cl_2]^{\frac{1}{2}}$; Rate reverse $= k_{-3} K_1^{\frac{1}{2}}[COCl_2][Cl_2]^{\frac{1}{2}}$ **(b)** Yes. **(c)** No, because different mechanisms can predict the same experimental results. **3. (a)** Rate $= K_1 k_2[O_3]^2/[O_2]$ **(b)** Rate $= k[O_3]^2$; Yes, this is different from the rate expression as shown in (a) where the rate is inversely proportional to $[O_2]$.
4. (a) Rate $= K_1 K_2 k_3[H_2][I_2]$ **(b)** No, both mechanisms predict the same rate expression. On the basis of the experimental rate expression, we cannot differentiate between the two mechanisms. Further information is needed. **5.** $l^{\frac{1}{2}}mole^{-\frac{1}{2}}sec^{-1}$
6. Rate $= k[NO_2][F_2]$; $k = 40$ l mole^{-1} sec^{-1}
7. $(2)^{\frac{3}{2}} = 2.83$ **8.** $E^* = 20$ kcal/mole; $A = 5.35 \times 10^{11}$ **9. (a)** $k = 7.6 \times 10^{-14}$ **(b)** $E_r^* = 85.7$ kcal/mole; $k_{-1} \approx 10^{-52}$ **10.** $NO_2 + SO_2 \longrightarrow NO + SO_3$

11. (a) $CH_2{=}CH{-}CH_2{-}CH_3 + {-}\overset{|}{Ni}{-}\overset{|}{Ni}{-} \longrightarrow$

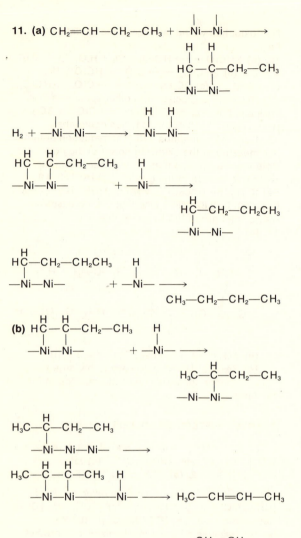

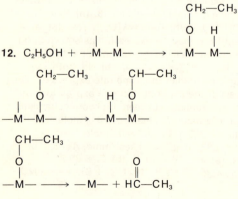

CHAPTER FIFTEEN EXERCISES

1. (a) Rate $= k[A]^\alpha[B]^\beta$ **(b)** Rate $= k[A]^3[B]$
(c) $K = 0.094\ l^3\ \text{mole}^{-3}\text{min}^{-1}$ **2. (a)** Increase
temperature to increase the rate constant.
(b) Choose proper catalysts.   **(c)** Completely stir the
reactor to eliminate slow diffusion effect.   **(d)** Rate
$= k[A]^\alpha[B]^\beta[C]^\gamma\cdots\cdot$ Increase the concentration
of those chemicals with positive reaction orders and
decrease the concentration of those chemicals
with negative reaction orders.   **3. (a)** $R_1 = 4.35 \times$
10^{-5} mole/min; $R_2 = 2.23 \times 10^{-5}$ mole/min; $R_3 =$
2.01×10^{-4} mole/min   **(b)** $R = k[H_2PO_2^-][OH^-]^2$
4. $k_2/k_1 = 58$ **5. (a)** $\Delta H° = -13.87$ kcal/mole
(b) $\Delta E° = -13.23$ kcal/mole   **(c)** $E_r^* = 48.56$

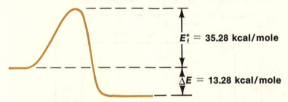

$E_f^* = 35.28$ **kcal/mole**

$\Delta E = 13.28$ **kcal/mole**

kcal/mole   **6. (a)** We could not tell because different mechanisms could give the same expression.

b.

The reaction in one step involves a single *O* atom
transfer. Furthermore, bond dissociation energy
D_{O_2-O} is low and the bond can be readily broken.
On the other hand, the large D_{ON-O} bond energy
would further facilitate the reaction.
(c) $K = k_f/k_r = [O_2][NO_2]/[O_3][NO]$
(d) Calculate $\Delta G°$ and K; $\Delta G° = \Delta G°_{f,O_2} + \Delta G°_{f,NO_2} -$
$\Delta G°_{f,O_3} - \Delta G°_{f,NO}$ ($\Delta G°_{f,i}$ from appendices); $\Delta G° =$
$-RT \ln K_p = -RT \ln K$; $k_r = k_f/K$ **7. (a)** Yes,
because the three molecules $2NO$, O_2 must collide
simultaneously to proceed in a single step, and
this probability is very small.   **(b)** I. Rate $= K_1 k_2$
$[NO]^2[O_2]$; Yes; II. Rate $= K_1 k_2[NO]^2[O_2]$, Yes;
III. Rate $= k_1[NO][O_2]$, No.   **(c)** Comparing
coefficients, we can obtain $k_f = K_1 k_2$ **(d)** $\Delta H° = 27.0$
kcal/mole; $\Delta E° = -26.4$ kcal/mole

(e)

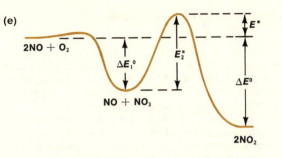

8. I. Rate $= k[H_2]^2[NO]^2$, No; II. Rate $= k_1[H_2][NO]$, No; III. Rate $= k_1[H_2][NO]^2$, Yes; IV. Rate $= k_2K_1[H_2][NO]^2$, Yes.   **9. (a)** $\Delta G° = -73.7$ kcal. The reaction is feasible so H_2O_2 is not stable. **(b)** Because thermodynamic information tells only the feasibility of the reaction and does not indicate the reaction rate. In this example the reaction rate could be very slow even though thermodynamically feasible.   **(c)** Catalyst   **(d)** $\mathcal{E}_{cell} = 1.09$ volts > 0, so the reaction is feasible.   **(e)** 3 bonds have to be broken and 2 bonds formed, so the reaction is slow.
(f) 1. $H_2O_2 + 2H^+ + 2Fe^{2+} \longrightarrow$
$\qquad 2H_2O + 2Fe^{3+}$ $\mathcal{E}° = 1.00$ volts > 0

2. $H_2O_2 + 2Fe^{3+} \longrightarrow$
$\qquad O_2 + 2H^+ + 2Fe^{2+}$ $\mathcal{E}° = 0.09$ volts > 0

$\qquad 2H_2O_2 \longrightarrow 2H_2O + O_2$ $\mathcal{E}_{cell}° = 1.09$ volts
Fe^{2+} is unchanged. The potential for the first step will be lowered and the potential for the second step will be increased when $[H^+] = 10^{-7}$ M is considered.   **10. (a)** Rate $= k[N_2O_5]^2$ for a single step reaction and this does not agree with the experimentally determined rate expression.   **(b)** The first step is the only slow reaction step, so it is the rate controlling step and Rate $= k_1[N_2O_5]$

(c)

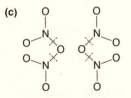

For a single-step reaction, the intermediate must be highly oriented as shown, and four N—O bonds must be broken simultaneously, so it is not likely to happen.   **(d)** Rate $= k[N_2O_5]$ **11.** The first step is the initiation step; the next two steps form a chain mechanism; reaction four is an inhibition step; formation of Br_2 is the termination step.

12. $Cl_2 \xrightarrow{h\nu} 2Cl$ $\qquad\qquad$ initiation
$\left.\begin{array}{l} Cl\cdot + CHCl_3 \longrightarrow CCl_3\cdot + HCl \\ CCl_3 + Cl_2 \longrightarrow CCl_4 + Cl\cdot \end{array}\right\}$ chain mechanism
$Cl\cdot + Cl\cdot \longrightarrow Cl_2$ $\qquad\qquad$ termination
13. (c) $y = [S]/R$; $x = [S]$; $m = 1/k_2[E]_0$;
$b = K_m/k_2[E]_0$

(d)

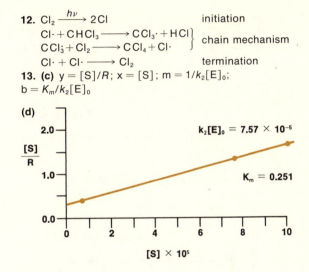

14. (a) $n_A° = 0.00312$ moles   **(b)** Graph of ln n_A versus t is linear
15. (a) $As(III) + I_2 \longrightarrow As(V) + 2I^-$ $\mathcal{E} = 0.43$

$\quad$ (B) $2Ce(IV) + 2I^- \longrightarrow$
$\qquad\qquad I_2 + 2Ce(III)$ $\mathcal{E} = 0.924$

$\qquad As(III) + 2Ce(IV) \longrightarrow$
$\qquad\qquad As(V) + 2Ce(III)$ $\mathcal{E}_{cell} = 1.35$ volts

If step (A) is slow compared to step (B), then
Rate $= k[As(III)][I_2]$
16. (A) $V^{3+} + H_2O + Cu^{2+} \longrightarrow$
$\qquad\qquad VO^{2+} + 2H^+ + Cu^+$ $\mathcal{E} = 0.228$

$\quad$ (B) $Cu^+ + Fe^{3+} \longrightarrow Cu^{2+} + Fe^{2+}$ $\mathcal{E} = 0.612$

$\qquad V^{3+} + H_2O + Fe^{3+} \longrightarrow$
$\qquad\qquad VO^{2+} + 2H^+ + Fe^{2+}$ $\mathcal{E}_{cell} = 0.840$

If step (A) is slow compared to step (B), then
[Rate] $= k[V^{3+}][Cu^{2+}]$

17.

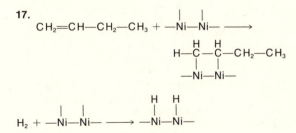

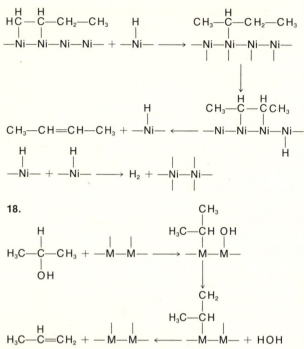

18.

CHAPTER SIXTEEN PROBLEMS

1. (a) Uni.; **(b)** Uni.; **(c)** bi.; **(d)** Uni.; **(e)** bi.;
(f) Uni.; **(g)** bi. **2. (a)** 6; **(b)** 6; **(c)** 4; **(d)** 2;
(e) 6; **(f)** 2 **3. (a)** dibromotetramminecobalt (III)
chloride; **(b)** potassium hexafluoroferrate (II);
(c) hexaaquochromium (III) ion
4. (a) $[Co(NH_3)_4Cl_2]$ Br; **(b)** $[Cr(en)_3]^{3+}$;
(c) $Li_2[Ti(CN)_6]$. **5.** 5.92 BM

6. (a) Ti^{3+}; $3d^1$

(b) Zn^{2+}; $4d^{10}$ **(c)** Cu^{2+}; $3d^9$

7. (a) 1.73 BM; **(b)** 0; **(c)** 1.73 BM
8.   high   spin   low

9. d^{10} case $- t^6e^4$ and crystal field stabilization energy
is $6(-\frac{2}{5}\Delta_0) + 4(\frac{3}{5}\Delta_0) = 0$. **10. (a)** high; **(b)** high
11. $-\frac{4}{5}\Delta_0$ **12.** -671 kcal/mole
13. (a) t paramagnetic
     e

(b) e or e
          t t
     low spin   high spin
     for octahedral case

CHAPTER SIXTEEN EXERCISES

1. (a) 4.90 BM; **(b)** 6.92 BM; **(c)** 7.94 BM
2. It is not possible because there are only five $3d$
orbitals, which can accommodate a maximum of five
unpaired electrons. **3.** Number of oxidation states
increases to a maximum half way through and then
decreases again. **4.** In the third series of the
transition metals, there is the interjection of the
Lanthanides or rare earth elements with atomic
number 57 (counting La) through 71. This interjection
is accompanied by a phenomenon known as the
Lanthanide contraction. In the Lanthanides, the
$4f$ orbitals are being filled, and the valence shell
configuration is $6s^2$ and remains so during the
addition of the fourteen $4f$ electrons. As the atomic
number increases, the electrons are all going into
the same subshell $(4f)$ and there is a progressive
decrease in ionic size. The same is true for the size
of the atoms for similar reasons. Because of the
presence of the lanthanide contraction in the third
transition metal series, the sizes of the ions of the
same charge in the second and third series are
very similar. **5. (a)** cesium pentachloroaquotitanate
(III) **(b)** bis (ethylenediamine) platinum (II) chloride
(c) hexammine palladium (IV) ion
6. (a) $K_3[Fe(C_2O_4)_3]$ **(b)** $[HgI_4]^{2-}$ **(c)** $[Zn(CN)_4]^{2-}$
7. (a) acid; **(b)** base; **(c)** base; **(d)** acid;
(e) acid; **(f)** acid; **(g)** base; **(h)** base
8. (a) $[TiCl_6]^{3-}$; **(b)** $[FeCl_4]^{2-}$; **(c)** $[CuCl_2]^{-1}$;
(d) $[VCl_6]^{3-}$ **(e)** $[CoCl_4]^{2-}$; **(f)** $[CoCl_6]^{3-}$
9. (b), (c) and (e)

10.

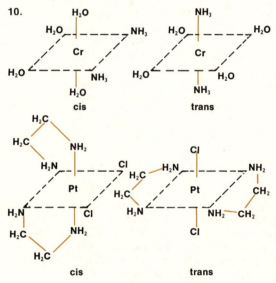

cis   trans

cis   trans

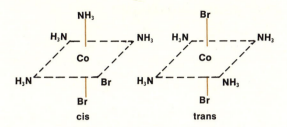

cis   trans

11. d^6 ↑↑ e —— e
　　 ⇈↑ t ⇈⇈ t
　　 high $- t^4e^2$ low $- t^6$

d^7 ↑↑ e ↑— e
　　 ⇈↑ t ⇈⇈ t
　　 high $- t^5e^2$ low $- t^6e^1$

12. (a) d^2 ——— t **(b)** d^3 —— e
　　　　 ↑↑ e　　　　 ↑↑↑ t

13. (a) 1.732; **(b)** high 4.9, low 0; **(c)** 0;
(d) 2.83; **(e)** high 3.464, low 1.732; **(f)** high 4.9, low 0. **14.** $M_s = \frac{4}{2}$ and four unpaired electrons exist giving a configuration ↑↑ e
　　　　　　　　　　　　　 ⇈↑↑ t

and the electrostatic field potential is relatively low. If it were high, the two electrons in the e orbital would pair with the two unpaired ones in the t orbital.

15. (a) $M_s = \frac{5}{2}$ and five unpaired electrons exist in an octahedral field: d^5: t^3e^2　　　↑↑ e
　　　　　　　　　　　　　　　↑↑↑ t
　(b) 1.732 BM; low spin state

16. (a) Octahedral **(b)** d^1: t^1 **(c)** 1.73 BM
(d) 8,330 Å **17. (a)** d^5 and in octahedral field

　　↑↑ e　　—— e
　 ↑↑↑ t ⇈⇈↑ t
 high spin$-t^3e^2$ low spin$-t^5$

(b) 0 **(c)** on the line

18. (a) Tetrahedral since four ligands **(b)** d^7 and based on magnetic moment, $M_s = \frac{3}{2}$

　　　　↑↑↑ t
　　　　⇈ ⇈ e

(c) $-\frac{6}{5}\Delta_t$

19. d^6 ↑↑↑ t ↑↑— t
　　 ⇈↑ e ⇈⇈ e
　　 high low spin

d^8 ⇈↑↑ t
　　 ⇈⇈ e

20. Since the complex is diamagnetic, a tetrahedral geometry is ruled out because paramagnetic behavior is always expected in this geometry (see solution to exercise 19, d^6 case).

CHAPTER SEVENTEEN PROBLEMS

1. (a) $R_{He} = 1.7 \times 10^{-13}$ cm **(b)** $R_{Ag} = 5.2 \times 10^{-13}$ cm
(c) $R_U = 6.8 \times 10^{-13}$ cm **2.** muon = 0.1134 amu; pion = 0.1498 amu **3.** $E = 2.15 \times 10^{13}$ cal
4. $E = 2.96 \times 10^{12}$ cal **5.** $^4_2$He 27.271 MeV. total; 6.883 MeV/nucleon. $^{12}_6$C 89.088 MeV. total; 7.423 MeV/nucleon. $^3_1$H 8.0166 MeV. total; 2.672 MeV/nucleon **6.** 18, 36, 54, 68 **7.** $^{40}_{20}$Ca, $^{288}_{82}$Pb, $^{16}_8$O = even protons and even neutrons, and therefore more stable; $^{45}_{20}$Ca, $^{211}_{82}$Pb, $^{15}_8$O = even protons and odd neutrons, less stable. **8. (a)** $^{140}_{57}$La **(b)** $^{13}_7$N
(c) $^1_0$n **(d)** $^4_2$He **(e)** $^4_2$He **(f)** $^{46}_{21}$Sc **(g)** $^{130}_{53}$I
9. $E = 2.39 \times 10^{10}$ cal/g **10.** $t = 15$ years
11. $t = 3.9 \times 10^9$ years

CHAPTER SEVENTEEN EXERCISES

1. The only correct statement is (d). **2.** Electron (β^-) and antielectron (β^+); proton (p^+) and antiproton (p^-)
3.

	total binding energy (MeV)	binding energy per nucleon (MeV/nucleon)
$^3_2$He	6.6958	2.232
$^7_3$Li	37.713	5.397
$^{31}_{15}$P	236.61	7.632
$^{127}_{53}$I	1049.2	8.262
$^{243}_{96}$Cm	1779.8	7.324

4. $^3_2$He is less stable than $^4_2$He because of the smaller binding energies. **5.** Rapid rise to a maximum around I and then slow decline (also see Fig. 17–2 in text). **6. (a)** $^3_2$He **(b)** $^{115}_{49}$In **(c)** $^{211}_{84}$Po **(d)** $^1_0$n
(e) γ **(f)** β^+ **(g)** $^{241}_{95}$Am **(h)** $^2_1$H **7.** $E = 1.274 \times 10^{11}$ cal/mole **8.** Rate = 8.76×10^5 counts/sec
9. ~10 g; essentially all $^{70}_{34}$Se disappeared. **10.** $k = 3.20 \times 10^{-2}$/min.; $t_{\frac{1}{2}} = 21.67$ min. **11.** After the equilibrium is reached, analyze the isotope composition (H_2O and D_2O) of the liquid water. Then inject a certain amount of D_2O vapor through the rubber. The existence of a greater amount of D_2O in the liquid water than that originally present will prove that a dynamic equilibrium is involved. **12.** $t = 9.0$ years **13.** $t = 12,300$ years **14.** $t = 3.2 \times 10^{11}$ years
15. 9.1×10^{-2} or 9.1% **16.** Half life of ^{3}H is too short, so there would be essentially no ^{3}H remaining and therefore no radioactivity would be observable.
17. 1.40×10^{-5} g **18.** Add radioactive silver isotope to a solution. Divide the solution into three parts and try the three methods of removing silver on each part. Determine the amount of radioactivity remaining in each solution and the one with the lowest activity has had the most silver removed.

CHAPTER NINETEEN PROBLEM

1. Rotational: 10^2 cm^{-1}; 0.286 kcal/mole;
Vibrational: 2×10^3 cm^{-1}, 5.728 kcal/mole;
Electronic: 2×10^4 cm^{-1}, 57.2 kcal/mole.

INDEX

The Elements

	Symbol	Atomic Number	Atomic Weight*
Actinium	Ac	89	(227)[b]
Aluminum	Al	13	26.9815
Americium	Am	95	(243)[b]
Antimony	Sb	51	121.75
Argon	Ar	18	39.948
Arsenic	As	33	74.9216
Astatine	At	85	(210)[b]
Barium	Ba	56	137.34
Berkelium	Bk	97	(247)[b]
Beryllium	Be	4	9.01218
Bismuth	Bi	83	208.9806
Boron	B	5	10.81
Bromine	Br	35	79.904
Cadmium	Cd	48	112.40
Calcium	Ca	20	40.08
Californium	Cf	98	(251)[b]
Carbon	C	6	12.011
Cerium	Ce	58	140.12
Cesium	Cs	55	132.9055
Chlorine	Cl	17	35.453
Chromium	Cr	24	51.996
Cobalt	Co	27	58.9332
Copper	Cu	29	63.546
Curium	Cm	96	(247)[b]
Dysprosium	Dy	66	162.50
Einsteinium	Es	99	(254)[b]
Erbium	Er	68	167.26
Europium	Eu	63	151.96
Fermium	Fm	100	(253)[b]
Fluorine	F	9	18.9984
Francium	Fr	87	(223)[b]
Gadolinium	Gd	64	157.25
Gallium	Ga	31	69.72
Germanium	Ge	32	72.59
Gold	Au	79	196.9665
Hafnium	Hf	72	178.49